T0135341

Algorithms for Intelligent Systems

This book series publishes research on the analysis and development of algorithms for intelligent systems with their applications to various real world problems. It covers research related to autonomous agents, multi-agent systems, behavioral modeling, reinforcement learning, game theory, mechanism design, machine learning, meta-heuristic search, optimization, planning and scheduling, artificial neural networks, evolutionary computation, swarm intelligence and other algorithms for intelligent systems.

The book series includes recent advancements, modification and applications of the artificial neural networks, evolutionary computation, swarm intelligence, artificial immune systems, fuzzy system, autonomous and multi agent systems, machine learning and other intelligent systems related areas. The material will be beneficial for the graduate students, post-graduate students as well as the researchers who want a broader view of advances in algorithms for intelligent systems. The contents will also be useful to the researchers from other fields who have no knowledge of the power of intelligent systems, e.g. the researchers in the field of bioinformatics, biochemists, mechanical and chemical engineers, economists, musicians and medical practitioners.

The series publishes monographs, edited volumes, advanced textbooks and selected proceedings.

Indexed by zbMATH.

All books published in the series are submitted for consideration in Web of Science.

Garima Mathur · Mahesh Bundele ·
Ashish Tripathi · Marcin Paprzycki
Editors

Proceedings of 3rd International Conference on Artificial Intelligence: Advances and Applications

ICAIAA 2022

Editors
Garima Mathur
Poornima College of Engineering
Jaipur, Rajasthan, India

Mahesh Bundele
Poornima College of Engineering
Jaipur, Rajasthan, India

Ashish Tripathi
Department of Computer Science
and Engineering
Malaviya National Institute of Technology
Jaipur
Jaipur, India

Marcin Paprzycki
Systems Research Institute
Polish Academy of Sciences
Warsaw, Poland

ISSN 2524-7565 ISSN 2524-7573 (electronic)
Algorithms for Intelligent Systems
ISBN 978-981-19-7043-6 ISBN 978-981-19-7041-2 (eBook)
https://doi.org/10.1007/978-981-19-7041-2

This Springer imprint is published by the registered company Springer Nature Singapore Pte Ltd.
The registered company address is: 152 Beach Road, #21-01/04 Gateway East, Singapore 189721, Singapore

Preface

This book gathers outstanding research papers presented at the 3rd International Conference on Artificial Intelligence: Advances and Applications (ICAIAA 2022), held on April 23–24, 2022, at Poornima College of Engineering, Jaipur, India, under the technical sponsorship of the Soft Computing Research Society, India. The conference is conceived as a platform for disseminating and exchanging ideas, concepts, and results of researchers from academia and industry to develop a comprehensive understanding of the challenges of the advancements of intelligence in computational viewpoints. This book will help in strengthening congenial networking between academia and industry. We have tried our best to enrich the quality of the ICAIAA 2022 through the stringent and careful peer-review process. This book presents novel contributions to advances in artificial intelligence and its applications and serves as reference material for advanced research.

We have tried our best to enrich the quality of the ICAIAA 2022 through a stringent and careful peer-review process. ICAIAA 2022 received many technical contributed articles from distinguished participants from home and abroad. ICAIAA 2022 received submissions from 18 different countries, viz., Algeria, Bangladesh, Bulgaria, Ethiopia, Iceland, India, Iran, Israel, Malaysia, Nigeria, Saudi Arabia, South Africa, South Korea, Spain, Taiwan, Thailand, United Arab Emirates, United Kingdom. After a very stringent peer-reviewing process, only 55 high-quality papers were finally accepted for presentation and the final proceedings.

This book presents 55 research papers related to advances in artificial intelligence and its applications serves as reference material for advanced research.

Jaipur, India — Garima Mathur
Jaipur, India — Mahesh Bundele
Jaipur, India — Ashish Tripathi
Warsaw, Poland — Marcin Paprzycki

Contents

About the Editors

Garima Mathur is currently working as the head of Department, Poornima College of Engineering, Jaipur, since July 1, 2016. Previously, she was working as the head of Department in Electronics and Communication Engineering at Jaipur Engineering College, Jaipur, Rajasthan, India, since October 2000 to December 2015. She has total 21 years of experience in teaching and research. She is working on various research projects sponsored by various agencies. She did her doctoral degree in Performance Evaluation of Modified Sphere Decoding Scheme for MIMO Systems. Dr. Mathur published and presented international, national journals, conferences, symposium, and seminar more than 30 research papers. She has guided more than 15 M.Tech. Dissertation theses. Her area of interest is wireless channel, channel modeling, and ad hoc networks. Dr. Mathur is a life member of the Institution of Electronics and Telecommunication Engineers (IETE), The Indian Society for Technical Education (ISTE), India, and many other professional bodies.

Mahesh Bundele is currently working as the principal and the director of Poornima College of Engineering, Jaipur, since September 1, 2018. He has total of 34 years of experience in teaching and research. He has developed many unique research methodology concepts and implemented. He is the mentor and controller of quality research and publications at the Poornima. He is also responsible for inculcation of innovative and critical analysis concepts across the university and across the Poornima Foundation involving three other campuses. He did his doctoral degree in Wearable Computing and guiding research in pervasive and ubiquitous computing, computer networks, and software-defined networking. His areas of interests are also wireless sensor networks, algorithmic research, mathematical modeling, and smart grids. He has more than 60 publications in reputed journals and conferences. He has been the reviewer of few IEEE Transactions. He is actively involved in IEEE activities in Rajasthan Subsection and Delhi Section and holding the responsibility on Standing Committee of IEEE Delhi Section for Technical and Professional activities for controlling quality of conferences and publications in IEEE. He is also a member of IEEE Delhi Section Execom. He has organized 5 editions of IEEE ICRAIE in India and Malaysia with conference position as the general co-chair. He has organized

PerCAA 2019 with Elsevier Computer Science Procedia as the program chair. He is the general chair and the advisor in many other international conferences including Springer ICAIAA 2019, 2021, IOP publication ICGSSET 2021 and an IEEE Delhi Section Oversight committee member of all IEEE International Conferences being organized in IEEE Delhi Section jurisdiction since last 3 years. He is the editorial manager for Elsevier's Computers and Electronics in Agriculture, Environmental Science and Pollution Research, Renewable and Sustainable Energy, Applied Mathematical Modeling, etc. He has received Excellence in Leadership award by ITSR foundation in 2020.

Ashish Tripathi (Member, IEEE) received his M.Tech. and Ph.D. degrees in Computer Science and Engineering from the Department of Computer Science and Engineering, Delhi Technological University, Delhi, India, in 2013 and 2019, respectively. He is currently working as an assistant professor with the Department of Computer Science and Engineering, Malviya National Institute of Technology (MNIT), Jaipur, India. His research interests include big data analytics, social media analytics, soft computing, image analysis, and natural language processing. Dr. Tripathi has published several papers in international journals and conferences including IEEE transactions. He is an active reviewer for several journals of repute.

Marcin Paprzycki is an associate professor at the Systems Research Institute, Polish Academy of Sciences. He has an MS from Adam Mickiewicz University in Poznan, Poland, a Ph.D. from Southern Methodist University in Dallas, Texas, and a D.Sc. Degree from the Bulgarian Academy of Sciences. He is a senior member of IEEE, a senior member of ACM, a senior fulbright lecturer, and an IEEE CS distinguished visitor. He has contributed to more than 450 publications and was invited to the program committees of over 500 international conferences. He is on the editorial boards of 12 journals and a book series.

Chapter 1
Prediction of Polycystic Ovary Syndrome (PCOS) Using Optimized Machine Learning Classifiers

Khushi Vora, Arya Shah, Nishant Shah, and Priyanka Verma

1 Introduction

Polycystic ovary syndrome is characterized by increased androgen levels, menstrual irregularities and/or cysts on one or both ovaries [1]. Women with PCOS may have infrequent or prolonged menstrual periods. There may be a marked increase in the amount of androgen, which is a male hormone, released. The ovaries may develop numerous small fluid-filled sacs (cysts) on the surface and fail to release eggs regularly [2]. PCOS may also cause skin changes such as increased facial and body hair and acne and infertility. Often, women with PCOS have metabolic issues too [3]. The disorder can be morphological (polycystic ovaries) or biochemical (hyperandrogenemia) in nature [4]. The clinical presentation of PCOS is variable as well. Patients may be asymptomatic or may have multiple metabolic, dermatologic or gynaecologic manifestations [5].

Limited research has been done in the field of PCOS detection using machine learning algorithms. The kind of system we propose in this paper could help people with understanding whether they are asymptomatic or actually have PCOS. It could also help them to determine the presence of PCOS by simply taking a look at their medical records, without a physical examination (eg. ultrasound) as it may not be a financially feasible option for all.

K. Vora (✉) · A. Shah · N. Shah · P. Verma
NMIMS University, Mumbai, India
e-mail: khushi.vora70@nmims.edu.in

A. Shah
e-mail: arya.shah82@nmims.edu.in

N. Shah
e-mail: nishant.shah55@nmims.edu.in

P. Verma
e-mail: priyanka.verma@nmims.edu

G. Mathur et al. (eds.), *Proceedings of 3rd International Conference on Artificial Intelligence: Advances and Applications*, Algorithms for Intelligent Systems,
https://doi.org/10.1007/978-981-19-7041-2_1

In our research paper, we have applied multiple ML algorithms for early detection of PCOS. We have used Optimizable Discriminant and Optimizable Naïve Bayes, whereas the non-linear models were Optimizable Tree, Optimizable SVM, Optimizable KNN, Optimizable Ensemble and Neural Networks.

The rest of the paper is divided as follows. Section 2 details a review of literature of the existing work in this field. It is followed by a detailed description of the methodology we used to build and test various machine learning models on our dataset. Section 4 discusses the results we obtained. It is followed by Sect. 5 which is the conclusion of the paper. In this section, we also discuss the ways in which we wish to expand this project in the future.

2 Literature Review

A literature review of the existing research in PCOS detection using machine learning is presented here. The techniques used range from basic to hybrid algorithms.

Chauhan et al. [6] used a split of 80:20 as the train to test the dataset. They used various classifiers and found out that the decision tree classifier gave an accuracy of 81% with precision of 70% and specificity of 94%. The scope of improvement in this paper can be tracking symptoms for a period of time and using the readings might increase the accuracy of the predictions.

Prapty and Shitu [7] used KNN, SVM, Naive Bayes classifier and Random Forest classifier for modelling their data. No Feature Engineering was done by them throughout the Machine Learning Process. The performance of KNN was not very excellent; however, the performance of SVM was strong enough to properly diagnose PCOS. Based on overall performance, the Random Forest Classifier is the best of these classifiers.

A decision tree is created using the Random Forest approach. Their study recommended that women should conduct regular home checks for observable changes such as BMI, hair growth, hair loss, weight increase and menstruation cycle. They should follow up with the remainder of the hormone testing and see a doctor based on the results. BMI, hair development, hair loss, weight increase and menstruation cycle are therefore derived as essential characteristics.

Abu Adla et al. [8] used hybrid feature selection. They employed the use of filters and wrappers to narrow down features to be selected by performing ANOVA test for feature significance and consulted a domain expert. Verify that results complied with medical standards. They also used K-Fold Cross Validation (KCV) to eliminate biased data splitting. The final model selected is SVM with Accuracy 91.6%, Precision 93.665% and Recall 80.6%.

Nabi et al. [9] conducted research on PCOS detection in Bangladeshi women. They collected 550 data points through a questionnaire in Bangla and English. They used a train-test split of 80–20 and achieved 99.09% using SVM technique. XGB Classifiers and Gaussian Naive Bayes both gave 98.18% accuracy. Logistic Regression (LR) and

KNN and both gave 97.97% accuracy. Decision Tree (DT) and Gradient Boosting Classifiers and Random Forest Classifier (RFC) gave the same accuracy of 97.27%.

Khan Inan et al. [10] proposed the use of Extreme Gradient Boosting along with XGBoost, for detection of PCOS at an early stage. They resampled their data using a combination of ENN (Edited Nearest Neighbour) and SMOTE (Synthetic Minority Oversampling Techniques). Using ANOVA Test and Chi-Square Test, they identified 23 significant parameters to best classify PCOS. They found that the Extreme Gradient Boosting classifier (XGBoost) performed best among all other classifiers with a Cross-validation score of 96.03% using 10-Folds. There was a 98% Recall in the detection of patients without PCOS.

Bharati et al. [11] ranked features in the training dataset according to p-value and selected features accordingly. They identified the significant feature to be the ratio of Follicle-stimulating hormone (FSH) and Luteinizing hormone (LH). Their results demonstrated that hybrid random forest and logistic regression (RFLR) on applying cross validation with 40 folds to the best 10 features demonstrated testing accuracy of 91.01% and recall of 90%.

Considering the scope of the literature review conducted, it was observed that of all the existing research conducted on the dataset, not many of the authors had emphasized on hyperparameter tuning during the training of models. In [12], the authors extracted 8 useful features and applied Principal Component Analysis (PCA) but no hyperparameter tuning was done. Similarly, in [7, 8] and [11], the authors have used various algorithms but little to no fine-tuning has been done. Only [10] has done a fair amount of hyperparameter tuning and used improved sampling techniques along with feature selection. Our paper adds onto their techniques by optimizing the hyperparameters to achieve better results overall. Sajwan and Ranjan [13] developed a new feature descriptor for image retrieval. Altaf et al. [14] deployed machine learning techniques for detecting liver functionalities. Oza and Bokhare [15] implemented logistic regression for predicting disease.

3 Methodology

To develop an effective ML-based diagnostic tool for PCOS, a comparison of evaluation metrics of multiple ML models applied to our dataset is necessary. Because it provides the foundation for the investigation, the model preparation is the most significant part in the research process. Figure 1 depicts the phases needed in establishing an apt model and fine-tuning it for the best potential outcome.

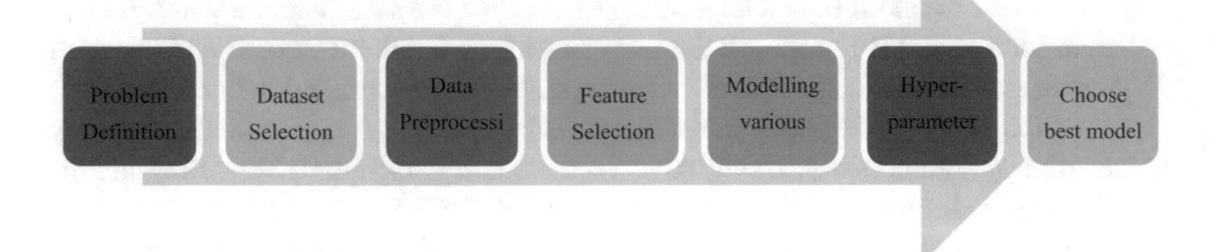

Fig. 1 Workflow for the machine learning model

3.1 Problem Definition

The first and most critical stage is to properly characterize the problem, including the model's inputs and the intended outcome. It is built on the principle that the outputs may be projected based on the inputs.

3.2 Dataset Selection

This research takes into account a dataset from Kaggle [16] named Polycystic Ovary Syndrome (PCOS). It consists of 541 samples spread across 45 physical and clinical parameters of various patients. The parameters of the dataset were finalized with the help of professional opinions and by taking into account recent studies that had an impact on PCOS in some manner. The original dataset was observed to be highly imbalanced with 362 instances of class: 0 and only 176 instances of class: 1 (see Fig. 2).

3.3 Data Cleaning

In order to prepare data for analysis, it is necessary to remove or change data that is inaccurate, incomplete, irrelevant, redundant, or poorly formatted. Steps taken were dropping the null values, solving the corrupted values and replacing the irrelevant values. To add the missing BMI values, the formula used was

$$\text{BMI} = \text{weight}/\text{height}^2 \tag{1}$$

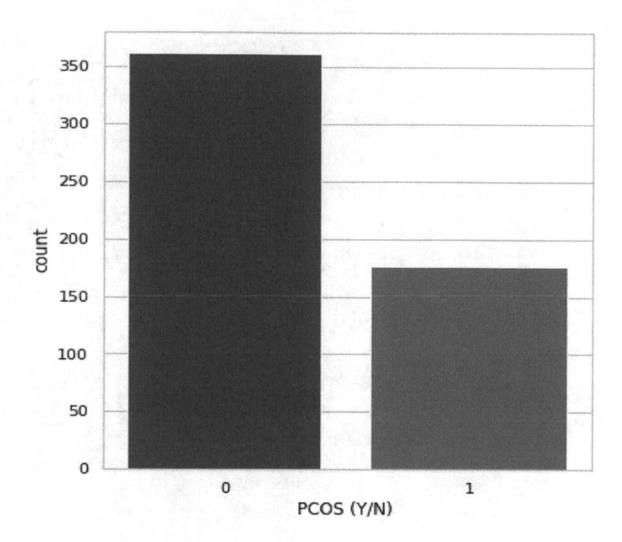

Fig. 2 Count plot of the original database

3.4 *Oversampling and Feature Selection*

Process of minimizing the usage of input variables when designing a predictive model is Feature Selection. Statistics are used to evaluate the relation between each input variable and the goal variable, input variable with strongest relationship with respect to target variable is identified. Features that we have extracted are through EDA by checking the heatmap and correlation matrix (see Fig. 3). The selected features are

1. Cycle (R/I)
2. Weight gain
3. Hair growth
4. Skin darkening
5. Follicle No. (L)
6. Follicle No. (R)

For an increasing amount of samples for our non-majority class, we used Edited Nearest Neighbours (ENN) along with Synthetic Minority Oversampling (SMOTE) criteria to eliminate outliers. SMOTE works by selecting close-together, creating a line between samples in subspace, then selecting a new sample along that line [17]. We utilize ENN to under sample our data after increasing the sample size in order to limit the impact of outliers. We apply ENN on two classes and discard all instances where K-nearest neighbours (KNN) prediction varies from that class. As a consequence, an instance will be deleted if it has more neighbours of a different class. The dataset after improved sampling was observed to have 296 instances of class: 0 and 268 instances of class: 1 (see Fig. 4).

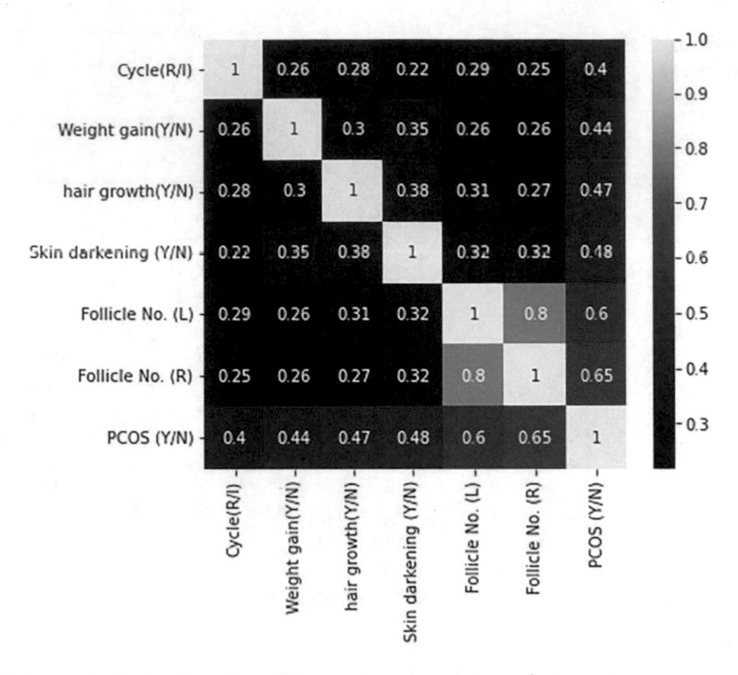

Fig. 3 Heatmap depicting the selected features

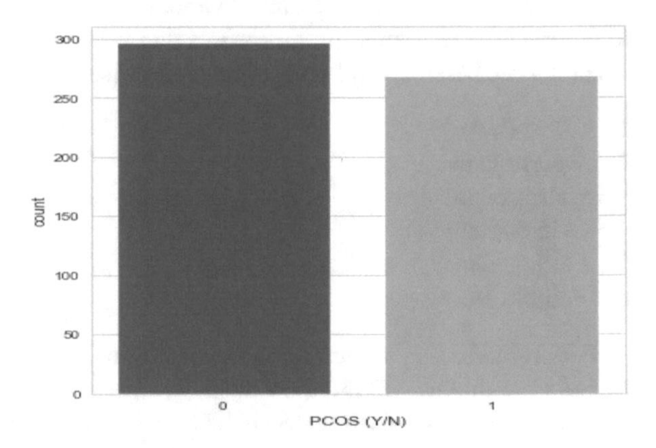

Fig. 4 Count plot of the dataset after improved sampling

3.5 Optimized Classifiers

One of the most difficult challenges in the application of machine learning systems is model optimization. Hyperparameter optimization in machine learning seeks to identify the hyperparameters of a particular machine learning algorithm that offer the greatest performance as assessed on a validation set. Using MATLAB 2021 we were able to take advantage of the inbuilt classifier which automatically optimizes the model. Table 1 presents the results summary in Fig. 5.

Table 1 Hyperparameters used for various optimizable models

Classifiers	Hyperparameters
Optimizable tree	Maximum number of splits: 66 with maximum deviance reduction
Optimizable discriminant	Discriminant type: Diagonal quadratic
Optimizable naïve bayes	Distribution name: Gaussian with box kernel
Optimizable SVM	Kernel function: Quadratic
Optimizable KNN	Number of neighbours: 1 with minkowski distance metric
Optimizable ensemble	Ensemble method: Bag with 459 maximum number of splits, 10 number of learners and 4 predictors to sample
Narrow neural network	First layer size: 10 with relu activation function in 1000 no of iterations
Medium neural network	First layer size: 25 with relu activation function in 1000 no of iterations
Wide neural network	First layer size: 100 with relu activation function in 1000 no of iterations
Bi-layered neural network	First layer size: 10, second layer size: 10 with relu activation function in 1000 no of iterations
Tri-layered neural network	First layer size: 10, second layer size: 10, third layer size: 10 with relu activation function in 1000 no of iterations

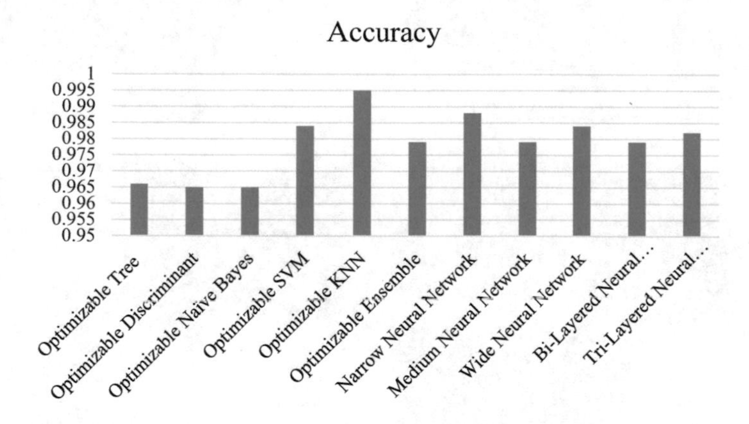

Fig. 5 Bar chart comparing model accuracies

3.6 Comparison of Various Models

We compare a series of ML models to determine the best one by analysing the confusion matrix of each algorithm, the accuracy, the precision, the sensitivity, the specificity and the F1 score and the ROC curve and AUC (Area under curve). See Figs. 6 and 7 for the same.

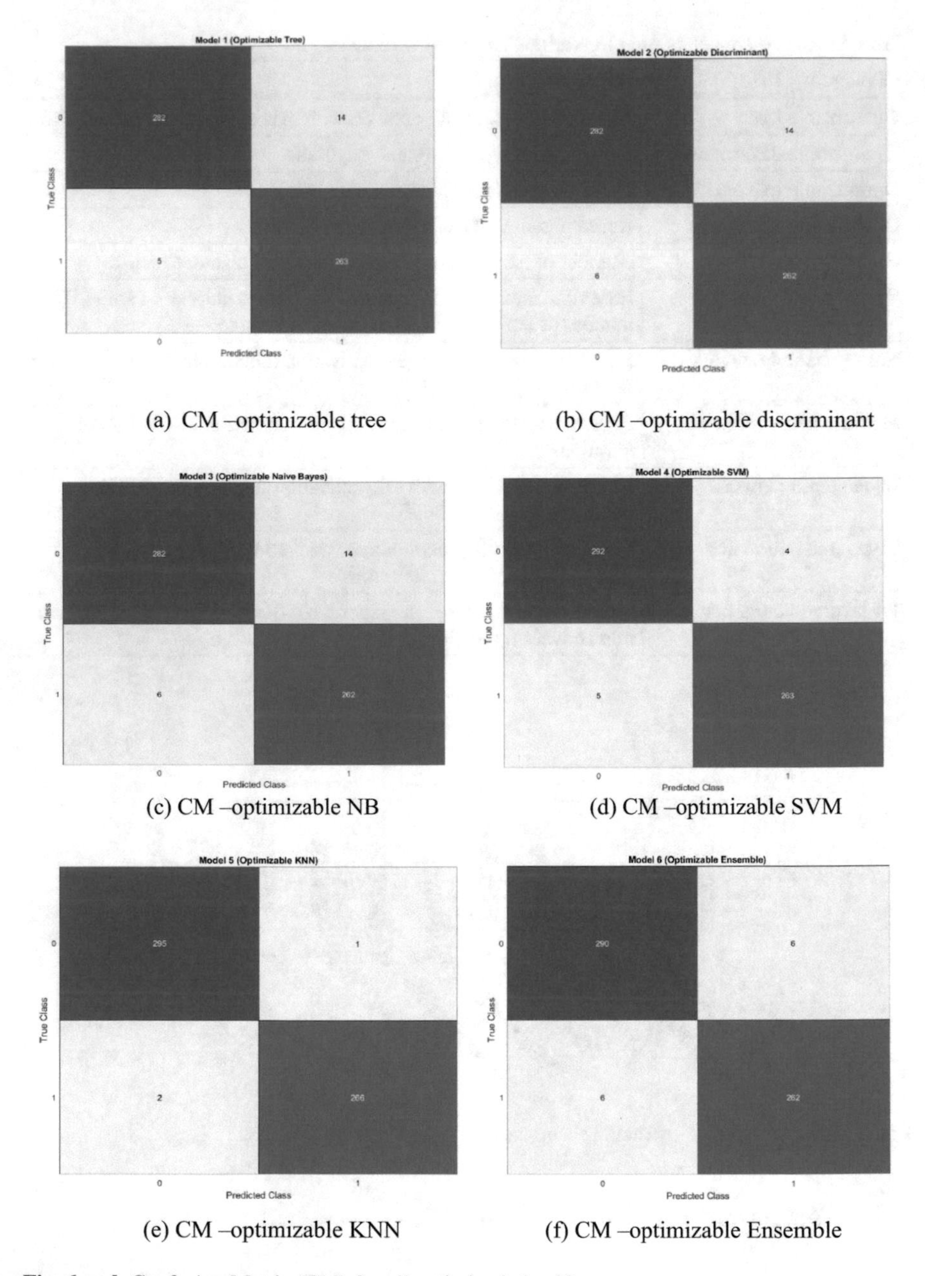

(a) CM –optimizable tree

(b) CM –optimizable discriminant

(c) CM –optimizable NB

(d) CM –optimizable SVM

(e) CM –optimizable KNN

(f) CM –optimizable Ensemble

Fig. 6 a-k Confusion Matrix (CM) for all optimized classifiers

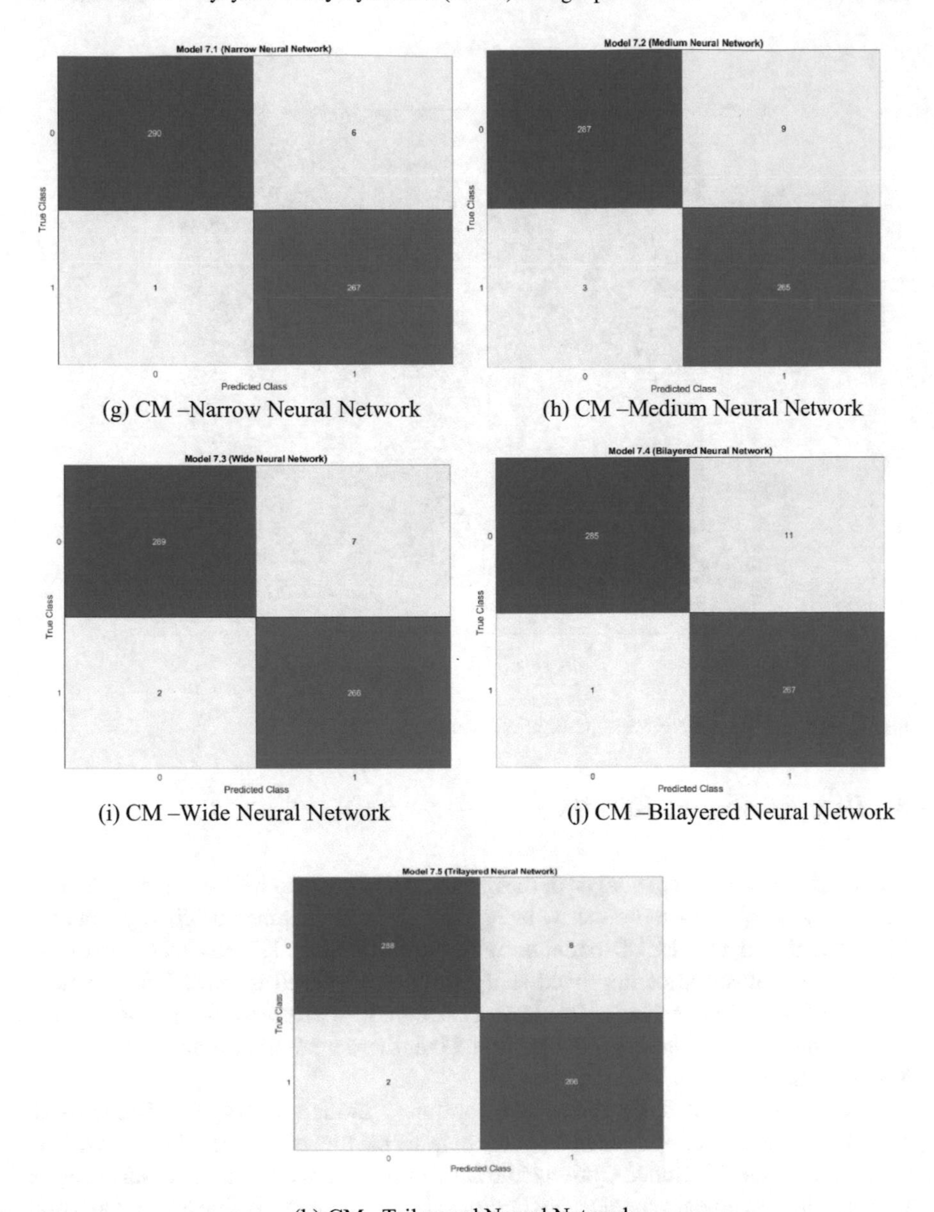

(g) CM –Narrow Neural Network

(h) CM –Medium Neural Network

(i) CM –Wide Neural Network

(j) CM –Bilayered Neural Network

(k) CM –Trilayered Neural Network

Fig. 6 (continued)

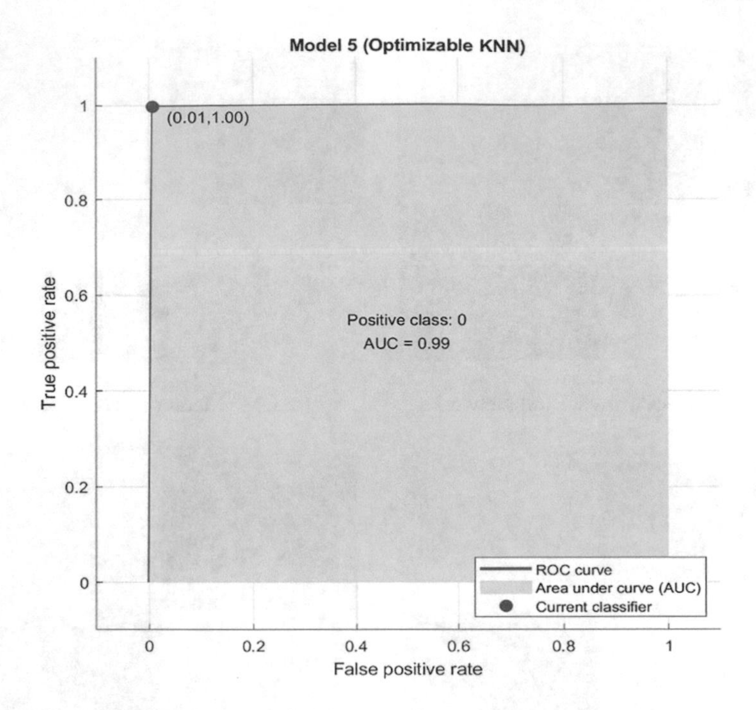

Fig. 7 ROC curve for KNN (Best performing model)

4 Discussion and Results

A total of 541 cases were analysed, the data included women between the ages of 18 and 40 who were in the reproductive age group. The data obtained originally included 364 normal and non-PCOS patients, with the remaining 177 instances reporting PCOS. The dataset after improved sampling was observed to have 296 instances of class: 0 and 268 instances of class: 1. Although there were 45 features, upon performing statistical analysis, it was found that there were only 6 major conditions affecting the outcome.

There is a mix of basic linear and non-linear models in the algorithms used. The models used an optimized version outputting the best possible results. The linear models used included Optimizable Discriminant and Optimizable Naïve Bayes whereas the non-linear models were Optimizable Tree, Optimizable SVM, Optimizable KNN, Optimizable Ensemble and Neural Networks. Each model's accuracy was calculated through hold out validation, and the estimated accuracy scores of each model were obtained. Each models' precision, sensitivity, specificity, F1 score, confusion matrix and AUC were recorded as presented in Table 2. The Optimizable KNN model presenting the highest accuracy of 99.5% is selected and deemed to be the best.

Table 2 Model performance metrics

Classifiers	Accuracy	Precision	Sensitivity	Specificity	F1-Score	AUC
Optimizable tree	0.966	0.9527	0.9826	0.9495	0.9674	0.99
Optimizable discriminant	0.965	0.9527	0.9792	0.9493	0.9658	0.99
Optimizable Naïve Bayes	0.965	0.9527	0.9792	0.9493	0.9658	0.99
Optimizable SVM	0.984	0.9865	0.9832	0.9850	0.9848	1.00
Optimizable KNN	0.995	0.9966	0.9933	0.9963	0.9949	0.99
Optimizable ensemble	0.979	0.9797	0.9797	0.9776	0.9797	0.99
Narrow neural network	0.988	0.9797	0.9966	0.9780	0.9881	0.99
Medium neural network	0.979	0.9696	0.9897	0.9672	0.9795	0.98
Wide neural network	0.984	0.9764	0.9931	0.9744	0.9847	0.98
Bi-layered neural network	0.979	0.9628	0.9965	0.9604	0.9794	0.98
Tri-layered neural network	0.982	0.9730	0.9931	0.9708	0.9829	0.99

5 Conclusion and Future Work

Polycystic Ovary Syndrome (PCOS) is a form of endocrine condition that affects women of reproductive age. Infertility and anovulation may arise as a result of this. Clinical and metabolic indicators, which are biomarkers for the illness, are included in the diagnostic criteria. We created a technique that detects PCOS automatically using a small number of possible indicators. The proposed model involves taking inputs of only Follicle No. (R), Follicle No. (L), Skin darkening, Hair growth, Weight gain, Cycle (R/I) for predicting the risk of PCOS. Among the various models tested, we found the Optimizable KNN Algorithm to perform the best. This automated technology can aid doctors in saving time when assessing patients and thereby minimizing the time it takes to diagnose PCOS risk.

We can pair up with healthcare facilities to gain real time data from these institutions based on the clinical and radiological investigations/check-ups and deploy the model as a web application for quick diagnosis, making it convenient for the doctors thereby saving time.

Further research can also be done to examine the link between patients' risk of PCOS diagnosis and their mental health.

References

1. Umland EM, Weinstein LC, Buchanan EM (2011) Pharmacotherapy: A pathophysiologic approach
2. Mayo Clinic PCOS webpage. https://www.mayoclinic.org/diseases-conditions/pcos/symptoms-causes/syc-20353439. Last Accessed on 16 Feb 22
3. NICHD PCOS Factsheet. https://www.nichd.nih.gov/health/topics/factsheets/pcos. Last Accessed on 16 Feb 22

4. Lin LH, Baracat MC, Maciel GA, Soares JM Jr, Baracat EC (2013) Androgen receptor gene polymorphism and polycystic ovary syndrome. Int J Gynecol Obstet 120(2):115–118
5. Williams T, Mortada R, Porter S (2016) Diagnosis and treatment of polycystic ovary syndrome. Am Fam Physician 94(2):106–113
6. Chauhan P, Patil P, Rane N, Raundale P, Kanakia H (2021) Comparative analysis of machine learning algorithms for prediction of PCOS. In: 2021 International conference on communication information and computing technology (ICCICT). pp 1–7. https://doi.org/10.1109/ICCICT50803.2021.9510101
7. Prapty AS, Shitu TT (2020) An efficient decision tree establishment and performance analysis with different machine learning approaches on polycystic ovary syndrome. In: 2020 23rd International conference on computer and information technology (ICCIT). pp 1–5. https://doi.org/10.1109/ICCIT51783.2020.9392666
8. Abu Adla YA, Raydan DG, Charaf MZJ, Saad RA, Nasreddine J, Diab MO (2021) Automated detection of polycystic ovary syndrome using machine learning techniques. In: 2021 Sixth international conference on advances in biomedical engineering (ICABME). pp 208–212. https://doi.org/10.1109/ICABME53305.2021.9604905
9. Nabi N, Islam S, Khushbu SA, Masum AKM (2021) Machine learning approach: detecting polycystic ovary syndrome & its impact on Bangladeshi women. In: 2021 12th International conference on computing communication and networking technologies (ICCCNT). pp 1–7. https://doi.org/10.1109/ICCCNT51525.2021.9580143
10. Khan Inan MS, Ulfath RE, Alam FI, Bappee FK, Hasan R (2021) Improved sampling and feature selection to support extreme gradient boosting for pcos diagnosis. In: 2021 IEEE 11th annual computing and communication workshop and conference (CCWC). pp 1046–1050. https://doi.org/10.1109/CCWC51732.2021.9375994
11. Bharati S, Podder P, Hossain Mondal MR (2020) Diagnosis of polycystic ovary syndrome using machine learning algorithms. In: 2020 IEEE region 10 symposium (TENSYMP). pp 1486–1489. https://doi.org/10.1109/TENSYMP50017.2020.9230932
12. Denny A, Raj A, Ashok A, Ram CM, George R (2019) i-hOPE: Detection and prediction system for polycystic ovary syndrome (pcos) using machine learning techniques. In: TENCON 2019–2019 IEEE region 10 conference (TENCON). pp 673–678. https://doi.org/10.1109/TENCON.2019.8929674
13. Sajwan V, Ranjan R (2022) A novel feature descriptor: color texture description with diagonal local binary patterns using new distance metric for image retrieval. In: Congress on intelligent systems. Springer, Singapore, pp 17–26
14. Altaf I, Butt MA, Zaman M (2022) Machine learning techniques on disease detection and prediction using the hepatic and lipid profile panel data. In: Congress on intelligent systems. Springer, Singapore, pp 189–203
15. Oza A, Bokhare A (2022) Diabetes prediction using logistic regression and k-nearest neighbor. In: Congress on intelligent systems. Springer, Singapore, pp 407–418
16. Kaggle Dataset. https://www.kaggle.com/prasoonkottarathil/polycystic-ovary-syndrome-pcos. Last Accessed on 22 Oct 21
17. He H, Ma Y (2013) Imbalanced learning: Foundations, algorithms, and applications, 1st edn. Wiley-IEEE Press

Chapter 2
Literature Review: A Comparative Study of Software Defect Prediction Techniques

Tarunim Sharma, Aman Jatain, Shalini Bhaskar, and Kavita Pabreja

1 Introduction

Over the previous few decades, developers have steadily shifted their attention to software-based systems, with software quality and dependability seen as the most important factor in user functionality. In recent years, there is an increased computerization that has resulted in the creation of a variety of different software; however, measures need to be taken to make sure that the produced software isn't defective. If the source code is complex, chances are there that software will show defects which further leads to failure in software. The scientifically based administration of the software testing phase necessitates the early and correct prediction of software problems. Defect prediction model development is ordinarily utilized in the field of industry and these kinds of models help in additional fault prediction, testing, calculating effort, reliability of software, software quality, assessment of hazards, etc., during the developmental stage. The future direction for research in the software engineering field is detecting faults in the software as it helps developers and testers for locating Software Defects with high accuracy and within time [1]. The most essential task in the software development life cycle model's testing phase is to establish a method for forecasting software failures so that testing and maintenance expenses can be lowered. It determines which modules are prone to errors and requires rigorous testing. For many years, regression techniques have been used to predict defective code snippets, and more recently, machines learning algorithms, both supervised

T. Sharma (✉) · A. Jatain · S. Bhaskar
Amity School of Engineering and Technology, Gurgaon, Haryana, India
e-mail: Tarunimsharma@msijanakpuri.com

K. Pabreja
Maharaja Surajmal Institute, Janakpuri, New Delhi, India
e-mail: kavitapabreja@msijanakpuri.com

G. Mathur et al. (eds.), *Proceedings of 3rd International Conference on Artificial Intelligence: Advances and Applications*, Algorithms for Intelligent Systems,
https://doi.org/10.1007/978-981-19-7041-2_2

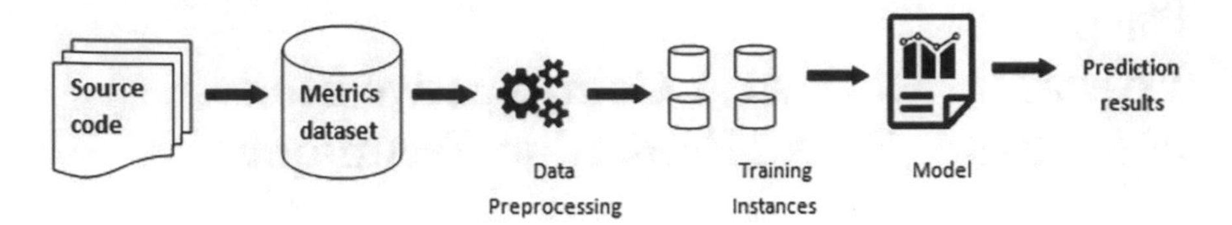

Fig. 1 Software defect prediction using machine learning

and unsupervised, have been used with a predefined collection of training data using historical data from software repositories, and forecasting problematic code snippets.

Machine learning techniques for automatic software failure prediction have gained a lot of attraction in recent times. They detect modules (modules) with a high risk of failure (that are more likely to have been mistaken) and then give top priority to more hazardous test cases to develop an efficient testing scheme with the least amount of work, time, and expense [2]. Approaches based on machine learning help in predicting the possible defects of the software modules as shown in Fig. 1. The algorithm then uses the training dataset to generate rules for a new set of data for predicting the class label. The predictor function is generated and strengthened using mathematical techniques in the learning phases. The process's training data has a defined output value and an attribute input value. The machine learning method's expected quality is compared to a usual output. This is repeated many times with different sets of training data until the best prediction accuracy is attained and the maximum number of iterations has been reached or not is checked. In the realm of unsupervised learning methods, the class label target value is unknown in the dataset. A cluster of data, on the other hand, is entered into the software, and the algorithm looks for patterns and correlations within it.

The primary focus is on the relationships between data attributes. Predicting faulty modules helps in improving software quality. In this research, several defect prediction models based on machine learning approaches have been suggested and verified on a variety of open-source datasets. "The method of designing models that are used in the early stages of the process to detect defecation is known as defect prediction". This can be achieved by categorizing the dataset into two classes or modules, i.e., defective and non-defective ones. Finding effective methods for assessing and forecasting problems in software modules is a key area of further research study. A variety of machine learning approaches, such as support vector machines, random forests, Artificial Neural Networks, K-Nearest Neighbor, Naive Bayes, Decision Trees, and Ensemble Techniques, have been used to conduct classification, grouping, and regression [3]. Although there are many additional algorithms, complicated data structures accuracy cannot be explored using only the above-mentioned algorithms.

These machine learning algorithms perform better when paired with strategies for reducing data imbalance as well as dimension reduction approaches. Some of the most common ways for determining the classification module are Support vector classifiers (SVC), Gradient boost, Random forests, Naive Bayes, Adaboost, etc. On the other hand, in order to forecast the quantity of software defects, various regression algorithms have been presented. It's a type of predictive modeling that looks at the relationship between a dependent (target) and independent variable (predictor). Some of the most frequent regression techniques employed in this field are linear regression, polynomial regression, logistic regression, multivariate regression, and lasso regression. During the diagnostic testing phases, the defect-prone modules are given high attention, while the non-defect-prone modules are evaluated as time and expense allow. Using equations for target classification, the classifier approach explores the classification feature known as the link between the attributes and the training dataset class label. These rules will also be required to determine the labels for future dataset classes. Thus, utilizing classification patterns and a classifier, unknown datasets can be classified. Because of the huge deployment of software, defining software problems, discovering the defect, and recognizing it is a recurring task for researchers. However, the development of algorithms and the prediction of accuracy remains a challenge.

However, these studies have various challenges yet to be met that includes an unbalanced class distribution, noisy data, an appropriate feature set selection, and proper machine learning model validation. An inadequate validation mechanism, less reliable research methodologies, difficulties in dataset class imbalance, coping with noisy data processes, and figuring out a way to combine feature selection and hybridization are all continuing research issues with existing frameworks advised for predicting defects in the software.

The class imbalance of samples is typical in many practical situations. As a matter of fact, most software modules in the majority class are defect-free, which is also true for the Software fault prediction task. There are very few numbers of software modules in the minority group having problems. The main objective of the SDP model is to correctly spot those software modules which are defective, hence accurately recognizing that minority classes are critical ones. As a result, the defect class is in the minority. It is very important to find out this minority class that further helps in detecting the defect in the software failing to do that will impact the quality of software drastically. Another issue is appropriate Feature Selection, which has become a necessary step in many applications as it is one of the data qualities challenges which includes identifying certain unnecessary items such as repetitive and irrelevant attributes. The prediction model's performance is unaffected by unrelated features, but the model's processing time is increased. The unessential sets of features add data which is not readable or outliers to the foretelling model, even if useful information is provided [4]. Furthermore, early research has also shown that presence of redundant features will decline the precision of these prediction models [5, 6]. As a result, selecting the appropriate features can aid data perception and visualization while also lowering test costs, storage needs, training time, and improving prediction model performance [7].

Various studies advocate the fact that the quality of the data used will strongly impact the performance of classifiers [8, 9]. As a result, corrupted data in real datasets makes decision-making difficult, and also the ensemble classifiers based on such data result in poor accuracy. In these situations, an ensemble-based framework with noise filtering, feature selection, and data balancing can be effective, and the authors in literature review present a framework that integrates Ensemble Learning Algorithm with feature selection, data balancing, and noise filtering. Feature selection in this case eliminates duplicate and insignificant features, leaving only the most critical attributes for further consideration. The use of essential features reduces the difficulty of the algorithm, resulting in faster processing and lower costs.

Several cost-based performance metrics have been proposed in the literature. "A cost metric considers particular aspects of expenses but does not attempt to quantify the total costs associated with the defect prediction model". It incorporates several, or each and every important feature of a defect prediction model's costs. Globally various studies have been proposed that are based on evaluation-based performance metrics. A cost metric considers particular aspects of expenses but does not attempt to quantify the total costs associated with the model for predicting defects. Numerous or every single relevant element of a model detecting defects in the software costs need to be incorporated into a model for calculating cost. The cost model includes post-release defects cost, quality assurance costs, as there is a possibility that quality assurance will miss expected flaws, and one important thing is how by-product of software development phase and defect prediction are related. The cost model is set up with various assumptions, and experiments are run to reveal trends in cost behavior before and after prediction on real-world projects. The major contribution of these studies is an analysis of various software defect prediction models and machine learning algorithms used in their implementation, as well as the proposal of an Ensemble SDP model to overcome all of the drawbacks present in these prediction models, such as imbalanced class distribution, noisy data, and appropriate feature set selection. Not a single machine learning technique has been found to be capable of tackling all of these difficulties, resulting in performance loss. As a result, Ensemble Modeling is commonly regarded as a way to overcome the constraints of individual ML classifiers [10, 11]. This assumption is supported by Ensemble Learning Algorithms (ELA) demonstrating competency in a variety of study domains [12, 13].

Ensemble Classifiers are also claimed to overcome the limitations of individual classifiers in literature. The suggested system is a hybrid and integrated approach that shows how class imbalance, ensemble learning, and selecting the best feature set can lead to improvement in the performance of fault classification. The ensemble learning approach is intended to be robust to data imbalance as well as feature redundancy as shown in Fig. 2.

The suggested system is a hybrid and integrated approach that shows how class imbalance, ensemble learning, and selecting the best feature set can lead to improvement in the performance of fault classification. The ensemble learning approach is intended to be robust to data imbalance as well as feature redundancy. As a result, experimentation needs to be done to see how ensemble modeling affects SFP. As a result, the conclusion reached is that the goal of this study is to propose developing a model for software defect prediction that will provide more accurate results in determining whether or not a software module is prone to errors, as well as assisting in the discovery of previously unknown problems [3]. This can be accomplished through a thorough examination of the program metrics as well as the dataset's characteristics and further verified on a variety of open-source datasets. To obtain clean data

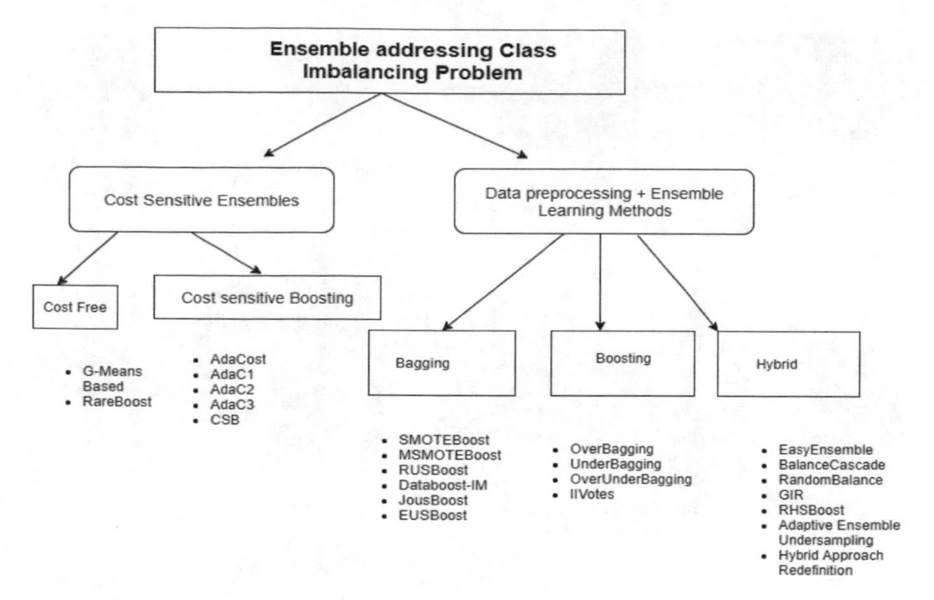

Fig. 2 Ensemble techniques for handling class imbalance problem

that can be adequately evaluated, appropriate features must be excluded from the feature collection. The goal of this study is to combine feature extraction and feature selection methods in order to obtain more accurate findings showing below the experimental framework in Fig. 3. In addition, the research would present cost model tests to reveal patterns in cost behavior on real projects before and after the completion. The proposed methodology may assess the model quality before utilizing it for the final software system.

2 Related Work

In today's time, researchers are persistently working in the direction of improving the performance of software defect prediction models. While implementing the machine learning algorithm its accuracy is dependent on the standard of the historical data which is composed of a number of classes, modules, files, etc., which is further categorized as faulty and non-faulty. Researchers have proposed a number of algorithms and techniques for selecting the optimal subset of features which is discussed in Table 1. They use software measurements to effectively identify problematic software modules; however, they don't consider the skewness in the dataset or other usual statistical properties. Ignoring such attributes will significantly affect the performance and accuracy of classifiers. Number of researches is also being done for ascertaining the supreme learning algorithm but datasets canted nature because of class imbalance problem makes these algorithms subprime.

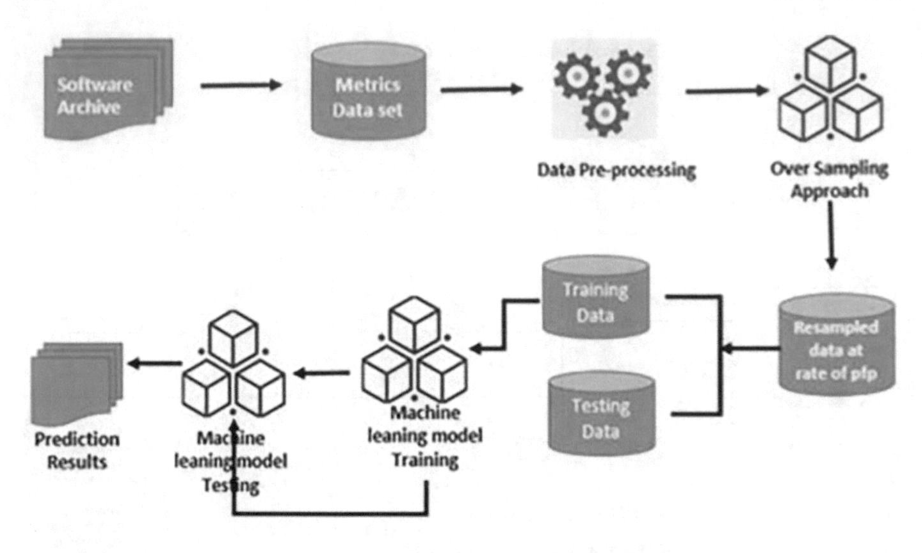

Fig. 3 Experimental framework for SDP

Aggarwal et al. [14] study gives a detailed analysis of metaheuristic algorithms and their binary variations that have been applied to feature selection problems from 2009 to 2019 time period. A complete description and mathematical model of the feature selection problem are provided, which may aid researchers in fully comprehending the topic. Furthermore, the methods for resolving feature selection problems have also been improved as shown in their study but class imbalancing was not present. Oluwagbemiga [15] presents a hybrid multi-filter wrapper technique for feature selection of relevant and irredundant features. They offer a method that takes advantage of the superiority of the relationship between the filter wrapper method and the filter method to produce an ideal subset of features, hence boosting the performance accuracy of software defect prediction models. But their performance surely would be affected as they missed dealing with one of main challenges present in the dataset, i.e., the problem of class imbalance which affects the accuracy, recall, precision of the SDP model so they suggested exploring this problem in future. Raheem et al. [16] study metaheuristic-based Correlation Feature Selection (CFS) techniques employed along with classification algorithms. The Firefly Algorithm (FA) and Wolf Search Algorithm (WSA) were used as metaheuristic search methods in the recommended Feature Selection approaches to evaluate and choose significant qualities. Though the results obtained are very promising, they did not consider class imbalance issues into consideration which further improves the accuracy of prediction. They further suggested experimenting with other metaheuristic algorithms to improve performance of the model. Alweshah et al. [17] introduced two techniques for improving the FS process these are in Exploratory phase mine blast algorithm is used for enhancing feature selection process and secondly in Exploitation phase Mine

Table 1 Literature review table

S. No	Author, year, pub	Title of paper	Contribution	Gaps for further improvements
1	Suresh Kumar, 2021, IEEE" [18]	"Bootstrap aggregation ensemble learning-based reliable approach for software defect prediction by using characterized code feature"	Paper implements bagging algorithm which avoids overfitting and reduces variance	In future they suggested implementing various deep learning algorithms to process the features set and uncovering all defects present in the product
2	Agrawal,2021, IEEE [14]	"Metaheuristic algorithms on feature selection: A survey of one decade of research (2009–2019)"	Paper examines metaheuristic algorithms and their fluctuating binary variations employed resolving the problem of selecting best feature sets by various authors	Gap is a classification task in which participants have to employ a variety of classifiers and compare them to the most often used ones
3	Oluwagbemiga, 2020, IEEE [15]	"A hybrid multi-filter wrapper feature selection method for software defect predictors"	Study presents using multi-filter wrapper feature selection hybrid framework for dealing with defects in the software for selecting features having irreducible and analytical properties	In future finding more good ways to combine filter and wrapper-based feature selection is still in progress and planning to investigate other data quality issues such as imbalance in class affecting SDP
4	Goyal, 2020, IEEE [19]	"Heterogeneous stacked ensemble classifier for software defect prediction"	Study shows how stacked ensemble improves the SDP classifier's accuracy and prediction capacity by addressing dataset class imbalance	Work can be reproduced for different datasets using Deep learning structures in future

(continued)

Table 1 (continued)

S. No	Author, year, pub	Title of paper	Contribution	Gaps for further improvements
5	Oloduowo, 2020, IEEE [16]	"Software defect prediction using metaheuristic-based feature selection and classification algorithms"	Study employed metaheuristic-based correlation feature selection techniques in conjunction with classification algorithms for evaluating and choosing best features metaheuristics search methods firefly algorithm and wolf search algorithm were used	Other feature selection metaheuristics, as well as other unique classification methods, could be researched in the future to improve the SDP models' performance
6	Alweshah, 2020, Springer [17]	"A hybrid mine blast algorithm for feature selection problems"	Study proposes the FS process in this research firstly exploratory phase mine blast algorithm and secondly in exploitation phase Mine Blast Algorithm (MBA) is blended with simulated annealing to improve solutions given by MBA	Future work could be to build a Hybridize model using MBA along with other metaheuristic algorithms and classifiers other than KNN like SVM, etc.
7	Alsawalqah et al.,2020, IEEE [20]	"Software defect prediction using heterogeneous ensemble classification based on segmented patterns"	Research proposes hybrid classification techniques for predicting failures. The suggested strategy is developed and tested in two ways, i.e., Simple classifiers first, while ensemble learning used in the second	Two areas where future work could be done to improve things are firstly to investigate more complex clustering methods, while the second is to find approaches for automatically finding the optimal number of clusters for each dataset

(continued)

Table 1 (continued)

S. No	Author, year, pub	Title of paper	Contribution	Gaps for further improvements
8	Jiang et al. 2020, MDPI [21]	"Heterogeneous defect prediction based on transfer learning to handle extreme imbalance"	Study shows the Grassmann manifold optimal transfer defect prediction technique (GMOTDP), which combines sampling with the majority method (swim) oversampling, feature selection	In future effects of high-class imbalance in semi-supervised software defect predictors on larger datasets will be seen, as well as observing how to combine other supervised learning approaches with sample and algorithm level methods and proposing transfer learning too
9	Chakraborty and Chakraborty, 2020, IEEE [22]	"Hellinger net: a hybrid imbalance learning model to improve software defect prediction"	Works on mapping of tree to network technique which is similar to deep feed forward neural network for which special distance measure is used in handling this problem known as Hellinger distance	Model is applied on Nasa Datasets only and they didn't take how cost get affected with so early defects detection Further they suggested extension of this work is to apply on other domains and to check semi-supervised settings also

(continued)

Table 1 (continued)

S. No	Author, year, pub	Title of paper	Contribution	Gaps for further improvements
10	Bejjanki et al. in 2020 [23]	"Class Imbalance Reduction (CIR): A novel approach to software defect prediction in the presence of class imbalance"	Proposes the class imbalance reduction algorithm to handle the imbalance between defective and non-defective records in the datasets which is further compared with smote and K-Means	Further accuracy can be improved by implementing optimization techniques like ant colony optimization, feature subset selection, and proposed work performance need to be experimented in cross project defect prediction too
11	Wei et al., 2019, Elsevier [24]	"Establishing a software defect prediction model via effective dimension reduction"	Proposed a unique model by using a "Local tangent space alignment support vector machine (LTSA-SVM)" technique for solving the defect prediction problem in software	Future involves using a more effective method, such as "Gaussian kernel or incremental LTSA" initiating the mapping between the high and low dimension space, improving validity of feature sample extraction, and consideration of the autocorrelation of process data during prediction

(continued)

Table 1 (continued)

S. No	Author, year, pub	Title of paper	Contribution	Gaps for further improvements
12	Khuat and Le, 2019, IEEE [25]	"Ensemble learning for software fault prediction problem with imbalanced data"	Research learns more about how the various techniques of sampling on getting imbalance data as input affects the accuracy of ensemble classifiers	····Future involves diversification of ensemble classifiers with a number of rules designed employing various sampling techniques and changing the strategies for sampling to use with binary classification models
13	Cai et al. 2019, IEEE [26]	"An under-sampled software defect prediction method based on hybrid multi-objective cuckoo search"	In this firstly a multi-objective under sampled technique cuckoo search based on support vector machine is suggested after that 3 under sampled methods were used for selecting non-defective modules	Algorithm they proposed is more compatible for software having limited resources and of high risks, one of their drawbacks. Comparison of the model with the most powerful XGBoost is also not being performed
14	NezhadShokouhi et. al in 2019, IEEE [27]	"Software defect prediction using over-sampling and feature extraction based on mahalanobis distance"	Proposes a method for handling class imbalance and feature selection problems known as Mahalanobis distance	Drawback is paper using traditional metrics like Halstead and Mc Cabe's cyclomatic complexity which are not based on Gaussian distribution so need for good metrics selection arises

(continued)

Table 1 (continued)

S. No	Author, year, pub	Title of paper	Contribution	Gaps for further improvements
15	Ghosha, 2018, IEEE [28]	"A nonlinear manifold detection based model for software defect prediction"	Study proposes a novel and dynamic methodology for dealing with datasets containing noisy attributes and huge dimensions based on "Nonlinear Manifold Detection Techniques"	In future suggested unique models will be compared to other Nonlinear MDTs to analyze performance and further check the effect of other Machine Learning methods
16	Miholca, 2018, IEEE [29]	"An improved approach to software defect prediction using hybrid machine learning model"	*HyGRAR*, a hybrid defect prediction model, integrating Gradual Relational Association Rules (GRARs) with ANNs	Model should be improved by leading the predictions of independent MLPs based on their performance on the data set being attested
17	Ali Nawaz et. al (2018) [30]	" A novel multiple ensemble learning models based on different datasets for software defect prediction"	Proposed an ensemble learning model on Promise Datasets showing an excellent classification accuracy in comparison with other single machine learning algorithms like SVM, KNN, Decision Tree, etc.	Future work should focus on developing a general ensemble learning model for the rest of the datasets accessible in the Promise repository, with 97% or higher accuracy
18	Tomar and Agarwal 2016, IEEE [31]	"Prediction of defective software modules using class imbalance learning"	An early prediction of a faulty software module can be generated using a weighted least squares twin support vector machine, according to studies	Selecting relevant features, class imbalance, parameter selection in future for improving accuracy

Blast Algorithm (MBA) is blended with simulated annealing to improve solutions given by MBA. They clearly show that MBA alone does not give good accuracy as it shows when it is hybridizing with other metaheuristic algorithms and strong classifiers. Moreover, the weakness in this implementation is the methods were applied to flat files but the real-world complex problems often handle gigabyte-sized databases.

Nawaz et al. [30] developed an ensemble learning model and compared them with other single machine learning algorithms like SVM, KNN, Decision Tree, Random Forest on Promise Datasets showing an excellent classification accuracy among all. They tested the accuracy on a few datasets and advised that in the future should focus on developing a general ensemble learning model for the rest of the datasets accessible in the Promise repository, with an accuracy of around 97% or higher. Goyal and Bhatia [19] implemented a Heterogeneous stacked ensemble method and compared it with a number of baseline models showing that the model they proposed performs better than that. They stacked three weak and two strong classifiers to improve the prediction but what is lacking in their model is noise filtering and feature selection is not being performed. They have also not shown how defect prediction in the early phase will affect the cost reduction in further phases of software development. Alsawalqah [20] implemented a hybrid classification technique for predicting failures in software. The suggested strategy is developed and tested in two ways. Simple classifiers are used in the first, while ensemble learning is used in the second. They concluded that the ensemble approach performs well as compared to the approach implemented by simple classifiers. Paper did not focus on class imbalance issues and feature subset selection which affects performance further. They further suggested exploring more advanced algorithms in clustering and determining how many clusters should be selected from the dataset for further improvement. Jiang et al. [21] show the Grassmann manifold optimal transfer defect prediction technique (GMOTDP), which combines Sampling with the Majority method (SWIM) oversampling, feature selection, and proposal of transfer learning is also being done in this study as a three-phase heterogeneous software prediction method. The method performs better no doubt but noise handling was missing in their study. They also throw light on class imbalance issues related to semi-supervised and supervised learning as it strengthens the accuracy of these algorithms.

NezhadShokouhi et al. [27] proposed a method for handling class imbalance and feature selection problems known as Mahalanobis distance. Class imbalancing is dealt with by producing diverse synthetic samples for minority class and feature subset selection using metric learning by Mahalanobis distance in which sample data is projected to a new space which is further evaluated with a series of experiments on 12 datasets from Nasa repository. Drawback in this paper is that researchers have used traditional metrics like Halstead and Mc Cabe's cyclomatic complexity which are not based on Gaussian distribution. So, there is a need for good metrics selection from the code itself. Bejjanki et al. in 2020 [23] suggested the class imbalance reduction algorithm to handle the imbalance between defective and non-defective records in the datasets which is further compared with Smote and K-Means Smote and tested on forty open-source software defect datasets using eight different classifiers and evaluated using six performance measures. But accuracy can be improved further as they suggested for further optimization techniques like ant colony optimization, feature subset selection is also not implemented and proposed work performance needs to be experimented in cross project defect prediction for proving its performance.

Weiad et al. [24] use Local tangent space alignment support vector machine (LTSA-SVM) technique for solving the defect prediction problem in software by proposing a unique model which improves data redundancy problem. But this study suffers from cost time problem as training of samples need to be done number of times. They further recommended finding the effective approach for creating mapping between low dimension and high-dimensional space and suggested considering autocorrelation among the data while predicting defects in the software. Khuat and Le [25] implement the integration of various sampling techniques and some of the popular classification techniques to form an ensemble model solving the issues of class imbalance to improve software defect prediction problem. But Feature set Selection algorithm are missing in this paper and further they leave research to apply different types of other oversampling techniques and new sampling strategies that can play an important role in improving the accuracy and performance of SDP model. Ghosha et al. [28] study a novel and dynamic methodology for dealing with datasets containing noisy attributes and huge dimensions based on "Nonlinear Manifold Detection Techniques (Nonlinear MDTs)" is proposed. How to deal with class imbalance problems is not taken into consideration in this study which makes this model more novel and in today's time many advanced machine learning techniques are present which reduces the time and cost further and handles problems of defects in software efficiently. Miholca [29] proposes *HyGRAR*, a hybrid defect prediction model, integrating "Gradual Relational Association Rules" with ANNs. The limitation of this approach is that it is not adaptive. Class imbalance and feature selection issues are not considered in this study which affects the performance and accuracy of the model. Tomar et al. [31] make an early prediction of a faulty software module generated using a Weighted Least Squares twin support vector machine, according to studies. They missed implementing feature selection techniques which also affects the accuracy of prediction models. Cai et al. [26] developed a hybrid model solving the problem of class imbalance and parameter selection to predict software defect prediction. In this firstly a multi-objective under sampled technique cuckoo search based on support vector machine is suggested after that 3 under sampled methods were used for selecting non-defective modules. In this study, they used three simulation indicators only false positive rate (pf), detection of probability (pd), G-mean leaving others like AUC, Log Loss, and F1-Score which gives a more real picture about the precision and accuracy. Moreover, the algorithm they proposed is more compatible for software having limited resources and of high risk though they are solving the problem synchronously which is one of their drawbacks. Comparison of the model with the most powerful XGBoost is also not being performed. Chakraborty et al. [22] proposed a novel hybrid model known as Hellinger net for handling class imbalance problems while predicting defects in the software. It works on mapping of tree to network technique which is similar to deep feed forward neural network. They used a special distance measure in handling this problem known as Hellinger distance. They missed focusing on dimension reduction. Model is applied on Nasa Datasets only and they didn't take into account how costs get affected if defects get predicted so early. The extension of this work that they suggested is to apply on other domains and to check semi-supervised settings also. Kumar et al. [18] showed how to improve software defect prediction using Bagging method which is an ensemble learning technique that proved out to be one of the most effective predictors. Their implementation technique avoids the problem of overfitting which leads to reduction in variance. As a research gap, they have not considered feature selection techniques while performing the prediction which can further enhance the capability of increasing the accuracy of this approach and more over there are many more advanced machine learning techniques like deep learning algorithms that can be implemented to analyze the attributes of software and detect faults with more accuracy in the future. Paper published by Bahaweres et al. [32] discussed the solution of the class imbalance problem, and the researchers used Smote along with Neural Networks in blended form, and further optimized each hyperparameter of these blended form frameworks by implementing the technique of searching randomly. However, there are some caveats: the hyperparameter optimization introduced in this study is random search, which does not allow an algorithm to reach an optimal state because of which the work gets further affected.

3 Conclusion and Future Work

Software defect prediction is the most critical and significant activity in the software development process for improving software quality. Timely identification of software defects in the initial phases of development results in enormous savings in terms of time, cost, and effort. As the complexity of the software continues to increase, a wide range of issues arise while performing software defect prediction, such as feature selection, noise filtering, and class imbalance to name a few. Traditionally, researchers have offered numerous ways for dealing with such issues by adopting machine learning techniques both in raw form and in ensemble form. There are still many challenges to be undertaken for improvement of various software performance metrics by adopting metaheuristic algorithms, combining different supervised learning approaches with sample and algorithm level methods; and applying optimization techniques like ant colony optimization. It is equally important to maintain the restrictions related to budget and time. This study is an attempt to summarize the defect prediction techniques used by developers and researchers globally and identify the research gaps for further improvement.

References

1. Dam HK, Pham T, Ng SW, Tran T, Grundy J, Ghose A, Kim T, Kim CJ (2019) Lessons learned from using a deep tree-based model for software defect prediction in practice. In: Proceedings of 16th international conference on mining software repositories (MSR). IEEE, Montreal, QC, Canada, pp 46–57
2. Jin C (2021) Software defect prediction model based on distance metric learning. Soft Comput:447–461. Springer
3. Shenvi AA (2021) Defect prevention with orthogonal defect classification. In: Proceedings of the 2nd India software engineering conference. pp 83–88
4. Yu L, Liu H (2004) Efficient feature selection via analysis of relevance and redundancy. J Mach Learn Res:1205–1224
5. Gao K, Khoshgoftaar TM, Wang H, Seliya N (2020) Choosing software metrics for defect prediction: an investigation on feature selection techniques. Softw: Pract Exp 41(5):579–606. Wiley Online Library
6. Rodriguez D, Ruiz R, Cuadrado-Gallego J, Aguilar-Ruiz J, Garre M (2007) Attribute selection in software engineering datasets for detecting fault modules. In: 33rd EUROMICRO Conference on software engineering and advanced applications, EUROMICRO 2007. IEEE, pp 418–423
7. Peng Y, Wu Z, Jiang J (2010) A novel feature selection approach for biomedical data classification. J Biomed Inform 43(1):15–23. Elsevier
8. Sharma D, Chandra P (2018) Software fault prediction using machine-learning techniques. In: Smart computing and informatics. Springer, Singapore, pp 541–549
9. Meiliana SK, Warnars HLHS, Gaol FL, Abdurachman E, Soewito B (2017) Software metrics for fault prediction using machine learning approaches: A literature review with PROMISE repository dataset, In: 2017 IEEE international conference on cybernetics and computational intelligence (CyberneticsCom). IEEE, pp 19–23
10. Ran L, Zhou L, Zhang S, Liu H, Huang X, Sun Z (2019) Software defect prediction based on ensemble learning. In: Proceedings of the 2019 2nd International conference on data science and information technology. pp 1–6
11. Matloob F, Aftab S, Iqbal A (2019). A framework for software defect prediction using feature selection and ensemble learning techniques. Int J Mod Educ & Comput Sci 11 (12)
12. Khan MZ (2020) Hybrid ensemble learning technique for software defect prediction. Int J Mod Educ & Comput Sci 12(1)
13. Matloob F, Ghazal TM, Taleb N, Aftab S, Ahmad M, Khan MA (2021) Software defect prediction using ensemble learning: a systematic literature review. IEEE Access:98754–98771
14. Agrawal P, Abutarboush HF, Ganesh T, Mohamed AW (2021) Metaheuristic algorithms on feature selection: a survey of one decade of research (2009-2019). IEEE Access 9:26766-26791
15. Oluwagbemiga BA, Shuib B, AbdulKadir SJ, Sobri AH (2019) A hybrid multi-filter wrapper feature selection method for software defect predictors
16. Raheem M, Ameen A, Ayinla F, Ayeyemi B (2020) Software defect prediction using metaheuristic algorithms and classification techniques. Ilorin J Comput Sci Inf Technol 3(1):23-39. Bennin KE, Keung J, Phannachitta P, Monden A, Mensah S (2017) Mahakil: Diversity based oversampling approach to alleviate the class imbalance issue in software defect prediction. IEEE Trans Softw Eng 44(6):534-550
17. Alweshah M, Alkhalaileh S, Albashish D, Mafarja M, Bsoul Q, Dorgham O (2021) A hybrid mine blast algorithm for feature selection problems. Soft Comput 25(1):517–534
18. Suresh Kumar P, Behera HS, Nayak J, Naik B (2021) Bootstrap aggregation ensemble learning-based reliable approach for software defect prediction by using characterized code feature. Innov Syst Softw Eng:1–25
19. Goyal S, Bhatia PK (2020) Heterogeneous stacked ensemble classifier for software defect prediction. In: 2020 Sixth International conference on parallel, distributed and grid computing (PDGC). pp 126–130. IEEE

20. Alsawalqah H, Hijazi NM, Eshtay M, Faris H, Radaideh AA, Aljarah I, Alshamaila Y (2020) Software defect prediction using heterogeneous ensemble classification based on segmented patterns. Appl.Sci 10(5):1745
21. Jiang K, Iwahori Y, Wang A, Wu H, Zhang Y (2020) Heterogeneous defect prediction based on transfer learning to handle extreme imbalance. Appl Sci 10(1):396
22. Chakraborty T, Chakraborty AK (2020) Hellinger net: A hybrid imbalance learning model to improve software defect prediction. IEEE Trans Reliab, IEEE 70(2):481-494
23. Bejjanki KK, Gyani J, Gugulothu N (2020) Class imbalance reduction (CIR): a novel approach to software defect prediction in the presence of class imbalance. MDPI, Symmetry 12(3):407
24. Wei H, Hu C, Chen S, Xue Y, Zhang Q (2009) Establishing a software defect prediction model via effective dimension reduction. Inf Sci 477:399-409
25. Khuat TH, Le MH (2019) Ensemble learning for software fault prediction problem with imbalanced data. Int J Electr Comput Eng 9(4):3241
26. Cai X, Niu Y, Geng S, Zhang J, Cui Z, Li J, Chen J (2019) An under-sampled software defect prediction method based on hybrid multi-objective cuckoo search. Concurr Comput: Pract Exp 32(5):e5478
27. NezhadShokouhi MM, Majidi MA, Rasoolzadegan A (2020) Software defect prediction using over-sampling and feature extraction based on Mahalanobis distance. J Supercomput 76(1):602-635
28. Ghosh S, Rana A, Kansal V (2018) A nonlinear manifold detection-based model for software defect prediction. Procedia Comput Sci 132:581–594. Elsevier
29. Miholca DL (2018) An improved approach to software defect prediction using a hybrid machine learning model. In: 2018 20th International symposium on symbolic and numeric algorithms for scientific computing (SYNASC), IEEE. pp 443–448
30. Nawaz A, Rehman AU, Abbas M (2020) A novel multiple ensemble learning models based on different datasets for software defect prediction. arXiv preprint arXiv: 2008.13114. Cornell University
31. Tomar D, Agarwal S (2016) Prediction of defective software modules using class imbalance learning, Applied Computational Intelligence and Soft Computing Volume 2016. Article ID 7658207:6
32. Bahaweres RB, Agustian F, Hermadi I, Suroso AI (2020) Software defect prediction using neural network based SMOTE. In: 7th international conference on electrical engineering, computer sciences and informatics (EECSI)

Chapter 3
Crop Recommendation Using Machine Learning Algorithms and Soil Attributes Data

Ritesh Ajoodha and Takalani Orifha Mufamadi

1 Introduction

Agricultural activity plays an important and significant role in the economies of most countries, especially in less developed countries where the contribution of agricultural activity towards the economy is substantial [1]. Hence, high levels of production in the agricultural sector is of utmost importance for the economic well-being of most countries and the welfare of their people [1–3]. This is indeed the case for most countries in Africa including South Africa and it's neighbouring countries.

Besides the need to increase (maximise) productivity, there is also a variety of factors such as climate change [3], reduction of farmland due to population growth [4], and globalisation [2, 4] that lead to the need for farmers to adapt by changing the crops they choose to grow. The greatest challenge for most farmers and in particular subsistence farmers whose contribution in the agricultural sector is a significant one is being able to decide what to grow and where [4–6].

In other words, farmers would be more productive if they knew what types of crops or livestock to grow depending on their location and its attributes, as well as other environmental factors that affect farm productivity. Which would result in the improvement of the food security of a country and it's overall economic well-being.

Our main aim is to come up with a crop recommendation (CR) system that will help farmers select crops that will produce the greatest yield for that particular soil type. This CR system should be able to recommend a crop based only on the attributes of the soil, which would make it more accessible to most farmers because of ease of access to such data. An inaccurate CR system would result in great losses for farmers, the agricultural sector, and the economy as a whole if the CR system is adopted by

R. Ajoodha · T. O. Mufamadi (✉)
University of the Witwatersrand, Johannesburg, RSA GP 2001, South Africa
e-mail: 1100162@students.wits.ac.za
URL: https://www.wits.ac.za/

G. Mathur et al. (eds.), *Proceedings of 3rd International Conference on Artificial Intelligence: Advances and Applications*, Algorithms for Intelligent Systems,
https://doi.org/10.1007/978-981-19-7041-2_3

a significant number of farmers. Therefore, it is crucial that the CR system is highly accurate to avoid such losses.

Other authors have incorporated more information about the soil (or land) and other environmental factors that affect crop growth the development of their CR systems, which yields higher accuracy than the model we develop. However, this also makes their models more complicated which makes it difficult for most farmers to adopt (use) these models. Therefore, our research aims to develop a CR system that can suggest crops to farmers using only the soil attributes.

Guided by the literature, we use machine learning algorithms, including SVM and RF, to develop a CR system with high accuracy by just considering the three main soil nutrients, namely, nitrogen, potassium, and phosphorus; as well as soil pH level. This would mean that farmers would be able to use the CR system with minimum resources required because this information is more readily available.

The main contributions of our study to the body of literature are: (a) we develop a high accuracy CR system that is more accessible to farmers using only soil attributes such as the three main soil nutrients, namely, nitrogen, potassium, and phosphorus; as well as soil pH level. (b) We compare various machine learning algorithms, including SVM and RF algorithms, to see which one is best suited for developing a CR system using only soil attribute data. (c) We compare the power of the features we use in this study with that of other previously used features to establish which features will yield higher accuracy.

This research report is structured as follows: we begin by introducing the topic in Sect. 1. In Sect. 2 we provide an overview of the literature related to our study. Following this, we highlight and discuss the methodology and data used to implement the our research in Sect. 3. Then in Sect. 4 we highlight and discuss the results obtained using the methodology highlighted in Sect. 3. Section 5 is the final chapter, wherein, we summarise the main points and findings discussed in this study as well as potential future work.

2 Background and Related Work

2.1 Precision Agriculture (PA)

Researchers have over the years attempted to come up with ways that can assist farmers to make better crop choices to maximise their yield using precision agriculture (PA) [1, 2].

The maximisation of farm productivity can be achieved by implementing PA, which could have a great impact on the well-being of farmers, the agricultural sector, and the economy as a whole. PA is a modern farming technique that uses research data and algorithms to suggest the most appropriate crop to grow and/or livestock to raise based on parameters that are specific to the farmer's location [2, 7, 8].

Many authors have developed various PA models that attempt to assist farmers in choosing the best crop for their soil based on the attributes of the soil [2, 7, 8]. The majority of these PA models that have been developed over the years recommend crops for particular land based on the attributes of the soil, and can be referred to as crop recommendation (CR) systems [9].

2.2 Recommendation Systems (RSs)

CR systems are part of a broader category of systems called recommendation systems (RSs), that are used to recommend items to users. RSs are programmes that are able to recommend or suggest the best items to specific users based on the history of the user, information regarding the items, and the interaction between the user and items [9]. Such systems use a variety of data processing methods to analyse the user's history and similar users' behaviour to be able to predict the interest of the user.

There are three categories of RSs, namely, content-based, collaborative filtering, and hybrid; CR systems belong to the content-based category of recommendation systems. RS that are content based suffer from the problem of new user cold-start. What this entails for CR systems is that they are unable to recommend a crop if the crop characteristics are not available [9]. Various authors have attempted to remedy this problem in order to yield better accuracy [10–13]. They use opinion mining methods as well as neural networks to circumvent the cold-start problem associated with content-based RS. For our future work we will consider using similar methods for our CR system to circumvent the cold-start problem.

RSs play a major role in multiple areas of application such as e-commerce wherein items that could be interesting to the target user are recommended. Perhaps one of the most important areas of application of RSs is in PA for the development of CR systems [9].

2.3 Crop Recommendation (CR) Systems

The development of a crop recommendation system at any level requires information and knowledge about the crops and the soil in which crops are grown as well as other environmental factors which affect crop growth such as temperature, precipitation and humidity [1, 2, 7, 9]. The soil attributes that can be considered are the soil nutrients, soil pH level and soil colour, to name a few [2, 6–8, 14]. When it comes to the sample of crops to be used, we can choose the crops that are most dominant in the location being studied [2, 7, 14].

Some CR systems are machine learning-based models [2, 7, 15, 16], some are probabilistic models [3], others are based on artificial intelligence [8], and so on. The most prevalent models amongst these are machine learning based models, which use machine learning algorithms to solve the classification task at hand, namely, soil classification.

The scientific field that gives learning ability to machines without the need for strict programming is referred to as machine learning [17]. Machine learning-based models are preferred for the development of CR systems because they provide the best (most accurate) soil classification regardless of the performance measures that are used to evaluate the performance of algorithms [14].

Naive Bayes (NB) and k-nearest neighbours (K-NN) can be used for soil classification [16] which allows them to predict crop yield using the data that is available. They also suggest that classification algorithms like support vector machines can be used to create models that are more efficient. This was confirmed by [15] who use kernel-based support vector machines (SVM) to develop their CR system and report an accuracy of 94.95%. However, [2] found that random forests (RF) based models performed better than their SVM counterparts when it comes to soil classification. They report an accuracy of 75.73 and 86.35% for SVM and RF, respectively.

2.4 Evaluation Criteria

Various evaluation parameters are used to rank the accuracy of the models and how effective they are. These performance measures can be categorised into two broad categories, namely, offline and online assessment [9]. Offline assessment uses offline data to assess the recommendation method's efficiency and reliability whereas online assessment uses real-time data for this purpose [9].

Examples of offline performance measures for the evaluation of CR systems include: standard deviation, standard error, root mean square error (RMSE), mean square error (MSE), recall and precision, cross-validation tests, and train-test analysis [9]. Conversion rate, A/B testing, and click-through rate (CTR) are examples of online performance measures [9].

For the purposes of this study, we will focus only on offline methods of model evaluation since offline data is more readily available than real-time data. Using this approach instead of the alternative should not have any significant impact on the results we obtain, and thus, our conclusions are not affected by the evaluation criteria we choose.

In this chapter we have gathered that we can be able to help farmers increase their productivity by implementing PA [1, 15]. The development of a CR system is one way of implementing PA that could help crop farmers increase their productivity [8, 9]. What we also learned from the above discussion is that information about the soil is the most important information we will need to develop our CR system [18–20].

We also gathered from the chapter that machine learning algorithms were better suited for the development of a CR system [9, 21, 22] and that, in particular, the RF algorithm is best suited for soil classification [2]. This insight was gained by ranking the models and algorithms based on the various evaluation parameters that are commonly used in the literature as discussed above. Another insight we obtained, is that, CR systems suffer from the cold-start problem of content-based RSs which can be solved using neural networks and opinion mining [9].

3 Methodology

The methodology used in this study is similar to that of [2, 6] who use machine learning algorithms for soil classification, which is the basis of our CR system. Reference [2] suggests that the RF machine learning algorithm, in particular, is best suited for soil classification. Following this, we develop our CR system using RF algorithm as well as other machine learning algorithms including SVM.

These algorithms are implemented on soil data that includes information about the nutrients in the soil and the pH level of the soil. The reason for this is that the development of a CR system requires information about factors that affect crop growth, for example, soil attributes. The various offline measures that we use to assess the accuracy and the performance of our CR system are borrowed from [9]

The rest of the section is structured as follows: in Sect. 3.1 we discuss the data and features that are required for the implementation of crop recommendation (CR) systems. In Sect. 3.2 we discuss the models that are used to develop the CR system. The performance measures that are used to evaluate the performance of the developed CR systems are discussed in Sect. 3.3.

3.1 Data and Features

To develop our crop recommendation system we use a cross-sectional data-set that was sourced from Kaggle [23] and consists of soil attributes as well as other environmental factors that affect crop growth, namely, temperature, humidity, and precipitation. There are no ethical considerations that we had to undertake when obtaining the crop and soil data used in this study.

The following soil attributes are included in the data-set: soil pH, phosphorus, potassium, nitrogen, and magnesium. Soil pH (level of acidity or alkalinity) is also included in the data-set because it affects the availability of nutrients in the soil. It is a master variable that can affect the level of exchangeable aluminium in the soil and also the activity of micro-organisms present in the soil.

Temperature, humidity, and precipitation were found to have the greatest power for crop recommendation and models that include these features yield higher accuracy. However, this also makes the model more complicated and less accessible to farmers. As a results we do not use these features in this research which means that our model can be adopted by most farmers. We also exclude magnesium since it does not have a great impact on plant growth for plants grown here in South Africa. Thus, for our final data-set we only remain with the soil pH as well as the three main soil nutrients.

Table 1 Definition or descriptions of the features

Variable	Definition or descriptions
1. Nitrogen (N)	The atomic element that is responsible for photosynthesis in the plant
2. Phosphorus (P)	The atomic element that plays a major role in a crop to store and transfer energy for growth of the crop
3. Potassium (K)	The atomic element that is required for reproduction of crops
4. pH level (pH)	The level of alkalinity or acidity

In Table 1, we provide the definition or descriptions of the features (soil attributes) we use for our proposed research. Most of these attributes are described in terms of the role they play in overall crop growth in order to highlight their importance in our research.

Our crop selection criteria is based on crops that are important in the location of study, namely, South Africa. The crops included in the original crop data are maize, rice, banana, mango, grapes, watermelon, apple, orange, papaya, coconut, cotton, jute, coffee, muskmelon, lentil, black-gram, kidney beans, pigeon beans, mung beans, moth beans, and pomegranate. Therefore, we select the following crops from the original crop data: maize, kidney beans, banana, mango, grapes, watermelon, apple, orange, papaya, and cotton. These crops are the ones that are commonly grown in South African and have the greatest contribution to the agricultural sector.

3.2 Machine Learning Algorithms

In the subsections below, we provide a brief overview of SVM and RF. We focus mainly on SVM and RF because the literature suggests that they are the best performing algorithms when it comes to soil classification [2, 15]. From our results we see that between these two algorithms, RF outperforms SVM when it comes to soil classification. Other algorithms that have been used in the literature include K-NN, K-Star, and NB to name a few. In this study we will compare the performances of some of these algorithms to establish which ones are best suited for soil classification. We also can see from our results that there exist other machine learning algorithms that perform better than those suggested by the literature which contradicts the finding suggested by the literature mentioned above.

SVM It uses a set of hyper-planes or a single hyper-plane for different errands, relapse, or characterisation. These hyper-planes are developed in boundless or high-dimension space. SVM calculation stays away from over-fitting by having a regularisation parameter. It also uses portion traps which allows us to develop master learning regarding the issue [2].

RF The RF algorithm belongs in the category of supervised machine learning algorithms and can be used for classification and regression problems. This algorithm is based on ensemble learning and one of its advantages is that it has multiple trees and a subset of the data is used for each tree which makes it unbiased. The advantage of RF algorithm is that the introduction of a new data point in the data-set does not affect the overall algorithm which makes it stable [2].

3.3 Evaluation Parameters

In the subsections below, we provide a quick overview of some of the online evaluation parameters that were mentioned earlier (Sect. 2). See [9] for the definitions of RMSE, MAE, and RAE.

Train-test analysis When it comes to train-test analysis the data is split into data for training and data for testing which are used separately. The training data (data used for training the model) will usually consist of between 70 and 80% of the original data. The testing data is used for model evaluation and would consist of the remaining 20–30% of the original data [9]. For this study we use 70% of the original data as our training data and use the remaining 30% for testing purposes.

Cross-validation tests Cross-validation tests can be thought of as an n-folds repetition of train-test analysis because the same process of splitting data into train and test data that is used in train-test analysis is repeated n-times. At each fold the accuracy of the is assessed for all the folding models and these accuracies are ultimately used to determine the final accuracy of the model [9]. In this study we use tenfold cross-validation.

Recall and precision Precision and recall are the other measures of accuracy that are widely used to evaluate the accuracy of crop recommendation systems by identifying the number of labels that are classified correctly compared to the actual class. The proportion of important labels that are in the top-k set is known as precision at k (nearest neighbours). The percentage of important labels that are in the list of recommendations out of total important labels is the recall at k [9].

4 Results and Discussion

In this section we highlight analyse discuss the results we obtained using the methodology and data discussed in the preceding section.

Table 2 shows the results of RF and SVM-based CR systems using the tenfold cross-validation test. We have included the results of other machine learning algorithms for comparison purposes, and we can see that neither the SVM algorithm nor the RF algorithm is the best algorithm for soil classification. The best performing machine learning algorithm is the K-Star algorithm whose accuracy is 91.3%, which is significantly greater than the 88.6% accuracy of SVM algorithm.

Table 2 Evaluation of the RF and SVM models

Variable	Precision	Recall	RMSE	MAE	RAE (%)
1. RF model	0.911	0.911	0.1044	0.0197	10.9522
2. SVM model	0.886	0.884	0.2731	0.1605	89.1778
3. K-NN model	0.910	0.910	0.1335	0.0198	10.989
4. K-Star model	0.913	0.913	0.1037	0.024	13.309
5. NB model	0.906	0.906	0.1028	0.0213	11.8521

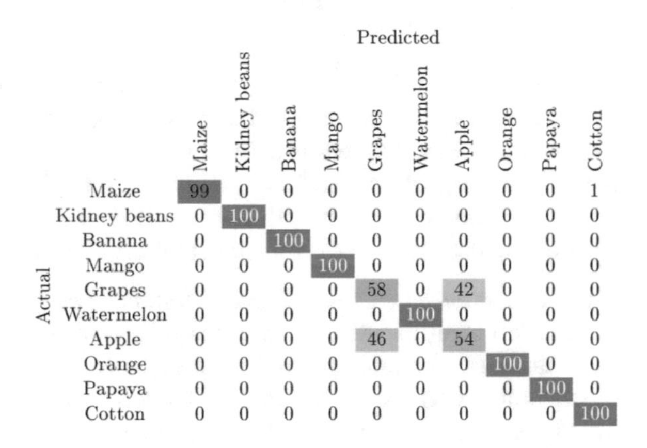

Fig. 1 Random forest (RF)—91.1% Accuracy, Correctly identified 911/1000 labels

However, it is very close to the 91.1% accuracy of RF algorithm which means that we can conclude that the RF algorithm performs just as well as the best performing algorithm. Therefore, RF algorithm is well suited for soil classification as suggested by the literature, and as such we can use it for the development of our CR system. Using precision and recall as our evaluation criteria we can see that the RF algorithm has a recall of 0.911 and a precision value of 0.911. On the other hand, SVM has a recall value of 0.884 and a precision value of 0.886.
The general theme across all evaluation parameters including RMSE and MAE is that RF-based models yield more favourable results compared to their SVM counterparts. In particular, when using RAE as the evaluation parameter, we see that the RF model (RAE value of 10.9522%) outperforms the SVM model (RAE value of 89.1778%) by a significant margin.

From the two confusion matrices above (i.e. Figs. 1 and 2) we can see that the RF-based model yields an accuracy of 91.1% whereas the SVM-based model yields an accuracy of 88.4%. Our results are in agreement with those of [2] who found that RF models are better than SVM models when it comes to soil classification.

The main purpose of our CR system is to classify soil based on its attributes and determine which crops are suitable for that particular soil. By using the RF algorithm, we were able to develop a model with an accuracy of 91.1% that only considers the

Fig. 2 Support vector machines (SVM)—91.1% Accuracy, Correctly identified 911/1000 labels

Actual \ Predicted	Maize	Kidney beans	Banana	Mango	Grapes	Watermelon	Apple	Orange	Papaya	Cotton
Maize	99	0	0	0	0	0	0	0	0	1
Kidney beans	0	100	0	0	0	0	0	0	0	0
Banana	0	0	100	0	0	0	0	0	0	0
Mango	0	0	0	100	0	0	0	0	0	0
Grapes	0	0	0	0	58	0	42	0	0	0
Watermelon	0	0	0	0	0	100	0	0	0	0
Apple	0	0	0	0	46	0	54	0	0	0
Orange	0	0	0	0	0	0	0	100	0	0
Papaya	0	0	0	0	0	0	0	0	100	0
Cotton	0	0	0	0	0	0	0	0	0	100

three main soil nutrients, namely, nitrogen, phosphorus, and potassium; as well as the pH level of the soil.

The RF algorithm-based CR system we have developed is of high accuracy relative to previous implementations of such a CR system, for example, [2]. Especially considering that we only considered the three main soil nutrients of the soil as well as the pH level in the development of our CR system. Incorporating more information about the soil or other environmental factors that affect the soil could increase the accuracy of our system, but it would also make our system more complicated and less accessible.

An example of a model that uses more features and yields better accuracy is that of [15] which considers all the features used in our model as well the following chemical attributes: zinc, sulphur, boron, calcium, magnesium, iron, copper, manganese, and organic matter %. This model yields a higher accuracy of 94.95% when comparing it to our model whose accuracy is 91.1%. However, we can conclude that our model is more efficient because it yields a comparable accuracy regardless of having used fewer features. This means that the features used in this study are better suited for the development of a CR system. We can also conclude that our model is more accessible to farmers because the data for the features we use for our model is more readily available.

5 Conclusion

In this study we were able to develop a crop recommendation system that can determine which crop to grow based on the attributes of the soil. We employed RF algorithm and SVM for the development of our CR system and found that RF algorithm works best for soil classification which is the basis of our CR system. Our final system uses RF algorithm and yields an accuracy of 91.1% as shown in the discussion of our results.

The soil attributes that we considered were the three main soil nutrients, namely, nitrogen, phosphorus, and potassium, as well as the soil pH level. Choosing to only consider these four soil attributes whose data is more readily available and inexpensive to attain, allows our system to be more accessible to most farmers. Therefore, our developed CR system would have a greater impact on the farming industry, the agricultural sector, and hence, the economy as a whole. This is especially the case for South Africa and other countries similar to it, whose agricultural sector plays a major role in the economy. The crop recommendation system developed in this study can be adopted by farms of all sizes across South Africa because it is easily accessible and not resource intensive, and if widely adopted, this system would have a great impact on the economic output of the agricultural sector.

Acknowledgements This work is based on the research supported in part by the National Research Foundation of South Africa (Grant numbers: 121835).

References

1. Babu S (Aug 2013) A software model for precision agriculture for small and marginal farmers. In: 2013 IEEE global humanitarian technology conference: South Asia satellite (GHTC-SAS). IEEE, pp 352–355
2. Bondre DA, Mahagaonkar S (2019) Prediction of crop yield and fertilizer recommendation using machine learning algorithms. Int J Eng Appl Sci Technol 4(5):371–376
3. Seo SN, Mendelsohn R (2008) An analysis of crop choice: adapting to climate change in South American farms. Ecol Econ 67(1):109–116
4. Bandara P, Weerasooriya T, Ruchirawya T, Nanayakkara W, Dimantha M, Pabasara M (2020) Crop recommendation system. Int J Comput Appl 975:8887
5. Shinde M, Ekbote K, Ghorpade S, Pawar S, Mone S (2016) Crop recommendation and fertilizer purchase system. Int J Comput Sci Inf Technol 7(2):665–667
6. Kumar A, Sarkar S, Pradhan C (Apr 2019) Recommendation system for crop identification and pest control technique in agriculture. In: 2019 international conference on communication and signal processing (ICCSP). IEEE, pp 0185–0189
7. Rajak RK, Pawar A, Pendke M, Shinde P, Rathod S, Devare A (2017) Crop recommendation system to maximize crop yield using machine learning technique. Int Res J Eng Technol 4(12):950–953
8. Suresh G, Kumar AS, Lekashri S, Manikandan R (2021) Efficient crop yield recommendation system using machine learning for digital farming. Int J Modern Agric 10(1):906–914
9. Patel K, Patel HB (2020) A state-of-the-art survey on recommendation system and prospective extensions. Comput Electron Agric 178:105779
10. Fletcher KK (Jun 2017) A method for dealing with data sparsity and cold-start limitations in service recommendation using personalized preferences. In: 2017 IEEE international conference on cognitive computing (ICCC). IEEE, pp 72–79
11. Kumbhar N, Belerao K (Aug 2017) Microblogging reviews based cross-lingual sentimental classification for cold-start product recommendation. In: 2017 international conference on computing, communication, control and automation (ICCUBEA). IEEE, pp 1–4
12. Sang A, Vishwakarma SK (Aug 2017) A ranking based recommender system for cold start & data sparsity problem. In: 2017 tenth international conference on contemporary computing (IC3). IEEE, pp 1–3
13. Wei J, He J, Chen K, Zhou Y, Tang Z (2017) Collaborative filtering and deep learning based recommendation system for cold start items. Expert Syst Appl 69:29–39

14. Liakos KG, Busato P, Moshou D, Pearson S, Bochtis D (2018) Machine learning in agriculture: a review. Sensors 18(8):2674
15. Rahman SAZ, Mitra KC, Islam SM (Dec 2018) Soil classification using machine learning methods and crop suggestion based on soil series. In: 2018 21st international conference of computer and information technology (ICCIT). IEEE, pp 1–4
16. Paul M, Vishwakarma SK, Verma A (Dec 2015) Analysis of soil behaviour and prediction of crop yield using data mining approach. In: 2015 international conference on computational intelligence and communication networks (CICN). IEEE, pp 766–771
17. Samuel AL (1959) Machine learning. Technol Rev 62(1):42–45
18. Pudumalar S, Ramanujam E, Rajashree RH, Kavya C, Kiruthika T, Nisha J (Jan 2017) Crop recommendation system for precision agriculture. In: 2016 eighth international conference on advanced computing (ICoAC). IEEE, pp 32–36
19. Lacasta J, Lopez-Pellicer FJ, Espejo-García B, Nogueras-Iso J, Zarazaga-Soria FJ (2018) Agricultural recommendation system for crop protection. Comput Electron Agric 152:82–89
20. Akshatha KR, Shreedhara KS (2018) Implementation of machine learning algorithms for crop recommendation using precision agriculture. Int J Res Eng Sci Manag (IJRESM) 1(6):58–60
21. Ghadge R, Kulkarni J, More P, Nene S, Priya RL (2018) Prediction of crop yield using machine learning. Int Res J Eng Technol (IRJET), 5
22. Chougule A, Jha VK, Mukhopadhyay D (2019) Crop suitability and fertilizers recommendation using data mining techniques. Progress in advanced computing and intelligent engineering. Springer, Singapore, pp 205–213
23. Ingle A (2021) Kaggle: crop recommendation dataset, viewed 13 June 2021
24. Kumar V, Dave V, Bhadauriya R, Chaudhary S (Jan 2013) Krishimantra: agricultural recommendation system. In: Proceedings of the 3rd ACM symposium on computing for development, pp 1–2

Chapter 4
Application of Machine Learning Models to Improve the Accuracy of Earnings Management Prediction

D. Kaviyameena, D. Kavitha, B. Uma Maheswari, and R. Sujatha

1 Introduction

An earnings management technique typically shows a positive view of the financial situation by manipulating financial reports. Earnings Management (EM) has been a major concern for most organizations for several decades now. Generally, earnings are viewed as a measure of an enterprise's past performance. Stakeholders consider corporate earnings as a vital indicator for a firm's operating performance. Chen et al. [1] proposed that the management of earnings has therefore become one of the major objectives of the firm. Companies often manipulate earnings through discretionary accruals. Such conduct and intent may lead to serious consequences for the stakeholders and create information asymmetry. Earnings management carried out continuously by a company can lower its credibility. Additionally, manipulating the cost with earnings information may undermine the organization's value. Studies made by Ariza et al. [2] show that earnings management methods have a negative impact on the brand's success. Kalbuana et al. [3] suggested that earnings management can affect the accuracy of financial statements by causing users to rely on profit figures from technical financial statements. Tahmina and Naima [4] put forward that earnings manipulation weakens investors' trust and has a negative impact on the country's economy.

Earnings management involves manipulating accruals through discretionary choices in accrual accounting in order to achieve specific targets according to Kliestik et al. [5]. Researchers have identified two major categories of earnings management research: accrual-based earnings management and real earnings management. Chen and Shen [6] presented that accrual-based earnings management is the most common method of managing earnings since it does not violate accounting principles, and

D. Kaviyameena (✉) · D. Kavitha · B. U. Maheswari · R. Sujatha
PSG Institute of Management, Coimbatore 641004, India
e-mail: kaviyameena@gmail.com

G. Mathur et al. (eds.), *Proceedings of 3rd International Conference on Artificial Intelligence: Advances and Applications*, Algorithms for Intelligent Systems,
https://doi.org/10.1007/978-981-19-7041-2_4

management can adjust the results according to its own discretion. Studies made by Mahmoudi et al. [7] suggest that one of the most common ways to manage profits is by manipulating accounting accruals, especially discretionary accruals. Accounting accruals are, therefore, often considered in studies of earnings management. Real earnings management is a strategy for assisting a corporation in departing from the traditional financial statement operation guideline. The drawback of real earnings management according to Chen and Howard [8] is that it necessitates earlier execution, which affects an enterprise's actual value and real economic activity.

A model to predict earnings management would be useful in finding out the extent of the earnings management process that is carried out in an organization. Linear Regression models have been widely used in predicting earnings management. To help analysts and decision-makers such as investors predict earnings, a variety of techniques have been created over time. Huang and Yen [9] came up with the proposal that Machine Learning approaches can extract useful information from any sort of data using supervised and unsupervised algorithms. Machine learning methods have the advantage of providing variable selection approaches that allow us to locate the most relevant predictors from vast collections of financial variables while avoiding the frequent overfitting concerns associated with models with many predictors as suggested by Zadeh et al. [10]. Using hybrid machine learning methods, this study aims to develop an efficient and accurate hybrid machine learning models for predicting earnings management by using accrual-based earnings management.

2 Literature Review

Several machine learning models were developed and tested in an attempt to establish a model of high accuracy to predict earnings management. The Linear Regression (LR) model has been used extensively in traditional approaches to study earnings management. This technique however, has several limitations, such as linearity and assumptions like non-flexibility of regression models, nonexistence of correlation and homoscedasticity. Hence, Namazi and Maharluie [11] proposed to use other machine learning techniques could be employed to predict the degree of earnings management carried out in an organization.

In a study conducted by Chen and Howard [6] in electronic companies listed in the Taiwan Stock Exchange, earnings management prediction model was developed based on elastic net and C5.0 algorithms. It was found that C5.0 has a better accuracy and provides the best classification performance. In an attempt to evaluate the predicting ability of neural networks, Iranian companies from five different industries listed on the Tehran Stock Exchange were studied by Mahmoudi et al. [7]. The model developed consisted of multilayer perceptron neural network with two hidden layers. The Multilayer perceptron neural network has the ability to predict earnings management at various levels in different industries. Stepwise logistic regression, elastic net and random forest algorithms used on a sample consisting of 1,16,904 firm year observations found that the random forest-based model is most effective.

The combination of Principal Component Analysis, LightGBM and Hyperopt method predicted earnings better than the traditional Logistic Regression when tested on 3000 companies by Xinyue et al. [12] over a 30-year period. It was also found that using neural networks and decision trees together provides not only better predictability but also important decision rules compared to using only neural network. Huang and Yen [9] found that the XGBoost algorithm provides the most accurate prediction among GA fuzzy clustering, XG Boost, HACT and contrastive divergence algorithm. Chen and Howard [8] attempted to detect earnings management practices by using soft computing methods. The hybrid model constructed using stepwise regression, decision trees and random forest is the best choice as it provided optimal results in terms of accuracy in predicting earnings manipulation of publicly listed electronic companies.

Financial statements from big corporations over a period of four years were studied using Bayesian Naive Classifier (BNC) by Zaarour [13] to enhance the decision-making process by identifying earnings manipulation. Qualitative and quantitative models to detect earnings manipulation were developed and tested on two private and state-owned entities in Fijian. The performance of both the discretionary accrual model and the qualitative model were not consistent in the measurement of earnings management when tested by Naidu and Patel [14].

Data relating to 117 firms from eight different industries were collected and machine learning techniques were used by Fischer et al. [15] to predict earnings management. Quarterly Earnings Prediction using epsilon support vector regression QEPSVR produced more significant results than the Brown and Rozeff Auto-Regressive Integrated Moving Average (BR ARIMA) model. Differences between the neural network models and regression model in detecting earnings manipulation of 94 listed companies in the Tehran Stock Exchange were analyzed by Namazi and Maharluie [11]. The results show that multi-layer perceptron and generalized regression neural networks (GRNN) are more accurate than linear regression. Out of the two models developed by Chen et al. [1] to predict earnings management of biotechnology industries, the GRNN-based model had the best performance and the linear regression-based model had the least performance and the Bayesian Network screening method, together with the C5.0 decision tree, yields the best results. Based on the review of literature, this study adapts a combination of Random Forest and C5.0 Decision tree. Random Forest is used for variable screening and decision tree algorithm is used to develop a prediction model.

2.1 Random Forest

A common goal in prediction modelling is to reduce the number of variables required to obtain a forecast in order to increase efficiency of the model. Random forests are tree-based prediction models in which each tree is based on a random vector sampled independently and whose distribution matches that of all other trees in the forest. As a result of accurate calculation, random forests can determine an important index of independent variables. The index can capture the interaction between the predictors and the response variable. Chen and Howard [8] in their study found that the random forest model was stable in terms of accuracy and chose the most relevant predictive variables in identifying earnings management. In the random forest architecture, variable selection is a key factor for many applications in expert systems. For datasets with known underlying relationships between predictors and outcomes, conditional random forest methods for variable selection may be preferred since conditional random forest is often better at correctly finding significant associations between predictors according to Speiser et al. [16]. This study adopted random forest as a vital indicator to evaluate the variables of earning management due to its high performance in carrying out classification tasks.

2.2 C5.0 Decision Tree

Decision trees can be used for both classification and regression problems. This algorithm is used to analyze large datasets based on division rule, resulting in the best predictions. In addition to not being susceptible to any statistical hypothesis of the sample data, decision tree's key characteristics are its capacity to treat partial data and examine the potential relationships among large and sophisticated input and output variables. As a result, C5.0 is one of the most widely utilized decision tree algorithms according to Chen and Shen [6]. When applied to detect earnings manipulation in the biotechnology industry, the C5.0 decision tree produced excellent outcomes with very high accuracy in the tests performed by Chen et al. [1]. In this study, the C5.0 method is used for prediction.

2.3 ANN

Mahmoudi et al. [7] proposed three significant advantages of neural networks in comparison with statistical approaches. For starters, neural networks can learn any complex design or nonlinear mapping. Second, neural networks take no defaults into account when distributing data, and third, neural networks are particularly flexible when dealing with partial, missing, or ambiguous data. An influential neural network model is the multilayer perceptron (MLP), which consists of many layers of nodes. From the review of the work done by Tsai and Chiou [17], it can be seen that the input, the output and the hidden layer consist of nodes corresponding to input nodes, output nodes, and hidden nodes. A good layering and input selection can approximate a good result. Input, hidden, summation, and output layers make up a generalized regression neural network. The parameters and structures of the neural networks are not clearly defined when using neural networks, and the way in which they are defined is often based on trial and error. Namazi and Maharluie [11] compared the neural network model with other machine learning models and neural network model provided more accuracy in prediction.

3 Research Gap

Traditional auditing systems, which are constrained by time, human resources, price, and effects, struggle to detect anomalous activities in vast and complicated financial data. As a result, developing a prediction model for the level of earnings management is quite useful for auditors and investors in determining the degree of financial statement manipulation. The purpose of this study is to develop a hybrid machine learning model to predict earnings management.

4 Methodology

In the research process, data was identified and pre-processed, a training set was defined, algorithms were selected, training parameters were identified, and the test set was evaluated. In order to predict earnings management, potential predictive variables were identified. Considering the fact that the study has many variables, Random Forest method was used for variable screening and arriving at final list of variables which was used as input for the decision tree model (C5.0) for predicting earnings management. Data was collected from NIFTY 500 companies from 2012 to 2021 using the Prowess database. Mean imputation was used to treat missing values during the pre-processing stage (Figs. 1 and 2).

4.1 Research Variables

Earnings Management Proxy Variables. Discretionary Accruals were employed as a proxy variable for Earnings Management in this study. Non-discretionary Accruals

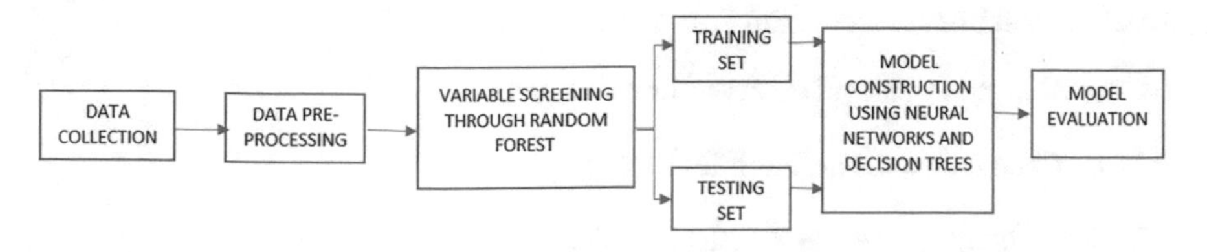

Fig. 1 Research methodology

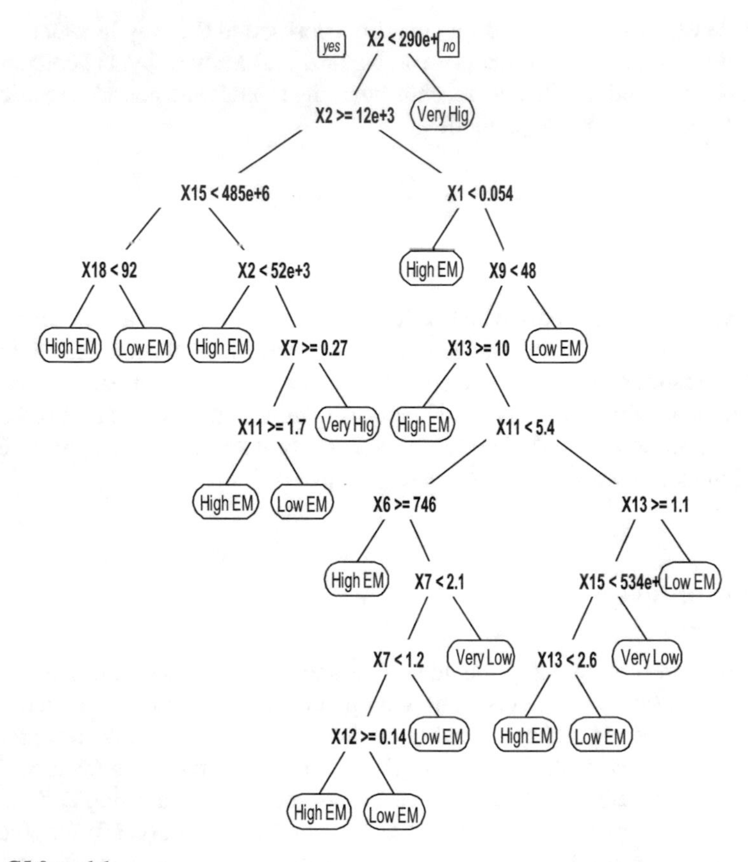

Fig. 2 C5.0 model

are subtracted from Total Accruals to arrive at Discretionary Accruals. The balance sheet approach and the statement of cash flows method are the two basic methods for calculating total accruals. This study uses the widely used balance sheet approach to calculate total accruals using the following formula.

$$TACC_{ft} = \Delta CA_{ft} - \Delta CL_{ft} - \Delta Cash_{ft} + \Delta STDEBT_{ft} - DEP_{ft} \quad (1)$$

where,

$TACC_{ft}$ Total Accruals of firm f on year t

ΔCA_{ft} Change in current assets of firm f on year t

ΔCL_{ft} Change in current liabilities of firm f on year t

$\Delta Cash_{ft}$ Change in cash and cash equivalents of firm f on year t

$\Delta STDEBT_{ft}$ Change in current maturities of long term debt and other short term debt included in current liabilities of firm f on year t

DEP_{ft} Depreciation and Amortization expenses of firm f on year t.

After calculating total accruals, the non-discretionary accruals were calculated using the modified Jones model. The formula to calculate non–discretionary accruals is given below.

$$NDA_{ft} = \alpha 1(1/TA_{ft-1}) + \alpha 2((\Delta REV_{ft} - \Delta REC_{ft})/TA_{ft-1}) + \alpha 3(PPE_{ft}/TA_{t-1}) \quad (2)$$

where,

NDA_{ft} Non-discretionary accruals of the firm f on year t

TA_{ft-1} Total Asset of the firm f on year t–1

ΔREV_{ft} Change in income of the firm f on year t

ΔREC_{ft} Change in accounts receivables of the firm f on year t

PPE_{ft} Gross Property, Plant and Equipment of the firm f on year t

$\alpha 1$, $\alpha 2$, and $\alpha 3$ are estimated using the below model.

$$TACC_{ft}/TA_{t-1} = \alpha 1\ (1/\ TA_{ft-1}) + \alpha 2\ ((\Delta REV_{ft} - \Delta REC_{ft})\ /\ TA_{it-1}) + \alpha 3\ (PPE_{ft}/\ TA_{t-1}) + \varepsilon_{ft} \quad (3)$$

where, ε_{ft}–Residuals of the firm f on year t.

After estimating the alpha values, the non-discretionary accruals were computed. Reducing the non-discretionary accruals from the total accruals gives the discretionary accruals. Natural logarithm values for the discretionary accruals were calculated and the degree of earnings management was classified based on those values. The average and standard deviation of the discretionary accruals were calculated and the ceiling and floor values were set by adding and subtracting the standard deviation from the average respectively. The observations had an average of 8.90 and standard deviation of 3.42. If the value is above the ceiling or below the floor, it is classified as Very High Earnings Management or Very Low Earnings Management. If the values fall between the average and floor, it is classified as Low Earnings Management and if it falls between average and ceiling, it is classified as High Earnings Management. This classification of firms based on discretionary accruals was adopted from the study of Chen et al. [1] (Tables 1 and 2).

Independent Variables. Based on previous studies potential predictive variables were identified. 19 variables were selected which could probably affect earnings management. The variable names and formulas are mentioned below.

Table 1 Earnings management classification

Classification	Description	No of samples
Very low earnings management	DA < 5.47	313
Low earnings management	5.47 < DA < 8.90	1282
High earnings management	12.33 > DA > 8.90	2635
Very high earnings management	DA > 12.33	435
Total		4665

5 Analysis and Interpretation

5.1 Variable Screening Using Random Forest

Given the fact that the study has many variables for the prediction of earnings management, random forest method was used to find out the significance of the variables before establishing decision tree model. The random forest packages were installed, and the data was split into training and testing sets in an 8:2 ratio. Thereafter, the random forest algorithm was used to determine the significance of the variables employed.

The random forest approach considers mean decrease gini as an important indicator in determining the significance of the variable in earnings management prediction. Higher the value of mean decrease gini, greater the influence of the variable on earnings management. Table 3 shows the mean decrease gini values of the variables used in this study. The order of importance of the variables are Operating Cash Flow, Corporate Size, Operating Profit Margin, Corporate Performance, Long Term Funds Appropriate Rate, Managerial Ownership, Total Assets Turnover, Return on Assets, Net Profit Margin, Leverage Coefficient, Quick Ratio, Return on Equity, Current Ratio, Operating Cash Flow Ratio, Sales to Equity Ratio, Employee Profitability. The variables Loss and Financing Activities do not have any significance and hence will be removed in the further study. The accuracy of the random forest model was found using the confusion matrix. The random forest model developed in the study to find out the significance of the variables had an accuracy of 75%. Mean decrease accuracy specifies the level of decrease in accuracy of the model when the specific variable is dropped. With three variables tried at each split and the number of trees being 500, the random forest model had an out of bag estimate of error rate 24.59% resulting in an accuracy of 75% (Table 4).

Table 2 Independent variables

Variable code	Variable name	Formula	Source
X1	Leverage coefficient	Total liabilities/Total assets	Mahmoudi et al. [7], Chen and Howard [8]
X2	Corporate size	Sales revenue of the firm	Mahmoudi et al. [7], Chen et al. [1]
X3	Corporate performance	Cash from operations	Chen and Howard [8]
X4	Return on equity	Net income/Shareholder's equity	Chen and Howard [8]
X5	Return on assets	Net income/Total assets	Chen and Howard [8]
X6	Operating cash flow	Cash flow from operating activities	Chen and Howard [8], Chen et al. [1]
X7	Total assets turnover	Net sales/Average total assets	Chen and Shen [6]
X8	Current ratio	Current assets/Current liabilities	Chen and Shen [6]
X9	Net profit margin	Revenue—Cost/Revenue	Chen et al. [1]
X10	Operating cash flow ratio	Operating cash flow/current liabilities	Chen and Shen [6]
X11	Long term funds appropriate rate	(Total stockholders' equity + Long term liabilities) ÷ Total fixed assets	Chen and Shen [6]
X12	Sales to equity ratio	Sales revenue/total equity	Chen and Shen [6]
X13	Employee profitability	Net profit before tax/total number of employees	Chen et al. [1]
X14	Quick ratio	(Current assets—Inventory)/current liabilities	Chen and Shen [6]
X15	Managerial ownership	Shares held by promoters	Chen et al. [1]
X16	Loss	1 if loss, else 0	Chen et al. [1]
X17	Financing activities	Change in outstanding shares more than 10%—1 else 0	Mahmoudi et al. [7], Chen and Howard [8]
X18	Operating profit margin	Operating Income/Sales revenue	Chen and Shen [6]

5.2 C5.0 Decision Tree Model

The decision tree model was built using discretionary accruals as the dependent variable, which is a proxy for earnings management and independent variables were chosen using random forest. The observations were divided into training and testing set at a ratio of 8–2. (80% to the training dataset and 20% to the testing dataset). Decision Trees can be used to analyze the link between variables and select the most

Table 3 Variable importance output from random forest

Variable	Mean decrease accuracy	Mean decrease gini
Operating cash flow	0.053	207.9973
Corporate size	0.068	175.3198
Operating profit margin	0.068	168.2167
Corporate performance	0.051	153.5326
Long term funds appropriate rate	0.054	147.8837
Managerial ownership	0.037	136.1943
Total assets turnover	0.031	112.1690
Return on assets	0.026	107.8518
Net profit margin	0.024	102.3296
Leverage coefficient	0.023	102.0554
Quick ratio	0.023	98.3986
Return on equity	0.012	97.1454
Current ratio	0.020	97.1209
Operating cash flow ratio	0.020	96.5214
Sales to equity ratio	0.022	94.1772
Employee profitability	0.018	93.9009
Loss	0.000	4.4778
Financing activities	0.000	0.1753

Table 4 Random forest confusion matrix

	High EM	Low EM	Very high EM	Very low EM
High EM	1655	150	21	9
Low EM	374	493	11	16
Very high EM	102	12	186	0
Very low EM	39	66	3	129
OOB estimate of error rate: 24.59% Accuracy: 0.75				

important factors for predicting a specific outcome. The variables Loss and Financing Activities were removed after variable screening through Random Forest and the study was carried out with 17 variables. Decision tree algorithm was employed and the variables used in the construction of the tree are Operating Profit Margin, Corporate Size, Long Term Funds Appropriate Rate, Corporate Performance, Managerial Ownership and Operating Cash Flow. Initially, the model's training set accuracy was 80%, while the testing set accuracy was 71.6%. The accuracy was improved by tuning the hyperparameters. Hyperparameter tuning was carried out using the grid search algorithm. The optimal values for the parameters were found by using grid search. RF + C5.0 The training set had 88% accuracy, whereas the testing set had 85.1% accuracy. The model had a type I error rate of 14.89% (Tables 5 and 6).

Table 5 Decision tree confusion matrix

	High EM	Low EM	Very high EM	Very low EM
High EM	505	24	1	6
Low EM	46	195	0	6
Very high EM	24	3	46	3
Very low EM	8	17	1	48

Table 6 Accuracy of RF+C5.0

	Overall accuracy rate (%)	Type I error (%)
Training set	88	12
Testing set	85.1	14.89

Figure 3 helps in determining the rules to predict the degree of earnings management carried out in a firm. Numerous rules can be formed using the model and few of the rules are mentioned below.

- When the Corporate Size is greater than Rs. 290000 million (or) the managerial ownership is greater than 485,000,000 and corporate size is greater than Rs. 520000 million and total assets turnover ratio is less than 0.27, the firm is said to have a very high degree of earnings management practices.
- Firms with corporate size less than Rs.12000 million and leverage coefficient less than 0.054 (or) firms with corporate size greater than 52,000, total assets turnover

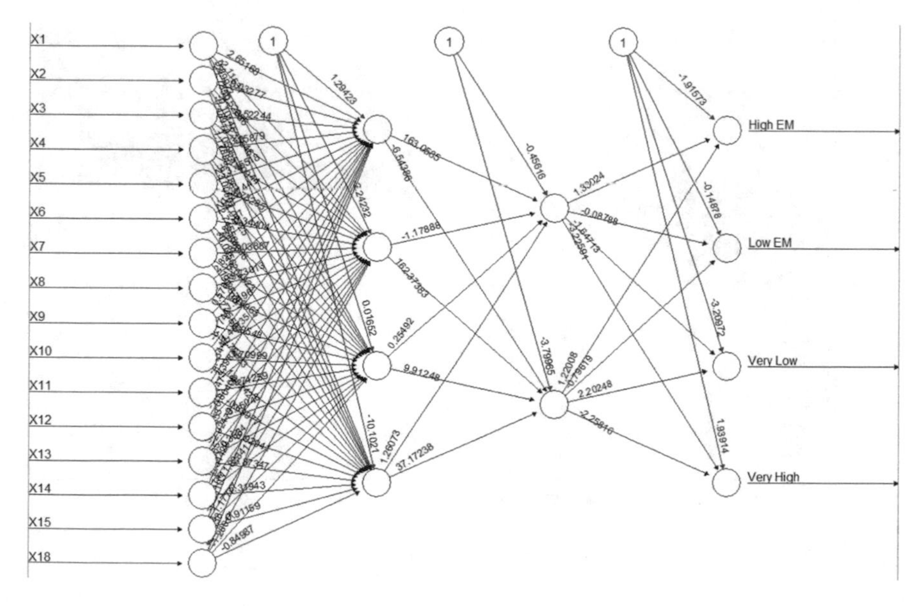

Fig. 3 Neural network

greater than 0.27 and long term funds appropriate rate greater than 1.7 are said to have a high degree of earnings management practices.

- Firms with corporate size less than Rs.12000 million, leverage coefficient greater than 0.054 and net profit margin greater than 48% are said to have a low degree of earnings management practices.
- If the corporate size is less than Rs.12000 million, leverage coefficient is greater than 0.054, net profit margin is less than 48%, employee profitability is less than 10, long term funds appropriate rate is less than 5.4, operating cash flow is less than Rs.746 million and total assets turnover ratio is greater than 2.1, the firm is said to have very low levels of earnings management practices.

5.3 Artificial Neural Network

The dataset was randomly split into training set (80%) and testing set (20%). Figure 4 represents the neural network created in the study. The neural network constructed has an input layer, two hidden layers and an output later. The input layer receives the variables whereas the output layer generates the prediction results. The hidden layers constitute neurons and serve to increase the complication of the network. Variable Importance for the neural networks was found using Olden's method. Figure 5 shows the variable importance plot for the neural network. The accuracy of the model was found to be 78.2%.

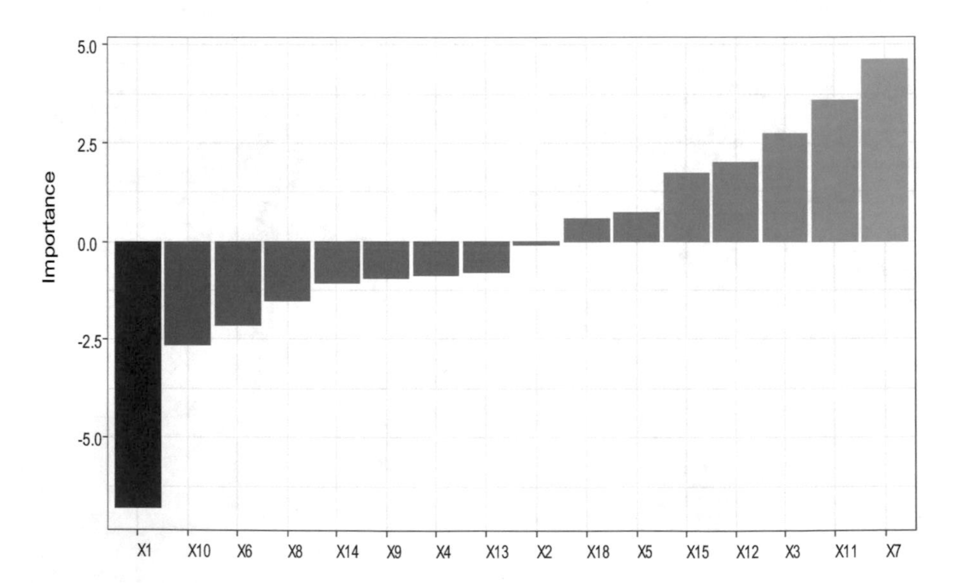

Fig. 4 Variable importance in neural network

Table 7 Accuracy of C5.0 and ANN

	Overall accuracy rate (%)	Type I error (%)
C5.0	85.1	14.89
ANN	78.2	21.5

6 Discussion and Implications

Finding important predictive variables would be critical, since it would affect the model's accuracy and classification. It could be seen that the usage of random forest for variable screening improves the predictive accuracy of the earnings management prediction models. Variables such as Corporate Size, Leverage Coefficient, Operating Profit Margin and Total Assets Turnover play a major role in determining the level of earnings management practices carried out by a firm as could be seen in the Fig. 3. The results of the experiments shed light on why the recommended model is optimal in this case. Variable Importance for Neural Networks found through Olden's method suggests that Corporate Size, Managerial Ownership, Return on Assets, Employee Profitability, Corporate Performance and Total Assets Turnover have a significant influence on the level of earnings management practices of a firm. The variable corporate size being an important parameter in earnings management prediction is obvious because it represents the sales revenue of the firm which can be easily manipulated according to the convenience of the firm. Operating profit shows a company's ability to manage its indirect costs. According to Rahimipour [18], Firms manipulate the operating profit by either accumulating the profit of all quarters in the final quarter or splitting the profit of the first quarter to all the other quarters to look more profitable. Hence operating profit margin is one of the vital factors in influencing earnings management. One issue with evaluating accrual-based earnings management models is the challenge of gauging the models' success because the exact amount of earnings management is unknown. To some extent, the use of discretionary accruals has assisted in overcoming this problem. The decision tree-based model is quite effective than the neural network model in terms of accuracy. Earnings management prediction models assist investors in making educated decisions by revealing the extent of earnings manipulation (Table 7).

7 Conclusion and Recommendations

This study is preliminary research focusing on the development of a machine learning model based on C5.0 decision tree to predict accrual-based earnings management. This study proposes a method for detecting earnings manipulation that can help stakeholders in corporations make more accurate judgments about their financial statements. The sample for this study was the financial data of NIFTY 500 companies from 2012 to 2020. Random Forest method was used to screen the variables and c5.0 decision tree model was developed for the prediction of earnings management. The findings show that the model constructed using a blend of random forest and c5.0 decision tree has a test group accuracy of 85.1 and a low rate of type I error. The neural network model had an accuracy of 78.2%. Though the performance of the hybrid model is good, the accuracy of the prediction model could be improved still. Further studies in this area could use more efficient machine learning methods to screen variables and could also consider using more financial variables other than the ones used in this paper. Future studies could consider the scenarios that the neural network model successfully predicts and use it to build a decision tree model that generates effective decision rules and compare the results with that of decision tree and neural network used separately to arrive at a more efficient prediction model.

References

1. Chen FH, Chi DJ, Wang YC (2015) Detecting biotechnology industry's earnings management using Bayesian network, principal component analysis, back propagation neural network, and decision tree. Econ Model 46:1–10
2. Rodriguez-Ariza L, Martínez-Ferrero J, Bermejo-Sánchez M (2016) Consequences of earnings management for corporate reputation: Evidence from family firms. Account Res J
3. Kalbuana N, Prasetyo B, Asih P, Arnas Y, Simbolon SL, Abdusshomad A, Mahdi FM (2021) Earnings management is affected by firm size, leverage and roa: Evidence from Indonesia. Acad Strat Manag J 20:1-12
4. Tahmina A, Naima J (2016) Detection and analysis of probable earnings manipulation by firms in a developing country. Asian J Bus Account 9(1):59–82
5. Kliestik T, Belas J, Valaskova K, Nica E, Durana P (2021) Earnings management in V4 countries: The evidence of earnings smoothing and inflating. Econ Res-Ekon Istraživanja 34(1):1452–1470
6. Chen S, Shen ZD (2020) An effective enterprise earnings management detection model for capital market development. J Econ, Manag Trade 26(4):77–91
7. Mahmoudi S, Mahmoudi S, Mahmoudi A (2017) Prediction of earnings management by use of multilayer perceptron neural networks with two hidden layers in various industries. J Entrep, Bus Econ 5(1):216–236
8. Chen FH, Howard H (2016) An alternative model for the analysis of detecting electronic industries earnings management using stepwise regression, random forest, and decision tree. Soft Comput 20(5):1945–1960
9. Huang YP, Yen MF (2019) A new perspective of performance comparison among machine learning algorithms for financial distress prediction. Appl Soft Comput 83:105663
10. Amel-Zadeh A, Calliess JP, Kaiser D, Roberts S (2020) Machine learning-based financial statement analysis. Available at SSRN 3520684
11. Namazi M, Maharluie MS (2015) Detecting earnings management via statistical and neural network techniques. Int J Econ Manag Eng 9(7):2520–2528
12. Xinyue C, Zhaoyu X, Yue Z (2020) Using machine learning to forecast future earnings. Atl Econ J 48(4):543–545
13. Zaarour BDI (2017) Financial statements earnings manipulation detection using a layer of machine learning. Int J Innov, Manag Technol 8(3)
14. Naidu D, Patel A (2013) A comparison of qualitative and quantitative methods of detecting earnings management: Evidence from two Fijian private and two Fijian state-owned entities. Australas Account, Bus Financ J 7(1):79–98
15. Fischer JA, Pohl P, Ratz D (2020) A machine learning approach to univariate time series forecasting of quarterly earnings. Rev Quant Financ Acc 55(4):1163–1179

16. Speiser JL, Miller ME, Tooze J, Ip E (2019) A comparison of random forest variable selection methods for classification prediction modeling. Expert Syst Appl 134:93–101
17. Tsai CF, Chiou YJ (2009) Earnings management prediction: A pilot study of combining neural networks and decision trees. Expert Syst Appl 36(3):7183–7191
18. Rahimipour A (2017) Investigating the validity of signaling theory in the Tehran stock exchange: Using real earnings management, accrual-based earnings management and firm growth. Int J Econ Perspect 11(3)

Chapter 5
Ensemble of Parametrized Quantum LSTM Neural Networks for Multimodal Stress Monitoring

Anupama Padha and Anita Sahoo

1 Introduction

The standard of living of a person is strongly affected by his emotional states such as stress and anxiety. Most of the work nowadays involves information processing and computer usage where the different levels of stress can be seen among the knowledge workers whose main task is to use and produce information [1]. They need to handle multiple deadlines to meet the targets, and a lot of stress is generated when the demands imposed on them are not met. According to Jerry Chen et al. [2], severe stress for a longer period of time can lead to muscle tension and depression among individuals. Due to this, there is a critical need for efficient, unobtrusive stress monitoring among individuals using daily life sensors or devices. The data from these sensors can then be integrated using modern computing techniques such as artificial intelligence (AI) and Internet of Things (IoT) [10] to handle stress management. Nath et al. [18] evaluate the effectiveness of wearable stress management systems by detecting the stressed and non-stressed states among old people. Among the different machine learning models, random forest gave the best performance. In order to preserve health data privacy, Can et al. [19] developed a federated deep learning algorithm on the heart data collected from smart bands worn by individuals for detecting stress levels. Zhang et al. [20] proposed a framework based on bi-directional long short-term memory networks (LSTM) and convolutional neural network (CNN) for detecting real-time stress among people using electrocardiogram (ECG) signals and achieved an accuracy of up to 0.865 on three levels of stress. Hence, the data gathered from multimodal sensors can be effectively analyzed by deep learning techniques for detecting stress among individuals.

A. Padha (✉) · A. Sahoo
Jaypee Institute of Information and Technology, Noida, India
e-mail: anupama.padha@gmail.com

G. Mathur et al. (eds.), *Proceedings of 3rd International Conference on Artificial Intelligence: Advances and Applications*, Algorithms for Intelligent Systems,
https://doi.org/10.1007/978-981-19-7041-2_5

With quantum computers becoming a reality [3, 4, 15], more and more researchers are trying to implement classical ML models in the quantum environment to see their effectiveness. As quantum computing is taking a technological leap and causing a paradigm shift in the behavior and performance of certain ML algorithms, it generates a strong need to effectively monitor stress using quantum enhanced machine learning techniques [5]. Many researchers are trying to implement hybrid quantum classical models to reap the quantum benefits. Li et al. [6] investigate quantum convolutional networks using quantum parametrized circuits and indicate that the model provides exponential acceleration on MNIST and GTSRB datasets. Yogesh Kumar et al. [7] proposed an ensemble of hybrid quantum deep NN and quantum inspired gravitational search algorithms for the application of facial expression recognition where the model outperformed the state-of-the-art techniques. In [8], Chengying Mao et al. use two backpropagation neural networks and two swarm intelligence algorithms (basic and quantum) to create an ensemble for addressing the trust problem for cloud services. The results outperformed the current state-of-the-art algorithms along with the basic backpropagation algorithms. Emanuela Paladini et al. [9] proposed two ensemble approaches in deep neural network which are Mean-Ensemble-CNN and NN-Ensemble-CNN in which four convolutional neural networks (ResNet-101, ResNeXt-50, Inception-v3, and DenseNet-161) were evaluated. These ensemble models also showed outperformed results compared to hand-crafted feature-based methods in detecting colorectal cancer tissue. Based on the study, we observe that quantum computing methods have been applied to traditional machine learning and deep learning models to enhance performance. The quantum ensemble LSTM approach is a novel attempt in this direction for improved stress monitoring among knowledge workers which effectively combines multiple predictions.

The objective of the paper is to develop an ensemble of quantum neural networks to monitor stress levels utilizing multimodal data. This paper presents a novel attempt to develop a PQLSTM model for stress management. In this regard, an ensemble PQLSTM (ENS_PQLSTM) has been developed that integrates the knowledge obtained by different PQLSTM models trained with different modality data.

The paper's organization is as follows. Section 2 discusses the architecture of Parametrized Quantum Long Short-Term Memory Network (PQLSTM). Section 3 explains the proposed ensemble of quantum LSTMSs (ENS_PQLSTM), while Sect. 4 gives the experimental results of the ENS_PQLSTM model. Finally, Sect. 5 gives the conclusion along with future work.

2 Parametrized Quantum LSTM

To benefit from the performance enhancement offered by quantum technologies, a parametrized quantum LSTM model (PQLSTM) is created with the help of a classical LSTM neural network and a quantum circuit. The fundamental idea is to split the issue into two parts, with the classical part preprocessing the data and determining the parameters of the quantum circuit, and the quantum part, which is composed of a

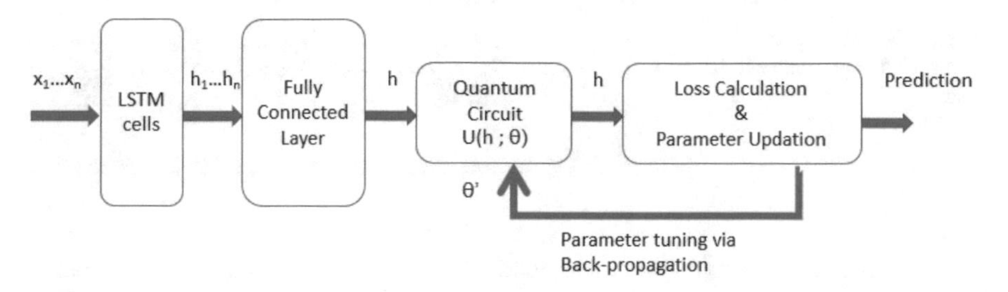

Fig. 1 Parameterized quantum LSTM (PQLSTM) model

parametrized quantum circuit (PQC), using the data to forecast measurements. The classical component refines the measuring result [11].

LSTMs can retain knowledge for a long time because they gradually develop long-term dependencies [17]. This characteristic makes them effective in processing data in sequences, such as time series data. As seen in Fig. 1, the input is delivered to the LSTM cells with the aid of a real vector. The data is preprocessed for effective data reduction and feature selection using input and forget gates. The fully connected layer receives the output that was obtained.

A single hidden layer is utilized in between the classical layers of the LSTM network to solve the issue of exploding and vanishing gradients. During backpropagation, this parametrized quantum circuit layer employs gradient descent. The layer consists of a single qubit and two gates, a Hadamard and a RY gate. A single qubit when affected by the Hadamard gate creates a superposition in which the qubit's quantum state can be either |0 > , |1 > , or a combination of both:

$$H = 1/\sqrt{2}\begin{bmatrix} 1 & 1 \\ 1 & -1 \end{bmatrix} \tag{1}$$

It makes the conversion given in Eq. (1) and determines the basis state $|0 > |0 > +1 > /\sqrt{2}$ and $1 > |0 > -1 > /\sqrt{2}$. After Hadamard, RY gate where a single qubit through angle θ is rotated within the y-axis is used:

$$Ry(\theta) = \begin{pmatrix} cos\frac{\theta}{2} & \cdots & -sin\frac{\theta}{2} \\ \vdots & \ddots & \vdots \\ sin\frac{\theta}{2} & \cdots & cos\frac{\theta}{2} \end{pmatrix} \tag{2}$$

As given in Eq. (2), the likelihood of measuring the qubit either as 1 or 0 signifying the condition of an individual, i.e. stressed or not stressed accordingly, is controlled by the RY gate, which sets the state vector angle θ of the qubit.

For prediction, the qubit value is measured after passing through two gates as shown in Fig. 2. A single qubit is rotated through an angle around the y-axis using the RY gate.

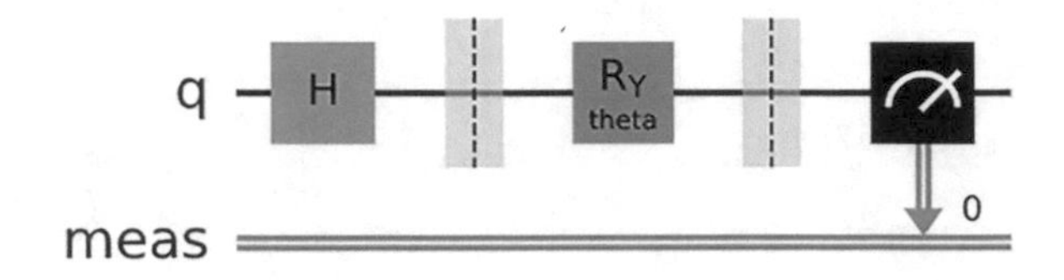

Fig. 2 Quantum circuit using Hadamard gate and RY gate

The qubit's angle of rotation θ is modified during backpropagation in order to maximize the neural network's measurement for precise prediction. The gradient descent using the parameter shift rule, a method for assessing the gradients of quantum circuits in relation to their parameters, is applied by the quantum layer. It entails making a minor adjustment and figuring out how the result will alter [16], i.e. we deduce the derivative's value by dividing the results of two linked circuit evaluations, as seen in Eq. (3). By adjusting the parameter θ, the measurement or expectation values are iteratively optimized:

$$\nabla\theta QC(\theta) = QC(\theta + s) - QC(\theta - s) \tag{3}$$

Finally, the parameters θ are updated, and the entire cycle is executed in a closed loop between the classical and quantum hardware multiple times.

3 Proposed Framework for Ensemble of PQLSTM Models

Ensemble learning [12, 13] is a distinctive and important machine learning method that uses an integrated learner consisting of multiple individual learners combined according to a specific combination strategy. In this section, a framework to create an ensemble of four different PQLSTM models trained with multimodal sensor datasets, namely facial expressions (FE_PQLSTM), computer interaction (CI_PQLSTM), body postures (BP_PQLSTM), and heart rate variability (HRV_PQLSTM), is elaborated.

Stress can be effectively detected from the data captured from multiple sensors as during the stressed state the person behaves differently and the postures, face movements, heart rate, etc. of a person change. The computer logging sensors capture the mouse and keyboard event patterns such as the number of left or right clicks, number of mouse wheel scrolling, shortcut keys, etc. The facial expressions such as head and facial movements capture different emotions of being sad, happy, or angry. The features of the eyes, mouth, and eyebrows used to detect the stressed state are captured through the video camera. Similarly, the Kinect SDK is used for capturing body postures such as bone orientations of the body w.r.t. different axis, changes in joint angles, average distance of the user from the image depth, etc. The physiological sensors capture heart rate variability and skin conductance of a stressed person measured through body sensors. Each PQLSTM model has learnt to detect stress by learning from specific data. The aim is to improve the overall performance

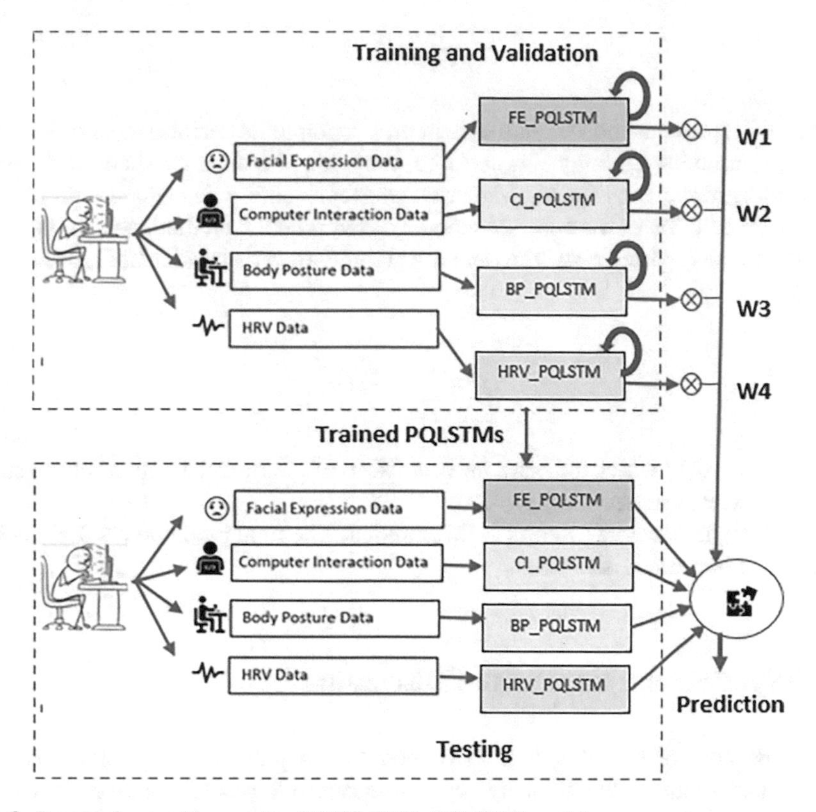

Fig. 3 Proposed ensemble quantum LSTM (ENS_PQLSTM) model

and stability in stress predictions by creating an ensemble of such diversified learning models. The structure of the proposed ensemble of PQLSTM(ENS_PQLSTM) is shown in Fig. 3. To generate an ensemble quantum model, the performance of all four classifiers is calculated by training them on respective datasets [14].

Each PQLSTM model is trained on different modality data from the sensors as discussed above. The performance of each model is calculated using F1 score. F1 score is a harmonic mean of precision and recalls providing a better measurement of stress prediction. The F1 scores of all the models are summed up as shown in Eq. (4):

$$F = \sum_{i=1}^{C} F_i \tag{4}$$

where i = 1,2…C, C representing the number of PQLSTM models. F_i represents the F1 score of ith model.

Weights are generated corresponding to each model based on their performance using Eq. (5):

$$W_i = \frac{F_i}{F} \tag{5}$$

where W_i represents the weight that is in proportion to its performance. A weighted average ensemble technique is used to calculate the final prediction. A better performing model is assigned a higher weight representing its major contribution to the final prediction of the ensemble model. Each model's prediction value is multiplied with the respective weights, and a weighted average is calculated as shown in Eq. (6):

$$y = \sum_{i=1}^{C} W_i D_i \tag{6}$$

where $D_i = \{0,1\}$ which indicates the prediction of i th model. 0 represents no stress and 1 represents stress.

The weighted average value y is thresholded. The final prediction is 0 if y is less than 0.5, otherwise it is 1.

4 Experimental Results and Discussion

The main goal of the ENS_PQLSTM model is to predict if knowledge workers are stressed or not in the office by combining multiple predictions from individual learners.

The Python implementations of the ENS PQLSTM and four parametrized quantum LSTM models make use of Pytorch, an open-source toolkit for the Deep Learning pipeline, and the Jupyter Notebook that Anaconda 3 makes available for algorithm development. Qiskit, an open-source quantum computer software development kit, provides the "qasm simulator," a noise-free simulator, which is used to simulate the single qubit quantum circuit developed utilizing Hadamard and RY gates for parameter adjustment. 100 shots are taken into consideration.

On the multimodal SWELL–KW [1] dataset with the two class labels stressed and not stressed, the ensemble model and the base learners are applied. High-quality data was obtained by doing data preprocessing on the Swell–KW files to eliminate redundant and missing NaN values. The mean values of the relevant feature were used to fill in the missing values during data preprocessing, and samples that included all NaN values were eliminated to enhance performance. This dataset's condition feature has four categories: relaxed, neutral, interrupted, and time pressure. Interruption and Time pressure are changed to 1 (stressed) and 0 (non-stressed) in place of relaxed and neutral. The four data sets have 12, 88, 40, and 3 features, respectively. For computer interaction, bodily positions, and facial expressions, the total number of samples is 3139 for each, whereas for HRV data, it is 3140. All four datasets were searched

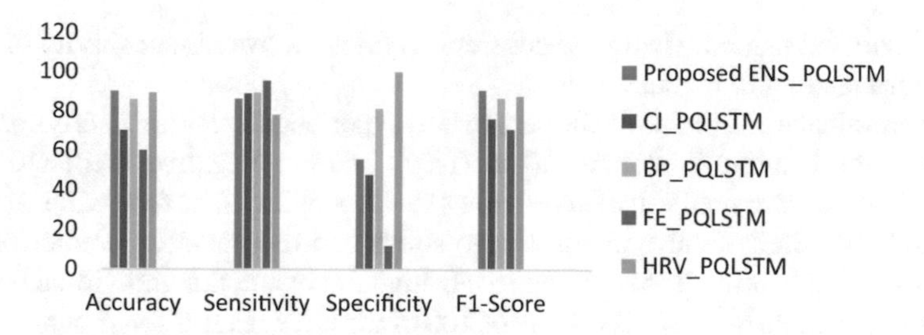

Fig. 4 Prediction performance of proposed ENS_PQLSTM model

for samples with similar timestamps for the same individuals which were analyzed using the ensemble model.

The model's learning process is controlled by hyper-parameters like the optimizer, learning rate, activation function, etc. Adam optimizer with extremely competitive performance is utilized. The training and test losses are calculated using a negative log-likelihood loss function, and the model was trained at a learning rate of 0.01. The dataset is divided in an 80:20 ratio for model training and model testing.

The proposed ENS PQLSTM model is assessed based on its F1 score, accuracy, specificity, and sensitivity. The accuracy numbers show how well the model can distinguish between stressed and non-stressed conditions. Figure 4 compares the performance of ensemble ENS PQLSTM and PQLSTM models on four multimodal datasets.

The results indicate that from among the base classifiers, a highest accuracy of 89.3% is achieved using the HRV_PQLSTM model. It shows that the heart rate variability features and skin conductance are the best features to monitor the stress of an individual. This is followed by the BP_PQLSTM with an accuracy of 85.9% showing the body posture features contribute second highest among the four classifiers in monitoring stress levels. The facial expressions model FE_PQLSTM having an accuracy of 60.1% indicates the facial expression features are not much indicative of the stress levels in an individual. By combining the predictions of four classifiers, it is observed that 90.2% accuracy is obtained using ensemble models. A high F1 score of 87.7 is obtained for the HRV_PQLSTM model contributing maximum for the higher F1 score of the ensemble model to be 90.6%. This shows that the model can successfully detect the cases where stress levels were incorrectly anticipated. A low F-value of 70.8 for the FE_PQLSTM model shows that facial features are not good at capturing wrongly predicted cases.

Sensitivity/Recall metric indicates the proportion of the positive cases that are predicted as positive, while the specificity metric indicates the proportion of negative values that are predicted as negative. From Fig. 4, it is observed that a highest sensitivity of 95.3% is obtained from facial expressions data indicating that facial expression features can best identify if a person is stressed, but the low sensitivity value of 11.9% indicates that these features are not efficient to detect a non-stressed person thus giving a low accuracy value. For the HRV model, it can be observed that a highest specificity value of 100 indicates that HRV features can effectively identify the non-stressed person. Using the ensemble model, an overall sensitivity of 86.2% is obtained leading to higher accuracy.

The prevalence of how likely the person is to experience stress is directly correlated with both the Positive Predicted Values (PPV) and the Negative Predicted Values (NPV). The positive predictive value shows the probability that following a positive test result, that individual will truly have stress and the negative predictive value indicates the probability that following a negative test result, that individual will truly not have stress. The highest PPV value of 100 of the HRV_PQLSTM model indicates that HRV features are the best to correctly indicate the stressed people from the predicted stressed people, while a high NPV value of 87.0 for BP_PQLSTM indicates the good features to correctly identify the non-stressed people from predicted ones.

5 Conclusion

Since Quantum Computing leverages the laws of quantum mechanics to build computers endowed with tremendous computing power, it is attracting more and more researchers trying to implement the classical algorithms in the quantum environment. In modern culture, stress has grown to be a significant and inescapable issue that requires careful monitoring. The proposed model seeks to predict knowledge workers' stress levels at the office using ensemble models with better performance. With the rise in the quantum computing world, the model can be implemented on quantum machines.

The PQLSTM model provides an improvement over the LSTM model during backpropagation by eliminating the vanishing gradient and exploding gradient problem. This is done using a parameterized quantum circuit that implements parameter tuning during backpropagation.

In this work, we proposed an ensemble quantum LSTM (ENS_PQLSTM) model which is a combination of four different hybrid quantum PQLSTM models implemented on four different modality datasets from the Swell-KW dataset. The proposed model uses a weighted averaging-based technique utilizing the F1 score of the individual learners to assign weights. The model is tested using the test data and achieved an accuracy of 90.2% which is higher than the individual predictive models indicating how effectively the model can capture the correct cases. A high F1 score of 90.6% indicates that the model can efficiently capture the wrongly predicted cases.

References

1. Koldijk S, Sappelli M, Verberne S, et al. (2014) The swell knowledge work dataset for stress and user modeling research. In: Proceedings of the 16th international conference on multimodal interaction. https://doi.org/10.1145/2663204.2663257

2. Chen J, Abbod M, Shieh J-S (2021) Pain and stress detection using wearable sensors and devices—a review. Sensors 21:1030. https://doi.org/10.3390/s21041030
3. Chakraborty S, Mandal SB, Shaikh SH (2018) Quantum image processing: challenges and future research issues. Int J Inf Technol. https://doi.org/10.1007/s41870-018-0227-8
4. Parameshwara MC, Nagabushanam M (2021) Novel low quantum cost reversible logic based full adders for DSP applications. Int J Inf Technol 13:1755–1761. https://doi.org/10.1007/s41870-021-00762-3
5. Sharma S (2020) QEML: (Quantum Enhanced Machine Learning) using quantum computation to implement a k-nearest neighbors algorithm in a quantum feature space on superconducting processors. arXiv
6. Li YC, Zhou R-G, Xu RQ et al (2020) A quantum deep convolutional neural network for image recognition. Quantum Sci Technol 5:044003. https://doi.org/10.1088/2058-9565/ab9f93
7. Kumar Y, Verma SK, Sharma S (2021) An ensemble approach of improved quantum inspired gravitational search algorithm and hybrid deep neural networks for computational optimization. Int J Mod Phys C 32:2150100. https://doi.org/10.1142/s012918312150100x
8. Mao C, Lin R, Towey D et al (2021) Trustworthiness prediction of cloud services based on selective neural network ensemble learning. Expert Syst Appl 168:114390. https://doi.org/10.1016/j.eswa.2020.114390
9. Paladini E, Vantaggiato E, Bougourzi F et al (2021) Two ensemble-CNN approaches for colorectal cancer tissue type classification. J Imaging 7:51. https://doi.org/10.3390/jimaging7030051
10. Raval D, Shukla A (2021) Stress detection using convolutional neural network and internet of things. Turk J Comput Math Educ Res Artic 12:975–978
11. Macaluso A, Clissa L, Lodi S, Sartori C (2020) A variational algorithm for Quantum Neural Networks. Lect Notes Comput Sci:591–604. https://doi.org/10.1007/978-3-030-50433-5_45
12. Sagi O, Rokach L (2018) Ensemble learning: A survey. Wiley Interdiscip Rev Data Min Knowl Discov 8:1–18. https://doi.org/10.1002/widm.1249
13. Kuncheva LI, Whitaker CJ (2003) Mach Learn 51:181–207. https://doi.org/10.1023/a:1022859003006
14. Araujo IC, da Silva AJ (2020) Quantum ensemble of trained classifiers. In: 2020 International Joint Conference on Neural Networks (IJCNN). https://doi.org/10.1109/ijcnn48605.2020.9207488
15. Schuld M, Petruccione F (2018) Quantum ensembles of quantum classifiers. Sci Rep. https://doi.org/10.1038/s41598-018-20403-3
16. Dimitriev DA, Saperova EV, Indeykina OS, Dimitriev AD (2019) Heart rate variability in mental stress: The data reveal regression to the mean. Data Brief 22:245–250. https://doi.org/10.1016/j.dib.2018.12.014
17. Choi JY, Lee B (2018) Combining LSTM network ensemble via adaptive weighting for improved time series forecasting. Math Probl Eng 2018:1–8. https://doi.org/10.1155/2018/2470171
18. Nath RK, Thapliyal H, Caban-Holt A (2021) Machine learning based stress monitoring in older adults using wearable sensors and cortisol as stress biomarker. J Signal Process Syst. https://doi.org/10.1007/s11265-020-01611-5
19. Can YS, Ersoy C (2021) Privacy-preserving federated deep learning for wearable IOT-based biomedical monitoring. ACM Trans Internet Technol 21:1–17. https://doi.org/10.1145/3428152
20. Zhang P, Li F, Zhao R et al (2021) Real-time psychological stress detection according to ECG using deep learning. Appl Sci 11:3838. https://doi.org/10.3390/app11093838

Chapter 6
Evolutionary Algorithms for Marine Dynamical Systems: Towards Sustainable Renewable Energy Harvesting from Offshore Environment

R. Manikandan

1 Introduction

Offshore electrical dynamical systems(OEDS) are cited in a severe environment and experienced large loads [1–3]. OEDS are offshore wind turbines, marine robots and underwater vehicles. These systems are using permanent magnet brushless motor [2, 4, 5]. Brushless direct current (BLDC) motors are popular device in oceanic applications like fixed and floating sustainable offshore wind turbines. The brushless DC motor is used to run the loads at the regular speed, by achieving the same a suitable control algorithm is mandatory [6–9]. The investigations of offshore electrical dynamical system associated with controls were carried out and the following things were located [1, 3].

- The Routh–Hurwitz criteria, pole placement methodology, Root locus approach, and Ziegler–Nichols (*ZN*) tuning formula are commonly used to produce traditional PID-controller gains.
- Concepts related with the optimal settings need dynamic weighting matrices.
- Advance control concepts based on artificial intelligence along with soft computing techniques are mandated for improving the overall performance of the system.

So, the outcome of the literature study and the research work focused on developing the suitable control technique to improve the performance of BLDC motor. The current research work speaks the controller concept for BLDC motors which are developed from GA and ACO-based PID mechanism. Numerical simulations are conducted and results were discussed. The paper is organized as follows: System descriptions are reported in Sect. 2, PID controller mechanism is described in Sect. 4, GA and PSO details are in Sect. 5, and the end-like discussions and results are in Sect. 6.

R. Manikandan (✉)
Department of Mechanical Engineering, Technion, Israel Institute of Technology, Haifa, Israel
e-mail: mkaucbe@gmail.com

G. Mathur et al. (eds.), *Proceedings of 3rd International Conference on Artificial Intelligence: Advances and Applications*, Algorithms for Intelligent Systems,
https://doi.org/10.1007/978-981-19-7041-2_6

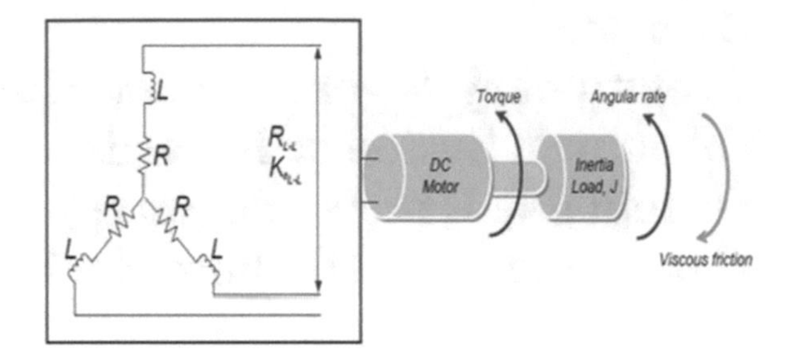

Fig. 1 Schematic illustration of a Brushless DC motor

2 Mathematical Model of BLDC Motors

A brushless DC motor's mathematical model is typically similar to that of a normal DC motor [10–12]. The phases involved in the BLDC motor have a significant impact on the overall results. The resistive and inductive properties of the BLDC configuration are affected by the phases' peculiarities. A basic design with a symmetrical three-phase and internal "wye" link, for example, might provide a quick overview of the complete phase idea in Fig. 1. The mechanical and electrical constants are, respectively, $\tau_m = \frac{J.\,3R}{K_e K_t}$ and $\tau_e = \frac{L}{3.\,R}$. By the introduction effects, one can derive the transfer function BLDC motor systems and same can be

$$G(s) = \frac{\frac{1}{K_e}}{\tau_m \tau_e . S^2 + \tau_m S + 1} \tag{1}$$

where J is the inertia of motor, R is the internal resistance, K_e and K_t are the back EMF and torque constant correspondingly.

3 PID Control Algorithm

The concept of controller variant is PID, which is a readily available industry-based controller and the same utilized in this study for achieving the response control of sustainable marine machine applications. In the vast majority of applications, the PID algorithm is employed to regulate feedback loops. The primary objective is to maintain the demand in highly stochastic loads in the offshore scenario and maintain the stable operation under Morison-type wave loads. The preliminary details along with operation are shown in Fig. 2.

The controller is responsible for providing the system with the necessary excitation and for controlling the system's overall behavior. The structural configurations of the PID controller are divided into numerous groups. The series and parallel structures are the most prevalent, while hybrid forms of the series and parallel structures exist in some circumstances.

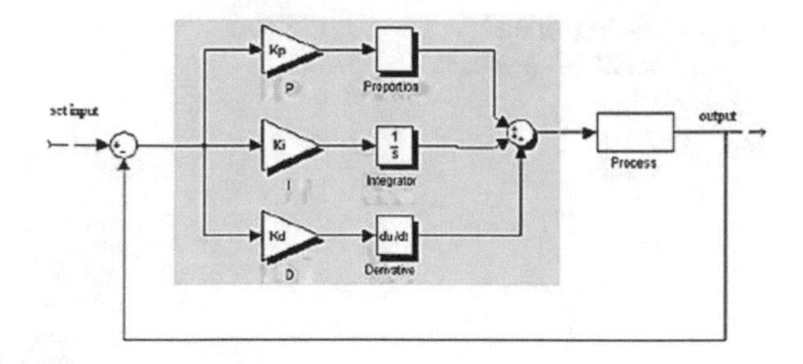

Fig. 2 PID control block

In Fig. 2, the error term e as required demand R versus output Y of the system in the offshore scenario. The addition of errors such as K_p proportional error signal, K_i the integral error and K_d the derivative error are given as an input to the controller module. That is,

$$u = K_p + K_i \int edt + K_d \frac{de}{dt} \tag{2}$$

4 Non-traditional Optimization Techniques

4.1 GA—Genetic Algorithm

This variant of AI uses approaches inspired by natural processes including selection, inheritance, mutation, and crossover to solve optimization issues. GA concept is used to achieve the best pair of optimized controller variables to engineering dynamical systems cited in the offshore scenario. John Holland deployed this variant in the early decades of 1970 and the reason is to locate the best pair solutions to unsolvable engineering and real-time problems. The variant has the theory, often known as the fundamental hypothesis of hereditary calculations, and is widely accepted as the foundation for elucidating the force of hereditary calculations. It claims that, in progressive periods, brief, low-request schemata with above-normal health increase rapidly.

Genetic algorithm is deployed to yield the best globally converged parameters for controller to improve the overall performance of the offshore sustainable systems. Figure 3 demonstrates the genetic algorithm process design flowchart. It populates a beginning set of PID controller settings at that point. The populace is hastily assembled, encompassing the whole spectrum of possible scenarios. Chromosomes make up the population. Every chromosome has a competing solution to the problem. The controller variables are packed into the fundamental chromosomal structure. In the offshore dynamical marine system, the chromosomes are linked, and the system's dynamic execution characteristics are resolved for each chromosome. After that, the target capacity is used to measure the wellness value of each chromosome. Based on

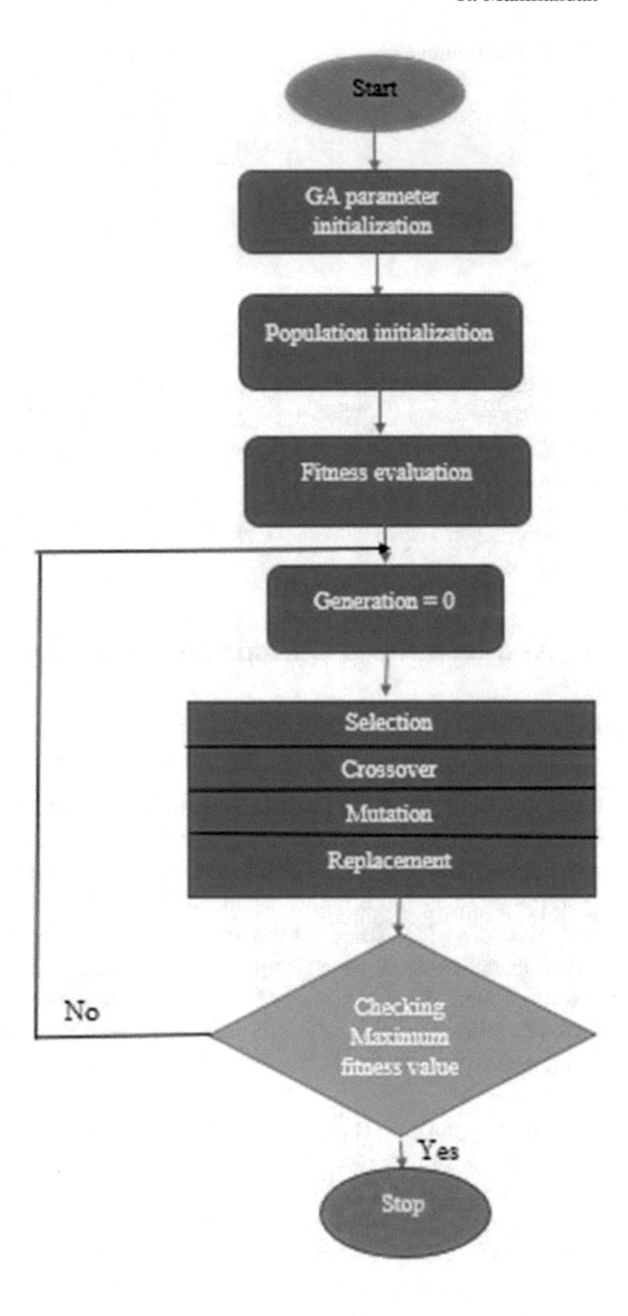

Fig. 3 A flow chart for Genetic algorithm operations

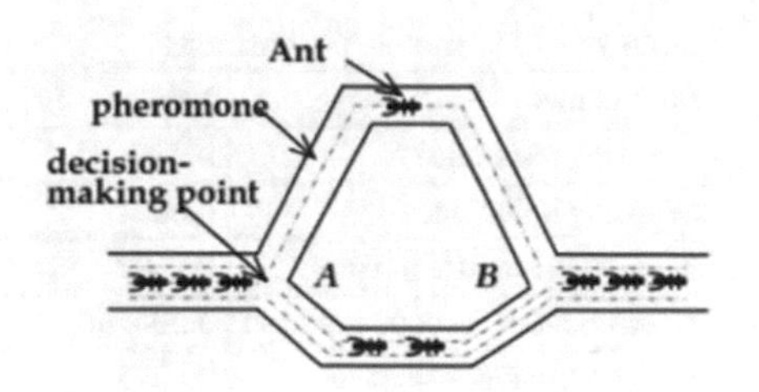

Fig. 4 Decision-making process of ACO

the original quality estimates, a group of the best chromosomes is picked to create the next population. The first stage in GAs is to create the first population. The population is made up of chromosomes that are connected together in a parallel pattern.

4.2 Ant Colony Optimization

ACO is a metaheuristic algorithm under development that is based on the combined activities arising from the hybrids of many search threads and is shown to be successful in handling hybrid optimization problems. The PID controller gains (G) are modified using a trial-and-error technique based on the prior experiment and system behavior. The gains are maximized using the ACO, and the results are sent to the system's controller. This algorithm's goal is to optimize the PID controller's gains for the provided systems. The controller responds to the loss by using the error proportional gain. The controller benefits from reducing steady-state error and minimizing overshooting of the integral derivative gain. ACO is a metaheuristic algorithm design approach for combinatorial optimization issues. Ant colonies' natural behaviors of searching out food sources and returning it to their nest by creating a one-of-a-kind path activate the ACO.

The distance between each city is adjusted based on its own pheromone level and the pheromone level of the other ant. The standard fixed gain PID controller is extensively used in various control processes. Proportional gain (KP), integral constant (Ki), and derivative constant (Kd) are the parameters used to develop the controller (Kd). The primary goal of ACO is to describe the issue as a search for the cheapest path in a graph using an evolutionary metaheuristic algorithm. Artificial ants' behavior is based on that of real ants. They leave pheromone trails and use transition probability to select their course. Ants like to go to nodes with short edges and a high concentration of pheromone. The pheromone matrix is essential for ACO.

Table 1 BLDC motor specifications

Parameters	Value	Units
Winding resistance	0.014	Ω
Winding inductance	0.0001	H
Moment of inertia of rotor	0.0002	
Coefficient of viscous function	0.004586	
Motor torque constant	0.0275	
Back EMF constant	0.0275	
Load torque	10	Nm
Reference speed	1500	rpm

Table 2 Input details of Genetic Algorithms

Parameters	Values
Population size	15
Maximum number of iterations	20
Selection rate	0.25
Mutation rate	0.1

Table 3 Input details of Ant Colony Optimization

Parameters	Values
Maximum number of iterations	20
Number of ants	40
Pheromone exponential weight	1
Number of nodes	500
Evaporation factor	0.7

5 Results and Discussions

The results of a typical PID controller for high-performance control system design of BLDC motor systems are not ideal. Similarly, the use of genetic algorithms with ACO-based PID controllers yields the most appropriate optimized coefficient values for the controller. The steady and dynamic performances of a certain BLDC motor control system will be linked using a simulation software. The model is used for building Simulink transfer functions, which makes simulation easier. The parameters used, are given in Table 1. Table 2 gives the genetic algorithm parameters and Table 3 gives the ACO parameters used.

The parameter seeding details of aenetic algorithm and ant colony optimizations are reported in Tables 2 and 3 correspondingly.

Table 4 Numerical results

Parameters	GA	ACO
Kp	5.9564	2.7856
Ki	95.4084	54.7094
Kd	0.1604	0.9579
Rise time	19.425 ms	17.969 ms
Settling time	19.2 ms	17.4 ms
Overshoot	48.529	22.619
Undershoot	1.99	0

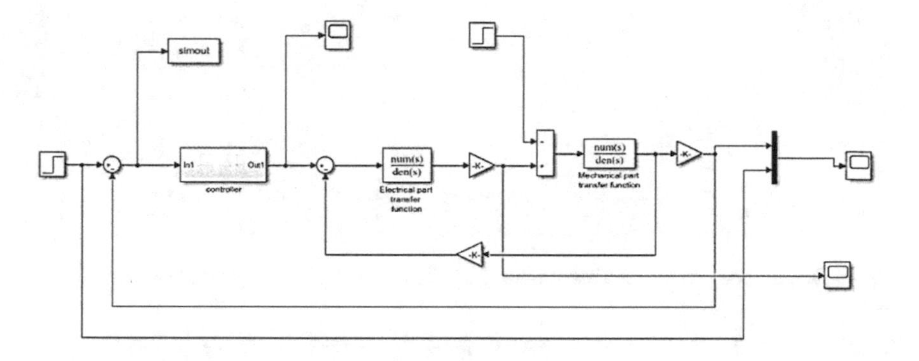

Fig. 5 MATLAB–Simulink diagram model of the BLDC motor

Figure 4 shows ant colony algorithm implementations. The model controller optimizes the PID gain parameters by minimizing the error and achieving the global minima of objective function designed. The numerical results are presented in Table 4 using various parameter-choosing ways for the controller variable.

The numerical simulations carried out for range of parameters and the best performance of the proposed algorithms are reported in the Figs. 5, 6, and 7.

6 Closure

In this work, we proposed the advanced artificial intelligence technique-based PID mechanism proposed for BLDC motors in marine green energy applications. The proposed methods were validated with readily available techniques such as conventional modeling for performance evaluation and the same was visualized via simulation outputs. From the results, the proposed method performed well in overall performances of the machine and the better time domain characteristics were also achieved simultaneously. The steady-state responsiveness and performance indices

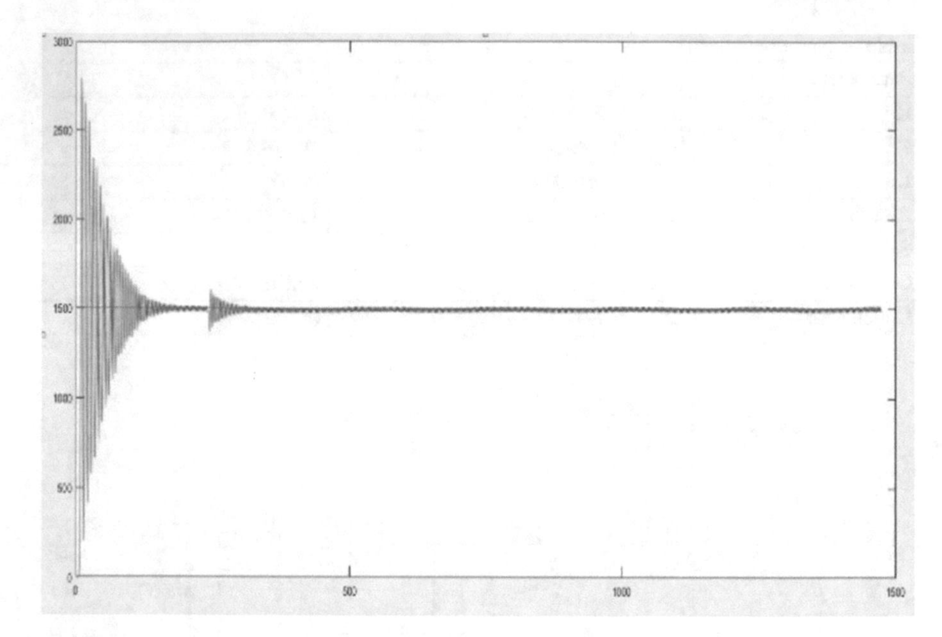

Fig. 6 Output speed versus time response of BLDC motor using PID controller without optimization

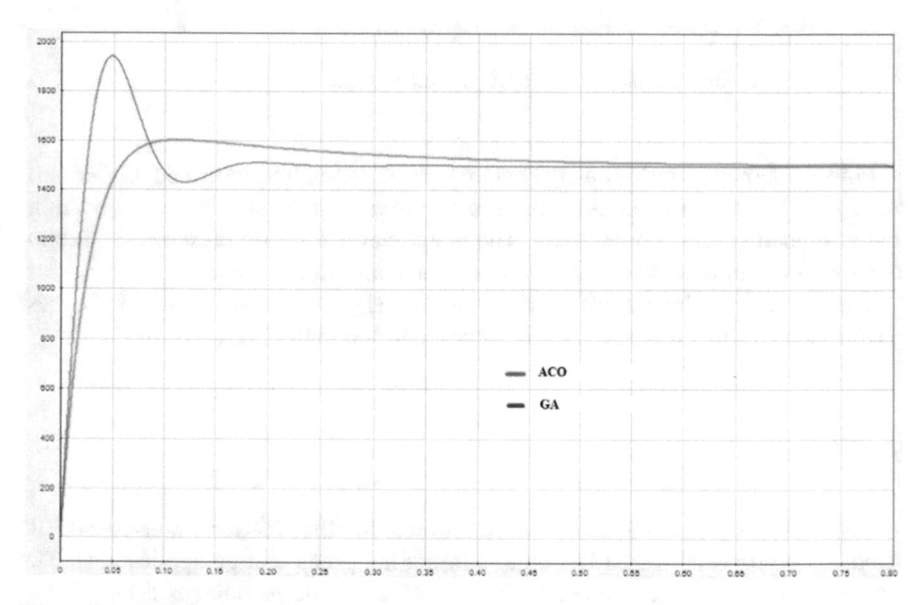

Fig. 7 Speed response of motor with GA and ACO-based PID controller

of the ACO-tuned machine are excellent. The same can be plugged into the real-time higher order experimental systems in ocean-based applications.

Acknowledgements The author would like to thank the University Grant Commission—New Delhi and Dr. D.S. Kothari Postdoctoral Fellowship Cell, Pune, for their support of his UGC-Dr. D.S. Kothari Post Doctoral Fellowship (Fellowship Award No: 202122-EN/20-21/0051).

References

1. Kim DH, Hoon JC (2006) Biologically inspired intelligent PID controller tuning for AVR systems. Int J Control Autom Sys 4(5):624–636
2. Passino KM (2002) Bio mimicry of bacterial foraging for distributed optimization and control. IEEE Control Syst Mag 17(08):52–67
3. Manikandan R, Saha N (2015) A control algorithm for nonlinear offshore structural dynamical systems. Proc R Soc Math
4. Kim DH, Abraham A (2007) A hybrid genetic algorithm and bacterial foraging approach for global optimization and robust tuning of PID controller with disturbance rejection. Stud Comput Intell 75:171–199
5. Ying S, Zengqiang C, Zhuzhi Y (2007) Adaptive constrained predictive PID controller via PSO. In: Proceedings of the 26th Chinese control conference, Zhangjiajie, Hunan, China, pp 729–733
6. Alrashidi MR (2009) A survey of PSO applications in electric power system. IEEE Trans Evol Comput 13(4):913–918
7. Ogata K (2010) Modern control systems engineering. Prentice Hall India
8. Manikandan R, Saha N (2011) Soft computing based optimum parameter design of PID controller in rotor speed control of wind turbines
9. Manikandan R, Saha N (Mar 2014) On the elimination of destabilizing motions of guyed offshore wind turbines using geometrical control mechanism. ICCES14-Changwon, Korea. Tech Science Press
10. Manikandan R, Saha N (2016) Modeling and PI control of spar offshore floating wind turbine. IFAC-PapersOnLine
11. Manikandan R, Saha N (2019) Dynamic modelling and non-linear control of TLP supported offshore wind turbine under environmental loads. Mar Struct
12. Vanchinathan R (2005) improvement of time response for sensorless control of BLDC motor drive using Ant Colony Optimization technique. Int J Appl Eng Res

Chapter 7
Comparison of Different Swarm Based Nature Inspired Algorithm for Solution of Optimization Problem

Kirti Pandey and C. K. Jha

1 Introduction

Optimization can be outlined as the skill of making something effective. This is done through searching for a substitute with good performance in less time span. Most of the Real-world problems are difficult to tackle and come in this category. Given the existence of many optimizations means, there is no certainty about how an optimal solution will be identified. Different algorithms have been used to assess their performance in producing an ideal solution is a wide topic of investigation. A wide classification of Optimization techniques divides them into Conventional and Unconventional methods. There are enormous numbers of Survey papers on Conventional techniques. Natural inspired algorithms are examples of unconventional algorithms, they exhibit a combination of artificial intelligence nature phenomena and computation, and therefore it is the main attractiveness point of nature inspired algorithms. Combination of technology with nature encouraged researchers to work in this field. We learn much from nature, so it seems as a coach for corps [1]. Unconventional algorithm is a capsule of explicit or uncertain algorithms. These algorithms are used in different domain areas like healthcare, engineering, marketing and many areas. This paper also shows the fine details of basic concepts in relevant areas.

Multiple Nature Inspired Optimization Algorithms (NIA) have been constructed in recent years by replicating the characteristics of various biological systems [1]. Nature inspired heuristic and metaheuristic algorithms are another term for such algorithms [1, 2]. Heuristic algorithms are using a trial-and-error algorithm to make new solutions, whereas Metaheuristic algorithms are higher-level heuristics that are using memory, a way to solve history, as well as other forms of 'learning' strategy to produce new solutions [1]. NIA is specifically built for global search, as opposed to

K. Pandey (✉) · C. K. Jha
Computer Science Department, Banasthali Vidyapith, Banasthali newai, Rajasthan, India
e-mail: pandey.kirti30@gmail.com

G. Mathur et al. (eds.), *Proceedings of 3rd International Conference on Artificial Intelligence: Advances and Applications*, Algorithms for Intelligent Systems,
https://doi.org/10.1007/978-981-19-7041-2_7

Fig. 1 Different swarms in nature

conventional algorithms, they see problems with opaqueness, and they might work on a wide range of problems not depending on specific knowledge. They are generally gradient-free approaches (Fig. 1).

Swarm Based NIA have balanced global (exploration) and local (exploitation) search features for choosing the best option. Local search (exploitation) describes the exploration of new locations in the near area of a frequently accessed location [37]. However, different tactics can be used to enhance the balance between exploration and exploitation. Swarm based is one of the categories of NIA, there are many categorizations proposed by different researches. These categories are like biological based, physics based and chemical concept based. Swarm based is one of the categories of NIA, there are many categorizations proposed by different researches. These categories are like biological based, physics based and chemical concept based (Fig. 2).

In this study, we give depth knowledge about some new ones: Ant Colony Optimization, PSO, Grasshopper, Grey Wolf, Butterfly Optimization Method, Spider Monkey and Cat Swarm Optimization have been studied. The goal of this revision is to explore the various NIA based on its source, basic function or characteristics parameters, types or variants and area of application where these algorithms have been successfully used. Additionally, it also supports the shorting of listing methodologies [3].

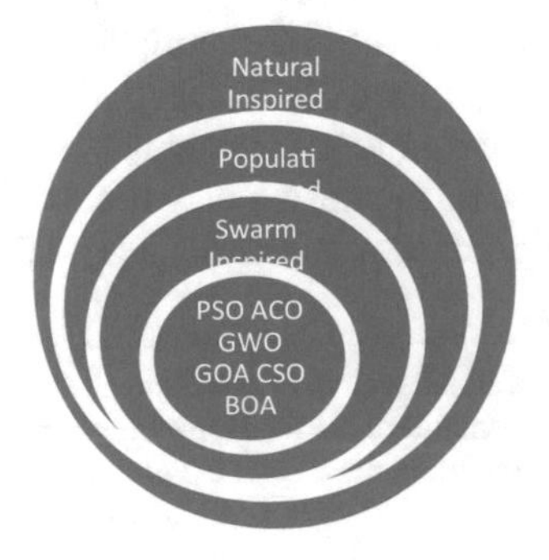

Fig. 2 General categorization of optimization algorithm

This study is based on other studies that have been proposed by other researchers. As far as research methodology, swarm based natural inspired algorithms may be heuristics algorithms and metaheuristics algorithms are concerned.

Firstly, this research presents an analysis of six approaches from oldest to the latest one. Several of these algorithms are very new, there has been little comparative information published on old one. Six optimization algorithms are explained in Sect. 2. The Comparative Study in Sect. 3 and comparison of algorithm with each other and limitations are discussed briefly in Sect. 4. Finally, the paper concludes In Sect. 5.

2 The Optimization Methods Studied

In this section we attempt to present, discuss, and analyze the state-of-the-art of the different Swarm based newly defined algorithms [4]. We started with first discussing the basic idea, steps of these algorithms and graphical representation which was originally the work of other researchers. NIA can deal with linear, non-linear, differential, non-differential, separable, non-separable, scalable, non-scalable, unimodal, and multimodal issues with continuous, discontinuities [8]. In terms of random numbers and random walks, these methods include stochastic components [9]. As a result, unlike typical deterministic algorithms, no identical solution will be found, even if the starting points are identical (Fig. 3).

Stochastic nature leads these algorithms to be able to escape local minima (and hence have a lower chance of being stuck or caught in local minima). Some of the past related works describes following.

2.1 *Ant Colony Optimization (ACO)*

The ACO was inspired by real ants food-finding activity and their ability to discover the best paths. It's a general search that's based on the population [10]. A method for resolving challenging combinatorial problems other ants follow one of the solutions to optimization problems Paths are chosen at random, and pheromone trails are left

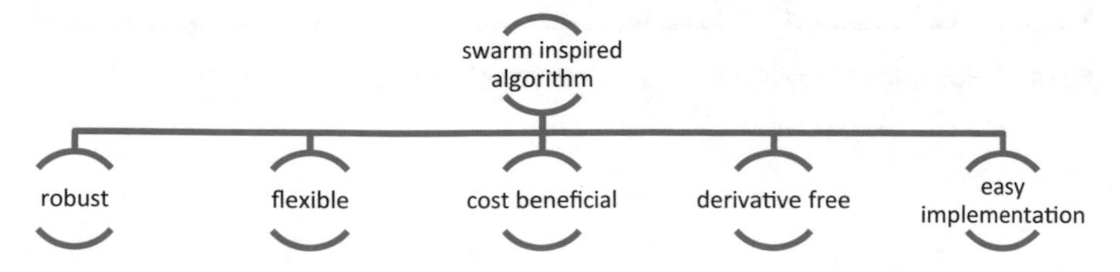

Fig. 3 Advantages of swarm based nature inspired algorithm

behind. Since Pheromone trails are left by the ants on the shortest way. This path is strengthened with increased speed as time goes on [11]. Future ants will find it more enticing because of the pheromone. As time goes on, the ants are more likely to follow the leader. Since it is always reinforced by a, it is the shortest path. a greater number of pheromones [12]. The pheromone is a scent that is produced by the human body. The longer pathways' trails vanish.

2.2 *Particle Swarm Optimization (PSO)*

PSO is a stochastic global optimization method based on the social behavior (bird flocking or fish schooling) and intelligence of swarm searching for the global optimal. It is developed by Edward and Kennedy in 1995. PSO algorithm [13] simulates animal's social behavior, including insects, herds, birds and fishes. These swarms contain a cooperative way to find food, and each member in the swarms keeps changing the search pattern according to the learning experiences of its own and other members. PSO exhibit the idea of evolutionary algorithm, and artificial life. Proximity, Quality, Diverse response, Stability and Adaptability are the 5 basic principles to establish the swarm artificial life system [14].

Figure 4 describes the typical flow, and how nature inspired metaheuristic algorithms works for finding solution to optimization problems. In this paper, we described the newly developed algorithms like GWO, GOA, BOA and CSO by using their flow charts.

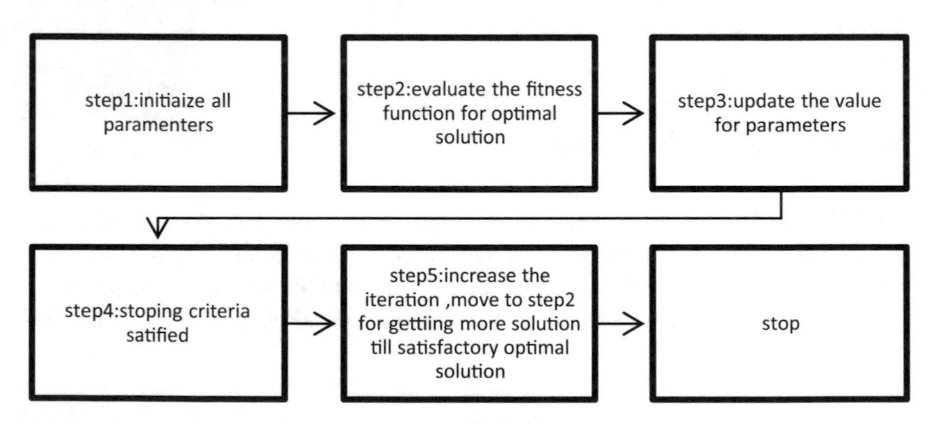

Fig. 4 Typical methodology of naturally inspired algorithm [41]

2.3 Grey Wolf Optimization (GWO)

GWO is a newly created population-based Meta heuristic swarm based nature inspired approach that has been successfully utilized to solve optimization issues [17]. It is inspired by the leadership hierarchy and hunting mechanism of grey wolves in nature. Faris et al. [16] created a nature inspired algorithm in 2014. Alpha, Beta, Omega and Delta [18] are the groups of agents GWO hierarchy. In addition, three primary processes of hunting are incorporated for optimization: seeking for prey, encircling prey, and attacking prey [19]. The grey wolf optimizer has gotten a lot of attention with its improved accuracy, and easy implementation, and needs a smaller number of control constants [20] (Figs. 5 and 6).

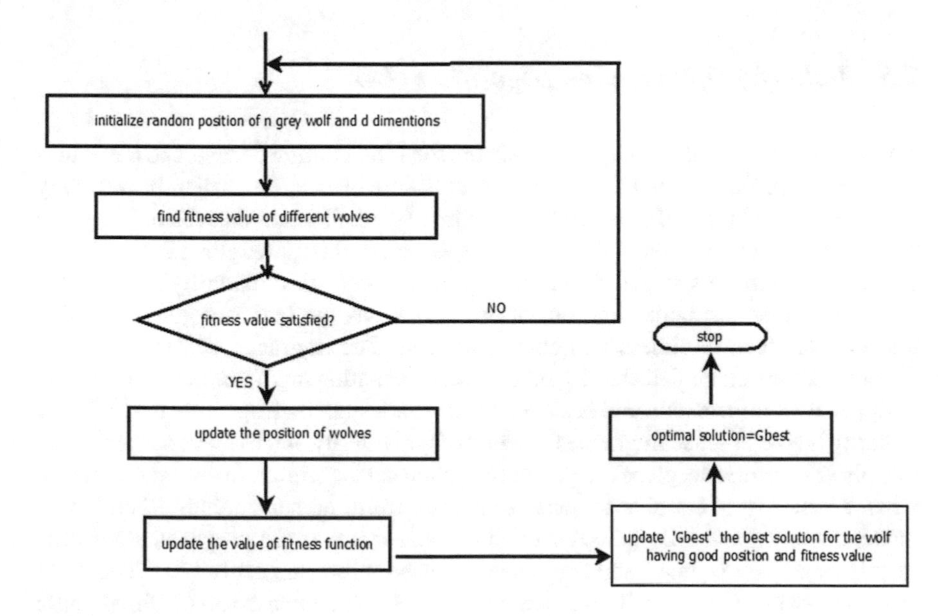

Fig. 5 Flow chart of GWO

Fig. 6 Hierarchy of golf in GWO

Alpha Wolf (Highest privileged dominant grey wolf)
Beta Wolf
Delta Wolf
omega Wolf (Lowest privileged dominant grey wolf)

2.4 Grasshopper Optimizations

It is population-based metaheuristic algorithm that replicates the behavior of grasshopper swarm. It is issued by Saremiet al. [39] in year 2017. It is basically working on social and hunting behavior of grasshoppers Fig. 9. Exploration and exploitation are the terms used in this. It gives the global solution to the real problems like engineering, in industries etc. which may be unconstraint and constraints optimization problems with high accuracy. But it is easy to fall in local optima [22]. Grasshopper search for their best position by using basic methodology of nature inspired algorithms, if find the best position then stop otherwise increases the number of iterations for getting the solution.

2.5 Butterfly Optimization Algorithm (BOA)

A unique optimization technique that mirrors the food hunting behavior of butterflies is proposed in this paper [23]. As per scientific evidence, butterflies have a very good understanding of, from where the source of smell is coming from. They can also differentiate between various scents and detect their strengths (Wyatt 2003). Butterflies are BOA's search agents for optimization [24]. A butterfly's smell will fluctuate in strength depending on its fitness, i.e., as the butterfly goes from one region to another, its fitness will change as well. The fragrance will travel a long distance before being detected by other butterflies, allowing them to communicate their personal information and build a collective social knowledge network. When a butterfly is able to sense fragrance from any other butterfly, it will move toward it and this phase is termed as global search in the proposed algorithm. In another scenario, when a butterfly is not able to sense fragrance from the surrounding, then it will move randomly, and this phase is termed as local search in the proposed algorithm. In this paper, terms smell and fragrance are used interchangeably [25]. There are three phases in BOA: (1) Initialization phase, (2) Iteration phase and (3) Final phase (Figs. 7 and 8).

2.6 Cat Swarm Optimization (CSO)

The original Cat Swarm Optimization Technique is a single-objective, continuous algorithm [15, 26]. It was inspired by a cat's resting and tracing activity. Cats appear to be inactive and spend most of their time sleeping. Their consciousness is very high during their rests, and they are very aware of what is going on around them. As a result, they are constantly and carefully studying their surroundings, and when they find a target, they immediately begin moving towards it. As a result, the CSO algorithm is built on merging these two key cat behaviors. The CSO algorithm is

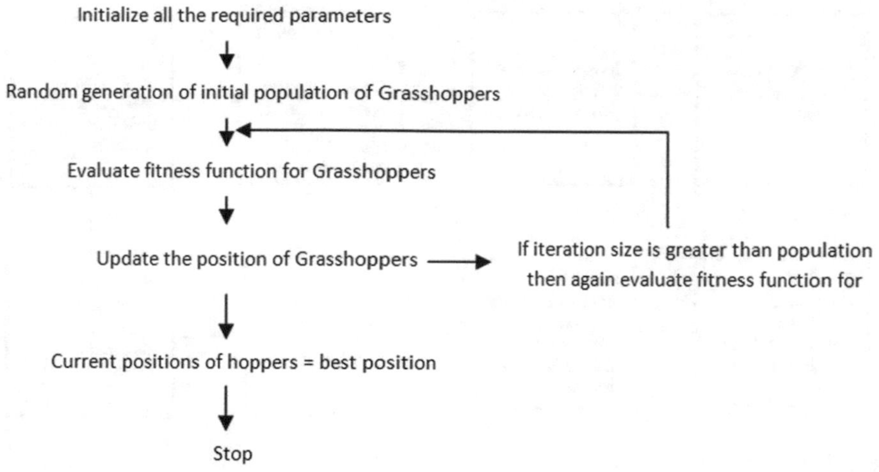

Initialize all the required parameters
Random generation of initial population of Grasshoppers
Evaluate fitness function for Grasshoppers
Update the position of Grasshoppers
If iteration size is greater than population then again evaluate fitness function for
Current positions of hoppers = best position
Stop

Fig. 7 Flow chart of GOA

Step1: set initial value of search agent, no. of generation, objective function, Iteration value
Step 2: generate initial random population of butterflies' value
Step 3: evaluate each solution of population to find best solution
Step 4: update position of butterfly with new value
Step 5: reset all the parameters if result violates the best criteria
Step 6: if all feasible solutions are not evaluated the go to step 3 for remaining populations.
Step 7: if step 3 gives best solution then assign best solution=current solution
Step 8: if termination criteria satisfied then stop finding solution, otherwise goto step 4

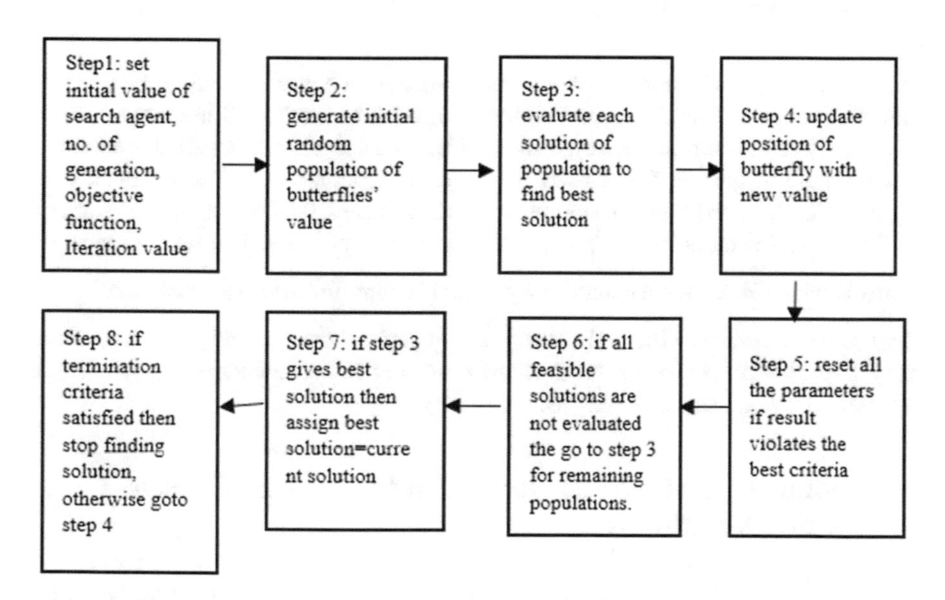

Fig. 8 Working flow of BOA

divided into two modes: tracing and seeking. The flag is used to classify the cats as either seeking or tracing. First, let's figure out how many cats there should be. We should first specify how many cats should be engaged in the iteration and run them through the algorithm best cat in each iteration is saved into memory, and the one at the final iteration will represent the final solution (Fig. 9).

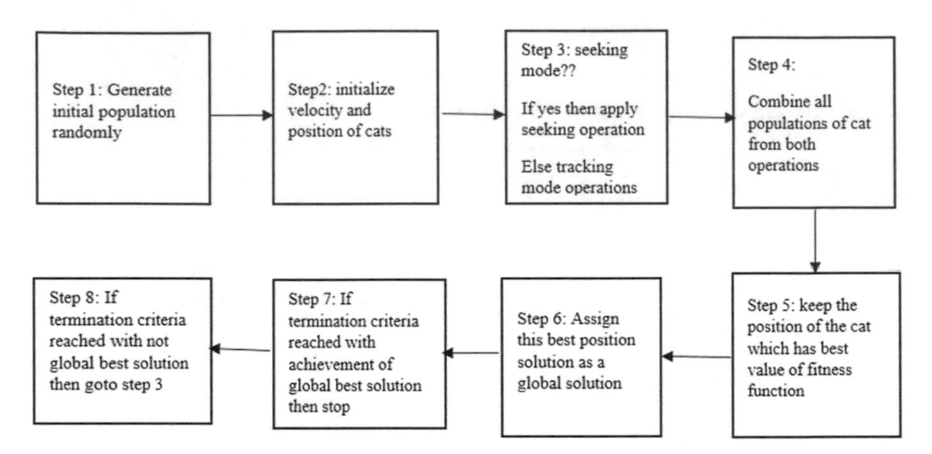

Fig. 9 Working flow of CSO

3 The Comparison of the Methods

In past days, most of the researcher worked on swarm based methods at huge level. Some of these proposed non-conventional algorithms work well in comparison to previously stated conventional algorithms. Here we compare the swarm based meta-heuristic algorithms on the basis of their nature, type, accuracy, convergence rate, optimal solution and types of optimization they solved. The terms "global search" or "exploration" describes the process of exploring a new area in a search space.

Comparison Of Above Algorithm by Considering Following Parameters

Any problems cannot efficiently be solved by each of the swarm based NIA, some may trap at a certain point, takes more time and increase complexity of simple problems or small size of problems.

4 Comparison with Each Other and Limitation of Above Studied Algorithms

1. Grasshopper Optimization Algorithm [39–41] has a strong capability of exploration, it is simple, Robust, Scalable and has good solution with Reasonable time for Execution. But it suffers from the problem of, balancing between exploration and exploitation.
2. Butterfly optimization algorithm avoids bad solution but consider every solution and has equal opportunity to improve [22]. It is totally performed random moves in updating the location of butterflies and in mating and local search phase. But it suffers the problem of diminished population that is poor exploitation ability and tendency to get trapped in local optimization. It is highly efficient in giving solution than other Metaheuristics like GWO, ACO and CSO.

3. PSO with high convergence rate gives a global solution. It is having excellent Robustness and parallel computing is possible in it but has complex Implementation. Easily combine with other algorithms to give a hybrid solution.
4. Grey wolf optimization algorithm is easy and simple to implement but has bad local search capability and trapped at local optimum. But it has a faster convergence rate which is less than PSO.
5. Ant colony optimization algorithm is best for distributed computing. It is a robust type of algorithm, and it is easily mixed with other algorithms to provide best solution. But sometimes it converges at local optimal solution. Convergence rate is also uncertain.
6. Cat Swarm optimization Algorithm gives local and global optimal solutions separately, but suffers from premature convergence hence trapped at local optima. But it outperforms PSO most of the times.

We can say not all the algorithms proved better for all optimization problems, their result may degrade in one and perform best in the other. Although Swarm based methods give the best result in comparison to conventional algorithms, in a given time.

5 Conclusion

Swarm based Natural Inspired algorithm [1] is growing vastly in various fields. This rapidly increases in non-conventional algorithms are the focus of today's research. The depth comparison of different algorithms has been done in this paper. This swarm based NIA covers all the areas like automation of real word problems, military, healthcare and data mining and network fields of engineering. A comparison among many swarm based algorithms has been made including Ant Colony Optimization, Particle Swarm Optimization, Grasshopper Optimization algorithm, Grey Wolf, Butterfly Optimization algorithm, and Cat Swarm Optimization. As seen in Table 1 of this paper, different algorithms are compared to benchmark the performance. According to the study mentioned above all algorithms are swarm based metaheuristic algorithms that is they are derivative free algorithms. They are probabilistic metaheuristic and some works on global optimization, and some are trapped at premature levels. Like ACO, PSO, BOA and CSO having fast convergence rates, GWO, and GOA have slow convergence, these algorithms [33, 41] solve nonlinear and linear nature problems, and we also gave which one is best for benchmark. Each algorithm which I studied in previous papers is either time efficient and gave the near best optimal solution. As PSO with high convergence rate gives a global solution, excellent Robustness and parallel computing is possible in it, so in future we use PSO in combination with any algorithms like cat swarm, firefly [36] or other swarm based algorithm which solve the vehicle routing problem in an efficient manner with a smaller number of iterations and give the best result.

Table 1 Comparison of different algorithms

Algorithm	Optimiza-Tion	Nature	Optimal solution	Implementation	Convergence rate	Accuracy	References
Ant colony	Combinatorial/continuous optimization	Population based	Local	Complex	High	High	[8, 9, 33–35]
Particle swarm	Continuous optimization	Population based	Global	Complex	High	High	[1, 2, 15, 29, 32, 33]
Grew wolf	Nonlinear/continuousoptimization	Population based	Global/local	Easy/simple	Low	Low accuracy	[3, 4, 12, 17]
Cat swarm	Combinatorial/continuous optimization	Population based	Local/global	Complex	High	Moderate	[27, 28, 30– 33]
Butterfly optimization algorithm	Combinatorial/continuous optimization	Population based	Global/local	Simple/easy	High	Best performance	[23– 26, ,37]
Grasshopper optimization algorithm	Continuous optimization	Population based	Global	Simple	Low	High	[7, 19–23, 38, 40, 41]

References

1. Yang XS (2020) Nature-inspired optimization algorithms. Academic Press
2. Hussain K, Mohd Salleh MN, Cheng S, Shi Y (2019) Metaheuristic research: a comprehensive survey. Artif Intell Rev 52(4):2191–2233
3. Dhal KG, Ray S, Das A, Das S (2019) A survey on nature-inspired optimization algorithms and their application in image enhancement domain. Arch Comput Methods Eng 26(5):1607–1638
4. Meraihi Y, Gabis AB, Mirjalili S, Ramdane-Cherif A (2021) Grasshopper optimization algorithm: theory, variants, and applications. IEEE Access 9:50001–50024
5. Pandey K, Jain P (2015) Comparison of different heuristic, Metaheuristic, nature based optimization algorithms for travelling salesman problem solution. Int J Manag Appl Sci 1(2):43-47
6. Dahl G, Mannino C (2009) Notes on combinatorial optimization
7. Cunningham WJCWH, Schrijver WRPA (1997) Combinatorial
8. Serani A, Leotardi C, Iemma U, Campana EF, Fasano G, Diez M (2016) Parameter selection in synchronous and asynchronous deterministic particle swarm optimization for ship hydrodynamics problems. Appl Soft Comput 49:313–334
9. Dhal KG, Das A, Ray S, Gálvez J, Das S (2021) Histogram equalization variants as optimization problems: a review. Arch Comput Methods Eng 28(3):1471–1496
10. Dorigo M, Stützle T (2003) The ant colony optimization metaheuristic: Algorithms, applications, and advances. In: Handbook of metaheuristics. Springer, Boston, MA, pp 250–285
11. Dorigo M, Birattari M, Stutzle T (2006) Ant colony optimization. IEEE Comput Intell Mag 1(4):28–39
12. Dorigo M, Socha K (2018) An introduction to ant colony optimization. In: Handbook of approximation algorithms and metaheuristics, 2nd Edn. Chapman and Hall/CRC, pp 395–408
13. Wang D, Tan D, Liu L (2018) Particle swarm optimization algorithm: an overview. Soft Comput 22(2):387–408
14. Yan D, Cao H, Yu Y, Wang Y, Yu X (2020) Single-objective/multiobjective cat swarm optimization clustering analysis for data partition. IEEE Trans Autom Sci Eng 17(3):1633–1646
15. Hu P, Chen S, Huang H, Xiao Z, Huang S (2018) Alpha guided grey wolf optimizer and its application in two stage operational amplifier design. In: 2018 13th World congress on intelligent control and automation (WCICA) IEEE, pp 560–565
16. Faris H, Aljarah I, Al-Betar MA, Mirjalili S (2018) Grey wolf optimizer: a review of recent variants and applications. Neural Comput Appl 30(2):413–435
17. Long W, Liang X, Cai S, Jiao J, Zhang W (2017) A modified augmented Lagrangian with improved grey wolf optimization to constrained optimization problems. Neural Comput Appl 28(1):421–438
18. Singh N (2020) A modified variant of grey wolf optimizer. Scientia Iranica 27(3):1450–1466
19. Panda M, Das B (2019) Grey wolf optimizer and its applications: a survey. In: Proceedings of the third international conference on microelectronics, computing and communication systems. Springer, Singapore, pp 179–194
20. Mirjalili S, Mirjalili SM, Lewis A (2014) Grey wolf optimizer. Adv Eng Softw 69:46–61
21. Sherry ST, Ward MH, Kholodov M, Baker J, Phan L, Smigielski EM, Sirotkin K (2001) dbSNP: the NCBI database of genetic variation. Nucleic Acids Res 29(1):308–311
22. Arora S, Singh S (2019) Butterfly optimization algorithm: a novel approach for global optimization. Soft Comput 23(3):715–734
23. Zhou H, Cheng HY, Wei ZL, Zhao X, Tang AD, Xie L (2021) A hybrid butterfly optimization algorithm for numerical optimization problems. Comput Intell Neurosci
24. Bidar M, Mouhoub M (2021) techniques for dynamic constraint satisfaction problems
25. Sharafi Y, Khanesar MA, Teshnehlab M (2013) Discrete binary cat swarm optimization algorithm. In: 2013 3rd IEEE international conference on computer, control and communication (IC4). IEEE, pp 1–6
26. Ahmed AM, Rashid TA, Saeed SAM (2020) Cat swarm optimization algorithm: a survey and performance evaluation. Comput Intell Neurosci

27. Ihsan RR, Almufti SM, Ormani BM, Asaad RR, Marqas RB (2021) A survey on cat swarm optimization algorithm. Asian J Res Comput Sci 10:22–32
28. Kraiem H, Aymen F, Yahya L, Triviño A, Alharthi M, Ghoneim SS (2021) A comparison between particle swarm and grey wolf optimization algorithms for improving the battery autonomy in a photovoltaic system. Appl Sci 11(16):7732
29. Abdulhussein KG, Yasin NM, Hasan IJ (2021) Comparison between butterfly optimization algorithm and particle swarm optimization for tuning cascade PID control system of PMDC motor. Int J Pow Elec & Dri Syst ISSN 2088(8694):8694
30. Assiri AS (2021) On the performance improvement of butterfly optimization approaches for global optimization and feature selection. PLoS ONE 16(1):e0242612
31. David D, Widayanti T, Khairuzzahman MQ (2019) Performance comparison of cat swarm optimization and genetic algorithm on optimizing functions. In: 2019 1st International conference on cybernetics and intelligent system (ICORIS), vol 1. IEEE, pp 35–39
32. Pratama DH, Suyanto S (2020) Comparison of PSO, FA, and BA for discrete optimization problems. In: 2020 3rd International seminar on research of information technology and intelligent systems (ISRITI). IEEE, pp 17–20
33. Saleh AA, Mohamed AAA, Hemeida AM, Ibrahim AA (2018) Comparison of different optimization techniques for optimal allocation of multiple distribution generation. In: 2018 International conference on innovative trends in computer engineering (ITCE). IEEE, pp 317–323
34. Dorigo M, Birattari M, Stutzle T (2006) Ant colony optimization: artificial ant as a computational intelligence technique. University libre de bruxelles. IRIDIA Technical report Series, Belgium, Tech. Rep
35. Ning J, Zhang C, Sun P, Feng Y (2019) Comparative study of ant colony algorithms for multi-objective optimization. Information 10(1):11
36. Fister I, Fister I Jr, Yang XS, Brest J (2013) A comprehensive review of firefly algorithms. Swarm Evol Comput 13:34–46
37. Custódio AL, Madeira JA (2015) GLODS: global and local optimization
38. Qin P, Hu H, Yang Z (2021) The improved grasshopper optimization algorithm and its applications. Sci Rep 11(1):1–14
39. Saremi S, Mirjalili S, Lewis A (2017) Grasshopper optimisation algorithm: theory and application. Adv Eng Softw 105:30–47
40. Mirjalili SZ, Mirjalili S, Saremi S, Faris H, Aljarah I (2018) Grasshopper optimization algorithm for multi-objective optimization problems. Appl Intell 48(4):805–820
41. Nature based optimization algorithms for travelling salesman problem solution. Int J Manag Appl Sci 1(2):43–47

Chapter 8
Assessment of firm's Performance by Employing Text Mining Techniques

N. Nowshith Parveen, D. Kavitha, B. Uma Maheswari, and R. Sujatha

1 Introduction

The proliferation of corporate information available to investors from various channels has been the subject of numerous studies. Chakraborty and Bhattacharjee [1] perceived that the limitation of the manual content analysis methodologies has been overcome by the usage of automated methods in disclosure studies. Chhatwani [2] determined that studies using automated methods for disclosure analysis have used various sources such as websites, analyst reports, annual reports, and news. Tran et al. [3] perceived that the firm has considerable flexibility in the content that it can convey through its narrative sections, and can possibly change the interpretation of the message that is conveyed through the quantitative information. However, as the information contained in the narrative sections of the annual report are voluntary in nature, the firm has the ability to change the content to suit its information requirement.

Yearly reports are documents that reflect a firm's operations and financial results. Financial statements are used by shareholders, lenders, and industry analysts to evaluate a firm and to arrive at conclusions. The financial data that is included in the yearly report is made up of both text and numbers. The textual data comprises Chairman's speech, Management discussion and analysis, Director's report, Notes to account, CSR report, etc., Pradeep et al. [4] suggested that the text information is as significant as the arithmetic information and they give end user of annual reports with useful insight for making wise choices. These reports reveal information regarding management's practices and how they are translated into behavior of the company.

Text analysis can be employed to assess this text data efficiently. Text mining is the process of drawing out valuable data from disorganized text. It entails discovering

N. N. Parveen (✉) · D. Kavitha · B. U. Maheswari · R. Sujatha
PSG Institute of Management, Coimbatore 641004, India
e-mail: parveenahamednp@gmail.com

G. Mathur et al. (eds.), *Proceedings of 3rd International Conference on Artificial Intelligence: Advances and Applications*, Algorithms for Intelligent Systems,
https://doi.org/10.1007/978-981-19-7041-2_8

and evaluating trends and patterns in order to derive useful information from data. Text analysis is a method of extracting subjective information from unstructured text. Allahyari et al. [5] proposed that the overview pertaining to the previous or current performance of a company is easily accessed through the annual report, which also provides insight into its future prospects. As of now, analysts have been analyzing financial data and making meaningful deductions. Owing to current fads and difficulties in financial statements, analysts must analyze disorganized text data to evaluate the efficiency of the company. Text mining methods can be used for collecting relevant information and to provide relevant stakeholders with useful data.

Mukhopadhyay [6] ascertained that analyzing and comprehending the annual report, as well as taking appropriate decisions, are emerging as core objectives of investors, analysts, shareholders, etc., that can make or break the future of any business. Several strategies for identifying, retrieving, and categorizing ideas, sentiments, and emotions of unstructured text data have emerged. This method is known as sentiment analysis. Andrea et al. [7] stated that an opinion "is essentially a favorable or unfavorable sentiment, perspective, outlook or emotion of an entity or an aspect of the entity" expressed by a person at a point in time. Sentiment analysis aims to discover individuality in content and gather and categorize views and sentiments. Cao et al. [8] proposed that emotion analysis is employed to capture the emotive aspects in the annual reports. Reading ease is the ability to understand the content of mandatory disclosures. Loughran and McDonald [9] perceived that the financial disclosures are the main source of data for investors and the textual properties have a significant bearing on the effectiveness with which valuable data is communicated between the firm and the market.

2 Literature Review

Earlier experts examined that textual analysis could be used as predictor for future financial performance and have found that current performance plays an important role in the emotions in the annual report.

The linguistic narrative accounts for the core of the reporting in a conventional company report. Text data contain a great deal of emotion when it comes to a company's performance. Lo et al. [10] proposed that reviewing of the core elements of financial reports is the focus of text analytics. Comprehending and deciphering the facts presented in the report is dependent on the accuracy of this substantial portion of obligatory disclosure. Firms that have managed earnings compared to the previous year have more complicated MD&A.

Yearly reports contain implied key data that can be extremely beneficial to stockholders. Text analytics refers to new approaches for converting contextual information into meaningful information about a company's sustainability efforts. Raghupathi et al. [11] proposed that by using text analytics, it is possible to glean valuable insight into company performance by analyzing the textual information in the Annual report. We can learn about the shareholders' opinions and goals by examining shareholder resolutions. Shareholders, for example, might have an impact on a company's long-term viability through resolutions. Text mining methods are used to provide understanding to scholars investigating large amounts of fragmented text, allowing for a better grasp of shareholder resolutions.

Text analysis can be performed to uncover content that illustrates how value relevance can be modified utilizing various automated approaches and source materials of information. Chakraborty and Bhattacharjee [1] performed research on the influence of such a tone on the economy. It aids experts in predicting earnings and returns, as well as stakeholders in making decisions.

NLP methods are widely employed in Text analysis, including the building of Semantic Networks and Emotion Mining techniques. Natural language processing algorithms allow investment companies to read news announcements in a split second. As a result, it's critical for public firms to arrange their interactions in such a manner that takes into consideration how well the economy processes relevant data and avoids excessive fluctuation. Sai et al. [4] performed text analytics on the annual reports for 12 selected companies from the Indian IT sector spanning from 2015 to 2018. Company types include both product and services-oriented companies. It was discovered that the future performance of the firm has a direct relationship with the emotions in annual reports.

Andrea et al. [7] proposed that sentiment analysis is a recent branch of Natural Language Processing (NLP) research that aims to discover subjectivity in content and/or retrieve and categorize attitudes and emotions. Lin and Mao [12] ascertained that various government agencies and public bodies analyze the information propagating in social networks using personality-based sentiment analysis as this is a way in which personality influences the way people write and talk. Prior to making a purchase decision, many consumers read reviews posted by other consumers. It is common for people to express their opinion on a number of entities. This has led to the growth of opinion mining. Bhonde et al. [13] suggested that an entity's expressed opinion about it is evaluated using Sentiment Analysis to determine whether it is positive or negative.

Hardeniya and Borikar [14] examined that there are two approaches to sentiment analysis: machine learning-based and dictionary-based. The sentiment analysis of documents is based on machine learning algorithms that include Support Vector Machine, Naive Bayes, and Neural Networks. An algorithm based on machine learning identifies text by analyzing it through classification techniques such as support vector machines or neural networks.

Machine learning techniques have had a significant effect on a variety of businesses, commencing during the last decade and continuing now, with the potential of becoming ubiquitous. Data contained in various quasi and fragmented sources of data could be extracted with the use of machine learning. Machine learning aids in financial fraud detection, stock price prediction, and market forecasting using online text mining.

Statistics are applied to text analytics to understand trends and patterns, which provides highly accurate information. It's accomplished by working on the unstructured information scattered across different sections in detail. This is accomplished by obtaining important data, translating it to quantitative form, and then applying appropriate machine learning algorithms. Bach et al. [15] suggested that analysis of vast amounts of data is conducted by companies, predominantly based on news and media posts on stock market predictions and fraud detections. Focus should be on external data sources, such as news and online media posts.

In lexicon-based approaches to sentiment analysis, individual words are polarized and their scores are aggregated to determine the polarity of the text. The further classification of Lexicon-based approaches can be done into two subcategories: (1) Corpus-based approach, (2) Dictionary-based approach. Dictionary-based methodology utilizes dictionaries that contain sentiment words which are matched with the textual data to determine polarity. Askerov et al. [16] assessed the emotions of communication contents and looked into distinct communication trends across prominent technological businesses. The stock performance of the firm has strong correlation with Loughran McDonald dictionary derived sentiment scores.

Todevski [17] proposed that for determining the probability of key phrases, the text analytics scenario can be illustrated using a bag of words and a word cloud. The goal of sentiment analysis is to establish the stated viewpoint that is obviously linked to the current and subsequent business periods.

Li et al. [18] inferred opinion indicators, such as tonal shifts in financial reports, that have been efficiently used to estimate share price in recent articles. In the long haul, shifts in the tonality of financial reports can forecast share price, implying relatively low market efficiency.

Chhatwani [2] examined the financial reports for the year 2018–19 and analyzing the range and existence of sentiment and readability across several different segments of the financial reports, including the Chairman's Speech, the Director's Report, Management Discussion & Analysis (MD&A), and Notes to Accounts, was done utilizing the NIFTY 100 companies as a sample. Despite the fact that such segments are all included in almost the same yearly statements, their comprehension and emotion values diverge. Chairman's speech had the lowest reading score, considering it the simplest to comprehend, and the highest sentiment score, indicating that it is written in a positive manner. The Director's Report, MD&A, and Notes to Account, on the other hand, have higher comprehension values but lesser sentiment values.

Significant emphasis has been placed on valuation tactics in the wake of the Initial Public Offering (IPO) process. Deokar and Tao [19] developed FOCAS-IE for extracting sentiments from MD&As by textual analysis of the prospectus parts containing MD&A. FOCAS-IE provided information that was incorporated into a predictive model to predict IPO closing prices.

Tao et al. [20] proposed that forward-looking statements (FLSs) provide valuable insight in applications such as stock price forecasting. FLSs are found in the Management Discussion & Analysis (MD&A) portions of IPO prospectuses as well as offer anticipated data on the firm's projected efficiency and advancement. FLS traits, as per the sentiment analysis, are often more accurate at forecasting before IPO valuations than after IPO valuations.

Gupta et al. [21] examined the significant impact of text-mining technology on the finance sector. Text mining has evolved as a vital phenomenon under investigation in the sphere of finances since the amount of information across every domain of finances has increased dramatically. As a result, analyzing relevant findings on text-mining applications in finances might aid in identifying areas where more study is needed. Text mining, which was initially assumed to be an irregular element in the securities industry, has cast doubt on this idea. The banking industry, on the other hand, has witnessed continual innovative developments. Text mining can be used to retrieve financial statements and corporate reports.

3 Research Gap

In the past, sentiment analysis and readability analysis were performed manually by humans, demanding more time, human resources, price, and effects. As a means of overcoming the above limitations, text analysis can serve as a useful tool for stakeholders and investors to determine the sentiment, emotions, and readability of texts. This study aims to provide useful information about a firm's profits by examining the text section of a yearly report. An analysis of emotions on future performance of the firm was conducted using a statistical model.

4 Research Methodology

The data used here has been gathered from annual reports of NIFTY 100 companies present in NSE of India. Chairman's speech and Management discussion and analysis from the year 2017 to the year 2021 were two different textual contents used in this study. Financial companies are excluded because their annual reports contain disclosures that differ from those of non-financial firms. Data was obtained from Prowess IQ and Capitaline databases. From Table 1, we can infer that a total of 385 MD&A reports and 385 chairman's speeches have been collected excluding the financial firms.

Table 1 Selection of samples

Selection of samples	Number of companies	Number of firm year observations	Percentage
Preliminary sample	100	500	100
Final sample after excluding financial firms	77	385	77

4.1 Sentiment Analysis

The management discussion and analysis & Chairman's speech portion of the annual report was subjected to sentiment analysis. Using R studio, this analysis was carried out. Chopra and Bhatia [22] analyzed the steps involved in sentiment analysis and put forth that it includes obtaining text data, preprocessing, analyzing, classification, and scoring of sentiment. Figure 1 depicts the steps involved in the process of sentiment analysis.

- The process of sentiment analysis begins with the collection of text, which will be used to generate data.
- Pre-processing turns the raw text into a more structured version. Commonly occurring words, such as articles, prepositions, conjunctions, and adverbs, may be eliminated through removal of stop words. Stemmed words are those that have been reduced to their root form for information retrieval purposes.
- The essence of sentiment analysis can be summed up by adding sentiment annotations to pre-processed records using linguistic tools.
- Text is classified based on its sentiment as positive, negative, or neutral by using Loughran and McDonald dictionaries.
- Scoring of sentiment: The L&M dictionary detects the emotions in the text data, resulting in the sentiment scores assigned to them.

The reports were assessed for sentiment and emotions. The Diction and Harvard GI word lists were commonly used by researchers. According to Henry and Leone [23], the language used in the accounting and finance domain tends to be very specific, meaning that general dictionaries do not have a great deal of predictive power in the financial domain. By examining word usage in 10Ks and creating a dictionary which is referred to as the L & M dictionary, Loughran and McDonald [24] designed a vocabulary that specifically addresses financial text. To determine the sentiment from the MD&A & Chairman's speech, Loughran and McDonald dictionary is used to categorize the narrative into either positive or negative. The sentiments include

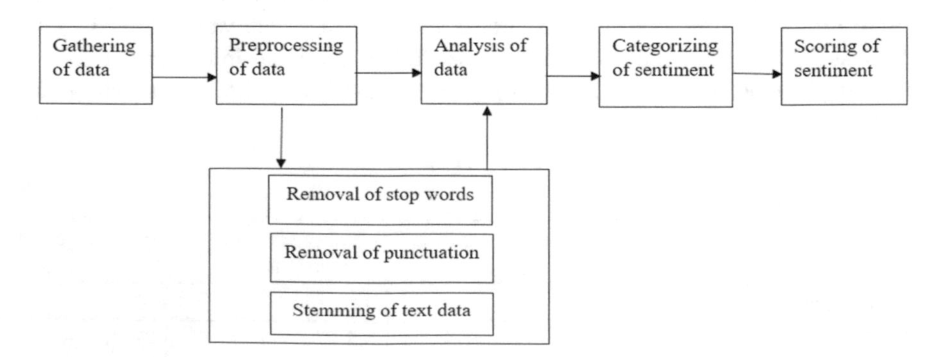

Fig. 1 Sentiment analysis process

both positive and negative elements. The NRC Word-Emotion Association Lexicon (aka EmoLex) is used to classify emotions. There are eight basic emotions listed in the NRC Emotion Lexicon. They are agitation, panic, excitement, faith, astonishment, grief, delight, and dismay are the emotions.

4.2 Readability

Understanding and interpreting the contents of the reports is not possible without clear content. Lo et al. [10] examined the content in the firms' disclosure and concluded that is becoming exceedingly difficult to interpret, which concerns investors. By definition, the Fog proposes that higher the number of words in the sentences or higher the number of syllables in a word imply a greater Fog, thus making a text harder to understand. In order to assess the readability of a text, the Gunning Fog Index is used. Table 2 represents the education level and grasping level required for respective fog scores.

4.3 Multiple Linear Regression

Yong and Awang [25] said that multiple Linear Regression provides a comprehensive explanation of the relationship between a response variable and its predictor variables. The general equation of multiple linear regression model is given below

$$y = \alpha 0 + \alpha 1x1 + \alpha 2x2 + \alpha 3x3 + \ldots. \quad (1)$$

where $\alpha 0$ is intercept

Table 2 Fog score classification

Fog score	Education level	Grasping level
Less than or equal to 5	Kindergarten through 5th grade	Simple and easy to read
6	6th grade	Easy to read
7	7th grade	Fairly easy to read
8	8th grade	Conversational English
9–12	9th through 12th grade	Fairly difficult to read
13–16	College	Hard to read
17	College graduate	Very hard to read
Greater than or equal to 18	Professional	Extremely hard to read

Table 3 Sentiment scores for chairman speech

Year	Sentiment score	Positive score	Negative score
2017	0.02888	0.05885	0.02997
2018	0.02397	0.05171	0.02773
2019	0.024327	0.054099	0.029772
2020	0.011844	0.05209	0.04025
2021	0.01553	0.05286	0.03734

$\alpha 1, \alpha 2, \alpha 3$ are the coefficients
$x1, x2, x3$ are the independent variables
y is the dependent variable.

A statistical model was constructed to assess the impact of emotions on the future performance of the firm.

5 Analysis and Interpretation

5.1 Sentiment Scores for Chairman's Speech

Table 3 consists of the respective sentiment scores obtained using Loughran and McDonald dictionary. From the table, we can infer that the positive sentiment score was the highest for the year 2017 while it was lower for the years 2018, 2020, and 2021. The year 2020 and 2021 exhibited more negative sentiment.

5.2 Sentiment Scores for Management Discussion and Analysis

Table 4 consists of the respective sentiment scores obtained using Loughran and McDonald dictionary. From the table, we can infer that the positive sentiment score was the highest for the year 2018 while it was lower for the year 2020 and 2021. Similar to the chairman's speech, the year 2020 and 2021 exhibited more negative sentiment.

5.3 Emotions for Chairman's Speech

From Fig. 2, we can see that words associated with the positive emotion "trust" occurred predominantly in the text for the year 2021. The underlying emotion of the Chairman's speech is positive, exhibiting a positive trend between 2017 and 2019. On

Table 4 Sentiment scores for MD&A

Year	Sentiment score	Positive score	Negative score
2017	0.0079487	0.03965	0.03171
2018	0.0093242	0.03948	0.03016
2019	0.012258	0.04109	0.02883
2020	0.0033010	0.03784	0.03454
2021	0.0082383	0.04003	0.03179

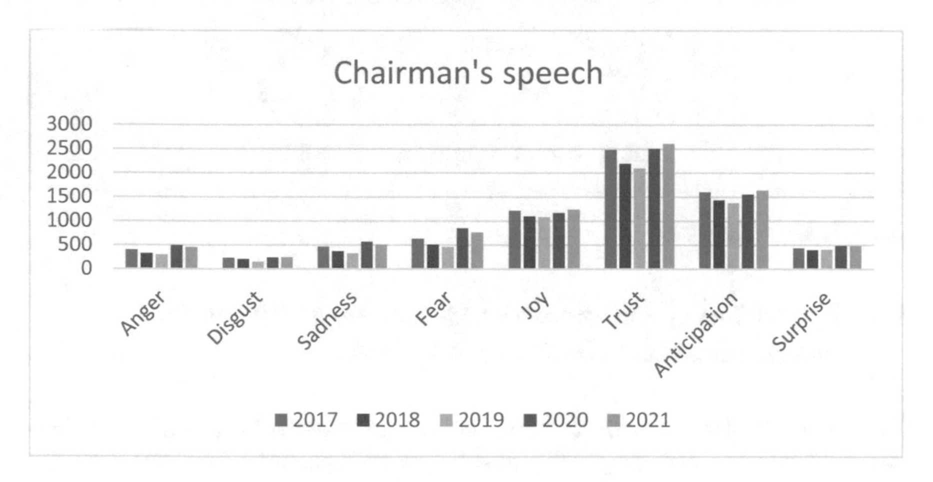

Fig. 2 Emotions for chairman speech

the other hand, negative emotions occurred predominantly between 2020 and 2021. Chairman's speech positivity showed a general upward trend in terms of negative emotions between 2020 and 2021.

5.4 *Emotions for Management Discussion and Analysis*

From Fig. 3, we can conclude that words linked with the positive emotion of "trust" are high between 2017 and 2019. 2019 was the best year yet in terms of returns. Positive outlook in the MD&A section of the annual report points to improved current performance. MD&A report positivity can be measured by emotions and revealed to be on the rise. In contrast, there was a considerable increase in negative emotions between 2020 and 2021. There was also a noticeable decrease in trust.

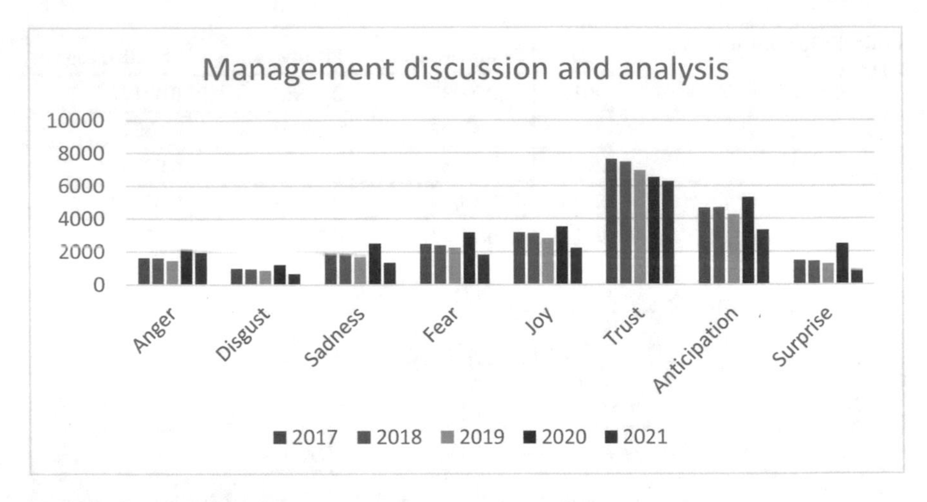

Fig. 3 Emotions for MD&A

5.5 Proportion of Emotions for Management Discussion and Analysis and Chairman's Speech

Figure 4 shows that the emotion "trust" appears to have the longest bar. This indicates that words contributing to this positive emotion comprise just over 25%. The bar corresponding to the emotion of "disgust" is short and shows that words that expressed this negative emotion accounted for less than 7.5%. The majority of the words in the text indicate that positive emotions account for more than 60% of the text, which can be seen as a good sign.

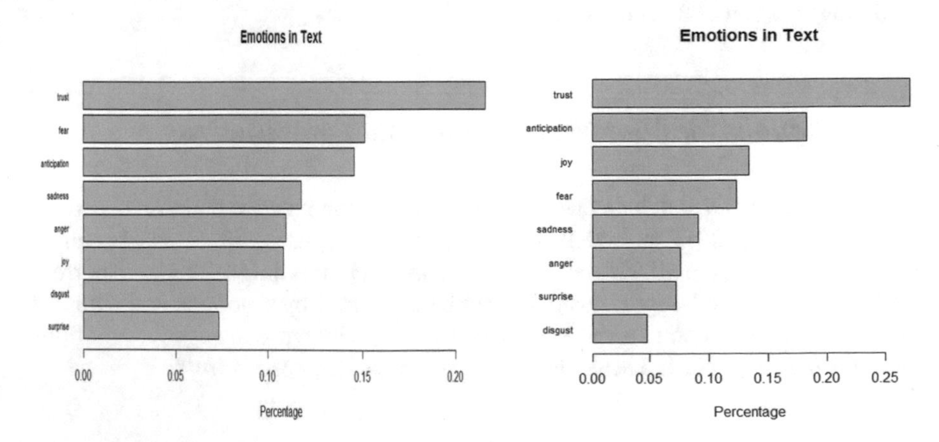

Fig. 4 Proportion of emotions in chairman's speech and MD&A

Table 5 Fog score

Chairman's speech		Management discussion and analysis	
Year	Fog score	Year	Fog score
2017	15.96	2017	15.91
2018	15.98	2018	16.14
2019	15.88	2019	15.99
2020	16.11	2020	16.19
2021	16.36	2021	16.40

5.6 Readability of Chairman's Speech & Management Discussion and Analysis

The chairman's speech and the MD&A section of the annual report were tested for readability. From Table 5, we can conclude that the fog score for the years 2020 and 2021 was high. This is high compared to the previous years. When we compare the readability with the sentiments, we can observe that the level of sentiment was low for 2020 and 2021 and the readability score was also high for the same period. The returns of the companies under study were not high during the year 2020 while it was recovering during 2021, which can also be considered as the post covid era. This could also be the reason for low sentiment score and increased fog score.

5.7 Statistical Model for Chairman's Speech

To establish a relationship between emotions and the subsequent years' return on assets, a multilinear regression model is developed.

Predictor variable: Anger + Fear + Trust + Sadness + Anticipation + Disgust + Surprise

Response variable: Return on assets

From Table 6, it can be inferred that the R square value is 0.74. It indicates that 74% of variability on average returns can be explained by independent variables namely anger, fear, trust, sadness, anticipation, disgust, and surprise. Adjusted R square is considered as an estimate for variability without bias which is (0.74) which is 74% of variation.

Table 6 Regression value for chairmans speech

Regression (Sig)	0.000
R squared	0.74
Adjusted R squared	0.74

Table 7 Significance values for chairmans speech

Emotion	Sig
Anger	0.033
Fear	0.004
Trust	0.000
Sadness	0.031
Anticipation	0.216
Disgust	0.358
Surprise	0.034

The significance value of the multiple linear regression model in Table 6 confirms the model built is statistically significant because significance level is less than 0.05. The overall model is proved to be statistically significant and it also makes sense to understand the importance of individual factors and its statistical significance towards the model of asset returns.

From Table 7, we can infer that emotions play a vital role in average returns. The significant values of anger, sadness, fear, Joy, trust, and surprise are 0.033, 0.031, 0.004, 0.016, 0.000, and 0.034 which is less than 0.05. Therefore, it is proved to be statistically significant. Whereas the significance values of disgust and anticipation are 0.358 and 0.216 which is above 0.05. Hence it is statistically insignificant.

The model can be characterized as follows:

Return on assets = 0.010 + 0.033 (anger) + 0.000 (trust) + 0.004 (fear) + 0.358 (disgust) + 0.216 (anticipation) + 0.034 (surprise) + 0.031 (sadness)

5.8 *Statistical Model for Management Discussion and Analysis*

Predictor variable: Anger + Trust + Sadness + Anticipation + Disgust + Surprise

Response variable: Return on assets

From Table 8, it can be inferred that the R square value is 0.73. It indicates that 73% of variability on average returns can be explained by independent variables namely anger, trust, sadness, anticipation, disgust, and surprise. Adjusted R square is considered as an estimate for variability without bias which is (0.72) which is 72% of variation.

Table 8 Regression value for MD&A

Regression (Sig)	0.000
R squared	0.73
Adjusted R squared	0.72

Table 9 Significance values for MD&A

Emotion	Sig
Anger	0.036
Fear	0.036
Trust	0.000
Sadness	0.094
Anticipation	0.024
Disgust	0.034
Surprise	0.466

The significance value of the multiple linear regression model in Table 8 confirms the model built is statistically significant because significance level is less than 0.05. The overall model is proved to be statistically significant and it also makes sense to understand the importance of individual factors and its statistical significance towards the model of asset returns.

From Table 9, we can infer that emotions play a vital role in average returns. The significant values of anger, disgust, fear, joy, trust, and anticipation are 0.036, 0.034, 0.036, 0.016, 0.000, and 0.024 which is less than 0.05. Therefore, it is proved to be statistically significant. Whereas the significance values of sadness and surprise are 0.094 and 0.466 which is above 0.05. Hence it is statistically insignificant.

The model can be characterized as follows:

Return on assets = 0.032 + 0.036 (anger) + 0.000 (trust) + 0.034 (disgust) + 0.024 (anticipation) + 0.036 (fear) + 0.016 (joy)

6 Discussion and Implications

The annual report is the conduit through which the management communicates information about the corporation to internal and external stakeholders. With the changes happening across the economic and environmental conditions, it is becoming increasingly important for stakeholders of financial statements, such as analysts, advisors, and investors to dive deeply into the enormous amount of unstructured data embedded in financial statements. Globally, firms' social responsibilities are under greater scrutiny as the business environment changes. Several national and international regulations protect corporate social responsibility. A company's annual report must include information about its activities in this field. The information is reported in text form. It is possible to use text analysis to extract the information that is not formally structured from annual reports, so that stakeholders can use the results to make informed decisions about companies according to their performance. There are no limitations to text analysis in terms of time, manpower, or cost. For large volumes of data, the computation is easy. This study has examined the performance of NIFTY

100 companies, with an emphasis on analyzing how emotions in the annual reports influence the company's performance presently and how future performance may be affected.

7 Conclusion and Recommendations

Text analysis is a new field of research. Although it is being used in a variety of fields, there are still other areas where Text Mining can make a significant difference. Source code, in association with the culture of continuous improvement, must be used to analyze and transform text information into organized information, from which analysts and others can draw useful conclusions.

Analyzing yearly reports to comprehend the firm's growth is a time-consuming effort for researchers and consultants. The shifts that occur throughout the period will exacerbate the problem. Text mining, when used effectively, can help to speed up these processes and give correct insights concealed in text material. This research focuses on the NIFTY100 firms, with the goal of determining how sentiments in financial reports as a whole influence the firm's future health. Greater sample should be considered, as well as delving deeper into other portions of the annual reports. Further research can also be done with respect to a specific sector.

References

1. Chakraborty B, Bhattacharjee T (2020) A review on textual analysis of corporate disclosure according to the evolution of different automated methods. J Financ Report Account
2. Chhatwani M (2021) Readability and sentiment analysis of financial statements: Evidence from India. SCMS J Indian Manag 18(3):87–94
3. Tran AD, Pallant JI, Johnson LW (2021) Exploring the impact of chatbots on consumer sentiment and expectations in retail. J Retail Consum Serv 63:102718
4. Sai PK, Gupta P, Fernandes SF (2019) Analysing performance of company through annual reports using text analytics. In: International conference on digitization (ICD). IEEE, pp 21–31
5. Allahyari M, Pouriyeh S, Assefi M, Safaei S, Trippe ED, Gutierrez JB, Kochut K (2017) A brief survey of text mining: Classification, clustering and extraction techniques. arXiv preprint arXiv:1707.02919
6. Mukhopadhyay S (2018) Opinion mining in management research: the state of the art and the way forward. Opsearch 55(2):221–250
7. Alessia D, Ferri F, Grifoni P, Guzzo T (2015) Approaches, tools and applications for sentiment analysis implementation. Int J Comput Appl 125(3)
8. Cao L, Peng S, Yin P, Zhou Y, Yang A, Li X (2020) A survey of emotion analysis in text based on deep learning. In: IEEE 8th International conference on smart city and informatization (iSCI). pp 81–88
9. Loughran T, McDonald B (2014) Measuring readability in financial disclosures. J Financ 69(4):1643–1671
10. Lo K, Ramos F, Rogo R (2017) Earnings management and annual report readability. J Account Econ 63(1):1–25

11. Raghupathi V, Ren J, Raghupathi W (2020) Identifying corporate sustainability issues by analyzing shareholder resolutions: A machine-learning text analytics approach. Sustain 12(11):4753
12. Lin J, Mao W (2015) Personality based public sentiment classification in microblog. In: IEEE International conference on intelligence and security informatics (ISI). pp 151–153
13. Bhonde R, Bhagwat B, Ingulkar S, Pande A (2015) Sentiment analysis based on dictionary approach. Int J Emerg Eng Res Technol 3(1):51–55
14. Hardeniya T, Borikar DA (2016) Dictionary based approach to sentiment analysis-a review. Int J Emerg Eng Res Technol 2(5):239438
15. Pejić Bach M, Krstić Ž, Seljan S, Turulja L (2019) Text mining for big data analysis in financial sector: A literature review. Sustain 11(5):1277
16. Askerov R, Kwon E, Song LM, Weber D, Schaer O, Dadgostari F, Adams S (2020) Natural language processing for company financial communication style. In: Systems and Information Engineering Design Symposium (SIEDS). IEEE, pp 1–6
17. Todevski D (2020) Text analytics on the case of Macedonian companies. J Econ 5(1):39–46
18. Li J, Luo W, Deng X (2019) The Effect of chairman's statement tone changes in annual reports from Hong Kong. In J Phys: Conf Ser 1168(3):032024
19. Deokar A, Tao J (2015) Text mining for studying management's confidence in IPO prospectuses and IPO valuations
20. Tao J, Deokar AV, Deshmukh A (2018) Analysing forward-looking statements in initial public offering prospectuses: a text analytics approach. J Bus Anal 1(1):54–70
21. Gupta A, Dengre V, Kheruwala HA, Shah M (2020) Comprehensive review of text-mining applications in finance. Financ Innov 6(1):1–25
22. Chopra FK, Bhatia R (2016) Sentiment analyzing by dictionary based approach. Int J Comput Appl 152(5):32–34
23. Henry E, Leone AJ (2016) Measuring qualitative information in capital markets research: Comparison of alternative methodologies to measure disclosure tone. Account Rev 91(1):153–178
24. Loughran T, McDonald B (2011) When is a liability not a liability? Textual analysis, dictionaries, and 10-Ks. J Financ 66(1):35–65
25. Ng KY, Awang N (2018) Multiple linear regression and regression with time series error models in forecasting PM10 concentrations in Peninsular Malaysia. Environ Monit Assess 190(2):1–11

Chapter 9
Decisions Prediction Techniques Using Language Processing and Learning Algorithms

Aastha Budhiraja and Kamlesh Sharma

1 Introduction

A large number of cases are filed daily in courts seeking a fair result, but only a small percentage of them are resolved in the shortest period of time. It is unquestionably critical to enhance the functioning of the Indian law in order to guarantee that the legal software judgments are correct and logical, and that they are made in a timely manner as discussed by Chih-Fong and Jung-Hsiang [1]. In addition, attorneys are needed to research such cases and present respective clients in court. It takes an inordinate amount of time and effort for a lawyer to do substantial legal research in favor of a client's case. In this contemporary era, if a lawyer is supported by a machine system to provide appropriate law pertinent to the matter, any lawyer's work would be decreased. This service's result may become the ability to forecast court decisions taking shorter time. In a specified time, every claimant may obtain the judgment. Legal cases will take less time and cost less money. Machine Learning is an application of AI Technologies that includes a variety of approaches to assist machines in understanding a variety of datasets. Language processor focuses on teaching machines how to handle and evaluate huge amounts of complex data. The suggested approach provided by Virtucio et al. [2] offers a blueprint using NLP to forecast strategic planning for the Indian law study. This issue is mostly viewed as a supervised learning explained by Katz et al. [3]. For predictive data mining problems, Sulea et al. [4] showed the Support Vector Machine (SVM) technique.

A. Budhiraja (✉)
Research Scholar, Manav Rachna International Institute of Research and Studies, Faridabad, Haryana, India
e-mail: aasthakohli0410@gmail.com

K. Sharma
Professor, Manav Rachna International Institute of Research and Studies, Faridabad, Haryana, India
e-mail: kamlesh.fet@miru.edu.in

G. Mathur et al. (eds.), *Proceedings of 3rd International Conference on Artificial Intelligence: Advances and Applications*, Algorithms for Intelligent Systems,
https://doi.org/10.1007/978-981-19-7041-2_9

SVM is a fascinating method with straightforward ideas. The classifier uses the hyper plane with the greatest margin to separate data points. Learning sets of data to support vectors that form the classified decision boundary. The hyper plane with the greatest distance from a nearby training data point of any functional margin achieves excellent classification. For text categorization, an SVM is employed. The SVM's job is to categorize data into a set of areas depending on structure. The automation of law procedure allows us to have a deeper understanding of the procedures that go into drafting and enforcing laws. In the duties of current humanities research, elements of predictive modelling are a vital instrument for its growth. The technical advantages of machine learning-based analysis are valuable and significant in concerned field, as governance procedures are confined to the implementation of laws. Data analysis and modelling of legal processes can be used to determine the structure and elements of the overall public complex process, as well as the connections between civil system elements and their interactions with scientific, technical, social, economic, political, and cultural aspects of the environment. The primary goal of the research is to create ground-breaking technologies, techniques, and tools to aid in the administration of law making, law enforcement, and legal judgments in order improve the legal system presented in detailed study given by Bielen et al. [5].

The next section of this article discusses a heterogeneous challenge in data processing. The following part gives an overview of the work done so far and the current status of research in this field. It is demonstrated that the field requires minor development. The final section discusses the methodologies, technologies, and tools for collecting and pre-processing tens of millions of court judgments, including the use of Apache Spark for distributed data processing.

The study also proposes a strategy for finding similarities in policy making and local authorities by examining a set of data from court judgments on punishment for code violations. An initial phase to decide whether a theory is required; a precise definition of the ontology's domain; a specification of the task to which the ontology is dedicated; identification of domain concepts and relationships among them; the collection of the concepts and relations in an ontology formalized in an appropriate language to make machine readable; the integrator. Concentrating on the stage of recognizing ideas and their relationships to believe that using NLP approaches by domain writers can enhance this phase. Ontologies are made up of ideas and their relationships, and they organize a list of entities for believing that words encapsulate ideas and that semantic connections between concepts are encoded in syntactical relationships between these terms. Law ideas have definitions which might vary depending on a variety of variables (context, source, etc.). There are a variety of legal ontologies that may be developed, with the aspects varying based on the goal for which the ontologies were created. The publication's main emphasis is a written document analysis, from which data was collected using natural language processing (NLP) techniques such as patterns, latent semantic analysis (LSA), and TF-IDF.NLP, on the other hand, is deserving of a mention.

2 Literature Review

Besides generations, constitutional experts have used several analytical methods, that include referencing policies, addressing practical issues, contributing evaluative remarks to case laws, as well as creative research design (systematization), with the more basic versions of that research serving as the required basic components for the more complex things. The research process "provides a methodical explanation of the rules controlling a given legal category, analyses the connection between rules, clarifies areas of ambiguity, and, maybe, anticipates future changes."

Judicial research is a method of knowledge extraction from past data in order to aid lawyers. It entails collecting data from court filings and compiling the data to produce newly discovered judgments. The subsection emphasis on research that has previously been done in this area. Though several approaches, including as classification, clustering, and regression techniques, have been used to discover and analyse legal allusions. Regression and neural network are more effective than any other functions in predicting legal information.

Natural language processing (NLP), Artificial Intelligence (AI), Machine Learning (ML), and Network Analysis are some of the key strategic planning techniques utilized in the public profession. Because law things are expressed in natural languages in numerous legal documents, NLP-based solutions are applicable in this sector. Past researchers have shown that named entity recognition and POS tagging may be used to extract legal entities, information, and identify actors, among other things. Aggregating all related documents to improve performance, wide-ranging machine learning techniques like as clustering and text classification, as in any other information retrieval system, can be utilized.

Virtual Reality (AI) optimizes the concern and logical reasoning. Court writings, papers, and standards each use their own conceptual characteristics similar to work expressed by Frankenreiter [6] as evidence using norms. Smart systems perform data gathering, knowledge model construction, and system implementation in the legal area. There are a variety of tools available for evaluating intelligent systems in the legal area. The categorization system sorts legal situations into the most relevant categories so that a knowledge model may be built if necessary.

The design of a concept key findings to classify the factors of lead time at multiple stages, including nation, justice, and instances, from the first and third levels receiving the most attention. This is the first study to use a simple regression technique to examine the duration of civil cases in Belgian First Instance courts. The markers were chosen based on the available research, and the author utilized a database of 174 instances. Following talks with a team of legal and academic experts, additional indications were developed Economides et al. [7], concentrating on cases that were appealed to the council evaluating construction and settlement issues.

The creation of theories and techniques to understand the factors that influence the length of civil litigation. This paper provides an analytical resource for upcoming academic research and sets the groundwork for the development of improved time regulations that will allow civil courts to be observed more reliably, supervised, and evaluated across countries. Comparing investigations of the topic of lead time in civil justice, the authors suggested a shared lexicon and experimental method.

The authors compared the courts using Data Envelopment Analysis (DEA) and utilized multiple regressions to address the causes of efficient disparities. The Conselho Nacional de Justiça (CNJ) in Brazil included these metrics comply with various parameters presented by Hanson et al. [8] as assessed by mean, median, and standard deviation in the Justice in Numbers Report: lead—times; unsettled case lead time; trial lead time; interruption lead time.

Time is tracked in these kinds of indexes using multiple metrics with distinct markings for the start and finish of the litigation. Deployment and courtroom latency are the key terms explored in the relational study of the judicial realm given in this article as measures of proceedings length.

Geist and Anton [9] presented ten Court Tools system indicators, including disposition time, which is the proportion of cases eliminated or solved within the period that may be used to assess court performance in a controllable fashion. The CPMS—Court Performance Measure System—at the European Court of Justice is based on the balanced scorecard and includes the following five dimensions described Chen et al. [10] client satisfaction, internal operation, financial characteristics, innovation and learning, and the success of information systems, as defined by the international laboratory for the study of judicial systems. One of the indications of internal operation is the time it takes to process a request.

The most popular method is to employ a text analysis model to identify the similarity between two texts and use it as the edge weight. Synonyms and homonyms are an issue when using text similarity measures like cosine similarity and vector space models. Furthermore, the document's whole text may not be an accurate measure of document similarity. In the case of papers with identical structures, various weight values for different portions of the document can be utilized as Hall et al. [11] explained to calculate the total weight of the link.

The quantitative study of case law has a longer history in the United States than in other areas of the world. Several quantitative analyses of datasets including case law from American courts have been conducted. The majority of these analyses rely on case law that has been painstakingly gathered and coded. Many researches rely on the Supreme Court Database, which provides data on the US Supreme Court's decisions over the previous two centuries that has been gathered and accurately coded as presented by Frankenreiter [6]. A substantial number of these studies look at the connection between a judge's gender or political background and their decision-making.

However, if applying the same methodology to the existing or bigger amounts of data, a continually inconsistent performance seen than what claimed in the study. As a result, a new study is employed for comparable approaches and examined how might gradually they can enhance by utilizing all of the available data.

3 Legal Concepts and Tools

The rule envisions the universe by attempting to control actions. As a result, the technical realm interacts with a variety of subjects such as medicine and science. As a result, many words, whether generic or domain-specific, can be annexed to terms of the law since they name objects or entities perceived by law. Since the law controls and conceptualizes things, which eventually transform into legal entities and legal ideas.

Judicial words are defined as terms that designate particular legal ideas like agreement or responsibility, as well as generic or specialized concepts like passenger, doctor, or weapon: all world things or artifacts seized by law. Words identifying world objects perceived by justice and objects generated by law are classified as legal terms. Advocates must identify applicable law, which includes legal system sources, in order to construct a winning case. Because legal concepts under the common law are created and embedded in prior decisions, attorneys must frequently select a group of factually comparable instances that may be referenced to support their arguments. Branting et al. [12] frequently conducted Judicial authority research with the use of a range of techniques, which divides among these groups: primary, secondary, and tertiary.

The legal system and bill documents are writings which are generated and transmitted by a variety of agents. A judicial report compiled by a lawyer is an example of this type of material.

Documents offering discussion on case law and legislation are referred to as secondary sources (e.g., law journals). Legal digests and citations are tertiary sources that give compilations of systemized case law knowledge. Legal practitioners typically use a combination of primary, secondary, and tertiary sources to gather information.

The phrase ‘Legal Tools’ allows for immediate support assistance created by the ICC Legal Tools Project from 2006, largely with European Union funding. The Legal Tools Database (‘LTD’), the Legal Tools Website, and the Case Matrix are the three major offerings.

The Legal Tools Database is the world’s most comprehensive online resource on international criminal law. It gives free access to all important legal sources in international criminal law, including information on the fundamental international crimes, to legal practitioners, researchers, students, and the general public. Several Legal tools have been used since which Economides et al. [13] explained below:

1. **LegalZoom:** LegalZoom links you with a legal person in the area who is familiar with the law as it applies in your state. A review of your small firm’s contracts and legal papers, as well as on-demand ad-hoc legal assistance, is included in the standard business plan. A specific package for small business start-ups is also available, which includes guidance on establishing your company’s legal structure as well as forms for various sorts of business contracts. Legal Zoom provides both entrepreneurs and business companies with a range of alternative tools and legal consultation plans.

2. **DocuSign:** Managing a business online has unique challenges that brick-and-mortar firms never face. Having legal paperwork signed is one of the most difficult aspects of ecommerce. Customers must print documents, sign them and courier the original copies to you if you don't want to lose business. Fortunately, legal solutions exist to handle this issue, and DocuSign is one of the most well-known digital signature providers. This mechanism ensures that agreements and legal papers are legally binding. Cloud storage and document delivery tracking are also available via the firm.
3. **Trademarkia:** Users could also be seeking for judicial tools to assist you check the authenticity of your trademark, logo, and company name. Trademarkia, a Legal Force service, makes it easy to find patents filed in a variety of countries, such as the United States and Europe. On the Trademarkia website, you may use the search function for free. You may also do a free search on the site for patents, online domain names, and trademarks.
4. **LawDepot:** LawDepot is further admin software for drafting legal papers. Small companies and individuals in the United States will benefit from this service. Unlike Docstoc, which provides downloadable templates, LawDepot walks you through the process of compiling your own papers on its homepage, after which you may obtain the final terms and conditions.

Showing various notations and terminology in this part before formulating the Legal Judgment Prediction job

Law Cases: A case summary and a number of decisions are included in each legal case. The fact description is represented by f, which is a text document as shown in Fig. 1 [14]. Articles of relevant legislation, charges, punishment periods, and so forth may be included in the judgment outcomes. Assume that there are t types of judgment results, and that the i-th judgment result is represented as a categorical variable yi that gets its value from the set Yi. A tuple can therefore be used to represent a law case (f, y1......., yt).

Law Articles: The law suit includes a description of the case as well as a number of decisions. F, which is a text document, represents the fact description. The results may contain articles of applicable law, charges, penalty terms, and so forth. Suppose there will be t kinds of judgment results, and the i-th judgment result is expressed as a binary classifier yi with a value derived from the set Yi. As a result, a tuple may be used to describe a legal situation (f, y1....., yt).

4 Methodology

Machine intelligence (AI) has got a huge amount of press in the legal community as a method to save expenses, improved justice, speed up case studies, and eventually replace all people with robotic doppelgangers. This approach has the potential to explain and simplify Legal AI so that it may be discussed as anything more than

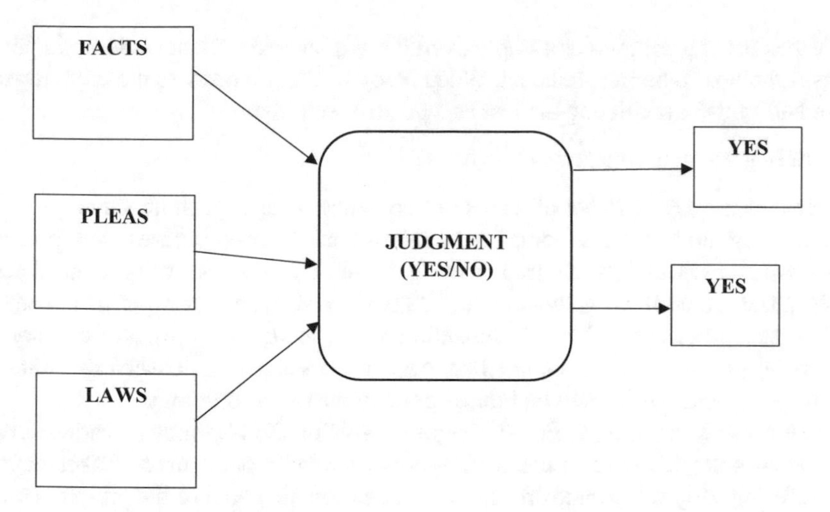

Fig. 1 Automatic judgment prediction

manual. Lawful AI is a catch-all term for a collection of machine learning techniques that are frequently used with NLP to decipher legal documents and text. Legal AI isn't truly AI, but it's a lot simpler to say than all of that. It's the concept of giving a computer the ability to reason, think, and learn about the world. The objective of true artificial intelligence research is to develop a generic algorithm that can learn to perform almost anything. That sort of AI is not available at any of the law tech firms. If they did, their robot attorneys would soon tire of examining legal paperwork and sign up for Pinterest and other social media accounts.

Linguistics, on the other hand, divides a phrase into nouns, verbs, and other speech components. The imprecise text in papers, contracts, spoken language, and stupid giraffe phrases is then transformed into a precise hierarchy of connected and tagged components using NLP technology. These phrase constructions can be utilized independently, but they're most commonly employed as inputs for machine learning algorithms to anticipate outcomes. NLP might be used by legal practitioners to speed up document evaluation. You might, for example, run an agreement across a natural language tool to help identify particular provisions, double-check the legal wording, and ensure the contract has all clauses necessary to comply with your safety protocols.

A continuous series of n elements from a given sample of text or voice is called an N-gram. The N-gram idea is frequently utilized in Computational Linguistics for information retrieval. A—unigram|| is a size 1 N-gram, a—bigram|| is a size 2, and a —trigram|| is a size 3 N-gram.

In Machine Learning, an N-gram plays an essential part in text analysis. A single word isn't always enough to understand the context of a document. Let's look at an example of how N-gram can be used for text analysis. For example, while anticipating the text's emotion, such as whether it is good or bad.

Text = "The pizza is not a bad taste."

When using a unigram or a single word for text analysis, the negative word "bad" causes the text to be mispredicted. When using bi gram, however, the bigram word "not bad" aids in predicting the text as a positive sentiment.

No. of N-gram in a sentence = X−(N−1)

where X is the total number of words in a sentence. K is an N-gram value

The program can detect complicated words such as noun phrases, verb phrases, adjective phrases, and so on based on these foundations and a set of syntactical principles. The program has gathered over 500,000 words from our experiment corpus. This list includes words from all semantic groups, including verbs, adverbs, nouns, and noun phrases. Using Natural Language Processing to discover legal phrases identifying ideas, which will be future components of our ontology.

The process depicted in Fig. 2. diagram explains the N-grams of various characteristics collected from numerous lawsuits serve as input sources. Tokenization, lowercasing, stop word removal, and lemmatization are some of the pre-processing methods employed here. The method's first phase eliminates several term classes from the initial list. First and foremost, only two syntactical categories were considered: nouns and noun phrases. This decision is based on the assumption that most ideas are contained within nouns. Aspects of law that are classified as adjectives or adverbs are excluded from our theory. Next, non-alphabetic words are eliminated from the initial list. The majority of the terms in our list are direct or indirect connections to documents like articles. Tokenization, lowercasing, stop word removal, and lemmatization are some of the pre-processing methods employed here.

1. **Tokenization**
   Metsker et al. [15] developed a method of condensing a sequence of strings data into the smallest unit feasible. Symbols, such as punctuation marks, are deleted during this procedure. The bits are then used as input in other processes such as parsing and text mining.
2. **Lowercasing**
   Metsker et al. [15] formatting data in lowercase characters. This method will make the search process easier. Although lowercasing has the potential to reduce.

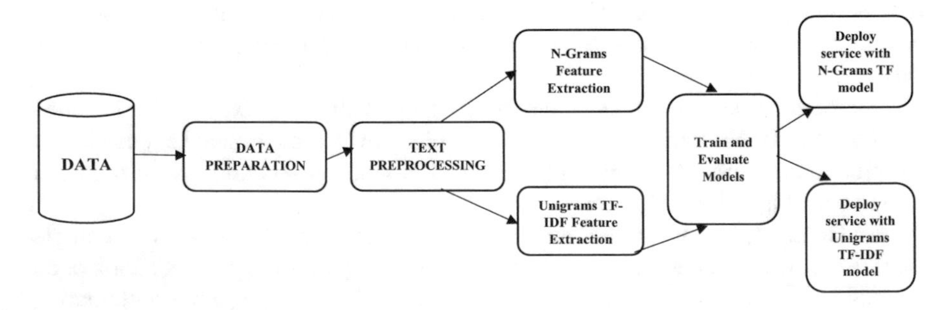

Fig. 2 Process flow for converting text into required target attribute

Table 1 Examples of lemmatization and stemming

Word	Stemming	Lemmatization
Information	Inform	Information
Informative	Inform	Informative
Computers	Comput	Computer
Feet	Feet	Foot

In some situations, the search engine's dependability. For example, MIT is an abbreviation that becomes "mit" in lowercase, which is a German word.

3. **Stop word Removal**
   The process of eliminating unnecessary words such as "the," "a," "an," and "in." Stop words are used in natural language processing to refer to worthless words (data). These words take up more room and take longer to digest. The software application has been set to ignore them while indexing entries for storing and retrieving them as a result of a search query.
4. **Lemmatization**
   Lemmatization is defined as "doing things properly with the use of vocabulary and morphological analysis of words, with the goal of removing only inflectional ends and returning the base or dictionary form of a word, which is called as the lemma". It is important to have thorough dictionaries for lemmatization so that the algorithm can relate the form back to its lemma.

The process of normalizing text materials shown in Table 1 is called stemming and lemmatization. The fundamental objective of text normalization is to keep the vocabulary limited, which helps many languages modelling tasks to be more accurate. Converting each word to lowercase, for example, the vocabulary size will be decreased. As a result, the distinction between now and how is overlooked. By eliminating inflectional forms, stemming and lemmatization enable us to standardize text and enhance vocabulary. It is the process of reducing a word to its most fundamental form.

5 Natural Language Processing

Known values can be used to eliminate full stops, numerals, and any other non-letter characters [2, 15]. The other tokens' words and phrases can be tagged using parts-of-speech (POS). Moreover, court rulings should be free of proper nouns and stop words. Collocations are popular words that contribute nothing to the sense of the text [2]. A corpus of proper nouns and English stop words is available in the NLTK library.

Tf-idf Vectorizer for text classification

At first, calculating the number of words in a text document is quite simple. However, a basic word count is insufficient for text processing since terms like “the,” “an,” and “your,” among others, are often used in text documents. Their huge word count has no impact on the text’s examination. Stop-words filtering from a text document may be done effectively with Tf-idf.

Another option for resolving this issue is to use word frequency. The TF-IDF technique is abbreviated as "Term Frequency—Inverse Document Frequency". The TF-IDF is a statistical metric for determining the importance of a term in a document.

The number of times a word appears in a text document is known as term frequency.

- Inverse Document Frequency: Determine if a word in a document is unusual or frequent. The formula that used to compute the IF-IDF is:

tf(t,d) = (Number of times term t appears in a document)/(Total number of terms in the document)

where tf(t,d)−Term Frequency, t = term, d = document

idf(t) = log [n / df(t)] + 1

where idf(t)−Inverse Document Frequency, n−Total number of documents, df(t) is the document frequency of term t;

tf-idf(t,d) = tf(t,d) * idf(t)

Count Vectorizer for text classification

The Scikit-Learn package includes a Count Vectorizer that transforms a group of word docs into a token count matrix. A Count Vectorizer makes tokenizing text input and building a vocabulary of recognized words straightforward. It also uses that established vocabulary to encode new text data. Because it contains so many zeros, the encoded vector is a sparse matrix.

To use Count Vectorizer, do the following steps: Create a Count Vectorizer class object.

To create a vocabulary of words from text input, use the fit () method.

To tokenize text data using constructed vocabulary, use the convert () function.

Metrics:

- Series strings as input
- Bool (lowercase) (Default-True). Before tokenizing, convert all characters to lowercase.
- Stop words: Remove the defined words from the vocabulary that results.
- Ngram range: For distinct n-grams, the lowest and highest boundaries of the collection of n-values to be returned.
- Max df: Ignore terms with a document frequency greater than or equal to a threshold.

- Ignore terms with a document frequency less than a threshold using min df.
- Max features are a variable that specifies the maximum number of features.

Create a vocabulary that only includes the most important words. max features are sorted by the number of times a word appears.

6 Machine Learning Techniques for Legal Analytics

Descriptive methods generate models and output without requiring human input, which may appear to be a significant benefit. Unsupervised techniques, on the other hand, always require event study expression of their models and output. There is no guarantee that the topics created by a topic model or the dimensions generated by an LSA model will be understandable. If unsupervised approaches are primarily employed for exploratory reasons, this is not always an issue. However, if an empirical claim is to be based on the findings of unsupervised techniques, verification of a minimum subset of these results may be necessary to show intersubjective vs. objective validity. There are several options for processing case law, and even while significant efforts have been done toward systematizing data and automating procedures, the number of options available might be overwhelming. As a result, one method of automatically digesting legal documents in this section is studied.

Legal information, in general, is written in a natural, but specialized, language. This material is, for the most part, unorganized. As a result, more techniques are need to be established in the field of natural language processing to interpret lawful large data efficiently.

In study presented by Katz et al. [16], a set of algorithms for information extraction was used in our data science studies. There has been no transition phase used due to the low quantity of annotated images accessible. Rather, choosing language patterns that would be relevant, pruned the feature set using automated feature selection, using cross-validation, designers tested a single machine learning algorithm with default settings and presented the results.:

- tf and idf were used to normalize attribute quantities.
- Across the dataset, feature picking (InfoGain, Attribute Eval in conjunction with Ranker (threshold = 0) search technique) was conducted.

Support Vector Machine

The goal is to determine the optimal data splitting border. The best fit line splits the sample in two dimensions. While dealing in vector space with a Support Vector Machine, so the dividing line is actually a separating hyper plane. The hyperplane with the “widest” margin between support vectors is designated as the best feature space. A training set is another name for the vector space.

Linear Support Vector Classifier

Given situation, the most appropriate learning algorithm is Generating a "best fit" hyper plane that splits or analyses the facts to suit the supplied data. After obtaining the hyper plane, use Linear SVC to get the "predicted" class.

Logistic Regression

Logistic regression model was applied for predicting a distinct result based on finite, linear, or hybrid data. Thus, logistic regression is a widely utilized approach is when indicator includes two or more distinct events. Given a collection of independent variables, the outcome may be Yes/No, 1/0, True / False, High/Low.

Random Forests

The Random Forest Method is a more polished version of the principles. We begin by gathering information. Second, in order to construct cross-validation, we split the data into many different testing sets, with some overlap between them. Third, these various testing sets result in various decision trees. Through Cross-validation, several decision trees are constructed, resulting in a more reliable model. Even if each decision tree created in this manner has flaws, the sheer number of trees ensures that the aggregate of each tree's different points of view will compensate for these flaws. As a result, issues like over fitting are reduced. This method is the underlying premise of crowd sourcing. The Random Forest Method, the antithesis of the Cult of the Expert, aggregates numerous decision trees to develop a prediction algorithm that suits the biggest available data environment.

Sequential Neural Networks

Supervised learning algorithms that additional control patterns of facts are known as sequence models. Content streams, audio snippets, video clips, time-series data, and other sequential data are examples. In models, recurrent neural networks (RNNs) are a common method.

A sequential approach is suited for a simple multistage process with one input tensor and one output tensor for each layer. Each of your layers can have numerous inputs and outputs. Surface share is required. You're looking for non-linear topology (e.g., a residual connection, a multi-branch model).

K-Fold Cross Validation

K-fold cross-validation is a technique for estimating the designer's skill on fresh data by dividing the dataset into smaller groups. Frequently used in pattern recognition applications to compare and choose a model for a specific predictive modelling issue. It entails splitting the collection of information into k subsets, or folds, of roughly similar size at random. The technique is fitted on the subsequent k–1 folds, with the initial fold serving as a validation set. By splitting the dataset into 10 folds, we utilized 10-Fold cross-validation.

Since many authorities follow the mandate to enhance public sector information accessibility and reuse by publishing considered cases online, the door for automated

Table 2 Performance of various techniques for high court judgments

Techniques used	Precision level
Logistic regression	0.90
Sequential neural networks	0.88
Random forest	0.85
SVM	0.78
K-Fold cross validation	0.75

legal data analysis is wide open. However, the concept of lawful automation and semi-automation is not new. Since the early 1990s, legal data search databases such as Westlaw and LexisNexis have existed.

Today, computers are striving to automate the summary and extraction of legal information (e.g., Decision Express), as well as the classification of legal materials. For a long time, the linguistic model has been applied in the legal and criminological fields. Text classification, for example, has been utilized in applied linguistics. Whereas in the past, such as in the case of the alleged identity, analysis was done by hand, now we can automate many of these activities.

7 Performance Metrics of Algorithms Trained to Predict the Presence of References

In the following student corpus of Belgian high court judgments, we developed various models to detect the categories "EU law" and "no EU law." We used a cross-validation technique because the data set was so tiny (1000 documents). Then we fitted tens of thousands of different versions of a few common methods, including logistic regression, support vector machine (SVM), random forest, and sequential neural network. While it is beyond the scope of this work to discuss the technical specifications of these algorithms, Table 2 shows the performance of the "best version" of each of these methods.

8 Conclusions

Looking at the legal judgments, as well as use of Language Classifiers on case facts from Judiciary System and Machine Learning methodologies to forecast the accuracy of constitutional violation/non-violation. In addition, semi-automated Verdict Prediction will assist the Judiciary System in carrying out its tasks. In this paper; a generic framework for assessing aspects (ideas and relationships between things) of a domain specialized to Information Retrieval through word-based NLP techniques. Language study is conducted out on a system of law texts known as the Codes. These papers were considered because they contain the following qualities: the Codes are rationally organized, so each law representatives are specified. As certain Codes represent a judicial project's conception. It's worth noting that the inaccuracy analysis is still hypothetical. Because the choice for a document is dependent on all n-grams in the text, it is hard to determine what has the most influence on the forecast. More advanced approaches, such as semantic analysis, should be employed in the future to not only anticipate decisions, but also to discover the variables that influence the machine learning algorithm's conclusion.

The above strategy mostly employs automated tools and methodologies, like speech syntax interpreters and analytics. Such automated approaches would not replace domain researchers; rather, they paid professionals in the practice for its creation, which involves discovering patterns and ideas. Of course, NLP approaches are useful for developing ontologies for IR. Furthermore, authors argue some of these approaches might be used to the creation of ontologies for other purposes, such as school systems, etc.

References

1. Tsai C-F, Tsai J-H (2010) Performance evaluation of the judicial system in taiwan using data envelopment analysis and decision trees. In 2010 Second international conference on computer engineering and applications, vol 2. pp. 290-294. https://doi.org/10.1109/ICCEA.2010.208
2. Virtucio MBL et al. (2018) Predicting decisions of the philippine supreme court using natural language processing and machine learning. In: 2018 IEEE 42nd Annual computer software and applications conference (COMPSAC). pp. 130–135. https://doi.org/10.1109/COMPSAC.2018.10348
3. Katz DM, Bommarito II MJ, Blackman J (2014) Predicting the behavior of the supreme court of the United States: A general approach. arXiv preprint arXiv:1407.6333
4. Sulea OM et al. (2017) Predicting the law area and decisions of French supreme court cases. arXiv preprint arXiv:1708.01681
5. Bielen S, Marneffe W, Vereeck L (2015) An empirical analysis of case disposition time in Belgium. Rev Law & Econ 11(2):293–316
6. Frankenreiter J (2018) Are advocates general political? Policy preferences of eu member state governments and the voting behavior of members of the European Court of Justice. Rev Law Econ, De Gruyter 14(1):1–43
7. Lepore L, Metallo C, Agrifoglio R (2012) Evaluating court performance: findings from two Italian courts. Int J Court Adm 4:82–93
8. Hanson R (2010) The pursuit of high performance. Int J Court Adm 3(1):2–12. https://doi.org/10.18352/ijca.50
9. Geist A (2009) Using citation analysis techniques for computer-assisted legal research in continental jurisdictions
10. Chen W, Wang G, Yin F (2014) Document similarity calculation model of CSLN. In: 2014 IEEE 5th international conference on software engineering and service science. pp 859–862
11. Hall M, Frank E, Holmes G, Pfahringer B, Reutemann P, Witten IH (2009) The WEKA data mining software: An update. ACM SIGKDD Explor Newsl 11:10–18. https://doi.org/10.1145/1656274.1656278
12. Branting LK (2017) Data-centric and logic-based models for automated legal problem solving. Artif Intell Law 25:5–27. https://doi.org/10.1007/s10506-017-9193-x
13. Economides, K., Haug, A.A., & McIntyre, J. (2016). Toward Timeliness in Civil Justice.

14. Aletras N, Tsarapatsanis D, Preotiuc-Pietro D, Lampos V (2016) Predicting judicial decisions of the european court of human rights: a natural language processing perspective. PeerJ Comput Sci. https://doi.org/10.7717/cs.93
15. Metsker O, Trofimov E, Sikorsky S, Kovalchuk S (2019) Text and data mining techniques in judgment open data analysis for administrative practice control. In: 5th International conference, EGOSE 2018. St. Petersburg, Russia, November 14–16, 2018, Revised selected papers. https://doi.org/10.1007/978-3-030-13283-5_13
16. Katz, DM, Bommarito MJ, Blackman J (2017) A general approach for predicting the behaviour of the supreme court of the United States

Chapter 10
Machine Learning-Based Tool to Classify Online Toxic Comments

Deepika Vodnala, Jadhav Shravya, Konagari Vishnupriya, and Vemuri Naga Sai Rohit

1 Introduction

In the initial days of the internet, people used to communicate with each other through email only and it was filled with spam emails. It was difficult to define emails as positive or negative, i.e., spam or not spam, back then. As time passed, communication and data flow through the internet altered dramatically, especially following the advent of social media sites.

Authorities have arrested many people in recent years as a result of their destructive and poisonous social media postings. Any form of societal harm must be prevented and also people's antisocial behavior should be controlled. It is necessary to detect such stuff before it is released since such harmful content is making the internet a dangerous place and negatively harming individuals.

The exponential progress of computer science and technology has resulted in one of the most significant developments of the "Internet" of the twenty-first century, in which one person may connect with another person anywhere in the world using only a smartphone and the internet. People used to communicate with one another via email exclusively in the early days of the internet, and it was clogged with spam.

It was difficult to define emails as positive or negative, i.e., spam or not spam, back then. As time passed, communication and data flow through the internet altered dramatically, particularly following the introduction of social media sites. With the growth of social media, it has become increasingly vital to categorize information as

D. Vodnala (✉)
Department of Emerging Technologies (CSE-Cyber Security), CVR College of Engineering, Hyderabad, India
e-mail: deepuvodnala19@gmail.com

J. Shravya · K. Vishnupriya · V. N. S. Rohit
Department of Information Technology, Vignana Bharathi Institute of Technology, Hyderabad, India

G. Mathur et al. (eds.), *Proceedings of 3rd International Conference on Artificial Intelligence: Advances and Applications*, Algorithms for Intelligent Systems,
https://doi.org/10.1007/978-981-19-7041-2_10

good or bad to prevent any type of harm to society and to govern people's antisocial behavior.

Many people have been arrested in recent years as a result of their destructive and toxic social media postings. However, the toxicity comment must go through a certain method before a classification algorithm is applied to it to evaluate the correctness of the resulting result.

Different machine learning methods will be utilized in the categorization of poisonous remarks on the Data set. To answer the problem of text categorization, this study provides seven machine learning algorithms: logistic regression method, Ensemble algorithm, SVM algorithm, Naive Bayes algorithm, decision tree algorithm, KNN algorithm, and hybrid algorithm. As a result, the model implements seven machine learning algorithms to the supplied data set and computes and evaluates their accuracy.

The organization of this paper is as follows: in Sect. 2, the issue of poisonous remarks in our everyday lives is examined. In Sect. 3, related work covers recent existing systems for the classification [1] of online toxic comments and their drawbacks. In Sect. 4, the overview of the proposed system and its methodology is discussed. In Sect. 5, results are provided and Sect. 6 concludes the paper.

2 Problem Statement

In general, Toxic comments are online statements that are offensive [2], abusive [3], or inappropriate, and they commonly force other users to leave a conversation. The danger of online bullying and harassment obstructs the free flow of ideas by restricting people's dissenting perspectives. Sites fail to effectively stimulate debate, prompting many communities to limit or delete user comments completely.

In this paper, the model classifies harmful comments on the Dataset using several machine learning methods [4]. To answer the problem of text classification, this comprises seven machine learning techniques: logistic regression, random forest, SVM classifier, Naive Bayes, decision tree, Hybrid, and KNN classification. As a result, we'll compute and compare the accuracy of all seven machine learning algorithms on the supplied data set.

3 Related Work

The proposed model for the Classification of Online toxic comments by Nobata et al. [5] is the detection of abusive language in user-generated online content has become an issue of increasing importance in recent years. The majority of existing commercial approaches rely on blacklists and regular expressions; however, these procedures fall short when dealing with more nuanced, less ham-fisted forms of hate speech. In this work, a machine learning-based approach is developed that outperforms a deep learning approach in detecting hate speech in online user comments from two domains. It also creates the first corpus of user comments that have been annotated for abusive language. Finally, it uses a detection tool to analyze abusive language over time and in different contexts to improve our understanding of this behavior.

Ikonomakis et al. [6] proposed that automated text categorization has long been regarded as a critical way of managing and processing a large number of digital documents that are widely distributed and growing. Text categorization is crucial for information extraction and summarization, text retrieval, and question answering in general. This study uses machine learning techniques to demonstrate the text categorization process. The referenced references address the key theoretical difficulties and point the researcher on the right path for further investigation.

Cheng et al. [7] proposed that the success of online communities depends on user contributions in the form of posts, comments, and votes. Allowing user interaction, on the other hand, fosters unwanted conduct such as trolling. In this work, the application analyzes individuals who were banned from three big online discussion forums to describe antisocial conduct. Examining the progression of these people from the time they first join a community until the time they are banned. Our research also identifies diverse groups of users with varying amounts of antisocial conduct, which can alter over time. It utilizes these data for identifying disruptive people early on is a crucial challenge for community administrators.

Toxic comments, according to Rahul et al. [8], are online statements that are disrespectful, aggressive, or inappropriate, and typically prompt other users to leave a conversation. The threat of online bullying and harassment obstructs the free flow of ideas by limiting people's dissenting viewpoints. Sites fail to properly encourage discussions, forcing many communities to limit or eliminate user comments. This article will explore the spectrum of online harassment and categorize the material so that the toxicity may be assessed as precisely as possible. The methodology employs six machine learning algorithms and applies them to our data to handle the challenge of text classification and to select the optimal machine learning method for categorizing damaging remarks based on our assessment metrics. It aims to assess toxicity with high accuracy to limit its negative consequences, which will serve as a motivation for enterprises to take the necessary changes.

Saleem et al. [9] proposed that the hateful speech material that shows hatred for a person or group of people abounds on social media platforms. Such content has the potential to terrify, intimidate, or silence platform users, and some of it has the potential to incite others to act violently. Despite the broad acknowledgment of the issues that such content causes, there are no viable ways for identifying hate speech. It shows why keyword-based techniques are insufficient for detection in this paper. Then, as training data, it presents a method for detecting hostile speech that employs material created by self-identified hateful communities. Our method avoids the time-consuming annotation procedure that is commonly necessary to train keyword systems and functions well on a variety of platforms, resulting in significant gains over current methods.

4 Proposed System

The key objective of the proposed system, i.e., Machine Learning-Based Tool to Classify Online Toxic Comments is to systematically examine the extent of online harassment and classify the content into labels to examine the toxicity as correctly as possible. Different machine learning algorithms can be used for the classification of toxic comments on the Dataset. The proposed system implements seven Machine Learning techniques, i.e., Logistic Regression (LR), Random Forest Classifier (RFC), Support Vector Machine (SVM) Classifier, Naive Bayes (NB), Decision Tree (DT), Hybrid, and K-Nearest Neighbor algorithm (KNN) Classification to solve the problem of text classification and calculates and compares their accuracy. The proposed system is used to identify the best machine learning algorithm based on evaluation metrics for toxic comments classification. The Architecture of the proposed system is shown in Fig. 1.

To carry out the objective of the proposed system, the tool is used Harmful Comments dataset, which comprises comments as well as labels indicating whether the contents are normal or toxic. This dataset is saved in the folder "Dataset". The dataset comprises column names, and rows that contain values and each comment will be labeled as 0 or 1, with 0 indicating a normal remark and 1 indicating a toxic comment.

Initially, the dataset is divided into two sections: train data and test data. The tool receives the test comments after training the model, and ML algorithms will predict if the uploaded test remark is normal or harmful.

The proposed system implemented the modules, i.e., upload input dataset, preprocessing, count vectorizer, create model, build models, plot graph, and prediction.

i. Upload input dataset: The dataset is uploaded in the model and based on the evaluation metrics it splits the records into train data and test data.

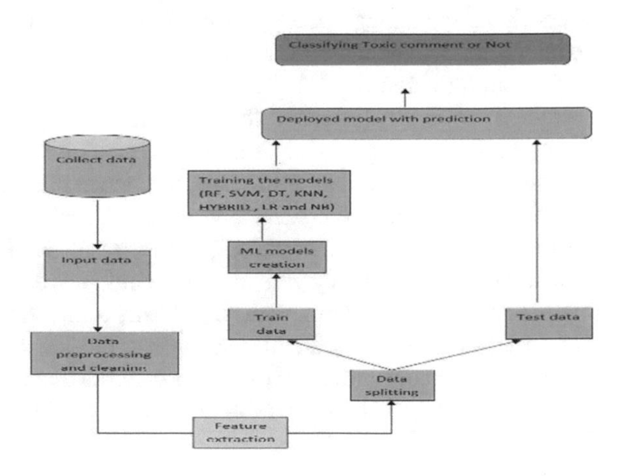

Fig. 1 The architecture of the proposed system

ii. Pre-processing: The model checks each comment and afterward eliminates stop words and special symbols, and makes all comments clean which enhances the analytical results of the training data and empowers accurate decision-making.
iii. Count vectorizer: first the model counts the occurrence of each word and calculates the average and generates a count vector and which will be trained with all algorithms.
iv. Create model: It builds a model by splitting the dataset into the train (80%) and test (20%). For instance, if the dataset contains 300 records, after splitting the train data includes 240 records and the test data contains 60 records.
v. Build models: The application constructs seven machine learning models such as NB, SVM, LR, RFC, DT, KNN, and Hybrid. Then the models get trained using training data. The seven ML algorithms are illustrated below.

   a. SVM: It's a prominent Supervised Learning technique that is used for both classification and regression and outlier identification. It's very useful in high-dimensional spaces and when the number of dimensions exceeds the number of samples, the method remains effective. It is depicted in Fig. 2. SVM is implemented in 5 phases that are (1) Collect the dataset which is split into two sub-sets, i.e., training and testing dataset, (2) Linear scaling of the train data set from 0 to 1, and calculation of the various parameters for establishing the classification function, (3) Estimation of the optimal model parameters (c, g) using the combined approach of k-fold cross-validation and grid search method, (4) Establishment of the final SVMs model with the help of the best parameters, and (5) Evaluation and validation of the SVMs model by evaluation with testing data.
   b. RFC: Random Forest is a supervised learning algorithm. It can be used for both classification and regression. It is also the most flexible and easy to use. A forest is comprised of trees. It is said that the more trees it has, the more robust a forest is. Random Forest creates decision trees on randomly selected data samples, gets a prediction from each tree, and selects the best solution through voting. It also provides a pretty good indicator of the

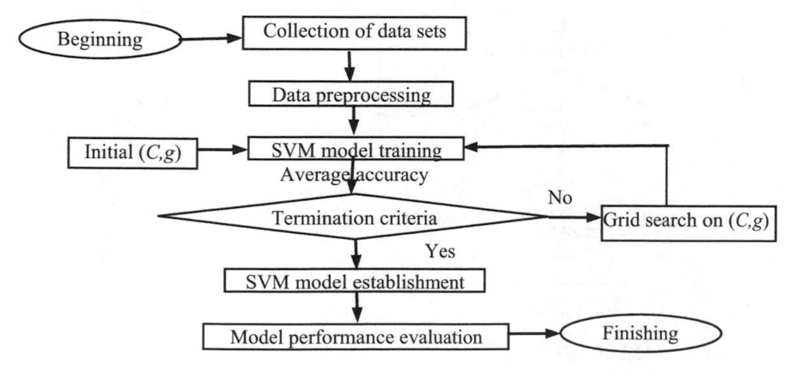

Fig. 2 Flowchart of support vector machine algorithm

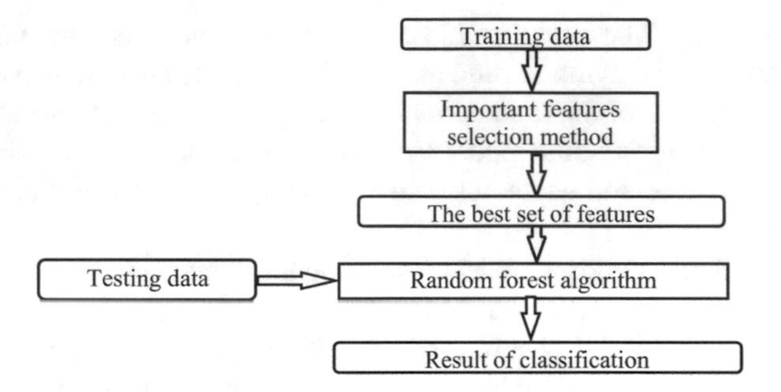

Fig. 3 Flowchart of RFC algorithm

feature's importance. A random forest algorithm can operate the task by constructing a multitude of regression trees at training time and outputting the mean prediction of the individual trees. It is depicted in Fig. 3.

c. NB: Naive Bayes algorithm is also referred to as the Bayes algorithm. The Naïve Bayes classifier algorithm classifies things by using Bayes' theorem. These classifiers believe in high, or naive, independence among data point properties. The major applications of these classifiers are text analysis, medical diagnosis, Spam filters, etc. These classifiers are generally utilized for ML as they are easy to carry out. A Naive Bayes classifier assumes that the presence of a particular feature in a class is unrelated to the presence of any other feature. It is depicted in Fig. 4.

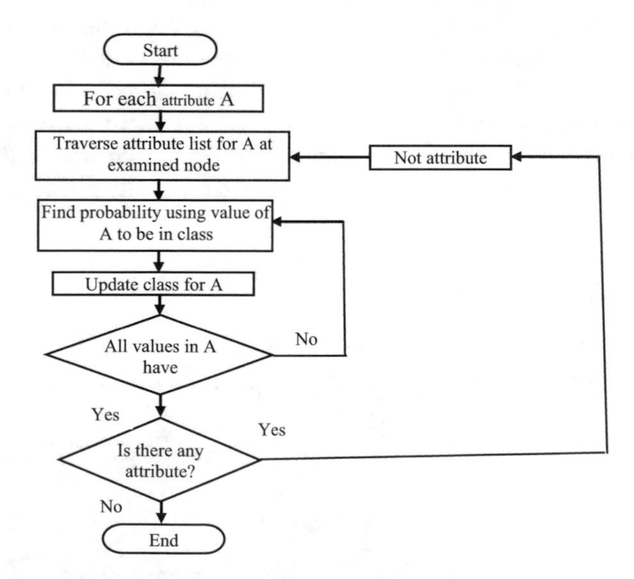

Fig. 4 Flowchart of NB algorithm

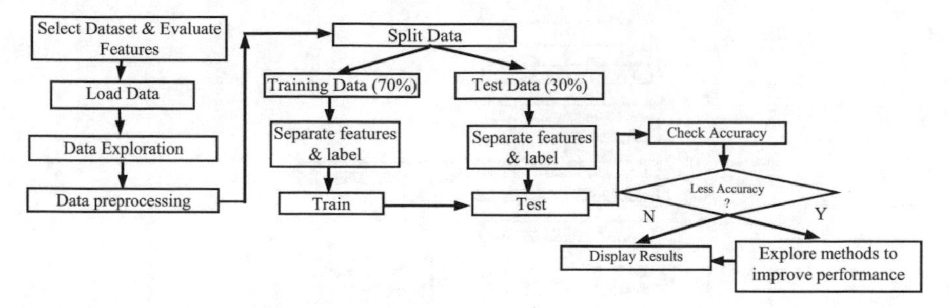

Fig. 5 Flowchart of LR algorithm

d. LR: The basic procedure uses probabilistic predictions to better clarify the correlation between the dependent variable and one or more independent variables in logistic regression. Explore several approaches to improve the performance and accuracy depending on certain diagnostic measurements used in the dataset. Unlike Linear Regression, which outputs continuous number values, Logistic Regression transforms its output using the Logistic Sigmoid function to return a probability value which can then be mapped to two or more discrete classes. It is depicted in Fig. 5.
e. KNN: For classification and regression, a non-parametric classification approach KNN can be utilized. In some scenarios, where the input is a dataset with k closest training samples, Whether KNN is being utilized for classification or regression determines the result. The working principle of the KNN is seeking the shortest distance between the data to be evaluated by the K neighbor closest to the training data. This technique is very simple and easy to implement. Figure 6 depicts KNN Algorithm.

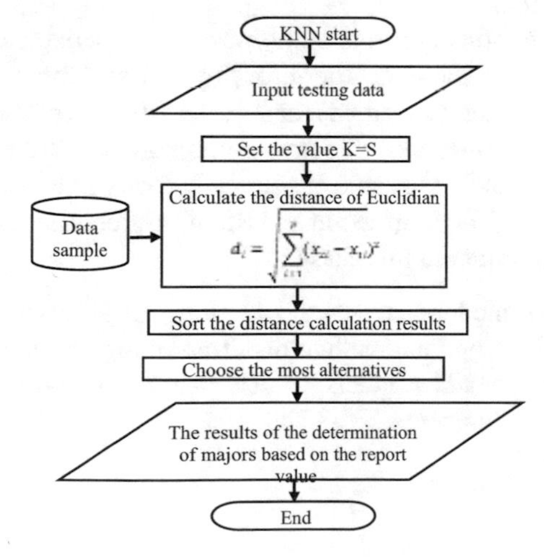

Fig. 6 Flowchart of KNN algorithm

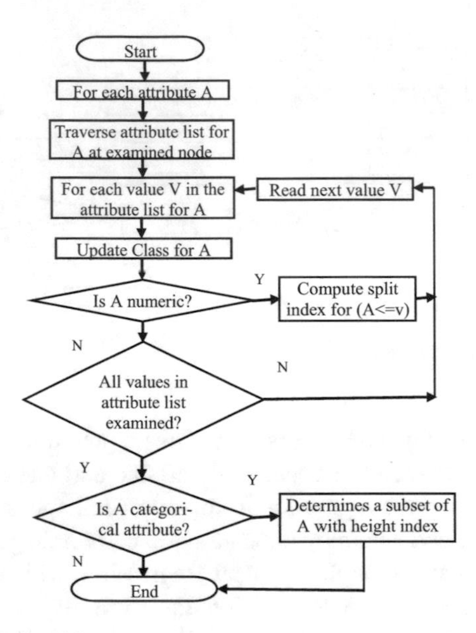

Fig. 7 Flowchart of DT algorithm

f. DT: In this algorithm, a tree-like model of operations and their resulting outcome including random occurrence results, resource costs, and effectiveness are served by a decision-support tool. The main objective of this technique is to develop a model which predicts the value of a target variable. To do this, tree representation is used by the decision tree to resolve the problem in which the leaf node resembles a class label and attributes are designated on the internal node of the tree. Figure 7 depicts the Decision Tree algorithm.
g. Hybrid: A hybrid algorithm combines two or perhaps more algorithms to find an optimal solution. The term "hybrid algorithm" may not relate to combining numerous methods for solving a distinct problem. Various strategies consider collections of discrete components. But rather to combining approaches that fix the same challenge but vary in other ways. A combination of the Logistic Regression and Random-Forest algorithm is considered a hybrid algorithm in this study.

vi. Plot Graph: The model analyzes the accuracy of all classification techniques and determines the optimal method resulting in high accuracy factors.
vii. Prediction: The model predicts if test results are normal or contain any potentially toxic comments.

5 Results

The proposed system, i.e., Machine Learning-Based Tool to Classify Online Toxic Comments works based on RFC, SVM, Naïve Bayes, Decision Tree, Logistic Regression, and Hybrid Algorithm. The proposed system is implemented using the toxic comments Dataset. The entire dataset is split into training and test datasets. Evaluation is done on the testing data set. Accuracy Comparison Graph will be plotted. And model predicts whether the given comment is toxic or non-toxic. The home page of the proposed system is shown in Fig. 8. The home page includes various kinds of labels such as Upload dataset, Data cleaning, Count vectorizer, Build ML Models such as SVM, RFC, NB, LR, KNN, Hybrid, and DT, Accuracy Graph, and Predict.

Once the toxic comments dataset is uploaded to the model, then it displays the total tweets found in the dataset, i.e., 300, which is shown in Fig. 9.

In the Data cleaning process, the application reads each comment and then removes stop words and special symbols and makes all comments clean which improves the quality of the training data for analytics and enables accurate decision-making which is shown in Fig. 10.

During the Count vectorization process, the application counts the occurrence of each word and then finds the average, and then builds a count vector and this vector will be trained with all algorithms. Total records were used to train machine algorithms using count vector 240 and Total records were used to test machine algorithms using count vector 60 as shown in Fig. 11.

Once the model selects Hybrid Algorithm, it predicts and displays the Accuracy, i.e., 98% as shown in Fig. 12.

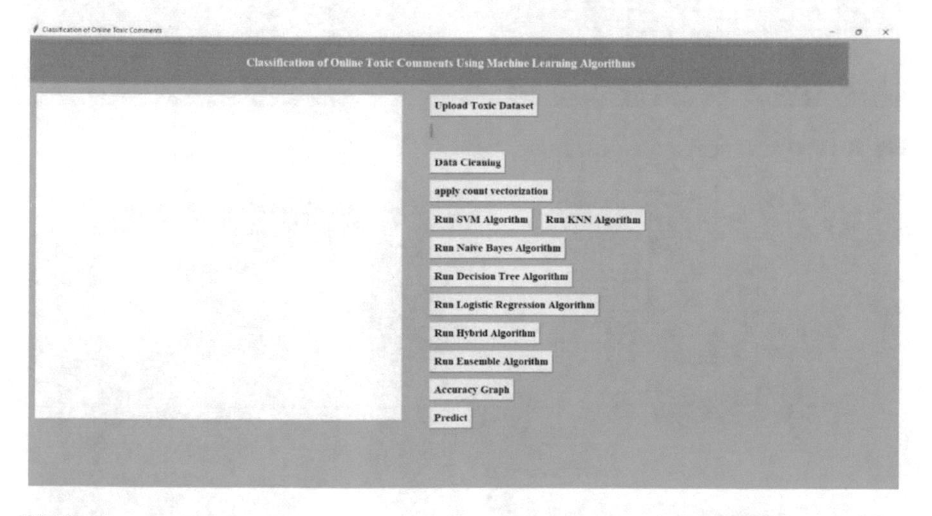

Fig. 8 Home page

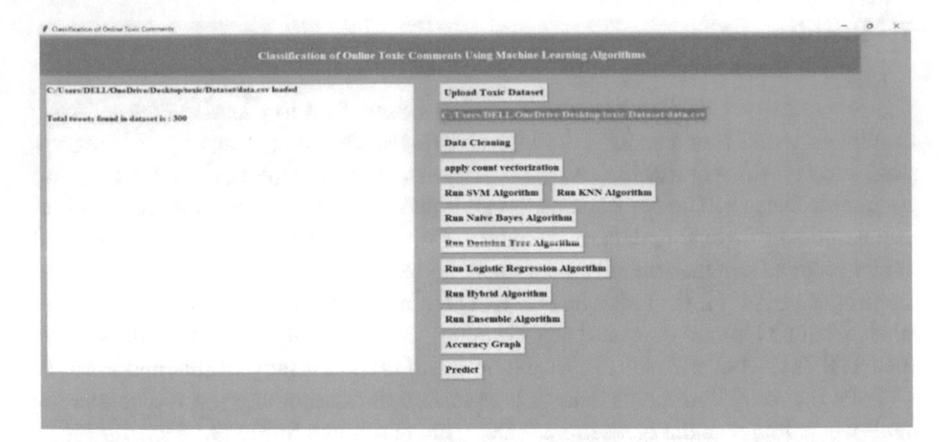

Fig. 9 Upload toxic dataset

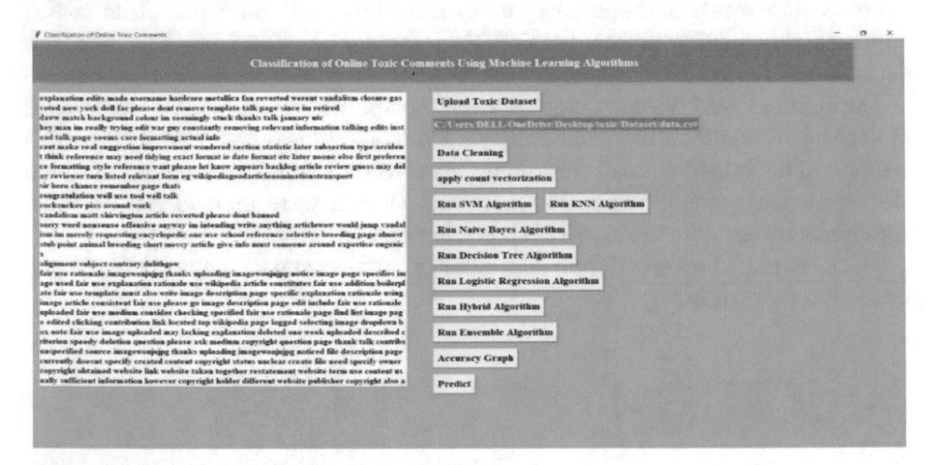

Fig. 10 Data cleaning

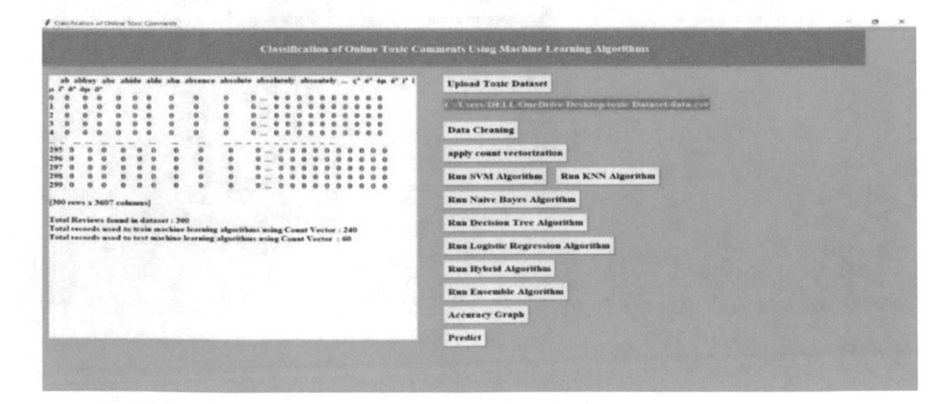

Fig. 11 Count vectorization

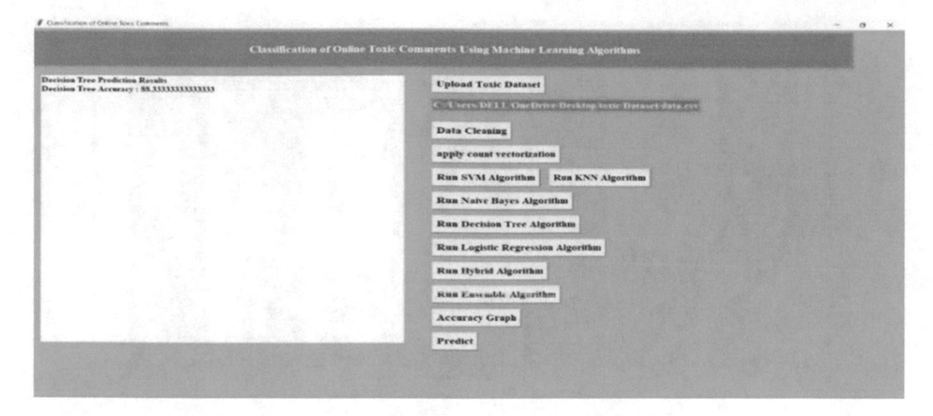

Fig. 12 Accuracy of hybrid algorithm

Once the model selects the Decision tree Algorithm, it predicts and displays the Accuracy, i.e., 90% as shown in Fig. 13. Similarly, the remaining five algorithms also predict and display the accuracy which is listed in Table 1.

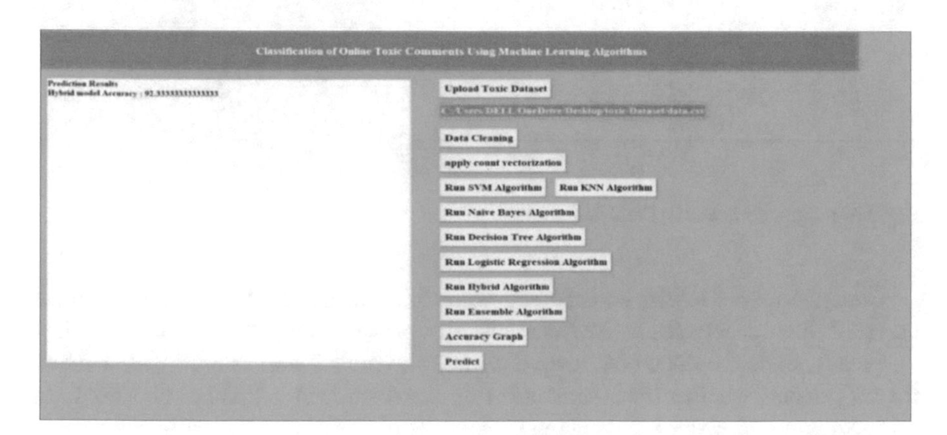

Fig. 13 Accuracy of decision tree algorithm

Table 1 Accuracy of seven algorithms

S.no.	Algorithms used	Accuracy (%)
1	SVM	83
2	KNN	86
3	Random forest	82
4	Logistic regression	92
5	Decision tree	90
6	Naïve Bayes	80
7	Hybrid	98

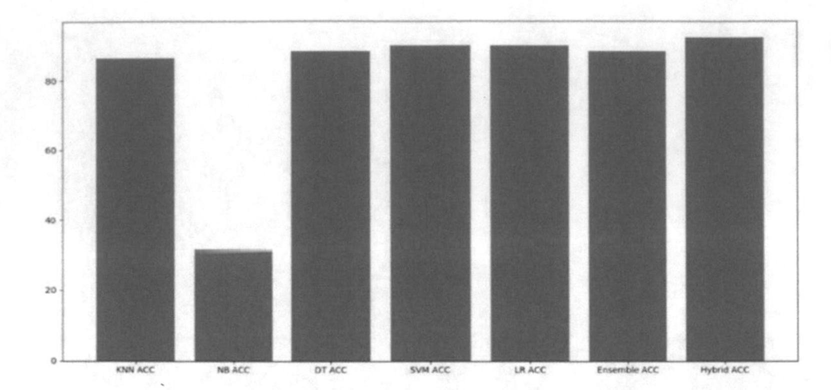

Fig. 14 Accuracy comparison graph

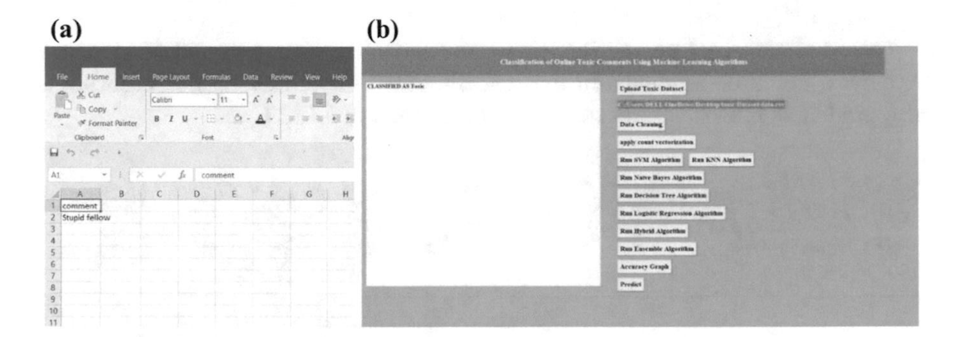

Fig. 15 **a** Input comment. **b** Predicted output

Once the model selects an accuracy graph it compares all the seven algorithms and plots their accuracies in the Accuracy comparison graph as shown in Fig. 14.

Finally, once the document is uploaded to the model, and selects predict label, then it predicts whether the comment inside the document is toxic or non-toxic. In this scenario, the given input comment, i.e., Stupid fellow as shown in Fig. 15a is predicted as toxic as shown in Fig. 15b.

6 Conclusion

The proposed system, i.e., Machine Learning-Based Tool to Classify Online Toxic Comments works on basis of Seven Machine learning techniques, i.e., logistic regression, Naive Bayes, decision tree, random forest, KNN classification, Hybrid Algorithm, SVM classifier, and compared their accuracy as a part of this research work. In the proposed system, Hybrid Algorithm yields better accuracy in comparison with other algorithms. And the proposed system predicts whether the given input comment is Toxic or Non-Toxic. In further research, effective machine learning models can be used to calculate accuracy, hamming loss, and log loss for better results. The researchers can also explore some effective deep learning algorithms such as LSTM (long short-term memory recurrent neural network) and multi-layer perceptron which improve the accuracy of prediction results.

References

1. Georgakopoulos SV, Tasoulis SK, Vrahatis AG, Plagianakos VP (2018) Convolutional neural networks for toxic comment classification. Cornell University. arXiv:1802.09957
2. Razavi AH, Inkpen D, Uritsky S, Matwin S (2010) Offensive language detection using multi-level classification. In: Canadian conference on AI
3. Nobata C, Tetreault J, Thomas A, Mehdad Y, Chang Y (2016) Abusive language detection in online user content. In: 25th international world wide web conference, WWW 2016, pp 145–153
4. Ikonomakis EK, Kotsiantis S, Tampakas V (2005) Text classification using machine learning techniques
5. Nobata C, Tetreault J, Thomas A, Mehdad Y, Chang Y (Apr 2016) Abusive language detection in online user content, yahoo, pp 11–15
6. Ikonomakis M, Kotsiantis S, Tampakas V (2005) Text classification using machine learning techniques. WSEAS Trans Comput (4)8:966–974
7. Cheng J, Danescu-Niculescu-Mizil C, Leskovec J (2015) Antisocial behavior in online discussion communities. In: Ninth international AAAI conference on web and social media
8. Kajla H, Hooda J, Saini G (2020) Classification of online toxic comments using machine learning algorithms. In: 4th international conference on intelligent computing and control systems (ICICCS). IEEE, pp 1119–1123
9. Saleem HM, Dillon KP, Benesch S, Ruths D (2017) A web of hate: tackling hateful speech in online social space. arXiv:1709.10159v1 [cs.CL]

Chapter 11
Pixel-Based Image Encryption Approaches: A Review

Asha J. Vithayathil and A. Sreekumar

1 Introduction

Rapid technological advancements have rendered the generation and transmission of data effortless, while, at the same time, security has become increasingly important. A cryptographic algorithm provides data security in the event that it is stored in a location and transmitted over an insecure network. The data is available in a variety of formats such as text, images, audio, video, etc., each requiring a specific method to deal with security issues. In multimedia communication, images play a significant role because of their high rate of availability, readability, and comprehensiveness, and are used in all fields, such as personal purposes, educational, medical, military, satellite communication, and so on. Image security can be achieved in either of the following ways: watermarking, steganography, or encryption [1]. Watermarking involves embedding a signature into a digital image that may be visible or hidden to determine ownership [2]. In steganography, the original image is hidden in the cover image so that the attacker cannot see it. With encryption, the plain image is transformed into a cipher image using a key, that is, encryption keeps a message's contents secret, while steganography and watermarking keep its existence secret [1, 3]. Below, we examine the encryption techniques for images.

In cryptography, the data is encoded by converting it into an unreadable form with the aid of a key, which is called encryption. Extracting the original data back from this unreadable form is called decryption. The original image is called plain image and the encrypted image is called cipher image. If the encryption–decryption process employs a single key for both operations, it is termed symmetric encryption, and if it uses a public key for encryption and a secret key for decryption, it is called asymmetric encryption. In contrast to textual data, the image needs special atten-

A. J. Vithayathil (✉) · A. Sreekumar
Department of Computer Applications, Cochin University of Science And Technology, Cochin, India
e-mail: ashavithayathil@cusat.ac.in

G. Mathur et al. (eds.), *Proceedings of 3rd International Conference on Artificial Intelligence: Advances and Applications*, Algorithms for Intelligent Systems,
https://doi.org/10.1007/978-981-19-7041-2_11

tion because of its attributes such as very large size, high redundancy, and highly correlated pixels and, therefore, most of the classical cryptographic algorithms like DES, AES, RSA, etc., are not useful for image security [4]. Hence, the encryption of images is being researched in order to solve these problems. An image encryption algorithm can be classified as a compressive sensing, spatial domain, optical domain, or transform domain approach depending on how it applies to an image [3]. Compressive sensing-based image encryption techniques enable compression and encryption simultaneously and thus enhance data security [5]. A spatial domain approach applies encryption techniques directly to the pixels of an image [3, 6–10], while a transform-based approach transforms the given image into a frequency domain by using a suitable transform model and then applies an encryption function to this newly transformed image [11–13]. Refregier and Javidi [14] first proposed optical image encryption using a double random phase encoding scheme (DRPE) based on Fourier transformation. Later, the use of optical methods for image encryption became a great attraction for many researchers [15].

In this paper, we review publications from 2018 onward on spatial domain-based image encryption techniques, which include image encryption using enhanced conventional algorithms, chaos maps, elliptic curves, and deep neural networks, and compare them using various performance evaluation metrics.

2 Evaluation Metrics

The evaluation metrics are very important to assess the performance of any technique. Various statistical, differential, and key analysis methods confirm the quality of an image encryption algorithm.

2.1 *Statistical Analysis*

Statistical analysis confirms the effectiveness of an encryption technique against statistical attacks by analyzing the pixel intensity distribution of an encrypted image. Histogram, correlation, and correlation coefficient are the commonly used statistical measures.

2.1.1 Histogram Evaluation

A histogram describes the appearance of an image by showing the frequency of its intensities by plotting the intensity values on the x-axis and the number of pixels on the y-axis. The histogram of the original image shows a pattern in its distribution, whereas that of the encrypted image should have a uniform intensity distribution without exhibiting any pattern.

2.1.2 Correlation Evaluation

In an image, each pixel intensity is almost the same as its neighboring pixels in horizontal, vertical, or diagonal direction; that is, each pixel is highly correlated with its neighboring pixels. The correlation graph shows the pixels' correlation of an image, with the x-axis representing the intensity value of the pixels and the y-axis representing either of its neighboring pixel values. The plain image correlation graph shows a pattern of points, whereas that of the encrypted image should have scattered points over the graph.

2.1.3 Correlation Coefficient

The correlation coefficient (CC) is a statistical measure that shows the strength of the correlation between two pixels, and zero is the ideal value for any image encryption technique. The computation [3] of the correlation coefficient is as follows:

$$\mathrm{CC} = \frac{\mathrm{cov}(x, y)}{\sigma_x . \sigma_y} \tag{1}$$

$$\mathrm{cov}(x, y) = \sum_{i=1}^{N} \frac{(x_i - E(x))(y_i - E(y))}{N} \tag{2}$$

$$\sigma_x = \sqrt{\sum_{i=1}^{N} \frac{(x_i - E(x))^2}{N}} \tag{3}$$

$$\sigma_y = \sqrt{\sum_{i=1}^{N} \frac{(y_i - E(y))^2}{N}} \tag{4}$$

2.2 *Differential Analysis*

Differential analysis is used to measure the sensitivity capability of the encryption technique, in which the same key is used to encrypt the plain image and the slightly modified plain image, and then compare the pixel values of these two encrypted images using the metrics Number of Pixel Change Rate (NPCR) and Unified Average Changing Intensity (UACI). The NPCR is the percentage of change in the number of pixels, and UACI is the average intensity difference of the encrypted images and can be computed [16, 17] using the following equations:

$$\begin{aligned}
\text{NPCR} &= \frac{\sum_{i,j} \text{dif}(i, j).100}{M.N} \\
\text{dif}(i, j) &= 0 \quad \text{if both pixels have same value} \\
&= 1 \quad \text{otherwise}
\end{aligned} \tag{5}$$

$$\text{UACI} = \frac{\sum_{i,j} I_1(i, j) - I_2(i, j)}{M.N.255}.100 \tag{6}$$

Here, M and N represent the dimension of an image, and dif(i,j) counts the number of different corresponding pixels between the encrypted image of the original $(I_1(i, j))$ and the encrypted image of the slightly modified image $(I_2(i, j))$.

2.3 Key Strength Evaluation

According to Kerckhoff's principle, all aspects of a cryptosystem are known to the attackers except the key [18]. As a result, an encryption technique's strength is primarily reliant on its encryption key and is achieved by a vast key space with high sensitivity [19]. The key space is determined by the key's length, and its sensitivity is measured by applying Eqs. 5 and 6 to two encrypted images of the same plain image, obtained with slightly different keys.

2.4 Information Entropy Analysis

Encryption renders images in a tangled format, while entropy is the measure of how tangled the data is and it determines [18] using the following equation:

$$\text{Entropy} = \sum_{i=0}^{255} p_i.\log_2 \frac{1}{p_i} \tag{7}$$

where p_i represents the probability of a particular pixel value i in the image. The ideal entropy value is 8 and any encrypted image with an entropy value near 8 shows high randomness [20].

2.5 NIST Tests for Randomness

The National Institute of Standards and Technology (NIST) has developed a standard test script that includes fifteen tests to validate the unpredictability of a PRNG used in cryptography applications [21, 22]. Each test, such as the frequency test and the Runs test, was briefly described in a toolbox [22] provided by NIST.

2.6 Encryption–Decryption Quality Measure

PSNR, peak signal–noise ratio is a metric used to measure the encryption–decryption grade quantitatively, and it can be calculated [23] as follows:

$$\text{PSNR} = 20\log_{10}\frac{255}{\sqrt{\text{MSE}}} \tag{8}$$

$$\text{MSE} = \frac{1}{\text{size}}\sum_{i,j}(I_{i,j} - E_{i.j})^2 \tag{9}$$

where MSE is the mean squared error, measures the difference between two images either plain image and cipher image or decipher image and plain image.

3 Image Encryption Approaches

Based on the above-mentioned performance evaluation metrics, various spatial domain-based image encryption techniques are discussed and analyzed in this section. The tables given provide a summary of whether the mentioned approaches accomplished or not the specified evaluation metrics, using yes and hyphen(−), respectively.

3.1 Image Encryption Using Enhanced Conventional Methods

Images are usually very large in size and redundant in nature, so the direct application of the conventional encryption algorithms like AES, DES, RSA, etc., on images will not result in good encryption [4, 9] and hence they need to be modified to improve the performance.

Ye et al. proposed image encryption using SHA 3 and RSA [24]. In this method, the plain image is disseminated with a fixed matrix to get a preprocessed image and SHA 3 is applied to this to get hash values which are later encrypted with RSA. Both the hash values and the ciphered hash values are used as initial parameters for a quantum logistic map to generate the random keystream. These random keys are then applied to the preprocessed image to employ two-level confusion and diffusion. In the decryption phase, hash values are obtained by RSA, and the inverse operations on the ciphered image generate the plain image. Performance metrics show better encryption, but the operations are tedious enough to increase the complexity and time. Reference [6] proposed Lorenz Hyperchaotic System with the RSA Algorithm for encryption. Four confidential numbers are encrypted using the RSA algorithm, and these numbers and their ciphered numbers are used as parameters of the Lorenz hyperchaotic system and further three pseudorandom sequences were generated. The first sequence is used to apply a diffusion by additive modular arithmetic, the second sequence is used to apply a nonrepetitive permutation, and the third is used for finite field diffusion on the image which result in a ciphered image. The inverse operations are performed at the decryption phase by utilizing the encrypted confidential numbers and corresponding decrypted numbers with RSA.

Gafsi et al. proposed [7] a combined method of both symmetric and asymmetric encryption approaches to provide image encryption and authentication. SHA-256 is used to generate two secret keys from a plain image and the owner's signature respectively. The first key with the Lorenz chaotic system generated a pseudo-random sequence and was used for symmetric encryption of the entire image by XOR diffusion, and then these two keys were combined and encrypted together with the RSA. The encryption was done with the receiver's public key, and at the receiver's end, the authenticated first secret key was decrypted using the receiver's secret key. The key randomness is tested with NIST; evaluation and analysis results show improved performance with less complexity and time than other reviewed methods.

For medical images, Shin and Choi [8] proposed a modified version of the RSA using three prime numbers, with one fixed Mersenne prime. The encryption is applied to DICOM images. The performance evaluations show improved encryption capability with a very slight improvement in time aspects as compared with the conventional RSA, but less time is taken than RSA with three prime numbers.

Reference [10] combined the RSA algorithm with a novel chaos-based RNG that encrypted both text and images. A three-dimensional chaotic map designed to construct random bit series at three phases. Random numbers generated in the first two phases are used to generate the first two prime numbers for RSA. The random bit series generated from the third phase had the same length as that of the data to be encrypted. The original image pixels are diffused with this random bit series, and RSA encryption is applied to the diffused data. In the decryption phase, after RSA decryption is applied to the received data, it undergoes an XOR operation with the same bit sequence constructed by the chaotic map, and thus the original image is recovered.

Reference [25] proposed an improved AES with ten rounds of encryption in which the encryption keys were constructed using Arnold's chaos sequence. The substituting and integrating column values are replaced with linear conversion and summation of pixel values. This method showed much-improved performance over the AES algorithm. Reference [26] combined AES with chaotic maps to encrypt satellite images. In this method, initially, a key stream was generated using AES-CTR, then chaotic sequences were produced using a combined logistic adjusted sine map, and they were further used to confuse the plain image and the keystream. Then the permuted image blocks are XOR with the confused keystream to generate the cipher image. This method shows high encryption quality and less encryption time compared with AES-CTR.

Zhang and Y conducted a study on AES used for image encryption [27] and concluded that AES in CBC mode is more efficient than many chaos-based methods,

Table 1 The performance evaluation tests conducted by enhanced conventional image encryption approaches

References	NPCR	UACI	HE	CE	CC	IE	PSNR	NIST	KA
[6]	Yes	Yes	Yes	Yes	Yes	Yes	–	–	Yes
[7]	Yes	Yes	Yes	–	Yes	Yes	Yes	Yes	Yes
[8]	–	–	–	–	–	–	Yes	–	–
[10]	Yes	Yes	Yes	Yes	Yes	Yes	–	Yes	Yes
[25]	Yes	Yes	Yes	Yes	Yes	Yes	–	–	–
[26]	Yes	Yes	Yes	Yes	Yes	Yes	–	–	Yes
[28]	Yes	Yes	Yes	Yes	Yes	Yes	–	–	–

unlike the existing viewpoint that AES cannot adequately secure images. Reference [28] proposed a technique with dynamic AES in which a dynamic S-Box was constructed through dynamic irreducible polynomials and affine constants. In correlation evaluation, the proposed method performed better than the standard AES.

Table 1 shows the comparison of enhanced conventional encryption techniques on images based on performance metrics. Almost all performance metrics are satisfied by these techniques; however, randomness tests and encryption–decryption quality tests are poorly conducted by these methods.

3.2 *Image Encryption Using Elliptic Curve*

Encryption using an elliptic curve offers high security and high speed, even with small key sizes, when compared to other algorithms such as DES and RSA [29], and is based on algebraic curve properties. The elliptic curve is mainly used as a pseudo-random generator in image encryption techniques.

Hayat and Azam studied [30], which is achieved in two phases. In the first phase, a substitution box (S-box) and a pseudo-random number generator are constructed using the elliptic curve points over a prime field. In the second phase, the image pixels were masked by PRN and then permuted by S-box. The various tests conducted confirm the strength of the S-box and PRN proposed in this paper. Dawahdeh et al. [31] combined elliptic curve cryptography (ECC) and the hill cipher, in which the encryption keys and a self-invertible key matrix were constructed with an elliptic curve. The encryption process was carried out using this self-invertible key, in which image pixels were divided into four-row column vectors, and modular multiplica-

Table 2 The performance evaluation tests conducted by image encryption approaches using elliptic curve

References	NPCR	UACI	HE	CE	CC	IE	PSNR	NIST	KA
[30]	Yes	Yes	Yes	Yes	Yes	Yes	–	–	Yes
[31]	–	Yes	Yes	–	–	Yes	Yes	–	–
[32]	Yes	Yes	Yes	Yes	Yes	Yes	Yes	–	Yes
[33]	Yes	Yes	Yes	Yes	Yes	Yes	Yes	–	Yes
[34]	–	Yes	Yes	Yes	Yes	Yes	Yes	–	Yes

tion was performed to get the ciphered vectors, from which the cipher image was constructed. In the decryption phase, the self-invertible key was constructed, and the inverse operation was performed to construct the original image. The elliptic curve-dependent self-invertible key made it really resistant to different attacks.

Reference [32] used an elliptic curve-based image encryption algorithm. Initially, the image pixels are grouped together and converted into big numbers, a chaotic system is then used to generate a chaotic image from the plain image and further chaotic numbers. Then these two numbers are encrypted with ECC, and then pixels are recovered from these encrypted numbers, and hence a cipher image is generated. Luo et al. studied a combination of elliptical and chaotic encryption on images [33]. Initially, SHA-512 was applied to the image to generate the hash values, which were used as the initial parameters to the chaotic system, and the chaotic sequence generated was used to scramble the given image. Then the generated scrambled image is embedded into the elliptic curve and encrypted by elliptic curve-ElGamal encryption. Finally, a chaotic DNA diffusion is applied to these encrypted values to get the final cipher image. Even though this method ensures security, it leads to a very high computational cost. In [34], Adhikari et al. proposed a combined elliptic curve pseudo-random number and chaotic random number sequence method for image encryption. An elliptic curve is used to generate the curve points for the given image and is used to create pseudo-random numbers. A random image is generated using these pseudo-random numbers and a map table. The map table is encrypted with the help of a chaotic number sequence to enhance security. In the decryption phase, the map table is used to decode the image data.

Table 2 shows the comparison of EC-based image encryption techniques, EC properties are mainly used to generate secure encryption keys and efficient random numbers; however, none of the methods checks the NIST-specified randomness test.

Table 3 The performance evaluation tests conducted by image encryption approaches using chaos

References	NPCR	UACI	HE	CE	CC	IE	PSNR	NIST	KA
[36]	Yes	Yes	Yes	Yes	Yes	Yes	Yes	Yes	Yes
[37]	Yes	Yes	Yes	Yes	Yes	Yes	–	–	Yes
[17]	Yes	Yes	Yes	Yes	Yes	Yes	–	–	Yes
[38]	Yes	Yes	Yes	Yes	Yes	Yes	Yes	Yes	Yes

3.3 Image Encryption Using Chaos

Chaos is a branch of mathematics that has various properties, such as ergodicity, sensitivity to initial conditions and control parameters, deterministic behavior, and so on [35]. This makes it possible to employ chaotic maps in cryptography, and is used in key stream generation, pseudo-random number generation, hash function design, and so on. A number of reviewed papers [6, 10, 24–26, 32–34, 40, 41, 46, 47] adopt chaotic maps to improve the encryption quality.

Pak et al. [36] designed a new two-dimensional map to generate keystreams, which were later used for pixel-level and bit-level permutation. The proposed method satisfied all the performance criteria. In [37], Noshadian et al. employed chaos to encrypt the images, where the image is jumbled and modified with a logistic map and Knuth shuffling algorithm respectively. The chaos parameters optimized with evolutionary algorithms are used as keys, that result in the least correlation and the highest entropy.

Shahna and Muhamed proposed [17] a novel chaotic map to generate two keystreams. Pixel-level scrambling is performed with the Hilbert curve and bit-level scrambling is performed using a cycle shift operation with the first generated keystream. The second keystream is used for diffusion of the scrambled image, which generates the final cipher image. Inverse operations are performed in the decryption phase. Reference [38] combined one-dimensional Logistic, Sine, and Tent maps to generate sine–tent maps and logistic–logistic chaotic maps. The sine–tent map is used for image scrambling, and the second map is used for zigzag transformation. Finally, the logistic–logistic chaotic map is again used for block-wise diffusion of the above-scrambled image and the ciphered image is generated. This method successfully tested all the mentioned metrics.

Based on the performance metrics, the comparison of chaos-based image encryption approaches is given in Table 3. Chaotic maps are mainly used as keystream generators and all of the reviewed methods satisfy the sensitivity and space necessities of keys.

3.4 Image Encryption Using Deep Neural Network

Deep learning or deep neural networks is a recent research trend that has applications in a wide range of fields [23, 35, 36]. The huge number of attributes, dense network topology, and an arbitrary training procedure characterize a deep learning model [39], and these allow deep learning to be used to encrypt images. In this section, image encryption with deep learning models is reviewed and analyzed, and it is seen that deep learning is primarily used for the generation of image-dependent one-time pad keys. The review shows it can be effectively used for image encryption, but the high level of complexity is a limiting factor.

A deep learning-based key generation network (DeepKeyGen) is suggested in [39]. Generative Adversarial Network (GAN) is the adopted deep neural network that uses image-to-image conversion to learn the desired style of the private key, which is also an image and is further used to encrypt the given medical image. Image-to-image translation is performed with the help of two domains, one is the source domain that generates the key and contains images of similar distribution. The next is a transformation domain that depicts the key's style, the generator transfers images from the first domain to the second domain, and the output is the encryption key. In the transformation domain, the discriminator distinguishes between the generated key and the data. The randomness of the training process makes it a one-time pad. This proposed method is secure in generating the secret key, but very complex in nature. Sang et al. used a logistic chaotic system in conjunction with a deep autoencoder [40]. At first, the image pixels are confused with the chaotic sequence of the logistic map and then the deep autoencoder incorporated with the uniform distribution constraint received this jumbled image and generated the cipher image. Various statistical and key analyses have been done and the differential analysis is not performed for this proposed method.

Reference [41] proposed a hybridization of deep neural network and chaotic map for image encryption, this method used stacked autoencoders to generate the secret key and an enhanced chaotic map to perform the encryption. The various result analysis techniques also ensure better decryption quality, security, and computational performance on real-time images. Bao et al. proposed an adversarial autoencoder to encrypt the color image [42]. It includes three parts, and they perform as follows: the encoder scrambles the image; the decoder unscrambles the encoded image; and the discriminator improves the encoding performance by differentiating between the scrambled and pseudo-random pixel images. The encryption and decryption keys are constructed from the final parameters of the encoder and decoder. This model shows poor performance in differential analysis.

In [43], Lekshmi et al. proposed a method in which encryption is achieved in three phases. In the first phase, a key is generated based on the features of the image with the help of a neural network. In the second phase, the image-specific key is given as a parameter to the Hopfield Neural Network, which generates a random sequence. In the last phase, the key and random sequence generated in the previous phases are used to perform the confusion and diffusion operations on the image. The unique key is generated for each image and hence highly resistant to chosen-plaintext attack. Ding et al. used deep learning for image security [44]. They used Cycle GAN as the learning network to perform the transformation of the image from the original form to the targeted form. The encryption key is constructed from the generator's final parameters. The decryption is achieved by the reconstruction network. They also proposed Region Of Interest (ROI) mining, by which the extraction of ROI from the encrypted image is possible. This method shows an enhanced performance in key analysis.

Li et al. studied biometric image encryption based on deep learning [45], in which the unique features of the biometric image are extracted with the help of a deep neural network and are used for key generation. Furthermore, the plain image is diffused with this key to generate the cipher image. Wu et al. proposed image encryption with adversarial neural cryptography [46]. In this approach, a GAN is used to get an intermediate image, and a pseudorandom matrix with the same image size is also generated using a SHA-256-constrained logistic map. Subsequently, the intermediate image performs XOR with a pseudorandom matrix that generates the final cipher image.

Wang et al. proposed [47]. In this method, first, the original image is converted into a binary image, then a permutation matrix is generated with the help of a chaos-constrained gray code bit-level permutation, and is converted into a diffusion matrix with a bit-level reverse order. A key stream is generated with the help of a chaos-based back propagation neural network, and this key stream is XOR with the diffusion matrix to generate the cipher image. This method can resist common attacks.

The performance evaluation tests conducted by image encryption approaches using deep neural networks are depicted in Table 4. It shows no single method satisfies all the metrics and the majority of the reviewed methods lack the encryption–decryption quality and randomness tests.

4 Conclusion

This paper reviews the present state of pixel-based or spatial domain-based image encryption, which includes upgraded conventional techniques, elliptic curves, chaos, and deep neural network-based solutions. We also discuss the statistical, differential, and key analysis metrics for the performance evaluation of these techniques. Algorithms with their key ideas and performance are discussed here. The performance comparison tables show that the majority of the reviewed techniques do not follow the complete evaluation criteria, which is in conflict with the assurances of security they provide. As shown in the reviewed articles, larger numbers of confusion diffusion operations lead to increased security, which, in turn, leads to greater complexity and inhibits their use in real-life applications. Developing an image encryption technique satisfying complete evaluation metrics, improved security, and minimum complexity remains challenging. And this area needs to be explored further by applying encryption techniques to different imaging systems such as satellites, remote sensing, three-dimensional imaging, and so on.

Table 4 The performance evaluation tests conducted by image encryption approaches using deep neural networks

References	NPCR	UACI	HE	CE	CC	IE	PSNR	NIST	KA
[39]	Yes	Yes	Yes	–	Yes	Yes	–	Yes	Yes
[40]	–	–	Yes	Yes	Yes	Yes	–	–	Yes
[41]	Yes	Yes	–	Yes	Yes	Yes	–	–	–
[42]	Yes	Yes	Yes	Yes	Yes	Yes	Yes	–	Yes
[43]	Yes	Yes	Yes	Yes	Yes	Yes	–	–	Yes
[44]	Yes	–	Yes	–	–	Yes	Yes	–	Yes
[45]	–	–	–	–	–	–	–	–	–
[46]	Yes	Yes	Yes	Yes	Yes	Yes	–	–	Yes
[47]	Yes	Yes	Yes	Yes	Yes	Yes	–	–	Yes

References

1. Kalaichelvi V, Meenakshi P, Vimala Devi P et al (2021) A stable image steganography: a novel approach based on modified RSA algorithm and 2–4 least significant bit (LSB) technique. J Ambient Intell Hum Comput 12:7235–7243. https://doi.org/10.1007/s12652-020-02398-w
2. Kumar C, Singh AK, Kumar P (2018) A recent survey on image watermarking techniques and its application in e-governance. Multimed Tools Appl 77:3597–3622. https://doi.org/10.1007/s11042-017-5222-8
3. Kaur M, Kumar V (2020) A comprehensive review on image encryption techniques. Arch Comput Methods Eng 27:15–43. https://doi.org/10.1007/s11831-018-9298-8
4. Chai X, Chen Y, Broyde L (2017) A novel chaos-based image encryption algorithm using DNA sequence operations. Opt Lasers Eng 88:197–213. https://doi.org/10.1016/j.optlaseng.2016.08.009. ISSN 0143-8166
5. Zhang Y, Zhang LY, Zhou J, Liu L, Chen F, He X (2016) A review of compressive sensing in information security field. IEEE Access 4:2507–2519. https://doi.org/10.1109/ACCESS.2016.2569421
6. Lin R, Li S (2021) An image encryption scheme based on Lorenz hyperchaotic system and RSA algorithm. Secur Commun Netw 2021:18. https://doi.org/10.1155/2021/5586959. Article ID 5586959
7. Gafsi M, Hajjaji MA, Malek J, Mtibaa A (2020) Efficient encryption system for numerical image safe transmission. J Electr Comput Eng 2020:12. https://doi.org/10.1155/2020/8937676. Article ID 8937676
8. Seung-Hyeok Shin, Won Sok Yoo, Hojong Choi (2019) Development of modified RSA algorithm using fixed mersenne prime numbers for medical ultrasound imaging instrumentation. Comput Assist Surg 24(sup2):73–78. https://doi.org/10.1080/24699322.2019.1649070

9. Kandar S, Chaudhuri D, Bhattacharjee A, Dhara BC (2019) Image encryption using sequence generated by cyclic group. J Inf Secur Appl 44:117–129. https://doi.org/10.1016/j.jisa.2018.12.003. ISSN 2214-2126
10. Çavuşoğlu Ü, Akgül A, Zengin A, Pehlivan I (2017) The design and implementation of hybrid RSA algorithm using a novel chaos based RNG. Chaos Solitons Fractals 104:655–667. https://doi.org/10.1016/j.chaos.2017.09.025. ISSN 0960-0779
11. Hua Z, Zhou Y, Huang H (2019) Cosine-transform-based chaotic system for image encryption. Inf Sci 480:403–419. https://doi.org/10.1016/j.ins.2018.12.048. ISSN 0020-0255
12. Wu X, Zhu B, Hu Y, Ran Y (2017) A novel color image encryption scheme using rectangular transform-enhanced chaotic tent maps. IEEE Access 5:6429–6436. https://doi.org/10.1109/ACCESS.2017.2692043
13. Yao L, Yuan C, Qiang J, Feng S, Nie S (2017) An asymmetric color image encryption method by using deduced gyrator transform. Opt Lasers Eng 89:72–79. https://doi.org/10.1016/j.optlaseng.2016.06.006. ISSN 0143-8166
14. Philippe Refregier, Bahram Javidi (1995) Optical image encryption based on input plane and Fourier plane random encoding. Opt Lett 20:767–769
15. Wang X, Zhou G, Dai C, Chen J (Apr2017) Optical image encryption with divergent illumination and asymmetric keys. IEEE Photonics J 9(2):1–8. https://doi.org/10.1109/JPHOT.2017.2684179. Art no. 7801908
16. Belazi A, Abd El-Latif AA, Belghith S (2016) A novel image encryption scheme based on substitution-permutation network and chaos. Signal Proc 128:155–170. https://doi.org/10.1016/j.sigpro.2016.03.021. ISSN 0165-1684
17. Shahna KU, Mohamed A (2020) A novel image encryption scheme using both pixel level and bit level permutation with chaotic map. Appl Soft Comput 90:106162. https://doi.org/10.1016/j.asoc.2020.106162. ISSN 1568-4946
18. Murillo-Escobar MA, Meranza-Castillón MO, López-Gutiérrez RM, Cruz-Hernández C (2019) Suggested integral analysis for chaos-based image cryptosystems. Entropy 21:815. https://doi.org/10.3390/e21080815
19. Ghebleh M, Kanso A, Noura H (2014) An image encryption scheme based on irregularly decimated chaotic maps. Signal Proc Image Commun 29(5):618–627. https://doi.org/10.1016/j.image.2013.09.009. ISSN 0923-5965
20. Sang Y, Sang J, Alam MS (2022) Image encryption based on logistic chaotic systems and deep autoencoder. Pattern Recognit Lett 153:59–66. https://doi.org/10.1016/j.patrec.2021.11.025. ISSN 0167-8655
21. Toughi S, Fathi MH, Sekhavat YA (2017) An image encryption scheme based on elliptic curve pseudo random and Advanced Encryption System. Signal Proc 141:217–227. https://doi.org/10.1016/j.sigpro.2017.06.010. ISSN 0165-1684
22. Nist randomness test suite. http://csrc.nist.gov/groups/ST/toolkit/rng/stats-tests.html
23. Murillo-Escobar MA, Meranza-Castillón MO, López-Gutiérrez RM, Cruz-Hernández C (2019) Suggested integral analysis for chaos-based image cryptosystems. Entropy 21:815. https://doi.org/10.3390/e21080815
24. Ye G, Jiao K, Huang X (2021) Quantum logistic image encryption algorithm based on SHA-3 and RSA. Nonlinear Dyn 104:2807–2827. https://doi.org/10.1007/s11071-021-06422-2
25. Arab A, Rostami MJ, Ghavami B (2019) An image encryption method based on chaos system and AES algorithm. J Supercomput 75:6663–6682. https://doi.org/10.1007/s11227-019-02878-7
26. Bentoutou Y, Bensikaddour E-H, Taleb N, Bounoua N (2020) An improved image encryption algorithm for satellite applications. Adv Space Res 66(1):176–192. https://doi.org/10.1016/j.asr.2019.09.027. ISSN 0273-1177
27. Zhang Y (2018) Test and verification of AES used for image encryption. 3D Res 9:3. https://doi.org/10.1007/s13319-017-0154-7
28. Singh A, Agarwal P, Chand M (2019) Image encryption and analysis using dynamic AES. In: 2019 5th international conference on optimization and applications (ICOA), pp 1–6. https://doi.org/10.1109/ICOA.2019.8727711

29. Koblitz N, Menezes A, Vanstone S (2000) The state of elliptic curve cryptography. Des Codes Cryptogr 19:173–193. https://doi.org/10.1023/A:1008354106356
30. Hayat U, Azam NA (2019) A novel image encryption scheme based on an elliptic curve. Signal Proc 155:391–402. https://doi.org/10.1016/j.sigpro.2018.10.011. ISSN 0165-1684
31. Dawahdeh ZE, Yaakob SN, bin Othman RR, (2018) A new image encryption technique combining Elliptic Curve Cryptosystem with Hill Cipher. J King Saud Univ-Comput Inf Sci 30(3):349–355. https://doi.org/10.1016/j.jksuci.2017.06.004. ISSN 1319-1578
32. Zhang X, Wang X (2018) Digital image encryption algorithm based on elliptic curve public cryptosystem. IEEE Access 6:70025–70034. https://doi.org/10.1109/ACCESS.2018.2879844
33. Luo Y, Ouyang X, Liu J, Cao L (2019) An image encryption method based on elliptic curve elgamal encryption and chaotic systems. IEEE Access 7:38507–38522. https://doi.org/10.1109/ACCESS.2019.2906052
34. Adhikari S, Karforma S (2022) A novel image encryption method for e-governance application using elliptic curve pseudo random number and chaotic random number sequence. Multimed Tools Appl 81:759–784. https://doi.org/10.1007/s11042-021-11323-y
35. Teh JS, Moatsum A, Sii YC (2020) Implementation and practical problems of chaos-based cryptography revisited. J Inf Secur Appl 50:102421. https://doi.org/10.1016/j.jisa.2019.102421. ISSN 2214-2126
36. Pak C, Kim J, Pang R et al (2021) A new color image encryption using 2D improved logistic coupling map. Multimed Tools Appl 80:25367–25387. https://doi.org/10.1007/s11042-021-10660-2
37. Noshadian S, Ebrahimzade A, Kazemitabar SJ (2018) Optimizing chaos based image encryption. Multimed Tools Appl 77:25569–25590. https://doi.org/10.1007/s11042-018-5807-x
38. Xingyuan W, Xuan C (2021) An image encryption algorithm based on dynamic row scrambling and Zigzag transformation. Chaos Solitons Fractals 147:110962. https://doi.org/10.1016/j.chaos.2021.110962. ISSN 0960-0779
39. Ding Y, Tan F, Qin Z, Cao M, Choo K-KR, Qin Z (2021) DeepKeyGen: a deep learning-based stream cipher generator for medical image encryption and decryption. IEEE Trans Neural Netw Learn Syst. https://doi.org/10.1109/TNNLS.2021.3062754
40. Sang Y, Sang J, Alam MS (2022) Image encryption based on logistic chaotic systems and deep autoencoder. Pattern Recognit Lett 153:59–66. https://doi.org/10.1016/j.patrec.2021.11.025. ISSN 0167-8655
41. Maniyath SR, Thanikaiselvan V (2020) An efficient image encryption using deep neural network and chaotic map, microprocessors and microsystems 77:103134. https://doi.org/10.1016/j.micpro.2020.103134. ISSN 0141-9331
42. Bao Z, Xue R, Jin Y (2021) Image scrambling adversarial autoencoder based on the asymmetric encryption. Multimed Tools Appl 80:28265–28301. https://doi.org/10.1007/s11042-021-11043-3
43. Lakshmi C, Thenmozhi K, Rayappan JBB et al (2021) Neural-assisted image-dependent encryption scheme for medical image cloud storage. Neural Comput Appl 33:6671–6684. https://doi.org/10.1007/s00521-020-05447-9
44. Ding Y et al (2021) DeepEDN: a deep-learning-based image encryption and decryption network for internet of medical things. IEEE Int Things J 8(3):1504–1518. https://doi.org/10.1109/JIOT.2020.3012452
45. Li X, Jiang Y, Chen M et al (2018) Research on iris image encryption based on deep learning. J Image Video Proc 2018:126. https://doi.org/10.1186/s13640-018-0358-7
46. Jianhua Wu, Weixia Xia, Gailin Zhu, Hai Liu, Lujuan Ma, Jianping Xiong (2021) Image encryption based on adversarial neural cryptography and SHA controlled chaos. J Modern Opt 68(8):409–418. https://doi.org/10.1080/09500340.2021.1900440
47. Wang X, Lin S, Li Y (2021) Bit-level image encryption algorithm based on BP neural network and gray code. Multimed Tools Appl 80:11655–11670. https://doi.org/10.1007/s11042-020-10202-2

Chapter 12
Predictive Model with Twitter Data for Predicting the Price Volatility of Cryptocurrencies Using Machine Learning Algorithms

R. Sujatha, B. Uma Maheswari, D. Kavitha, and A. R. Subash

1 Introduction

Cryptocurrency is a new and potentially revolutionary technology where the social indicators and trading behaviour that surround them are highly volatile and change frequently. Cryptocurrency trading is gaining momentum and is one of the fastest growing trading segments in the world (Delfabbro et al. [4]). Thousands of crypto traders are signing up to trade cryptocurrencies every day. Bitcoin is the most widely used cryptocurrency, and it is the one that laid the groundwork for other cryptocurrencies (www.coinmarketcap.com). Cryptocurrency is a decentralised medium of payment that signifies there is no individual who can acquire the ownership of the cryptocurrency. The number of transactions done with the cryptocurrency is increasing as they are easy to transfer and not claimable by any of the countries. Recently, there has been an increase in the number of countries using cryptocurrency, and future predictions show that there is a possibility of developing a common cryptocurrency all over the globe. Even though the cryptocurrency market is on a rising trend, the price is highly volatile. The price volatility is a concern as there is uncertainty for investors who want to use cryptocurrency as an investment opportunity (Mohapatra et al. [11]). Cryptocurrency prices differ from traditional currency prices, and it is difficult to predict the reason for the volatility (Liu and Serletis [10]; Akyildirim et al. [2]). In order to predict the price volatility, large volumes of data from publicly used platforms such as blogs and social media sites are required. The most popular social media website, Twitter, provides the sentiments of the most influential personalities and the common man that impact the cryptocurrency market. Opinion mining is the process of analysing and extracting people's thoughts about a

R. Sujatha (✉) · B. U. Maheswari · D. Kavitha · A. R. Subash
PSG Institute of Management, Tamil Nadu, Coimbatore 641 004, India
e-mail: sujatha@psgim.ac.in

G. Mathur et al. (eds.), *Proceedings of 3rd International Conference on Artificial Intelligence: Advances and Applications*, Algorithms for Intelligent Systems,
https://doi.org/10.1007/978-981-19-7041-2_12

specific item, such as a product, an event, etc. This study aims to develop a predictive model to predict the price of cryptocurrency using machine learning algorithms by analysing Twitter sentiments.

2 Literature Review

Cryptocurrency prediction has been a niche area of research in the past few years. Cryptocurrency prices and public sentiment are interrelated, and many techniques are applied to predict the volatility of the price. Abraham et al. [1] predicted the cryptocurrency price using tweet volumes and sentiment analysis. Huang et al. [6] developed a prediction model using a long short-term memory algorithm based on sentiment analysis. Valencia et al. [15] researched the prediction of cryptocurrency price movement using sentiment analysis and machine learning. Jain et al. [8] developed models to forecast the price of cryptocurrency using sentiment analysis of tweets. Hassan et al. [5] used emotion theory and lexicon sentiments to explore opinions on cryptocurrency. Kim et al. [9] used a hidden Markov model to examine the cryptocurrency market shifts. Pano and Kashef [13] created a valence aware dictionary for sentiment reasoning models for bitcoin price prediction. Inamdar et al. [7] predicted the value of cryptocurrency using sentiment analysis. There are a number of studies done to analyse the relationship between the price volatility of cryptocurrency and social media sentiments (Naeem et al. [12]; Shen et al. [14]; Wu et al. [16]). But this study differs from other studies by providing the opinions (tweets) as input to the machine learning algorithms and analysing the important words in the tweets that influence the price volatility of cryptocurrencies.

3 Methodology

Figure 1 shows the steps involved in the machine learning model building process.

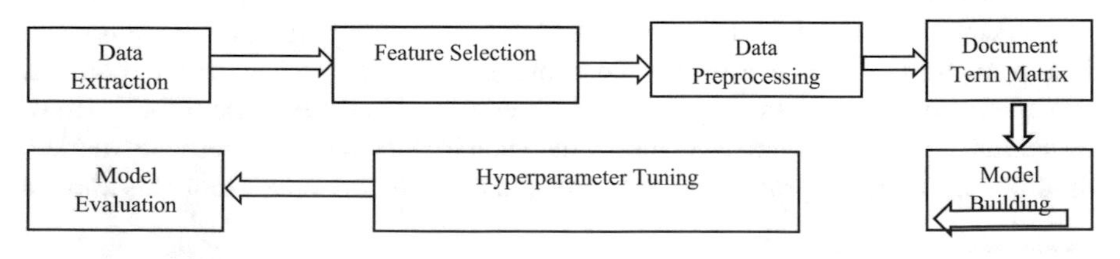

Fig. 1 Model building process

3.1 Data Extraction

There are a number of cryptocurrencies that are traded daily. Bitcoin (BTC) is the largest cryptocurrency by market cap in the cryptocurrency market. For this study, the price of bitcoin is considered. The data from 4th April 2021 to 3rd May 2021 from the website https://coinmarketcap.com/ is extracted for model building. Coinmarketcap is ‘the’ go-to price checker available in the market that provides cryptocurrency prices in the real time (Bitcoin.com [3]). This site provides a listing of top 100 cryptocurrencies on the basis of market capitalization. This site provides information about the current market cap, price of the cryptocurrency, 24 h trading volume, the percentage change in price of cryptocurrency over the last 24 h and the circulating supply of cryptocurrencies. To understand people’s sentiments about cryptocurrency trading, tweets are extracted from Twitter. Tweets are extracted during the same time span from April 4th, 2021 to May 3rd, 2021. Tweets were collected independently for each cryptocurrency, resulting in a dataset comprising 3,589,657 public tweets. Using the Twitter API, a live stream crawler was created that continually saved tweets as and when the tweets were posted in real time. This method ensures that a broader range of Tweets are obtained. The Twitter community generally uses themes prefixed with hashtags (#) and comments about financial goods such as cryptocurrencies or stocks prefixed dollar symbol ($). Various permutations of names of the cryptocurrency in combination with tickers were utilised to generate the Twitter search phrases. Non-English tweets were omitted during extraction.

3.2 Feature Selection

All cryptocurrencies are volatile in nature. In this study, the opinion of crypto traders is the predictor variable, and an increase or decrease in price, i.e. price volatility, is the response variable. In order to predict the increase or decrease in the price of the cryptocurrency, the first frequency price value of the cryptocurrency is compared with the next-frequency price. If the first frequency price is greater, then 1 is assigned. Otherwise, 0 is assigned for the decrease.

3.3 Data Processing

There is always an excessive amount of noise in the Twitter data. Therefore, considerable preprocessing is required before the Twitter data can be used for sentiment analysis. A total of 18 sentiment preprocessing methods are used, together with specially built approaches for removing noise from Tweet messages. To begin with, the Tweets are tokenized and normalised by eliminating URLs, superfluous white spaces, and user references (e.g. @account). If a tweet is retweeted or not, the alphabet RT at the beginning of the tweets is erased. Tweets containing fewer than five tokens are not included since they are unsuitable for performing sentiment analysis at the sentence level. In addition, a novel method for extracting hashtags' potential linguistic significance is offered. If the token's hashtag prefix is found in the NLTK Reuters English lexicon, it is removed. The whole hashtag is stripped from the text if the token is not found in the dictionary. Consider the following Tweet as an illustration: #really #good #buy #buynow #btc #cryptocurrency. If all the hashtags were removed from the tweet, it would lose a lot of its significant sentiments. This statement would then be processed in the same way as in the previous example. So, token contractions are enlarged (e.g. there's into there is), tokens containing number characters (e.g. 3rd or 456) are eliminated, and negations (e.g. didn't and did not) are handled. The ticker symbols (e.g. $BTC) that are used to gather Tweets are deleted from the sentiment analysis because they constitute noise. A carefully produced list of slang abbreviations and acronyms such as LOL or BTW is used, as well as case-folding of words is also done (e.g. purchase to buy). A list of words for cryptocurrency is compiled from various online articles and blogs related to cryptocurrency and other terms that are known to be common in the cryptocurrency area. To decrease noise, tokens with character sequences with more than the same three alphabets are reduced to three alphabets (for e.g. heeeellllloo to heeellloo). The punctuations are eliminated. Stop words that convey no meaning (e.g. is, has, me) are removed using NLTK's English stop word list. After preprocessing, the total number of Tweets was 1,568,698.

3.4 Database Creation

Word clouds are essentially a technique for visualising qualitative information. The preprocessed tweets are converted to a term document matrix with each term listed in the first column and their frequency listed in the second. The word cloud (Fig. 2) showed the frequently occurring words. "Bitcoin" and "btc" are the most frequently occurring words, indicating that bitcoin is the most traded cryptocurrency. After removing the words, bitcoin and btc, the word cloud shows that "igaming', "xrp", "follow", "doge", "usd" as the most frequently occurring words in the tweets.

The term document matrix is converted to a document term matrix so that the text responses can be converted into variable names and be used for model building. The rows of the Document Term Matrix represent the text responses to be analysed, and the columns of the matrix represent the words from the text that are to be used in the analysis. The document term matrix is converted to a data frame and then the price volatility is added as the response variable.

Fig. 2 Word cloud of tweets

4 Model Building

The supervised machine learning algorithms that can be applied for classification are used for building prediction models. Three algorithms, such as support vector machines (SVM), decision trees, and random forest, are used for model building. The hyperparameters of all three algorithms were tuned using the grid search technique to make the most accurate predictions.

4.1 Support Vector Machine

SVM algorithm finds a hyperplane that creates a boundary between the types of data and uses the boundary for classification. The best hyperplane is the one that has the maximum distance from the observations. The performance of the SVM classifier depends on the choice of the regularisation parameters C and gamma. When these hyperparameters are tuned, considerable non-linear classification hyperplanes can be achieved with more accuracy. The C parameter is used for controlling the cost of misclassification of the model. Gamma parameter indicates the influence of a single training of the model on the training data. The hyperparameters in the model are tuned using *tune()* function. The kernel trick used is radial. The parameters c and gamma are tuned by providing a list of values. The C parameter is provided with values ranging from 1/10 to 10 to the power of 6 and gamma is provided with values ranging from 1/10 to 10. Finally, the best parameters identified are cost = 10 and gamma = 1. The model is built again with the parameters cost = 10 and gamma = 1. The number of support vectors identified by the tuned model is 105. The output of tuning is provided in Fig. 3. After tuning, the accuracy of the model is 69.73% and Area under the Curve (AUC) is 0.4613. The AUC curve after tuning is given in Fig. 4.

```
Parameter tuning of 'svm':

- sampling method: 10-fold cross validation

- best parameters:
 cost gamma
   10     1

- best performance: 0.07362637
```

Fig. 3 SVM hyperparameter tuning

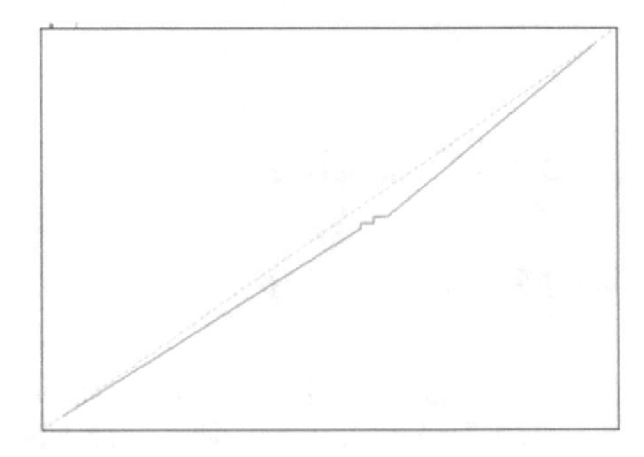

Fig. 4 AUC of support vector machine

4.2 *Decision Tree*

Classification and Regression Trees (CART) algorithm is used for building the decision tree. CART can be used for both regression and classification. The model was built using "rpart" computation engine. Ten-fold cross-validation is used for building the model. For the CART algorithm, three hyperparameters were considered. They are minsplit, minbucket and cost_complexity. The minsplit is used as a criteria for splitting the nodes and is set to 200. The minbucket represents the minimum number of observations in any terminal node and is set as 20. The cost_complexity parameter is used for pruning the tree so that tree has optimal size. The cross-validation error "xerror" is initially 1.000 when the nodes are split and the tree is growing and the error is decreasing to a point and then increasing. The corresponding complexity parameter is 0.0038 and the value is considered for pruning the tree. The complexity parameter and the cross-validation error is given in Fig. 5. After pruning the accuracy of the model is 68.55% and the AUC value is 0.5252. The AUC plot is given in Fig. 6.

4.3 *Random Forest*

A random forest classifier creates a number of decision trees by taking subsets of the dataset. The random forest model was built on a split rule using "gini". The random forest model was built using the parameters mtry = 3, number of trees = 501 and node size = 10. Out of bag (OOB) error rate was used for measuring the prediction error. The initial OOB estimate before tuning is 36.58% (Fig. 7). The random forest

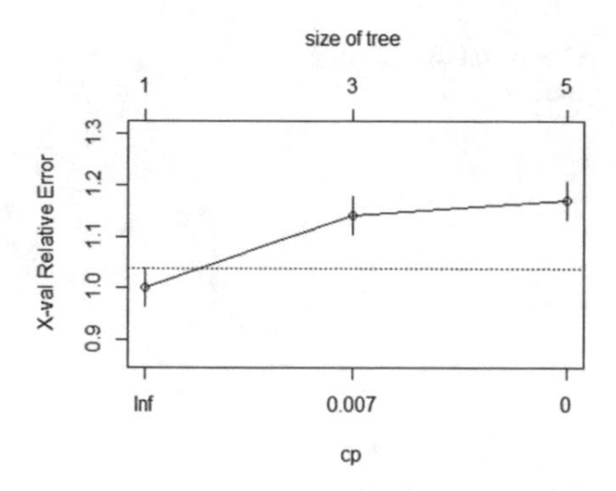

Fig. 5 Complexity parameter of decision tree

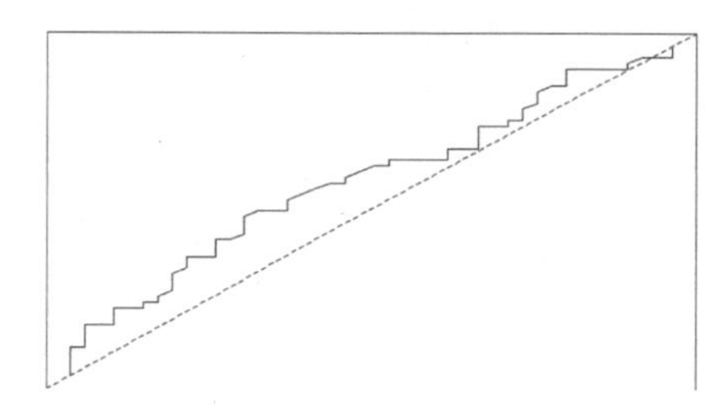

Fig. 6 AUC of decision tree

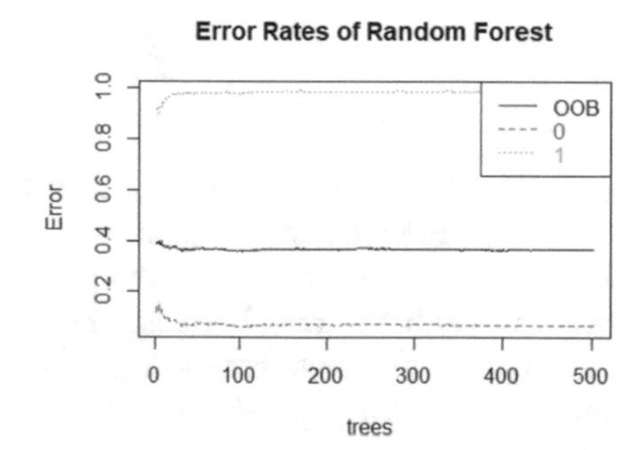

Fig. 7 OOB error before tuning

is tuned with mtryStart = 3, ntreeTry = 100, step factor = 1.5, improve = 0.0001, and node size = 100. After tuning the OOB error rate decreased to 32.33.

Figure 8 shows the OOB error at mtry = 3 is 32.4%, decreases to 32.33% at mtry = 2 and remains at 32.33% at mtry = 1. The final model is created with mtry = 2 and OOB error at 32.33%. The accuracy of the final model is 71.37% and the AUC is 0.6268. The AUC curve for tuned random forest model is given in Fig. 9.

4.4 Model Evaluation

Accuracy, sensitivity, specificity, area under the curve, Kolmogorov Smirnov (KS) parameter and gini coefficient are used for evaluating the model. The comparison

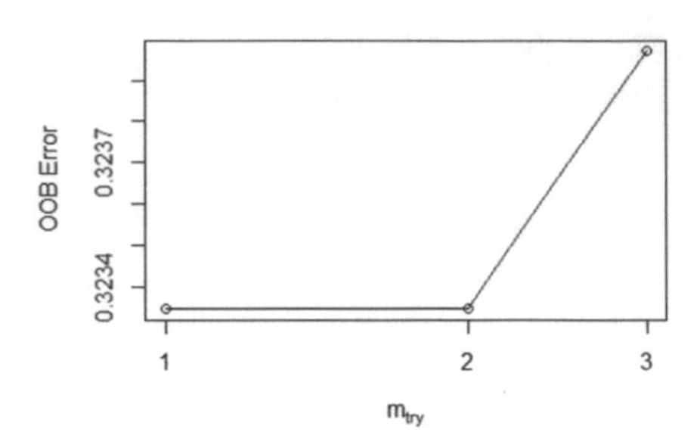

Fig. 8 OOB error after tuning

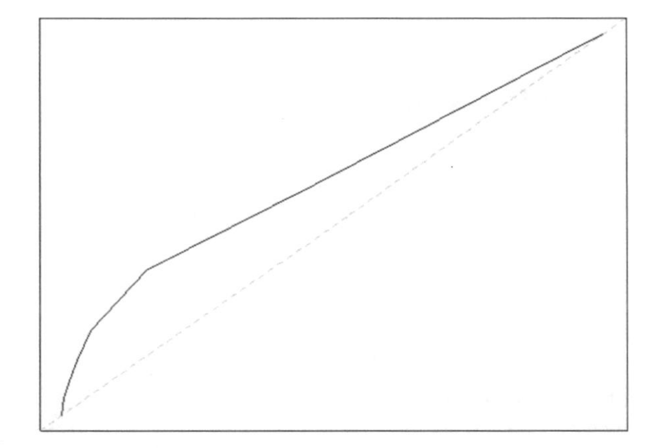

Fig. 9 AUC of random forest

of the model performance measures of the three algorithms are presented in Table 1. AUC is the measure of the classifier model to differentiate between the classes in the response variable. Sensitivity is the true positive rate and specificity is the negative predictive rate. KS measures the degree of separation between the positive and negative distributions. The higher the value, the better the model is at separating the positive from negative cases. Among the three models developed for the study, the random forest classifier has the highest accuracy of 71.3%, AUC value of 0.6268 and KS measure of 0.583. By comparing the model performance measures, it is found that random forest classifiers score high when compared to support vector machines and decision tree algorithms.

Table 1 Model performance measures

Measure	Accuracy	Sensitivity	Specificity	AUC	KS
Support vector machine	0.6973	0.9482	0.1943	0.4613	0.251
Decision tree	0.6855	0.8723	0.2144	0.5252	0.189
Random forest	0.7137	0.9613	0.1139	0.6268	0.583

5 Analysis and Discussion

After model building, further analysis was done to understand the words in the tweets that highly impact the price volatility of the cryptocurrencies. The variable importance plot (Fig. 10) or random forest classifier provides the important words. The variable that contributes to the maximum decrease in Gini is the most important variable. In this model, **excitement, experience, together, power, and hit** contributes to the mean decrease in Gini. Similarly, **hunt, excitement, platform and pump together** contribute more to MeanDecreaseAccuracy and are considered important for the model. "Exciting" is an important variable as it gives enthusiasm for the crypto traders to invest in the currency. Whenever the word "Exciting" is tweeted, there is an increasing trend in the cryptocurrency price. "Hunt" can be taken as a key-search word to invest money by an individual in order to get more profit. The use of the word "Hunt" in the tweets results in more people getting into the trade and also getting profit from their investments. "Experience" is a word often used by the crypto traders indicating their experience in trading cryptocurrency. Since cryptocurrencies are traded based on block chain technology, the investors have a secure environment which makes them tweet with the word "Experience". This word also adds to the influence in price. The word "Platform" is based on the block chain technology that is making the crypto traders often describe the platform in their tweets.

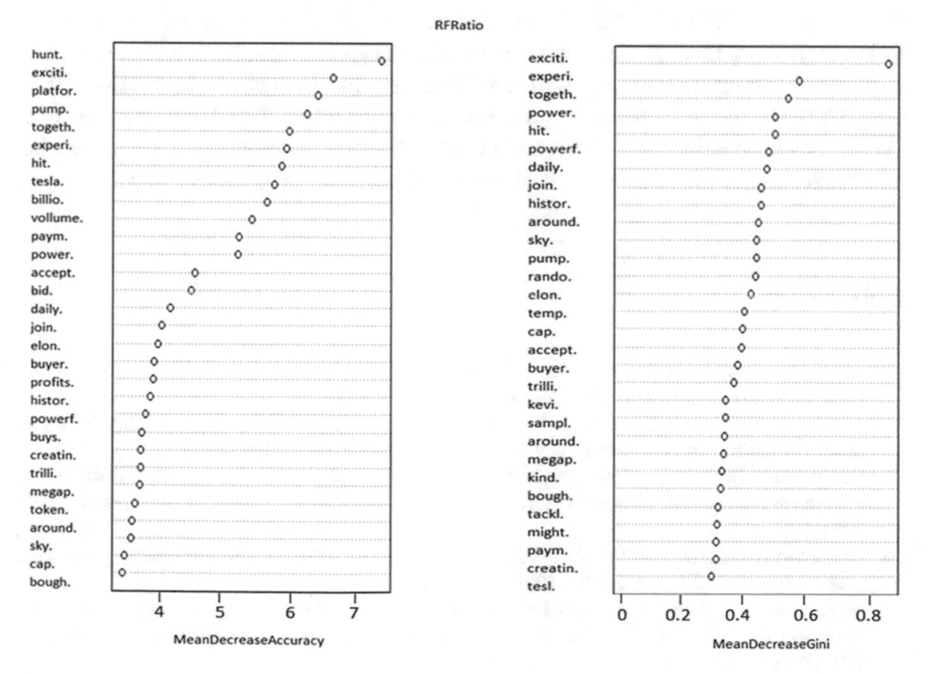

Fig. 10 Important variables influencing price volatility

6 Conclusion and Scope for Further Research

Price volatility is a major problem in cryptocurrency trading, and this study attempted to design a predictive model using machine learning algorithms. Random forest algorithms provided better accuracy after tuning the hyperparameters when compared to other algorithms. Further, the important words in the tweets influencing the price volatility provide some important insights about cryptocurrency trading. The design of new cryptocurrency and its service (or current cryptocurrency innovation activities) should prioritise performance as the most important aspect in determining acceptance. Consumers must see cryptocurrency as a high-value-added offer, and significant marketing efforts should be done so that the potential customers foresee that the cryptocurrency provides high returns. The more value a cryptocurrency provides, the more likely it is to be used. The suggested strategy in the bitcoin market is focusing on utility. The enabling conditions for trading are very important for the cryptocurrency market. The decision to buy or sell an existing or new cryptocurrency is highly influenced by the terms under which potential consumers benefit from it. Many factors influence the cryptocurrency adoption, but the important factors are technical resources and knowledge required to operate a cryptocurrency; the compatibility of a customer's usage of technology with cryptocurrency's technical requirements; the existence of widely accepted norms for operating with the cryptocurrency; and full time availability of a helpdesk that is easily accessible to the customers in case of any problems. The next point to be considered is how much work an investor must put out in order to use a cryptocurrency. Any improvement in the usage of cryptocurrency will have a favourable impact on the trading. This research has some limitations as the study only concentrated on Bitcoin prices and Twitter (social media data). It can be extended to other cryptocurrencies and social media sites to get a holistic picture of the cryptocurrency trading practices.

References

1. Abraham J, Higdon D, Nelson J, Ibarra J (2018) Cryptocurrency price prediction using tweet volumes and sentiment analysis. SMU Data Sci Rev 1(3):1
2. Akyildirim E, Corbet S, Lucey B, Sensoy A, Yarovaya L (2020) The relationship between implied volatility and cryptocurrency returns. Financ Res Lett 33:101212
3. Bitcoin.com. Here are 7 crypto comparison sites chasing coinmarketcap's crown. https://news.bitcoin.com/seven-crypto-comparison-sites-chasing-coinmarketcaps-crown/, Accessed 3 Oct 2021
4. Delfabbro P, King DL, Williams J (2021) The psychology of cryptocurrency trading: risk and protective factors. J Behav Addict 10(2):201–207
5. Hassan MK, Hudaefi FA, Caraka RE (2021) Mining netizen's opinion on cryptocurrency: sentiment analysis of Twitter data. Stud Econ Financ 39(3):365–385.https://doi.org/10.1108/SEF-06-2021-0237
6. Huang X, Zhang W, Tang X, Zhang M, Surbiryala J, Iosifidis V, Zhang J et al (Apr 2021) Lstm based sentiment analysis for cryptocurrency prediction. In: International conference on database systems for advanced applications. Springer, Cham, pp 617–621

7. Inamdar A, Bhagtani A, Bhatt S, Shetty PM (May 2019) Predicting cryptocurrency value using sentiment analysis. In 2019 international conference on intelligent computing and control systems (ICCS). IEEE, pp 932–934
8. Jain A, Tripathi S, Dwivedi HD, Saxena P (Aug 2018) Forecasting price of cryptocurrencies using tweets sentiment analysis. In: 2018 eleventh international conference on contemporary computing (IC3). IEEE, pp 1–7
9. Kim K, Lee S-YT, Assar S (2022) The dynamics of cryptocurrency market behavior: sentiment analysis using Markov chains. Ind Manag Data Syst 122(2):365–395. https://doi.org/10.1108/IMDS-04-2021-0232
10. Liu J, Serletis A (2019) Volatility in the cryptocurrency market. Open Econ Rev 30(4):779–811
11. Mohapatra S, Ahmed N, Alencar P (Dec 2019). Kryptooracle: a real-time cryptocurrency price prediction platform using twitter sentiments. In: 2019 IEEE international conference on big data (Big Data), IEEE, pp 5544–5551
12. Naeem MA, Mbarki I, Suleman MT, Vo XV, Shahzad SJH (2021) Does Twitter happiness sentiment predict cryptocurrency? Int Rev Financ 21(4):1529–1538
13. Pano T, Kashef R (2020) A complete VADER-based sentiment analysis of bitcoin (BTC) tweets during the era of COVID-19. Big Data Cognit Comput 4(4):33
14. Shen D, Urquhart A, Wang P (2019) Does twitter predict Bitcoin? Econ Lett 174:118–122
15. Valencia F, Gómez-Espinosa A, Valdés-Aguirre B (2019) Price movement prediction of cryptocurrencies using sentiment analysis and machine learning. Entropy 21(6):589
16. Wu W, Tiwari AK, Gozgor G, Leping H (2021) Does economic policy uncertainty affect cryptocurrency markets? Evidence from Twitter-based uncertainty measures. Res Int Bus Financ 58:101478

Chapter 13
Recognition of Indian Sign Language Characters Using Convolutional Neural Network

Siddhesh Gadge, Kedar Kharde, Rohit Jadhav, Siddhesh Bhere, and Indu Dokare

1 Introduction

Communication is one of the most important skills you require for a successful life. Communication works for those who work at it. The primary source of communication is speech but what about hearing-impaired people? Deaf people use Sign language to communicate with one another. Sign language is a kind of nonverbal communication that includes hand gestures, movements, finger alignment, and, in some cases, facial expressions. There are 143 existing different sign languages all over the world, mainly the American Sign Language (ASL), British Sign Language, French sign language, Japanese Sign Language, and Indian Sign Language (ISL). The American sign language uses one hand for communication. On the other hand, the Indian Sign language uses both hands for communication. In comparison to ISL, ASL uses a single hand to depict gestures, making it simpler. ISL is more difficult than ASL since it uses both hands to represent gestures. Figure 1 shows ISL signs of 10 digits and 26 alphabets.

People who are used to communicating through speech find it challenging to communicate using sign language. Additionally, some parents of deaf children are

S. Gadge · K. Kharde · R. Jadhav · S. Bhere · I. Dokare (✉)
Department of Computer Engineering, VES Institute of Technology, Mumbai, India
e-mail: indu.dokare@ves.ac.in

S. Gadge
e-mail: 2018siddhesh.gadge@ves.ac.in

K. Kharde
e-mail: 2018.kedar.kharde@ves.ac.in

R. Jadhav
e-mail: 2018.rohit.jadhav@ves.ac.in

S. Bhere
e-mail: 2018.siddhesh.bhere@ves.ac.in

G. Mathur et al. (eds.), *Proceedings of 3rd International Conference on Artificial Intelligence: Advances and Applications*, Algorithms for Intelligent Systems,
https://doi.org/10.1007/978-981-19-7041-2_13

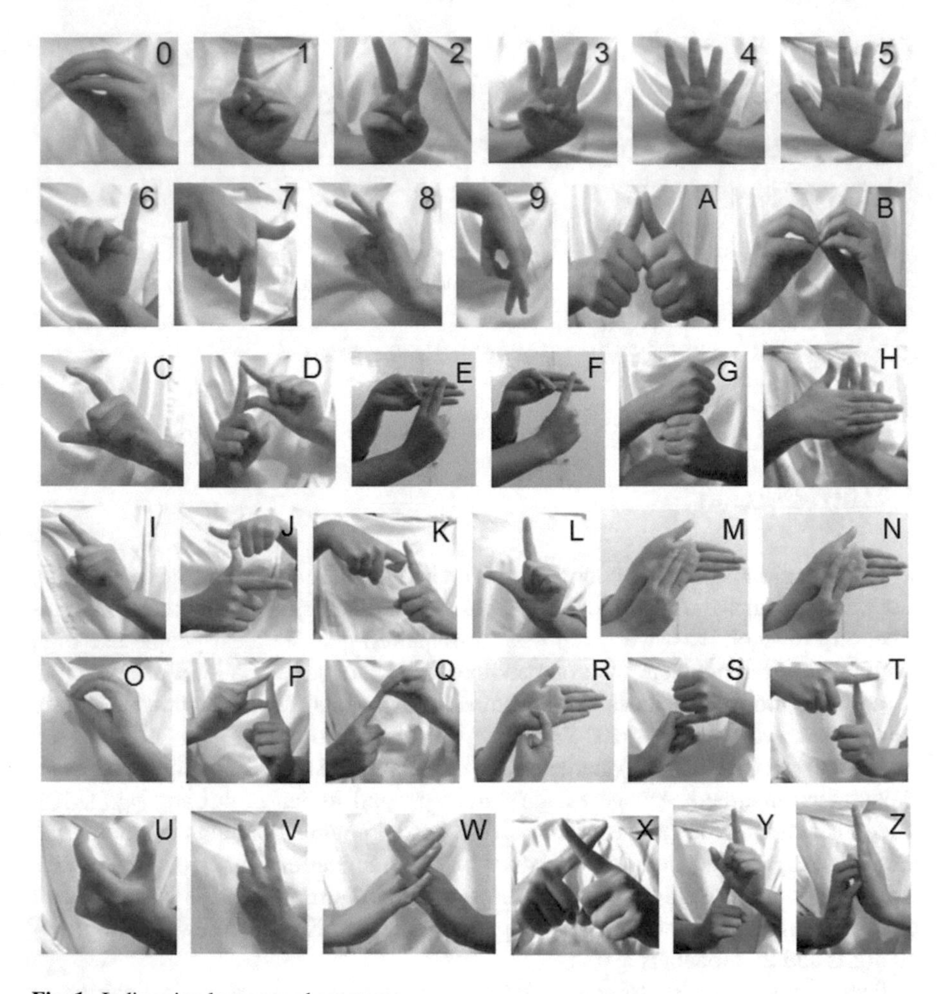

Fig. 1 Indian sign language characters

unaware of sign language's ability to bridge communication gaps. The major goal of the ISL system discussed in this paper is to eliminate this communication barrier and create a simple bridge between ordinary people and impaired persons.

The fundamental purpose of this research is to provide a custom CNN architecture for Indian sign language recognition. This research examines a self-compiled dataset of the Indian sign language characters using this architecture. This paper is divided into several sections. The survey of the existing systems and approaches is covered in the first section. The ISL dataset and its production are described in the second section. In the third section, the proposed architecture is further explained. In the fourth section, the results obtained are compared and examined. Section five brings the paper to a close.

2 Related Work

In recent times various methods have been implemented for Indian Sign Language Recognition. Patel et al. [10] discussed various methods of preprocessing, feature extraction, and classification that have been used mainly for ISL recognition thus analyzing them overall. Raghuveera et al. [11] introduced an algorithm that translates ISL signs to meaningful English sentences by using image denoizing for preprocessing, k-means clustering for segmentation, histogram of oriented gradients for feature extraction, and SVM for classification which was performed over the self-extended peasant dataset. Adithya et al. [4] presented a neural network-based method for automatically recognizing the fingerspelling in Indian sign language. A new shape feature based on the distance transform of the image is proposed in this work. Gangrade et al. [12] proposed that Oriented FAST and Rotated BRIEF (ORB) performs better by comparing it with other methods, and it is also used with KNN classification achieving 93.26% recognition accuracy with the ISL dataset. Deora et al. [3] presented a recognition system using principal component analysis with 94% recognition accuracy and later it was proposed that the distance of fingertips from the center of the hand and the number of fingertips can be used to improve the performance. Mali et al. [8] presented an ISL recognition model where they made their own dataset using PCA for feature extraction, and SVM for classification and achieved 95.31% accuracy. Mangamuri et al. [9] worked on making a standard dataset of ISL freely available and verified it on 6 different classification algorithms with very good accuracy.

Rajam et al. [2] proposed a method for identifying the ISL signs using gesture dataset. The results show that the system is able to recognize images with 98.125% accuracy. Euclidean distance, Canny Edge Detection was used for feature extraction. Shenoy et al. [5] introduced an android application that uses the smartphone's camera to capture the sign language used by the person. The frames were taken at 5 frames per second place. The majority of these photos were taken with a regular webcam, although a handful was taken with a smartphone camera. Kaur et al. [7] proposed an Indian sign language recognition system with an accuracy of 99.43%. The dataset utilized was built from the ground up and made available to the public. Shravani et al. [13] presented an Indian sign language recognition using the Bag of words model (BOW). Not only the recognition of static images is discussed but also the real-time recognition of gestures is proposed. Mistree et al. [15] discussed the first Indian Sign Language video dataset, INSIGNVID, and utilizing this dataset as input, a unique approach is presented that uses transfer learning to translate video of ISL sentences into acceptable English sentences. Bhagat et al. [6] proposed depth perception techniques that enabled effective real-time background subtraction. A one-to-one mapping between the depth and the RGB pixels was achieved using computer vision techniques. Patil et al. [1] presented a Convolutional neural network to train the model and recognize the images. The accuracy achieved is about 95%. Gedam et al. [14] introduced a framework that has been prepared using 3000 static images of RGB photographs captured using a standard camera. The accuracy achieved is 99.56%.

Table 1 Comparative analysis of literature survey

Ref. no	Author/year	Methods	Results—accuracy
[11]	Raghuveera et al. (2019)	SVM, HOG	Avg = 71.85%
[4]	Adithya et al. (2013)	ANN, Euclidean, city block, and Chessboard, Fourier Descriptors	91.11%
[12]	Gangrade et al. (2020)	SIFT, k mean, SVM	93.26%
[8]	Mali et al. (2019)	PCA, morphological processes, SVM classifier	95.31%
[2]	Rajam et al. (2011)	Euclidean distance, Canny Edge Detection	98.125%
[5]	Shenoy et al. (2018)	K-Nearest Neighbors, Hidden Markov Model, Histogram of Oriented Gradients, Kernelized Correlation Filter (KCF) Tracker	Static hand poses = 99.7, Gesture recognition = 97.23%
[7]	Kaur et al. (2019)	SIFT and FFBPNN	99.43%
[13]	Shravani et al. (2020)	SVM, KNN, CNN, LR	99%
[15]	Mistree et al. (2021)	MovileNetV2	93.89%
[1]	Patil et al. (2021)	CNN	95%
[14]	Gedam et al. (2021)	CNN	99.56%

Overall, different techniques for image preprocessing and feature extraction had been explored in the past which gave a good performance with common classification algorithms, later researchers shifted to using Convolutional Neural networks. More effort needs to be taken for developing a system that is completely capable of recognizing ISL with as high performance as possible (see Table 1).

3 Materials and Methods Used

3.1 Dataset Used

There are various datasets available for different sign languages, but there is a lot of research pending for the Indian sign language. The dataset used in this study was built from scratch. This custom dataset was utilized for this study. The Indian Sign Language Recognition dataset has a total of 142,765 images taken by four distinct people. There are a total of 36 classes in the dataset, including 10 for numerals (0–9) and 26 for alphabets (A–Z). This dataset has roughly 4000 images in each class. Figure 2 shows sample images from a dataset of four different persons.

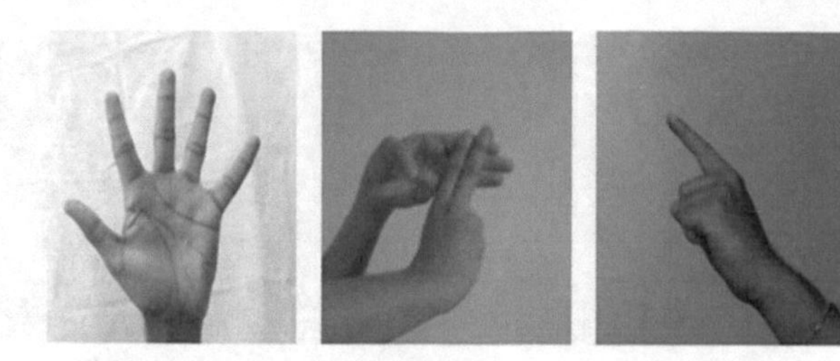
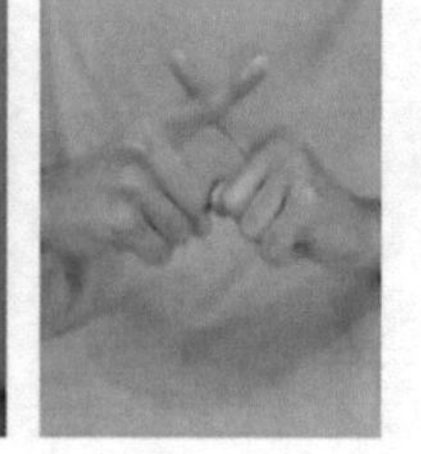

Fig. 2 Sample Images from the dataset

3.2 Image Processing

The primary purpose of this study is to identify the sign. For good architectural performance, several Image Processing steps were performed on the dataset. The first step of Image processing is to convert the image into grayscale. Figure 3a shows the grayscale image of the original image. The extraction of the hand from the image is the next stage. It begins by transforming the original image into an HSV image, which stands for Hue, Saturation, and Value model. This model can detect certain colors and can reduce the intensity of light coming in from the outside. The image in Fig. 3b has been converted to HSV. The next stage is to create a mask for the extraction of the hand. Firstly, skin color of the hand was defined and then a skin mask was created for extraction purposes. The image in Fig. 3c shows the image with a mask produced in the shape of a hand. The mask produced can contain some noise which we can see in Fig. 3c. As a result, the next stage in image processing is to eliminate the noise. Figure 3d shows the image after removing the noise. The next step is a Morphological operation on skin masks. Morphological operation is used to structure the mask created according to the correct shape. Basically, Morphology is a broad set of processing operations based on the shape of the image. The image in Fig. 3e depicts the image after the morphological operation. After this the skin mask is applied on a grayscale image for extraction of the hand. The image in Fig. 3f shows the hand-extract from the original image.

This research includes two different forms of Image Processing for study. One form of Image processing is done till the extraction of the hand in the grayscale image as shown in Fig. 3f. In the second form, the image is further processed. Canny edge detection is applied on grayscale images to detect the sharp edges of the hand. The canny edges of the extracted hand are shown in Fig. 3g. Because of the significant disparity in camera quality, two separate dataset variants are constructed. So, even if the design is well-suited, when it comes time to anticipate a sign, it may not forecast as desired. At this point, a grayscale image can be useful.

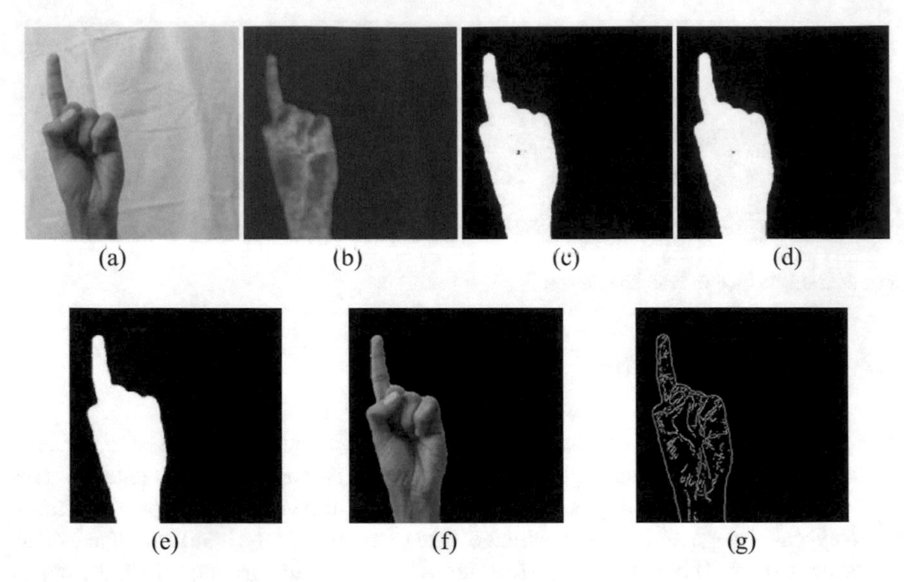

Fig. 3 Image processing

3.3 Deep Convolutional Neural Network

The main aim of this research is to identify the Indian sign language. We have used Convolutional Neural Network for training and recognition purposes. A previous study says that Convolutional Neural Networks are best suited for 2D data like Image classification, Object detection. The preprocessed images will be trained and validated on a low-cost Convolutional Neural Network model and approximately 8000–10,000 images are tested on the same Convolutional Neural Network model. Also, approximately 1200–1500 images were tested separately on the same Convolutional Neural Network model.

4 Proposed Architecture

The proposed architecture of convolutional neural networks classifies Indian sign language characters. The architecture is trained and tested for numerals and alphabets separately. The numerals consist of 10 classes and alphabets of 26 classes. The proposed architecture consists of two tasks: feature extraction (through convolution utilizing a succession of convolution layers) and feature mapping (by fully connected layers) onto the specified classification labels.

The convolutional part does the feature extraction which contains 2 groups of a sequence of two convolutional layers with 64 filters & kernel size 3 followed by a max-pooling layer. The output from the convolution block is flattened. The output

Table 2 Layer Description, Output Shape, Number of Parameters

Layer (Type)	Output shape	Parameters
conv2d (Conv2D)	(None, 30, 30, 64)	640
conv2d (Conv2D)	(None, 28, 28, 64)	36,928
max_pooling2d (MaxPooling2D)	(None, 14, 14, 64)	0
conv2d_2 (Conv2D)	(None, 12, 12, 64)	73,856
conv2d_2 (Conv2D)	(None, 10, 10, 64)	147,584
max_pooling2d (MaxPooling2D)	(None, 5, 5, 64)	0
flatten (Flatten)	(None, 3200)	0
dense (Dense)	(None, 512)	1,638,912
dropout (Dropout)	(None, 512)	0
dense_1 (Dense)	(None, Number of classes)	5130
Total parameters: 1,903,050 Trainable parameters: 1,903,050 Non-Trainable parameters: 0		

from flattening into a 1d array is fed into a dense layer that has 512 nodes with ReLu activation and is regularized with a 20% dropout layer. The final layer contains softmax activation for classification. In Table 2, the layer-wise description of the proposed architecture is given.

5 Results and Discussion

The proposed architecture gives good performance in validation and testing with accuracies greater than 98%. Below is a detailed discussion on the performance of the proposed architecture.

5.1 *Train, Test, and Validation Sets*

The dataset used for numerals which are preprocessed till canny edges and the dataset for numerals which is preprocessed till grayscale hand extraction each consists of 40,000 images where each class has 4000 images with 10 classes. Similarly, the dataset used for alphabets consists of 65,000 images and each class has 2500 images with 26 classes. 20% of images are used for testing purposes. From the remaining 80% of the training set 20% of images are used for validation. Table 3 shows how the dataset is split.

Table 3 Number of classes & Splitting of datasets

Dataset	No. of classes	Total Input Images	Train size after splitting	Validation size after splitting	Testing samples
Numeral dataset with canny edges	10	40,000	32,000	6400	8000
Alphabet dataset with canny edges	26	65,000	52,000	10,400	13,000
Numeral dataset with grayscale	10	40,000	32,000	6400	8000
Alphabet dataset with grayscale	26	65,000	52,000	10,400	13,000

5.2 *Model Performance*

For training, validation, and testing, the proposed architecture achieves accuracies of better than 98%. The training and validation testing for the model for numerals is performed for 15 epochs and the model for alphabets is run for 10 epochs. Figure 4 depicts the accuracy curves for four different dataset types.

Figure 4 shows the accuracy curve of training and validation of the grayscale dataset for numerals for 15 epochs. Till 4 epochs, training accuracy increases rapidly and then stays between 98 and 100%, whereas validation accuracy is always more than 98%.

Figure 5 shows the accuracy curve of training and validation of the grayscale dataset of alphabets for 10 epochs. In this training and validation accuracy fluctuates till 4 epochs but always stays greater than 95%. Validation accuracy stays between 98 and 100% from the first epoch to the last.

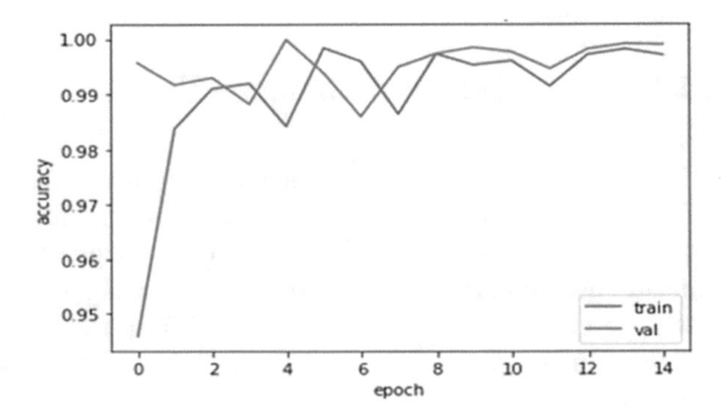

Fig. 4 Numeral dataset with grayscale

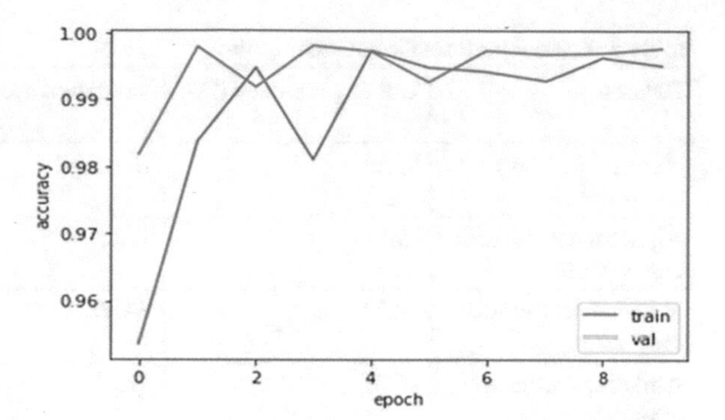

Fig. 5 Alphabet dataset with grayscale

Figure 6 shows that validation accuracy fluctuates continuously till 8 epochs but still gives above 96% accuracy which is a good fit. Here training accuracy significantly climbs to 98–99% in the first two epochs and then remains stable.

Figure 7 shows the accuracy curve for the alphabet dataset which includes canny edges which run for 10 epochs. Here training accuracy increases rapidly in the first two epochs and remains stable for the rest of the epochs. Validation accuracy shows a good fit for the model by staying stable from 98 to 100%.

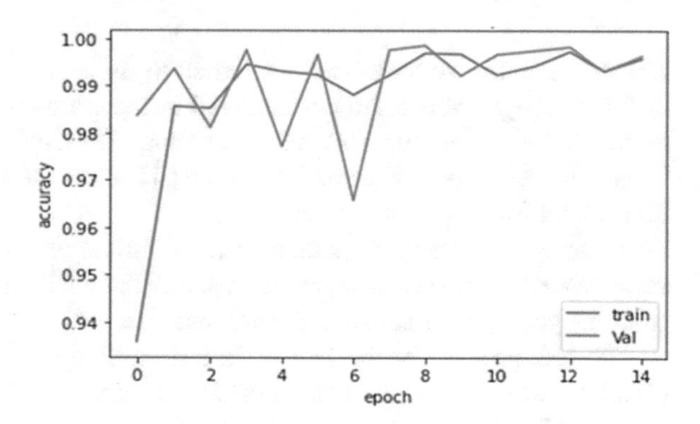

Fig. 6 Numeral dataset with canny edge

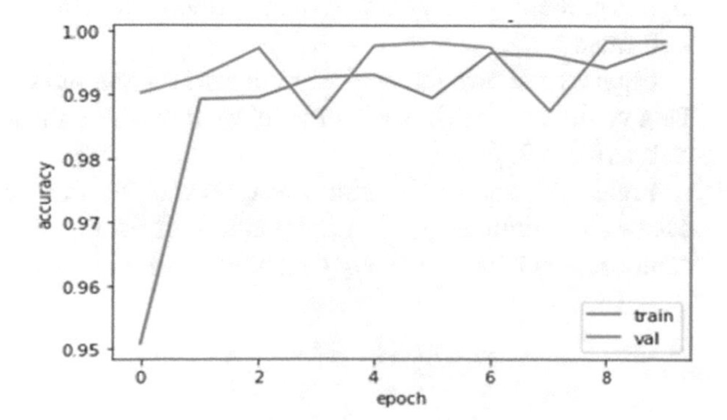

Fig. 7 Alphabet dataset with canny edge

Table 4 Overall performance of architecture

Dataset	Training accuracy (%)	Validation accuracy (%)	Testing accuracy (%)
Numeral dataset with canny edges	99.58	99.63	99.82
Alphabet dataset with canny edges	99.76	99.85	99.82
Numeral dataset with grayscale	99.70	99.89	99.87
Alphabet dataset with grayscale	99.51	99.77	99.76

Table 4 shows the performance of architecture on all four differently preprocessed datasets. For all four dataset types, the accuracy of training, testing, and validation remained more than 99%. This demonstrates how precise and well-suited this architecture is to the dataset.

5.3 Testing

The architecture we have used for Indian Sign Language Recognition is tested on different images which are not included in the training dataset. For alphabets and numerals, each class has 700 images totaling 7000 images for numerals and 18200 images for Alphabets. Figures 8, 9, 10 and 11 show the Confusion matrix of testing done on all four types of datasets.

Figure 8 shows the confusion matrix for Numerals grayscale dataset. The darkest color shows the most correct predictions. Most of the images were predicted correctly ensuring good performance of the architecture.

The background for the Numerals testing sample images had some obstacles, therefore some of the classes are unable to predict the correct alphabet.

Figure 9 shows the confusion matrix for the alphabet grayscale dataset. For the alphabet, there are very few wrong predictions. This shows that the architecture is well fitted.

Figure 10 shows the confusion matrix for the numerals dataset of canny edges. This confusion matrix shows that class 0 predicts most of the images correctly as well as class 9.

Figure 11 shows the confusion matrix for the alphabet of the canny edge dataset. The number of incorrect predictions for the alphabet is extremely low. This demonstrates how effectively the architecture fits well.

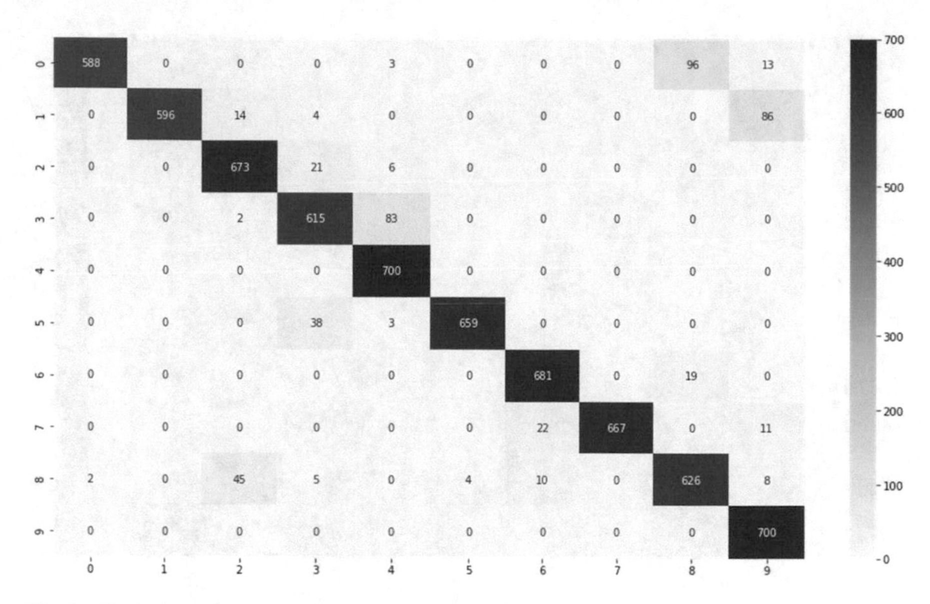

Fig. 8 Confusion matrix for numerals grayscale dataset

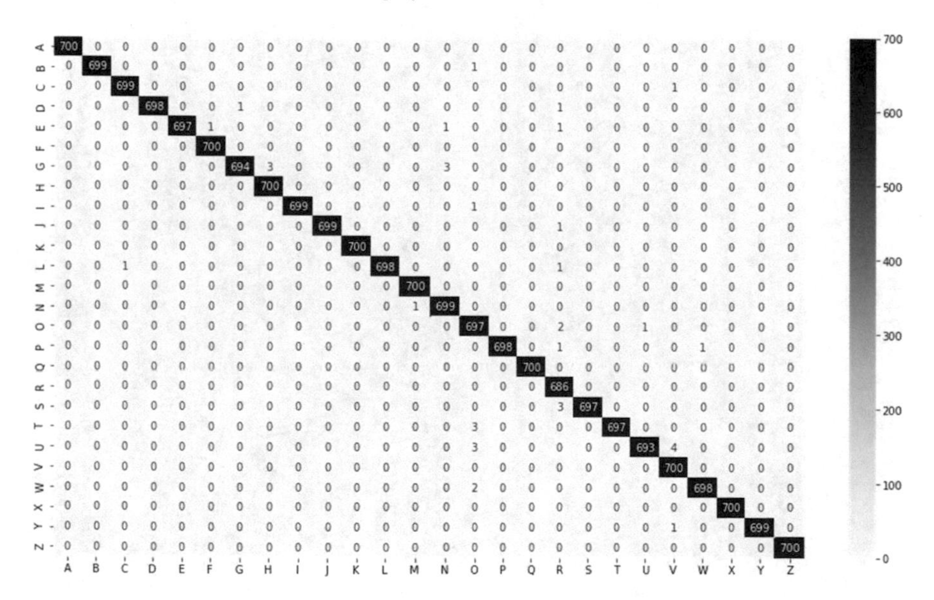

Fig. 9 Confusion matrix for alphabet grayscale dataset

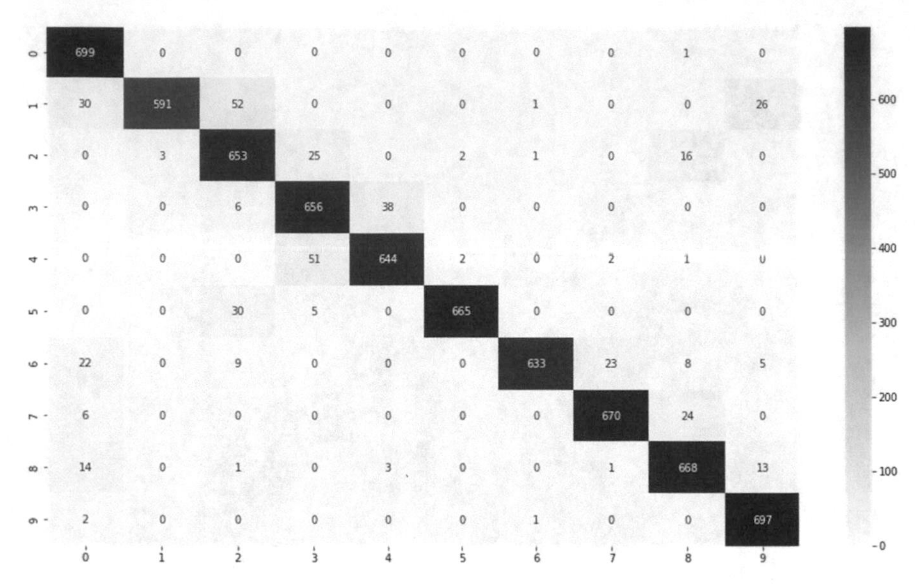

Fig. 10 Confusion matrix for numerals canny edge dataset

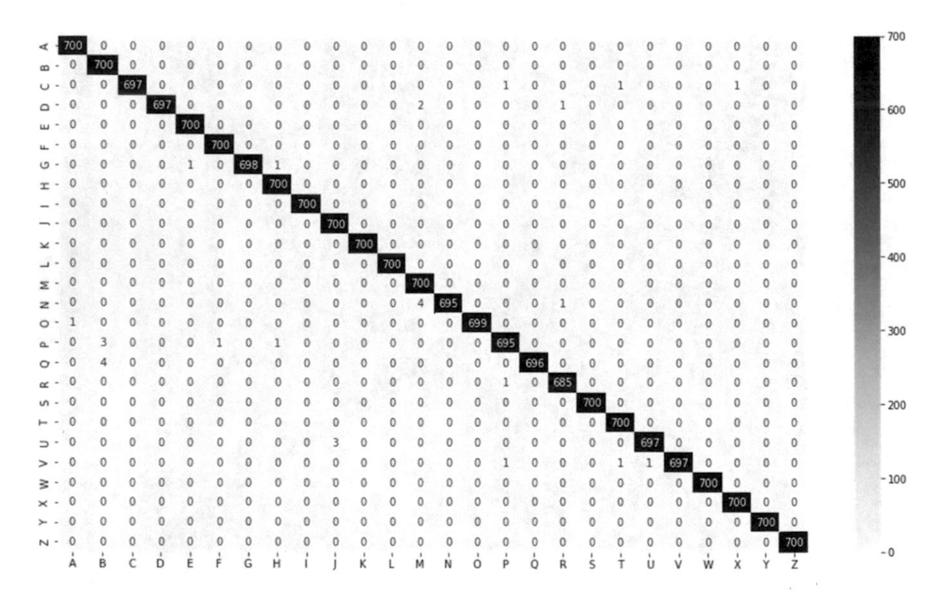

Fig. 11 Confusion matrix for alphabets canny edge dataset

6 Conclusion

In this paper, we have portrayed the utilization of custom architectures of Convolutional Neural networks for Indian sign language recognition. The dataset used is self-compiled and preprocessing is done on images for better accuracy. CNN gives 99.82% accuracy for grayscale digits and 99.88% accuracy for grayscale alphabets. In the case of canny edges, the given architecture gives an accuracy of 99.78 and 99.74% for digits and alphabets respectively. Results indicate that our proposed model achieved performance better than previous works and can be integrated further with any application related to ISL. The future scope for this system is that it can be used to translate normal language into sign language and vice versa.

References

1. Patil R, Patil V, Bahugunal A, Datkhile G (2021) Indian sign language recognition using convolutional neural network. ITM web of conferences 40, 03004, ICACC-2021
2. Rajam PS, Balakrishnan G (2011) Real time Indian sign language recognition system to aid deaf-dumb people. In: 2011 IEEE 13th international conference on communication technology, pp 737–742
3. Deora D, Bajaj N (2012) Indian sign language recognition. In: 2012 1st international conference on emerging technology trends in electronics, communication & networking
4. Adithya V, Vinod PR, Gopalakrishnan U (2013) Artificial neural network based method for Indian sign language recognition. 2013 IEEE conference on information & communication technologies, pp 1080–1085
5. Shenoy K, Dastane T, Rao V, Vyavaharkar D (2018) Real-time Indian sign language (ISL) recognition. In: 2018 9th international conference on computing, communication and networking technologies (ICCCNT), pp 1–9
6. Bhagat NK, Vishnusai Y, Rathna GN (2019) Indian sign language gesture recognition using image processing and deep learning. In: 2019 digital image computing: techniques and applications (DICTA), pp 1–8
7. Kaur J, Krishna C (Aug 2019) An efficient Indian sign language recognition system using sift descriptor. Int J Eng Adv Technol (IJEAT) 8(6)
8. Mali D, Limkar N, Mali S (18 May 2019) Indian sign language recognition using SVM classifier. In: Proceedings of international conference on communication and information processing (ICCIP)
9. Mangamuri L, Jain L, Sharmay A (2019) Two hand Indian sign language dataset for benchmarking classification models of machine learning. In: 2019 international conference on issues and challenges in intelligent computing techniques (ICICT), pp 1–5
10. Patel P, Patel N (Feb 2019) A survey on state of the art methods for Indian sign language recognition. In: J Appl Sci Comput (JASC) VI(II)
11. Raghuveera T, Deepthi R, Mangalashri R et al (2020) A depth-based Indian sign language recognition using Microsoft Kinect. Sādhanā 45:34
12. Gangrade J, Bharti J, Mulye A (2020) Recognition of Indian sign language using ORB with bag of visual words by Kinect sensor. IETE J Res

13. Shravani K, Lakshmi A, Geethika M, Kulkarni S (May–Jun 2020) Indian sign language character recognition. IOSR J Comput Eng (IOSR-JCE) 22(3), Ser I
14. Gedam R, Soni A, Kharat K, Mulmule V, Sood S (2021) Indian sign language recognition. Int J Res Eng Sci (IJRES) 9(5)
15. Mistree K, Thakor D, Bhatt B (2021) Towards Indian sign language sentence recognition using INSIGNVID: Indian sign language video dataset. Int J Adv Comput Sci Appl (IJACSA) 12(8)

Chapter 14
Effective Vehicle Classification and Re-identification on Stanford Cars Dataset Using Convolutional Neural Networks

B. Cynthia Sherin and Kayalvizhi Jayavel

1 Introduction

Video surveillance has a vital role in modern transportation systems for security and traffic control [1]. Videos provide abundant information about the vehicle specifications such as colour, make, model, the licence number plate and traffic density, vehicle count, and spatial locations of the vehicles etc. The development in image processing techniques and contributions of intelligent transportation systems has paved the way to extract all this information from the surveillance videos and perform vehicle detection, categorization, pose assessment and re-identification. Vehicle re-identification (Re-ID) has various applications such as city traffic flow monitoring, tracking anomalous vehicles, estimating the trajectory of the vehicle [2], maintaining the safety of automated vehicles, and reduction of traffic energy consumption [3]. Vehicle Re-ID is the process of capturing the same vehicle from multiple cameras which generally has non-overlapping views. Most of the vehicles appear to possess similar appearance and visual hints based on their local appearance as shown in Figures 1 and 2. Cars being 3D objects, the angles from which they are viewed also accounts to the difference in leading to the viewpoint variations and also poses challenges in re-identification. So these attributes exhibit visually discriminative features by connecting low level appearance features with high level properties to classify the vehicle image, where automatic selection of attributes have to be made for better discriminative results [4]. So the first step in the re-identification process [5, 6] is performed by the CNN [7, 8] where the features of the vehicles are determined by

B. Cynthia Sherin (✉) · K. Jayavel
School of Computing, Department of Networking and Communications, SRM Institute of Science and Technology, Kattankulathur, Chennai 603203, India
e-mail: cb4881@srmist.edu.in

K. Jayavel
e-mail: kayalvij@srmist.edu.in

G. Mathur et al. (eds.), *Proceedings of 3rd International Conference on Artificial Intelligence: Advances and Applications*, Algorithms for Intelligent Systems,
https://doi.org/10.1007/978-981-19-7041-2_14

the similarity metric. After the feature extraction, comparison occurs to classify the images into the right class. This includes comparison of vehicle image from multiple angles in the absence of stringent resolution needs for every compared image. Eventually the effect of over-fitting has to be avoided as in certain cases best results are obtained from the training images, but do not perform well on the test image [9].The proposed approach extracts the features from the existing Stanford cars dataset [10] based on different CNN models namely ResNet152, ResNet50, DenseNet201. The pre-trained CNN models are trained on the specific dataset to extract the features and classify the vehicle images based on the brand which forms the classes.

Along with these discriminative features, the influence of varied camera angles also creates different styles on the captured images. Some research studies included the spatio-temporal information and license plate data along with the extracted features. Since all datasets do not possess this spatial information, the requirement of high resolution front and rear images for license plate identification is quite challenging in the real world scenario.

The paper is organized as follows, Sect. 2 reviews the related works in Vehicle Re-identification. Section 3 presents the outline of the methodology. Section 4 gives the clear picture about the experimentation results and discussions on robustness of each

Fig. 1 Intra-class variability Eg: Land Rover Range Rover SUV 2012

Fig. 2 Inter-class similarity Eg: Honda Accord Sedan 2012 and Hyundai Elantra Sedan 2007

CNN model based on the experiments conducted. Section 5 provides the conclusions drawn out of the work done and the enhancements which can be implemented in future work.

2 Related Works

To improve the generalization ability, deep convolutional neural networks (CNNs) were employed with multiple hidden layers to understand the high level features of the image and to attain good performance in re-identification. Deep learning also provides better generalization to other computer vision tasks such as object identification, image classification, video monitoring, semantic segmentation, etc when compared to traditional machine learning methods.

Key point positioning and region segmentation are the commonly used methods for extracting the **local features** from the vehicle image for re-identification. Lin et al. [11] gave a complete comparison on the features extracted from the key parts of the object and the key points. Liu et al. [12] employed fine grained recognition of regions based on self adaptive reinforcement learning with low supervision. Point Pair Feature Network (PPFNet) by Deng et al. [13] presents the three dimensional (3D) local features which are obtained geometrically and are aware of the global content.

Changes in the capturing angle of the camera can influence the local areas used for vehicle re-identification. Accuracy in re-identification is not just achieved by extracting the local features but by feature learning. In **representation learning** Bengio et al. [14] proposed multiple non-linear transformations that are made as the input to produce better representations for classification, prediction and other tasks. Representation learning trains huge amounts of data in order to obtain valid representations with automatic classification and recognition using CNNs. Li et al. [15] presented that representation learning is highly stable and robust and also used for person re-identification. Hongchao et al. [16] proposed DF-CVTC to understand the discriminative representations from the appearance of the vehicle. Hou et al. [17] proposed a vehicle re-identification technique based on random occlusion assisted deep representation learning, where original training images were occluded and simulated real world situations to a definite degree.

Metric learning which is also known as distance learning or similarity learning is the process of mapping into feature space by feature transformation followed by cluster formation in the feature space. Xing et al. [18] proposed this method which is mainly applied in the field of person re-identification, face recognition and vehicle re-identification. Metric learning works on the similarity of images where the distance between the similar images are closer and those of the different images are farther in a network. When it comes to vehicle re-identification, vehicles with similar IDs are closer to each other than those vehicles with different IDs. Hence metric learning relies on the key features of learning objectives such as the distinguishing individual features.

In **unsupervised learning** the training is carried out on the unlabelled input data with a two step cascaded architecture by Bashir et al. [19] when combined with CNN architecture for feature extraction. This technique can also be effectively employed for person re-identification proposed by Zhao et al. [20]. Marin-Reyes et al. [21] applied this unsupervised training method for vehicle re-identification for interpreting a weakly labelled training dataset. Bashir et al. [22] proposed that the base network architecture is trained with unsupervised learning architecture which transferred the deep learning representations on an unlabelled dataset.

Visual **attention techniques** in combination with deep learning concentrated on the usage of masks to structure the attention mechanism. Masks identify the key features of the image and attention is created to learn about the areas which need to be focussed by training the deep neural networks. There are two distinct attention mechanisms, soft attention and hard attention. Soft attention gives more attention to areas by Jaderberg et al. [23] and channels by Hu et al. [24] and it is more differentiable as it calculates the gradient through neural network and learns the weight of attention through forward propagation and backward feedback by Zhao et al. [25], Attention learning mechanism is also deterministic because it is directly obtained from the network after learning

3 Methodology

3.1 Dataset

The Stanford cars dataset comprises 16,186 images in 196 classes. The data in each class is approximately split into 75–25 divide ratio with 12,309 images in the training set and 3877 images in the testing set as in Table 1. The classes in the dataset are categorised based on the brand, model and year of release. (Eg: Land Rover Range Rover SUV 2012). The distribution of various classes in the Stanford cars dataset is given in Fig. 3.

Table 1 Train and test split ratio for the Stanford cars dataset

Stanford cars vehicle re-ID dataset		
Split Set	Ratio	Image count
Gallery	100	16,186
Train	75	12,309
Test	25	3877

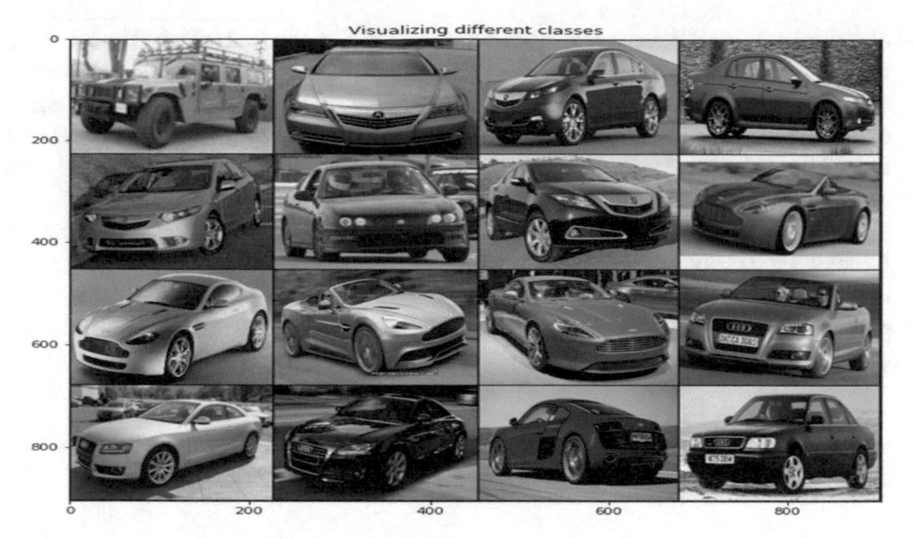

Fig. 3 Distribution of classes in Stanford cars dataset

3.2 CNN Architecture for Vehicle Re-identification

Neural Networks are used extensively in the field of computer vision, the progress in hardware utilization and growth in the availability of wide-ranging training data has made deep neural networks successful in computer vision jobs. They have surpassed the human accuracy levels in many image recognition tasks. Convolutional neural networks (CNN) are unique deep learning architectures applied on spatial data such as images and give elevated performance on image classification, segmentation and detection. Different variants of CNN architectures are developed for various real world scenarios. These CNN architectures are used to train the deep neural network which then incorporates transfer learning techniques for complex models. In deep learning, transfer learning is a techniquc of applying the knowledge obtained from one problem and thereby applying it with the others when there are inadequate data resources. In our work the pre-trained parameters of ImageNet are used along with certain modifications on the structure of the network to make a comparative study between the CNN models: ResNet152, ResNet50 and DenseNet201.

3.3 Training Procedure

In this paper, from the 196 classes in the Stanford cars dataset 9925 images are taken to the train set, 708 images belong to the validation set and 3877 images to the test set. These vehicle images are trained on three CNN models namely ResNet152, ResNet50, DenseNet201 using Keras and ImageDataGenerator for classifying the algorithms. For all these CNN models the methods in the ImageDataGenerator are used to load and transform the images in the dataset under various specifications thus producing different angles of the same image, preventing the need to load the entire dataset into the memory. The dimensions of the images in the dataset are resized to 224 × 224. The computation speed is based on the batch size and for the 196 classes, the batch size is set to 32 with 311 instances per id. The network is trained for 100 epochs with the learning rate adjusted to 0.001 with Stochastic Gradient Descent (SGD) optimizer. As there are 196 classes total in this Stanford cars dataset the loss in the network is set to categorical Cross-Entropy loss to alleviate the training process. The final layers make use of Relu activation function and the output layer uses Softmax.

3.4 Testing Procedure

During the testing process the image size is reduced in the dimension 224 × 224 with the various data augmentation specifications which are mentioned before the training stage. After training the model, the training accuracy and loss along with the validation accuracy and loss are generated for each CNN model from which the performance metrics are calculated.

4 Experimental Results and Discussion

The entire experimentation is carried out in Dell Workstation15 in the system model Precision Tower 5810. Intel(R) Xeon(R) CPU E5-2630 V4 processor is used with a processing speed of 2.20 GHz, 32 GB RAM and GPU Nvdia XP environment. Google Colabs is used to perform the execution of the deep learning models using Python 3.7.12. Keras and Tensorflow 2.7.0 python packages which are imported for executing the different CNN models. The following sections discuss the hyperparameter metrics and the performance study on three CNN models ResNet152, ResNet50 and DenseNet201.

4.1 Hyper Parameter Metrics

In CNNs the hyper parameters are the variables which are related to the network structure and the variables based on which the network is trained. These hyper parameters are set prior to the training phase before the optimization of weights and bias. The hyper parameters are related to the network structure and training. The selection of learning rate and minimization of losses incurred are imposed on the dataset based on its performance to the various metrics. Table 2 illustrates the network performance on

Table 2 Hyper Parameter metrics applied on Stanford cars dataset

Hyperparameter metrics						
Loss	Optimizer	Activation function	Learning rate	Momentum	Number of epochs	Batch size
Categorical cross entropy	Stochastic gradient descent (SGD)	ReLU Softmax (output layer)	0.001	0.9	100	32

Stanford Cars dataset when trained with varied hyper parameter settings. From the experimental results, it is observed that the CNN models ResNet152, ResNet50, DenseNet201 trained based on these configurations provided good results when compared with the configurations (see Table 2).

4.2 *Performance Analysis on Stanford Cars Dataset*

The trained network is tested for its robustness on the Stanford cars dataset and the performance study on the CNN models ResNet152, ResNet50, DenseNet201 models are analysed. Accuracy can be given as the ratio of observations which are correctly predicted to the total number of observations. The value of accuracy for each epoch is obtained based on the training and validation accuracy respectively, followed by the loss which is calculated depending on the mean square error that occurs during the training and validation for all the 100 epochs. From the experimentation it is found that ResNet152 gives a highest accuracy of 89% followed by ResNet50 and DenseNet201 producing 86% and 50% accuracy. The comparison Table 3 given below represents the loss and accuracy rates in each of the CNN models for an average of 10 epochs in each series.

The plots for accuracy and losses for the three CNN models are given in the Figs. 4 and 5 respectively where ResNet152 shows the highest accuracy of 0.89 which is highlighted in the overall accuracy chart in Fig. 6.

The other performance metrics are precision, recall and F1 score. These can be explained with the help of four major terms, True positive (TP), False Negative (FN), True Negative (TN), False Positive (FP). Precision is given as the ratio of the correct positive predictions to the total number of positively predicted values. High precision explains that there is a low false positive rate.

$$\text{Precision} = \frac{\text{True Positive}}{\text{True positive} + \text{False Positive}} \tag{1}$$

Recall is given as the ratio of the correct positive predictions to the overall correct observations in the actual class. F1 Score gives the weighted average of both precision and recall.

Table 3 Average Accuracy and Loss comparison for ResNet152, ResNet50, DenseNet201

Model	ResNet152		ResNet50		DenseNet201	
Epochs	Loss	Accuracy	Loss	Accuracy	Loss	Accuracy
10	4.657	0.223	3.895	0.211	4.285	0.186
20	1.502	0.35	1.784	0.309	3.799	0.267
30	1.74	0.474	1.589	0.352	3.811	0.291
40	1.531	0.557	0.699	0.471	3.368	0.346
50	0.288	0.658	0.58	0.586	3.127	0.388
60	0.517	0.681	0.532	0.612	2.549	0.408
70	0.182	0.786	0.436	0.685	2.488	0.457
80	0.319	0.834	0.397	0.743	2.297	0.443
90	0.231	0.875	0.351	0.821	1.327	0.495
100	0.078	0.891	0.196	0.865	1.522	0.507

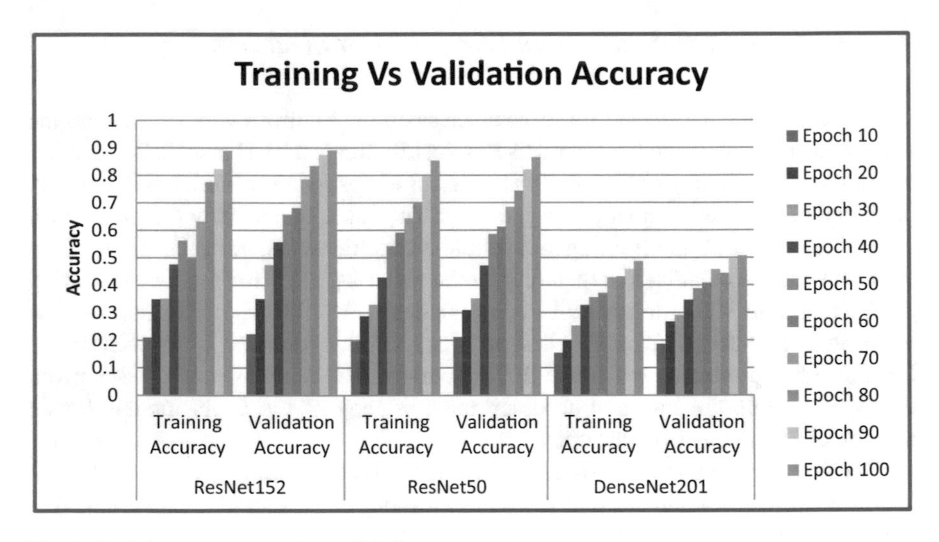

Fig. 4 Training accuracy versus validation accuracy

$$\text{Recall} = \frac{\text{True Positive}}{\text{True positive} + \text{False Negative}} \tag{2}$$

The values of false positives and false negatives are considered for calculating the F1 score. The overall values obtained for these performance metrics are tabulated and presented in Table 4. The detailed classification report on each class for the first 25 classes in the Stanford dataset is given in Table 5.

$$\text{F1 Score} = 2x\,\frac{\text{Precision} * \text{Recall}}{\text{Precision} + \text{Recall}} \tag{3}$$

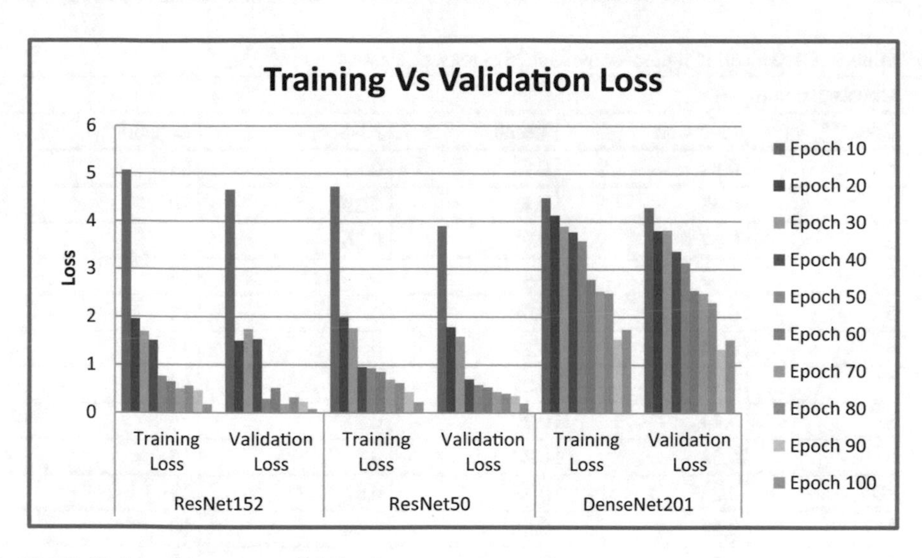

Fig. 5 Training loss versus validation loss

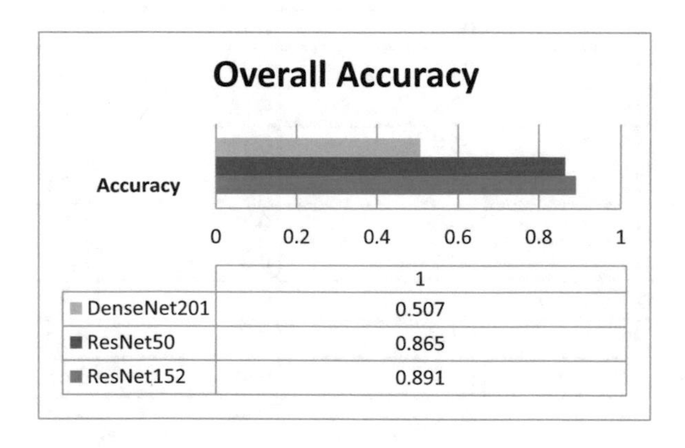

Fig. 6 Overall accuracy chart

Table 4 Performance metrics for the CNN models

CNN model	Accuracy	Precision	Recall	F1 score
ResNet152	0.891	0.893	0.891	0.881
ResNet 50	0.865	0.869	0.865	0.85
DenseNet201	0.507	0.512	0.492	0.486

The prediction result for vehicle identification is done with the BMW X5 SUV 2007 model in Fig. 7 using ResNet152 model and obtained the highest accurate prediction when compared with the other models. BMW X5 SUV 2007 shows the highest similarity percentage when compared with the other models such as Daewoo Nubira Wagon 2002, Buick Rainer SUV 2007, BMW X5 SUV 2012 and BMW 3 Series Sedan 2012.

Table 5 Classification report for the first 25 classes of ResNet152

Classification report				
Class	Precision	Recall	F1-Score	Support
1	0.82	0.93	0.87	15
2	0.35	0.45	0.39	20
3	0.57	0.27	0.36	15
4	0.33	0.47	0.15	15
5	0.12	0.08	0.12	15
6	0.44	0.27	0.33	15
7	0.44	0.27	0.28	20
8	0.44	0.20	0.25	20
9	0.33	0.20	0.21	15
10	0.23	0.78	0.62	18
11	0.52	0.35	0.44	20
12	0.58	0.40	0.52	20
13	0.73	0.92	0.37	20
14	0.24	0.15	0.12	20
15	0.23	0.30	0.21	20
16	0.75	0.78	0.62	20
17	0.19	0.12	0.18	20
18	0.31	0.35	0.33	20
19	0.32	0.35	0.33	20
20	0.10	0.05	0.07	20
21	0.28	0.55	0.37	15
22	0.50	0.27	0.35	15
23	0.33	0.53	0.41	15
24	0.11	0.14	0.09	20
25	0.53	0.69	0.56	15

4.3 Confusion Matrix

Confusion matrix determines the performance of the classification model with the actual values on the rows and predicted values on the columns with the help of the values TP, FN, TN, and FP. A confusion matrix is generally created for the number of classes in the dataset. This is represented as an n × n matrix where n represents the total number of classes. In case of a dataset with 2 classes the confusion matrix is generated as a 2 × 2 matrix. The Stanford cars dataset comprises 196 classes for which a 196 × 196 matrix is obtained. Figure 8 shows the confusion matrix and balanced confusion matrix obtained for the best result yielding the CNN model ResNet152 in our paper. Figure 9 shows the heat map generated for ResNet152.

Fig. 7 Prediction results on BMW X5 SUV 2007 from image classification performed on Stanford Cars dataset

```
Confusion matrix
[[ 14  0  0 ...  0  0  0]
 [  0  9  0 ...  0  0  0]
 [  0  0  4 ...  0  0  0]
 ...
 [  0  0  0 ... 10  0  0]
 [  0  0  0 ...  0  9  0]
 [  0  0  1 ...  0  0 16]]


confusion matrix balanced
[[0.93333333 0.         0.         ... 0.         0.         0.        ]
 [0.         0.45       0.         ... 0.         0.         0.        ]
 [0.         0.         0.26666667 ... 0.         0.         0.        ]
 ...
 [0.         0.         0.         ... 0.45454545 0.         0.        ]
 [0.         0.         0.         ... 0.         0.375      0.        ]
 [0.         0.         0.04347826 ... 0.         0.         0.69565217]]
```

Fig. 8 Sample snapshot of confusion matrix generated for ResNet152

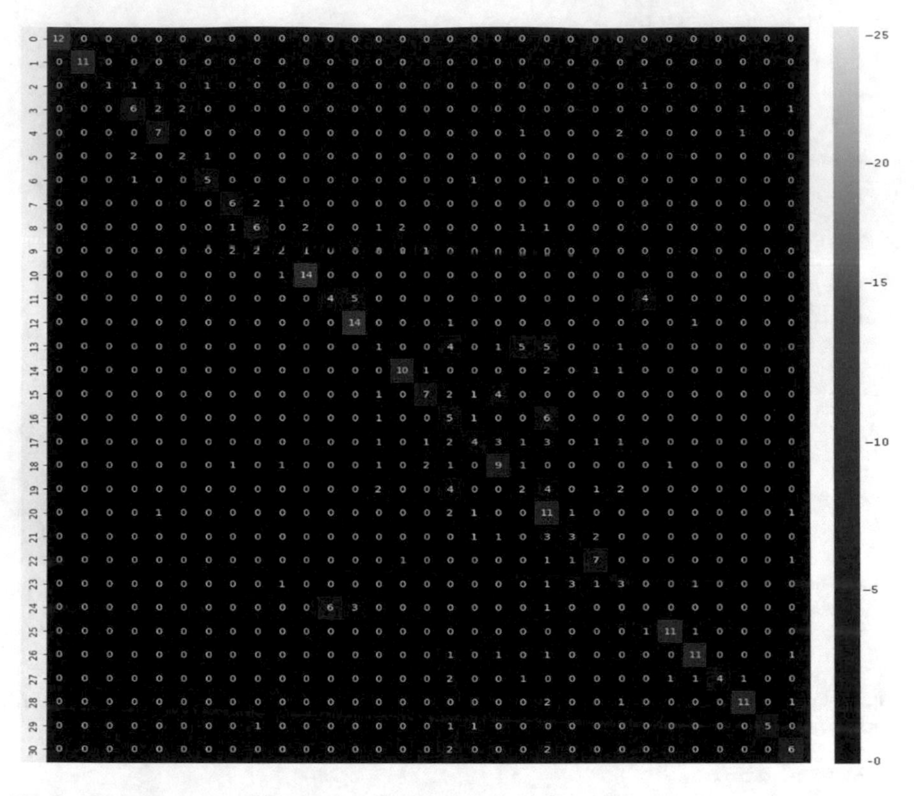

Fig. 9 Heatmap generated for the first 30 classes using ResNet152

5 Conclusion

This paper examines the vehicle classification and identification problem on the Stanford cars dataset which possesses around 16,000 images using three convolution neural network models ResNet152, Resnet50, DenseNet201. From the experimental observations, ResNet152 gives a higher accuracy of 89% by its fine tuned extraction of features from the sample images in each class when compared with the other two CNN models. Moreover implementation of data augmentation has also enlarged the train images by adding flip and rotations thereby improving the robustness of the model. As a future enhancement we would implement changes in the loss functions and incorporate Adam optimizer which has a quick convergence rate on various other datasets which possess real traffic image data to analyse the vehicle re-identification performance. In future we would also try to implement the different CNN models in Amazon EC2 instance from Amazon Web Services (AWS) platform which has higher GPU optimization and varied configuration levels when compared to Google Colabs.

References

1. Lv Z, Zhang S, Xiu W (2021) Solving the security problem of intelligent transportation system with deep learning. IEEE Trans Intell Trans Syst 22:4281–4290
2. Bui N, Yi H, Cho J (2020) A vehicle counts by class framework using distinguished regions tracking at multiple intersections. In: Proceedings of the IEEE/CVF conference on computer vision and pattern recognition (CVPR) workshops
3. Arvin R, Khattak AJ, Kamrani M, Rio-Torres J (2021) Safety evaluation of connected and automated vehicles in mixed traffic with conventional vehicles at intersections. J Intell Trans Syst 25(2)
4. Duan K, Parikh D, Crandall D, Grauman K (2012) Discovering localized attributes for fine-grained recognition. In: Proceedings of the IEEE conference on computer vision and pattern recognition (CVPR)
5. Rublee E, Rabaud V, Rabaud V, Bradski G (2011) ORB: an efficient alternative to SIFT or SURF. In: IEEE international conference on computer vision, ICCV, Barcelona, Spain
6. Pan X, Lyu S (2010) Region duplication detection using image feature matching. IEEE Trans Inf For Secur 5:857–867
7. Liu H, Tian Y, Wang Y, Pang L, Huang T (2016) Deep relative distance learning: tell the difference between similar vehicles. In: Proceedings of the IEEE conference on computer vision and pattern recognition, Las Vegas, NV, USA
8. Zapletal D, Herout A (2016) Vehicle re-identification for automatic video traffic surveillance. In: Proceedings of the IEEE conference on computer vision and pattern recognition workshops
9. Xu Y, Jiang N, Zhang L, Zhou Z, Wu W (2019) Multi-scale vehicle re-identification using self-adapting label smoothing regularization. In: ICASSP 2019-2019 IEEE international conference on acoustics, speech and signal processing (ICASSP), Brighton, UK
10. Krause J, Stark M, Deng J, Fei-Fei L (2013) 3D object representations for fine-grained categorization. In: 4th IEEE workshop on 3D representation and recognition at ICCV, Sydney, Australia
11. Lin D, Shen X, Lu C, Jia J (2015) Deep LAC: deep localization, alignment and classification for fine-grained recognition. In: IEEE conference on computer vision and pattern recognition (CVPR), Boston, MA, USA
12. Liu X, Xia T, Wang J, Yang Y, Zhou F, Lin Y (2017) Fully convolutional attention networks for fine-grained recognition. In: Computer Vision and Pattern Recognition
13. Deng H, Birdal T, Ilic S (2018) PPFNet: global context aware local features for robust 3D point matching. In: Proceedings of computer vision and pattern recognition, Lake City, UT, USA
14. Bengio Y, Courville A, Vincent P (2013) Representation learning: a review and new perspectives. IEEE Trans Pattern Anal Mach Intell 35:1798–1828
15. Li Y, Zhuo L, Hu X, Zhang J (Dec 2016) A combined feature representation of deep feature and hand-crafted features for person re-identification. In: International conference on progress in informatics and computing (PIC), pp 224–227
16. Li H, Lin X, Zheng A, Li C, Luo B, He R, Hussain A (2019) Attributes guided feature learning for vehicle re-identification. In: Proceedings of computer vision and pattern recognition
17. Hou J-H, Zeng H-Q, Cai L, Zhu J-Q, Chen J (2018) Random occlusion assisted deep representation learning for vehicle re-identification. In: Control theory applications
18. Xing EP, Ng AY, Jordan MI, Russell S (2002) Distance metric learning, with application to clustering with side-information. In: Proceedings of advances in neural information processing systems, Vancouver, BC, Canada
19. Bashir RMS, Shahzad M, Fraz MM (2019) VR-PROUD: vehicle reidentification using progressive unsupervised deep architecture. Pattern Recognit
20. Zhao R, Ouyang W, Wang X (2013) Unsupervised salience learning for person re-identification. In: Proceedings of computer vision and pattern recognition, Portland, OR, USA
21. Marín-Reyes PA, Palazzi A, Bergamini L, Calderara S, Lorenzo-Navarro J, Cucchiara R (2018) Unsupervised vehicle reidentification using triplet networks. In Proceedings computer vision and pattern recognitions workshops, Salt Lake City, UT, USA

22. Bashir RMS, Shahzad M, Fraz MM (2018) DUPL-VR: deep unsupervised progressive learning for vehicle re-identification. In: Proceedings of ISVC pattern recognition, Las Vegas, NV, USA
23. Jaderberg M, Simonyan K, Zisserman A, Kavukcuoglu K (2015) Spatial transformer networks. In: Advances in neural information processing systems, Montreal, QC, Canada
24. Hu J, Shen L, Sun G (2019) Squeeze-and-excitation networks. IEEE Trans Pattern Anal Mach Intell 42(8)
25. Zhao B, Wu X, Feng J, Peng Q, Yan S (2017) Diversified visual attention networks for fine-grained object classification. IEEE Trans Multimed 19(6):1245–1256

Chapter 15
A Literature Review on Sentiment Analysis Using Machine Learning in Education Domain

Bhavana P. Bhagat and Sheetal S. Dhande-Dandge

1 Introduction

Teachers are the backbone of the education system. The performance of teachers is a central point of attention of foremost educational researchers. Teachers' effectiveness is judged not only by their qualifications and knowledge, but also by their style of teaching, dedication, commitment, and use of new tools to stay on the cutting edge in the classroom. To address the changing demands of the classroom, effective teachers with a broad repertoire and the ability to employ various strategies skillfully are required. Knowing what students think of teaching is one of the most effective strategies for a teacher to improve teaching methodology. Thus, gaining access to students' opinions in the educational process entails allowing them to provide feedback on their teacher's performance, including their perspective on instruction, organization, classroom environment, and amount learned quality.

One of the most essential strategies for assessing the quality of the educational process is teacher evaluation. It is mostly used in colleges and universities to assess teacher effectiveness and course delivery in higher education. An evaluation questionnaire can be used together information. Quantitative data can be collected using closed-ended questions like MCQs, and qualitative data can be collected with open-ended questions like comments and suggestions from students in textual form. Instructors frequently struggle to draw conclusions from such open-ended comments because they are usually loaded with observations and insight.

B. P. Bhagat (✉)
Department of Computer Engineering Government Polytechnic, Yavatmal (M.S), India
e-mail: bhavana.bhagat30475@gmail.com

S. S. Dhande-Dandge
Department of Computer Science and Engg Sipna College of Engg and Technology, Amravati (M.S), India

G. Mathur et al. (eds.), *Proceedings of 3rd International Conference on Artificial Intelligence: Advances and Applications*, Algorithms for Intelligent Systems,
https://doi.org/10.1007/978-981-19-7041-2_15

Quality education plays a very vital role nowadays in the growth of educational institutions. Also, success of any educational institution depends on quality education, good academic performance, and retention of students. Students and their families do extensive online research by searching data to gain better knowledge of prospective institutions. Academic quality, campus placement, financial aid, campus facility, infrastructure, socialization, and educational policies are some of the key factors that students focus on before the admission process.

Thus, Students are one of the most important key stakeholders for every educational institute. Enhancing knowledge at the student intellectual level is the responsibility of the teacher. Using innovative techniques in teaching improves the performance of students, teachers, and academic institutions. The primary need of today is to monitor students' understanding about learning and improving or changing teaching methodology in the field of education domain. To ensure an ongoing development in teaching and learning experience, it is essential to assure that students' thoughts and feedback are taken seriously [1].

Sentiment analysis is gaining popularity in various areas of text mining and natural language processing these days. Many industries including education, consumer information, marketing, literature, applications, online review websites, and social media have begun to analyze sentiments and opinions. Because of its widespread use, it has attracted the attention of various stakeholders, including customers, organizations, and governments. As a result, one of the most important functions of sentiment classification is to assess student feedback in the education area, as well as online documents such as blogs, comments, reviews, and new products as a whole, and categorize them as positive, negative, or neutral. The study of sentimental analysis has recently gained popularity among academics, and a number of research projects have been done.

2 Related Work

The education sector is undergoing a revolution to live up to standard in today's competitive world. With the generation of huge data, technologies have been developed to store and process data much easier. Thus, the use of sentiment analysis in the education domain plays a very important role for large users, namely teachers, students, and educational institutes. For analyzing the feedback, different machine learning languages and deep learning models will be used [1]. In this section, we are going to discuss existing work in the area of sentiment analysis in the education domain using machine learning and deep learning methods.

2.1 Methods

We started our survey by searching for relevant research studies on internet websites, i.e., digital libraries like IEEE Xplore, Google Scholar, Researchgate, Springer, ACM library, etc. The web search was conducted by using other search engines to trawl the digital libraries and databases. The key terms or search strings that we used were "Sentiment Analysis in education domain", "Effectiveness of teacher's performance using sentiment analysis", "sentiment analysis of student feedback", "sentiment analysis using machine learning". The above different terms are mainly used to search for research study. Also, many articles were identified through scanning the reference list of each one of these articles. There is a lot of literature on the use of sentiment analysis in the education domain. Thus, studies published during the year 2014–2022 related to use of sentiment analysis in education sectors, tools and techniques related to it were surveyed. Nearly 96 different papers have been referred, out of which nearly 80 papers were related to use of sentiment analysis in different domains. Thus, 29 studies related to sentiment analysis using different methods specifically in the education domain have been included for further studies. Some of the important work in this area has been discussed as follow:

In this paper 2014, Altrabsheh et al. [2], a true time student feedback is analyzed using sentiment analysis. The comments were acquired in real time during lectures at the University of Portsmouth's Computing department. The total number of data points collected was 1036, with 641, positive, 292 negative, and 103 neutral opinions. Inter-rater reliability was measured to ensure the accuracy of labels. The Fleiss-kappa was 0.625, and Krippendor's alpha was 0.626. The percent agreement was 80.6%. Naïve Bayes, Complement Naïve Bayes, Maximum Entropy, and Support Vector Machines were used in their investigations. The tenfold cross-validation was used for measuring accuracy, precision, recall, and F-score of all the models. When pre-processing was used, almost all models performed better. There are, however, some exceptions, including: Unigrams performed well in a variety of models. Unigrams and bigrams did well with CNB, but trigrams did significantly better with ME. Except for NB, all of the methods are fairly accurate, and the SVM linear kernel was outperforming the others by 95% and the SVM radial basis kernel performs 88% as a second position. SVM and CNB models have good precision, recall, and F-score, whereas NB and ME models have low precision, recall, and F-score; nevertheless, SVM and CNB models perform better when the neutral class is taken into account. According to the results, the simplest models were SVM and CNB. CNB was also a better choice for training classes with uneven distribution. Future work could include pos-tagging as a feature, as well as additional pre-processing techniques like negation and the preservation of numerals and punctuation. They also made a decision to broaden their study by looking for precise emotions related to education.

Pong-Inwong [3] offers a new approach for building a Teaching Sentilexicon and automating sentiment polarity definition for teaching assessment in 2014. The method is divided into two parts: data preparation and modeling and evaluation. For the teaching assessment method, 175 instances of student opinions from the second semester of 2013 were chosen. After the text has been parsed, it is filtered to remove any extraneous terms. The Weka tool was utilized for modeling and evaluation, and the ID3, Naive Bayes, and SMO algorithms were employed to categories the teaching sentiment, and the model was evaluated with tenfold cross-validation. In comparison to the others, the Support Vector Machine has the highest level of correctness at 97.71%. In the future, a feature selection approach for teacher evaluation can be extracted from online teaching reviews, such as those found on Twitter or Micro blogging, which are widely used in all sectors of the educational arena.

In 2016, Rajput et al. [4] proposed sentiment analysis on faculty evaluations based on student comments. Knime, an open-source data analytics tool, is used to do sentiment analysis. As a consequence, the proposed approach achieved an accuracy of 91.2%. The system's performance is also measured using the sentiment classification technique, recall, precision, and F-measure. With the highest recall and precision rates, positive sentiments performed the best. In the future, sentiment scores will be computed using bigram and higher order n-grams. The sentiment lexicon would also be broadened to include more words that are regularly used in academic settings. Researchers have used unsupervised dictionary methods or supervised learning models such as Naive Bayes, Complement Naive Bayes, Multinomial Naive Bayes, Support Vector Machines, Decision Trees, Conditional Random Forests, Maximum Entropy, Logistic Regression, and a modern method, namely Deep Learning, such as Convolutional Neural Networks, Long Short Term Memory, bidirectional LSTM, Recursive Neural Networks, N-Grams, etc.

Balahadia et al. [5] proposed a sentiment analysis and opinion mining-based teacher performance rating technique in 2016. Opinion mining and a sentiment analysis engine are used to analyze the qualitative ratings. Students' opinions were gathered via a database for data processing. After that, the data is pre-processed to exclude any undesirable characters. The Naive Bayes algorithm is used to analyze the retrieved opinions. The major challenge of reading and comprehending all of the written opinions is one of the most important obstacles. Furthermore, designing an assessment index system for evaluating teachers from an index system with various sub-scores and obtaining a general valuation from the entire scores is difficult. Administrators will be able to combine results, produce statistics and reports based on a evaluation form. Also, student opinions written in free form natural language, which will aid in analyzing individual teachers' performance and identifying their strengths, potential, and weaknesses. Moreover, other algorithms can be applied to improve system performance accuracy.

In 2016, Sindhu et al. [6] employed deep learning approaches to perform aspect-based sentiment analysis on student opinions for faculty teaching performance evaluation. This is one of the most important aspects of the annual appraisal process. A two layered LSTM model was used for aspect extraction and sentiment classification in this system. The first layer categorized review sentences into six categories: teaching pedagogy, behavior, knowledge, assessment, experience, and general. The second layer LSTM is used to predict if a particular aspect has a positive, negative, or neutral sentiment orientation. They used five years of student feedback from Sukkur IBA University for this study. For sentiment analysis, more than 5000 comments were included. Finally, four matrices were used to evaluate the model's performance: precision, recall, F-score, and accuracy. In both tasks, the system achieves greater accuracy with the domain embedding layer: 91% in the aspect extraction band and 93% in sentiment polarity detection. One of the primary issues is that due to the greater number of aspects in each review phrase, a few comments were misclassified. As a result, such comments were misclassified, which is a significant setback when it comes to calculating accuracy.

Rani and Kumar [7] discuss challenging issues in teaching and learning in their study. Student's comments were given to machine learning in order to assist university administrators and educators in taking appropriate action. The system analyzed student's opinions from course surveys and online platforms to determine sentiment polarity (positive, negative, and neutral), sentiments impressed (e.g., eight classes-anger, fair, joy, anticipation, disgust, sadness, surprise, and trust), and satisfaction and dissatisfaction with teacher performance and course satisfaction. Sentiment Analysis system API will be integrated with Student Response Systems and online education portals in the future, allowing for real time analysis of student response and extending the system's multilingual capabilities.

Esparza et al. [8] suggest that the opinion of students is the main factor to evaluate the performance of the teacher. The study of a Social Mining Model is based on a corpus of real Spanish language comments. To identify positive, negative, and neutral comments, the model consists of three phases: comment extraction, feature selection, and classification using the Support Vector Machines algorithm with linear, radial, and polynomial kernels. The dataset used in their research consists of 1040 Spanish comments from Polytechnic University of Aguascalientes systems engineering students. The study involves the evaluation of 21 teachers in the first grade. Sensitivity, specificity, and predictive values were used to calculate the results. With the linear kernel, they were able to achieve a balanced accuracy of above 0.80. The results of this study may aid in the enhancement of the comment classification process as well as teacher development. Also, implementation of other methods will improve the performance in the classification process.

Sivakumar et al. [9] focus on use of different machine learning methods for analyzing feedback collected from Twitter API. The data is pre-processed for word classification. The Naive Bayes classification and K-means algorithm are used to calculate semantic relatedness between an aspect and an opinion sentence. Finally, sentiment analysis is performed on the classified sentences using an open-source technology called SentiWordNet. Accuracy, Precision, Recall, and F-Score measures all yielded positive results. Preprocessing will need to improve in the future to acquire more accurate results from student feedback. In addition, the online application must establish a realistic work scenario.

Nasim et al. [10] proposed two strategies in their study. Machine learning and lexicon-based approaches were used to analyze the sentiment of student's responses. Positive, negative, and neutral sentiment polarities are classified using the methodology described. The dataset included 1230 comments taken from the IBA website. To begin, the Python NLTK package was used to preprocess the data. The data set is randomly split into train and test sets for training and evaluation purposes. The following phase is feature extraction, which uses N-gram and TF-IDF to convert these processes into numerical feature vectors. The training module used the machine learning techniques, Random Forest, and Support Vector Machine after extracting features from the train and test data sets. Finally, test data is used to evaluate the learned model. The result shows that the proposed hybrid approach TF-IDF with a domain-specific lexicon has accuracy as 0.93 and F-score as 0.92. Future studies may include a fine-grained analysis of student comments at the aspect level, such as the instructor's teaching style, lecture planning, competency, and timeliness.

Aung et al. [11], this paper, evaluate the different levels of teaching performance. Students' feedback comments are analyzed using lexicon-based sentiment analysis. A database of English sentiment words is created to determine the polarity of words. Every opinion word in the database is assigned a value ranging from −3 to +3. By matching opinions in a sentiment dictionary, the lexicon-based method was able to determine polarity. In the sentiment word database, there are 745 words, a total of 24 intensifier words, 448 positive words, 263 negative words, and the remaining are neutral words. The lexicon-based approach has the drawback of being domain-specific, and the contextual semantic orientation of words is ignored.

Gutiérrez et al. [12] explain the design and development of a model for analyzing reviews of students to measure teacher performance. Two techniques were used to get the opinions of the students. The first set of opinions were gathered in 2016, while the second set of comments were gathered through student interaction on Twitter. This model comprises three stages: comment extraction and cleaning, feature selection, and a SVM-based classification of positive, negative, and neutral comments. They choose three Kernels: linear, polynomial, and radial basis functions, which are often used in SVM and have good text categorization performance. The ROC, Sensitivity, and Specificity are the performance measurements used. The Linear kernel has a balanced accuracy of above 0.80 according to the results. Additionally, in all kernels, Sensitivity values are higher than Specificity, indicating that negative comments are correctly classified. They plan to use other machine learning techniques in the future to compare the performance of other comment classification algorithms. Students' opinions are usually ignored when determining whether or not to recommend a course to a teacher. As a result, teaching performance will improve.

Atif et al. [13] in their work present an enhanced framework to increase a student's retention, teaching, and facilities in university business courses. Their work targets student's opinions with review and enhances decision making capability of university administrators. The process of prediction of sentiment is that the sentiment comes from review opinion on individual courses and the classifier classifies the document based on a positive and negative sentiment written in natural language. They found that four grams is best in terms of performance and accuracy. Some programs in which comments content other languages (Arabic opinion) were not studied and analyzed in the above system.

In 2018, Newman and Joyner [14] examined Student Evaluations of Teaching (SET) of one course from three sources: official feedback, forum opinion, and an unofficial reviews site monitored by students using a sentiment analysis tool called Valence Aware Dictionary and Sentiment Reasoner. They studied at positive and negative opinions, and evaluated the impact on the positivity or negativity of comments, and skewed positive or negative values by asking course SET opinion questions. The investigation is useful for those evaluating opinion at many universities as related questions are posed at a variety of universities.

In 2018, Nguyen et al. [15] proposed the LSTM model and dependency Tree LSTM for analyzing the sentiments of student opinion in their work. The UIT-VSFC corpus is available publicly and consists of 16,175 sentences from Student feedback. The focus in this paper is on sentiment-based tasks. For the SVM classifier they combine the final hidden states of LSTM and Dependency LSTM model. The results show that, the LD-SVM model obtained F1-score 90.2% and accuracy 90.7%, i.e., highest performance. In the future, they plan to use Vietnamese pre-trained word vectors for deep neural network on UIT-VSFC corpus. Also, plan to carry experiments with a different Tree-structured LSTM network and use other corpus followed by Linear SVM, Long Short Term Memory, DT-LSTM, DT-SVM, Bi-SVM, and Bi-NB.

In 2018, Cabada et al. [16] in this paper proposed that deep learning techniques were used to predict a positive and negative polarity of students' opinion related to exercise or Java language in an intelligent learning environment It also detects learning center emotions like engagement, boredom, and frustration. In this work dataset for training use the Corpus yelp which is an open data Corpus consisting of more than 158,000 reviews. Additionally, to build opinions related to Programming language the Corpus, SentiText was used. The corpus consists of 13,010 phrases along with some artificially added data. Another Corpus used is Edu Era (education resource assessment system) which has 7386 comments with a learning centered sentiment. The pre-processing is cleaning and preparing data for later use. After extensive testing the best model found was one that combined a CNN network with an LSTM. Accuracy is 88.26% in SentiText and 90.30% in Edu Eras. Future work required is to test opinion mining with other learning environments. Improving Corpus to include more phrases, implementing emotion recognition.

Lalata et al. [17] focus on teacher performance in educational institutions in 2019. Sentiment analysis or opinion mining were employed to evaluate teacher to analyze the comments of the students. There are 1822 feedbacks in the student's dataset. Data collection, data preparation, data categorization, and data summarizing and visualization are all phases in measuring the sentiment of student opinions. The technique used for feature extraction is N-gram analysis. The parameters used to determine how important a word is in a text were identified using a metric known as the Term Frequency-Inverse document frequency (TF-IDF) technique, which is used in statistics. In this work, supervised learning techniques such as Naive Bayes, Logistic Regression, Support Vector Machine, Decision Tree, and Random Forest were used to classify the opinions based on the Majority Voting Principle. The accuracy, F-score, and recall of individual classifiers are all evaluated. With NB, accuracy is 90.26%, 90.20% accuracy with SVM, and 89% accuracy with DT. Use of multiclass and multi-label sentiment analysis to identify a teacher's strengths and weaknesses for multilevel decision-making for course preparation and faculty development is one of the institute's limits or future proposals.

Chauhan et al. [18] presented that machine learning and lexicon-based approaches were used to perform aspect-based sentiment analysis. The methodology followed is collection of data, i.e., they gathered 1000 students' comments from social media and pre-processing is done in order to remove unwanted data and other irrelevant comments. To extract students' comments POS tagger was used. Naïve Bayes classifier and Online sentiment analyzers were used for predicting sentiment polarity. The polarity of multiple aspects based on different aspect categories is aggregated once the polarity of an aspect category has been discovered. Precision, recall, and F-score as matrices are used to evaluate performance. Positive: 0.75 and negative: 0.25 are the averaged polarities, i.e., for teaching evaluation and course evaluation. In the future, machine learning with an aspect-based technique to solve aspect level tasks such as finding aspects and their polarities could be considered.

In 2020, Kaur et al. [19] conducted a study in analyzing emotions and sentiments of student feedback using supervised learning techniques. The study conducted by analyzing feedback of students for "Thermal Engineering course from Mechanical Engineering faculty of University Teknologi Mara (UiTM)". For sentiment classification Machine learning algorithms, i.e., Naïve Bayes and Support Vector Machine were used and for emotion analysis NRC Emotion lexicon was used using Weka 3.7 platform. F-measure scores of SVM and NB produces average results as 84.33 and 85.53, respectively, for all features, i.e., Chapter Modules, Assessment Measures, and Teaching Evaluation. As a result, the findings can be used to provide insight into how to enhance teaching and learning process, hence improving overall student quality. The future recommendation is to expand study for other courses.

Chandra et al. [20] stated that data generated at a rapid rate by internet users on different platforms is critical to evaluate and use it for defense and government sectors, to take relevant actions in appropriate time. Datasets used were First GOP Debate, Bitcoin tweets and IMDB Movie reviews from Kaggle for experiments. They used various methods like a voting-based classification system and polarity-based sentiment analysis to perform sentiment analysis and results were generated for Machine learning classifiers in range of 81.00–97 accuracy percentage for training and testing data. The deep learning models like CNN-RNN, LSTM and their combination shows better performance in the range of 85–97 accuracy percentage. Further work needs to find out which architecture can be designed in order to achieve accuracy percent as good as human accuracy.

Multi head attention, Embedding, Dropout, LSTM, and Dense layers were employed in this model by Sangeetha [21]. In this study, the student feedback corpus [15] was employed. For pre-processing and feature extraction, they employed deep learning approaches. Multiple heads with embedding and LSTM layers are concatenated in this model. The results of their fusion model outperform those of the baseline models. They also intend to use the fusion model to undertake tests on various student feedback datasets.

Qaiser et al. [22] focus on Sentiment analysis using different machine learning algorithms. The study analyzes various machine learning approaches, including Naive Bayes, Support Vector Machines, Decision Trees, and a more recent method, Deep Learning (DL). Between February and March 2019, the ML approaches were applied to a single dataset (a total 4289 rows) in English, and their performance was compared in terms of accuracy. Deep Learning achieved the highest accuracy of 96.41%, followed by Naive Bayes and Support Vector Machines with 87.18% and 82.05%, respectively. The Decision Tree has worst performer, which had an accuracy of 68.21%. The limitation is the use of a small dataset. It has the potential to grow into a big dataset. More parameters can be added to Deep Learning as well.

Mabunda et al. [23] developed a sentiment analysis model to analyze students' feedback to evaluate the efficiency of teaching and learning. They used a dataset of 185 student feedback from Kaggle to classify sentiment classes as negative, positive, or neutral. They use different machine learning models. Among these models, the K-Nearest Neighbors and Neural Network were hypothesized to work better than the other three models. This paper used a small dataset for experimental purposes. All models are biased towards the negative and neutral classes in terms of accuracy. Thus, further work suggested use of large datasets to avoid biases and use of Neural Networks as it gives efficient accuracy in prediction.

Dhanalakshmi et al. [24] used supervised learning algorithms to determine the polarity of student's comments based on teaching and learning features. The Rapid Miner Tool was used to categorize student responses. Students' input from six programs was collected using the Survey Monkey platform and a manual Blitz Survey. In this research, English comments were retrieved using Rapid Miner's relevant operators. Support Vector Machines, Naïve Bayes, K-Nearest Neighbor, and Neural Network are supervised learning techniques used to train the model. Precision and Recall values were determined for each classifier algorithm using the Performance Operator of RapidMiner Accuracy, and the comparative performance of the four algorithms was stated. The outcome demonstrates the best precision result of 100% for KNN, the best recall for NB, and accuracy of 97.07 and 99.11%, respectively.

Sultana et al. [25] proposed model based on deep learning method to do sentiment analysis on education data. The dataset was used from the Kalboard 360 dataset source. Weka tool was used to work with data as it is capable of performing a wide range of data mining activities. MLP, a deep learning technique, and other classifiers such as Support Vector Machines, Decision Tree, Bayes Net, K-star, Simple Logistics, and Random Forest are used to classify the data. Support Vector Machines and MLP-deep learning methods predict student performance well, with maximum accuracies of 78.75 and 78.33%, respectively, whereas Bayes Net has a minimum accuracy of 70.20%.

In 2020, Kastrati et al. [26] proposed a framework to automatically analyze emotions of students presents in reviews. It is based on aspect-level sentiment analysis and seeks to detect sentiment polarity stated in relation to the MOOC automatically. The tests were carried out using a large education dataset, namely 104,999 from Coursera. The suggested Aspect Based Sentiment Analysis (ABSA) framework's performance was investigated for two primary tasks: aspect category identification and sentiment classification. For aspect category identification the achieved performance for F1 score is 86.13 and 82.10% for aspect sentiment classification

(CNN+FastText). This framework was developed for the education domain, but it can be used for other domains with minor changes to the input parameters. In the future, word embeddings models like BERT or ELMo, which handle both out-of-vocabulary and in-vocabulary words, could be investigated.

In 2021, ONAN [27] uses ensemble learning and deep learning to provide an effective sentiment classification system with high prediction performance in MOOC ratings. The corpus of 66,000 MOOC reviews was analyzed. Several preprocessing activities were completed with the help of the NLP toolbox. In a machine learning-based approach, three term weighting schemes (TP, TF, and TF-IDF) were used, and the representation schemes were evaluated in conjunction with five supervised learners (Naive Bayes, support vector machines, logistic regression, k-nearest neighbor, and random forest) and five ensemble learning methods (i.e., AdaBoost, Bagging, Random Subspace, voting, and Stacking). In comparison to traditional approaches, the results of the empirical evaluation show that ensemble learning methods produce better prediction performance. As per results, deep learning approaches outperform traditional supervised learning methods and ensemble learning approaches. Long short term memory networks (LSTM) offer the best predictive performance among deep learning models. The second and third best predictive performance was achieved by the recurrent neural network with attention mechanism (RNN-AM) and gated recurrent units (GRU), respectively. The empirical research shows that the GloVe word embedding scheme outperforms the other word embedding schemes. The fastText CBOW model came in second with the best predictive results, followed by the fastText skip gram model.

In 2020, Lwin et al. [28] proposed a feedback analysis method based on rating score and textual comments. Nearly 3000 university students were polled for information. The language of the comments is either English or Myanmar. The data was pre-processed using the Weka tool. The dataset was clustered and labeled using the K-mean clustering algorithm with the Euclidean Distance Function. It has 1104 instances and 15 attributes. The labeled dataset was utilized as the training dataset for the classification model after the clustering stage. Logistic Regression, Multilayer Perceptron, Simple Logistic Regression, Support Vector Machine, LMT, and Random Forest were chosen as effective classifiers. To assess classification performance, four quality measures were used. In terms of performance, Support Vector Machine outperforms the other classifiers in terms of precision and F-score.

2.2 *Recent Advances*

Sentiment analysis is a technique for obtaining information about an entity and identifying whether or not that entity has any subjectivities. The main purpose is to determine whether user-generated language generates positive, negative, or neutral feelings. The document level, phrase level, and aspect level are the three phases of sentiment categorization extraction. There are three approaches to dealing with the problem of sentiment analysis: lexicon-based techniques, machine learning-based techniques, and hybrid approaches. The original sentiment analysis methodology was lexicon-based, which was later split into two methods: dictionary-based and corpus-based. In the first method, sentiment classification is performed using a dictionary of terms, such as those found in SentiWord Net and WordNet, whereas in the second method, corpus-based sentiment analysis is performed using different methods such as K-Nearest Neighbor, conditional random fields, and hidden Markov models among others.

The two types of machine learning-based approaches provided for sentiment analysis challenges are traditional models and deep learning models. Traditional models include machine learning algorithms such as the Naive Bayes, Support Vector Machines, and Maximum Entropy Classifier. These algorithms take a variety of inputs, including lexical features, sentiment lexicon-based features, part of speech, and adjectives and adverbs. When compared to traditional models, deep learning models can produce better results. Convolutional Neural Networks, Deep Neural Networks, and Recursive Neural Networks are examples of deep learning models that can be used for sentiment analysis. These techniques identify classification problems at the document, phrase, or aspect level. The hybrid technique includes lexical and machine learning methodologies.In the majority of these systems, sentiment lexicons are widely used [29]. In recent years, machine learning approaches have been used for sentiment analysis to achieve classification using either supervised or unsupervised algorithms. On few occasions, hybrid approaches were used. Machine learning or lexical approaches are used in the majority of studies. Furthermore, based on the above-mentioned reviews, it appears that neural networks are now the preferred approach of most authors in their educational work. Deep network models such as Long Short Term Memory, bidirectional LSTM, Recursive Neural Networks, and Convolutional Neural Networks have been used in machine learning solutions.

2.3 Summary and Discussion

In Table 1, the summary is given for the above reviewed techniques. From the above literature review, we can say that most sentiment analysis have been extensively applied in the education domain in the last five years. In the early years, the lexicon-based techniques have been used for evaluating the sentiments of students. Later different machine learning algorithms have been used for sentiment analysis. In recent years, the use of deep learning in the education domain has increased. Due to the use of sentiment analysis in the education domain improved learning process in course performance, improved reduction in course retention, improved teaching process and fulfillment with the course of students increased. Also, according to the above literature, the most used techniques under the supervised machine learning approach are Naive Bayes, Support Vector Machine, which provides higher precedence than other methods. Further, recently the use of deep learning approaches such as LSTM, bidirectional LSTM, RNN, and CNN provide higher precedence over machine learning approaches. Moreover from the above reviews, the dataset used in most of the studies is from online comments or from datasets available online on different repositories like Kaggle, etc.

Research Gaps: After reviewing previous work related to the education domain, some limitations were found as follows:

- Many of these works, such as Rajput et al. [4], Rani [7], and Aung et al. [11] used a lexicon-based approach that is less accurate than machine learning.
- Some of these researches, including Balahadia [5], Rani et al. [7], ShivKumar et al. [9], Aung et al. [11] and Newman et al. [14], were not evaluated using common sentiment analysis evaluation metrics, such as accuracy, precision, recall, and F-score and also they have not included any details of evaluation or testing the performance of the model.
- Naive Bayes and Support Vector Machine are mostly used techniques for the Sentiment Analysis. Also trends to combine Machine Learning and Lexicon based or Machine and Deep Learning to perform better for the Sentiment Analysis process.
- There is scope to explore for newer algorithms or optimized the existing ones for obtaining better accuracy.
- Some models do not provide visual results for sentiment analysis, nor do they provide any data or research to evaluate the visualization's usefulness and usability.
- The majority of research in this area focuses on e-learning [26]. Very few models use primary data sets and use the machine and deep learning approaches along with evaluation metrics.

3 Conclusion

In this paper, we provide studies from various researchers on the use of sentiment analysis in the educational domain to measure teacher's performance and improve teaching quality and overall institution performance. For various levels and scenarios, we studied numerous ways to sentiment analysis. We also showed some deep learning work in the educational domain for sentiment analysis based on aspects. As a result, this information could aid researchers in identifying solutions to educational issues and limitations. Finally, we examined some of the limitations of past educational research.

Table 1 Summary of Reviewed Papers. *NP-Not Provided

S. no.	Year	Study	Method/algorithm	Data corpus	Performance
1	2014 Nov	Altrabsheh [2]	Naïve Bayes, CNB SVM ME	1036 commens	F-Score = 0.94, P = 0.94, R = 0.94, A = 0.94 (For SVM Linear Kernel Unigram)
2	2014	Pong-Inwong [3]	SVM, ID3 and Naïve Bayes	175 commens	Support Vector Machine Accuracy-0.98, Prcision, Recall, F-score-0.98
3	2016	Rajput [4]	Lexicon based method	1748 comments	A = 91.2, P = 0.94, R = 0.97, F-score = 0.95
4	2016	Baladia [5]	Naïve Bayes	Comments	NP
5	2016	Irum [6]	LSTM Model for layer 1 and 2	5000 comments	F-Score = 0.86, P = 0.88, R = 0.85, A = 0.93
6	2017	Rani [7]	NRC, Emotion Lexicon	4000 comments during course and 1700 after course	NP
7	Jun-17	Esparza [8]	SVM-Linear, SVM Radial, SVM-Polynomial	1040 comments in Spanish	B.A = 0.81, A = 0.80, Sensitivity = 0.74 (For SVM Linear), Specificity = 0.74 (SVM Radial)
8	2017	Shivkumar [9]	Naive Bayes and K-means algorithm	Online student comments	NP
9	2017	Nasim [10]	Random Forest and SVM	1230 comments	Hybrid Approach Accuracy = 0.93, F-Measure = 0.92
10	2017	Aung [11]	Lexicon based method	745 words	NP
11	2018	Gutiérrez [12]	SVM-Linear, SVM Radial, SVM-Polynomial	1040 comments in Spanish	SVM linear kernel-above 0.80
12	2018	Atif [13]	N-gram,	Students responses	N-gram Accuracy = 0.80

(continued)

Table 1 (continued)

S. no.	Year	Study	Method/algorithm	Data corpus	Performance
13	2018	Newman [14]	VADER as SA tool	Students comments	NP
14	2018	Nguyen [15]	LSTM and DT-LSTM Models	UIT-VSFC corpus 16,175 sentences	LD-SVM F1-score 90.2% and accuracy 90.7%
15	2018	Cabada [16]	Convolutional Neural Networks, Long Short Term Memory	Yelp-147,672 SentiText-10,834 EduSere-4300 comment	CNN+LSTM A = 88. 26% in SentiText and A = 90.30% in EDuEras
16	2019	Lalata [17]	Naïve Bayes, SVM DT RF	1822 comments	Best Model NB with A = 90.26 SVM with A = 90.20, DT A = 89
17	2019	Chauhan [18]	Naive Bayes classifier and online sentiment analyzer	1000 students comments	Performance using sentiment score F-Score = 0.72, P = 0.69, R = 0.76 Performance teaching aspects F-Score = 0.80, P = 0.76, R = 0.84
18	2020	Kaur [19]	Naïve Bayes, SVM, Emotion lexicon	4289 comments	SVM F-score = 84.33, NB F-score = 85.53
19	2020	Chandra [20]	Convolutional Neural Networks, Recursive Neural Networks, Long Short Term Memory	First GOP Debate, Bitcoin tweets and IMDB Movie reviews	For ML classifiers Accuracy range 81.00 to 97, For CNN-RNN, LSTM and their combination-Accuracy range 85–97
20	2020	Sangeetha [21]	LSTM, LSTM+ATT, Muitihaed ATT and FUSION	16,175 students feedback sentences	FUSION Model (multi-head attention+Embedding+LSTM) A = 94.13, R = 88.72, P = 97.89
21	2021	Qaiser [22]	Naïve Bayes, SVM DT DL	5000 comments	DL-A = 96.41 (Best), NB-A = 87.18, SVM-A = 82.05, DT-68.21 (Poor)

(continued)

Table 1 (continued)

S. no.	Year	Study	Method/algorithm	Data corpus	Performance
22	2021	Mabunda [23]	Support Vector Machines, Multinomial Naive Bayes, Random Forests, K-Nearest Neighbours and Neural Networks	Students feedback from Kaggle dataset 185 records	SVM-81% Multinomial NB-81% Random Forest-81% K-NN-78% Neural Network-84%
23	2016	Dhanalakshmi [24]	Support Vector Machines, Naive Bayes, K-Nearest Neighbours and Neural Networks	Total number of responses overall-6433	Precision result of 100% for KNN, Recall for NB-97.07% and Accuracy for NB 99.11%
24	2018	Sultana [25]	SVM, DT, Bayes Net, K-star, Simple Logistics and Random Forest	Kalboard 360 dataset repository	SVM Acc-78.75% (max) and MLP Acc-78.33% (max) and BayesNet Acc-70.20% (min)
25	2020	Kastrati [26]	Convolutional Neural Networks, Long Short Term Memory	105 k reviews from Coursera and a 5989 feedback in traditional classroom	F1 score-86.13% for aspect category identification and F1 score-82.10% for aspect sentiment classification
26	2021	Onan [27]	Five Supervised learners and Five ensemble learning methods	66,000 MOOCs reviews	LSTM with GloVe classification accuracy-95.80%
27	2020	Lwin [28]	Logistic Regression, Multiplayer Perceptron, Simple Logistic Regression, SVM, LMT and Random Forest	1104 instances	Precision and F-score for SVM-0.972

References

1. Archana Rao PN, Baglodi K (2017) Role of sentiment analysis in education sector in the era of big data: a survey. Int J Latest Trends Eng Technol, 022–024
2. Altrabsheh N, Cocea M, Fallahkhair S (Nov 2014) Sentiment analysis: towards a tool for analysing real-time students feedback. In 2014 IEEE 26th international conference on tools with artificial intelligence. IEEE, pp 419–423
3. Pong Inwong C, Rungworawut WS (Aug 2014) Teaching senti-lexicon for automated sentiment polarity definition in teaching evaluation. In: 2014 10th international conference on semantics, knowledge and grids. IEEE, pp 84–91
4. Rajput Q, Haider S, Ghani S (2016) Lexicon-based sentiment analysis of teachers'evaluation. Appl Comput Intell SoftComput
5. Balahadia FF, Fernando MCG, Juanatas IC (May 2016) Teacher's performance evaluation tool using opinion mining with sentiment analysis. In: 2016 IEEE region 10 symposium (TENSYMP). IEEE, pp 95–98
6. Sindhu I, Daudpota SM, Badar K, Bakhtyar M, Baber J, Nurunnabi M (2019) Aspect-based opinion mining on student's feedback for faculty teaching performance evaluation. IEEE Access 7:108729–108741
7. Rani S, Kumar P (2017) A sentiment analysis system to improve teaching and learning. Computer 50(5):36–43
8. Esparza GG, de-Luna A, Zezzatti AO, Hernandez A, Ponce J, Álvarez M, Cossio E, de Jesus Nava J (Jun 2017) A sentiment analysis model to analyze students reviews of teacher performance using support vector machines. In: International symposium on distributed computing and artificial intelligence. Springer, Cham, pp 157–164
9. Sivakumar M, Reddy US (Nov 2017) Aspect based sentiment analysis of students opinion using machine learning techniques. In: 2017 international conference on inventive computing and informatics (ICICI). IEEE, pp 726–731
10. Nasim Z, Rajput Q, Haider S (Jul 2017) Sentiment analysis of student feedback using machine learning and lexicon based approaches. In: 2017 international conference on research and innovation in information systems (ICRIIS). IEEE, pp 1–6
11. Aung KZ, Myo NN (May 2017) Sentiment analysis of students' comment using lexicon based approach. In: 2017 IEEE/ACIS 16th international conference on computer and information science (ICIS). IEEE, pp 149–154
12. Gutiérrez G, Ponce J, Ochoa A, Álvarez M (Mar 2018) Analyzing students reviews of teacher performance using support vector machines by a proposed model. In: International symposium on intelligent computing systems. Springer, Cham, pp 113–122
13. Atif M (2018) An enhanced framework for sentiment analysis of students' surveys: Arab open university business program courses case study. Bus Econ J 9(2018):337
14. Newman H, Joyner D (Jun 2018) Sentiment analysis of student evaluations of teaching. In: International conference on artificial intelligence in education. Springer, Cham, pp 246–250
15. Nguyen VD, Van Nguyen K, Nguyen NLT (Nov 2018) Variants of long short-term memory for sentiment analysis on Vietnamese students' feedback corpus. In: 2018 10th international conference on knowledge and systems engineering (KSE). IEEE, pp 306–311
16. Cabada RZ, Estrada MLB, Bustillos RO (2018) Mining of educational opinions with deep learning. J Univ Comput Sci 24(11):1604–1626
17. Lalata JAP, Gerardo B, Medina R (Jun 2019) A sentiment analysis model for faculty comment evaluation using ensemble machine learning algorithms. In: Proceeding of 2019 international conference on big data engineering, pp 68–73
18. Chauhan GS, Agrawal P, Meena YK (2019) Aspect-based sentiment analysis of students' feedback to improve teaching–learning process. In: Information & communication technology for intelligent systems. Springer, Singapore, pp 259–266
19. Kaur W, Balakrishnan V, Singh B (Apr 2020) Improving teaching and learning experience in engineering education using sentiment analysis techniques. In: IOP conference series: materials science and engineering, vol 834, no 1. IOP Publishing, p 012026

20. Chandra Y, Jana A (Mar 2020) Sentiment analysis using machine learning and deep learning. In: 2020 7th international conference on computing for sustainable global development (INDIACom). IEEE, pp 1–4
21. Sangeetha K, Prabha D (2021) Sentiment analysis of student feedback using multi-head attention fusion model of word and context embedding for LSTM. J Ambient Intell Humaniz Comput 12(3):4117–4126
22. Qaiser S (2021) A comparison of machine learning techniques for sentiment analysis. Turkish J Comput Math Educ (TURCOMAT) 12(3):1738–1744
23. Mabunda JGK, Jadhav A, Ajoodha R (2021) Sentiment analysis of student textual feedback to improve teaching. In: interdisciplinary research in technology and management. CRC Press, pp 643–651
24. Dhanalakshmi V, Bino D, Saravanan AM (Mar 2016) Opinion mining from student feedback data using supervised learning algorithms. In: 2016 3rd MEC international conference on big data and smart city (ICBDSC). IEEE, pp 1–5
25. Sultana J, Sultana N, Yadav K, AlFayez F (Apr 2018) Prediction of sentiment analysis on educational data based on deep learning approach. In: 2018 21st Saudi computer society national computer conference (NCC). IEEE, pp 1–5
26. Kastrati Z, Imran AS, Kurti A (2020) Weakly supervised framework for aspect-based sentiment analysis on students' reviews of MOOCs. IEEE Access 8:106799–106810
27. Onan A (2021) Sentiment analysis on massive open online course evaluations: a text mining and deep learning approach. Comput Appl Eng Educ 29(3):572–589
28. Lwin HH, Oo S, Ye KZ, Lin KK, Aung WP, Ko PP (Jun 2020) Feedback analysis in outcome base education using machine learning. In: 2020 17th international conference on electrical engineering/electronics, computer, telecommunications and information technology (ECTI-CON). IEEE, pp 767–770
29. Dang NC, Moreno-García MN, De la Prieta F (2020) Sentiment analysis based on deep learning: a comparative study. Electronics 9(3):483

Chapter 16
Deep Learning Approach for Segmenting and Classifying Knee Osteoarthritis Using MR Images

S. A. Revathi, B. Sathish Babu, and K. N. Subramanya

1 Introduction

A degenerative disease that causes pain, inflammation, and dysfunction in joints is Osteoarthritis. It generally affects the knees, hips, and hands. The main concern for this disease is immobility and physical disability at 50 years or more. Knee OA is characterized by degeneration in cartilage, decreased joint space width, formation of osteophytes, and variations in the bone structure.

1.1 Segmentation

Analysis of Knee MR Image segmentation via human–computer interaction by Doctors and Radiologists is done using semi-automatic methods. State-of-art algorithms for semi-automatic methods are live wire, region growing, active contours, watershed and ray casting. Few deep-learning bone segmentation methods used for the post-processing of tibia and femur bones are statistical shape models. Later a deep CNN and 3D simplex deformable modelling called SegNet was developed to segment cartilage along with bones.

S. A. Revathi (✉)
Department of CSE, RV College of Engineering, Bangalore, India
e-mail: revathisa@rvce.edu.in

B. Sathish Babu
Department of AIML, RV College of Engineering, Bangalore, India

K. N. Subramanya
Department of IEM, RV College of Engineering, Bangalore, India

G. Mathur et al. (eds.), *Proceedings of 3rd International Conference on Artificial Intelligence: Advances and Applications*, Algorithms for Intelligent Systems,
https://doi.org/10.1007/978-981-19-7041-2_16

1.2 Classification

Classification model which uses Kellgren and Lawrence (K&L) scale, showing five stages of severity.

Grade 0 is normal with the absence of OA, Grade 1 is doubtful with joint cartilage narrowing, Grade 2 is minimal with the presence of osteophytes and joint cartilage narrowing, Grade 3 is moderate with osteophytes, joint cartilage narrowing, and few sclerosis, Grade 4 is severe with large osteophytes, broad joint cartilage spacing, and many sclerosis.

The above classification is grounded on radiological features, namely joint space reduction and osteophytes. Various imaging modalities are available to acquire the knee images, namely MRI, X-Ray, CT scan, and scintigraphy. For diagnosis of knee O.A., the gold standard is M.R. Imaging. Due to wide availability, differentiation in cartilage and bone in acquired images are easily understood. However, assessing the cartilage defect varies from one radiologist to another. To get rid of this discrepancy and evaluate the severity of knee cartilage defects, many grading techniques are implemented, e.g., Kellgren Lawrence grading [1], MRI Osteoarthritis Knee Score (OKS) [2], and Whole-Organ MRI Score (WORMS) [3]. These grading systems permit computer-aided diagnosis (CAD) to assess the cartilage defect, delivering precise and complete results and increasing the efficacy of diagnosis. Though various CNN-based models for automatic assessment of cartilage degeneration which are done using deep learning methods. These methods can be classified as follows:

1. **Patch-based methods:** Liu et al. [4], femoral cartilage was cropped along the patches. To classify defects in cartilage, the patches are used as encoders in the U-Net architecture.
2. **Slice-based methods:** Huo et al. [5] trained CNNs can concentrate on OA-affected cartilage regions with the introduction of a self-ensembling framework with dual consistency loss in the MRI Slice.
3. **Subject-based methods**: Guida et al. [6] 3D CNN model was implemented to extract the 3D features of the whole knee MRI. It performed better than the slice-based method.

Research based on CNN is observed as a state-of-the-art tool for the assessment of automatic knee cartilage defect, but many issues need to be considered.

1. **Information Dilution**: Large background is considered for computation in subject-based and slice-based methods for diluting the portion of cartilage information. Knee cartilages are usually thin layers in structure wrapping the heads of bones. However, in CNNs, the receptive cartilage field cannot be simply adjusted with the target anatomy's shape and size.
2. **Structure Misidentification**: Irrelevant background is reduced by cropping patches along the cartilage in Patch-based methods resulting in loss of the global shape information of knee cartilages [7]. The appearance of individual patches at different positions of the cartilage is diverse. This confusion in local appearance

can reduce the capability of patch-based methods to find defects in the cartilage for knee Osteoarthritis [8].

1.3 Proposed Technique

U-Net architecture has a skip connection to move the output from the encoder to decoder. For the up-sampling operation, the feature maps are concatenated with the output, and the concatenated feature maps are passed on to successive layers.

MultiResU-Net provides different-scale features to distinguish an object from the whole image. It doesn't perform well because of fuzzy objects and background interference. Hence it was modified to provide more effective features. This idea led to the development of the DC U-Net model.

AlexNet CNN is the simplest method in CNN architecture. AlexNet uses Local Response Normalization (LRN) to maximize the activation of adjacent neurons. LRN is not widely used as other methods are more effective for normalization.

He et. al. introduced 34-layer ResNets (Residual Network), which gave competitive accuracy, and then 152-layer ResNet has a 4.49% of a validation error.

This paper has a Dual Channel U-Net Model for segmentation and AlexNet, ResNet V2 Model for classification.

2 Proposed Technique for Segmenting and Classification of Knee OA

The sagittal MR Image data was collected from diagnostic centres in Bangalore which are used to segment and classify the knee health in patients for OA. Images were divided based on K&L grades. Figure 1 represents the working model.

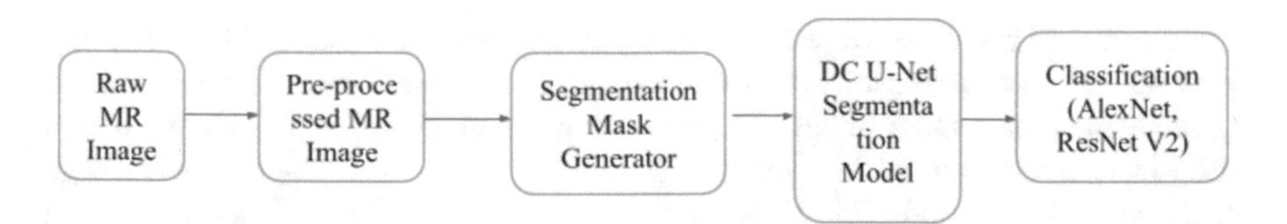

Fig. 1 Working model

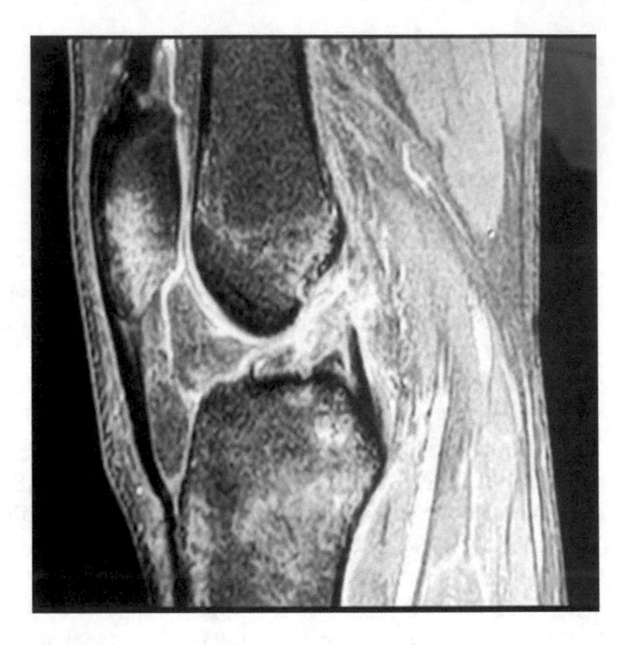

Fig. 2 Pre-processed image

2.1 Pre-processing of MR Image

Initially, the sagittal view of the raw M.R. knee osteoarthritis Image was cropped to obtain the Region of Interest (ROI) and then dimensions are reduced to 384*384-pixel dimension for the DC U-Net segmentation model, and bone annotation was performed [9]. To reduce training time the model images were converted to greyscale.

2.2 Segmentation Mask Generator

Annotations of the image were done to generate segmentation masks for the training network. Labelbox was used to annotate the femur and tibia. Then it was exported to generate the binary mask image. Ground truth was considered as the original image and binary mask to train the model [10]. Figure 2 shows the pre-processed image of the knee. Femur and Tibia Bones are segmented using a label box and act as a training image in Fig. 3.

2.3 Segmentation Model

U-NET is the most widely used model for medical image segmentation, due to several limitations, it was modified: (1) Replacement of the encoder and decoder with

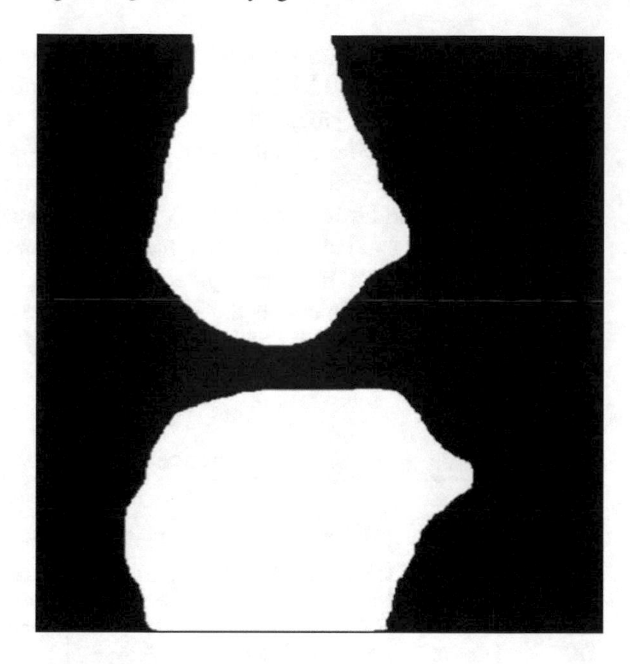

Fig. 3 Segmentation mask

efficient CNN architecture, (2) Residual module was applied to replace connection between encoder and decoder. This modification improved the Dual Channel U-Net model compared to the U-Net version. The below figure depicts the DC U-Net Architecture [14–16] (Fig. 4).

The U-Net model does not perform well in challenging medical image cases such as noisy objects and interference of backgrounds, and overlap of borders. To solve images having small bordered objects, spatial features are of utmost importance. Therefore, to avoid the problem of lack of spatial features, a sequence of three 3 × 3 Conv layers was used to change the residual connection in the multi-Res (modified U-Net for different scale features) block called the Dual Channel Block [17, 18].

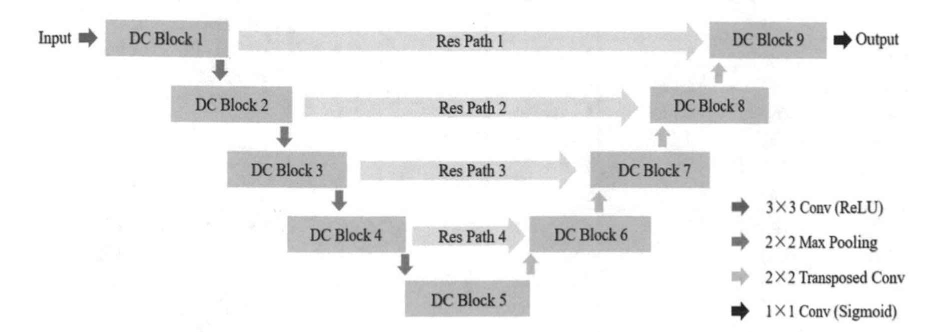

Fig. 4 DC U-Net model architecture

Every channel in the Dual Channel block has exactly half the filter numbers of the multi-Res block: [32, 64, 128, 256, 512] And, the number of filters in the Residual-path are [32, 64, 128, 256] respectively, and the number of each layers' filter as shown in Table 1. Every Conv layer in the Dual Channel U-Net [20–22] is activated by the Rectified Linear Unit (ReLU) function. To avoid overfitting Batch normalization is used. The last output layer is triggered by the sigmoid activation function.

Most frequently applied loss function for Image segmentation is pixel-wise cross-entropy loss. It examines individual pixels with class predictions to a target vector.

Pixel-wise loss is evaluated as the log loss summed over input images as shown in Eq. (1). Cross entropy loss is calculated as class predictions for each pixel over an average of all pixels as in Eq. (2).

Table 1 Shows the various hyperparameters of the DC-UNet

DC-UNet							
Block	Layer (left)	#Filters	Layer (right)	#Filters	Path	Layer	#Filters
DC Block 1 DC Block 9	Conv2D (3,3) Conv2D (3,3) Conv2D (3,3)	8 17 26	Conv2D (3,3) Conv2D (3,3) Conv2D (3,3)	8 17 26	Res Path 1	Conv2D(3,3) Conv2D(1,1) Conv2D(3,3) Conv2D(1,1) Conv2D(3,3) Conv2D(1,1) Conv2D(3,3) Conv2D(1,1)	32 32 32 32 32 32 32 32
DC Block 2 DC Block 8	Conv2D (3,3) Conv2D (3,3) Conv2D (3,3)	17 35 53	Conv2D (3,3) Conv2D (3,3) Conv2D (3,3)	17 35 53			
DC Block 3 DC Block 7	Conv2D (3,3) Conv2D (3,3) Conv2D (3,3)	35 71 106	Conv2D (3,3) Conv2D (3,3) Conv2D (3,3)	35 71 106	Res Path 2	Conv2D(3,3) Conv2D(1,1) Conv2D(3,3) Conv2D(1,1) Conv2D(3,3) Conv2D(1,1) Conv2D(3,3) Conv2D(1,1)	64 64 64 64 64 64 64 64
DC Block 4 DC Block 6	Conv2D (3,3) Conv2D (3,3) Conv2D (3,3)	71 142 213	Conv2D (3,3) Conv2D (3,3) Conv2D (3,3)	71 142 213			
DC Block 5	Conv2D (3,3) Conv2D (3,3) Conv2D (3,3)	142 284 427	Conv2D (3,3) Conv2D (3,3) Conv2D (3,3)	142 284 427	Res Path 3	Conv2D(3,3) Conv2D(1,1) Conv2D(3,3) Conv2D(1,1)	128 128 128 128
					Res Path 4	Conv2D(3,3) Conv2D(1,1)	256 256

Consider every input image as X, assuming the prediction of model output is ŷ, and the ground truth is y. Hence, the binary cross-entropy is given as:

$$Cross\ Entropy\,(\mathrm{y}, \hat{\mathrm{y}}) = \sum_{\mathrm{x} \in \mathrm{X}} (-\mathrm{y}(1 - \mathrm{y})\log(1 - \hat{\mathrm{y}})) \tag{1}$$

Loss function j for a batch of n images is given as:

$$J = \frac{1}{n}\sum_{j=1}^{n} \text{Cross Entropy}\,(\mathrm{y}, \hat{\mathrm{y}}) \tag{2}$$

2.4 *Knee OsteoArthritis Classification*

Knee Osteoarthritis is classified based on the cartilage thickness to know the severity of the disease [11–13].

2.4.1 AlexNet

The classifier model consists of 8 layers with weights of 5 fully convolutional layers and 3 fully connected layers. The convolutional layer is followed by max-pooling layers. ReLU Activation Function is applied to improve the network's performance over sigmoid and tanh functions and to add non-linearity. It is one of the simplest methods to implement among the CNN architecture [19] (Fig. 5).

The AlexNet makes use of transfer learning and has employed weights of the pre-trained network on a classified dataset of knee O.A. and has shown better performance. The image was resized to 224, 224, and the trained dataset was given as input to the AlexNet Model, which generated an accuracy of 80% and cross-entropy of 57.62%.

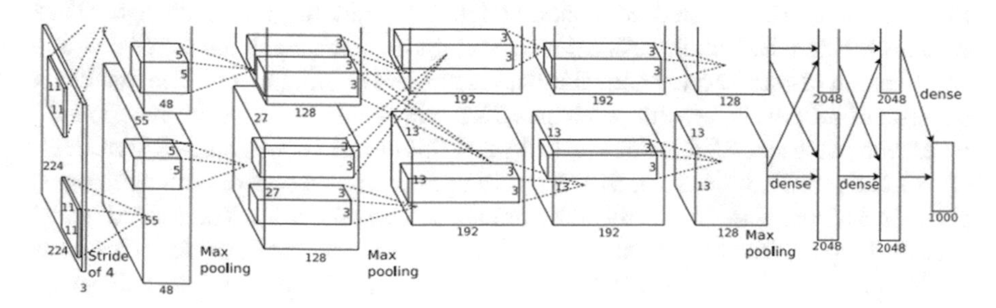

Fig. 5 AlexNet model architecture

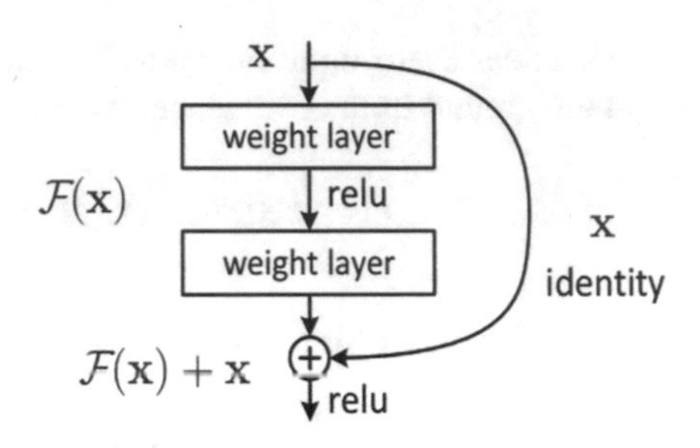

Fig. 6 Building block of residual network

2.4.2 ResNet Version 2

The segmentation outcome is given as input to the classification model with actual parameters used for the multi-classification model [23, 24]. Here ResNet version 2 with 152 layers is used (Fig. 6).

Microsoft developed a new architecture called ResNet (Residual Network) to overcome the drawback of plain network vanishing/exploding gradients. A skip/shortcut connection is directly added to the input x to the output after a few weights' layers, as shown above. In deep neural networks, a concept of Residual blocks was used. The residual block consists of skip connections that skip the training from a few underperforming layers or some layers that can harm the performance of the entire network.

Before multiplying input to the weight matrix (Conv operation) Residual Network Version-2(V2) applies Batch-Normalization & ReLU activation function on input. i.e., V2 performs pre-activation instead of post-activation used in V1. The output of the addition operation between the identity mapping and the residual mapping should be provided as is to the next block for further operations, according to Residual Network Version-2. In the identity function 'f', the signal will be transferred straight between any two units. As a result, the gradient value calculated by the output layer will readily reach the starting layers even if the signal does not change.

The model uses manually generated segmentation masks of size 384X384 as input for training. The input is augmented using various parameters to increase the data folds. The augmentations applied include zoom, shear, vertical and horizontal flips, rotations, vertical and horizontal shifts, and some brightness changes. Data then flows through multiple residual blocks of the network. Each block consists of a combination of zero or more convolutional layers and pooling layers. The first block consists of a Conv layer with patch size 7X7 with stride 2, followed by a pooling layer with patch size 3X3, stride 2. Blocks that follow have filters ranging from 1X1, 64, 1X1, 2048, 3X3, 64, 3X3, 512. Input Image undergoes convolution operations, pooling, and reshaping to finally arrive at the categorical output. The trained model uses the output of the Segmentation model as input to give one of the four classes as the final output. This output can be mapped to K&L grades as mentioned above.

3 Results

Trained Data is labelled manually. Then the model is fed with knee MR Image as input, next pretrained models are used for segmentation [25–27] and classification which generates the output of segmented bone part and K&L grade severity level of knee OA respectively.

DC U-Net Model is evaluated using a dice coefficient calculated by twice the overlapped area divided by the over-all pixels in both images. Another easiest way is to calculate pixel error by measuring the pixel difference between labelled and segmented images.

133 Images have been collected from diagnostic centre, used as training data and 11 Dicom images as testing data.

Segmentation of the femur and tibia with the two models U-Net and DC U-Net resulted in a pixel error of 0.0611 and 0.0184, respectively. The evaluation metric for the DC U-Net segmentation model was the Dice coefficient which resulted in 80%. Table 2 shows the performance of the Segmentation Model. The outcome of the model is shown in Fig. 7.

Accuracy is the most frequently used metric for classification evaluated as the mean of correct predictions.

The Classification model AlexNet validation accuracy was 80% and ResNet 152-V2 validation accuracy was 76.67%. The classification Precision value was 0.5204 (Table 3).

Along with the Dice score the segmentation model is evaluated using pixel error, calculated by measuring the difference in labelled and segmented images. The value

Table 2 Performance of segmentation model

Model	Epoch val	Dice score	Loss	Training accuracy	Training loss
DC U-Net	20	0.80	0.1391	0.80	0.0335

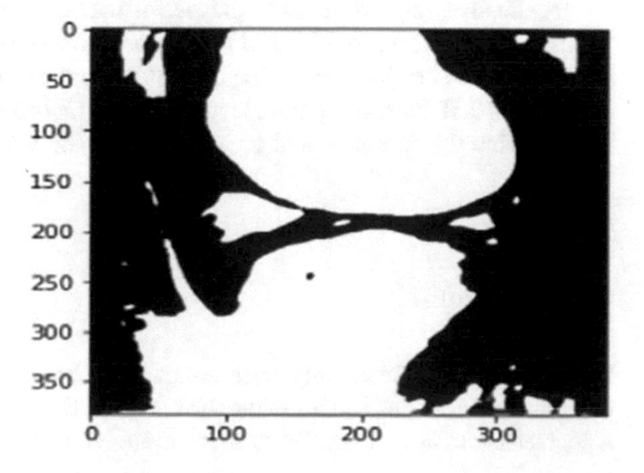

Fig. 7 Segmented image

Table 3 Performance of classification model AlexNet and ResNet 152-V2

Model	Epoch val	Accuracy value	Training accuracy
AlexNet	6	0.8000	0.7541
ResNet 152-V2	5	0.7667	0.6391

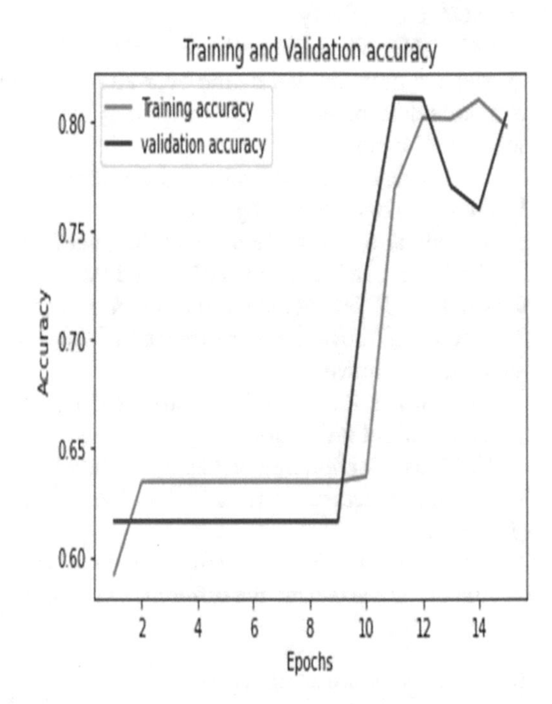

Fig. 8 Training versus validation accuracy

of the Pixel error was 0.018, comparatively less compared to the U-Net Pixel Error of 0.0611.

An effective tool to monitor the performance of segmentation is done using learning curves. The score is used to minimise loss or error, in Fig. 8. Performance Learning Curve helps in evaluating and selecting the model. Training v/s Validation Accuracy is 0.80 indicating more learning Fig.9. Optimisation Learning Curves help in optimising the parameters of the model. Training v/s Validation Loss is less than 0.70.

4 Conclusion

Here, the model detects only knee bones in an M.R. image, thereby reducing the non-essential features and allowing the classification model to extract features from a segmented image. Since K&L score is based on the relative space between the

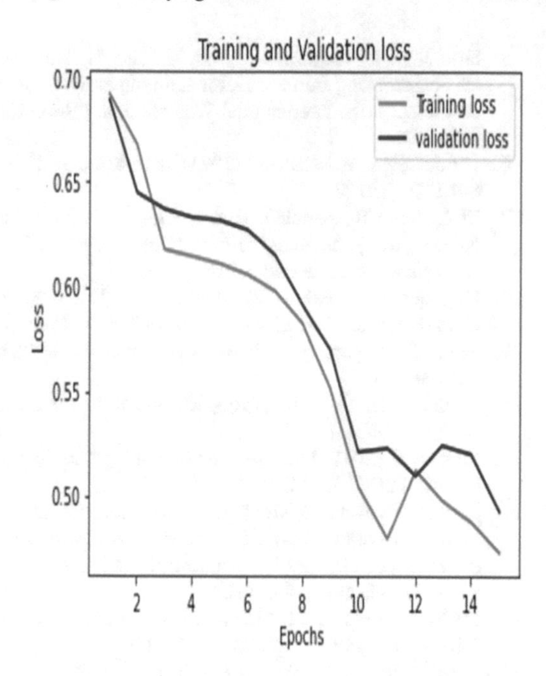

Fig. 9 Training versus validation loss

femur and tibia. We have made sure that the approach will give the best results in real-time and provide a research base for other researchers to carry out further work in the field of Medical Image Processing and Deep Learning. The constraint in the model was the quality of the dataset, which can be improved, allowing us to segment other tissues around the bone, further improving our classification accuracy. This model can be used for other medical image classification that has limited data and as a diagnosis model for knee osteoarthritis.

References

1. Emrani PS, Katz JN, Kessler CL, Reichmann WM, Wright EA, McAlindon TE, Losina A (2008) Joint space narrowing and Kellgren-Lawrence progression in knee osteoarthritis: an analytic literature synthesis. Osteoarthr Cartil 16:873–882
2. Hunter DJ, Guermazi A, Lo GH, Grainger AJ, Conaghan PG, Boudreau RM, Roemer FW (2011) Evolution of semi-quantitative whole joint assessment of knee oa: Moaks (mri osteoarthritis knee score). Osteoarthr Cartil 19(8):990–1002
3. Alizai H, Virayavanich W, Joseph GB, Nardo L, Liu F, Liebl H, Nevitt MC, Lynch JA, McCulloch CE, Link TM (2014) Cartilage lesion score: comparison of a quantitative assessment score with established semiquantitative mr scoring systems. Radiology 271(2):479–487
4. Liu F, Zhou Z, Samsonov A, Blankenbaker D, Larison W, Kanarek A, Lian K, Kambhampati S, Kijowski R (2018) Deep learning approach for evaluating knee mr images: achieving high diagnostic performance for cartilage lesion detection. Radiology 289(1):160–169

5. Huo J, Si L, Ouyang X, Xuan K, Yao W, Xue Z, Wang Q, Shen D, Zhang L (2020) A self-ensembling framework for semi-supervised knee cartilage defects assessment with dual-consistency. In: International Workshop on PRedictive Intelligence In MEdicine. Springer, pp 200–209
6. Guida C, Zhang M, Shan J (2021) Knee osteoarthritis classification using 3d cnn and mri. Appl Sci 11(11):5196
7. Si L, Xuan K, Zhong J, Huo J, Xing Y, Geng J, Hu Y, Zhang H, Wang Q, Yao W (2020) Knee cartilage thickness differs alongside ages: a 3-t magnetic resonance research upon 2,481 subjects via deep learning. Front Med 7
8. Glyn-Jones S, Palmer A, Agricola R, Price A, Vincent T, Weinans H, Carr A (2015) Osteoarthritis. The Lancet 386(9991):376–387
9. Maini R, Aggarwal H (2010) A comprehensive review of image enhancement techniques. J Comput 2:8–13
10. Arya RK, Jain V (2013) Osteoarthritis of the knee joint: an overview. J Indian Acad Clin Med 14(2):154–62
11. Sanjeev K (2013) Measurement of cartilage thickness for early detection of knee osteoarthritis (KOA). IEEE Trans 208–211
12. Pedoia V, Norman B, Mehany SN, Bucknor MD, Link TM, Majumdar S (2019) 3d convolutional neural networks for detection and severity staging of meniscus and pfj cartilage morphological degenerative changes in osteoarthritis and anterior cruciate ligament subjects. J Magn Reson Imag (JMRI) 49(2):400–410
13. Sinha A, Dolz J (2021) Multi-scale self-guided attention for medical image segmentation. IEEE J Biomed Health Inform 25(1):121–130
14. Bien N, Rajpurkar P, Ball RL, Irvin J, Park A, Jones E, Bereket M, Patel BN, Yeom KW, Shpanskaya K, Halabi S, Zucker E, Fanton G, Amanatullah DF, Beaulieu CF, Riley GM, Stewart RJ, Blankenberg FG, Larson DB, Jones RH, Langlotz CP, Ng AY, Lungren MP (2018) Deep-learning-assisted diagnosis for knee magnetic resonance imaging: development and retrospective validation of MRNet. PLoS Med 15(11):e1002699
15. Roth HR, Shen C, Oda H, Oda M, Hayashi Y, Misawa K, Mori K (2020) Deep learning and its application to medical image segmentation. Med Imaging Technol 36(2):63–67
16. Hesamian MH, Jia W, He X, Kennedy P (2019) Deep learning techniques for medical image segmentation: achievements and challenges. J Dig Imaging 32:582–596
17. Merkely G, Borjali A, Zgoda M, Farina EM, Gortz S, Muratoglu O, Lattermann C, Varadarajan KM (2021) Improved diagnosis of tibiofemoral cartilage defects on MRI images using deep learning. J Cartil Joint Preserv 1(2):100009
18. Antony J, McGuinness K, O'Connor NE, Moran K (2016) Quantifying radiographic knee osteoarthritis severity using deep convolutional neural networks. In 2016 23rd International conference on pattern recognition (ICPR). IEEE, pp 1195–1200
19. Ciresan D, Giusti A, Gambardella L, Schmidhuber J (2012) Deep neural networks segment neuronal membranes in electron microscopy images. In: NIPS. pp 2852–2860
20. Zhou Z, Zhao G, Kijowski R, Liu F (2018) Deep convolutional neural network for segmentation of knee joint anatomy. In: 2018 International Society for Magnetic Resonance in Medicine
21. Patil1 S, Udupi VR (2012) Preprocessing to be considered for MR and C.T. Images containing tumors. IOSR J Electr Electron Eng (IOSRJEEE) 1(4):54–57 ISSN: 2278–1676
22. Gornale SS, Patravali PU, Uppin AM (2019) Study of segmentation techniques for assessment of osteoarthritis in knee X-ray images. Image Graph Signal Proc 2:48–57
23. Tiulpin A, Thevenot J, Rahtu E, Lehenkari P, Saarakkala S (2018) Automatic knee osteoarthritis diagnosis from plain radiographs: a deep learning-based approach. Sci Rep 8(1):1727
24. Swiecicki A, Li N, O'Donnell J, Said N, Yang J, Mather RC, Jiranek WA, Mazurowski MA (2021) Deep learning-based algorithm for assessment of knee osteoarthritis severity in radiographs matches performance of radiologists. Comput Biol Med 133:104334

25. More S, Singla J, Abugabah A, AlZoubi AA (2020) Machine learning techniques for quantification of knee segmentation from MRI Hindawi. Rev Article 6613191
26. Gu Z, Cheng J, Fu H, Zhou K, Hao H, Zhao Y, Zhang T, Liu J (2019) CE-Net: context encoder network for 2D medical image segmentation. IEEE Trans Med Imaging 38(10)
27. Mallikarjuna Swamy MS, Holi MS (2012) Knee joint articular cartilage segmentation, visualization and quantification using image processing techniques: a review. Int J Comp Appl 42

Chapter 17
Recognition of Handwritten Assamese Characters

Olimpia Borgohain, Pramod Kumar, and Saurabh Sutradhar

1 Introduction

Handwritten character recognition is a technique of detecting, segmenting, and identifying characters from images. The main objective of handwritten character recognition is to replicate the human reading capabilities so that the computer can understand, read, and work as humans do with text. It has been one of the most interesting and laborious research areas in the field of image processing and pattern recognition lately. Several research works have been focusing on new methods that would reduce the processing time without compromising high recognition accuracy.

A lot of work for handwritten character recognition for languages like European, Chinese, and Arabic has been done. Genetic algorithms and Artificial Neural networks (ANN) were used as a classifier to recognize English digits and alphabets [6]. Deep-Learning was also used to recognize English digits and alphabets [10]. Even for domestic Indian languages like Hindi, Bangla [4], Oriya [2], Malayalam [5], Tamil [11], etc. a lot of research has been done till now. Indic scripts like Oriya, Telugu, Bangla, and Roman have been extensively explored [2, 4].

The Assamese language has not been much explored due to its less usage. The Assamese language is a part of the Assamese-Bengali Script prominently used in the North-Eastern region of India. The Assamese script consists of 11 vowels and 40 consonants [15]. It also has 122 conjunct characters [15]. Recognition of Assamese Characters is a difficult task because of the complex features of the characters like convoluted edges (ক্ষ, শ, ঋ,etc.), similarity in appearance (অ, আ, য, য়,etc.), presence of loops (ভ, ঞ,etc.), etc. Unlike English script, the Assamese script does not have the concept of capitalization of first characters. Also, different people have

O. Borgohain (✉) · P. Kumar · S. Sutradhar
Department of Computer Science and Engineering, Royal Global University, Guwahati, India
e-mail: olympiaborgohain@gmail.com

G. Mathur et al. (eds.), *Proceedings of 3rd International Conference on Artificial Intelligence: Advances and Applications*, Algorithms for Intelligent Systems,
https://doi.org/10.1007/978-981-19-7041-2_17

different handwriting styles therefore it becomes a very laborious task of determining the characters.

Nowadays, CNNs have been effectively applied to pattern recognition, image classification, and forecast studies to name a few. For the classification of Assamese digits and characters, a few versions of CNN have been used so far [13, 14]. This research area of recognizing Assamese characters has not been explored much. The potential of this field is immense considering the popularity of the language in its native state.

We have used the performance metrics of precision, recall, and F1 score, for our calculation. The formulas of the respective performance metrics are given below:

$$\text{Precision} = Tp/(Tp + Fp)$$

$$\text{Recall} = Tp/(Tp + Fp)$$

$$\text{F1} = 2 \text{ x } [(\text{P x R}) /(\text{P} + \text{R})]$$

where

Tp = Number of true positives.

Fp = Number of false positives.

Fn = Number of true positives (Tp) + Number of false positives (Fp).

Additionally, this paper is divided into five sections. The first is the Related Work section, where we have given a detailed survey in this area. Next is the Methodology section where we have explained the workflow of our system. We have also discussed our dataset and explained each of the steps we have thoroughly used in our system. It is followed by the Results and Discussion section which focuses on the results obtained in our research. In the end, we have the conclusion and future scope section where we have underlined the future breadth of this research area.

2 Related Work

Medhi et al. [1] compared printed and handwritten Assamese digits. Part feed-forward neural networks and a tree-based classifier were used for the feature extraction and classification. Digits were scanned from the document after pre-processed images were cropped with a bounding box to obtain individual digits. Features were extracted from each bounded area of the image. The grayscale documents were converted into a binary image and the background noises were removed with the help of linear filtering, medial filtering, and adaptive filtering. After that skew detection and correction were performed, following line, word, and character segmentation.

Word-level script identification for six handwritten Indic scripts (Bangla, Devanagari, Gurmukhi, Malayalam, Oriya Telugu, and Roman) was proposed by Singh et al. [2]. This paper proposed the elliptical and approximation approach to design features. The original images are in grayscale and Otsu's Global thresholding method is used to convert them into binary images, removing the noisy pixels from the binarized image using the Gaussian Filter. For the classification part, the feature sets have been applied to 7 different classifiers namely-Naïve Bayes, Bayes Net, MLP, SVM, Random Forest, Bagging, and Multi-Class Classifiers. The multi-layer perceptron (MLP) achieved the highest accuracy of 94.35%.

Bania et al. [3] proposed a computational model for Handwritten Assamese to perform the preprocessing, text segmentation, and then extraction of different features from individual characters. To segment, a document image into various parts, a global projection profile approach of a word was used to identify the upper, lower, and middle areas of a word. A combination of diagonal features using zoning concept and texture features via GLCM (Grey Level Co-occurrence Matrix) were computed for extracting various features of individual characters. The mean and standard median filters (SMF) were used to clean noises from an input image. Random transform-based techniques were used for skew detection and correction. An artificial neural network as a backend was used to perform classification and recognition tasks.

Alif et al. [4] proposed a modified ResNet-18 architecture (a convolution neural network architecture) to recognize Bangla handwritten characters. This paper considered three main challenges: Recognizing convoluted edges to distinguish between the repetitions of the same pattern in different characters and different handwritten patterns for the same characters. To have wider input for the generalized performance of the network, input images are preprocessed through the removal of noise with the median filter, and edge thickening filter, and the image is resized to a square shape with appropriate paddings by default. The dataset used for this paper is the two recently introduced datasets Bangla Lekha-Isolated dataset and the CMATERdb dataset.

Sujala et al. [5] proposed a hybrid approach for recognizing Malayalam handwritten characters. It takes into account both the dependent and independent features of the language. MATLAB is used to implement the proposed OCR system. The proposed method of OCR is a hybrid approach for feature extraction combining both structural and statistical features. In this paper, the curve features are extracted using the water reservoir principle and a decision tree classifier is used for classification.

Bhattacharya et al. [7] used an Artificial Neural Network (ANN) based approach to segment handwritten text in Assamese. After the feature extraction part, the input was fed to an ANN model, and the similarity measure was found. Gopinath et al. [8] used a text-to-speech synthesizer to facilitate English text reading. The text was manually typed into the screen and the system was made using MATLAB. Sanjrani et al. [9] recognized handwritten Sindhi numerals using K-NN and SVM. They evaluated their system using the correlation coefficient. The Sindhi numeral 0 achieved an accuracy of 100% whereas the numeral 3 achieved 63%. Then for Tamil script, Wahi et al. [11] considered features like character height, width, slope, etc. Zernike moments were used for feature extraction and then fed to a backpropagation model.

Similarly, for the classification of Assamese handwritten digits, vowels, and consonants a zoning feature was used by Medhi et al. [12]. A feed-forward neural network model was used with a sigmoid function at every neuron to calculate the output. For digits, vowels, and consonants a recognition accuracy of 70.6, 69.62, and 71.23% was achieved respectively. Dutta et al. [14] recognized Assamese handwritten digits using CNN (DigiNet model). This paper used six alternative Convolution and Max-Pooling layers which were later followed by a Fully Connected layer and a Softmax classifier. CNN was further used by using LeNet-5, ResNet-50, Inception V3, and DenseNet-201 model by Yadav et al. [13].

3 Methodology and Dataset

The proposed system consists of the following phases:

- Data Collection
- Digitization and Normalization
- Preprocessing
- Convolution neural network/Classification.

The block diagram in Fig. 1. Depicts the workflow of our system.

3.1 Data Collection

The collection of datasets is the first and most necessary part of the work. For our research work, we went to a local school to ask small children to write the Assamese characters in different font sizes. People from 7 years and above were asked to fill in the boxes of 0.75, 1.0, and 1.25 cm^2. We wanted to include diversity and variation in our dataset as much as possible.

We have included the snapshot of a page of the dataset in Fig. 2.

3.2 Digitalization and Normalization

After the collection of datasets, the image of every character was taken by our phone cameras. The images were then manually cropped. For every character, 225 images were collected. So, our total dataset consists of 9,225 images. Each image of the dataset is fixed to a size of 50 × 50.

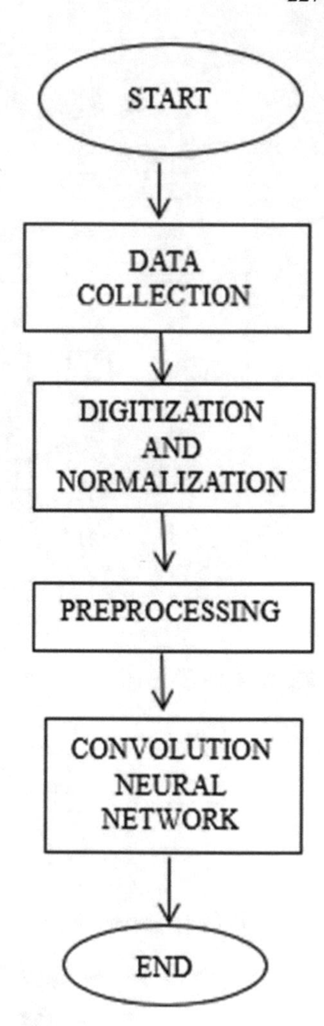

Fig. 1 Workflow diagram of our system

3.3 Preprocessing

The images were converted into a grayscale image. The color of the images was then inverted. For noise removal, we used Gaussian blur and Otsu's thresholding. Gaussian blur was used for smoothening and Otsu's thresholding was used for filtering in our system. Figure 3 shows the first character of the Assamese script from our image dataset before preprocessing. Then Fig. 4 shows that image after all the preprocessing work.

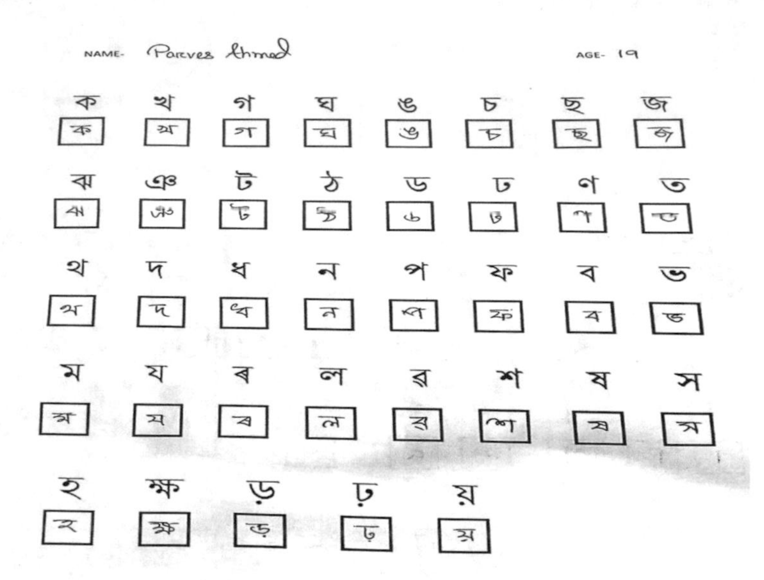

Fig. 2 A sample page of our collected dataset

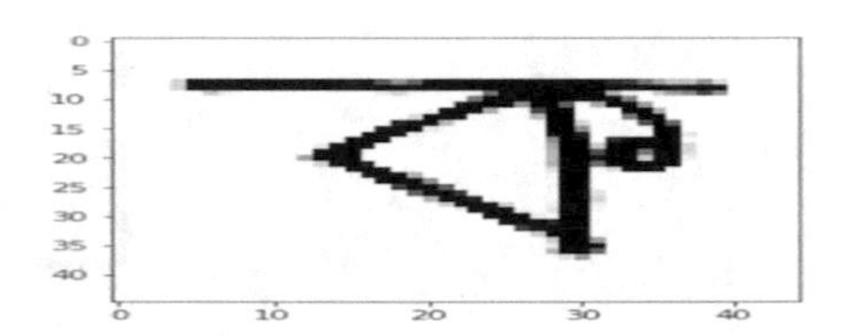

Fig. 3 Before preprocessing

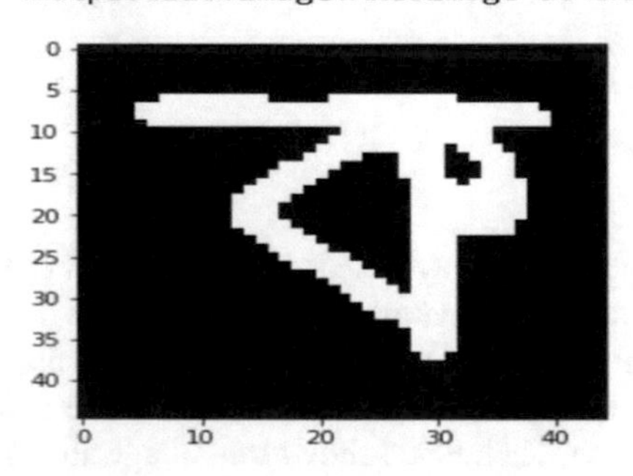

Fig. 4 After preprocessing

3.4 CNN/Classification

We divided our dataset into two parts in an 80:20 ratio, namely training and testing. The training dataset consists of 7,380 images and the testing dataset consists of 1,845 images. After dividing the datasets, the training images are fed to Convolution Neural Network.

A CNN pipeline model usually is made of a feature learning phase and a classification phase. The feature learning phase consists of the Convolution and Pooling layers. It learns the high-level features of our pre-processed input image. The Convolution layer performs linear operations in the input image for feature extraction. Then we used the relu and softmax functions as the activation function to ensure non-linearity in our network. Then for the pooling layer, we used the max-pooling function to reduce the dimensionality of the network and detect the strong features. The convolution and pooling operations are performed on every pixel of the input image independently. The output of the feature learning phase is a flattened column vector.

Since we have to classify 41 characters (consonants), our output layer consists of 41 nodes. Then for the classification phase, a feed-forward network must be established. It is achieved by mapping the flattened column vector to the output layer with the aid of a fully connected layer. This fully connected layer learns the logic behind the feature learning phase and performs the classification of Assamese characters. We have used five layers in our CNN network, the dropout and dense layers being alternatives. We went for categorical_crossentropy and RMSprop() for the loss function as the optimizer.

The testing dataset (1845 images from the test folder) was compared with the training dataset in the testing phase. We tested for 41 Assamese alphabets.

4 Results and Discussions

We evaluated the performance of the classification by the metrics accuracy, recall, precision, and F1 score.

We went for a batch size of 128 with epochs of 15. The result of our system has been shown in Table 1.

Table 1 Performance measure of our system

Accuracy	Precision	Recall	F1 Score
70.02%	68.5%	69.23%	70.71%

5 Conclusion and Future Scope

We have discussed the progress in the field of Assamese handwritten character recognition in this paper. Our research work achieved an accuracy of 70.02%, precision of 68.5%, recall of 69.23%, and F1 score of 70.71% by using a convolution neural network. The accuracy can be further improved by increasing the size of the dataset, adding more pre-processing steps, or by using a hybrid model.

In the future, the Assamese handwritten character recognition system can further be extended by adding a voice feature in the post-processing part. The recognition part could also be used to identify conjunct characters shortly. After successful recognition of conjunct characters, words and sentences can be targeted next.

References

1. Medhi K, Kalita S (2015) Assamese digit recognition with feed forward neural network. Int J Comp Appl 109(1)
2. Singh P, Doermann D (2015) Word-level script identification for handwritten indic scripts. In: 13th International conference on document analysis and recognition (ICDAR). IEEE
3. Bania R (2018) Handwritten assamese character recognition using texture and diagonal orientation features with artificial neural network. Int J Appl Eng Res 13(10):7797–7805
4. Alif M, Ahmed S, Hasan M (2017) Isolated Bangla handwritten character recognition with convolutional neural network. In: 20th International conference of computer and information technology (ICCIT). IEEE.
5. Sujala K, James A, Saravanan C (2017) A hybrid approach for feature extraction in Malayalam handwritten character recognition. In: Second international conference on electrical, computer and communication technologies (ICECCT). IEEE
6. Agarwal M, Kaushik B (2015) Text recognition from image using artificial neural network and genetic algorithm. IEEE
7. Bhattacharya K, Sarma K (2009) ANN-based innovative segmentation method for handwritten text in assamese. IJCSI Int J Comp Sci Issues 5
8. Gopinath J, Aravind S, Chandran P, Saranya S (2015) Text to speech conversion system using OCR. Int J Emerg Technol Adv Eng 5(1)
9. Sanjrani A, Baber J, Bakhtyar M, Noor W, Khalid M (2015) Handwritten optical character recognition system for Sindhi numerals. IEEE
10. Vaidya R, Trivedi D, Satra S (2018) Handwritten character recognition using deep-learning. In: 2nd International conference on inventive communication technologies (ICICCT)
11. Wahi A, Sundaramurthy S, Poovizhi P (2013) Handwritten Tamil character recognition. In: Fifth international conference on advanced computing (ICoAC)
12. Medhi K, Kalita S (2018) Assamese character recognition using zoning feature. Adv Electr Commun Comp
13. Yadav M, Mangal D, Srinivasan N, Ganzha M (2021) Assamese character recognition using convolutional neural networks
14. Dutta P, Muppalaneni N (2021) DigiNet prediction of assamese handwritten digits using convolutional neural network. In: Concurrency and computation: practice and experience
15. Wikipedia (2021) https://en.wikipedia.org/wiki/Assamese_alphabet. Accessed on 12 10 2021

Chapter 18
Frequency Reconfigurable C Shape Substrate Integrated Waveguide Antenna for Cognitive Radio Application

Sandeep Pratap Narayan Mishra, Sarwang Patel, Vidya Sagar Singh, and Ruchi Agrawal

1 Introduction

For wireless communication, Afzal et al. [1] show that one of the most important resources is the radio spectrum. David et al. [2] tells that the spectrum of radio is mostly underutilized, according to recent research. Fadhali et al. [3] explain that the technology of Cognitive radio provides a solution of the problem of spectrum shortage and ineffectiveness in networks of wireless. The new paradigm Peng et al. [4] explains about Internet of things that brings together a variety of technologies, including sensor with wire and without wire(wireless) as well as actuator networks, better communication concord, distribution of intelligence for smart devices, smartphones, and, of course, Internet. The basic thought behind the Internet of Things is to link things in order to improve many elements of prospective users' everyday lives and behaviour.

There are numerous Shah et al. [5] demonstrate that tough concerns must be solved for to establish a union as cognitive radio for IoT. Interoperability of interconnected gadgets, expanded adroitness, further developed versatility dependent on the current climate, industrialization and normalization, and, above all, security and trust in the unavoidable climate are only a couple of instances of main points of contention that should be painstakingly thought of assuming the two advances are to assume a main part soon.

Song et al. [6] proposed that the developed system supports the multiple number of frequency bands for application of CR-IoT based is a critical aspect of modern technology of wireless, which necessitates systematic spectrum exploitation and therefore no interference occurs between its neighboring devices, most probably in

S. P. N. Mishra · S. Patel · V. S. Singh · R. Agrawal (✉)
Galgotia's College of Engineering and Technology, Knowledge Park II, Greater Noida, Uttar Pradesh 201310, India
e-mail: ruchi.agrawal@galgotiacollege.edu

G. Mathur et al. (eds.), *Proceedings of 3rd International Conference on Artificial Intelligence: Advances and Applications*, Algorithms for Intelligent Systems,
https://doi.org/10.1007/978-981-19-7041-2_18

the IoT, which combines lots of devices in our lifestyle. Due to its tiny size, small amount of volume, simplicity integration with system, and underweight, frequency reconfigurable antennas are better suited for supplying many wireless technologies in one antenna unit. In 1930s, Tawk et al. [7] explains about one of the important concepts that came to mind which was the reconfigurable antenna. In 1979, FR-Antenna was constructed for satellite communication. The proposed antenna has six distinct beam angles Chaudhary et al. [8] show that it may be switched off and also mentioned a multi-beam reconfigurable antenna for satellite communications. Tawk et al. [9] tells that re-configurable PIN diode leaky patch antenna was invented in 1999. From 1999 to the present, Tawk et al. [10] explain that reconfigurable antennas have been designed using microstrip and SIW antennas as a platform. The effective radiator length of so many antennae such as microstrip, slot, loop, dipole, and monopoly antenna Prasad et al. [11] describe that they are typically defined by antenna resonance, and the operational antenna length takes part in a vital role for defining the operating frequency. In result, Mansoul et al. [12] tell that frequency may be changed in a way by adjusting the effective length of all proposed antenna. In such antenna configuration allows for operation at 2.59–3.40 GHz and four distinct bands generated from main bands. Ghosh et al. [13] describe about another FR antenna system which is a wireless communication network that supports many standards. One central microstrip is coupled to four neighboring microstrip through copper strips to achieve frequency reconfigurability. There were 16 distinct states that might be produced throughout the bandwidth of 0.65–2.95 GHz. As this is not necessary to match with those networks. In recent years, so many great research has been done. Karmokar et al. [14] tell how to improve the Microwave Wave guide efficiency and wave element in millimetre which are made with minimum effort and at a cheap cost, SIW (Substrate Integrated Waveguide) is one of them. The SIW approach for wireless components, micro and millimetre waves is a novel developing application and integration technology. The utilization of sophisticated SIW structures, as well as the modern way of research in future related to networks of wireless sensor, are friendly to environmental technologies and materials. Jafari et al. [15] illustrate that SIW fills electric field distribution and spreads the surface currents over a wider total area on the waveguide walls, resulting in losses of lower connection and application of multiband which is more demanding and integrated electronic systems. For providing different functions of wireless communication systems, a multiband FR antenna employing a proposed SIW antenna. At the same time, Lou et al. [16] demonstrate that frequency reconfigurable multi-band antennas are required for smart to cognitive radio broadcasting, where only one antenna may be statically hanged to transmit or receive different bands of frequency. Single-antenna systems have shown unsuitable response for cognitive radio applications, mainly for large systems like IoT. Mansoul et al. [17] disclose that according to the FR antenna has been widely used in the application of wireless communication systems particularly in the Cognitive Radio (CR).

This research aims to create a multiband FR antenna designed on layer of SIW that may be used in the application of cognitive radio. Fadhali et al. [18] demonstrate about various approaches, plus SIW slot supported proposed antenna, are recommended as for overcome the above-mentioned difficulties came in microstrip, co-planar as well as a classic waveguide. SIW is a dependable approach for creating narrowband of low-frequency that aids in addressability, discernment, and integration. Thus, SIW fits best for certain data of wireless networks, Alnahwi et al. [19] explain that Internet of Things needs minimum bandwidth approximately 186 kHz. In this paper the return loss (−13.394, −16.586, −17.483, −15.406, −17.606, −22.184, −18.635, −26.480 dB) obtained at respective frequencies (2.592, 2.764, 2.848, 2.872, 5.240, 5.308, 5.332, 5.388 GHz). All the return loss in this design are less than −10 dB. Hence the obtained multiband frequency reconfigurable for cognitive radio and IoT application.

2 Antenna Design

The diagrammatic representation of C–C shaped frequency reconfigurable SIW antenna supported with strips of copper schematic is the evolution and Fadhali et al. [3] show the improvement compare to earlier design. In this design, C–C slots were created using a parameterized + research that looked at the maximum concentration of surface current on the interior part of the cavity. Computer Simulation Technology (CST) software was used in this study's simulation. The parameter and material which are used in this design are shown in Table 1. The SIW structure, which is made up of metallic via-hole arrays, is arranged as illustrated in Fig. 1.

SIW is made up of parallel variants with PEC via holes defining the spread zone of wave TE10, and cutoff frequencies are simply branded with waveguide width until the thickness of substrate 'h' is smaller than "w". Between the two arrays, parameter "w" determines the steady spread of central mode's, 'd' and 'q' are the via through holes parameters set to control the central mode's spread. Now both radiation and return loss occurs.

Table 1 Parameter used in this design

Parameter name	Material used	Measurement(mm) l = length, w = width, h = height
1. Ground	PEC	l = 36, w = 41,h = 0.035
2. Substrate	RogersRT5880lz(lossy)	l = 36, w = 41,h = 1.27
3. Top Layer	PEC	l = 36, w = 36, h = 0.035
4. Strap line	PEC	l = 9.40, w = 3.50, h = 0.035
5. Via diameter	PEC	d = 1
6. Distance between two via	–	q = 1.9
7. Small c slot	–	l = 12, w = 2
8. Large C slot	–	l = 20.6, w = 2
9. Copper strip(present in top layer)	PEC	l = 2, w = 0.8, h = 0.035

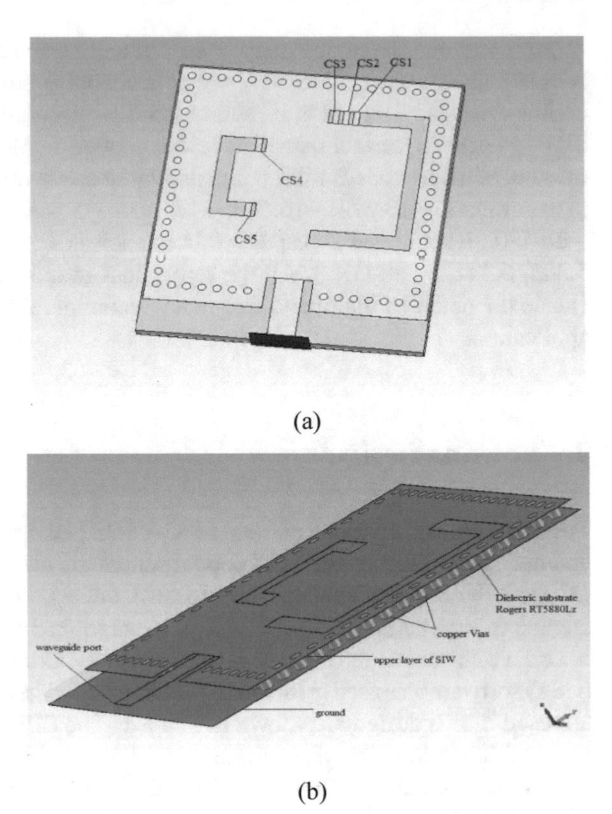

Fig. 1 **a.** top view of the SIW along with Via structures and copper strips. **b.** 3D SIW configuration synthesized by PEC metallic via-hole array

First, substrate thickness 'h' is manufactured due to the presence of physical constants, the geometrical parameter of SIW "w" is significantly larger than "b". Secondly the width of SIW 'w' is not the same as the width between the centre of first and the last via hole array 'W_{eff}'. In a similar vein, several trials and changes have led to the confirmation of the estimation of the W_{eff} given is one empirical criteria for determining W_{eff}. The length, width, height and diameter of top layer, bottom layer, substrate layer, C–C slot, vias, copper strips are shown in Fig. 2. All the vias and copper strips dimensions are symmetrical. Dimensions of the SIW structure can be calculated using the following formula,

$$W_{SIW} = \frac{a_W}{\sqrt{\epsilon_r}} \tag{1}$$

SIW width is computed using the rectangular waveguide width aw as a reference. SIW cavity size is defined by the appropriate resonant frequency as given initial dimension using the equation below.

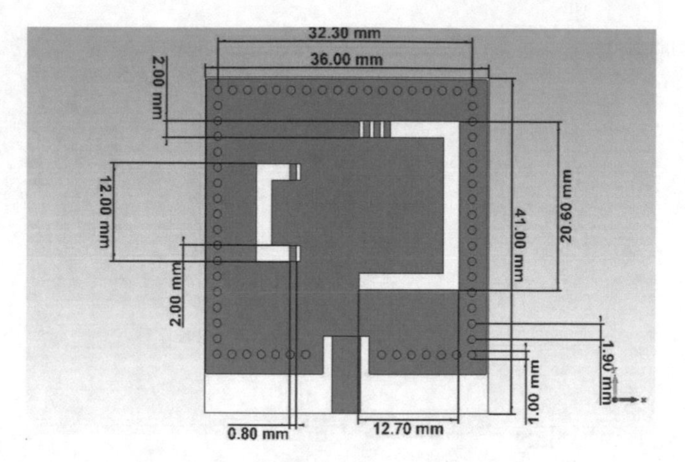

Fig. 2 Dimension of Frequency reconfigurable SIW C–C shape slot antenna

$$f_{101} = \frac{c}{2\pi\sqrt{\mu_r \epsilon_r}} \sqrt{\left(\frac{\pi}{W_{eff}}\right) + \left(\frac{\pi}{l_{eff}}\right)^2} \tag{2}$$

where W_{eff} and l_{eff} are equivalent width and length respectively that can be mentioned as

$$d = \frac{\lambda_g}{5} \tag{3}$$

$$q < 2d \tag{4}$$

where λ_g is wavelength guide which can be determined from below equation

$$\lambda_g = \frac{2\pi}{\sqrt{\left(\frac{\varepsilon_r (2\pi f)^2}{c^2}\right) - \left(\frac{\pi}{W}\right)^2}} \tag{5}$$

$$W_{eff} = W - \frac{d^2}{0.95q}, l_{eff} = l - \frac{d^2}{0.95q} \tag{6}$$

where

l = length of SIW

W = width of SIW

d = metallic via hole diameter

q = distance between center of two adjacent metallic via holes

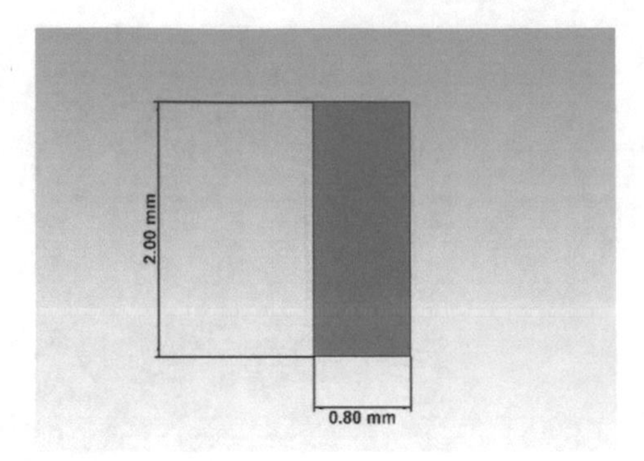

Fig. 3 Copper strip

μ = relative permeability

ϵ_r = relative permittivity

c = speed of light in vacuum

SIW leakage loss decreases by taking the correct dimension of vias and the distance of two neighboring vias. Hence, for low-loss output substrate thickness should be high. The proposed antenna is built upon Rogers RT5880lz (lossy) substrate and its thickness is 1.27 mm and its dielectric constant is 1.96 and its tangent loss is of 0.0009.

Here, copper strips are used instead of pin diode, copper strips denote the ON state (as the pin diode work as forward biasing). The length and width of copper strips are 2 and 0.8 mm which are inserted in the cavity of the C–C top layer.

All five strips of copper are used to analyse the basics of frequency reconfigurability. These copper strips sets on the edges of C–C shape slots which are shown in Fig. 3. In this antenna by using five copper strip (2^5 = 32) electronic switches status are generated. Out of which only five states are selected to analyse the reconfigurability concept in the simulation result and all other states are deleted because they are redundant that is overlap each other during resonance and those states which are having maximum coefficient of reflection.

3 Simulation Result and Discussion

All the designing and simulation part was done with CST studio software. For all states of switches, the proposed antenna performance is lower than −10 dB from 2 to 6 GHz. The proposed antenna also shows radiation properties. The radiating element which shows radiation in the antenna is C–C cavity slots. The fed is given by a microstrip patch line for impedance matching to produce good return loss at required frequencies. Due to a suitable change of copper strips position, it can produce less impact on the characteristic of radiation. In this geometry the structure position of the substrate and cavity joins in order to produce less loss and high productivity.

The resonance frequency and return loss of the metallic strip are operated in two ranges, one range is 2–3 GHz and another range is 5–6 GHz. Figure 4 displays the result of the combination of all five copper strips from 2 to 3 GHz. Figure 5 shows the result of the combination of all five copper strips from 5 to 6 GHz. The result is simulated for the cases which are (2.592, 2.764, 2.848, 2.872, 5.24, 5.308, 5.332, 5.388 GHz) for required multiband in order to analyse the frequency reconfigurability concept. But due to the effect of overlapping of copper strips on each other all other cases were ignored, coefficient of reflection was also not achieved and therefore these strips are deleted.

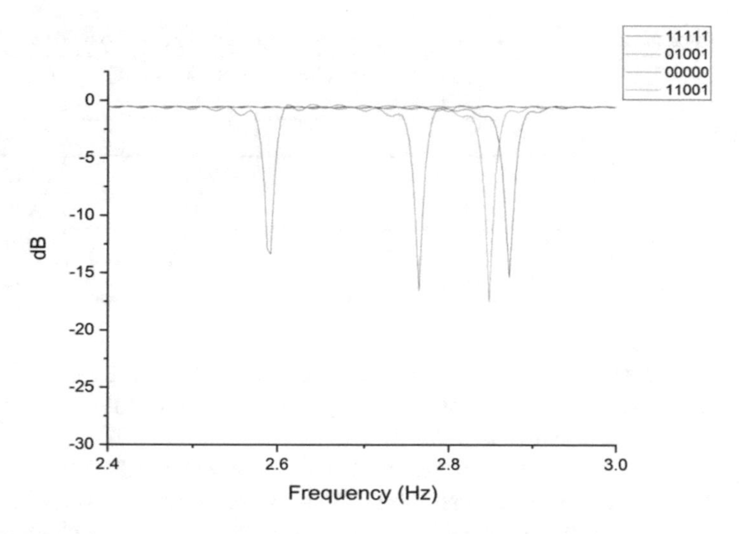

Fig. 4 Resonant frequencies and return loss of copper strips states from 2 to 3 GHz

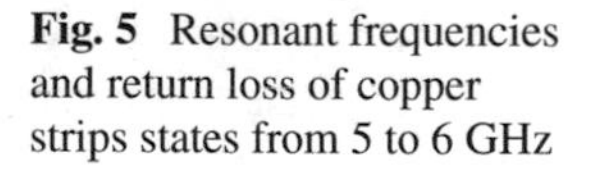

Fig. 5 Resonant frequencies and return loss of copper strips states from 5 to 6 GHz

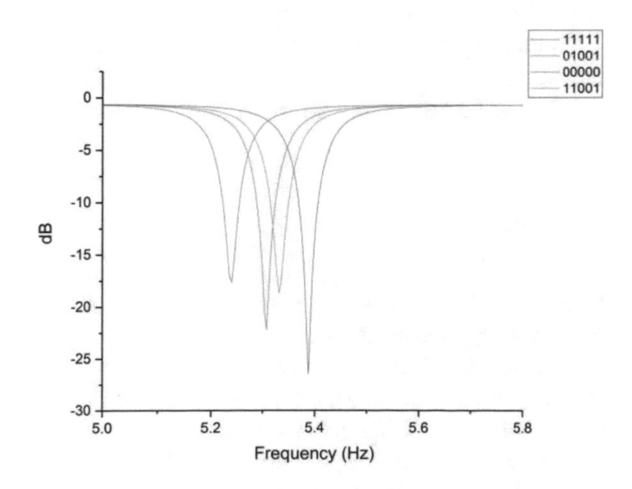

Table 2 Resonant frequencies and return loss of copper strips states from 2 to 3 GHz

States					Resonant frequency (GHz)	Return loss (dB)	Gain (dBi)
C1	C2	C3	C4	C5			
0	0	0	0	0	2.592	−13.394	3.779
0	1	0	0	1	2.764	−16.586	4.210
1	1	0	0	1	2.848	−17.483	4.360
1	1	1	1	1	2.872	−15.406	4.520

Table 3 Resonant frequencies and return loss of copper strips states from 5 to 6 GHz

States					Resonant frequency (GHz)	Return loss (dB)	Gain (dBi)
C1	C2	C3	C4	C5			
0	0	0	0	0	5.24	−17.606	7.447
0	1	0	0	1	5.308	−22.184	7.471
1	1	0	0	1	5.332	−18.635	7.537
1	1	1	1	1	5.388	−26.480	7.464

The resonant frequency and return loss (S11) result from the range 2–3 GHz shown in Table 2. The result of resonant frequency and return loss (S11) from the range of 5 to 6 GHz is shown in Table 3.

The result which is shown in Tables 2 and 3 showed a reflection coefficient below –10 dB. Resonant frequencies which are obtained from 2 to 6 GHz, include at (00,000, 01,001, 11,001, 11,111) these are the resonant frequencies which supports the cognitive radio systems. The good result is achieved in the paper which is less than –10 dB return loss in every case.

In this paper, at each frequency the directional radiation pattern is generated. At copper strip state 11,111 the radiation pattern of the E- plane and H-plane resonates at 2.872 and 5.388 GHz respectively as shown in below Fig. 6.

4 Conclusion

In future, CR as well as IoT play a vital role in the research field. In this paper, CR functionality such as innovative, closed packed, and vigorous SIW antenna allows a proper communication between wide bands which are interconnected to each other. For so many applications, the proposed antenna can produce self-reconfigurability to IoT solutions. A frequency reconfigurability SIW C–C slot antenna is presented which consist of five copper strips. During simulation, communication band was obtained by changing the positions of copper strips which are present in the cavity according to the states condition. All the resonant frequencies include (2.592, 2.764,

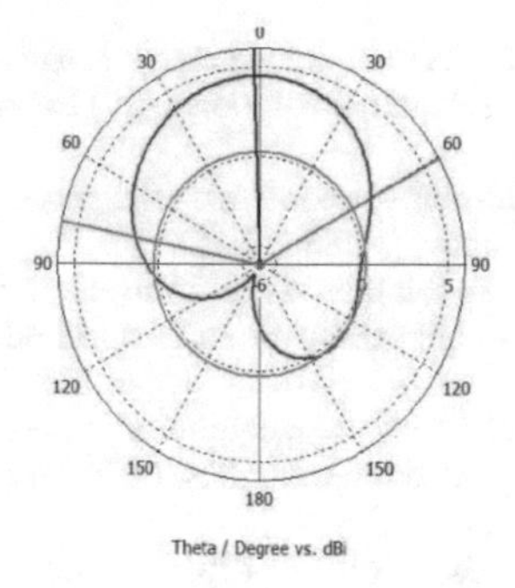

E-plane (11111) at 2.872GHz

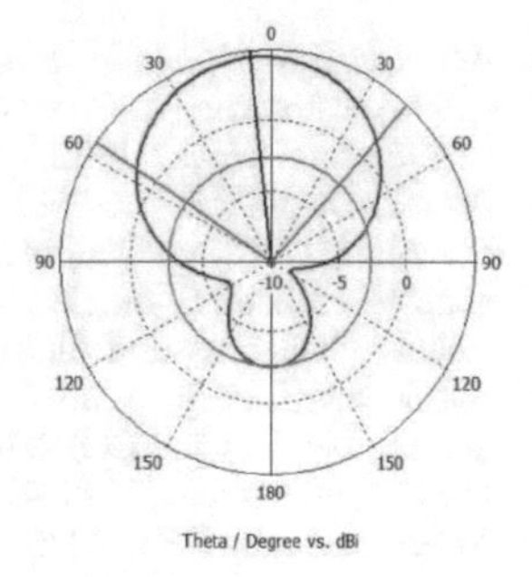

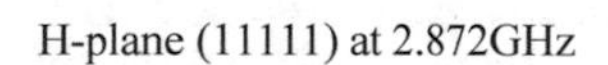

H-plane (11111) at 2.872GHz

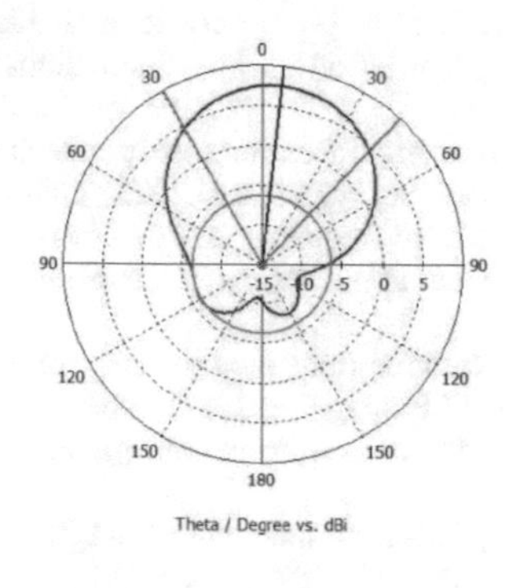

E-plane (11111) at 5.388GHz

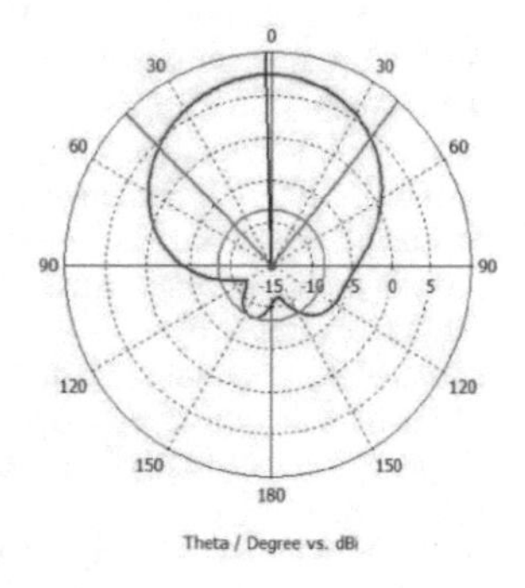

H-plane (11111) at 5.388GHz

Fig. 6 Radiation pattern of resonant frequency by copper strips at (11,111)

2.848, 2.872, 5.24, 5.308, 5.332, 5.388 GHz) at which the reconfigurability occurs. This design supports the CR system and generates the decent performance. In this paper, the proposed antenna gives a decent performance. In future, the frequency reconfigurable SIW antenna of multiband frequency can be designed by using various types of slots which are embedded on the top layer which are placed above SIW. This shows reconfigurability and generates the decent performance of that design.

References

1. Afzal H, Mufti MR, Awan I, Yousaf M (2019) Performance analysis of radio spectrum for cognitive radio wireless networks using discrete time Markov chain. J Syst Softw 151:1–7
2. Tuberquia-David L, Hernandez C (2018) Multifractal modeling of the radio electric spectrum applied in cognitive radio networks. In: 2018 ITU Kaleidoscope: machine learning for a 5G future (ITU K). IEEE, pp 1–6

3. AL-Fadhali NMA, Majidd HA, Omar R, Mokhtar MH, Mosali NA (2019) Frequency reconfigurable substrate integrated waveguide (SIW) cavity F-shaped slot antenna. Indonesian J Electr Eng Inform (IJEEI) 7(1):135–142
4. Peng L, Dhaini AR, Ho PH (2018) Toward integrated Cloud-Fog networks for efficient IoT provisioning: Key challenges and solutions. Futur Gener Comput Syst 88:606–613
5. Shah MA, Zhang S, Maple C (2013) Cognitive radio networks for Internet of Things: applications, challenges and future. In: 2013 19th International conference on automation and computing. IEEE, pp 1–6
6. Song L, Gao W, Chui CO, Rahmat-Samii Y (2019) Wideband frequency reconfigurable patch antenna with switchable slots based on liquid metal and 3-D printed microfluidics. IEEE Trans Antennas Propag 67(5):2886–2895
7. Tawk Y, Albrecht AR, Hemmady S, Balakrishnan G, Christodoulou CG (2010) Optically pumped frequency reconfigurable antenna design. IEEE Antennas Wirel Propag Lett 9:280–283
8. Chaudhary P, Verma S (2018) A swtichable frequency reconfigurable UWB antenna for cognitive radio application. In: 2018 second international conference on electronics, communication and aerospace technology (ICECA). IEEE, pp 1174–1177
9. Tawk Y, Costantine J, Christodoulou CG (2010) A frequency reconfigurable rotatable microstrip antenna design. In: 2010 IEEE antennas and propagation society international symposium. IEEE, pp 1–4
10. Tawk Y, Christodoulou CG (2009) A cellular automata reconfigurable microstrip antenna design. In: 2009 IEEE antennas and propagation society international symposium. IEEE, pp 1–4
11. Prasad BS, Rao PM, Madhav BTP (2018) Trapezoidal notch band frequency and polarization reconfigurable antenna for medical and wireless communication applications. Indian J Public Health Res Dev 9(6)
12. Mansoul A, Ghanem F, Hamid MR, Trabelsi M (2014) A selective frequency-reconfigurable antenna for cognitive radio applications. IEEE Antennas Wirel Propag Lett 13:515–518
13. Ghosh S, Lim S (2018) Fluidically switchable metasurface for wide spectrum absorption. Sci Rep 8(1):1–9
14. Karmokar DK, Guo YJ, Qin PY, Chen SL, Bird TS (2018) Substrate integrated waveguide-based periodic backward-to-forward scanning leaky-wave antenna with low cross-polarization. IEEE Trans Antennas Propag 66(8):3846–3856
15. Jafari M, Moradi G, Shirazi RS, Mirzavand R (2017) Design and implementation of a six-port junction based on substrate integrated waveguide. Turk J Electr Eng Comput Sci 25(3):2547–2553
16. Lou Q, Wu RX, Meng FG, Poo Y (2017) Realizing frequency reconfigurable antenna by ferrite-loaded half-mode SIW. Microw Opt Technol Lett 59(6):1365–1371
17. Mansoul A, Ghanem F (2018) Frequency reconfigurable antenna for cognitive radios with sequential UWB mode of perception and multiband mode of operation. Int J Microw Wirel Technol 10(9):1096–1102
18. AL-Fadhali N, Majid HA, Omar R, Dahlan SH, Ashyap AY, Shah SM, Rahim MK, Esmail BA (2020) Substrate integrated waveguide cavity backed frequency reconfigurable antenna for cognitive radio applies to internet of things applications. Int J RF Microwave Comput Aided Eng 30(1):e22020
19. Alnahwi F, Abdulhameed A, Swadi H, Abdullah A (2019) A compact wide-slot UWB antenna with reconfigurable and sharp dual-band notches for underlay cognitive radio applications. Turk J Electr Eng Comput Sci 27(1):94–105

Chapter 19
Handwritten Text Recognition for Regional Languages of Indian Subcontinent

Jagdish Kumar and Apash Roy

1 Introduction

The handwriting text recognition initially started way back in the sixties and continued its journey with real time application in postal address and bank cheque writing recognition systems. With increasing data volume, the need for automation was felt in the 20 s. Handwritten text or data could not prove its utility in digital applications. Simply storage of handwritten data by scanning or in jpeg format is costs exorbitant and time consuming on one hand and often purposeless on the other. Ancient documents, museums, libraries and offices are contributors of much of handwritten data which deteriorates with the passage of time and other reasons leading to severe data loss. Handwritten text recognition (HTR) system is therefore a blessing to those working with data as it converts handwritten data to editable digital form which could be used in important applications. Digital data so produced can easily be stored and retrieved and is compatible for searching. Enhanced Data Security is also an important aspect of digital data so produced from handwritten form. With increasing amount of data and need for accuracy, automation in HTR is a demanding area. Continues research in this area is very much required as much of the data in different languages and scripts is still untouched.

Statistical features or structural features of an image of contribute to Feature extraction techniques in handwritten character Recognition. The recognition or classification process of symbols, characters or words is normally done using a template or feature-based approach. In template-based approach, comparison of a new test pattern is done with stored pattern and percentage of similarity between both is used for classification. In Feature-based approach, features are

J. Kumar (✉) · A. Roy
Lovely Professional University, Jalandhar, Punjab, India
e-mail: jagdishkkvk@gmail.com

G. Mathur et al. (eds.), *Proceedings of 3rd International Conference on Artificial Intelligence: Advances and Applications*, Algorithms for Intelligent Systems,
https://doi.org/10.1007/978-981-19-7041-2_19

taken out from test patterns and further used in classification models namely Artificial neural networks(ANN), Convolution Neural network(CNN), hidden Markov models(HMM), K-NN, support vector machines(SVM), Decision trees, Fuzzy Classification and modified quadratic discriminant function etc., for recognition. The process of Handwritten Text recognition is a step wise process of Digitization/Image Acquisition, Pre-processing, Segmentation, Feature extraction, Classification and recognition. Brindha [1] explained that to process the huge data manually and classify is difficult which forced the researchers to develop some tools for text processing to analyze linguistic and lexical patterns. According to Vasa [2] Classification is an important phase in which various classification techniques/models are employed based on certain parameters such as volume, nature and type of data. Clustering followed by classification and finally categorization are major techniques used in text analytics. Kumar et al. [3] have done a review for character recognition of Indic and non-Indic scripts in which major issues and challenges for numerals and character numeral recognition are examined. Roy [4] elaborated Bangla Character recognition process and study in the recent decade. Roy and Ghosh [5] have presented different achievements in handwritten character recognition in Bangla script and also future tasks to be done.

From last many decades various algorithms, models, feature extraction and classification techniques are being used across the world by many researchers and in regional languages also but a comparative study of the above would give a clear idea of its advantages and disadvantages for better accuracy and also it will serve as a blueprint for the researchers to get a better understanding and decision to proceed further in this field. Efficient classifier for one script can work well with other similar scripts also. Different classification algorithm approaches need to be tried for better accuracy. Roy et al. [6–13] have been working with Bangla script mainly on segmentation and classification algorithm/approaches and have portrayed variable results and accuracy according to different patterns.

2 Methodology

India is a country of mixed religions, faith and languages. Researchers have tried to develop an HTR system for different languages. Many methods/algorithms/techniques have been used for the recognition of handwritten text which mainly depends on the data type. There are ten prominent scripts in India such as Devanagari, Gurumukhi, Bangla, Gujarati, Oriya, Tamil, Malayalam, Kannada, Telugu and Urdu. Many Indian scripts originated from Brahmi which was an ancient script. Many languages like Hindi, Nepali, Marathi, Konkani, Sanskrit, Santali, Maithili, Sindhi, and Kashmiri are written using Devanagari script. Hindi, the national language of India and written in Devanagari and that is why Devanagari is not considered as a regional language. For this study, 50 papers of high rated journals and technical blogs in different Indian Languages preferably Gurumukhi, a north Indian script, Tamil, Gujarati, Kannada, Malayalam and Devanagari were

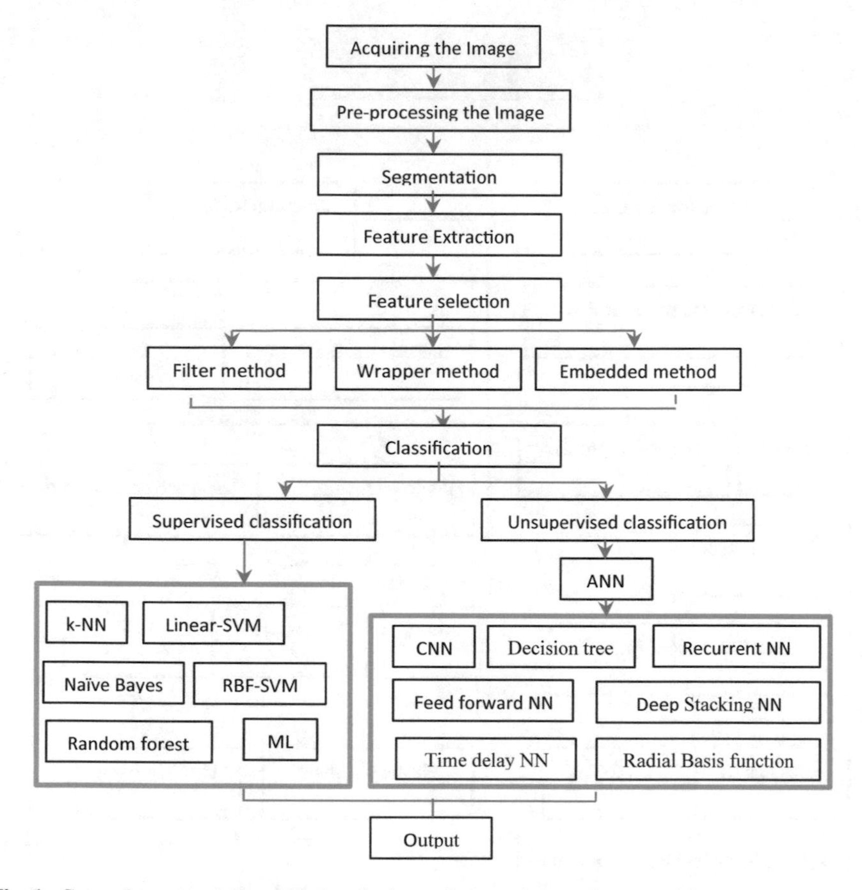

Fig. 1 General process followed in handwritten character/numerals recognition

downloaded and analyzed as much of the work is carried out in these languages. Each article was thoroughly read. The research problems along with techniques of feature extraction, classification and algorithms involved were analyzed. A comparative table for each script under study was drawn with respect to feature extraction, classification techniques and accuracy obtained. General procedure adopted in the handwritten text recognition process is presented in Fig. 1.

2.1 *Features Extraction/Feature Selection Techniques*

Fundamental components of character are features. Main aim of feature extraction and selection technique is to seek/track out the most relevant and effective features for improving accuracy in classification. Feature Extraction reduces the number of features in a dataset by developing new features from existing features. Many feature

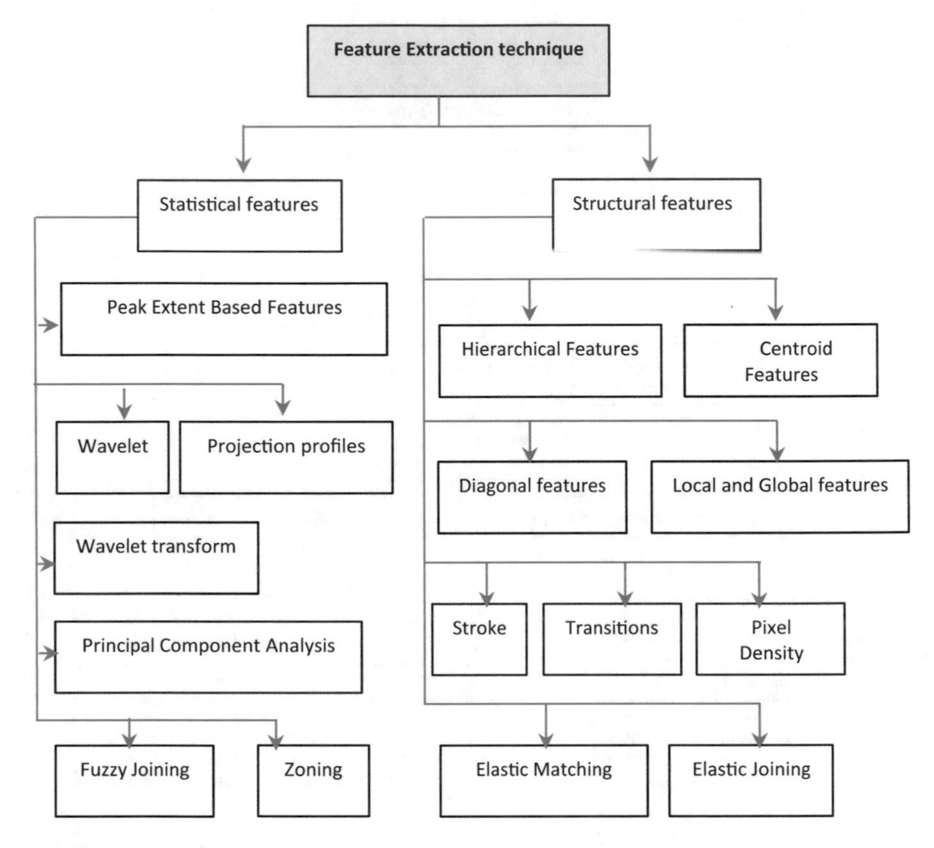

Fig. 2 Prominent feature extraction techniques

extraction techniques are available out of which some prominent are presented in Fig. 2. Feature selection aims to select important features in the dataset and discard the least important ones without creating any new features. Filter, Wrapper and Embedded are some methods that are used in feature selection. General approach and algorithms adopted in feature selection is depicted in Fig. 3.

2.2 Population Statistics of Speakers in India and World

Population of speakers of the languages under study worldwide and in India is presented below in Table 1.

The data depicted in Table 1 above was presented in the form of a chart to give a comparative visualization of the speakers of languages under study in India in Fig. 4 and at the world level in Fig. 5. Worthwhile to mention here that we have a reasonable number of speakers of said languages at world level also.

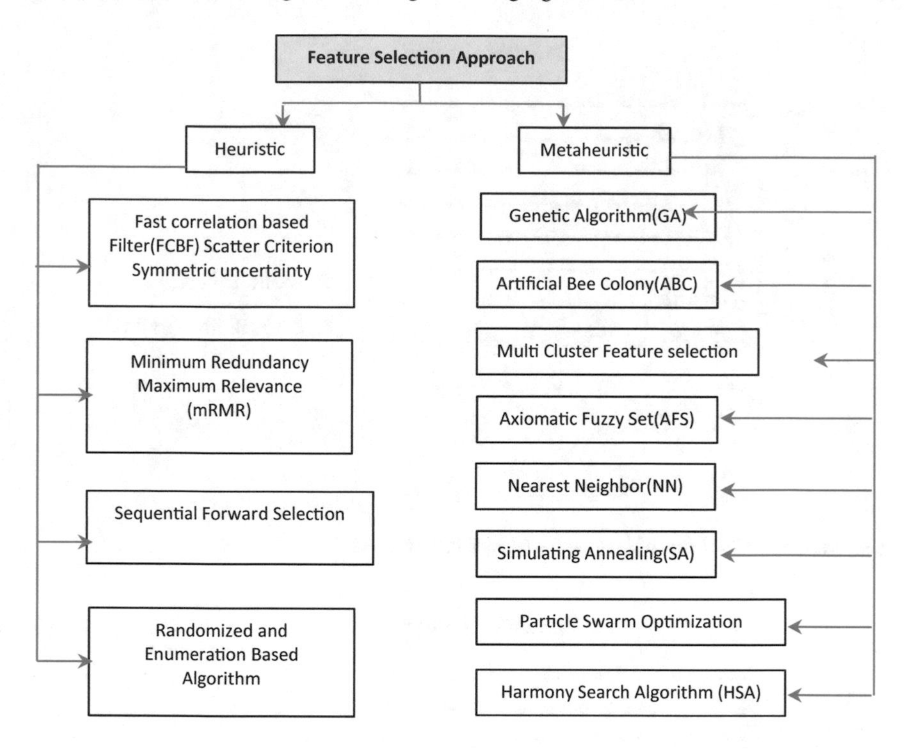

Fig. 3 General approach and algorithms adopted in feature selection

Table 1 Number of speakers in India and in world

Language	Nos. in India (Cr)	% population in India[a]	Nos. in World (Cr)	%population in World[b]
Hindi	32	26.6	34.1	4.429
Bengali	9.7	8.85	30	4
Gujarati	5.6	4.75	5.64	0.732
Kannada	4.4	3.59	4.36	0.566
Malayalam	3.5	2.87	3.71	0.482
Punjabi	3.11	2.57	3.26	0.423

[a] As per 2011 census India
[b] As per 2019 edition of Ethnologue, a language reference published by SIL International

3 Gurumukhi Script

Gurumukhi script is used for writing in Punjabi language. It is the official language of the Punjab state. The word “Gurumukhi” means “from the mouth of the Guru”. Among different language writing scripts of the world, Gurumukhi is the eighteenth

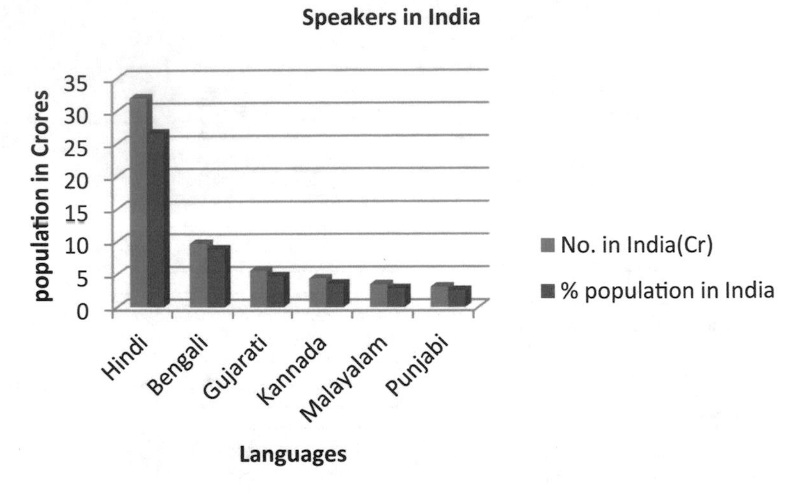

Fig. 4 Population statistics of language speakers in India

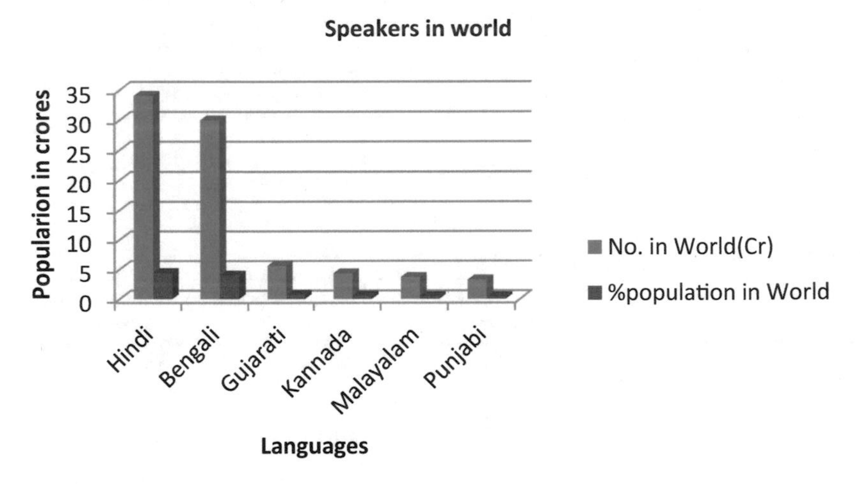

Fig. 5 Population statistics of language speakers in the world

most popular writing script. The Gurumukhi script has 32 consonants with additional 6 consonants, 3 vowel bearers and 9 vowel modifiers, 3 auxiliary signs and 3 half characters. According to Kumar [14] a Gurumukhi word can be divided into 3 zones: upper, middle and lower zone. Sharma and Jain [15] have proposed use of neocognitron technique for extracting local features in Gurumukhi character recognition system. The overall recognition accuracy for both learned and unlearned Gurumukhi characters was 92.78% using the neocognitron neural network approach. Lehal and Singh [16] using a hybrid classification scheme consisting of nearest neighbors (kNN) and binary decision trees achieved a recognition rate of 96.6%. Kumar [17] using three types of feature extraction techniques, Directional features, LBP features and regional features with deep neural network extracted 117 features using which achieved an accuracy of 99.3%. Kumar et al. [18] did a comparison of Nearest Neighbors (NN) and Support Vector Machine (SVM) classifiers using various classification schemes for performance checks of various feature selection techniques. Results outcome clarified that performance of Chi Squared Attribute feature selection technique gave better results than other feature selection techniques using Neural Network and Support Vector Machine. Recognition accuracy of 91.3, 95.2 and 88.3% for lower zone, middle zone and upper zone respectively were obtained. Kumar et al. [19] evaluated various classifiers such as k-nearest neighbors, Naive Bayes, linear-support vector machine (SVM), CNN, decision tree, RBF-SVM, and random forest classifier using 13,000 samples data set consisting of 7000 characters and 6000 numerals and results figures out that Random Forest classifier performance is better compared to other classifiers both for handwritten Gurumukhi character and numeral and achieved 87.9% recognition accuracy with 13,000 samples. Kumar et al. [19] explained that classifiers have reciprocal and interdependent quality and implementing them in a hybrid format could give better accuracy rates. Kumar et al. [20] have noticed that few classifiers perform better with increasing the training data set samples irrespective of the features. Accuracy achieved with respect to combination of features and classifiers used in Handwritten Text Recognition of Gurumukhi script is presented in Table 2.

Table 2 Gurumukhi script recognition performance with respect to feature and classifier selection

Methodology	Features	Classifier	Accuracy (%)
Sharma and Jain [15]	Local features	Neocognitron Neural Network approach	92.78
Lehal and Singh [21]	Zoning Local and global features	Hybrid classification scheme using binary decision trees and nearest neighbors	96.6
Sharma et al. [22]	Elastic matching	K-means	87.4
Sharma and Jhajj [23]	Zoning	SVM	72
Kumar [17]	Local binary pattern features, regional features and directional features	Deep neural network	99.1
Kumar et al. [24]	General	Random forest	87.9
Kumar et al. [24]	Hierarchical	SVM	91.8
Agarwal et al. [25]	8-directional gradients	SVM classifier equipped with RBF kernel	97

4 Tamil

Tamil is a popular southern Indian script mostly spoken in the southern state of Tamil Nadu and has a large number of speakers in and outside India. It is also used as the administrative language of Tamil Nadu. It is a member of the Dravidian Language family. Elakkiya et al. [26] developed a handwritten Tamil character recognition (offline) system using k-nearest neighbor (k-NN) which is a method used for classifying numerals/characters based on neighboring sample's similarity in the training data space. The accuracy of work was about 91%. Kowsalya et al. [27] detailed how the traditional neural network was modified utilizing optimization algorithm. The weights were optimized utilizing Elephant Herding Optimization (EHO) for recognition of Tamil characters. Based on recognition rate the usefulness of the proposed method was measured. The comparative results imply that the proposed MNN (NN + EHO) method gives an enhanced recognition rate of 93% when compared with existing classifiers such as SVM, NN, FNN, RBF, SOM and Quad tree method. Deepa and Rao [28] proposed a simple Nearest Interest Point (NIP) classifier, like Neural Network (NN), which yielded better results in Handwritten Character Recognition. Raj and Abirami [29] explained that challenges in Tamil handwritten characters are usually irrelevant character portions, discontinuation structure and structure over looping etc. The character portions were chosen in the research by implementing the Junction Point Elimination algorithm after meticulous analysis on algorithms in use such as Zoning and Junction Point. They introduced an algorithm to reduce problems in some existing feature selection and pre-extraction algorithms. Suitable classification algorithm and feature extraction techniques were applied on JPE features extracted to analyze success. The final analysis and experiments depicted that JPE performs better than others.

Accuracy achieved with respect to combination of features and classifiers used in Handwritten Text Recognition of Tamil script is presented in Table 3.

5 Gujarati

Gujarati belongs to a Devanagari family of languages; it has 12 vowels and 34 consonants. 55 million people speak Gujarati in Gujarat state and 65 million all over the world. Gujarati script is similar to other Indo-Aryan scripts. Research done for Gujarati script is less than other scripts. Chaudhary et al. [36] explained that few Gujarati characters have significant similarities because of which a little change in writing style can lead to wrong classification of characters because of which handwritten Gujarati character recognition is a challenging problem to researchers. Antani and Agnihotri started a research in Gujarati OCR in early 1999 [37]. They focused on printed text only by using K-Nearest Neighbor (KNN) and minimum hamming distance classifiers and realized 67 and 48% accuracy respectively. Shah and Sharma [38] implemented a template matching based system for Gujarati printed character

Table 3 Tamil script recognition performance with respect to feature and classifier selection

Methodology	Features	Classifier	Accuracy (%)
Sigappi [30]	Projection profile and transition	HMM Hidden Markov Model	90
Rajashekararadhya and Vanaja Ranjan [31]	Zoe based approach	SVM, NN	93.9
			94.0
Sarkhel et al. [32]	LGH feature, MartiBunke feature, pyramid histogram of oriented gradient	Multi-column multi-scale convolutional neural network (MMCNN)	98.12
Kavitha et al. [33]	Weights and bias using CNN	CNN	95.16
Sornum and Priya [34]	Feature extraction done by CNN	PCA and CNN	85.05
Ren et al. [35]	8 directional feature	RNN	97.60

recognition and achieved 72.30% accuracy using Fringe distance comparison of input images with template, though a dataset of 1375 images only was used which was very small. Goswami and Mitra [39] selected a dataset of 12,000-character dataset and utilized low-level stroke features, like junction points, endpoints, curves and line elements and realized 98.13% Gujarati characters recognition accuracy using KNN. Thaker and Kumbharana [40] selected 750 character dataset and used connected components, disconnected components, diagonal lines, horizontal and vertical lines, endpoints, crosspoints, curves types, and closed loops numbers as structural features for only 5 Gujarati handwritten characters recognition. 750 characters dataset was used and 88.78% recognition accuracy was realized by using Decision tree classifier. Pareek et al. [41] collected 10,000 images of Gujarati Characters from 250 persons and created a dataset from which training and testing data were selected and recognition success rate of 97.21% was achieved using CNN and 64, 28% was achieved using MLP. Patcl and Desai [42] achieved an accuracy of 63.1% by using kNN classifier for recognition of Gujarati handwritten numerals and characters. Statistical and structural features such as centroid distance and moment-based features along was used. Sharma et al. [43] proposed three new features to represent Gujarati handwritten characters such as features on normalized cross correlation, structural decomposition and zone pattern matching basis. Proposed features were applied on Naive Bayes and Support Vector Machine classifiers using 20,500 handwritten Gujarati characters dataset. Accuracy achieved with respect to combination of features and classifiers used in Handwritten Text Recognition of Gujarati script is presented in Table 4.

Table 4 Gujarati script recognition performance with respect to feature and classifier selection

Methodology	Features	Classifier	Accuracy (%)
Sharma et al. [43]	Feature extraction based 1. Structural decomposition 2. Zone pattern matching 3. Normalized cross correlation	SVM and NB	98.7 79.84 62.69
Hassan et al. [44]	Fringe distance map, shape descriptor, and Histogram of Oriented Gradients (HoG.)	Multiple Kernel Learning-based SVM (MKL-SVM)	95–99
Goswami and Mitra [45]	Low-level stroke features	Support Vector Machine with Radial Basis Function	98.46
Joshi et al. [46]	Morphological transformation	k-NN	82.03
Paneri et al. [47]	Histogram of Oriented Gradients (HoG)	SVM k-NN	85.87
Shirke et al. [48]	Darknet-53	Yolo Algorithm and deep learning techniques	Not quantified results ok

6 Kannada

Kannada, one of the 22 Indian languages, is mostly spoken by people in the south Indian state of Karnataka. The Kannada character set comprises of 16 vowels and 35 consonants. Maximum Kannada characters have similar orientation with minor differences. Sah and Indira [49] proposed recognizing online type of input for Kannada character recognition by using different types of features and obtained a variable rate of accuracy having an average of 97% accuracy. They used SVM classifier for the purpose. Ramappa et al. [50] proposed handwritten numerals recognition system and obtained an accuracy of 91% by using feature fusion for feature selection. Pasha and Padma [51] explains recognizing Kannada handwritten characters and numerals and achieved recognition accuracy of 97%. By using ANN classifier and wavelet transform feature extraction technique. Pereira et al. [52] explains online and offline characters recognition for Kannada including Hindi and English by using discrete artificial bee colony feature extraction algorithms. Rajput and Ummapure [53] also proposed a recognition system for Kannada, Hindi and English handwritten text by using kNN classifier and feature extraction using scale invariant feature transformation approach. Sushma and Veena [54] achieved an accuracy of 75% by using HMM model and scale invariant feature transformation for Kannada handwritten

Table 5 Performance detail with respect to feature and classifier selection in Kannada script

Methodology	Features	Classifier	Accuracy (%)
Ramesh et al. [59]	Hybrid	SVM, CNN	> 85
Pereira et al. [60]	Artificial bee colony	CNN	
Rajput and Ummaoure [53]	Scale invariant feature transformation (SIFT)	KNN	97.65
Sushma and Veena [54]	Scale invariant feature transformation (SIFT)	HMM	75
Pasha and Padma [51]	Wavelet transform for global feature extraction	ANN	97
Hebii et al. [61]	LBP (Local Binary Pattern, RLC (Run Length Count, CC (Chain Count), HOG (Histogram of Oriented Gradient)	K-Means clustering algorithm, Ranom Forest	89.92 highest

text. Patel and Reddy [55] implemented Handwritten Optical Character Recognition restricting to names in Karnataka states and obtained 68% accuracy by making use of principal component analysis (PCA) method for dimensionality reduction and classification. Aravinda and Prakash [56] proposed a system by using correlation technique and template matching for text classification in handwritten text of Kannada language. Pasha and Padma [57] proposed the use of KNN classifiers on hybrid features extracted and achieved an accuracy of 87% for Kannada HTR. Tushar et al. [58] proposed a model depicting knowledge transfer from one recognition system to another for written number recognition in scripts including Bangla, Arabic, and Hindi because of similarity in numerals writing and the best among the results was saved. CNN has been used for the transfer of learning yielding to less time consumption and enhanced accuracy and non-dependence of model-particular training decreases time of training again and again within a particular task. Accuracy achieved with respect to combination of features and classifiers used in Handwritten Text Recognition of Kannada script is presented in Table 5.

7 Malayalam

Malayalam is the official language in southern Indian states of Kerala, Puducherry and Lakshadweep, and have more than 34 million speakers worldwide (Wikipedia). The Malayalam language script belongs to the Dravidian Language family having a close proximity with Sanskrit and Tamil and thus having an effect on Malayalam vocabulary and grammar Govindaraju and Setlur [62]. Malayalam script in writing syllables uses symbols from Vatteluttu and Grantha script. As per Obaidullah et al. [63], in Malayalam, all characters are of same case rather than lower or upper case. Majority of Malayalam characters' shape is round unlike other Indian language

Table 6 Malayalam script recognition performance with respect to feature and classifier selection

Methodology	Features	Classifier	Accuracy (%)
Raju et al. [72]	Combination of gradient-based features and run length count i.e., GBF–RLC	Simplified Quadratic Classifier (SQDF) and Multi-Layer Perceptron (MLP)	99
Chacko et al. [74]	Histogram of Oriented Gradients	Feed forward neural network	94.23
James et al. [75]	Hybrid (Statistical and Structural) SSF	Decision Tree	97.87
John [76]	Topographical features	SVM-RBF	86.15
	Topographical & Distribution		87.44
	Transition count features		74.6
Manjusha et al. [77]	Scattering convolutional network	SVM	91.05

scripts. Only 27% of the basic character set are straight and most of the characters are having downward concavities. The basic characters set usually called as aksharmala of Malayalam script consist of 15 vowels also called as swarangal and 36 consonants also called as vyanganagal symbols making. As Malayalam language contains both old and new lip characters the total of character glyphs thus reached around 250. Pal and Choudhury [64] explained that Handwritten Malayalam character recognition work started after Bangla and Devnagari which were very popular languages in India. Because of large character classes and structural resemblance in shapes of characters, Malayalam character recognition has always been a challenging problem. Majority of works undertaken in Malayalam handwritten character recognition is structure based John et al. [65], statistical based Mani and Raju [66], Kuma et al. [67], transform domain based Chacko et al. [68], John et al. [69], Manjusha et al. [70] and gradient Jomy et al. [71], Raju et al. [72] features. SVM and NN based classifiers gave good performance in Malayalam character classification Jawahar [73]. Accuracy achieved with respect to combination of features and classifiers used in Handwritten Text Recognition of Malayalam script is presented in Table 6.

8 Devanagari

Devanagari is one of the most extensively used ancient scripts in India. It has 58 characters which includes 36 consonants, 12 vowels and 10 numerals. It also has composite characters consisting of two or more basic characters. Research initiatives started a few years back in recognition of handwritten Devanagari characters. Arora et al. [78] by using structural features and feed forward neural network achieved an accuracy of 89.12%. Hanmandlu et al. [79] achieved an accuracy of 90.65% by using a dataset of 4750 characters. Vector distance feature extraction method and fuzzy sets classifiers were used for handwritten Devanagari characters recognition. Pal et al. [80] achieved 94.24% accuracy by using a dataset of 36,172 characters for handwritten Devanagari characters used and Quadratic classifier for classification and Gaussian and gradient filters for feature extraction. Pal et al. [81] in their system used SVM and MQDF classifiers with gradient features and obtained 95.13% recognition accuracy. Deshpande et al. [82] obtained 82% accuracy by using dataset of 5000 handwritten Devanagari characters in their proposed system employing regular expressions techniques to perform operations such as searching, manipulation, validation, formatting of string and minimum edit distance technique. Mane and Ragha [83] realized an accuracy of 94.91% by using a 3600 characters dataset. They used feature extraction method which was eigen deformation-based and elastic matching technique for classification. Khanduja et al. [84] realized a recognition accuracy of 93.4% when applying a quadratic curve fitting (QCF) model on character images zone wise and coefficients of reasonably fitted curve utilizing feature vector. Aggarwal et al. [85] in a review study infers that both CNN and SVM classifier gives better results in handwritten Devnagari character recognition with an accuracy of 98.4 and 99.6% respectively. Khandukar et al. [86] has explained that if training data is increased better recognition rate is achieved. Singh et al. [87] evaluated and explained the effect of variable parameters on individual learning algorithm SVM and k-like using Keras, Accuracy achieved with respect to the combination of features and classifier used in Handwritten Text Recognition of Devanagari is presented in Table 7.

Table 7 Performance detail with respect to feature and classifier selection in Devanagari script

Methodology	Features	Classifier	Accuracy (%)
Singh and Lehri [88]	Binary vector Image	Backpropagation Neural Network	93
Indian et al. [89]	Feature extraction method "TARANG", based on natural wave movement	Backpropagation learning algorithm	96.2
Tawde [90]	Using Wavelets Features	Backpropagation Neural Network	60–70
Indian and Bhatia [91]	Combining feature vector of Zernike complex moments, directional gradient histogram, and Wave features	Back-propagation based Neural Network	96.4
Khanduja et al. [92]	Quadratic polynomial curve fitting	Neural Network	93.4
Kumar [93]	Multiple features like neighbor pixel weight, horizontal and vertical histogram, crossover points, gradients	Multi-layer Perceptron classifier (MLP)	72–80.8

9 Conclusion

The study reveals that different classification techniques and algorithms yield variable results. Nature and type of data is one the prime parameters to be considered while selecting the techniques and algorithm for handwritten text recognition. This paper explores the limitations, success and research trends in handwritten text classification in fields such as Artificial Intelligence. The automation in the process of Handwritten Text Recognition is the need of the hour with increasing size of data and accuracy requirements. Current trends call for further research opportunities which lie in simple algorithms in coding and also in implementation with deep learning and building sophisticated text data models. Comparison table presented above will act as a guiding line for new researchers in the field. Convolutional Neural Network and SVM is a more preferred machine learning algorithms used for classification than others. Use of Hybrid Neural networks has also shown a better performance in many cases, so future researchers may explore more on hybrid approaches. It is also inferred that more the training data the better is the recognition accuracy.

References

1. Brindha S, Sukumaran S, Prabha K (2016) A survey on classification techniques for text mining. In Proceedings of the 3rd International Conference on Advanced Computing and Communication Systems. IEEE. Coimbatore, India. https://doi.org/10.1109/ICACCS.2016.7586371
2. Vasa K (2016) Text classification through statistical and machine learning methods: a survey. Int J Eng Dev Res 4:655–658
3. Kumar M, Jindal MK, Sharma RK, Jindal SR (2018) Character and numeral than recognition for non-indic and indic scripts: a survey. Artif Intell Rev. https://doi.org/10.1007/s10462-017-9607-x
4. Roy A (2019) Handwritten Bengali character recognition-a study of works during current decade. Adv Appl Math Sci 18(9):867–875
5. Roy A, Ghosh D (2021) Pattern recognition based tasks and achievements on handwritten Bengali character recognition. In: 6th International conference on inventive computation technologies (ICICT). IEEE, Coimbatore, India, pp 1260–1265. https://doi.org/10.1109/ICICT50816.2021.9358783
6. Roy A, Manna NR (2015) An approach towards segmentation of real time handwritten text. Int J Adv Innov Res (2278–7844) 4(5)
7. Roy A, Manna NR (2014) Handwritten character recognition with feedback neural network. Int J Comp Sci Eng Technol (IJCSET 2229–3345) 5(1)
8. Roy A, Manna NR (2013) Recognition of handwritten text: artificial neural network approach. Int J Adv Innov Res (2278–7844) 2(9)
9. Roy A, Manna NR (2012) Handwritten character recognition using mask vector input(MVI)in neural network. Int J Adv Sci Technol (2229 5216) 4(4)
10. Roy A, Manna NR (2012) Handwritten character recognition using mask vector in competitive neural network with multi-scale training. Int J Adv Innov Res (2278–7844) 1(2)
11. Roy A, Manna NR (2012) Competitive neural network as applied for character recognition. Int J Adv Res Comp Sci Softw Eng (2277 128X) 2(3)

12. Roy A, Manna NR (2012) Handwritten character recognition using block wise segmentation technique (BST) in neural network. In: Proceedings of First International Conference on Intelligent Infrastructure, held during 1–2 December at Science City, Kolkata
13. Roy A, Manna NR (2012) Character Recognition with multi scale training. In: UGC Sponsored National Symposium on Emerging Trends in Computer Science (ETCS 2012) on 20–21 January
14. Kumar M (2018) Offline handwritten Gurmukhi script recognition. PhD Thesis, Thapar University, Patiala, 2014 Engineering (RICE), pp 1–6. https://doi.org/10.1109/RICE.2018.8509076
15. Sharma D, Jain U (2010) Recognition of isolated handwritten characters of Gurumukhi script using neocognitron. Int J Comput Appl 10(8):975–8887
16. Lehal G, Singh C (2000) A Gurumukhi script recognition system. Int Conf Pattern Recognit 2(2):557–560
17. Kumar N, Gupta S, Pradesh H (2017) A novel handwritten Gurumukhi character recognition system based on deep neural networks. Int J Pure Appl Math 117(21):663–678
18. Kumar M, Jindal MK, Sharma RK, Jindal SR (2018) Performance comparison of several feature selection techniques for offline handwritten character recognition. International Conference on Research in Intelligent and Computing in Engineering (RICE)
19. Kumar M, Sharma R, Jindal M, Jindal S (2020) Performance evaluation of classifiers for the recognition of offline handwritten Gurmukhi characters and numerals: a study. Artif Intell Rev 53:2075–2097. https://doi.org/10.1007/s10462-019-09727-2
20. Kumar M, Sharma RK, Jindal MK (2013) Size of training set vis-a-vis recognition accuracy of handwritten character recognition system. J Emerg Technol Web Intell 5(4):380–384
21. Lehal GS, Singh C, Lehal R (2001) A shape based post processor for Gurmukhi OCR. In: Proceedings of the 6th international conference on document analysis and recognition (ICDAR). pp 1105–1109
22. Sharma A, Kumar R, Sharma RK (2008) Online handwritten Gurmukhi character recognition using elastic matching. In: Proceedings of the congress on image and signal processing. pp 391–396
23. Sharma D, Puneet J (2010) Recognition of isolated handwritten characters in Gurmukhi Script. Int J Comp Appl 4. https://doi.org/10.5120/850-1188
24. Kumar M, Sharma RK, Jindal MK (2014) Efficient feature extraction techniques for offline handwritten Gurmukhi character recognition. Natl Acad Sci Lett 37(4):381–391
25. Aggarwal A, Singh K, Singh K (2014) Use of gradient technique for extracting features from handwritten gurumukhi characters and numerals. In: International conference of information and communication technologies. Elsevier, pp 1716–1723
26. Elakkiya V, Muthumani I, Jegajothi M (2017) Tamil text recognition using KNN classifier. Adv Nat Appl Sci 11(7):41–45
27. Kowsalya S, Periasamy PS (2019) Recognition of Tamil handwritten character using modified neural network with aid of elephant herding optimization. Multimed Tools Appl 78:25043–25061. https://doi.org/10.1007/s11042-019-7624-2
28. Deepa A, Rajeswara Rao RN (2020) A novel nearest interest point classifier for offline Tamil handwritten character recognition. Pattern Anal Appl 23:199–212. https://doi.org/10.1007/s10044-018-00776-x
29. Raj MAR, Abirami S (2020) Junction point elimination based tamil handwritten character recognition: an experimental analysis. J Syst Sci Syst Eng 29:100–123. https://doi.org/10.1007/s11518-019-5436-6
30. Sigappi AN, Palanivel S, Ramalingam V (2013) Handwritten document retrieval system for Tamil language. Int J Comp Appl 31(4):42–47
31. Rajashekararadhya SV, Vanaja Ranjan P (2009) Zone-Based hybrid feature extraction algorithmfor handwritten numeral recognition of two popular Indian script. In: World congress on nature and biologically inspired computing. pp 526–530
32. Sarkhel R, Das N, Das A, Kundu M, Nasipuri M (2017) A multi-scale deep quad tree based feature extraction method for the recognition of isolated handwritten characters of popular indic scripts. Pattern Recogn 71:78–93

33. Kavitha BR, Srimathi C (2019) Benchmarking on offline handwritten tamil character recognition using convolutional neural networks. J King Saud Univ-Comp Info Sci Commun
34. Sornam M, Priya CV (2018) Deep convolutional neural network for handwritten tamil character recognition using principal component analysis. In: Smart and innovative trends in next generation computing technologies. pp 778–787
35. Ren H, Wang W, Liu C (2019) Recognizing online handwritten Chinese characters using RNNs with new computing architectures. Pattern Recogn 93:179–192
36. Chaudhary M, Shikkenawis G, Mitra SK, Goswami M (2012) Similar looking Gujarati printed character recognition using Locality Preserving Projection and artificial neural networks. In: Proceedings of the 2012 Third International Conference on Emerging Applications of Information Technology (Kolkata, India). IEEE, pp 153–156
37. Antani S, Agnihotri L (2019) Gujarati character recognition. In Proceedings of the Fifth International Conference on Document Analysis and Recognition. ICDAR'99 (Cat. No. PR00318). IEEE, pp 418–421
38. Shah SK, Sharma A (2006) Design and implementation of optical character recognition system to recognize Gujarati script using template matching. J Inst Eng India Part ET Electron Telecommun Eng Division 86:44–49
39. Goswami MM, Mitra SK (2016) Classification of printed Gujarati characters using low-level stroke features. ACM Transactions on Asian and Low-Resource Language Information Processing (TALLIP) 15(4):25
40. Thaker HR, Kumbharana C (2014) Structural feature extraction to recognize some of the offline isolated handwritten Gujarati characters using decision tree classifier. Int J Comput Appl 99(15):46–50
41. Pareek J, Singahania D, Rekha E, Purohit S (2020) Gujarati handwritten character recognition from text images. Proc Comp Sci 171:514–523
42. Patel C, Desai A (2013) Gujarati handwritten character recognition using hybrid method based on binary tree classifier and k-Nearest Neighbour. Int J Eng Res Technol 2(6):2337–2345
43. Sharma A, Thakkar P, Adhyaru DM, Zaveri TH (2019) Handwritten Gujarati character recognition using structural decomposition technique. Pattern Recognit Image Anal 29:325–338
44. Hassan E, Chaudhury S, Gopal M (2014) Feature combination for binary pattern classification. Int J Doc Anal Recogn (IJDAR) 17(4):375–392
45. Gohel CC, Goswami MM, Prajapati YK (2015) On-line handwritten Gujarati character recognition using low-level stroke. In: Third international conference on image infonnation processing, December
46. Joshi DS, Risodkar YR (2018) Deep learning based Gujarati handwritten character recognition. In: IEEE 2018 International Conference On Advances in Communication and Computing Technology (ICACCT). pp 563–566. https://doi.org/10.1109/ICACCT.2018.8529410
47. Paneri PR, Narang R, Goswami MM (2017) Offline handwritten Gujarati word recognition. In: Fourth international conference on image information processing (ICIIP). pp 1–5. https://doi.org/10.1109/ICIIP.2017.8313708
48. Shirke A, Gaonkar N, Pandit P, Parab K (2021) Handwritten Gujarati script recognition. In: 2021 7th international conference on advanced computing and communication systems (ICACCS). https://doi.org/10.1109/icaccs51430.2021.9441811
49. Sah RK, Indira K (2017) Online Kannada character recognition using SVM classifier. In: IEEE international conference on computational intelligence and computing research (ICCIC). pp 1–6. https://doi.org/10.1109/ICCIC.2017.8524435
50. Ramappa MH, Srirangaprasad S, Krishnamurthy S (2014) An approach based on feature fusion for the recognition of isolated handwritten Kannada numerals. In International conference on circuits, power and computing technologies [ICCPCT-2014]. pp 1496–1502. https://doi.org/10.1109/ICCPCT.2014.7054777
51. Pasha S, Padma MC (2015) Handwritten Kannada character recognition using wavelet transform and structural features. In: International conference on emerging research in electronics, computer science and technology (ICERECT). pp 346–351. https://doi.org/10.1109/ERECT.2015.7499039

52. Pereira NA, Rao B, Kallianpur AK, Srinivasa KG (2017) Discrete artificial bee colony algorithm based optical character recognition. IEEE
53. Rajput GG, Ummapure SB (2017) Script identification from handwritten documents using SIFT method. In: IEEE international conference on power, control, signals and instrumentation engineering (ICPCSI). pp. 520–526. https://doi.org/10.1109/ICPCSI.2017.8392348
54. Sushma A, Veena GS (2016) Kannada handwritten word conversion to electronic textual format using HMM model. In: Proceedings of international conference on computational systems and information systems for sustainable solutions
55. Patel MS, Reddy SL (2014) An impact of grid based approach in offline handwritten Kannada word recognition. In: International conference on contemporary computing and informatics (IC3I). pp. 630–633. https://doi.org/10.1109/IC3I.2014.7019825
56. Aravinda CV, Prakash HN (2014) Template matching method for Kannada handwritten recognition based on correlation analysis. In: International conference on contemporary computing and informatics (IC3I). IEEE, pp 857–861
57. Pasha S, Padma MC (2013) Recognition of handwritten Kannada characters using hybrid features. In: proceedings of IET in ARTcom
58. Tushar AK, Ashiquzzaman A, Afrin A, Islam MR (2018) A novel transfer learning approach upon Hindi, Arabic, and Bangla numerals using convolutional neural networks. In: Hemanth D, Smys S (eds) Computational vision and bio inspired computing. Lecture notes in computational vision and biomechanics, vol 28. Springer, Cham. https://doi.org/10.1007/978-3-319-71767-8_83
59. Ramesh G, Kumar N, Sandeep, Champa, H.N. (2020). Recognition of Kannada handwritten words using SVM classifier with convolutional neural network. In: IEEE region 10 symposium (TENSYMP). Dhaka, Bangladesh
60. Pereira NA, Rao P, Kallianpur AK, Srinivasa KG (2017) Discrete artificial bee colony algorithm based optical character recognition. In: 14th IEEE India council international conference (INDICON). pp 1–6. https://doi.org/10.1109/INDICON.2017.8487826
61. Hebii C, Metri O, Bhadrannavar M, Mamtha HR (2021) Dataset building for handwritten Kannada vowel using unsupervised and supervised learning methods. In: SIRS 20 CCIS 1365. pp 75–89
62. Govindaraju V, Setlur S (2009) Guide to OCR for Indic scripts. Springer
63. Obaidullah SM, Halder C, Santosh K, Das N, Roy K (2017) Page-level handwritten document image dataset of 11 official indic scripts for script identification. Multimedia Tools 1–36
64. Pal U, Chaudhuri B (2004) Indian script character recognition: a survey. Pattern Recogn 37(9):1887–1899
65. John J, Pramod K, Balakrishnan K (2011) Offline handwritten Malayalam character recognition based on chain code histogram. In: International conference on emerging trends in electrical and computer technology, ICETECT. IEEE, pp 736–741
66. Moni BS, Raju G (2011) Modified quadratic classifier for handwritten Malayalam character recognition using run length count. In: International conference on emerging trends in electrical and computer technology (ICETECT). IEEE, pp 600–604
67. Kumar SS, Manjusha K, Soman K (2014) Novel SVD based character recognition approach for Malayalam language script. Recent Adv. Springer, Intell. Inf., pp 435–442
68. Chacko BP, Krishnan VV, Raju G, Anto PB (2012) Handwritten character recognition using wavelet energy and extreme learning machine. Int J Mach Learn Cybern 3(2):149–161
69. John J, Pramod K, Balakrishnan K (2012) Unconstrained handwritten Malayalam character recognition using wavelet transform and support vector machine classifier. Procedia Eng. 30:598–605
70. Manjusha K, Kumar MA, Soman K (2017) Reduced scattering representation for Malayalam character recognition. Arab J Sci Eng 1–12
71. Jomy J, Balakrishnan K, Pramod K (2013) A system for offline recognition of handwritten characters in Malayalam script. Int J Image Graph Signal Proc 5(4):53
72. Raju G, Moni BS, Nair MS (2014) A novel handwritten character recognition system using gradient-based features and run length count. Indian Acad Sci 39(6):1333–1355

73. Neeba NV, Jawahar CV (2009) Empirical evaluation of character classification schemes. In: Seventh international conference on advances in pattern recognition. IEEE Computer Society, pp 310–313
74. Chacko AMMO, Dhanya PM (2015) A comparative study of different feature extraction techniques for offline malayalam character recognition. In: Computational intelligence in data mining, vol 2. https://doi.org/10.1007/978-81-322-2208-8(Chapter2)
75. James A, Saravanan SKC (2018) A Novel hybrid approach for feature extraction in Malayalam handwritten character recognition. J Theoret Appl Inform Technol 96(13)
76. John J (2018) Spatial domain feature extraction methods for unconstrained handwritten Malayalam character recognition. In: International Conference on Machine Learning & Neural Information Systems (ICMLNIS 2021)
77. Manjusha K, Kumar MA, Soman KP (2019) On developing handwritten character image database for Malayalam language script. Eng Sci Technol Int J 22(2):637–645. ISSN 2215–0986
78. Arora S, Bhatcharjee D, Nasipuri M, Malik L (2008) A two stage classification approach for handwritten Devanagari characters. Proc Int Conf Comput Intell Multimed Appl ICCIMA 2:399–403
79. Hanmandlu M, Murthy OR, Madasu VK (2007) Fuzzy model based recognition of handwritten Hindi characters. Digit Image Comput Tech Appl 454–461
80. Pal U, Sharma N, Wakabayashi T, Kimura F (2007) Off-Line handwritten character recognition of devnagari script. Int Conf Doc Anal Recognit (ICDAR 2007) 1–5
81. Pal U, Chanda S, Wakabayashi T, Kimura F (2008) Accuracy improvement of Devnagari character recognition combining SVM and MQDF. ICFHR 367–372
82. Deshpande PS, Malik L, Arora S (2008) Fine classification and recognition of hand written Devnagari characters with regular expressions & minimum edit distance method. J Comput 3(5):11–17
83. Mane V, Ragha L (2009) Handwritten character recognition using elastic matching and PCA. In: International Conference on Advances in Computing, Communication and Control—ICAC3 '09. pp 410–415
84. Khanduja D, Nain N, Panwar S (2015) A hybrid feature extraction algorithm for devanagari script. 15(1)
85. Agrawal M, Chauhan B, Agrawal T (2022) Machine learning algorithms for handwritten Devanagari character recognition: a systematic review. J Sci Technol 07(01)
86. Khandokar I, Hasan M, Ernawan F, Islam S, Kabir MN (2021) Handwritten character recognition using convolutional neural network. J Phys Conf Ser 1918:042152. https://doi.org/10.1088/1742-6596/1918/4/042152
87. Singh R, Shukla AK, Mishra RK, Bedi SS (2022) An improved approach for Devanagari handwritten characters recognition system. In: Iyer B, Ghosh D, Balas VE (eds) Applied information processing systems. advances in intelligent systems and computing, vol 1354. Singapore. https://doi.org/10.1007/978-981-16-2008-9_20
88. Singh G, Lehri S (2012) Recognition of handwritten Hindi characters using backpropagation neural network. Int J Comp Sci Inform Technol 3(4):4892–4895
89. Indian A, Bhatia K (2017) Offline Handwritten Hindi "SWARs" recognition using a novel wave based feature extraction method. Int J Comp Sci Issues 14(4). ISSN 1694–0814 08 14
90. Tawde GY (2014) Optical character recognition for isolated offline handwritten devanagari numerals using wavelets,. Int J Eng Res Appl 4(2):605–611
91. Indian A, Bhatia K (2018) Off-line handwritten Hindi consonants recognition system using Zemike moments and genetic algorithm. 10–16. https://doi.org/10.1109/SYSMART.2018.8746934
92. Khanduja D, Nain N, Panwar P (2015) A hybrid feature extraction algorithm for Devanagari script. ACM Trans Asian Low-Resour Lang Inf Process 15(1):2
93. Kumar S (2016) A study for handwritten Devanagari word recognition. In: 2016 international conference on communication and signal processing (ICCSP). IEEEE, pp 1009–1014

Chapter 20
Image Denoising Using DT-CWT Combined with ANN and Grey Wolf Optimization Algorithm

P. Venkata Lavanya, C. Venkata Narasimhulu, and K. Satya Prasad

1 Introduction

A major operation achieved on noisy images is de-noising that is performed before the preprocessing methods like texture analysis, segmentation and extraction of the feature. By utilizing some predefined data on the process of degradation, the image de-noising process restores the original image from the noisy image. Some unwanted abbreviations such as noise are accompanied with received images in the communication systems. For further processing these noises must be removed. Awad [1] presented a detailed study on different types of noise occurring in digital images. The whole image might collapse and lose the image details due to the noise in the image. When the signal in analog systems is transmitted through a linear dispersive channel then noise occurs.

There are several works related to image denoising, some are related to image denoising by different filters, some methods related to removing noise by using optimization techniques, some methods using wavelet transform. Hassan and Saparon [2] developed a new method for Still image denoising by Discrete Wavelet Transform (DWT). Wavelet thresholding is one of the latest image de-noising techniques that utilize DWT. However, more attention is required for de-noising of edges. Moreover, because of the lack of shift invariance, wavelet coefficients are not suitable for realistic environments. To compute a probabilistic pixel value of the original image from noisy pixel values, the Bayesian network is another choice. To get a structure closer

P. V. Lavanya (✉)
ECE, TKR College of Engineering and Technology, Hyderabad, Telangana, India
e-mail: venkatalavanya@tkrcet.com

C. V. Narasimhulu
ECE, Lords Institute of Engineering and Technology, Hyderabad, Telangana, India

K. S. Prasad
ECE, JNTUK, Kakinada, Andhra Pradesh, India

G. Mathur et al. (eds.), *Proceedings of 3rd International Conference on Artificial Intelligence: Advances and Applications*, Algorithms for Intelligent Systems,
https://doi.org/10.1007/978-981-19-7041-2_20

to the original model, reliable priors are wanted by some of the sophisticated scoring functions and computational complexity is the drawback of this approach which is indicated by Sahu et al. [3]. In a DT-CWT, a pair of DWT trees are utilized in this scenario. The transforms of real and imaginary segments are represented by every tree. Into the dual trees the input image is given for disintegration and the results of both recreated trees are connected at the end step of construction that considers the higher redundancy, lack of shift invariance, directional selectivity of DWT and improper edge de-noising.

The coefficients of the non-zero filter under the particular threshold are set to zero in DT-CWT based de-noising. After attaining the rate-distortion trade-off iteratively, the input image is reconstructed on its convergence with inverse DT-CWT. The low-quality reconstructed image will be the result if we select improper threshold values. So, the proper threshold value selection is important. Based on this condition various wavelet thresholding techniques have been implemented. They are Visu Shrink, Sure Shrink and Bayes Shrink methods which are used for denoising and compared by Wang [4]. Some denoising methods related to the use of Neural networks and fuzzy logic to denoise underwater images were proposed by Zhang and Salari [5]. Some denoising methods focus on the optimization of threshold value to derive the best threshold value which is used in wavelet thresholding to get high PSNR. Some methods proposed by Gadekallu et al. [6, 7], Srinivasu et al. [8] will combine optimization algorithms with convolution neural networks(CNN) and with deep neural networks for feature extraction and classification before performing denoising. Some methods will optimize the DT-CWT filter weights to get a high PSNR of the denoised image rather than optimizing the threshold value. In methods developed by Lavanya et al. [9, 10] DT-CWT is applied to images to get the decomposed image into different sub-bands and before applying DT-CWT, its filter weights are optimized by optimization techniques so that these works provide accurate wavelet coefficients after image decomposition by DT-CWT to enhance PSNR. But the proposed work is novel compared to these methods because to improve PSNR of denoised images it is combining DTCWT with Artificial Neural Networks and optimization techniques.

For image denoising, DT-CWT is utilized in this paper along with Artificial Neural Network (ANN) and Grey Wolf optimization Algorithm (GWA). The image sub-bands are generated after applying DT-CWT and then properly trained ANN is applied to obtain noise-free coefficients. GWA-based optimized value of the threshold is utilized to implement the thresholding operation. The performance of the suggested approach is compared with other approaches like DWT, DT-CWT, Genetic algorithm-Bayesian regularisation based DT-CWT (GA-BR based DT-CWT) and combined Earthworm Grey wolf optimization algorithm based ANN (EWGW-ANN based DT-CWT) in terms of MSE and PSNR. This article is arranged as follows: The DT-CWT, GWA and ANN methodology is described in sect. 2. The proposed method is detailed in sect. 3. The results are discussed in sect. 4, and the paper is concluded in sect. 5.

2 Methods

2.1 DT-CWT's Mathematical Model

By DT-CWT, denoising is accomplished as discussed earlier. $LP_0(n)$ and $LP_1(n)$ represent low-pass and High-pass filters of Tree a of DT-CWT used by Naimi et al. [11]. $HP_0(n)$ and $HP_1(n)$ represent the low-pass and high-pass filters of Tree b of DT-CWT. Based on the theory of wavelets, imaginary and real parts of the scaling and wavelet coefficients are represented by $sc^{imaginary}{}_j(q)$, $sc^{real}{}_j(q)$ $wc^{imaginary}{}_i(q)$ and $wc^{real}{}_i(q)$. By the inner products they are obtained as given in Eqs. (1), (2), (3) and (4).

$$wc_i^{real}(q) = 2^{\frac{i}{2}} \int_{-\infty}^{\infty} V(u)\rho_h(2^i u - z)du, i = 1, \ldots j \tag{1}$$

$$sc_j^{real}(q) = 2^{\frac{j}{2}} \int_{-\infty}^{\infty} V(u)\phi_h(2^j u - z)du,\ j = 1, \ldots .i \tag{2}$$

$$wc_i^{imaginary}(q) = 2^{\frac{i}{2}} \int_{-\infty}^{\infty} V(u)\rho_g(2^i u - z)du, i = 1, \ldots j \tag{3}$$

$$sc_j^{imaginary}(q) = 2^{\frac{j}{2}} \int_{-\infty}^{\infty} V(u)\phi_g(2^j u - z)du \tag{4}$$

where the j represents the level of decomposition, the scale-factor is denoted by i, dual trees imaginary and real valued wavelet-scaling functions are represented by ϕ_h and ϕ_g and ρ_h and ρ_g and the finite energy signal is denoted by V(u). The DT-CWT's scaling coefficient can be obtained using Eq. (5)

$$sc_j^s(q) = sc_i^{real}(q) + jsc_j^{imaginary}(q) \tag{5}$$

The wavelet coefficient can also be derived as above. So, the scaling and wavelet coefficients can be obtained using Eqs. (6) and (7).

$$sc_j(u) = 2^{\frac{(j-1)}{2}} \left[\sum_n sc_j^{real}(K)\phi_h(2^j u - n) + \sum_m sc_j^{imaginary}(K)\phi_g(2^j u - m) \right] \tag{6}$$

$$wc_i(u) = 2^{\frac{(i-1)}{2}} \left[\sum_n wc_i^{real}(q)\rho_h(2^i u - x) + \sum_m wc_i^{imaginary}(q)\rho_g(2^i u - y) \right] \quad i = 1 \ldots j \tag{7}$$

By using Eq. (8) the decomposed signal by DT-CWT can be obtained.

$$V(u) = sc_j(u) + \sum_{i=1}^{j} wc_i(u) \quad (8)$$

where $wc_i(u)$ $i = 1,2,3,\ldots j$ and from higher to lower frequencies the sub-band signals are arranged and denoted by $sc_j(u)$.

Till a certain range the input image is decomposed via branches of two 2D-DWT, in 2D DT-CWT. Six sub-bands of high pass HL_g, HL_h, LH_g, LH_h, HH_g, HH_g are included in this process. Through differencing or averaging operations, the process is repeated linearly with the combination of sub-bands that have similar pass bands. At all levels, the sub-bands of 2D DT-CWT are given in Eq. (9).

$$\frac{(LH_g + LH_h\ )}{\sqrt{2}}, \frac{(LH_g - LH_h)}{\sqrt{2}}, \frac{(HL_g + HL_h)}{\sqrt{2}}, \frac{(HL_g - HL_h)}{\sqrt{2}}, \frac{(HH_g + HH_h)}{\sqrt{2}}, \frac{(HH_g - HH_h)}{\sqrt{2}} \quad (9)$$

2.2 *Artificial Neural Network*

The neurological process of the brain is the main idea of ANN. They have many neurons as linked processing components. To train the ANN, training sets are used. Data categorization and pattern identification are the main functions in the learning process of ANN. Conventional neural networks (CNN) are utilized in Deep learning, and ANNs are utilized in conventional neural networks proposed by Bashar [12]. Among the neurons, the biological process of absorption happens. This method is followed by ANN, and there are two network types. They are feed-forward networks (FFN) and feed-backward networks (FBN). The flow of signal is from input to output in feed-forward networks. Since there is no feedback loop, the input will not be affected by the output. The bidirectional signal flow is in Feed-backward networks. Thus the output affects the input and it is a strong and complex network. These networks are also called recurrent or interactive. Until this dynamic system reaches the equilibrium position the feedback networks' state differs constantly. Until the input locates another equilibrium, the network keeps a steady state. The proposed three-layer FFN is shown in Fig. 1.

The 6 sub-bands of complex wavelet coefficients are decomposed from the noisy image by applying DT-CWT. By individual ANN is utilized to train every sub-band. Thus six ANNs are totally used by the proposed approach. From sub-band coefficients' PXP neighbourhood area of DTCWT each ANN inputs are obtained. The variance between the corresponding pixel value and the center pixel value is represented by $P^2 - 1$ inputs, and the center of the PXP processing window is one

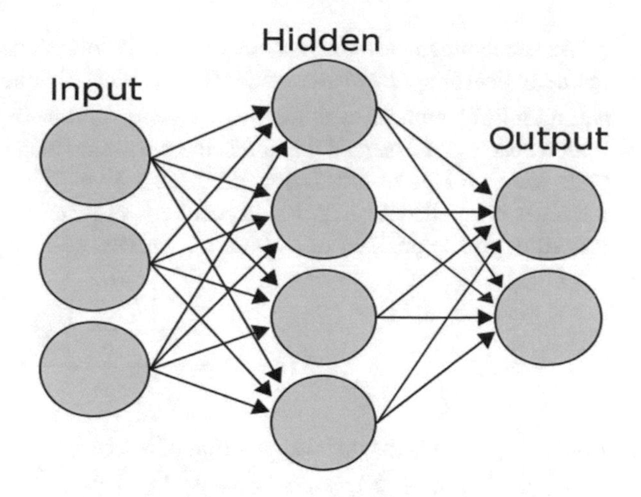

Fig. 1 Basic Neural Network. *Note* This image from wikimedia commons, the free media repository, File: Artificial neural networl.svg-wikimedia Commons

input. P^2 neurons are there in Input layers and Hidden layers. Sigmoid activation function is used by the output layer and hidden layer. For six corresponding DT-CWT sub-bands, six trained ANNs are applied after the training process and the output is noise-free coefficients. The denoised image can be restored by taking inverse DT-CWT.

2.3 Grey Wolf Optimization Algorithm (GWA)

GWA is an optimization technique founded by Lu et al. [13] on behaviour of Grey Wolves in hunting preys. A typical grey wolf pack consists of approximately 5–12 members. These wolves are categorized into 4 groups namely alpha (α), beta (β), delta (δ) and omega (ω). α wolves are leaders of the pack. It comprises a male and a female wolf. The decision of α wolves is followed by the entire pack. α wolves are responsible for decisions like time to wake up, walking, hunting, time and place to sleep, etc., and β wolves are subordinate wolves that assist α wolves in decision-making. They advise αwolves but command other wolves and are responsible for carrying out the decisions of α wolves in the entire pack. They replace α wolves if any of the α wolves become old or pass away. ω are the lowest-ranking wolves and obey all other wolves. If any of the wolves do not belong to α, β or ω categories, they are called δ wolves. They submit to α and β wolves but dominate ω wolves. δ wolves are subdivided into scouts; which are responsible for watching boundaries and were previously α or β wolves and caretakers; which are responsible for taking care of old, ill and wounded issue warning in case of danger, sentinels; which guard the pack, hunters; which assist α and β in hunting, elders; and which were wolves. The mathematical modelling of hunting behaviour of Grey Wolves is described as follows.

As stated previously, the hierarchy of dominance starts from α and ends in ω. For mathematical simplicity we consider that there is only one α, β and δ wolf in the pack and an infinite number of ω wolves. Consider that there is a prey at position $\overrightarrow{X}$ in search space. It is assumed that α has the best knowledge of prey position, followed by β and then δ. ω wolves follow all these wolves. The initial position of α, β and δ wolves is labelled $\vec{X}_1$, $\vec{X}_2$, $\vec{X}_3$ respectively. Now ω wolves change their positions according to the position of α, β and δ wolves. It is expressed mathematically in Eq. (10).

$$\vec{X}(t+1) = \frac{\overrightarrow{X_1} + \overrightarrow{X_2} + \overrightarrow{X_3}}{3} \tag{10}$$

where $\overrightarrow{X}(t+1)$ is the updated position of ωwolves,
$\overrightarrow{X_1} = \overrightarrow{X} - \overrightarrow{A_1}.\left(\overrightarrow{D}\right)$, $\overrightarrow{X_2} = \overrightarrow{X} - \overrightarrow{A_2}.\left(\overrightarrow{D}\right)\overrightarrow{X_3} = \overrightarrow{X} - \overrightarrow{A_3}.\left(\overrightarrow{D}\right)$ and
$\vec{A} = 2\vec{a}.\overrightarrow{r_1} - \vec{a}$, $\overrightarrow{D} = \left|\overrightarrow{C_1}.\overrightarrow{X} - \vec{X}\right|$, $\overrightarrow{D} = \left|\overrightarrow{C_2}.\overrightarrow{X} - \vec{X}\right|$, $\overrightarrow{D} = \left|\overrightarrow{C_3}.\overrightarrow{X} - \vec{X}\right|$ and $\vec{C} = 2.\overrightarrow{r_2}$.

Here t denotes iteration (hunting steps) and for the initial case its value is 0. Random vectors are represented by $\overrightarrow{r_1}$ and $\overrightarrow{r_2}$ with the range between 0 and 1. The $\overrightarrow{a}$ components are linearly reduced from 2 to 0 in the time of iteration as the distance between prey and wolves minimizes in each step. When Eq. (10) is substituted with values it becomes evident that ω wolves occupy positions within a circular region with respect to positions of α, β and δ values. The arrow mark above symbols represents vectors as movement includes both magnitude and direction. It is to be noted that the linear decrease of $\overrightarrow{a}$ along with the random number $\overrightarrow{r_1}$ result in a range of $\overrightarrow{A}$ values. The effect of this is that, when $\left|\overrightarrow{A}\right| > 1$, ω wolves position $\overrightarrow{X}(t+1)$ diverges from prey and when $\left|\overrightarrow{A}\right| < 1$ it converges towards prey. With random numbers of $\overrightarrow{r_2}$, value of $\overrightarrow{C}$ falls inrange [0 2]. This is modelled to represent obstacles in the path which wolves may experience in real situations.

Each time α, β and δ wolves advance towards prey,ω wolves update their positions and finally all wolves jointly attack the prey and eventually consume it. From the model it is clear that hunting works purely based on the position of α, β and δ wolves and the objective function is the minimisation of the distance between wolves and prey. However one of the main aspects of the hunting process i.e., advice of β to α wolves is not considered, as the model concentrates only on the position of ω wolves with respect to other dominant wolves. When we apply this process to optimization problems α, β and δ correspond to first, second and third best solutions respectively, satisfying objective functions of that particular optimization problem. In our case α, β and δ are first, second and third best groups of filter coefficients that maximize PSNR (our objective function) of reconstructed images. It is to be noted that in our approach only the first group of filter coefficients is significant as it is the group which provides maximum PSNR from so far concluded iterations.

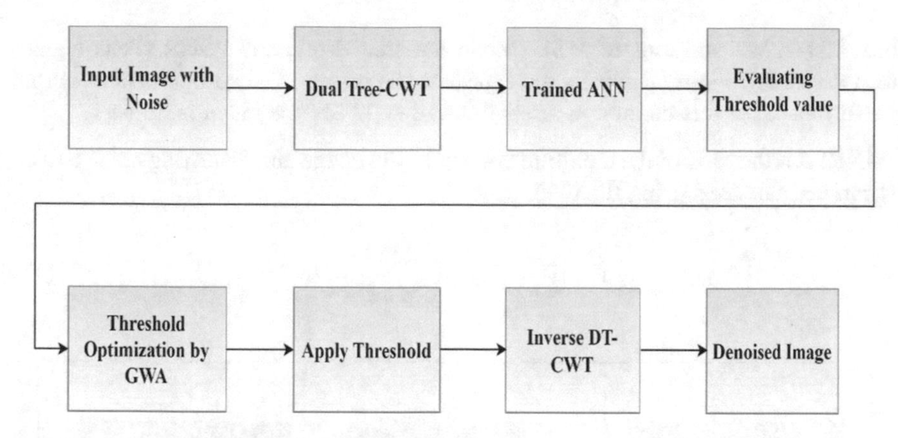

Fig. 2 Block Diagram of proposed method

3 Proposed Method

In this paper a novel image-denoising method is introduced with the use of GWA and ANN. Various frequency subbands are obtained from the noisy image by applying DTCWT. To get noise-free coefficients the ANN is properly trained. To the ANN, the correlation between the noisy image's wavelet coefficients is given. Using sure shrink, the computation of threshold value is performed and then GWA optimizes the threshold value. Soft thresholding is performed using the GWA-based optimized threshold value. At last, the denoised image as shown in Fig. 2 can be restored by taking inverse DT-CWT. The algorithm for wavelet-based denoising of the image is given below:

Step 1: Noisy image is given as the input.
Step 2: On noisy images apply DT-CWT and obtain frequency Subbands.
Step 3: To get noise-free coefficients trained ANN is applied on subbands.
Step 4: Using SURE shrinkage function, calculate the threshold value.
Step 5: Using GWA optimize the Threshold value for the population size of 200.
Step 6: With the optimized threshold value, on the subbands of DTCWT apply soft thresholding.
Step 7: Apply inverse DT-CWT.
Step 8: The denoised image is the output.
Step 9: In terms of MSE and PSNR, evaluate the quality of the denoised image.

4 Results

The Matlab 2016a is utilized for implementing the proposed method on nature Images [14] of 256 × 256 sizc. The performance of the suggested technique is compared with other technique such as DWT, DT-CWT, GA-BR based DT-CWT and EWGW-NN based DT-CWT in terms of MSE (Mean Squared Error) and PSNR (Peak-signal-to-noise-ratio). Figure 3 depicts the experimental results. The comparison of output performance for this method in terms of MSE and PSNR is given in Table 1.

PSNR: It is the ratio of the maximum power of the image and distorting noise power. It can be calculated using Eq. (11).

$$\text{PSNR} = log_{10}\left(\frac{255^2}{MSE}\right)dB \tag{11}$$

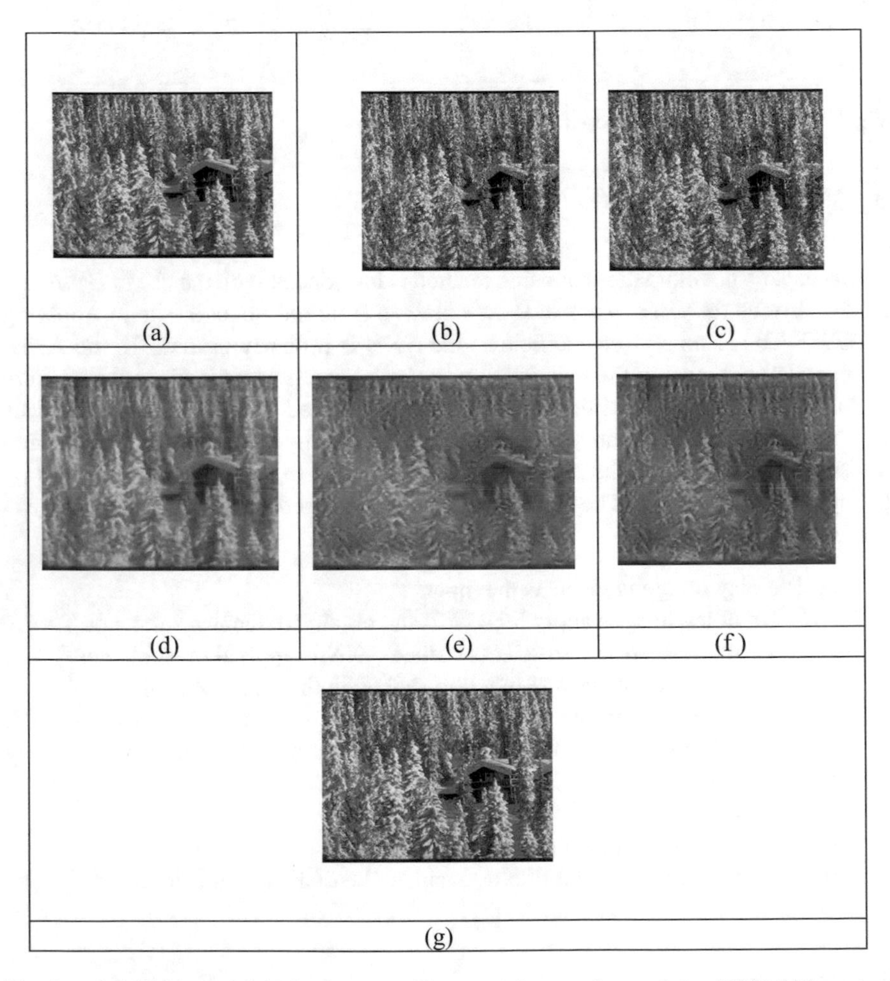

Fig. 3 **a** Original image. **b** Noisy image. **c** The output image after applying DWT **d** The output image after applying DT-CWT **e** The output image after applying. GA-BR based DT-CWT **f** The output image after applying EWGW-ANN based DT-CWT **g** The output image of proposed DT-CWT combined with ANN and GWA

Table 1 MSE and PSNR value comparison for different techniques of denoising in comparison to the proposed approach on 256 × 256 nature image with Gaussian noise variance $\sigma = 20, 40$ and 60

Method	$\sigma = 20$		$\sigma = 40$		$\sigma = 60$	
	PSNR in dB	MSE	PSNR in dB	MSE	PSNR in dB	MSE
DWT	21.80	449.37	20.223	469.23	19.13	476.35
DT-CWT	23.63	223.91	22.521	250.85	20.25	269.32
GA-BR based DT-CWT	24.58	226.50	23.85	245.82	21.78	275.12
EWGW-ANN based DT-CWT	23.83	269.82	22.691	286.32	21.75	295.87
Proposed ANN and GWA Based DTCWT Algorithm	26.69	139.24	25.582	146.54	24.452	189.58

MSE: It is the mean squared error among input and denoised image. It can be calculated using Eq. (12).

$$\text{MSE} = \frac{1}{\hat{N}} \sum_{\hat{i},\hat{j}} \left(\hat{X}_{\hat{i},\hat{j}} - \hat{Y}_{\hat{i},\hat{j}} \right)^2 \tag{12}$$

where the image size is represented by $\hat{N}$ and the distorted or modified and original images are depicted as $\hat{Y}\&\hat{X}$ respectively. The $\hat{i},\ \hat{j}$ represents the $\hat{N}$ pixel position.

The denoised image can be obtained as depicted in Fig. 3 by applying the proposed method and other existing methods with variance 20, 40 and 60 noise levels. Experimental results of various noise levels are tabulated in Table 1.

From these results it can be known that PSNR is improved in this suggested approach after applying ANN and thresholding function by optimized threshold value which is derived from GWA on DT-CWT subbands. This proposed technique outperforms other existing approaches such as DWT, DT-CWT, DT-CWT filter weights optimization using GA and DT-CWT filter weights optimization using EWGW in terms of MSE and PSNR.

5 Conclusion

A novel image-denoising process is proposed in this paper by using DT-CWT with GWA and ANN. ANN has trained to output the noise-free coefficients. The GWA is used to optimize the threshold value. The proposed approach performs well when compared with existing approaches in terms of MSE and PSNR. The experimental results show that the method proposed in this paper has the capability to denoise the images with higher visual quality. This method can be applied for different types of images in the future like medical images and satellite images also.

References

1. Awad A (2019) Denoising images corrupted with impulse, Gaussian, or a mixture of impulse and Gaussian noise. Eng Sci Technol Int J 22(3):746–753
2. Hassan H, Saparon A (2011) Still image denoising based on discrete wavelet transform. In: IEEE international conference on system engineering and technology. Shah Alam, Malaysia, pp. 188–191
3. Sahu S, Singh HV, Kumar B, Singh AK (2019) De-noising of ultrasound image using Bayesian approached heavy-tailed Cauchy distribution. Multimedia Tools Appl 78(4):4089–4106
4. Wang X (2000) Nonlinear multiwavelet transform based soft thresholding. In: IEEE APCCAS 2000. 2000 IEEE Asia-Pacific Conference on Circuits and Systems. Electronic Communication Systems. (Cat. No. 00EX394). IEEE, pp 775–778
5. Zhang S, Salari E (2005) Image denoising using a neural network based non-linear filter in wavelet domain. In: ICASSP 05, IEEE international conference on acoustics, speech, and signal processing
6. Gadekallu TR, Rajput DS, Reddy MPK, Lakshmanna K, Bhattacharya S, Singh S, Jolfaei A, Alazab M (2021) A novel PCA–whale optimization-based deep neural network model for classification of tomato plant diseases using GPU. J Real–Time Image Proc 18(4):1383–1396
7. Gadekallu TR, Alazab M, Kaluri R, Maddikunta PKR, Bhattacharya S, Lakshmanna K, Parimala M (2021) Hand gesture classification using a novel CNN-crow search algorithm. Complex Intel Syst 7:1855–1868
8. Srinivasu PN, Bhoi AK, Jhaveri RH, Reddy GT, Bilal M (2021) Probabilistic deep Q network for real-time path planning in censorious robotic procedures using force sensors. J Real-Time Image Proc 18(5):1773–1785
9. Lavanya PV, Narasimhulu CV, Prasad KS (2020) Dual stage Bayesian network with dual–tree complex wavelet transformation for image denoising. J Eng Res 8(1):154–178. ISSN:2307–1877
10. Lavanya PV, Narasimhulu CV, Prasad KS (2020) Modified Euler Frobenius half band polynomial filter based dual tree complex wavelet transform and metaheuristic optimization algorithms for image denoising with artificial neural networks. Int J Adv Sci Technol 29(6):3475–3496
11. Naimi H, Mitiche ABHA, Mitiche L (2015) Medical image denoising using dual tree complex thresholding wavelet transform and Wiener filter. J King Saud Univ Comp Inform Sci 27(1):40–45
12. Bashar A (2019) Survey on evolving deep learning neural network architectures. J Artif Intell 1(02):73–82
13. Lu C, Gao L, Yi J (2018) Grey wolf optimizer with cellular topological structure. Expert Syst Appl 107:89–114
14. http://www.ultrasoundcases.info/case-list.aspx?cat=26. Accessed on 15 02 2019

Chapter 21
Automated Detection of Multi Class Lung Diseases Using Deep Learning with the Help of X-ray Chest Images

S. R. Likhith and Salma Itagi

1 Introduction

By 2020, a newly imported virus known as coronavirus will have wreaked havoc on every industrialized and developing country on the planet [1]. The Covid-19 coronavirus is a brand-new strain that has never been seen in people previously. Corona is a virus that can infect humans through animals (perhaps mammals and birds). It was suggested in 2019 that bats and pangolins were the carriers of Covid-19 [2]. Covid-19, on the other hand, is presently rapidly spreading from human to human. By January 2022, it had afflicted 4.29 billion individuals worldwide, with the number of cases linked to mortality continuing to rise on a daily basis [3]. Rather than different diseases, the Covid has a long incubating season of 4–14 days. Heartbreakingly, the secondary effect simply appears following 5–6 days. Because they keep meeting and occasionally the symptoms do not show the longer the time span, the greater the risk of Covid-19 spreading to others. Consequently, both the time period and the patients with no symptoms make Covid acknowledgment essentially more irksome, and the conceivable outcomes of transmission spread grow radically. To prevent the growth of the virus illness, it is critical to identify people who are infected on a regular basis. Coronavirus is frequently focused on using a methodology, for instance, switch record polymerase chain reaction (RT-PCR) [4]. However, one disadvantage of this technology is that it takes a long time to test and is expensive. These defects can be overpowered by using radiographic methodologies, for instance, chest x-bar

S. R. Likhith (✉)
Data Science, Nagarjuna College of Engineering & Technology, Bangalore, India
e-mail: likhit.k@ncetmail.com

S. Itagi
CSE, Sai Vidya Institute of Technology, Bangalore, India
e-mail: salma.itagi@saividya.ac.in

G. Mathur et al. (eds.), *Proceedings of 3rd International Conference on Artificial Intelligence: Advances and Applications*, Algorithms for Intelligent Systems,
https://doi.org/10.1007/978-981-19-7041-2_21

pictures, which have exhibited to be a fruitful strategy for recognizing Covid impacts on bronchi on schedule. The chest X-shaft is an essential and shrewd procedure.

The capacities of a couple of advanced pre-arranged CNNs were perused up for the perpetual assessment of Covid-19 including chest radiographs in this audit. By using a little plan of photos, move learning (TL) was picked to achieve high accuracy results. To perceive the Covid-19 patient X–ray of lung from sound chest x-shaft analyzes, different CNN models were attempted. Moreover, perceive individuals who have pneumonia and the people who have Covid-19.

The rest of the paper is organized accordingly. The most current works for Health care detection are described in Sect. 2. In Sect. 3, the proposed methodology is explained. We can see the methodology for growing the system in Area IV. The effects obtained from the developed system are examined in Sect. 5. In Sect. 6, the conclusion is formed about the future scope.

2 Related Work

The making COVID-19 pandemic is tracking down a way or ways of bringing clinical advantages structures starting with one side of the planet and then onto the next to their knees. As of April 2021, a more prominent number of than 140 million admitted cases had been seen around the world, as exhibited by most recent figures. How much events keep on rising, individuals can be more careful expecting that they are given an accurate and lucky examination. Radha [1] utilizes X-Rays, one of the clinical imaging types used to perceive a patient's lung bothering, to see COVID-19 and Pneumonia patients. The CNN (Convolutional Neural Network) model sees Patients with COVID-19 and Pneumonia were seen as using a genuine world dataset of lung X-bar information. Pictures are pre-dealt with and prepared for a course of action of classes, including Normal, COVID-19, and Pneumonia. After pre-dealing with, it's an ideal chance to gather everything, burden ID is achieved by picking sensible elements from each dataset's photographs. The outcome shows the COVID versus Normal and COVID versus Pneumonia conspicuous verification accuracy. Coronavirus versus Normal has a more basic level of accuracy than COVID versus Pneumonia. With an exactness of 80 and 91.46%, this approach COVID and Pneumonia are recognized, as well as subtypes of Pneumonia like bacterial or viral Pneumonia. The proposed model for seeing COVID, Bacterial Pneumonia, and Viral Pneumonia maintains the fast confirmation and social occasion of COVID from Pneumonia and its developments, thinking about the utilization of reasonable and accommodating meds.

Pneumonia is a bacterial or viral lung illness that can be lethal. The early discovery of pneumonia anomalies in the lungs can save a child's or an elderly lady's life, especially if caught early. The size of these divergences is really small. Youssef et al. [2] suggested a new architecture based on ResNet50 and medical pictures of the chest to identify infectious pneumonia cases. When compared to various prior scientific studies and radiologists' hopes, the anticipated results are highly interesting (97, 65%).

This survey uses the textural attributes Researchers used the Gray Level Occurrence Matrix to endeavor to recognize and orchestrate pneumonia into three stages: delicate, moderate, and outrageous (GLCM). Deepika et al. [3] joins 30 individuals with delicate, moderate, and outrageous lobar pneumonia. (10 delicate, 10 moderate, and 10 limit). A CT result of the lungs was performed with respect to all matters using a Philips MX8000 IDT 16 cut CT scanner and saved in DICOM plan. Center filtering was used to take out upheaval from the data, and contrast stretching out was used to normalize it. The GLCM was used to remove textural contrast, energy, most prominent probability, change, mean, skewness, entropy, standard deviation, autocorrelation, center, mode, bunch observable quality, pack shadow, homogeneity, and kurtosis are among the properties that can be found. A neural association was used to arrange the data. The textural features contrast, energy, mean, standard deviation, autocorrelation, and center have quantifiably colossal differentiations at the level p 0.05, as shown by the revelations of the post hoc test (tukey HSD).

For diagnosing pneumonia, chest radiography has become the tool of choice. Analyzing chest X-ray images, on the other hand, can be difficult, time-consuming, and require professional knowledge that isn't always available in less-developed areas. As a result, computer-assisted diagnostic systems are required. Deep transfer learning provides the advantage of lowering development costs by borrowing structures from trained models and fine-tuning a few layers. However, in many situations, whether deep transfer learning is more effective than training from scratch in the healthcare context is still a research subject. Irfan et al. [4] look at how deep transfer learning can be used to classify pneumonia in chest X-ray pictures. Deep transfer learning, with minor fine-tuning, provides a performance advantage over training from scratch, according to experimental results. ResNet-50, Inception V3, and DensetNet121 were all trained separately from scratch and through transfer learning. The former can produce a 4.1–52.5% greater area under the curve (AUC) than the latter, implying that deep transfer learning is useful for diagnosing pneumonia in chest X-ray data.

Another method of detecting Covid-19 disease is to use deep learning models based on Convolutional Neural Networks to distinguish between Covid-19 and other lung ailments like various types of Viral Pneumonia or Bacterial Pneumonia using pictures from chest X-Rays. On a test dataset of 300 photos, the models constructed as part of this paper[5] achieved roughly 98% accuracy. Both AlexNet and 4 Layer CNN are 98% accurate.

The most common symptom of Mycoplasma pneumonia, a rare bacterial pneumonia, is a dry cough. It's conceivable that it'll coexist with COVID-19, which was identified recently. It's tough to tell the difference between COVID-19 and Mycoplasma pneumonia, a condition that can cause pneumonia. Due to the similar symptoms, distinguishing between mycoplasma pneumonia and typical viral pneumonia can be challenging. The purpose of [6] is to make diagnosing more straightforward. You can apply a variety of deep learning algorithms on computed tomography (CT) scans to categorize them as having mycoplasma pneumonia, viral pneumonia, or COVID-19. According to the outcomes of this study, the ResNet-18 and MobileNet-v2 designs function effectively during sickness discrimination.

The World Health Organization (WHO) declared this virus a global pandemic after it infected millions of people and claimed countless lives around the world. An infection caused by Covid-19 illness wreaks havoc on the human pulmonary system, leading to several organ failures and, in the worst-case scenario, death. For the aim of feature extraction, chest radiographs were fed into multiple deep learning CNN architectures in [7]. The dataset images were fed into the VGG16, VGG19, and Inception v3 models for feature extraction. The data was loaded into a succession of machine learning classifiers that categorized the chest radiographs as Covid-19 positive, pneumonia infection, or healthy scans after feature extraction.

Pneumonia is an irresistible lung disorder that essentially impacts individuals. Untreated pneumonia can make certifiable intricacies in the more established (north of 65 years old) and youths (under 5 years old), subsequently early acknowledgment and treatment are consistently urged to help sullied patients with recovering. Considering its clinical importance, lungs sicknesses are routinely perceived through chest radiography (X-shaft). The significant learning (DL) method for managing and recognize pneumonia is portrayed in [8]. The DL plan's ailment area execution is attested by a twofold course of action result using the SoftMax classifier unit. During this assessment, 2000 photographs (1000 sound and 1000 pneumonia) are penniless down, and execution estimations are made to confirm the revelations. The accuracy of AlexNet's preliminary outcomes on the considered photo database was >98%.

Pneumonia can cause lung tissue rottenness and a lung bubble. The spread of a pneumonia sickness could achieve different results. Pneumonia can loosen up to the pleural cavity, achieving empyema. Pleurisy, endocarditis, pericarditis, joint agony, and various hardships can result. It is fundamental to see pneumonia precisely. This investigation offers Depth-Wise Convolution Consideration of neural association (DWA) and an arrangement of modified pneumonia affirmation estimations considering significant learning advancement and the morphology of the lungs. Target disclosure models like Vgg16, DenseNet, and ResNet can be taught to perform course of action endeavors. During the planning time span, [9] uses comparable educational assortment and uses comparable readiness limits. The VGG16, DenseNet, ResNet, and DWA models were evaluated using a uniform appraisal rule. The test revelations suggest that DWA has the best request execution (planning accuracy of 97.5%, testing precision of 96% and endorsement accuracy of 79%).

In the human respiratory structure, the lungs are the most crucial organs. Our lungs, of course, are delicate organs that are instantly wounded by disturbance or impact wounds in our everyday schedules. Due to the growing advancement of certified and connected events with COVID-19 pneumonia, which consistently outperforms the limits of clinical associations, speedy and careful finding for patients has transformed into a first concern. Hence, ultrasound pictures have since they are more invaluable, adaptable, more reasonable, and don't make ionizing radiation than CT and CXR, they have begun to be used in lung finding. This paper attempts to use VGG, ResNet, and EfficientNet associations to unequivocally group Lung Ultrasound pictures of pneumonia according to specific clinical stages using free LUS datasets. The three associations' hyperparameters were changed, and their results were carefully checked out. Zhang et al. [10] show that the EfficientNet model defeated the others, with portrayal correctnesses of 94.62 and 91.18% for clinical stages 3 and 4 of pneumonia, independently. The best request accuracy for eight pneumonia imagological features is 82.75%. This review shows the LUS device's promising potential for use in pneumonia assurance, as well as the good judgment of significant learning for LUS characterization of pneumonia.

Significant learning strategies are frequently used in clinical medications and diagnostics, and they have applications in an arrangement of disciplines. Chest X-shaft pictures are routinely assessed to perceive infections like pneumonia, and the adequacy of findings can be amazingly updated with the use of PC upheld demonstrative devices. Srivastav et al. [11] uses significant learning systems to orchestrate chest X-pillar pictures to examine pneumonia. To deal with the model's display, significant convolutional generative hostile associations were ready to improve fake pictures and oversample the dataset. Then, including VGG16 as the foundation model for picture portrayal, move learning was applied with convolutional neural associations. On the endorsement set, the model had a 94.5% accuracy rate. The precision of the proposed model was considered to be significantly more essential than that of the unsophisticated models.

The photos reveal pathology of pneumonia conveyed by microorganisms and contaminations, according to the investigation of picture gathering computations. The proposed system in [12] relies upon isolating data credits and setting up the model course of action using the VGG16, VGG19, and DenseNet169 associations. Common individuals, patients with viral pneumonia, and patients with bacterial pneumonia are completely gathered in the X-radiates. Clinical data from chest X-pillar pictures of patients that were truly gathered by specialists was given as the source. Regardless, the portrayal accuracy is by and largely dependent upon how much photographs, picture objective, and accepting the X-shaft picture is unequivocally seen. The computations in this review produce commonly certain characterization disclosures, with a precision of around 85%.

In [13], an exceptionally gathered significant learning model for diagnosing pneumonia conditions by exploring radiographs is promoted. The hybrid CNN model has been arranged to sort unmistakable evidence of pneumonia into three (3) groupings: bacterial, normal, and viral pneumonia x-bar pictures. The proposed hybrid CNN procedure, which involves various convolution blocks with unequivocal burdens and different totally associated layers for careful request, was used in the preliminaries. In stood out from various models in this survey, the proposed significant learning model achieved an accuracy of 92.9%, making it the most raised situating model.

Artificial intelligence is becoming more prevalent in our daily lives. Convolutional neural networks (CNN) are a promising and emerging technique in the field of medical image processing, with the potential to make diagnoses easier and more accurate. Accurate diagnosis is critical in determining the most appropriate and effective treatment. This research presents a self-constructed convolutional neural network for lung X-ray image classification that was trained on a short dataset. This CNN allows patients to be classified into one of three groups: healthy, bacterial pneumonia, or viral pneumonia. Such a categorization, which takes pneumonia into account, is unusual in scholarly papers. A comparative investigation of the degree of impact of data augmentation on the model's performance and overfitting prevention was also carried out. The category classification accuracy was found to be 85%, with a sensitivity of 0.95. Such outcomes bode well for future research and development.

Due to the COVID-19 Pandemic, experts should make clinical decisions for their patients considering a combination of tests. Transfer learning, of course, has been used in different assessments and spotlights on only one kind of biomarker (e.g., CT-Scan or X-Ray) for Pneumonia assurance. As shown by the assessment, each procedure has its own gathering accuracy, and each biomarker for COVID-19 Pneumonia recognizable proof could give additional information. A CT result or X-shaft imaging of the chest can reveal the COVID-19 disease. Hilmizen et al. [14] merges two separate trade learning procedures. Models were worked to arrange CT-Scan and X-pillar pictures into two groupings: common and COVID-19 Pneumonia, using an open-source dataset containing 2500 CT-Scan and 2500 X-bar pictures. The going with associations were used: DenseNet121, MobileNet, Xception, and InceptionV3. In our investigation, we used the ResNet50 and VGG16 picture affirmation models. As a result, the connection of ResNet50 and VGG16 networks achieves the best course of action precision of 99.87%. We moreover had the most critical game plan precision of 98.00% while using a singular system of CT-Scan ResNet50 associations and 98.93% while using X-Ray VGG16 associations. When stood out from the philosophy of using a single technique of biomarkers, our multimodal mix methodology has a higher portrayal precision.

A get-together of viral pneumonia cases commonly through a short period of time could hail the start of a discharge or pandemic. Expedient and dependable assessment of viral pneumonia using chest X-bars can be of colossal benefit for enormous extension screening and scourge balance when other further evolved imaging modalities are not speedily open. The move of novel changed contaminations, clearly, causes a fundamental dataset shift, which can truly confine the show of requesting based approaches [15]. To deal with this, we convert the task of withdrawing viral pneumonia from non-viral pneumonia and sound controls into a one-class gathering based abnormality insistence issue. A typical component extractor, an anomaly revelation module, and a sureness assumption module are proposed in the conviction careful abnormality area (CAAD) model. The information will be considered as an odd case accepting the irregularity score obtained by the idiosyncrasy revelation module is adequately tremendous, or the sureness score surveyed by the conviction assumption module is pretty much nothing (i.e., viral pneumonia). We make the important strides not to unequivocally show express popular pneumonia classes and treat all recommended viral pneumonia cases as whimsies, which is a massive advantage of our technique over twofold solicitation. On the clinical X-VIRAL dataset, which contains 5,977 viral pneumonias (no COVID-19) cases, 37,393 non-viral pneumonia or strong cases, the supported model outfoxes twofold portrayal ways to deal with managing further develop the one-class mode. Besides, with close to no changing, our model achieves an AUC of 83.61% and responsiveness of 71.70% when tried clearly on the X-COVID dataset, which contains 106 COVID-19 cases and 107 standard controls, which looks like the presentation of radiologists portrayed in the design.

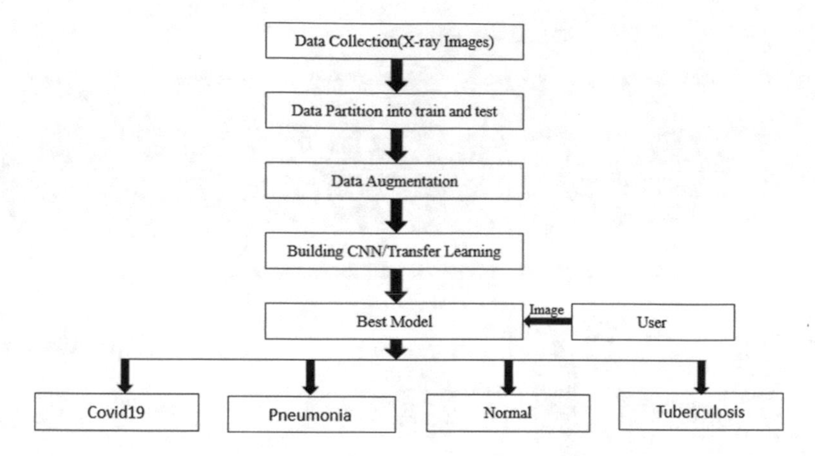

Fig. 1 Phases of proposed model

3 Proposed System

Proposing a Five stage structure to classify the X-ray image of lung as shown in Fig. 1. The first stage is Data collection during which the dataset with four classes is collected form the kaggel repository. In second stage the dataset is partitioned into 3 sub folders namely Train, Test and Validation. In third stage Data augmentation is done on the dataset to improve dataset number and to consider all possibility of data to extract even complex feature from the input images. In fourth stage ResNet, InceptionNet and XceptionNet were trained and new simple CNN architecture called CPT Net is built and trained on the dataset. In Fifth stage, the best architecture which is achieving high accuracy from ResNet, InceptionNet, XceptionNet and CPT Net is selected and the user given input X-image will be passed to the model to classify the image into one of the four classes.

4 Methodolgy

4.1 Dataset

Dataset for this study is collected from the kaggel repository. And the dataset which is considered for this study is X-ray images of patients. X-ray images plays a very important role in the medical field to identify any kind of lung diseases. The dataset is partitioned into three sub folders which are training, testing and validation. In each and every sub folder there are four sub folders which represents four classes of the lung disease namely Covid-19, Pneumonia, Tuberculosis and Normal. Covid-19, Pneumonia, Tuberculosis and Normal are four categories of the dataset which need

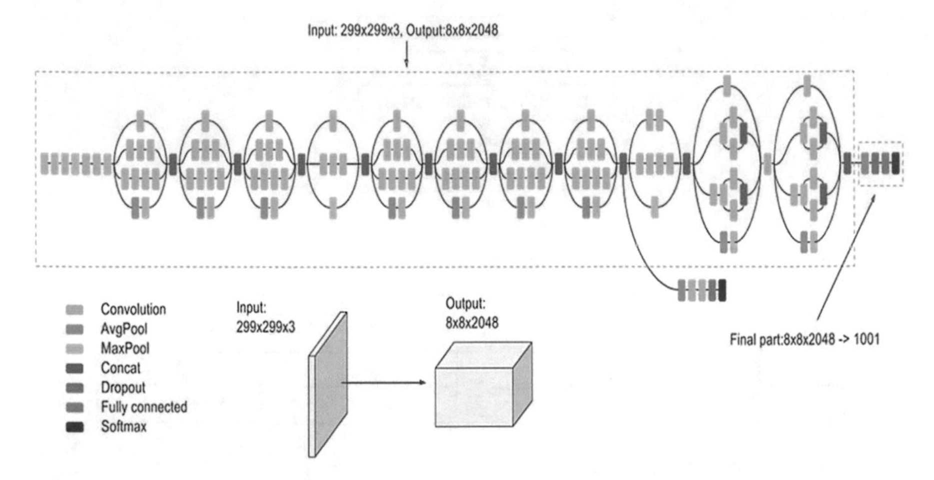

Fig. 2 Architecture of InceptionNet

to be classified. The dataset contains a total of 7135 X-ray images which belongs to all four categories of lung disease mentioned above.

4.2 *InceptionNet*

An inception block [4] in InceptionNet performs convolution utilizing three different filter sizes on an input (1 × 1, 3 × 3, 5 × 5). Additionally, max pooling is carried out. Following that, the results are pooled and forwarded to the inception module. Before the 3 × 3 and 5 × 5 convolutions, a 1 × 1 convolution is employed to limit the number of enter channels. Despite the fact that including a larger operation may appear to be more difficult, 1 × 1 convolutions are significantly less expensive than 5 × 5 convolutions, as they reduce the number of channel inputs. Keep in mind, though, that the 1 × 1 convolution is done after the max pooling layer, rather than before it. As seen in Figs. 2 and 9 inception blocks are stacked in a linear fashion, resulting in a depth of 22 layers.

4.3 *XceptionNet*

XceptionNet is a severe model of Inception, with a modified depth wise separable convolution [5]. It makes use of Depthwise Convolution, in which alternatively of making use of convolution of dimension d × d × Cd × d × C, we observe a convolution of dimension d × d × 1d × d × 1. In different words, we don't make the convolution computation over all the channels, however solely 1 by using 1. And a

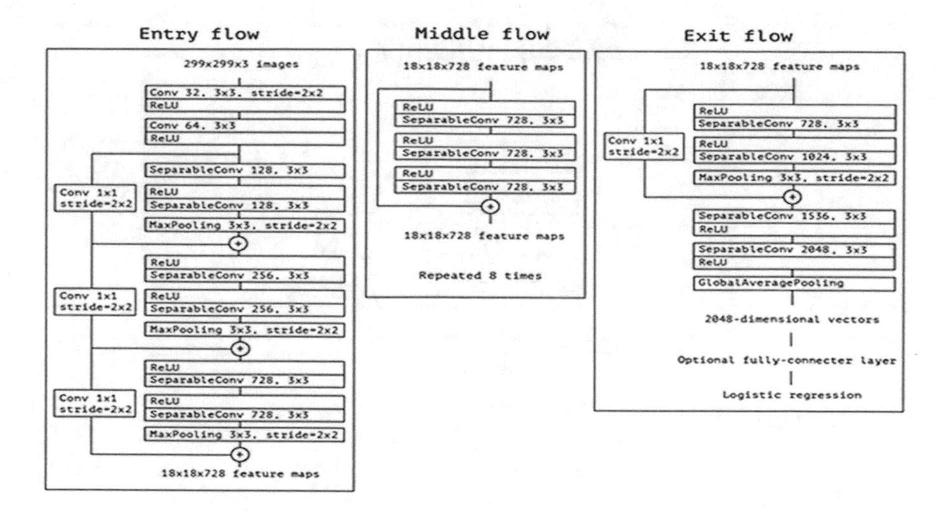

Fig. 3 Architecture of XceptionNet

Pointwise convolution, with measurement 1 × 1 × N over the K × K × C volume. This permits the growth of a KKN-like extent. As found in Fig. 3, the records starting pass through the entering stream, then, through the center float, which is reiterated on numerous occasions, and last through the leave float. Cluster screens all Convolution and Separable Convolution layers. In all Separable Convolution layers, a significance multiplier of 1 is applied.

4.4 ResNet

ResNet is proposed by Microsoft research in 2015. Gradient disappearing is one of the problems when we deal with the convolutional neural networks. To resolve this problem residual blocks were introduced in ResNet which uses skip connections which is as shown in the Fig. 4. In convolutional neural network each and every layer is expected to learn some features, in this process some of the layers may not learn well or might not extract the feature properly which will lead to low performance or low accuracy of the model. To overcome this problem skip connection are introduced in the ResNet which allows the model to skip those layers with did not learn well and pass the previous layer's output as input to the next layer by doing this accuracy of the model r the architecture will increase.

Residual Networks

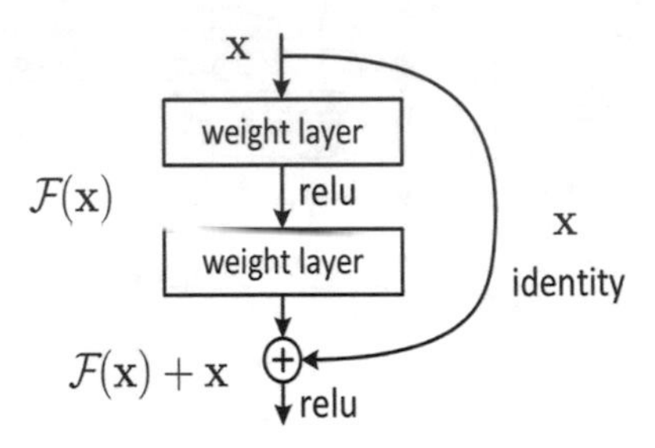

Fig. 4 Residual blocks

4.5 CPT Net

CPT Net architecture is shown in the Fig. 5. CPT Net have 10 layers. First two layer are convolutional layer with kernel size 3*3 and the activation function used is relu where the first convolutional layer takes the X-ray image of size 150*150 as input. The first two layers are followed by a maxpool layer with the kernel size 2*2. The fourth layer is a convolutional layer with same hyper parameters that were used in 1st and second layers. Fifth layer is maxpool layer with kernel size 2*2 followed with a convolutional layer. Seventh layer is where all the extracted features from the convolution will be flattened namely called flatten layer. The output of the flattened layer will be passed to dense layer. There are three dense layers in this architecture where the last dense layer has 4 units or neurons with softmax as an activation function since the model or the architecture has to classify 4 classes.

4.6 Transfer Leraing

Transfer studying is a computer gaining knowledge of approach, the place a mannequin skilled on one project is re-purposed on a 2nd associated challenge [8]. We pick a pre-trained supply mannequin which is chosen from reachable models. Many lookup establishments launch fashions on massive and difficult datasets that may also be protected in the pool of candidate fashions from which to select from. The mannequin pre-trained mannequin can then be used as the beginning factor for a mannequin on the 2d project of interest. This can also contain the usage of all or components of the model, relying on the modeling approach used. Optionally, the mannequin may also want to be tailored or sophisticated on the input–output pair information on hand for the assignment of interest.

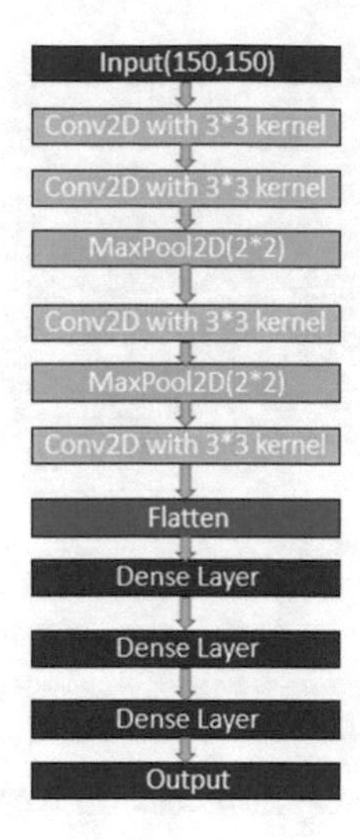

Fig. 5 CPT net architecture

5 Results and Analysis

Figure 6 show the number of instances in each class from the training data set.

Figure 7 show the number of instances in each class from the testing data set.

Below table will give us the training accuracy and testing accuracy of three architectures namely Resnet, InceptionNet and XceptionNet which are performed with the help of transfer learning.

Architecture	Training_accuracy	Val_accuracy
ResNet	94.15	92.19
InceptionNet	98.52	88.35
XceptionNet	93.09	87.07

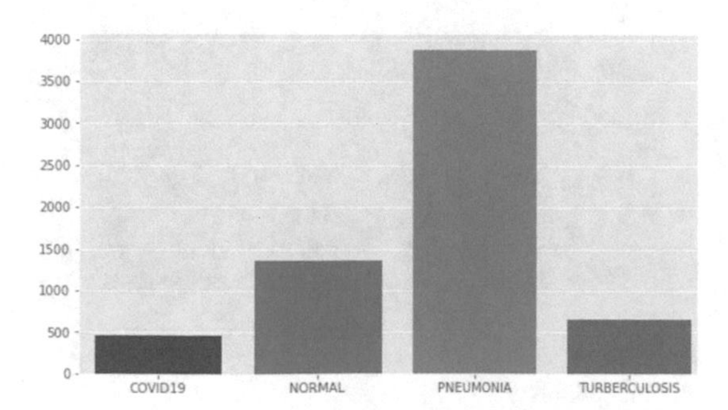

Fig. 6 Number of instances in training data set

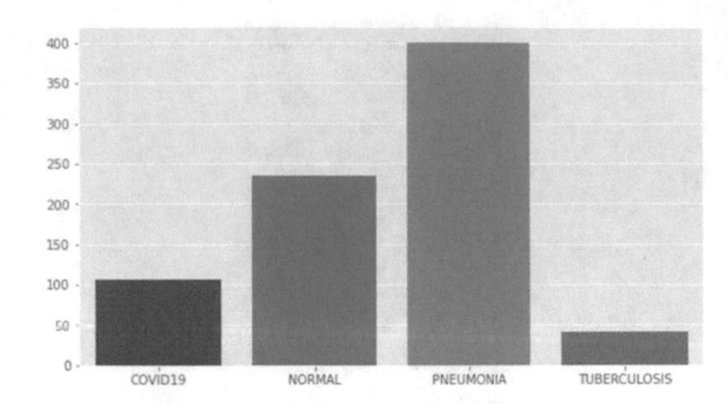

Fig. 7 Number of instances in test data set

Loss and accuracy plots for InceptionNet and XceptionNet.

InceptionNet achieved a Training and Validation accuracy of 98.52 and 88.35%, respectively, as shown in Figs. 8 and 9, and XceptionNet achieved a Training and Validation accuracy of 93.09 and 87.07%, respectively, as shown in Figs. 10 and 11 and ResNet achieved a Training and Validation accuracy of 94.15 and 92.19%, respectively, as shown in Figs. 12 and 13. The data collection is split into 90:10 proportions of training and test data with Adam optimizer and Cross specific entropy loss feature.

Confusion matrix of InceptionNet, XceptionNet and ResNet

In the following Confusion matrix 0 represents Covid-19, 1 represents Normal, 2 represents Pneumonia and 3 represents Tuberculosis. Figure 14 shows the confusion matrix of InceptionNet where 99 are correctly classified as Covid-19, 82 are correctly classified as normal, 400 are correctly classified as pneumonia and 41 are

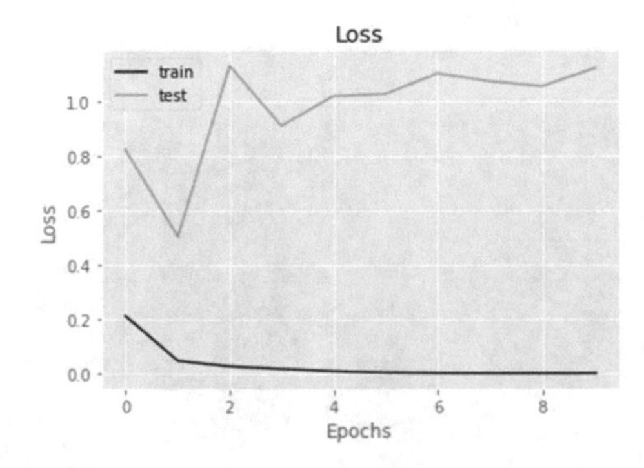

Fig. 8 Plots of loss for InceptionNet

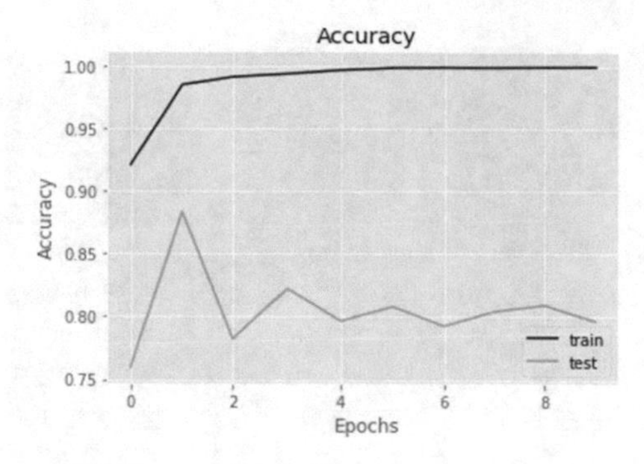

Fig. 9 Plots of accuracy for InceptionNet

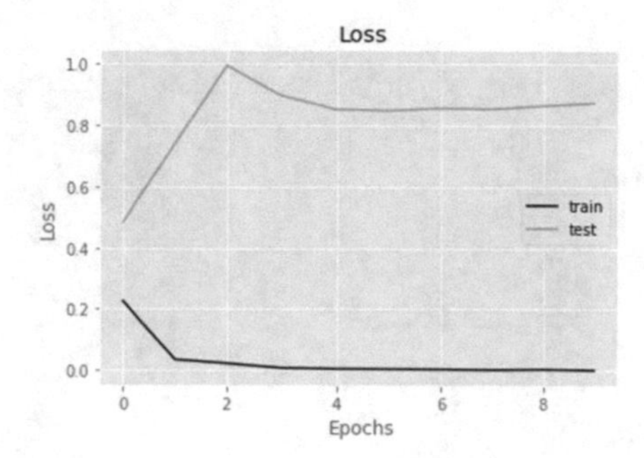

Fig. 10 Plots of loss for XceptionNet

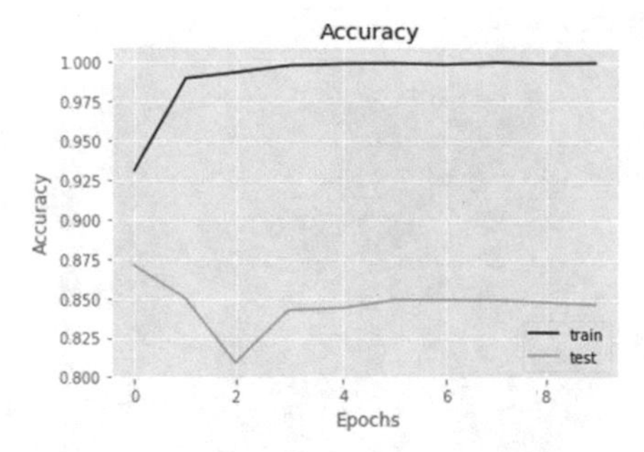

Fig. 11 Plots of accuracy for XceptionNet

correctly classified as tuberculosis. Figure 15 shows the confusion matrix of XceptionNet where 106 are correctly classified as Covid-19, 114 are correctly classified as normal, 400 are correctly classified as pneumonia and 41 are correctly classified as tuberculosis. Figure 16 shows the confusion matrix of ResNet where 106 are correctly classified as Covid-19, 107 are correctly classified as normal, 399 are correctly classified as pneumonia and 40 are correctly classified as tuberculosis.

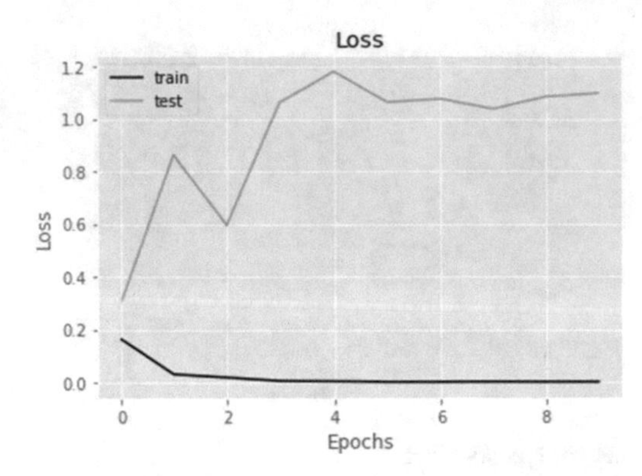

Fig. 12 Plots of loss for ResNet

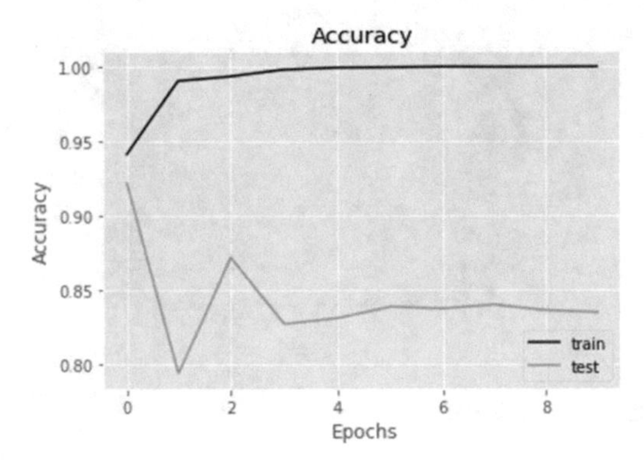

Fig. 13 Plots of accuracy for ResNet

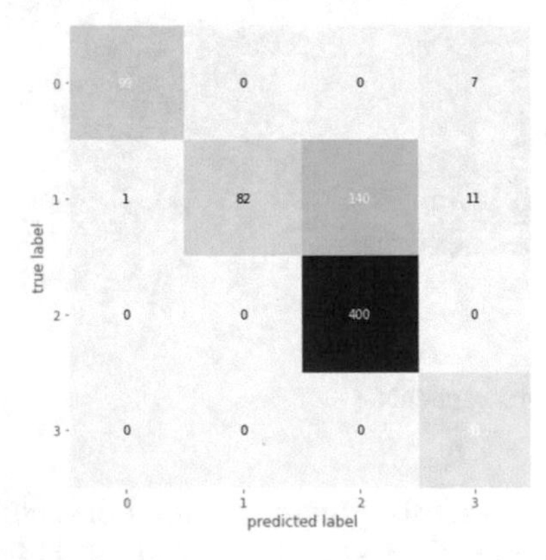

Fig. 14 Confusion matrix of InceptionNet

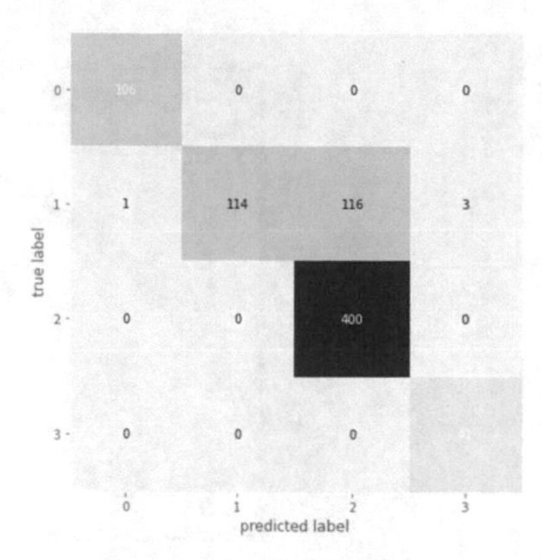

Fig. 15 Confusion matrix of XceptionNet

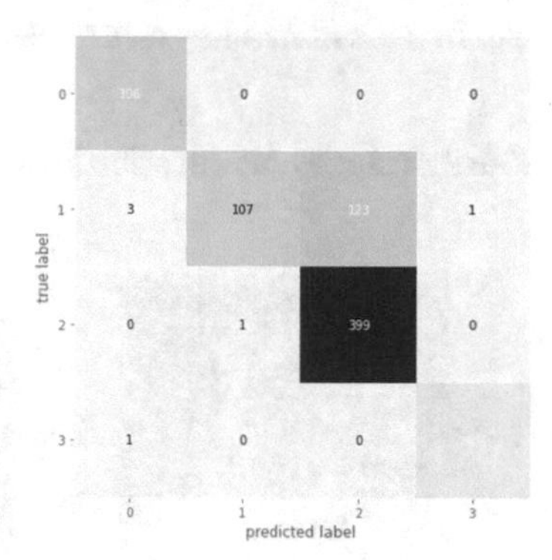

Fig. 16 Confusion matrix of ResNet

Loss and accuracy plots for CPT Net

InceptionNet achieved a Training and Validation accuracy of 91.09 and 76.36%, respectively, as shown in Figs. 17 and 18.

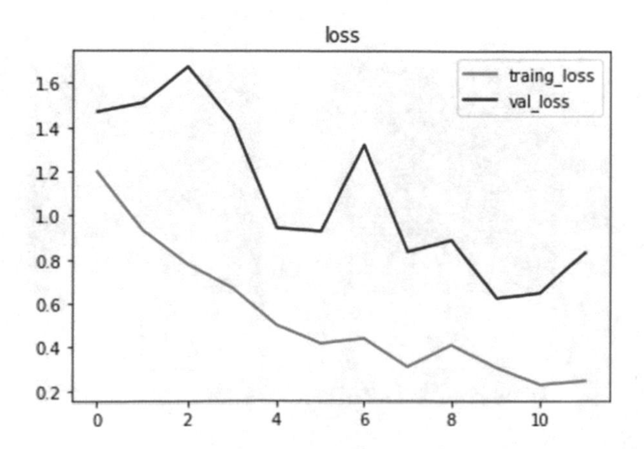

Fig. 17 Plots of loss for CPT Net

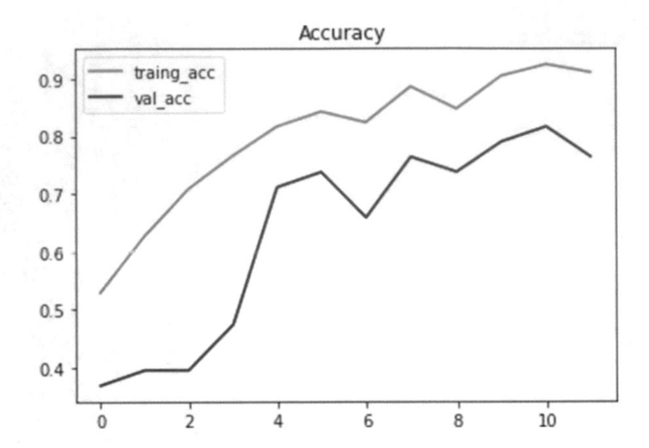

Fig. 18 Plots of Accuracy for CPT Net

X-Ray Image classification results

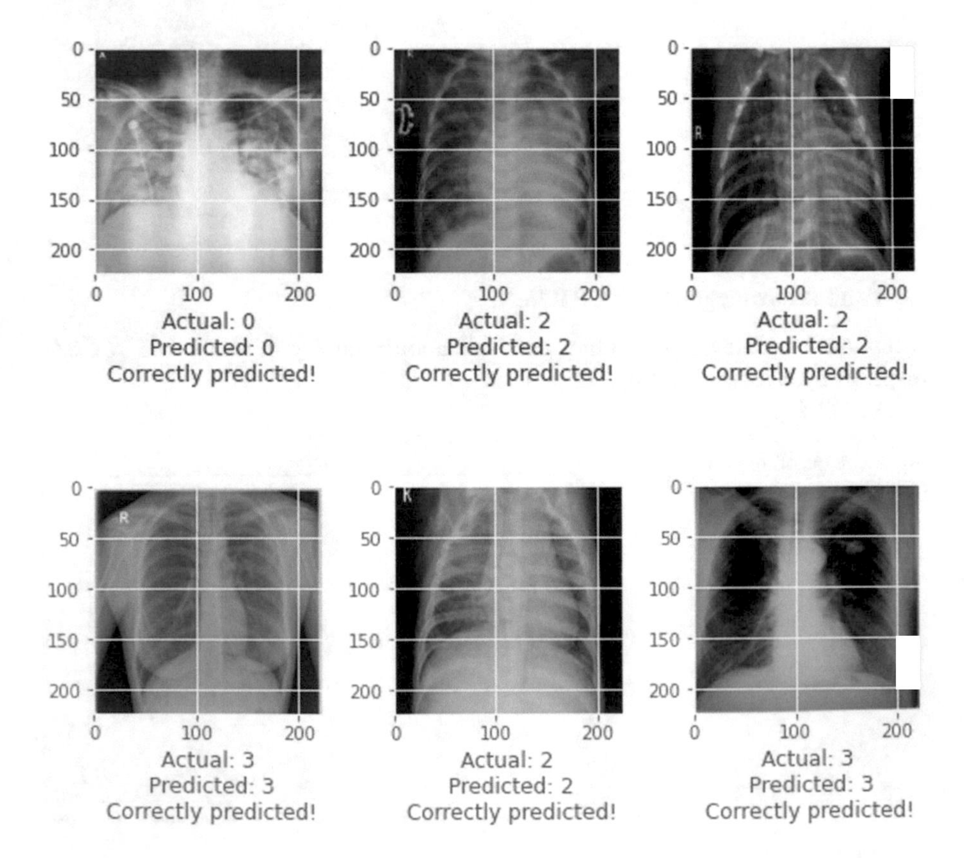

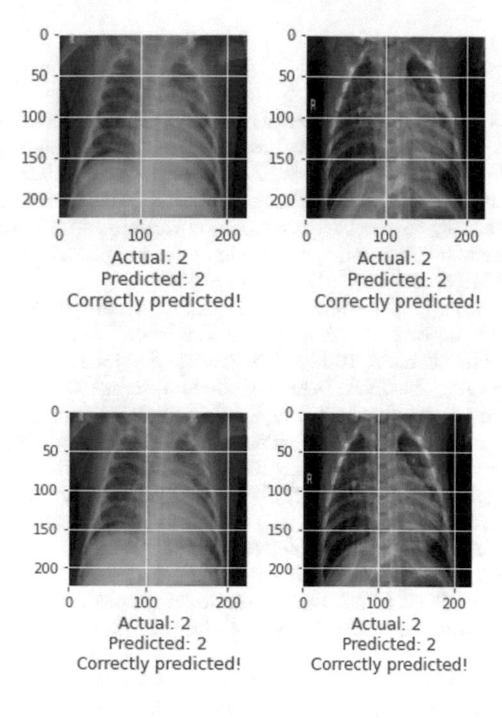

6 Conclusion and Future Scope

Different deep learning models were exhibited for categorizing the X-Ray images into classes such as Covid-19, Pneumonia, Tuberculosis or Normal. The images from the dataset were fed into the model for feature extraction. These findings show that when a fresh image is an input to the Network, it can correctly identify data as Covid-19, normal, tuberculosis, or pneumonia patients with an accuracy of 94.15, 98.52, 93.09, and 91.09%, respectively; for architectural designs Inception v3, Xception, ResNet, and CPNet are the different versions utilized. These models, however, cannot be implemented directly. They necessitate a correct clinical diagnosis. To improve the accuracy of the dataset, we will upload more data in the future. In addition, the current dataset can be used to train multiple deep learning models to improve the identification of Covid-19 infections.

References

1. Radha D (2021) Analysis of COVID-19 and pneumonia detection in chest X-ray images using deep learning. In: 2021 International Conference on Communication, Control and Information Sciences (ICCISc. pp 1–6. https://doi.org/10.1109/ICCISc52257.2021.9484888
2. Youssef TA, Aissam B, Khalid D, Imane B, Miloud JE (2020) Classification of chest pneumonia from x-ray images using new architecture based on ResNet. In: 2020 IEEE 2nd international conference on electronics, control, optimization and computer science (ICECOCS). pp 1–5. https://doi.org/10.1109/ICECOCS50124.2020.9314567
3. Deepika N, Vinupritha P, Kathirvelu D (2018) Classification of lobar pneumonia by two different classifiers in lung CT images. Int Confr on Commun Signal Proc (ICCSP) 2018:0552–0556. https://doi.org/10.1109/ICCSP.2018.8524384
4. Irfan A, Adivishnu AL, Sze-To A, Dehkharghanian T, Rahnamayan S, Tizhoosh HR (2020) Classifying pneumonia among chest X-rays using transfer learning. In: 2020 42nd Annual International Conference of the IEEE Engineering in Medicine & Biology Society (EMBC). pp 2186–2189. https://doi.org/10.1109/EMBC44109.2020.9175594
5. Panwar A, Dagar A, Pal V, Kumar V (2021) COVID 19, Pneumonia and other disease classification using chest X-ray images. In: 2021 2nd International Conference for Emerging Technology (INCET). pp 1–4. https://doi.org/10.1109/INCET51464.2021.9456192
6. Serener A, Serte S (2020) Deep learning for mycoplasma pneumonia discrimination from pneumonias like COVID-19. In: 2020 4th international symposium on multidisciplinary studies and innovative technologies (ISMSIT). pp 1–5. https://doi.org/10.1109/ISMSIT50672.2020.9254561
7. Panwar A, Yadav R, Mishra K, Gupta S (2021) Deep learning techniques for the real time detection of Covid19 and pneumonia using chest radiographs. In: IEEE EUROCON 2021–19th international conference on smart technologies. pp 250–253. https://doi.org/10.1109/EUROCON52738.2021.9535604
8. Arunmozhi S, Rajinikanth V, Rajakumar MP (2021) Deep-learning based automated detection of pneumonia in chest radiographs. In: 2021 International conference on system, computation, automation and networking (ICSCAN). pp 1–4. https://doi.org/10.1109/ICSCAN53069.2021.9526482
9. Wan S, Hsu C.-Y, Li J, Zhao M (2020) Depth-wise convolution with attention neural network (DWA) for pneumonia detection. In: 2020 International conference on intelligent computing, automation and systems (ICICAS). pp 136–140. https://doi.org/10.1109/ICICAS51530.2020.00035
10. Zhang J, et al. (2020) Detection and classification of pneumonia from lung ultrasound images. In: 2020 5th international conference on communication, image and signal processing (CCISP). pp 294–298. https://doi.org/10.1109/CCISP51026.2020.9273469
11. Srivastav D, Bajpai A, Srivastava P (2021) Improved classification for pneumonia detection using transfer learning with GAN based synthetic image augmentation. In: 2021 11th International Conference on Cloud Computing, Data Science & Engineering (Confluence). pp 433–437. https://doi.org/10.1109/Confluence51648.2021.9377062
12. Thanh HT, Yen PH, Ngoc TB (2020) Pneumonia classification in X-ray images using artificial intelligence technology. Appl New Technol Green Build (ATiGB) 2021:25–30. https://doi.org/10.1109/ATiGB50996.2021.9423017
13. Abubakar MM, Adamu BZ, Abubakar MZ (2021) Pneumonia classification using hybrid CNN architecture. International Conference on Data Analytics for Business and Industry (ICDABI). pp 520–522. https://doi.org/10.1109/ICDABI53623.2021.9655918
14. Hilmizen N, Bustamam A, Sarwinda D (2020) The multimodal deep learning for diagnosing COVID-19 pneumonia from chest CT-scan and X-ray images. In: 2020 3rd international seminar on research of information technology and intelligent systems (ISRITI). pp 26–31. https://doi.org/10.1109/ISRITI51436.2020.9315478

15. Zhang J et al (2021) Viral pneumonia screening on chest X-rays using confidence-aware anomaly detection. IEEE Trans Med Imaging 40(3):879–890. https://doi.org/10.1109/TMI.2020.3040950
16. Szegedy C, Liu W, Jia Y (2015) Going deeper with convolutions. 978–1–4673–6964–0/15/$31.00
17. Chollet F (2017) Xception: deep learning with depthwise separable convolutions. In: 2017 IEEE conference on computer vision and pattern recognition. Google, Inc.
18. Han X, Jin R (2020) A small sample image recognition method based on resnet and transfer learning. 978–1–7281–6042–9/20/$31.00. https://doi.org/10.1109/ICCIA49625.2020.00022

Chapter 22
Classification of Electrocardiogram Using Color Images with Pixel Method by Deep CNN

A. H. M. Zadidul Karim, Md. Badeuzzamal Sarker, Md. Rafiqul Alam Rejon, Md. Saimun Islam, Md. Rafatul Alam Fahima, and Md. Sazal Miah

1 Introduction

For a long time, electrocardiograms (ECGs) have been used to think about heart conditions. Exploring their patterns has been found to help diagnose heart diseases (Chng et al. [1]; Lee et al. [2]). The use of computer-based technologies for automated ECG signal processing may aid in diagnosis. Many approaches for detecting ARR, NSR, and CHF have been presented during the last few decades (Llamedo and Martínez [3]; de Chazal et al. [4]; Kutlu et al. [5]; de Chazal and Reilly [6]). Categorizing heartbeats using ECG data is a critical boosting technology for wearable health solutions.

It's been a long time since many ECG-based methods for figuring out what kind of ECG it is have been proposed (Islam et al. [7]). Prepossessing, feature extraction, and classification are the three key processes of a model in this category. The techniques for feature extraction and classification used by these models vary from one another. Morphology, temporal information, high-order statistics, Hermite basis functions, and convolutional neural networks are all examples of feature extraction models. On the PTB Diagnostic ECG Database, a model with 99.3% accuracy (Strodthoff et al. [8]) was described, in which CNN was utilized for feature extraction, and a support vector machine was employed for classification.

The appropriate selection of fiduciary factors like QRS complex and ST segment is essential for ECG analysis and categorization. Feature engineering is the technique

A. H. M. Z. Karim · Md. B. Sarker (✉) · Md. R. A. Rejon · Md. S. Islam · Md. R. A. Fahima
Department of Electrical and Electronic Engineering, University of Asia Pacific, Dhaka, Bangladesh
e-mail: 17208018@uap-bd.edu

Md. S. Miah
School of Engineering and Technology, Asian Institute of Technology, Pathum Thani 12120, Thailand

G. Mathur et al. (eds.), *Proceedings of 3rd International Conference on Artificial Intelligence: Advances and Applications*, Algorithms for Intelligent Systems,
https://doi.org/10.1007/978-981-19-7041-2_22

of detecting fiducial points. Traditionally, Feature Engineering is done by a doctor personally viewing the ECG graph and making a diagnosis.

Machine learning techniques can be used to find T wave anomalies and ST segment in ECGs with the sensitivity of 86% (Zhang et al. [9]). ST-deviation was detected by backpropagation neural network. The deviation is calculated by deducting the identified ST segment from the beat's iso-electric level. They accomplished a sensitivity of 90.75% (Smisek et al. [10]). The basic support vector machine, sometimes known as a binary classifier, is a kind of administered learning algorithm that divides data into two categories using isolating hyperplanes.

As used by the authors (Smisek et al. [10]; Xu et al. [11]), Support Vector Machines like Complex Support Vector Machine (CSVM) and Multi-class Support Vector Machine (MSVM) may be utilized to categorize ECG arrhythmia varieties into numerous groups. On the MITBIG and ESCDB databases, the authors used the Multi-Module NN (MMNNS) to identify S and V heart beats.

Many ways to use a CNN to classify heartbeats with arrhythmias (Ye et al. [12]; de Chazal et al. [4]) have been suggested. The authors refined the two-dimensional (2D) CNN utilized for pictures and introduced the first 1D CNN for automated heartbeat categorization. Oversampling was used to balance the five kinds of heartbeats, and a nine-layer CNN structure was developed. Another author created a dual heartbeat coupling matrix by connecting two additional ECG heartbeats (Xu et al. [11]) data, then utilized as an input sample and passed to a 7-layer 2D CNN for sorting.

We came up with a novel neural architecture depending on the current rise in acceptance of CNN to make an automatic heart-related disease treatment system that uses ECG signal to make the diagnosis. ECG signals, in particular, were studied and sent on to a CNN network that has been appropriately educated. The ECG is analyzed. ECG may give useful diagnostic information for detecting several sorts of heart problems, arrhythmia. ECG monitoring systems that use powerful machine learning technologies in real-time give real-time health information and have boosted the user's confidence. As a three-layer ECG signal analysis, this suggested pattern is shown. The pattern may be used to monitor real-time portable and wearable devices. We also employed the CWT. CWT is employed. CNN is used to break down ECG data into distinct time–frequency components to obtain characteristics from the 2D scalogram made up of the time–frequency data above factors. Taking into account, R peck interval is also important. Four RR interval characteristics are attached and deleted with the CNN, which is important for the ARR. ECG classification characteristics to feed into a fully linked layer We will discuss this study in detail to investigate the usefulness of CNN in detecting and classifying ARR, CHF, and NSR. Using the ECG dataset as a starting point, this section explains the goals and structures of this dissertation.

Use a Deep CNN and CWT to make a scalogram of an ECG signal from an ECG dataset; this study is about. These numbers are from the MIT-BIH datasets (ARR, CHF, and NSR) (Github).

2 Methodology

2.1 Types of ECG Signal for Classification

Here, we take three kinds of ECG signals for classification. Figure 1 represents the signal of Arrhythmia (ARR), Congestive Heart Failure (CHF), and Normal Sinus Rhythm (NSR).

The purpose is to train a CNN to tell the difference among CHF, ARR, and NSR so that it can do this.

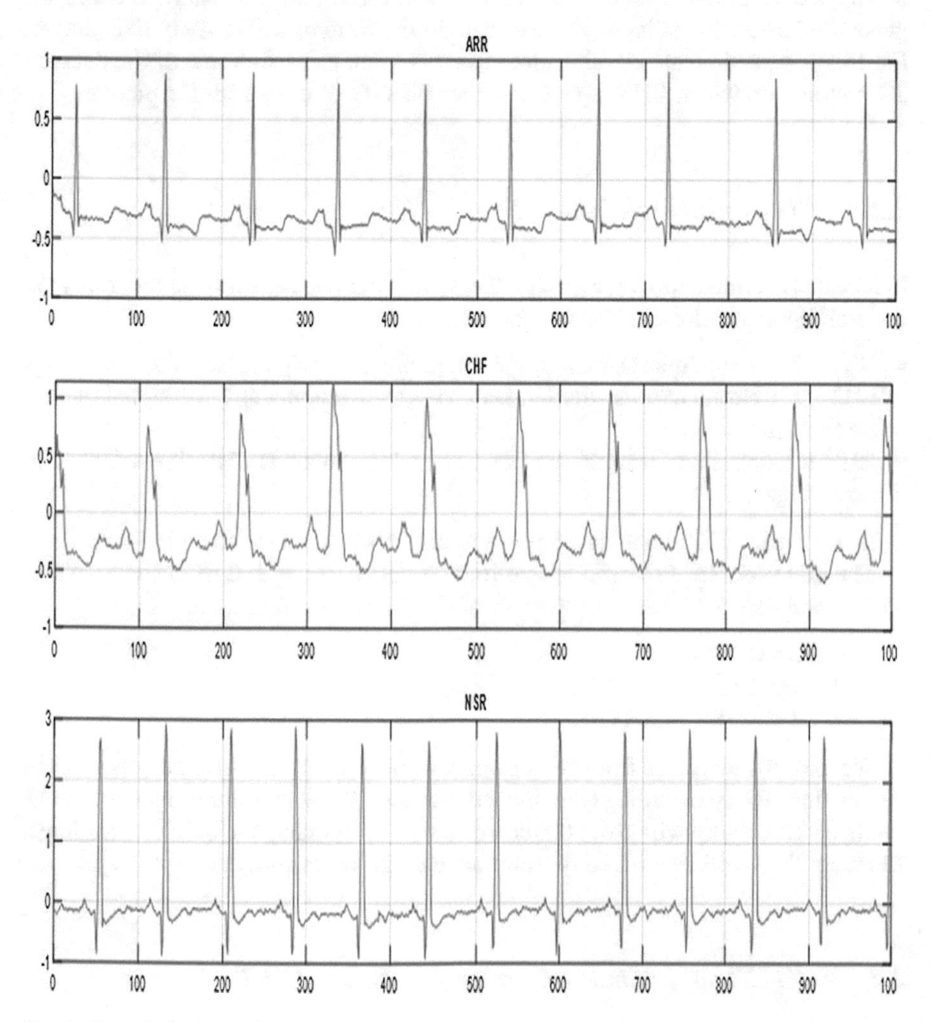

Fig. 1 Signal of ARR, CHF, and NSR

Table 1 ECG dataset

1	2	3	4	...	...	...	...	...	65,536
2									
3									
...									
162									

A CNN is a sort of ANN especially built to analyze the pixel input and is utilized in image detection and processing. It is a specialized artificial neural network for generating data with an input shape or data in the form of a 2D matrix-like picture. For image detection and classification, CNN is often utilized. Some of the data are 2D matrices on which CNN is used to either identify or categorize the picture.

2.2 *ECG Signal Database*

ECG data is divided into three groups. These numbers come from 162 recordings in three different PhysioNet datasets. There are

- MIT-BTH Arrhythmia Database, which contains ARR signal data of 96 recordings
- MIT-BTH Normal Sinus Rhythm Database, which contains NSR signal data of 30 recordings
- BIDMC congestive Heart failure Database, which contains CHF signal data of 36 recordings

In total, 162 ECG signal data that we took, whose demo is shown in Table 1.

The data variable is a matrix of size 162*65,536, a sum of 162 ECG signal of size 65,536 sample from labels we get types of ECG signals information.

1:96 are ARR signal (96)
97:126 are CHF signal (30) &
129:162 are NSR signal (36)

We took 30 recordings from every category (NSR, CHF, and ARR) to ensure even distribution. Every recordings are divided into ten 500-sample-long segments. As a result, each category will give 300 recordings with 500 samples for 900 recordings. 750 recordings will be utilized for training and 150 for testing out of 900 total.

2.3 *ECG Signal to Image Conversion Using CWT*

We will now use the CWT to turn all of the 1D signals or data sets into pictures to be given as input to a CNN for classification. For this, we will take the CWT of each 1D

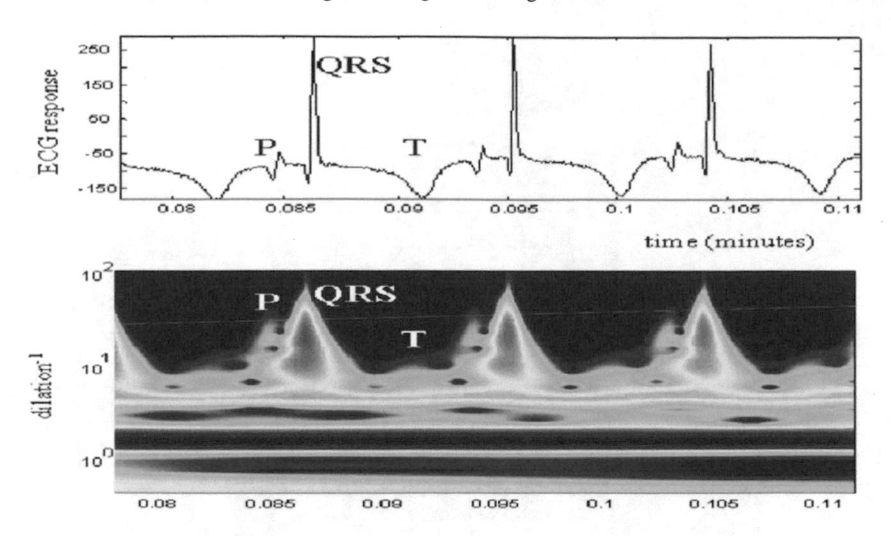

Fig. 2 Converting ECG to scalogram

signal or data set and arrange all of the coefficients to generate a CWT scalogram. Each scalogram is presented by a 128-color color map of a typical jet. Each class's scalogram is transformed into a picture and stored in its folder.

Each picture is 227*227 pixels in size (to be used for AlexNet) and is in RGB color format. Following conversion, we have 900 scalogram photos stored in three files, one for each of the three above-mentioned categories.

Analysis wavelet transform (AWT) is a wavelet transform that gives information about the magnitude and phase of a signal at a time scale or a frequency level. An Analytic wavelet ("amor") is employed in this case. The time and frequency variances in this wavelet are both equal. In the time domain, these wavelets have one-sided spectra and are complex-valued. The continuous wavelet transform is a superior alternative for determining a time–frequency analysis using these wavelets (CWT). In addition, for CWT, we employed 12 wavelet bandpass filters per octave (12 voices per octave). Scalogram is used to classify the EEG signals. Figure 2 shows the conversion of ECG to Scalogram.

We did not receive the specific frequency range to cover whether we used a high pass or low-pass filter. Filtering with a bandpass covering (1 Hz to 4 kHz).

2.4 *Transfer Learning via AlexNet*

A CNN called AlexNet had a big impact on the field of ML, especially in the use of DL for machine vision.

We utilized AlexNet's pre-trained deep CNN for ECG signal categorization. AlexNet was trained on over 1 million photos, which can categorize them into 1000

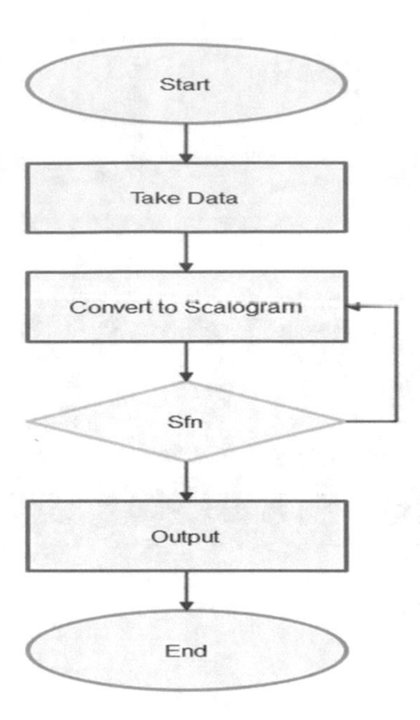

Fig. 3 Flowchart

individual clusters. Transfer learning is the process of fine-tuning a previously trained CNN to achieve categorization on a novel set of pictures. Transfer learning is rapid and straightforward rather than training the CNN from the scratch, takes millions of input pictures, a lot of training time, and high-speed efficient hardware. Figure 3 represents the working diagram.

3 Result and Discussion

3.1 Dataset

It indicates that the heart is depolarized. The P wave arises when the sinus node, known as sinoatrial node, which depolarizes the atria. The PR interval is the period from the beginning of the P wave to the starting of the QRS complex. The PR section of the ECG that runs from the end of the P wave to the starting of the QRS complex. The PR interval, which is calculated in minutes, is not to be confused with the PR segment, which is shown in Fig. 4.

The QRS complex begins with the Q wave, which is the first downward deflection following the P wave. There is no Q wave when the first deflection of the QRS complex is upright. In many, but not all, ECG leads, a typical person will have a modest Q wave.

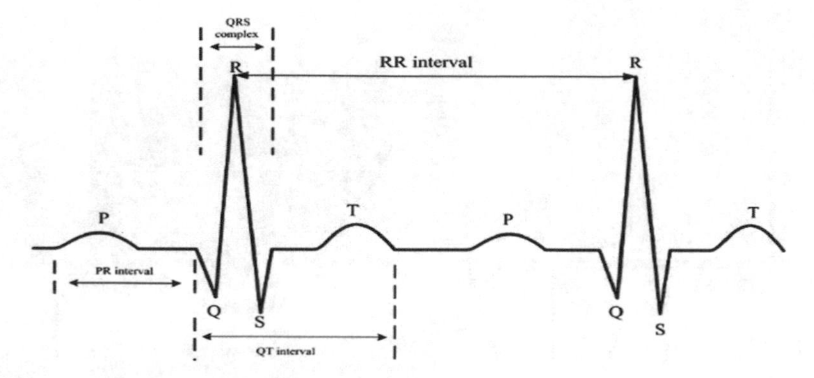

Fig. 4 RR intervals

The "QRS complex" combines the Q, R, and S waves that reflect ventricular depolarization. Although not all ECG leads include all three of these waves, the term "QRS complex" is used to describe the presence of a "QRS complex." The R wave, which follows the P wave and is part of the QRS complex, is the first upward refraction after the P wave. The R wave shape isn't very clinically significant, although it may change. After the R wave, the S wave is the first downward deflection of the QRS complex. However, an S wave may not appear in all of a patient's ECG leads. The section of the ECG from the conclusion of the QRS complex to the beginning of the T wave is known as the ST segment. Because the ST segment is generally isoelectric, ST segment depression or elevation might suggest heart disease. The T wave is a product of ventricular repolarization and occurs after the QRS complex. The section of the ECG from the end of the T wave to the starting of the P wave is known as the TP segment.

3.2 Image Conversion from Dataset by CWT

ECG heart beat signals and scalograms of regular heart beat and PVC heart beat are shown in Fig. 5. The PVC pulse seems to be distinct from the typical heartbeat on the scalogram, indicating that the scalogram may be used to classify heartbeats. However, establishing a direct link between scalogram and aberrant states is challenging. CNN was created to overcome this issue by automatically extracting the probable link between various arrhythmias and regular heartbeats.

3.3 Trained and Calculation

Figures 6 and 7 represent the progress of the training and validation loss and the training and validation accuracy and loss process.

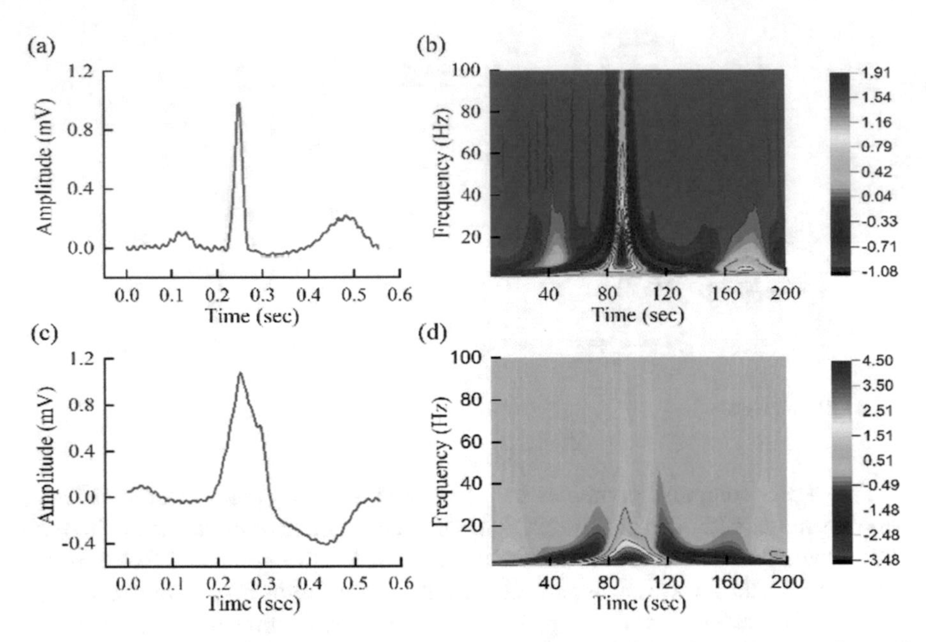

Fig. 5 Raw ECG heart beat signals & CWT scalogram, **a** the ECG, **b** CWT scalogram of normal heart beat, **c** ECG heart beat signal, and **d** CWT scalogram of irregular heart beat

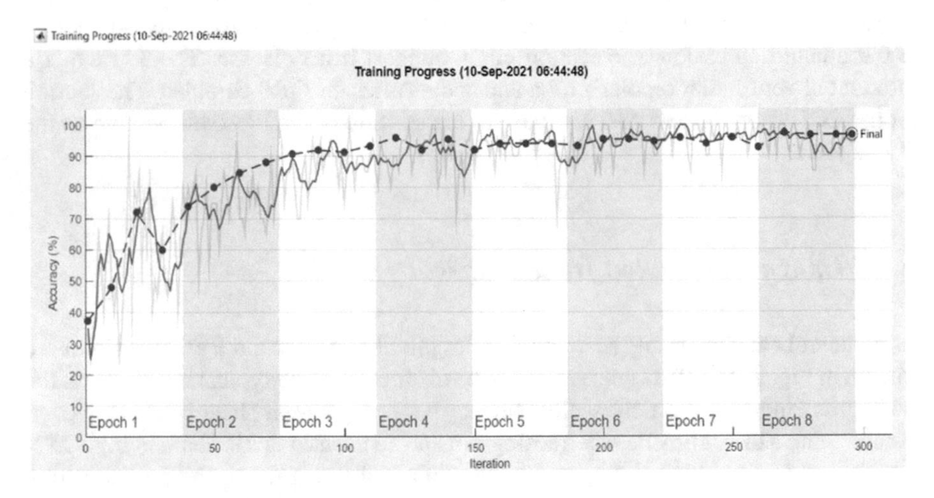

Fig. 6 Training and validation accuracy

In Fig. 8, we can see that training has come to the end of its cycle. The overall validation rate is 96.67%. And accomplished epoch 8 of a total of 8 epochs. There were 296 iterations in all, with 296 iterations completed. 37th iteration per epoch. The maximum number of iterations is 296. The validation frequency is 10 iterations.

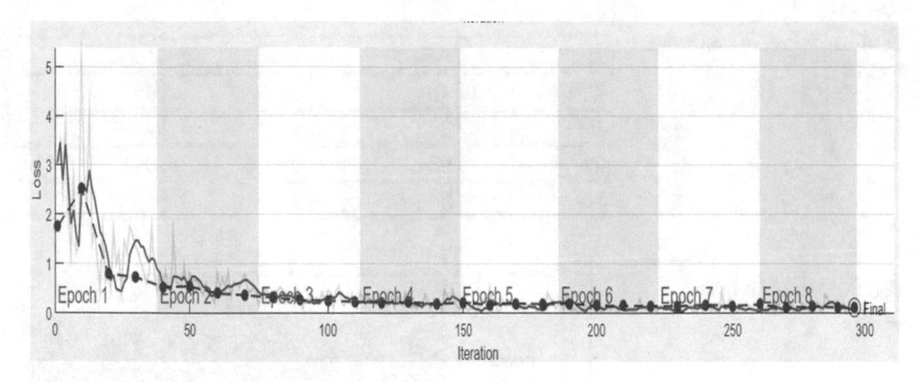

Fig. 7 Training and validation loss

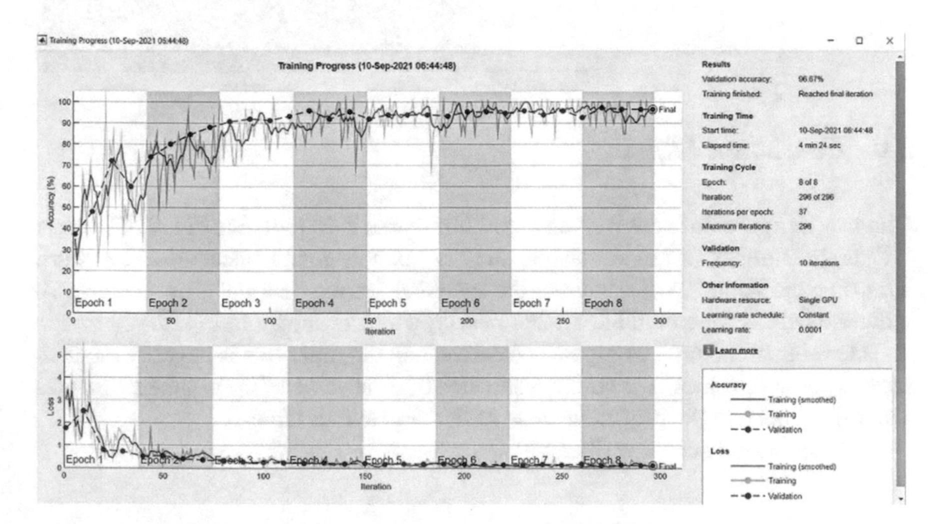

Fig. 8 Total training progress

Such information was quite promising since it was recognized that the categorization had a high degree of accuracy. The relative confusion matrix and statistical characteristics reported in Table 2 are shown in Fig. 10. From Fig. 10, we have calculated the confusion matrix, which is shown in Table 3.

Table 2 Calculation of confusion matrix

	ARR	CHF	NSR
TP =	Cell1	Cell5	Cell9
FP =	Cell2 + Cell3	Cell4 + Cell6	Cell7 + Cell7
TN =	Cell5 + Cell6 + Cell8 + Cell9	Cell1 + Cell3 + Cell7 + Cell9	Cell1 + Cell2 + Cell4 + Cell5
FN =	Cell4 + Cell7	Cell2 + Cell8	Cell3 + Cell6

Table 3 Result of confusion matrix

	ARR	CHF	NSR
TP =	48	48	49
FP =	2	1	2
TN =	98	99	98
FN =	2	2	1

	Positive	Negative
Positive	TP	FP
Negative	FN	TN

Fig. 9 Confusion matrix (2*2)

3.4 *Confusion Matrix*

One way to figure out how well an algorithm works is to look at a "N X N" matrix. "N" is the number of target classes, and "N" is how many there are. The matrix compares the actual goal values to the machine learning model's predictions. As indicated below, we would have a 2*2 matrix, which is shown in Fig. 9.

One value of the target variable is positive, and the other one is negative, so there are two possible values. The column indicates the true value of the targeted variables. The rows represent the projected value of the targeted variables.

Let's see where accuracy filters;

$$\text{Accuracy} = \frac{TP + TN}{TP + FP + TN + FN}$$

Precision tells us how many of the correctly predicted cases turned out to be positive.

$$\text{Precision} = \frac{TP}{TP + FP}$$

This value helps to calculate whether the proposed strategy is reliable or not.

Sensitive (TPR), Recall indicates that the number of model's real positive cases can be properly anticipated.

$$\text{Recall} = \frac{TP}{TP + FN}$$

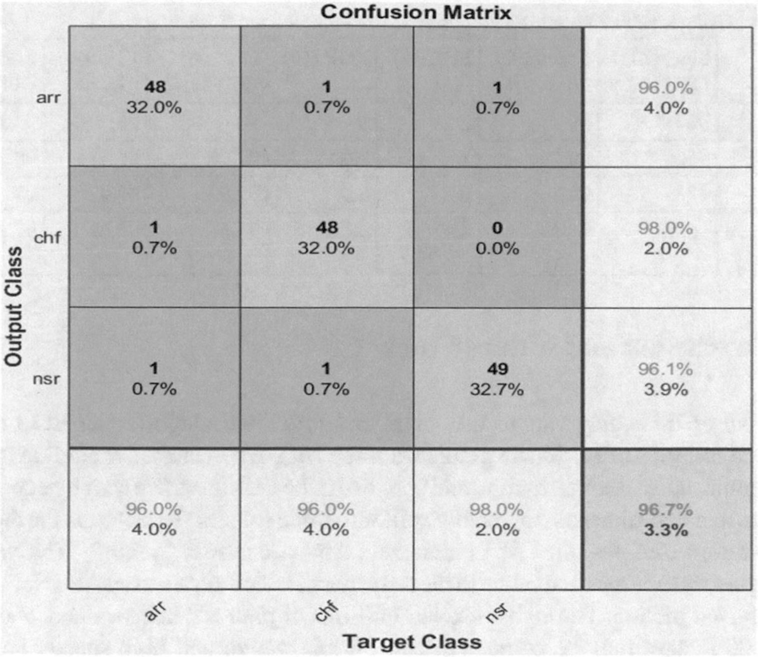

Fig. 10 Confusion matrix (3*3)

$$\text{Specificity or True Negative Ratio (TNR); TNR} = \frac{TN}{FP + TN}$$

$$\text{Fall out, False Positive Ratio (FPR); FPR} = 1 - \text{TNR}$$

$$\text{False Negative Ratio (FNR); FNR} = \frac{FP}{FP + TP}$$

The F1-score is determined by taking the harmonic mean of recall and precision. As a result, it provides a synthesis of these two measurements. In actuality, when we tried to improve our model's precision, the recall suffers, and vice versa.

$$\text{F1} - \text{score} = \frac{2TP}{2TP + FP + FN}$$

We combine it with other evaluation criteria to get a full picture of the outcome. So, from Fig. 10 we get

From Table 4, we can say that the mean term of those classifications are 97.77% Accuracy; 96.67% Sensitivity; 98.33% specificity; 1.67% false-positive ratio; 3.32% false-negative ratio; 96.68% precision; and 96.67% F1-score.

Table 4 Using a validation set the proposed method yielded the following result

	Sensitivity (TPR) (%)	Specificity (TNR) (%)	FPR (%)	FNR (%)	Precision (PPV) (%)	F1-score (%)	Accuracy (%)
ARR	96	98	2	4	96	96	97.33
CHF	96	99	1	2.04	97.96	96.97	98
NSR	98	98	2	3.92	96.08	97.03	98
Mean accuracy	96.67	98.33	1.67	3.32	96.68	96.67	97.77

4 Conclusion and Future Work

The goal of this study was to use CNN and CWT to categorize an ECG signal. We used millions of ECG data points from the MIT-BIH Dataset, including data on arrhythmia, normal sinus rhythm, and congestive heart failure. We start by expressing the data as a two-dimensional picture split into three folders. To describe the data in a 2D picture, we employ the CWT parameters Analytic morlet ("amor"). This wavelet distinguishes time and frequency in the same way. After that, we used AlexNet to train the received picture. Through AlexNet, millions of photos can be trained, and more than 100 images may be compared. It is simple, rapid, and high-speed hardware. It was proved experimentally that the overall accuracy of the classification could reach 96.67% validation accuracy. By comparing and contrasting different methods, we were able to ensure that the method used in this paper is a lot better than the best that has been done before. We may also classify ECGs in a variety of ways using this approach. For example, first-degree heart block, ischemia, and myocardial infarction (MI), to name a few. Since usage, we think our trained classifier model may be utilized for real-life and real-time applications with a little extra modification.

In the future study, we will analyze our model against numerous datasets and categorize additional illness kinds using an inter-patient division technique. In addition, we are attempting to gather fresh ECG pictures, construct our dataset to represent real-time circumstances, and test our approach in the actual world as a real-time system given in this work. We tried to transfer the trained network in a microcontroller-based system. Subsequently, train it for real-time monitoring and classification on a portable and wearable device. We are also attempting to create software that can scan ECG pictures, diagnose cardiac ailments, and offer ECG categorization results.

References

1. Chang KC, Hsieh PH, Wu MY et al (2021) Usefulness of machine learning-based detection and classification of cardiac arrhythmias with 12-lead electrocardiograms. Can J Cardiol 37:94–104. https://doi.org/10.1016/J.CJCA.2020.02.096

2. Lee H, Shin M, Zhu Y, Zhu X (2021) Learning explainable time-morphology patterns for automatic arrhythmia classification from short single-lead ECGs. Sensors 21:4331. https://doi.org/10.3390/S21134331
3. Llamedo M, Martínez JP (2011) Heartbeat classification using feature selection driven by database generalization criteria. IEEE Trans Biomed Eng 58:616–625. https://doi.org/10.1109/TBME.2010.2068048
4. de Chazal P, O'Dwyer M, Reilly RB (2004) Automatic classification of heartbeats using ECG morphology and heartbeat interval features. IEEE Trans Biomed Eng 51:1196–1206. https://doi.org/10.1109/TBME.2004.827359
5. Kutlu Y, Kuntalp D (2012) Feature extraction for ECG heartbeats using higher order statistics of WPD coefficients. Comput Methods Prog Biomed 105:257–267. https://doi.org/10.1016/J.CMPB.2011.10.002
6. de Chazal P, Reilly RB (2006) A patient-adapting heartbeat classifier using ECG morphology and heartbeat interval features. IEEE Trans Biomed Eng 53:2535–2543. https://doi.org/10.1109/TBME.2006.883802
7. Islam SS, Miah MS, Zadidul Karim AHM, Bashar SS (2020) Seizure detection and classification using different decomposition methods and robust statistical analysis from EEG signals. In: 2020 International symposium on advanced electrical and communication technologies ISAECT. https://doi.org/10.1109/ISAECT50560.2020.9523685
8. Strodthoff N, Wagner P, Schaeffter T, Samek W (2021) Deep learning for ECG analysis: benchmarks and insights from PTB-XL. IEEE J Biomed Health Inform 25:1519–1528. https://doi.org/10.1109/JBHI.2020.3022989
9. Zhang D, Yang S, Yuan X, Zhang P (2021) Interpretable deep learning for automatic diagnosis of 12-lead electrocardiogram. iScience 24:102373. https://doi.org/10.1016/J.ISCI.2021.102373
10. Smisek R, Nemcova A, Marsanova L, et al (2020) Cardiac pathologies detection and classification in 12-lead ECG. Computing in Cardiology. https://doi.org/10.22489/CINC.2020.171
11. Xu SS, Mak MW, Cheung CC (2019) Towards end-to-end ECG classification with raw signal extraction and deep neural networks. IEEE J Biomed Health Inform 23:1574–1584. https://doi.org/10.1109/JBHI.2018.2871510
12. Ye C, Vijaya Kumar BVK, Coimbra MT (2012) Heartbeat classification using morphological and dynamic features of ECG signals. IEEE Trans Biomed Eng 59:2930–2941. https://doi.org/10.1109/TBME.2012.2213253

Chapter 23
Semi-automatic Vehicle Detection System for Road Traffic Management

Manipriya Sankaranarayanan, Madhav Aggarwal, and C. Mala

1 Introduction

The lack of an accurate system constantly hinders vehicle detection for counting on Indian roads; there exists a need to help engineers, researchers, and surveyors. Further, a lack of manpower puts a constraint on the amount of CCTV footage that can be reviewed at a time. As a consequence of the lack of resources, it becomes impossible to analyze CCTV footage to detect vehicles, let alone analyze any attention to the direction of motion or classification. Despite this, CCTV footage is needed for security purposes and future development in terms of crowd and traffic analysis. Automatic analysis of videos is essential for reducing the workload of operators [1]. Traffic flow needs to be analyzed for a variety of factors, such as the direction of motion and traffic density, and all these studies should be consistent with any external factors such as climatic changes. This traffic data is essential to analyze the need to eliminate traffic congestion and effectively plan routes to reduce chaos. The existing solutions work well for streamlined motions but fail where the traffic density is high. They fail to perform accurately, specifically in regions where the directions are abnormal, and fail to follow any predictable patterns [2, 3].

Detection and Background Subtraction is accompanied by a lot of noise, distortion, low illumination, poor resolution, elimination of shadows, nightlight illumination, and many other problems. While CNNs can overcome these limitations, they are expensive, time-consuming, and memory inefficient. It is essential to strike a balance between accuracy and resource constraints. The proposed system can work well with limited computational power. The following sections showcase the areas

M. Sankaranarayanan (✉)
IIIT Sri City, Chittoor, Andhra Pradesh 517646, India
e-mail: manipriya.s@iiits.in

M. Aggarwal · C. Mala
National Institute of Technology, Tiruchirappalli 620015, India
e-mail: cmala@nitt.edu

G. Mathur et al. (eds.), *Proceedings of 3rd International Conference on Artificial Intelligence: Advances and Applications*, Algorithms for Intelligent Systems,
https://doi.org/10.1007/978-981-19-7041-2_23

which need extensive work. With the availability of more data, the system will be able to operate on the data to produce better results.

The proposed (Semi-)Automatic Vehicle Detection System (SAVDS) can rapidly recognize moving objects. Background subtraction methods are fast and with the recent developments in detection methods such as YOLOv3 [4], and RESNET [5], the gap has been amplified since they operate smoothly on real-time videos. Measuring volumes and types of vehicles operating on each road at a high rate is the prime goal of the work. Apart from this, the performance of the study should remain similar at both day and night time and in the case of rare climatic occurrences.

The organization of the paper is as follows. The existing work related to the proposed SAVDS is introduced in Sect. 2. It also comprises related works in four aspects: video surveillance, modified background subtraction, object detection, and vehicle detection and classification. Section 3 presents a framework consisting of multiple steps with a detailed description of the extracted images from the first step. The result section presents the results and reports the experimental result, and the last section presents the conclusion and discussion.

1.1 Related Work

The origins of data-driven real-time traffic analysis can be traced back to the advent of the GPS [6]. After several iterations of the technology, it was observed that Computer Vision technologies [7] are capable of retrieving and processing a lot more data. With Computer Vision, it became possible to retrieve spatial and visual feature information such as size, color, and text. The advent of the Convolutional Neural Network [8] and its usage in traffic analysis [9, 10] greatly increased the accuracy and analytical power of computer-based techniques. Modern improvements to CNN architectures including YOLO [11], RESNET [5] and MobileNet [12] have made great strides in increasing efficiency and moving ever closer toward achieving real-time detection systems.

Current vehicle detection techniques are based on a combination of Deep Learning and Computer Vision techniques. Zhang et al. [13] propose the SlimYOLOv3 architecture to perform real-time detection on embedded computers. Lin et al. [14] utilize Feature Pyramids to extract additional information from the input frames. Bansal et al. [15] improve upon Attention-based mechanisms [16] to enhance the Re-Identification of objects in a scene. However, a major downfall of these techniques is the expensive training step which requires considerable resources. While a degradation in speed does reduce expenses, it also hampers performance. Additionally, CNNs also suffer from significant over-fitting due to the lack of a sufficiently large dataset. Thus, it is required to generate a dataset of images based on different cameras taken at different angles in various conditions. In turn, this leads to an increase in expenses. Besides, there is a need for a model that can recognize not only cars and vehicles but also bicycles and people on roads owing to the traffic and lack of planning. Further, the system needs to be effective in poor lighting and adverse weather conditions.

This is not the only problem needing to be solved. Most of the time, CCTV footage consists of regions that are of no use to the individual in detecting the incident being observed or the analysis. Hence there is a need to shorten the area of focus and only focus on the essential part. In outdoor environments, the moving objects and noise in the background increases [2, 3, 17]. Thus, recognizing the class of the detected object is of utmost importance.

To summarize, this paper works toward the existing drawbacks such as (1) Rapid Detection of moving objects, (2) Identifying the direction of motion of vehicles, (3) Counting the number of vehicles on the road, (4) Classifying the vehicles, (5) Zooming in and observing the enlarged region of interest. The details of the proposed SAVDS are discussed in the subsequent sections.

1.1.1 Proposed Semi-automatic Vehicle Detection System (SAVDS)

The proposed Semi-Automatic Vehicle Detection System (SAVDS) is designed to use visual and spatial information to build an effective framework for traffic analysis. This leads the system to work in real-time and in challenging scenarios such as at night and in adverse climates with very high accuracy. This paper proposes a framework to recognize moving objects following a multi-step approach. These steps are applied sequentially to the video frames. Computer Vision allows the system to effectively detect count, classify vehicles, find the direction of motions, and identify lanes in all conditions. The strategies are non-invasive, cost-effective, and automated. A variety of color models are used, and an assessment is done to determine which one gives the best results. Different background subtraction strategies are employed and also different classifiers, all of which have been compiled and compared in this work. The overall process of the proposed SAVDS is presented in Fig. 1.

The steps are briefly described as follows:

- The first step involves drawing rectangular boundaries on the region to be detected.
- This is followed by a contour detection step performed on the background-subtracted frame.
- The detected regions are then fed into a classifier: a CNN, which outputs the detected object.

The application that uses SAVDS primarily assists traffic surveillance experts, traffic police, city planners, researchers, and surveyors in properly monitoring traffic.

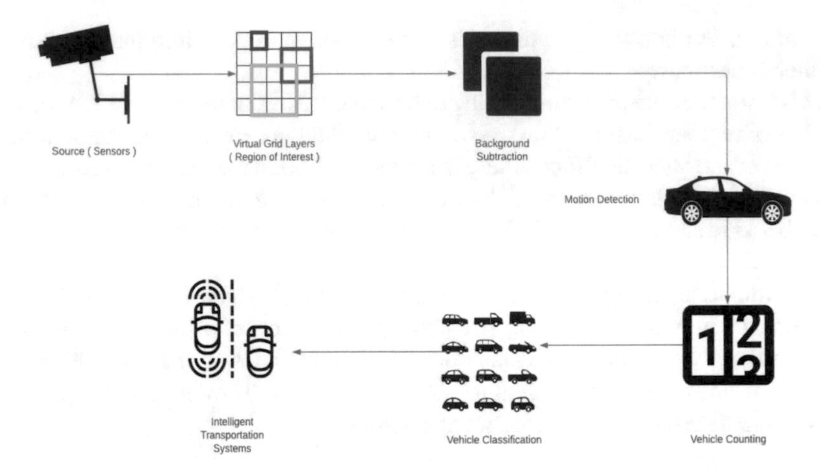

Fig. 1 Semi-automatic vehicle detection system

1.2 Selection of Virtual Grid Layers

As the first step in the system, the user selects the region of interest. The most innovative part of the proposed framework is that the selected region of interest is converted into virtual grids of layers, as shown in Fig. 2. This significantly enhances the computational power of the framework as compared to other existing works. This selection eliminates the regions in the field of vision that the user does not want to observe [2, 17]. This system uses a multi-level background subtraction strategy which effectively removes shadows and irregular headlight illuminations. A tracking module is in place to detect and count the number of vehicles. It works efficiently owing to the right choice of parameters and remains consistent for all models.

Fig. 2 Virtual grid layers for SAVDS

1.3 Background Subtraction

The detection of vehicles is done using background subtraction. The video frames are used to identify the background values for the virtual grid layers using the selection of a region of interest. SADVS uses a multi-level background subtraction strategy which effectively removes shadows, and irregular headlight illuminations [1, 3, 17]. In SADVS, background values are obtained using HSV Filtering in which H, S, and V are processed in parallel. The reliability of object detection is due to a balance between luminance and chrominance that is used to obtain high accuracy. The approach uses a multi-stage background subtraction model, which uses the disparity maps for chrominance and luminance and then executes algorithms on the frame CLAHE [18], Watershed [19], and Sobel [20] Algorithms. In the model, the background must be constantly updated over time to adapt to the scene, affecting the chromaticity and luminance values. This also aids in removing the shadow regions of the video. In order to initially set the values for the tuning parameters of background subtraction, track bars are used to set the threshold of background subtraction for the Hue, and the Value [2].

The selection of a region of interest and the selection of a threshold are the only two tasks the user is involved in, making this a Semi-Automatic system. This is mainly due to the disparity of the surveillance camera. With this selection, the proposed system can be used for any type of video frame irrespective of its angle, illumination, etc. Apart from these two selections, all the processing is done automatically. In the future scope, the automation of the region and threshold selection process are explored. A series of steps are followed to detect the contours in the background subtraction model. Contours care curves that join continuous points having the same color or intensity. Thus, contours are used for shape and object detection. Contour detection is useful for object detection and recognition. Contours are approximated by removing all redundant points and compressing the contour to save memory. Background subtracted image is obtained with which the vehicle contours are detected. Centroids of the contours are then obtained and appended to a list to further use for vehicle counting. The width and height of the contours are made static so that they can later be used for enumeration in classification and counting. As the contours are detected in the virtual grid layers, it is visualized by highlighting the grids of contours. The next step of SAVDS is to track the number of vehicles effectively

The algorithm for contour and object detection is described in Algorithm 1. The algorithm helps us get the contour boxes from the image frames effectively even in low light and extreme weather conditions.

1.4 Direction of Motion Detection

The directions of motion are detected for the vehicles using the Optical Flow method. Oriented FAST and Rotated BRIEF (ORB) are used to get the corner points in the

Algorithm 1 Object Detection

Input: p: number of frames, $\mathcal{F}(\mathcal{P})$: Video, h: hue, s: value, v: saturation
Output: Contour List C from the image
1: $average \leftarrow$ (F/p) : $Average; Background$
2: $DisparityMap(D) =$
3: **for** average[i] **do**
4: **for** average[i][j] **do**
5: $D[i][j] \leftarrow \sum((average[i][j])^2 - (padded[i][j])^2)/\sum(average[i][j] - padded[i][j])$
6: **end for**
7: **end for**
$D_L, D_C = Disparity\ Maps\ for\ Lumninance\ and\ Chrominance$
8: $t = WhiteThreshold - DifferentiateObjectorNoObjectforheadlightdetection$
9: **for** average[i] **do**
10: **for** average[i][j] **do**
11: **if** average[i][j]>[i][j] **then**
12: $final[i][j] = 1$
13: **else**
14: $final[i][j] = 0$
15: **end if**
16: **end for**
17: **end for**
Apply Clahe(final) and Obtain Foreground and Shadow using F(P) and average.
Update final and average by applying Watershed(Sobel(Foreground, Shadow)).
Identify Contours on the final frame and Store them in C.

frames. Lucas Kanade Optical Flow [21] is used to calculate the directions in which vehicles are moving and then predict the primary direction of motion of the vehicles using a formula. This is followed by a strategy that is similar to SIFT tracking [22] to find the number of vehicles based on whether the centroid has crossed a central of the virtual grid layers. A Series of filters is applied to the frames to fill holes. This method works by obtaining a list of trackable points which are compared across frames. This aspect was also used to predict the lanes in which vehicles are moving accurately. Lanes are differentiated based on abrupt changes in the direction of vehicles. Thus, it is considered an aberration when a vehicle changes lanes to overtake another. This is recorded in the system, which can later be used to detect all the different directions in the frame of view and then develop a strategy to counter too many vehicles moving abnormally in a space.

1.5 Vehicle Counting

Based on the contour identified in the previous section with the direction of motion, the vehicle moving across the virtual grid layers is identified. The details of the detection are described in Algorithm 2. At the same time, the features that were extracted are used to mark the different/variety of directions of motion that exist in the video.

Any abnormal directions in the vehicles that move in are marked on a frame and can be analyzed.

Algorithm 2 Vehicle Counting

Input: N: number of bounding boxes, BB(N): Set of Bounding Boxes
Output: Number of vehicles N
1: $Centroid \leftarrow length_{box}/2, breadth_{box}/2$
2: $UseORBAlgorithmtotracksameboxinconsecutiveframes$
3: **if** Contour Point p appears in x frames > threshold **then**
4: $\quad Contour = Valid$
5: **end if**
6: $DividerLine(0, Height_{frame}/2), (Length_{frame}, Height_{frame})/2$
7: **if** $Contour = Valid\ and\ Centroid_{box} > Divider\ Line$ **then**
8: $\quad VehicleCount = VehicleCount + 1$
9: **end if**

1.6 Classification of Vehicles

Following motion detection, classification is performed using a simple CNN framework using a self-built database initially inherited from the MIO-TCD [23] study. The proposed SADVS are trained on 32,500 images and six classes of vehicles. Most of the computation time is spent on the CNN stage while maintaining a fast computation speed that matches real-time detectors. The mini-batch size is set to 50 images per batch and 500 epochs for the CNN framework. For the convolutional layers, 5×5 windows are used. The dropout rate is kept at 10%. The input size of all training images is set to 64×64 to ensure uniformity. The aim was to make a computationally light system for detection. The focus is on the virtual grid layers for this step which are passed into the classifier. The input images are RGB images (64×64), categorized into six classes. The CNN of SADVS has two convolutional layers, which are followed by one pooling layer, each followed by two fully connected layers at the end, before feeding the final result into the Softmax activation layer. The convolutional layer is used along with the ReLU activation function. Each convolutional layer has 64 filters and a 5-unit patch with stride 1. In the pooling layer, 2×2 filters are used with a stride of 2, which reduces its size to half. Thus, the final size obtained is 16×16. Dropout is applied at the end of every layer. The number of fully connected points is 384 in the first connected layer and 192 in the second layer. Results from SADVS are evaluated using benchmark datasets in the next section.

2 Results and Performance Evaluation

The detection of vehicles using the proposed SAVDS is analyzed using the videos from DETRAC datasets. There are 100 video sequences in the DETRAC [24] dataset, which is used for calculating the accuracy of the model. The frame size of images in the DETRAC [24] dataset is 940 × 560, which is standard. In the DETRAC [24] dataset for testing, 23 Night video sequences contain cars with headlights on and 13 rainy day sequences among the 100 video sequences. The frames are all high-resolution video sequences. Videos are run at 25 fps using an NVIDIA GTX-940 2GB Graphics Processor and Intel i5 8750H with 8GB RAM. The screenshot of the videos is shown in Fig. 3.

The overall GUI of SAVDS is shown in Fig. 4

Directions of motion were successfully obtained using optical flow in SAVDS and correctly detected the lanes and the directions of motions according to the average slope calculation ORB. The vehicles were effectively counted while keeping track of the processed frames. The screenshot of the vehicle detection by virtual grid layers, background subtraction, and detection direction of motion are shown in Fig. 5. However, there are instances where huge vehicles have big holes, leading to the vehicle being counted twice. This can be improved by building a backpropaga-

Fig. 3 Screenshot of DETRAC datasets

Fig. 4 Screenshot of SAVDS user interface

(a) Detection in SADVS with Virtual Grid Layers

(b) Vehicle Tracking for Motion Detection

(c) Vehicle Centroid Tracking

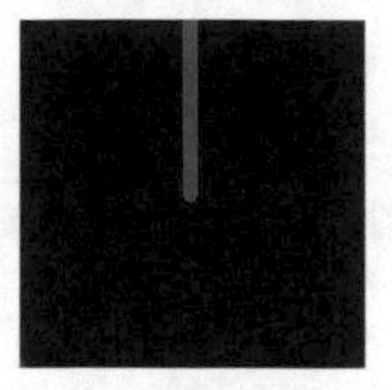

(d) Computed Direction of Motion in SADVS

Fig. 5 Screenshot of SADVS

Table 1 Detection accuracy by weather

Name	Ground truth	Detected	Accuracy (%)
Sunny	84	97	86.59
Rainy	42	77	54.54
Snowy	26	30	86.66
Night	75	140	53.57
Night + Rainy	36	75	48
Night + Snowy	18	45	40

tion system from the classifier to the detector, although this might reduce the model's speed. The detection accuracy of the model decreases significantly as the weather gets worse. Rainy weather obscures vision, due to which the accuracy drastically drops. The model fails to capture white cars when snowing and black cars at night. Headlights tend to be calculated as extra vehicles. The number of vehicles detected is more significant due to the different contours that are captured. The results are shown in Table 1.

The accuracy increases with the number of layers to a point after it saturates. Thus, having 3–4 layers on an average proves to be beneficial, as in Fig. 6. However, the same result varies when the weather conditions vary. In rainy weather, the accuracy increases until a layer count of 10 and in snowy weather until a count of 8.

The CNN with virtual grid layers of the SAVDS approach offers an advantage in processing, training, and manual classification speed. The approach comes very close to matching YOLO detection accuracy. YOLO gives a SOTA accuracy of 93% in test conditions. CNN in SAVDS gives a test accuracy of 88% but a massive reduction in complexity. The computational complexity was much lesser in the case of this model. This model used only 209K parameters, whereas a standard YOLO model would use over 7,00,000 K parameters. The processing speed of the framework is equal to the YOLOv2 framework and is very fast, with a speed of 20 fps at least. The training speed is also breakneck considering it is a reasonably simple CNN network. For a YOLO network, it was much longer. The manual annotation speed of the model is fast because there is no need to specify locations while training. Labeling of data

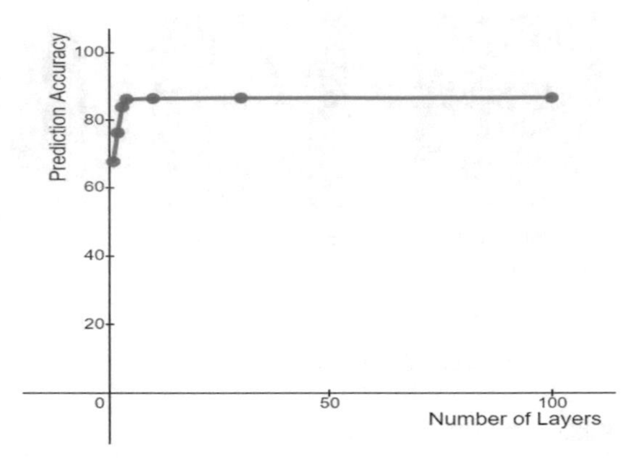

Fig. 6 Variation of prediction accuracy of SAVDS with number of layers under sunny weather conditions

is done for the portions extracted after background subtraction. The model works well on zoomed-out videos, which are used for testing.

A dataset of 32,500 images is more than enough to get good accuracy on the data. However, shadows and headlight illumination do affect final accuracy and results. Included background images in the training data ensures that they are categorized as part of the background when detected, which helps the model improve its accuracy.

The model can be made more accurate by training on the videos; however, that would involve labeling complexities.

3 Conclusion

Several applications aid in controlling the ever-growing vehicle density. The core requirement of any such application is to detect the vehicle's presence and provide the current traffic characteristics. Among several sources to enumerate traffic characteristics CCTV based video footage is the prime contributor. This paper proposes an innovative Semi-Automatic Vehicle Detection System (SAVDS) to detect vehicles from such traffic videos. The uniqueness of the proposed work is to use virtual grid layers over the user-selected region of interest in the CCTV videos for detecting and classifying vehicles. This system uses background subtraction and optical flow technique in the virtual grid layers to detect the vehicle contours and tracking. The detected vehicles are classified using Convolution Neural Network (CNN). The performance of SAVDS is evaluated in benchmark datasets such as DETRAC. The vehicle detection and tracking show 85% accuracy, whereas the classification shows 88% accuracy. This accuracy is evaluated in videos of various illumination such as night, rainy, and summer. This detection system contributes to improving several traffic management strategies and applications utilizing it.

References

1. Sankaranarayanan M, Mala C, Mathew S (2021) Pre-processing framework with virtual mono-layer sequence of boxes for video based vehicle detection applications. Multimedia Tools Appl 80(1):1095–1122
2. Manipriya S, Mala C, Mathew S (2014) Performance analysis of spatial color information for object detection using background subtraction. In: IERI procedia, international conference on future information engineering (FIE 2014), vol 10, pp 63–69
3. Manipriya S, Ramadurai G, Reddy VB (2015) Grid-based real-time image processing (grip) algorithm for heterogeneous traffic. In: 2015 7th international conference on communication systems and networks (COMSNETS), pp 1–6
4. Redmon J, Farhadi A (2018) Yolov3: an incremental improvement
5. He K, Zhang X, Ren S, Sun J (2015) Deep residual learning for image recognition
6. Taylor MA, Woolley JE, Zito R (2000) Integration of the global positioning system and geographical information systems for traffic congestion studies. Transp Res Part C: Emerg Technol 8(1):257–285
7. Beymer D, McLauchlan P, Coifman B, Malik J (1997) A real-time computer vision system for measuring traffic parameters. In: Proceedings of IEEE computer society conference on computer vision and pattern recognition, pp 495–501
8. LeCun Y, Boser B, Denker JS, Henderson D, Howard RE, Hubbard W, Jackel LD (1989) Backpropagation applied to handwritten zip code recognition. Neural Comput 1(4):541–551
9. Spampinato C, Faro A, Giordano D (2008) Evaluation of the traffic parameters in a metropolitan area by fusing visual perceptions and cnn processing of webcam images. EEE Trans Neural Netw 19(6):1108–1129
10. Sun Z, Bebis G, Miller R (2004) On-road vehicle detection using optical sensors: a review. In: Proceedings. The 7th international IEEE conference on intelligent transportation systems (IEEE Cat. No.04TH8749), pp 585–590
11. Redmon J, Divvala S, Girshick R, Farhadi A (2015) Unified, real-time object detection, You only look once
12. Howard AG, Zhu M, Chen B, Kalenichenko D, Wang W, Weyand T, Andreetto M, Adam H (2017) Efficient convolutional neural networks for mobile vision applications. Mobilenets
13. Zhang P, Zhong Y, Li X (2019) SlimYOLOv3: narrower, faster and better for real-time UAV applications. In: 2019 IEEE/CVF international conference on computer vision workshop (ICCVW). IEEE
14. Lin TY, Dollr P, Girshick R, He K, Hariharan B, Belongie S (2016) Feature pyramid networks for object detection
15. Bansal V, Foresti GL, Martinel N (2021) Where did i see it? object instance re-identification with attention. In: 2021 IEEE/CVF international conference on computer vision workshops (ICCVW), pp 298–306
16. Vaswani A, Shazeer N, Parmar N, Uszkoreit J, Jones L, Lukasz Kaiser AN, Polosukhin I (2017) Attention is all you need, Gomez
17. Sankaranarayanan M, Mala C, Mathew S (2020) Virtual mono-layered continuous containers for vehicle detection applications in intelligent transportation systems. J Discret Math Sci Cryptography 23:321–328
18. Yadav G, Maheshwari S, Agarwal A (2014) Contrast limited adaptive histogram equalization based enhancement for real time video system. In: 2014 international conference on advances in computing, communications and informatics (ICACCI) (2014), pp 2392–2397
19. Roerdink J, Meijster A (2003) The watershed transform: definitions, algorithms and parallelization strategies. Fundam Inf 41:10
20. Kanopoulos N, Vasanthavada N, Baker RL (1988) Design of an image edge detection filter using the sobel operator. IEEE J Solid-State Circuits 23(2):358–367
21. Lucas B, Kanade T (1981) An iterative image registration technique with an application to stereo vision (ijcai) 81:04

22. Zhou H, Yuan Y, Shi C (2009) Object tracking using sift features and mean shift. Comput Vis Image Understand 113(3):345–352. Special issue on video analysis
23. Luo Z, Branchaud-Charron F, Lemaire C, Konrad J, Li S, Mishra A, Achkar A, Eichel J, Jodoin P-M (2018) Mio-tcd: a new benchmark dataset for vehicle classification and localization. IEEE Trans Image Process 27(10):5129–5141
24. Wen L, Dawei D, Cai Z, Lei Z, Chang M-C, Qi H, Lim J, Yang M-H, Lyu S (2015) A new benchmark and protocol for multi-object detection and tracking, Ua-detrac

Chapter 24
FTIR-Based Characterization and Classification of Various Indian Monofloral Honey Samples

S. M. Annapurna, Sunil Rajora, Yoginder Kumar, V. Sai Krishna, and Navjot Kumar

1 Introduction

Honey is a sweet, fragrant, natural material. It is produced by honeybees from nectar of plants or from honeydew [1]. Since ancient times, honey has been consumed by humans without processing. It can be used as a dietary supplement due to its nutritional values. It also has medicinal, dermatological, and industrial uses. Honey has antioxidant and antibacterial properties that can contribute positively to the immune system and can even help with cardiovascular protection. Honey is effective for wound treatment and its intake can increase serum antioxidant potential [2]. Honey is a natural preservative. It can be stored, closed in a cool place, and it never gets spoiled. It is used in beverages due to its preservative properties [3, 4]. Antioxidants in honey can aid in reducing the risk of cardiovascular diseases, cancer, and inflammatory diseases [5]. Honey has nearly 200 components, of which sugar, glucose, and fructose are the major constituents. The other components in honey include proteins, minerals, flavonoids, phenolic acids, phytochemicals, enzymes, and vitamins. The therapeutic effects of honey mainly arise from the presence of phytochemicals, which in turn originate from the floral nectar and pollen collection by bees. There are variations among different plants, which contribute to the wide variety of bioactive substances in honey. However, flavonoids and phenolic acids are abundantly found [3]. Flavonoids are a large subset of phytochemicals and have a wide range of chemoprotective effects [6].

S. M. Annapurna (✉) · S. Rajora · Y. Kumar · V. Sai Krishna · N. Kumar
CSIR-Central Electronics Engineering Research Institute, Pilani, Jhunjhunu 333031, India
e-mail: smannapurna@gmail.com

V. Sai Krishna
e-mail: sai@ceeri.res.in

N. Kumar
e-mail: navjot@ceeri.res.in

G. Mathur et al. (eds.), *Proceedings of 3rd International Conference on Artificial Intelligence: Advances and Applications*, Algorithms for Intelligent Systems,
https://doi.org/10.1007/978-981-19-7041-2_24

The nature of substances used by bees greatly influences the characteristics of honey, followed by processes employed for its extraction, preparation for market, and conservation [7]. The floral source of honey is one of the key parameters in determining its quality. The composition and antioxidant property of honey are dependent on the flora source from which nectar is collected, climatic variations, and environmental factors [8]. These sets of parameters are further used to determine the price of the product and its certification. Certification of honey depends on its production mechanisms as well as on its botanical and geographical origin [9]. Knowledge of the botanical origin of a honey sample can increase its medicinal utility and can aid in further technological developments. The relative frequencies of pollen types of floral species could be used to determine the biological origin of honey [10]. Honey can be categorized as monofloral or multifloral depending on the botanical origin. Monofloral honey is defined as a sample that primarily comes from a dominant floral source and demonstrates the usual flowing characteristic of the associated type of honey. The relative frequency of that taxon's pollen in a sample must exceed 45% to be deemed monofloral [10]. The flavor, taste, and medicinal attributes make monofloral honey a better choice when compared to multifloral honey. Also, monofloral honey has the potential to compete with low-cost polyfloral honey.

Certain sensorial features, like color, aroma, texture, and flavor are characteristic of the floral source. Physicochemical properties such as ash, acidity, protein and sugar content, moisture, electrical conductivity, pH value, etc., are also influenced by the source. Furthermore, some plant species produce toxic nectar that is not fit for consumption [11]. Therefore, determination of the botanical origin of honey is an area of growing interest.

The honey industry, like any other food industry, is not free from quality fraud. In spite of strict administrative rules, adulteration is a common issue in the honey market. Samples can be adulterated by using syrups of sugar, jaggery, corn, caramel, glucose, etc. Monofloral honey can be mixed with other samples from different botanical origins and then be sold under a false name. Pollen analysis is a common practice to determine the origin of the honey sample [10]. It includes steps like identification of the floral calendar, density of the honey plant, and sensorial detection prior to pollen analysis. By using liquid chromatography, the variations in sugar content in honey samples can be expressed as a function of their floral origin [12]. Electronic tongue is an alternative to traditional methods which are time-consuming. It also involves less complicated sample preparation and can be used for multicomponent analysis [13, 14]. Cyclic voltammetry electric tongue can also be used for the classification of monofloral honey [9].

Chromatography-based measurements are time-consuming and require the use of chemicals. Also, they are destructive measurement strategies. Hence, FTIR spectroscopy integrated with chemometrics-based multivariate data analysis is an alternative for the detection of adulteration in honey. Lately, infrared (IR) spectroscopy is gaining more acceptance in the food industry. With this method, solid and liquid spectra can be recorded without any sample preparation. It is a low-cost method involving portable instrumentation that can be used for fast scanning.

This work aims at using an FTIR spectrometer with an ATR sampling accessory along with chemometric-based methods for the characterization of honey samples from different botanical origins. Classification models have been developed by using K-means, K-medians, and SVM.

2 Materials and Methods

2.1 Preparation of Samples

For this study, seven monofloral honey samples and four adulterant samples were collected from different parts of India. The monofloral honey samples analyzed are Neem (Azadirachta indica), Mustard (Brassica nigra), Tulsi (Ocimum tenuiflorum), Ajwain (Trachyspermum ammi), Rosewood (Dalbergia sissoo), Ginger (Zingiber officinale), and Litchi (Litchi chinensis). The honey and adulterant samples used for this work are listed in Table 1.

The samples are analyzed using the FTIR spectrometer. The block diagram of the equipment is shown in Fig. 1.

2.2 Data Acquisition Using NIR Spectrophotometer

The FTIR spectrophotometer has a spectral range of 4000–600 cm^{-1} and a resolution of 1.5 cm^{-1}. This instrument was used for the acquisition of spectra from monofloral honey and adulterant syrup samples. The equipment employs an Attenuated Total

Table 1 List of monofloral honey and adulterant syrup samples

S. no.	Name of the sample	Category
1	Neem honey	Monofloral
2	Mustard honey	Monofloral
3	Tulsi honey	Monofloral
4	Ajwain honey	Monofloral
5	Rosewood honey	Monofloral
6	Ginger honey	Monofloral
7	Litchi honey	Monofloral
8	Sugar syrup	Adulterant
9	Jaggery syrup	Adulterant
10	Glucose syrup	Adulterant
11	Rice syrup	Adulterant

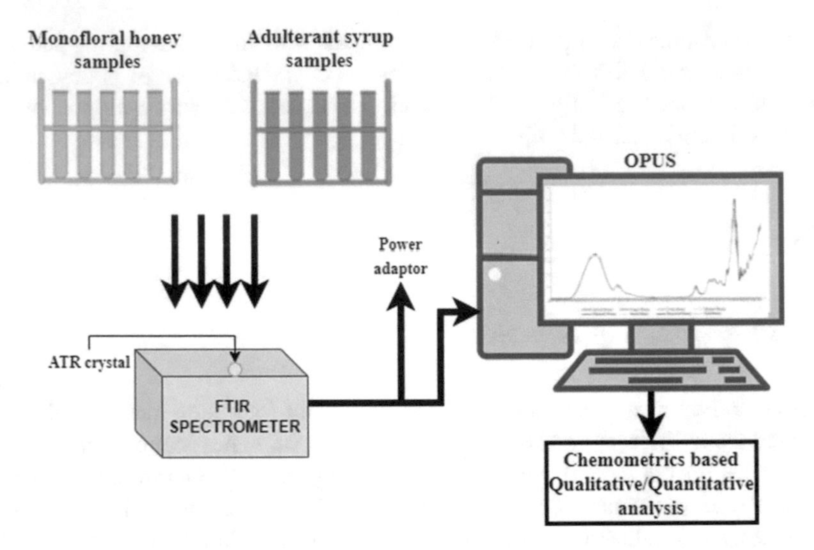

Fig. 1 Equipment for recording NIR spectra of monofloral honey and adulterant syrup samples

Reflectance (ATR)-based sampling technique using diamond ATR crystal. The penetration depth of infrared is in μm range. Hence, a very thin sampling path is required. This can be achieved by using an ATR-based sampling technique [15].

The samples are scanned one after another by dropping them over the ATR crystal. The crystal must be covered completely while placing the sample. Also, bubble formation should not happen. Absorbance mode was enabled while recording the spectrum from the spectrometer. The entire spectral range of 4000–600 cm^{-1} was utilized using OPUS software. Background subtraction was enabled while recording the spectrum. The spectrum was preprocessed by using techniques such as EMSC and second-order derivative preprocessing. The acquired spectra are as shown in Fig. 2.

3 Data Pre-treatment and Chemometrics

Scattering effects can cause unwanted redundancy while recording NIR spectra. Environmental conditions, functional defects of the electronic component, and surface roughness of the sample can induce scattering. Additive, multiplicative, and wavelength-specific scattering effects will arise due to the effect of these factors. Scattering results in baseline shift, slope variation, or unwanted peaks in the recorded spectra [16]. Also, the NIR spectra recorded are not free of noise and background variations. These are not related to the variables or properties of one's interest while analyzing the spectra. Therefore, application of preprocessing techniques becomes a very important step before qualitative or quantitative analysis. The sequence of steps involved in chemometrics-based qualitative analysis is illustrated in Fig. 3.

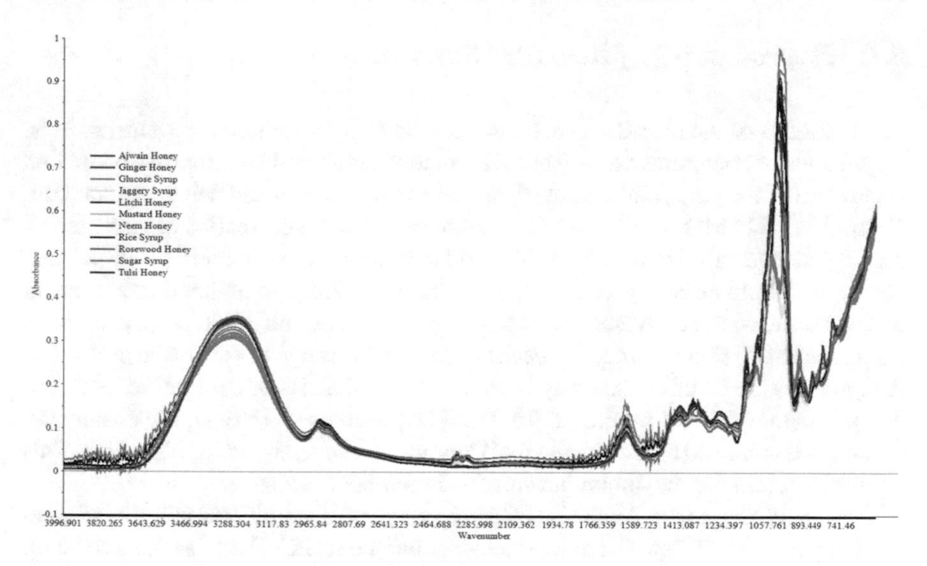

Fig. 2 Spectrum of monofloral honey and adulterant syrup samples

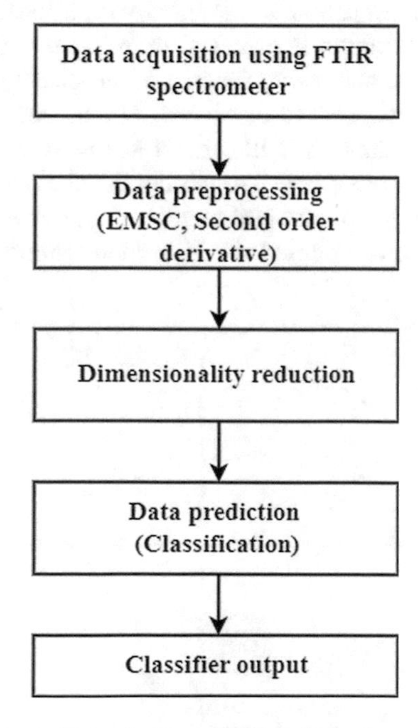

Fig. 3 Steps for data preprocessing and chemometric analysis

3.1 Preprocessing of Recorded Spectra

Preprocessing of the recorded spectra is a crucial step in chemometric data analysis. Application of chemometrics will result in robust models and will improve prediction accuracy. In this work, multivariate data analysis was conducted using Unscrambler X, from M/S CAMO. EMSC and second-order derivative correction are the preprocessing methods used to rectify additive and multiplicative scatter effects [16]. Visualization of multivariate spectral data is a complex task due to higher dimensions. In this work, a total of seven monofloral honey samples and four adulterant syrups have been scanned 20 times over 2656 spectral points in the range of 4000–600 cm^{-1}. The data matrix size in this case study becomes [220 * 2656] for calibration. For validation, another data set of size [220 * 2656] has been used. Further, dimensionality reduction is achieved by using Principal Component Analysis (PCA) algorithm. This step helps retain the maximum information in terms of variance.

Principal Component Analysis (PCA) represents the actual data in terms of principal components (PCs). The first principal component (PC-1) attains the maximum variance when compared to the other principal components (PC). Significant information in the actual data set can be represented with a lower number of PCs. This is how dimensionality reduction is achieved in PCA. It was possible to represent 99% of the variance in the data using the first seven principal components. Therefore, the remaining 1% data can be discarded. The score plot of PC1 versus PC2 for the monofloral honey samples is illustrated in Fig. 4. From Fig. 4, all the pure monofloral honey samples are grouped as per botanical source along PC1. Hence, it can be concluded that all the monofloral honey samples are uniquely characterized by using the FTIR spectrometer coupled with a chemometrics-based approach.

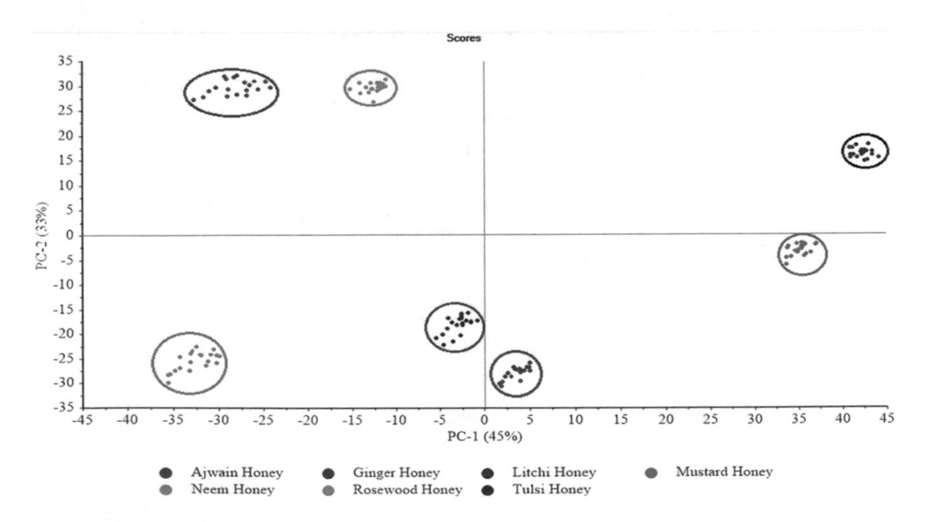

Fig. 4 PC1–PC2 score plot

Table 2 Classification algorithms and their prediction accuracy

Model	Total sample	True positive	True negative	Accuracy (%)
K-means	220	220	0	100
K-medians	220	220	0	100
SVM	220	220	0	100

3.2 Sample Classification

Classification is the process of assigning a predefined class to a given entity. Seven different varieties of pure honey were used for this study; hence, there are seven classes. Each FTIR spectrometer reading must be grouped into one of these seven categories. A classification model has to be built for each sample. While trying to classify a sample, there are three probable outcomes. The sample can belong to a single class, it can belong to more than one class, or it might not belong to any class.

During the spectral data analysis, it is found that the peaks are moderately overlapping. There are no signature traces for any honey sample over the entire spectral range. However, the monofloral samples are found to have some relevant bands in terms of unique peak width and peak shape. It can be observed that all the monofloral honey samples have a signature band. Mustard honey, neem honey, and litchi honey have distinct curves in the range of 1458–1170 cm^{-1}. Rosewood honey has a significant peak in the range of 2824–1499 cm^{-1}. Ajwain honey is distinguishable in the range of 1152–934 cm^{-1}. Ginger honey has a distinct peak at 1018–985 cm^{-1}. Tulsi honey is detected in the range of 2796–1743 cm^{-1}.

In this work, data classification is done by using three classification models. The models used are K-means, K-medians, and SVM. First, the models are built using the training data set [220 × 2656]. The model is tested using the test data set [220 × 2656]. Preprocessing (EMSC, second-order derivative-based) and PCA were done prior to building these models. The comparison of the performance of these classifiers is listed in Table 2.

4 Conclusion

In this work, FTIR spectroscopy with an ATR sampling accessory was utilized for the classification of pure honey samples. A total of seven varieties of monofloral honey samples were directly collected from farmers. Also, four adulterant syrups are analyzed. Preprocessing techniques have been applied to the data. Three chemometric models, namely K-means, K-medians, and SVM have been developed for classifying the data. All the models are successful in predicting the pure monofloral honey samples and adulterant syrups in the data sets. Therefore, it can be concluded that FTIR spectroscopy together with chemometrics strategy is efficient in identifying various monofloral honey samples and adulterant syrups.

Acknowledgements The authors would like to thank all the members of the Dairy and food instrumentation division, CSIR-CEERI, IIH, Jaipur Centre and the Director, CSIR-CEERI, Pilani, for this opportunity as well as their cooperation and support throughout this research work.

References

1. Codex Alimentarius Commission (1981) Revised Codex Standard for Honey Codex Stan 12. Codex Standard, vol 12, pp 1–7
2. Gheldof N, Wang XH, Engeseth NJ (2003) Buckwheat honey increases serum antioxidant capacity in humans. J Agric Food Chem 51(5):1500–1505
3. Naliganti C, Akkinepally RR, Valupadas C (2019) Association of quality of life with severity, nutrition and medication adherence in heart failure. Asian J Pharm Pharmacol 5:373–377
4. Nguyen H, Panyoyai N, Paramita V, Mantri N, Kasapis S (2017) Physicochemical and viscoelastic properties of honey from medicinal plants. Food Chem 241
5. Kuś PM, Congiu F, Teper D, Sroka Z, Jerković I, Tuberoso CI (2014) Antioxidant activity, color characteristics, total phenol content and general HPLC fingerprints of six Polish unifloral honey types. LWT Food Sci Technol 55:124–130
6. Mukisa IM, Muyanja C, Byaruhanga Y, Langsrud T, Narvhus J (2012) Physicochemical properties, sugars, organic acids and volatile organic compounds of different types of Obushera during natural fermentation. Afr J Food Agric Nutr Dev 12:6655–6685
7. Marcazzan GL, Magli M, Piana L, Savino A, Steffano MA (2014) Sensory profile research on the main Italian typologies of monofloral honey: possible developments and applications. J Apic Res 53(4):426–437
8. Gül A, Pehlivan T (2018) Antioxidant activities of some monofloral honey types produced across Turkey. Saudi J Biol Sci 25(6)
9. Tiwari K, Tudu B, Bandyopadhyay R, Chatterjee A (2012) Discrimination of monofloral honey using cyclic voltammetry. In: Proceedings—2012 3rd national conference on emerging trends and applications in computer science, pp 132–136
10. Belay A, Haki DG, Birringer M, Borck H, Addi A, Baye K, Melaku S (2016) Rheology and botanical origin of Ethiopian monofloral honey. LWT—Food Sci Technol 75(4)
11. Gupta A, Gill JPS, Bedi J, Manav M, Ansari M, Walia GS (2018) Sensorial and physicochemical analysis of Indian honeys for assessment of quality and floral origins. Food Res Int 108
12. Cotte JF, Casabianca H, Chardon S, Lheritier J, Grenier-Loustalot M (2004) Chromatographic analysis of sugars applied to the characterisation of monofloral honey. Anal Bioanal Chem 380:698–705
13. Tiwari K, Tudu B, Bandyopadhyay R, Chatterjee A (2013) Identification of monofloral honey using voltammetric electronic tongue. J Food Eng 117:205–210
14. Sousa M, Dias LG, Veloso ACA, Estevinho L, Peres AM, Machado A (2014) Practical procedure for discriminating monofloral honeys with a broad pollen profile variability using an electronic tongue. Talanta 128:284–292
15. Kumar N, Ranjan R, Kumar Y, Patel S, Krishna V, Appaiah A, Gupta K, Panchariya PC (2021) Discrimination of various pure honey samples and its adulterants using FTIR spectroscopy coupled with chemometrics, 808–811
16. Kumar N, Panchariya PC, Patel S, Kiranmayee AH, Ranjan R (2018) Application of various pre-processing techniques on infrared (IR) spectroscopy data for classification of different ghee samples, 1–6

Chapter 25
A Surveillance Mobile Robot Based on Low-Cost Embedded Computers

Krisanth Tharmalingam and Emanuele Lindo Secco

1 Introduction

Use of robots for the purpose of surveillance has been increasing. Companies are expressing their interests in implementing robots for their business. There are some benefits in using robots for this task, such as, for example, the fact that the robot can work continuously, the maintenance cost could be reasonable, the communication can be embedded in the system and allows the human operator to remotely monitor the plant [1].

In some countries like Asia, companies mainly hire humans for security purpose as they can't afford to buy a robot. Securities come and secure the property. If that is a large place, they hire multiple securities. They make a schedule between them and walk around the factory or warehouse according to their schedule. For example, if a person walks around the factory this time, another one goes after some minutes and hours. There are clearly some problems that security systems have to face, such as, for example:

1. Most of the time they walk around the warehouse lonely which is not safe for them.
2. It is very hard for them to walk continuously as they need some rest.
3. Even if they encounter any suspicious activities, it is very hard for them to alert others. Before they alert others, possibility to be attacked.

According to these scenarios, we want to develop a security robotic system, given the demand to make this kind of robot is high, especially in some countries. A

K. Tharmalingam
Azenta Life Sciences, Manchester, UK

K. Tharmalingam · E. L. Secco (✉)
Robotics Lab, School of Mathematics, Computer Science and Engineering, Liverpool Hope University, Liverpool, UK
e-mail: seccoe@hope.ac.uk

G. Mathur et al. (eds.), *Proceedings of 3rd International Conference on Artificial Intelligence: Advances and Applications*, Algorithms for Intelligent Systems,
https://doi.org/10.1007/978-981-19-7041-2_25

literature review has been also performed in order to analyze some existing similar systems which are currently available in the market. The following section presents some of these systems.

1.1 Security Robotic Systems

There are different systems for security and surveillance which are currently available in the market: here we present some details of the most significant ones.

Leo-Rover Robot

The Leo-Rover has a cost of £ 3024.79, it is not autonomous vehicle that can be controlled remotely. Its main software is based on Robotics Operating System (ROS) with video streaming feature, thanks to a camera at the front of the vehicle. It is a device suitable for outdoor explorations with a built-in Raspberry PI controller, 4 DC motors and Wi-Fi connectivity and an overall weight 6.5 kg [2].

The robot is quite big in size. Wheels are properly suitable for outdoor activity; however, the camera is fixed and there is no option to turn the camera position and monitor what is happening around the vehicle.

Si-Surveillance Robot

This robot provides video surveillance remotely with a panoramic camera at 360°. It is capable of detecting people up to 100 m in any direction; it also embeds an Artificial Intelligence based system to analyze the contents of the video. The device is suitable for indoor and outdoor explorations with a total weight of up to 185 kg, and facial recognition of up to 50 m. The robot can also detect weapons and identify who has the permission to stay inside the premise that you are going to monitor [3].

Some of the features of this robot are interesting with respect to the system which we are going to develop and present here. For example, one of the main purposes of the device could be the surveillance after the opening hours of a warehouse or a factory. Therefore, it is of interest to have the possibility to detect persons and/or objects in that practical scenario. Nevertheless, the weight of the SI robot is very high and it maybe more interesting to approach a design which involve a lower weight of equipment.

ID-2868 Robot

This device has a weatherproof chassis for outdoor activities, a camera that covers 360°, infrared cameras for night activities, object detection capability, as well as it can detect number plates and human beings. The robot is autonomous and can plan its path automatically. The robot can also send notifications with photos.

The robot's navigation system is equipped with a LIDAR which helps to adjust the path when it detects obstacles on its way. The main software is based on ROS. One of the main features of this robot is its autonomous function and navigation capability [4].

ID-2729 Robot

The ID-2729 is a small size device, which can easily reach areas that are not easily accessible to people. A remote access to its camera allows taking pictures and videos while the robot moves. It has a weight of 3.63 kg and a wheelie bar which provides the ability to climb obstacles without rolling over. A built-in SD card reader allows saving videos while the camera keeps recording even if the wireless communication drops down. Interestingly, it also embeds a swappable battery that can run the robot continuously for about 2 h [5].

According to the aforementioned characteristics of some surveillance robots, this paper presents the development of a mobile robot for security and surveillance which will be designed to perform its activity on daily time and, moreover, on night time.

The proposed system will be characterized by four wheels and a camera in the front. We would also like to integrate a buzzer on the top of the robot (Fig. 1).

In order to keep the cost of the system as low as possible, an Arduino Uno will be used to control the wheels by means of a low-cost L298N motor driver.

An Ultrasonic sensor—which is connected to the Arduino board—will also be included, since we want the robot to be able to explore the environment using obstacle avoidance technology.

Finally, a PIR Sensor—which will be connected to a further control board, namely a Raspberry PI, will be embedded. Whether the PIR sensor will perform a motion

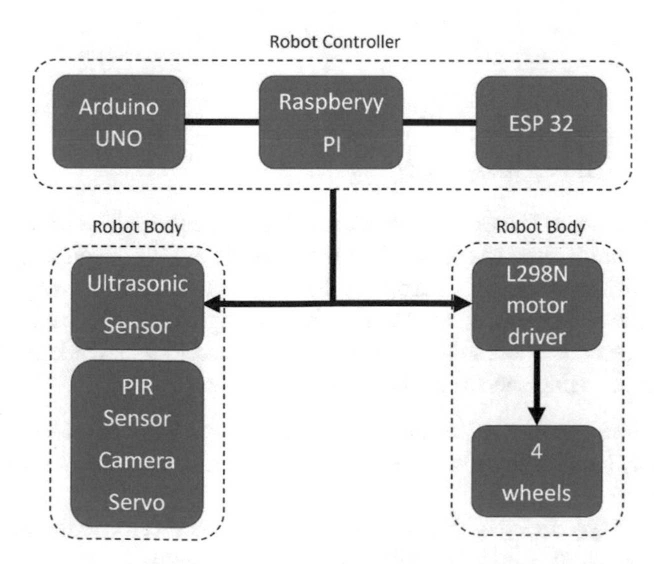

Fig. 1 Overall architecture of the robot

detection, then the system will trigger the camera (to capture a picture of the scene) and the buzzer (to provide an acoustic alarm feedback).

The end-user will then be able to wirelessly receive the picture over an IEEE 802.11 communication protocol (e.g., wi-fi). Moreover, the camera will be connected to a servo motor which is wirelessly controlled by an ESP32 board. By changing the position of the servo, the orientation of the camera will be changed (Fig. 1).

2 Materials and Methods

This section refers to the materials that were used in the project, and how the hardware and parts were assembled and organized. A summary about the Programming Code which has been integrated in the system is also reported.

The following parts were adopted for the project:

- Arduino Uno
- ESP32
- Raspberry PI
- Raspberry PI camera module
- 16 GB Micro SD card
- Servo Motor
- 4 DC Motors
- L298N Motor Driver
- HC-SR04 Ultrasonic Sensor
- PIR Sensor (Passive Infrared Sensor)
- LED light
- 4 Wheels
- Buzzer
- Router
- Jumper wires
- Other consumables/lab materials.

The Arduino board is used to control the vehicle wheels and the ultrasonic sensor. The ESP32 board is used to control the servo motor position wirelessly over a wi-fi communication protocol. The Raspberry PI is used to detect the motion and then capture a frame and trigger an alert to the end-user by means of a beeping buzzer.

All the microcontrollers in the vehicle are designed to work independently. The main reasons for using such a few microcontrollers are:

- It makes the system reactive (in case of many components connected to a single microcontroller, it may take more time to read all the sensors and then react accordingly)
- It makes the system versatile and it will be easy to add some extra features in the future (such as, for example, an object detection system)
- It simplifies the debugging process and enhances modularity.

The development of the project is divided into three main stages, namely

- Step 01—Communicate to the motors using the Arduino Uno and the Ultrasonic sensor
- Step 02—Communicate to the servo motor by using the ESP32 board
- Step 03—Communicate to the Raspberry PI and to the camera by using the Raspberry PI board

All these steps are developed and tested individually, then all the systems are connected to the vehicle and fully integrated at the end. The following paragraphs detail the above-mentioned three integration steps.

2.1 Controlling the Motors and The Ultrasonic Sensor (i.e., The Arduino Board)

Motors should not be connected to the microcontroller directly. They are then connected to a L298N motor driver. An HC-SR04 Ultrasonic sensor to avoid obstacles is also incorporated. The ENA, ENB, IN1, IN2, IN3, and IN4 are the outputs and Motor Terminal 1, whereas Motor Terminal 2 are inputs in the L298N motor controller. Therefore, the ENA, ENB, IN1, IN2, IN3, and IN4 are connected to the Arduino Uno board, while the Motor Terminal 1 and Motor Terminal 2 are connected to the motors.

The design foresees two motors in the left-hand side of the robot—which are connected to the Motor 1 Terminal—and two motors in the right-hand side—which are connected to the Motor 2 Terminal [6].

To turn the vehicle to the right or to the left, there are two basic approaches. One is based on turning off one side motors and activating the other side motors to go forward. A second method is based on turning both the side motors in the opposite side (i.e., the tank mode). In this project, second method is used. Table 1 summarizes the motor strategy according to this approach.

The ultrasonic sensor is placed at the front of the robot to scan the environment. The sensor is directly connected to the Arduino Uno board in order to measure how far are the obstacles from the front of the robot. In this project, a threshold distance is fixed at 20 cm. Whenever the distance is more than 20 cm between the robot and any

Table 1 Motor control strategy

Robot movement	Left motors/wheels	Right motors/wheels
Forward	Forward	Forward
Backward	Backward	Backward
Stop	Stop	Stop
Turn right	Forward	Backward/stop
Turn left	Backward/stop	Forward

objects, then the robot will keep moving forward. On the contrary, the vehicle will stop moving, it will then go back slightly and then it will randomly choose whether to turn right or left. This approach allows the robot to be unpredictable vs any possible presence which is trying to avoid the robot itself. The following structured code summaries the rational of the approach.

```
Int randNum = rand() % 2; //Generate number 0 or 1
  if (randNum == 0) {    //To turn Right
    turnRight();
    delay(1000);
  }
  else {
    turnLeft();        //To turn Left
}
```

2.2 Controlling the Servo Motor of the Camera (i.e., The ESP32 Board)

A servo motor is connected to the ESP32 microcontroller directly. The reason for using an ESP32 for this step is that the Arduino Uno board does not straightforwardly integrate a Wi-Fi communication module.

A bracket is also mounted at the end of the servo motor in order to position the camera and the PIR Sensor.

A webserver is available for the end-user by entering the IP address of the ESP32 in the Internet Browser. From the mask, the user can choose whether to turn the camera to the RIGHT or to the LEFT. A static IP address of ESP32 is also set on the board.

Then by means of the following code on the serial monitor, the IP address of the ESP32 board is communicated [7]:

```
#include <WiFi.h>
const char* ssid = "Your Router name";
const char* password = "Your Router's passward";
WiFiServer server(80);

void setup() {
  Serial.begin(1prachi200);
  Serial.print("Connecting to ");
  Serial.println(ssid);
  WiFi.begin(ssid, password);
  while (WiFi.status() != WL_CONNECTED) {
```

```
    delay(400);
    Serial.print(".");
  }
  // Print local IP address
  Serial.println("");
  Serial.println("WiFi is connected.");
  Serial.println("IP address;: ");
  Serial.println(WiFi.localIP());
  server.begin();
}
void loop(){
}
```

2.3 Raspberry PI, Buzzer and Camera

The Raspberry PI camera module is connected to a Raspberry PI Zero W board. The PIR Sensor and the buzzer are also connected to the Raspberry PI, such as, in case the PIR sensor will detect any movements, the camera will take a photo and the buzzer will start beeping. Camera and PIR sensor are mounted on a bracket which is connected to servo motor, whereas the Buzzer is placed on top of the vehicle. It is possible to access the Raspberry PI by using a VNC Viewer or PUTTY directly from the computer.

After logging into the board, the user needs to type on the terminal "*sudo raspi-config*", then go to the interface option and enable VNC. A reboot of the Raspberry PI by using the "*sudo reboot*" command enables then the camera. A reboot will be needed every time changes are applied (Fig. 2).

Finally, we need to connect one wire of the buzzer to the Raspberry PI (at the pin 17) and connect the other wire to the ground. Connect the middle wire of the PIR Sensor at the pin number 4 of the board, the red wire to the 5 V pin, and the black one to the ground pin as well [8, 9]. The following code need also to be implemented.

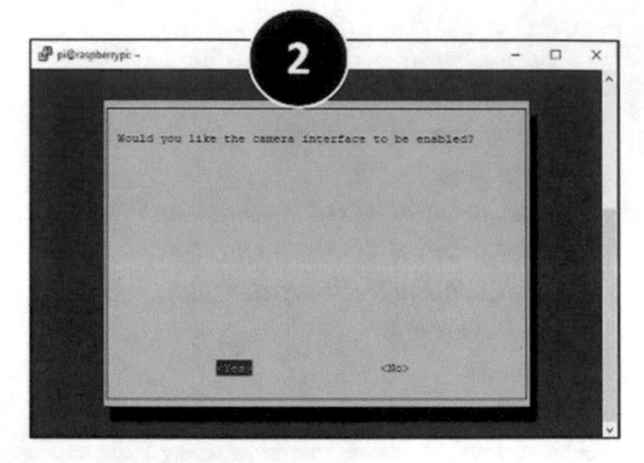

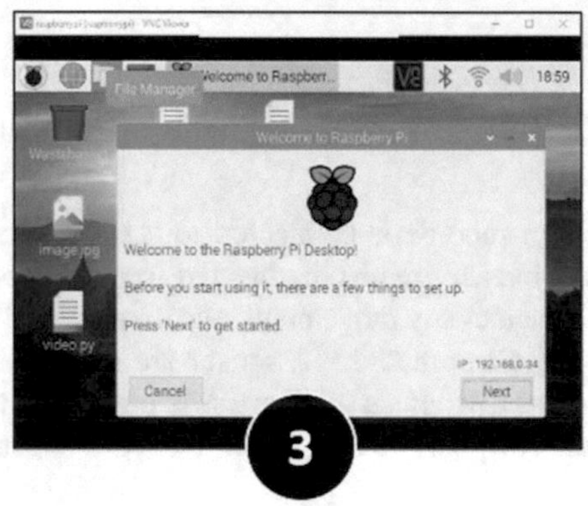

Fig. 2 Connecting to the Raspberry PI via Putty (1), enabling the camera (2) and logging into the Raspberry PI board (3)

```
from gpiozero import MotionSensor
from picamera import PiCamera
import RPi.GPIO as GPIO
import time

buzz = 17
motion = MotionSensor(4)
GPIO.output(buzz, GPIO.OUT)
cam = PiCamera()

while True:
    motion.wait_for_motion()        //Wait for motion
    GPIO.output(buzz, GPIO.HIGH) //Start the buzzer after detect the motion
    time.sleep(1)
    GPIO.output(buzz, GPIO.LOW)   //Off the buzzer
    time.sleep(1)
    cam.capture("/home/pi/image2.jpg")     //Capture the photo and save
    time.sleep(10)
```

3 Preliminary Tests and Results

This section reports all the preliminary trials which were run in order to evaluate the final performance of the robotic system. Initially, the testing was performed on single components, then tests were performed on the overall system. Each session of tests was made of 10 trials.

Test 1

In the current version of the system, the servo motor of the camera has been connected to the ESP32 board. Nevertheless, initially, this servo was connected to the Arduino Uno board. This latter board was then taking care of the servo motor, the Ultrasonic sensor, the L298N Motor driver, and, finally, the 4 DC Motors. This configuration made the system pretty low on reacting and, as soon as we performed 10 trials, we always noticed that the system was slow (4/10) if not very slow (6/10). As soon as the servo motor was removed from the Arduino board and connected to the ESP32, performances of the system changed (10/10).

Test 2

We evaluated the performance of Ultrasonic Sensor and whether it works well in order to detect obstacles and allow the robot to avoid them. On 10 trials, 9 times (over 10) the system was detecting the obstacle. The failure was due to a blind spot, since the Ultrasonic sensor was positioned at the front of the vehicle. There is a possibility to avoid this issue by using a minimum of 3 sensors. In future development, we should adopt 3 Ultrasonic sensors in the vehicle: one at the front, the second sensor should be 45° to the left with respect to the middle sensor, while the third one should be 45° to the right.

Test 3

In this set of trials, the position of the servo motor was controlled by the ESP32 wireless communication by means of the webserver. The main purpose of this test was to find out whether the servo was reacting as it was instructed to do. For example, when the "LEFT" button in the webserver was pressed, we checked whether the servo was turning to the right direction. Communication and servo behavior were observed as consistent and reliable over all the trials (10/10).

Test 4

In this section, some tests evaluated the performance of the Raspberry PI board. The main purpose of this test is to find out whether it captures a photo and beeps the buzzer when it detected a motion. It was necessary for the system to both capture the frame and beep the buzzer in case of the motion detection event. Otherwise, the test was reported as a failure. Table 2 reports the result of a set of 10 trials.

When repeating the test quickly, the buzzer and camera did not respond properly on some tests: this may require some changes in the coding when where the "time.

Table 2 Testing the camera and the buzzer versus motion detection

Trial	Camera performance	Buzzer performance
1	Success	Success
2	Success	Failure
3	Success	Success
4	Success	Success
5	Success	Success
6	Success	Success
7	Failure	Success
8	Success	Failure
9	Success	Success
10	Success	Success

sleep" function is set. Future investigation should be done in order to improve the system on this aspect.

Test 5

In the previous test, the system was checked by taking photo frames. In this test, a live video streaming function was added to the system and tested. During these trials, there was a latency in the live streaming. The latency was pretty relevant in the order of 5 s. Because of this latency, the live streaming function was removed from the system. This should be reconsidered in the future development of the system.

Test 6

A final session of trials was focusing on the performance of the whole integrated system. At this stage, all the components were installed into the vehicle and monitored.

The robot was observed in action while detecting obstacles, performing image capturing and beeping in case of motion detection. It was successfully performed 8 times out of 10: in two trials, problems were encountered while the robot was capturing the frame and beeping. In all the trials, the robot was properly avoiding the obstacles.

4 Discussion and Conclusion

The proposed system has multiple microcontrollers and therefore it will be very easy to add extra features to the system. The system also finds its path by scanning the environment through its Ultrasonic sensor and the vehicle is substantially moving randomly. In the future it would be of interest to use a better navigation strategy, such as for example *Simultaneous Localization and Mapping* (SLAM) technology. SLAM lets the vehicle to map the unknown environment and acts accordingly [10,

11]. SLAM may also use some information from the sensors of the vehicle, such as, for example, the data from the frontal camera, the wheel revolutions to determine its current location and so on [12]. Clearly there are also potentials in order to increment the visual recognition system of the robot and make it more interactive [13].

The proposed Raspberry PI camera captures a photo when it detects a motion and alerts the end-user by beeping the buzzer. At present this buzzer is placed on board of the vehicle, but—in the future—the buzzer maybe also integrated in the device of the end-user.

At present we are providing separate power supplies for each microcontroller, however, in a further step, it would be nice to use one integrated power source for the whole system.

Acknowledgements This work was presented in dissertation form in fulfilment of the requirements for the M.Sc. in Robotics Engineering for the student Krisanth Tharmalingam under the supervision of E.L. Secco from the Robotics Laboratory, School of Mathematics, Computer Science and Engineering, Liverpool Hope University.

References

1. Guest P (2021) Advantages & disadvantages of robots|PlastikMedia News. Plastikmedia.co.uk. https://www.plastikmedia.co.uk/advantages-disadvantages-of-industrial-robots/. Accessed 10 Oct 2021
2. Robotshop.com (2021) Leo Rover 4WD Developer Kit (Assembled). https://www.robotshop.com/uk/leo-rover-4wd-developer-kit-assembled.html. Accessed 9 Oct 2021
3. Series S (2021) Deep learning surveillance|Artificial intelligence of a robot|Big data for law enforcement and security|SMP robotics®. Smpsecurityrobot.com. https://smpsecurityrobot.com/products/ai-surveillance-robot/. Accessed 9 Oct 2021
4. Superdroidrobots.com (2021) SPAR—Autonomous Indoor/Outdoor Security Robot. https://www.superdroidrobots.com/index.php?route=product/product&product_id=2858. Accessed 9 Oct 2021
5. Superdroidrobots.com (2021) Security robots. https://www.superdroidrobots.com/shop/custom.aspx/security-robots/93/. Accessed 9 Oct 2021
6. Teachmemicro.com (2021). https://www.teachmemicro.com/use-l298n-motor-driver/. Accessed 10 Oct 2021
7. Randomnerdtutorials.com (2021) ESP32 Servo Motor Web Server with Arduino IDE|Random Nerd tutorials. https://randomnerdtutorials.com/esp32-servo-motor-web-server-arduino-ide/. Accessed 10 Oct 2021
8. Arduino Project Hub (2021) Use a Buzzer Module (Piezo Speaker) using Arduino UNO. https://create.arduino.cc/projecthub/SURYATEJA/use-a-buzzer-module-piezo-speaker-using-arduino-uno-89df45. Accessed 10 Oct 2021
9. W, T (2021) Time Lapse Raspberry Pi Zero W. Hackster.io. https://www.hackster.io/rjconcepcion/time-lapse-raspberry-pi-zero-w-25da6b. Accessed 10 Oct 2021
10. Resources A (2021) What is visual SLAM technology and What is it used for? Automate. https://www.automate.org/blogs/what-is-visual-slam-technology-and-what-is-it-used-for. Accessed 10 Oct 2021
11. Germanos V, Secco EL (2016) Formal verification of robotics navigation algorithms. In: 19th IEEE international conference on computational science & engineering, Paris, pp 177–180

12. Baxendale M, Pearson MJ, Nibouche M, Secco EL (2017) AG pipe, self-adaptive context aware audio localization for robots using parallel cerebellar models. Lecture notes in computer sciences, Chapter 6, pp 1–13. https://doi.org/10.1007/978-3-319-64107-2
13. McHugh D, Buckley N, Secco EL (2020) A low-cost visual sensor for gesture recognition via AI CNNS. In: Intelligent systems conference (IntelliSys), Amsterdam, The Netherlands

Chapter 26
Advance Plant Health Monitoring and Forecasting System Using Edge-Fog-Cloud Computing and LSTM Networks

Rugved Sanjay Chavan, Gaurav Srivastava, and Nitesh Pradhan

1 Introduction

Traditionally, all living creatures were classified as either plants or animals. Food production is a key challenge in developing nations such as India. To overcome this challenge, a suitable monitoring and forecasting system is required to maximize plant yield while maintaining soil health consistently. Soil quality or fertility is one of the most critical factors impacting crop yield. Aside from agricultural production, soil quality determines the cost for a farmer to produce one or more plants, as some of them require precise ratios of specific components in the soil, such as water, sunlight intensity, air, temperature, humidity, and so on. Farmers must regularly monitor soil conditions on their land plots to enhance agricultural production. Crop yield forecasting is critical for agricultural production. Governments all across the globe utilize analytical data on crop production projections to make sound judgments about their national import/export activities. In recent years, it has been nearly difficult not to come across the term "Internet of Things" (IoT) in some form or another. Particularly over the last years, there has been a great spike in interest in the Internet of Things. Consortia have been created to provide frameworks and standards for the Internet of Things. Companies have begun to launch a slew of IoT-based goods and services. And a number of IoT-related purchases have made news, including

R. S. Chavan
Department of Computer and Communication Engineering, Manipal University Jaipur, Jaipur, Rajasthan, India

G. Srivastava · N. Pradhan (✉)
Department of Computer Science and Engineering, Manipal University Jaipur, Jaipur, Rajasthan, India
e-mail: nitesh.pradhan@jaipur.manipal.edu

G. Srivastava
e-mail: mailto.gaurav2001@gmail.com

G. Mathur et al. (eds.), *Proceedings of 3rd International Conference on Artificial Intelligence: Advances and Applications*, Algorithms for Intelligent Systems,
https://doi.org/10.1007/978-981-19-7041-2_26

Google's $3.2 billion takeover of Nest and subsequent acquisitions of Dropcam by Nest and SmartThings by Samsung [1]. This paper investigates and proposes an IoT architecture of Edge-Cloud computing for monitoring and predicting plants' health.

The following is a complete breakdown of the paper's structure: Sect. 2 discusses past literature work, whereas Sect. 3 presents the proposed system architecture. Section 4 delves into the embedded system, while Sect. 5 focuses on the deep learning technique deployed for prediction. Section 6 covers all the results and finally, Sect. 7 concludes this paper.

2 Related Works

Siddagangaiah [2] discusses about plant health monitoring systems which will examine several environmental variables like temperature, humidity, and light intensity, all of which have an impact on plants. Also, they get the moisture out of the soil. All of this data is transferred to the Ubidots IoT (Internet of Things) cloud platform through Arduino Uno dev boards. However, future predictions and actions after analysis of anonymous data generated were not motioned. Liu et al. [3] present a remote monitoring and control system that is particular to plant walls in this research. To simplify the administration method, increase scalability, improve user experiences of plant walls, and contribute to a green interior environment, the system makes use of Internet of Things technologies and the Azure public cloud platform. The purpose of the article [4] is to highlight several Internet of Things applications that play an important part in people's daily lives which highlights the idea of smart agriculture. Arathi Reghukumara et al. [5] consider the moisture level and temperature of the environment in which the plant develops when deciding whether or not to release water from the electric motor. The data from the sensors will be shown in graphical form on an Adafruit cloud page, which is an IoT platform (hardware and software interface), and will be used to analyze the plant health and send an email alert to the farmer or person concerned. As a result, the technology saves water while irrigating the plants and avoids the need for constant human supervision.

The literature survey states that there was a lot of work to be done in terms of user interface, high-accuracy prediction, and a fully deployable architecture in plant health monitoring systems.

3 System Architecture

We propose edge-fog-cloud computing as a novel paradigm for arranging data pipeline operations in the IoT systems, such as acquisition, analytics, and processing. The system architecture is deployed in a distributed manner as indicated in Fig. 1.

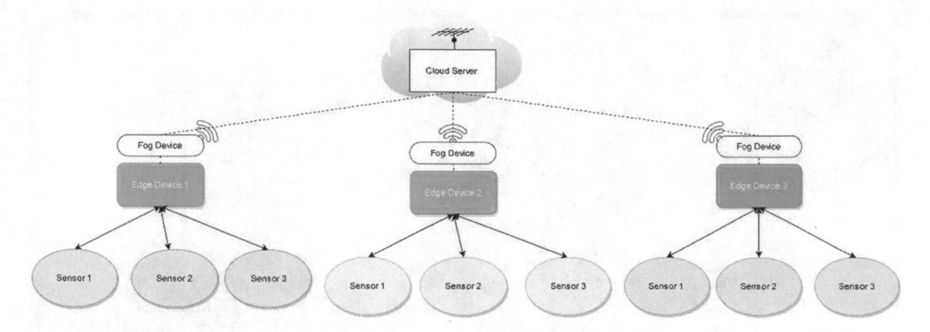

Fig. 1 Distributed architecture

The data intensiveness and on-site resource limits of IoT devices are addressed by cloud computing [6]. In this case, the cloud server is utilized to handle massive amounts of data generated by edge devices and convert it into valuable information. Furthermore, it is not just limited, but also incorporates computationally intensive tasks such as training of initial deep learning models, live forecasts, periodic backups, and backend for web apps. Putting all computing activities on the cloud has shown to be an efficient method for data processing because the computational power on the cloud outperforms the capabilities of items at the edge. However, in comparison to the rapidly increasing data processing speed, the network's capacity has come to a halt. With the increasing amount of data generated at the edge, data transmission speed is becoming a hurdle for the cloud-based computing paradigm [7].

Temperature and humidity sensors in agricultural fields, for example, provide vital data, but that data does not have to be evaluated or stored in real time. Edge devices can gather, sort, and do early data analysis before sending it to centralized apps or long-term storage, which can be on-premises or in the cloud. Because this traffic may not be time-sensitive, slower, less expensive internet connections may be employed. Furthermore, because the data is presorted, the amount of traffic that must be delivered may be decreased. Edge computing has the advantage of providing faster reaction time for applications that demand it while also reducing the expansion of expensive long-distance connections to processing and storage hubs.

Another example, the soil moisture is constantly monitored by the edge device, so maintaining a constant connection with the cloud server is a bottleneck for this system. To overcome this, a small amount of data collection and preprocessing is performed on the edge device itself, and the processed data is sent to the cloud after a specific time period. This will avoid a continuous connection to the cloud, as bandwidth is the bottleneck of the edge-cloud architecture.

The complete workflow of the plant health monitoring and the predicting system is depicted in Fig. 2.

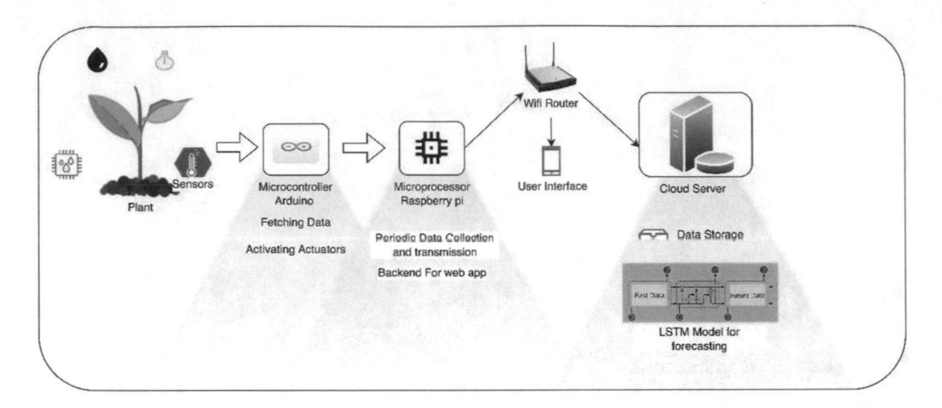

Fig. 2 Graphical abstract of proposed work

4 Hardware and Embedded System

An embedded system is a type of technical artifact that involves computing while being physically constrained as discussed in Henzinger and Sifakis [8]. Embedded systems are made up of Hardware, Software, and Real-Time Operating Systems (RTOS). Table 1 shows the hardware and RTOS configuration of cloud servers.

Table 1 Hardware configuration of cloud server

Hardware	Configuration
PROCESSOR I	Intel Core i9-10850 K @ 5.20 GHz
Core count	10
Thread count	20
Cache size	20 MB
Graphics	Gigabyte NVIDIA GeForce RTX3080 10 GB
Frequency	210/405 MHz
BAR1/visible vRAM	256 MiB
Display driver	NVIDIA 495.44
Motherboard	Gigabyte Z590 AORUS ELITE AX
BIOS version	F5
Chipset	Intel Tiger Lake-H
Audio	Realtek ALC1220
Network	Realtek RTL8125 2.5GbE + Intel Tiger Lake PCH CNVi WiFi
RAM	64 GB

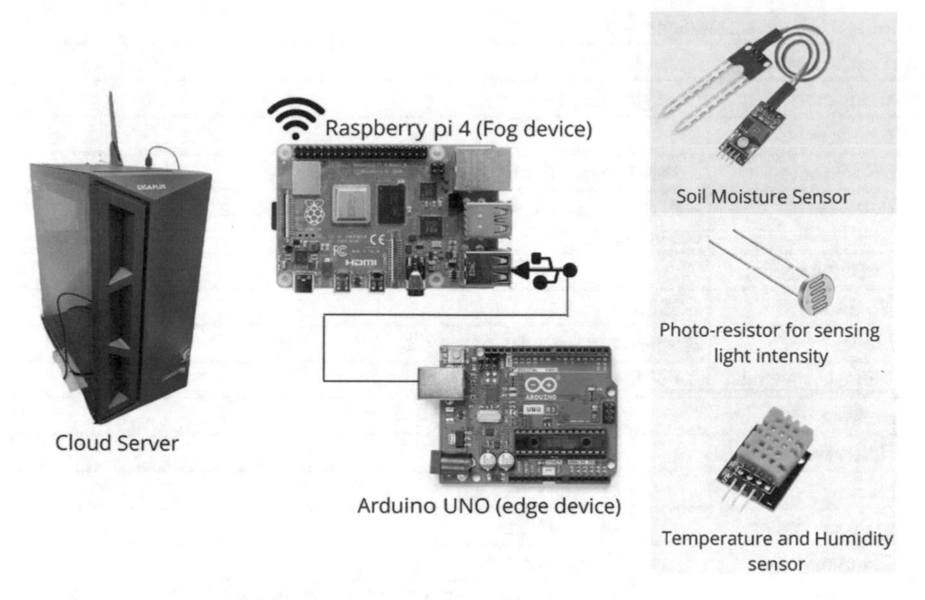

Fig. 3 Soil moisture, temperature, humidity, and light intensity sensors

4.1 Micro-processor and Micro-controllers

The Raspberry Pi 4 is the edge device in this system since it is a microprocessor that can handle light data processing operations [11]. However, due to its inability to receive analogue data, the Arduino Uno is linked to the Raspberry Pi 4 via a Universal Serial Bus (USB). As a result, the Raspberry Pi serves as a partial fog device. Furthermore, all the sensors are connected to the Arduino Uno via General Purpose Input Output (GPIO) pins as shown in Fig. 3. Hardware configuration of fog and edge devices is shown in Table 2.

4.2 Sensors and Actuators

A sensor is an electronic device that converts physical events or properties into electrical signals [12]. This is a hardware device that converts environmental input into information for the system. A thermometer, for example, uses temperature as a physical property before converting it into electrical signals. An actuator is a device that converts electrical impulses into physical events or characteristics [12]. It takes the system's input and outputs it to the environment. Actuators that are frequently used include motors and heaters. The proposed plant monitoring system is focused on sensing four parameters, which include soil moisture, temperature, humidity, and light intensity sensors as shown in Fig. 3.

Table 2 Hardware configuration

Raspberry 4

Hardware	Configuration
Processor	3
Model name	ARMv7 processor rev 3 (v7l)
Features	Half thumb fastmult vfp edsp neon vfpv3 tls vfpv4 idiva idivt vfpd32 lpae evtstrm crc32
CPU architecture	7
Hardware	BCM2711
Model	Raspberry Pi 4 Model B Rev 1.2
Wifi and Bluetooth	2.4 GHz 802.11n (150 Mbit/s)

Arduino UNO

Hardware	Configuration
Name	Arduino UNO R3
Microcontroller	ATmega328P
Communication	UART, 12c, SPI
Memory	ATmega328P- 2 KB SRAM, 32 KB FLASH, 1 KB EEPROM
I/O voltage	5 V
Clock speed	Main Processor—ATmega 328P 16 MHz USB-Serial Processor—ATmega16U2 16 MHz

4.3 Working

To monitor the content of Soil Moisture, the soil moisture sensor includes two probes. The resistance value is determined by passing a current between the probes. This DHT11 Digital Relative Humidity and Temperature Sensor Module is pre-calibrated with resistive sense innovation and an NTC thermistor for precise reading of relative humidity and surrounding temperature [13]. The DHT11 module communicates serially, i.e., via a single cable. This module transmits data in the form of a pulse train with a particular time period. Before transferring data to Arduino, some initialized instruction with a time delay is required. And the entire procedure takes around 4 ms. The single-wire serial interface speeds up and simplifies the system integration [9]. A photoresistor is used to detect the intensity of light [14]. The term photoresistor is derived from the terms photon (light particles) and resistor. A photoresistor is a type of resistor whose resistance lowers as the light intensity rises. In other words, as the intensity of light increases, so does the flow of electric current via the photoresistor [10].

All of the sensors and actuators are linked to Arduino via GPIO pins. Because the main advantage of the edge device is its fast response time [15], Arduino is used to executing lightweight programs, such as when the soil moisture level lowers, the actuator, such as a water pump, is turned on. Further, the Arduino is connected to

Raspberry pi 4 with a USB connection for sending serial data, this data is collected in the form of tables which is stored in Comma-separated values (CSV) file. The attributes of the table are TIMESTAMP, TEMPERATURE, HUMIDITY, LIGHT INTENSITY, SOIL MOISTURE. After every specific time-stamp, this data is sent to the cloud server for further analysis.

5 LSTM for Forecasting

Deep learning technologies, such as automated learning of temporal dependency and automatic handling of temporal structures like trends and seasonality, hold a lot of promise for time series forecasting [16]. Long short-term memory (LSTM) is an artificial recurrent neural network (RNN) architecture used in the field of deep learning [17]. Because there might be lags of undetermined duration between critical occurrences in a time series, LSTM networks are well-suited to classifying, processing, and making predictions based on time series data [18, 20]. LSTMs were created to solve the problem of vanishing gradients that can occur when training traditional RNNs [19].

The input **x(t)** of the LSTM can be the output of a CNN or the input sequence itself. The inputs from the previous timestep LSTM are **h(t−1)** and **c(t−1)**. The output of the LSTM for this timestep is **o(t)**. The LSTM also generates the **c(t)** and **h(t)** for the next time step LSTM consumes.

$$f_t = \sigma_g(W_f \times x_t + U_f \times h_{t-1} + b_f) \tag{1}$$

$$i_t = \sigma_g(W_i \times x_t + U_i \times h_{t-1} + b_i) \tag{2}$$

$$o_t = \sigma_g(W_o \times x_t + U_o \times h_{t-1} + b_o) \tag{3}$$

$$c'_t = \sigma_c(W_c \times x_t + U_c \times h_{t-1} + b_c) \tag{4}$$

$$c_t = f_t \cdot c_{t-1} + i_t \cdot c'_t \tag{5}$$

$$h_t = o_t \cdot \sigma_c(c_t) \tag{6}$$

where f_t is forget gate, i_t is input gate, o_t is output gate, c_t is cell state, h_t is hidden state, σ_g: sigmoid, σ_c: tanh.

6 Result

Arduino is used as an edge device to collect data from sensors and activate actuators in a timely manner. For light processing, a Raspberry Pi is utilized as a fog device, while for heavy processing of data, a cloud device is employed.

The authors were able to obtain real-time data on temperature, humidity, light intensity, and soil moisture using the architecture they presented. The authors produced 1 year of data and trained our LSTM network on it to verify the overall functioning of the proposed architecture, and for proper assessment of deep learning models. Proposed LSTM models showed a trailblazing performance on produced dataset with a mean squared error of 2.077 on the training set and 2.303 on the testing set as shown in Table 3. Figure 4 represents the LSTM predictions of temperature over a year for both the train and test data. Figure 5 depicts the loss curve for training of the LSTM model.

The system's real-time architecture is depicted in Fig. 6. Furthermore, a user interface was developed on the Flask framework as a wrap up backend to serve the entire architecture and display fetched and forecasted data.

Table 3 Evaluation metrics of proposed LSTM model

	Mean squared error	Mean absolute error	Mean absolute percentage error
Train set	2.077	1.693	0.059
Test set	2.303	1.966	0.105

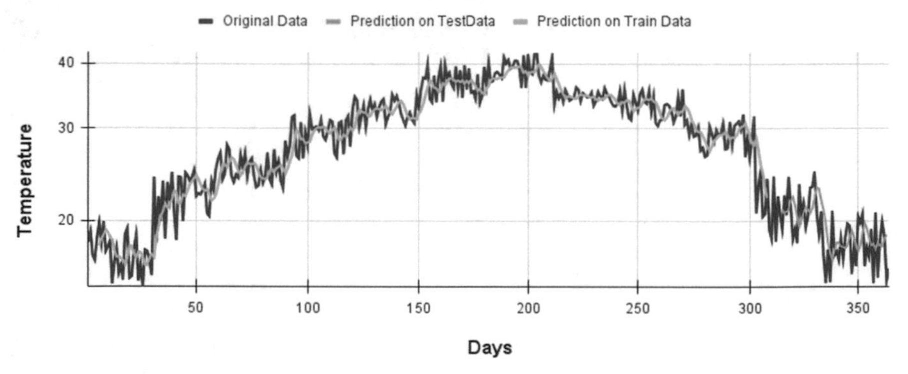

Fig. 4 LSTM predictions of temperature over a year

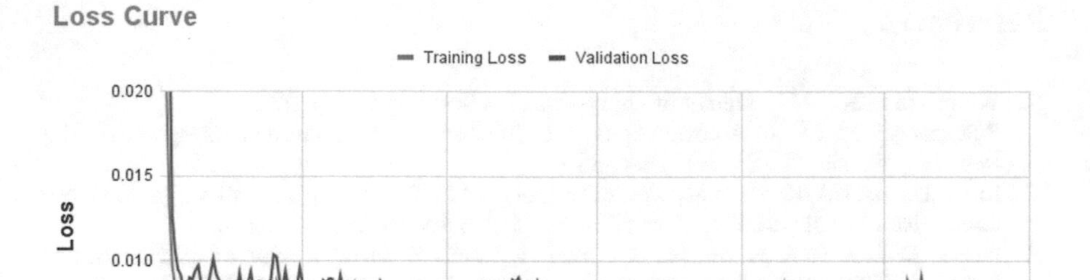

Fig. 5 Loss curve of the trained LSTM model

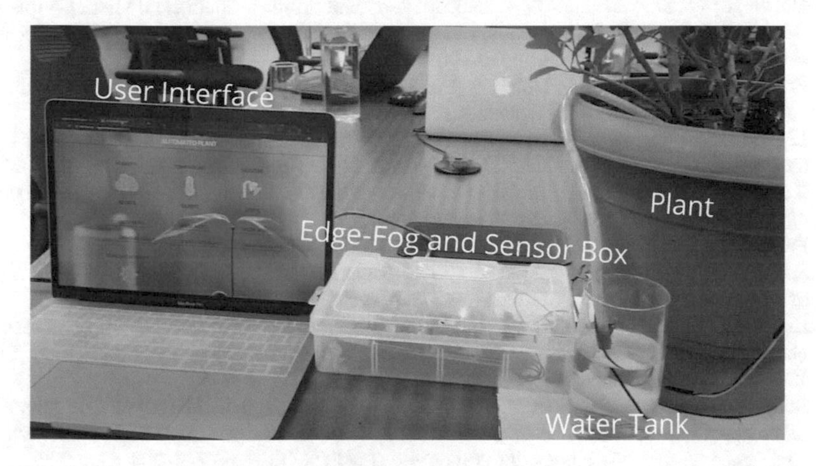

Fig. 6 Realtime working of proposed architecture

7 Conclusion

This design incorporates a cutting-edge Plant monitoring and forecasting system that is unique, reliable, robust, and user-friendly, as well as more efficient than currently available solutions. The completely working device, including software and hardware, along with edge-fog-cloud architecture was successfully implanted. The elegant user interface made all of the retrieved parameters, as well as notifications alert, available remotely. In addition, the cloud device forecasts critical data, which will aid in the planning process for the forthcoming parameter change. Users, such as farmers, will be able to monitor and enhance crop yields and overall production by implementing this system.

References

1. Wortmann FFK (2015) Internet of Things. Bus Inf Syst Eng 57:221–224
2. Siddagangaiah S (2016) A novel approach to IoT based plant health monitoring system. Int Res J Eng Technol (IRJET) 3(11):880–886
3. Liu Y, Hassan KA, Karlsson M, Weister O, Gong S (2018) Active plant wall for green indoor climate based on cloud and Internet of Things. IEEE Access 6:33631–33644
4. Farooq MU, Waseem M, Mazhar S, Khairi A, Kamal T (2018) Review on applications of Internet of Things (IoT). Int J Adv Res Comput Eng Technol (IJARCET) 7(12):841–845
5. Arathi Reghukumara VV (2019) Smart plant watering system with cloud analysis and plant health prediction. In: International conference on recent trends in advanced computing, vol 165, pp 127–135
6. Tianfield H (2018) Towards edge-cloud computing. In: IEEE international conference on big data (big data), Seattle, WA, USA
7. Shi W, Cao J, Zhang Q, Li Y, Xu L (2016) Edge computing: vision and challenges. IEEE Internet of Things J 3(5):637–646
8. Henzinger TA, Sifakis J (2006) The embedded systems design challenge. In: Misra J, Nipkow T, Sekerinski E (eds) FM: formal methods. Lecture notes in computer science, vol 4085. Springer, Berlin, Heidelberg. https://doi.org/10.1007/11813040
9. Srivastava D, Kesarwani A, Dubey S (2018) Measurement of temperature and humidity by using arduino tool and DHT11. Int Res J Eng Technol (IRJET) 5(12):876–878
10. Liu J, Liang Y, Wang L, Wang B, Zhang T, Yi F (2016) Fabrication and photosensitivity of CdS photoresistor on silica nanopillars substrate. Mater Sci Semicond Process 56:217–221
11. Kamath R, Balachandra M, Prabhu S (2019) Raspberry Pi as visual sensor nodes in precision agriculture: a study. IEEE Access 7:45110–45122
12. Pasquale M (2003) Mechanical sensors and actuators. Sens Actuat, A 106(1–3):142–148
13. Srivastava D, Kesarwani A, Dubey S (2018) Measurement of temperature and humidity by using Arduino tool and DHT11. Intl Res J Eng Technol (IRJET) 5(12):876–878
14. Lin C-H, Chen C-H (2008) Sensitivity enhancement of capacitive-type photoresistor-based humidity sensors using deliquescent salt diffusion method. Sens Actuat, B Chem 129(2):531–537
15. Singh R, Armour S, Khan A, Sooriyabandara M, Oikonomou G (2019) The advantage of computation offloading in multi-access edge computing. In: 2019 fourth international conference on fog and mobile edge computing (FMEC). IEEE, pp 289–294
16. Chatfield C (2000) Time-series forecasting. Chapman and Hall/CRC
17. Staudemeyer RC, Morris ER (2019) Understanding LSTM—a tutorial into long short-term memory recurrent neural networks (2019). arXiv preprint arXiv:1909.09586
18. Yu Y, Si X, Hu C, Zhang J (2019) A review of recurrent neural networks: LSTM cells and network architectures. Neural Comput 31(7):1235–1270
19. Ribeiro AH, Tiels K, Aguirre LA, Schön T (2020) Beyond exploding and vanishing gradients: analysing RNN training using attractors and smoothness. In: International conference on artificial intelligence and statistics. PMLR, pp 2370–2380
20. Park I, Kim HS, Lee J, Kim JH, Song CH, Kim HK (2019) Temperature prediction using the missing data refinement model based on a long short-term memory neural network. Atmosphere 10(11):718

Chapter 27
A Framework for Smart Agriculture System to Monitor the Crop Stress and Drought Stress Using Sentinel-2 Satellite Image

Tasneem Ahmed, Nashra Javed, Mohammad Faisal, and Halima Sadia

1 Introduction

Saving money by reducing inputs like crops, water, and resources while also protecting the environment is known as a precision agriculture management system. Precision agriculture makes use of cutting-edge technologies such as geographic information systems, positioning *technology,* optical and microwave remote sensing, and satellite navigation systems, while farm size, legal issues, and social contact are all variables that influence precision agriculture adoption. Precision agriculture may help farmers meet a variety of obstacles such as achieving ideal productivity and optimizing profitability, as a result of climate change and the rising global population [1]. The bedrock of the precision agriculture system is information technology, and management, and integrating these components decreases inputs while increasing efficiency. Precision agriculture was targeted by a variety of challenges, including threats to secrecy, reputation, affordability, and spectrum signal crowding [2]. Nowadays, many researchers have introduced the usage of artificial intelligence (AI) and

T. Ahmed (✉) · N. Javed · M. Faisal
Advanced Computing Research Laboratory, Department of Computer Application, Integral University, IndiaLucknow
e-mail: tasneemfca@iul.ac.in

N. Javed
e-mail: nashra@iul.ac.in

M. Faisal
e-mail: mdfaisal@iul.ac.in

H. Sadia
Department of Computer Science and Engineering, Integral University, Uttar Pradesh, Lucknow, India
e-mail: halima@iul.ac.in

G. Mathur et al. (eds.), *Proceedings of 3rd International Conference on Artificial Intelligence: Advances and Applications*, Algorithms for Intelligent Systems,
https://doi.org/10.1007/978-981-19-7041-2_27

Internet of Things (IoT) to develop analytical data-based smart agriculture systems for crop monitoring [3, 3].

Yield monitoring and mapping, field and crop scouting, soil sensing and tracking, variable rate application (VRA), satellite-based navigation, automated steering, geographical information systems (GIS), and remote sensing are all part of precision agriculture, also known as Precision Farming (PF) or Satellite Farming (SF). Precision agriculture has as its key goal to assist farmers by providing customized information and technology services that improve production, enhance profitability, and minimize emissions. Precision agriculture provides an opportunity to improve crop productivity and livestock, as well as the role of various factors such as soil fertility, water, and pest control in growing farmer income while also protecting the environment [5, 5]. Precision agriculture is a key component of developed nations' smart agriculture systems that offer farmers contextually meaningful and customized agricultural information through their mobile phones. Precision agriculture tests and improves current agricultural processes, making these consulting programs more customizable and intelligent over time. Precision agriculture is becoming an interesting field for controlling natural resources such as water, land, and plants, as well as implementing modern sustainable agricultural growth [2].

Multispectral satellites are playing a very crucial role in gathering high-resolution data for agricultural activities. Farmers can better control seeds, soil, pests, fertilizer, and water demand with the help of multispectral imaging camera sensors installed in agriculture drones. As a result, drones have proved to be advantageous in terms of rising yields and other advantages. Multispectral sensors collect images of crops and plants in visible band and unseen regions using four bands: red, green, red edge, and near-infrared (NIR) bands [7]. Multispectral sensors may provide information on plants, such as coverage area, crop greenness, and vegetation structure due to their wide spectral channels. However, the spectral resolution and bandwidth of these sensors are limited. Various multispectral satellites are available such as Landsat and Sentinel-2 with high spatial as well as temporal resolution, in which Sentinel-2A satellite provides better spatial and temporal resolution [8].

Sentinel-2A is equipped with a multispectral sensor (MSI), which is having 13 spectral bands with a spatial resolution ranging from 10 to 60 m (depending on the band) and a current temporal resolution of around 10 days, and is publicly available for free. Sentinel-2's high spatial and temporal resolutions, as well as the free availability, make it ideal for crop tracking in the context of precision agriculture [7]. Food protection preparation, temporal change analysis to detect changes in cropping trends over time, impact evaluation studies, and many other applications benefit from high spatial resolution crop style maps developed from Sentinel-2 satellite data. Crop style charts are often used at different stages of crop insurance [8]. Climatic agricultural mapping and high-accuracy statistics are critical sources for determining the effects of abiotic stresses like heat and drought stress, which occur frequently in the area and are expected to rise regularly and in severity as the environment changes [9, 10]. Sentinel-2 images, when used in the context of precision agriculture, will make significant contributions to agriculture tracking and increase our understanding of the effects of environmental variations. Indices like the Normalized Difference Vegetation Index (NDVI), which measures the amount of greenness in the vegetation, and the Normalized Difference Water Index (NDWI) index, which measures the amount of water in the vegetation or the saturation level of moisture in the soil are very useful for the monitoring of crop and drought stress. The NDVI, which is calculated from high spatial resolution images, is a useful tool for determining vegetative vigor and water status [11, 12].

This paper examines the role of precision agriculture in increasing agricultural productivity and farmer viability, preserving natural resources, and protecting the environment, as well as the various techniques used in precision agriculture and the challenges that have been directed at the precision agriculture system. The findings of the analysis can be used as a scientific guideline for extracting crop data in other diverse planting areas, as well as advice for related departments to understand the planting situation of major crops ahead of time and change crop planting structures appropriately. To utilize the analyzed information, machine learning (ML)-based smart agriculture system could be developed for effective and efficient crop monitoring. Hence, in this paper, a framework of a machine learning-based smart agriculture system has been proposed to monitor crop stress and drought stress using Sentinel-2 satellite images.

2 Study Area and Data Used

2.1 Study Area

In this study, Roorkee and its nearby areas have been taken as study regions, which are located in Haridwar District, Uttarakhand, India. The center latitude 29.85 N and longitude 77.88 E compose the study region, and it contains significant land covers urban (built-up areas in Roorkee town and its nearby villages), agricultural areas (crop and grassland nearby Roorkee town and its nearby villages), and water (due to Ganga Canal and Solani River, a seasonal river that normally remains dry in summers). Figure 1a and b shows the true color composite (TCC) image and Google Earth image of the study region. An image showing a mixture of the apparent red, green, and blue bands with the matching red, green, and blue channels of the satellite image on the screen monitor is a natural or true color composite image. The resulting composite is identical to what the human eye will normally observe: grass is green, water is dark blue to black, and bare grounds look brown and light gray. As colors look normal to our eyes, many people choose true color composites, but even small variations in characteristics are difficult to identify [13].

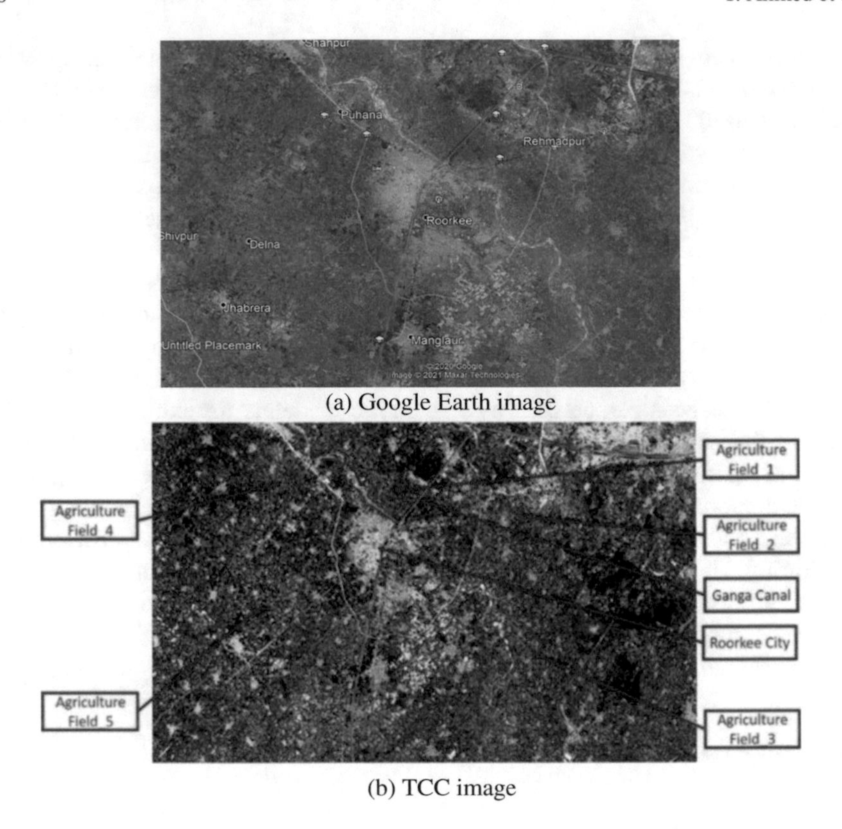

(a) Google Earth image

(b) TCC image

Fig. 1 Study region images. **a** Google Earth image and **b** TCC image

2.2 Data Used

Sentinel-2A satellite images have been used in this study. Sentinel-2A is intended to measure the reflectance at 10 m resolution of the blue, green, red, and near-infrared-1 bands; red edge 1–3, near-infrared-2 and short-wave infrared 1 and 2 at 20 m; and 3 atmospheric bands (Band 1, Band 9, and Band 10) at 60 m. However, Sentinel-2 images showed extensive action in the agricultural field, including chlorophyll and nitrogen content of plants and grasses, and vegetation status in pastures and savannas [14]. To monitor the agriculture fields, the main focus is to extract complete information on the cultivation and harvesting process of wheat crops. The growing time of the wheat crop is rabi season or during the winters in North India. Usually, the crop is shown in November–December and harvested in April. To monitor the status of wheat crop cultivation, harvesting, and post-harvesting cycle, a total five number of Sentinel-2A images from January 2020 to May 2020 have been downloaded and processed; the image IDs along with acquisition dates are given in Table 1.

Table 1 Details of satellite data used

S. no.	Dataset details	Acquisition date	Image_ID
1	S2A_MSIL2A_20200105T053221_N0213_R105_T43RGP_20200105T075223	05 January 2020	Image_01
2	S2A_MSIL2A_20200224T052811_N0214_R105_T43RGP_20200224T081949	24 February 2020	Image_02
3	S2A_MSIL2A_20200325T052641_N0214_R105_T43RGP_20200325T095908	25 March 2020	Image_03
4	S2A_MSIL2A_20200424T052651_N0214_R105_T43RGP_20200424T094235	24 April 2020	Image_04
5	S2A_MSIL2A_20200524T052651_N0214_R105_T43RGP_20200524T094211	24 May 2020	Image_05

Table 2 Sentinel-2 band details

S. no.	Band name	Resolution (m)	Central wavelength (nm)	Bandwidth (nm)
1	Coastal Aerosol	60	443.9	27
2	Blue	10	496.6	98
3	Green	10	560	45
4	Red	10	664.5	38
5	Vegetation red edge	20	703.9	19
6	Vegetation red edge	20	740.2	18
7	Vegetation red edge	20	782.5	28
8	Near-infrared	10	835.1	145
9	Narrow NIR	20	864.8	33
10	Water vapor	60	945	26
11	SWIR–Cirrus	60	1373.5	75
12	SWIR	20	1613.7	143
13	SWIR	20	2202.4	242

3 Methods and Approaches

3.1 Pre-processing of Sentinel-2A Images

Sentinel-2A/MSI acquires Earth's surface images in 13 spectral bands with a spatial resolution of 10, 20, and 60 m, and a current temporal resolution of about 10 days [8]. The complete bands' details along with spatial resolution, central wavelength, and bandwidths of each band are given in Table 2.

Sentinel-2A images have been pre-processed by using SNAP 8.0; all the bands are resampled at 10 m spatial resolution for unique pixel size into a raster stack, and Sentinel-2A image projection has been changed into Geographic Latitude/Longitude format. The main bands that were used in the study are bands 4 (Red), 8A (NIR), and bands 11 (SWIR1) from Sentinel-2A. The spectral response of band 8A is identical to band 5 of Landsat-8 data, hence, band 8A of Sentinel-2A is used instead of band 8 [15].

3.2 Monitoring of Crop Stress and Drought Stress

Crop identification and mapping are critical for managing agricultural fields, as well as for potential yield assessment. Some vegetation indices, which are combinations of spectral bands at different wavelengths, were utilized to analyze phenology and quantify biophysical parameters. As a result, crop maps in earlier studies have also been prepared more precisely, and the competencies of optical satellite images for agricultural field monitoring have been improved [16]. To measure the quantity of green biomass in vegetation, NDVI is used and it is the most common way to track vegetation phenology and to monitor drought stress NDWI index is used, which are as follows.

Normalized Difference Vegetation Index (NDVI). NDVI is the oldest and most widely used vegetation index for the successful monitoring of seasonal variability of crop phenology due to temperature changes and different precipitation regimes, so it is suitable for determining agricultural field variability [17]. Therefore, the NDVI index has been in the current study to monitor crop stress. The mathematical formulation of NDVI is given in Eq. (1):

$$\text{NDVI} = \frac{(\rho_{\text{NIR}} - \rho_{\text{RED}})}{(\rho_{\text{NIR}} + \rho_{\text{RED}})} \tag{1}$$

where ρ_{NIR} is Near-Infrared (0.865 μm) Band and ρ_{RED} is Red (0.665 μm). Band.

Around -1 and 1, NDVI index values vary, where values around 0.5 denote dense vegetation, while 0 represents no vegetation. Higher values indicate stable photosynthetic vegetation, while lower numbers indicate stressed or non-existent vegetation (bare soil). To monitor the amount of biomass, hydrologic stress, and vegetation health, NDVI has been widely used and appears to be the most efficient index for agriculture monitoring [18]. To monitor the crop stress, NDVI images have been computed by using Eq. (1), and obtained NDVI images from Image-01 to Image-05 are shown in Fig. 2a–e.

The bright regions in the NDVI images above are vegetated, while the dark areas (urban, water bodies, etc.) belong to non-vegetated areas. The trees that line the roads stand out against the dark background as gray linear features. The crop pattern can visually be analyzed from Fig. 2a–e, as the grayscale tone changes from January to May month, which usually belongs to the Rabi crop sowing and harvesting period.

Normalized Difference Water Index (NDWI). By using green band wavelengths, the NDWI maximizes water reflectance while minimizing low NIR reflectance by absorbing a maximum wavelength. As a consequence, positive values improve water features, while zero or negative values suppress vegetation and soil [19]. The mathematical formulation of NDWI is given in Eq. (2):

$$\text{NDWI} = \frac{(\rho_{\text{NIR}} - \rho_{\text{SWIR}})}{(\rho_{\text{NIR}} + \rho_{\text{SWIR}})} \tag{2}$$

where ρ_{NIR} is Near-Infrared (0.865 μm) Band and ρ_{SWIR} is Short-Wave Infrared (1.610 μm) Band.

To monitor the drought stress, NDWI images have been computed by using Eq. (2), and obtained NDWI images from Image-01 to Image-05 are shown in Fig. 3a–e.

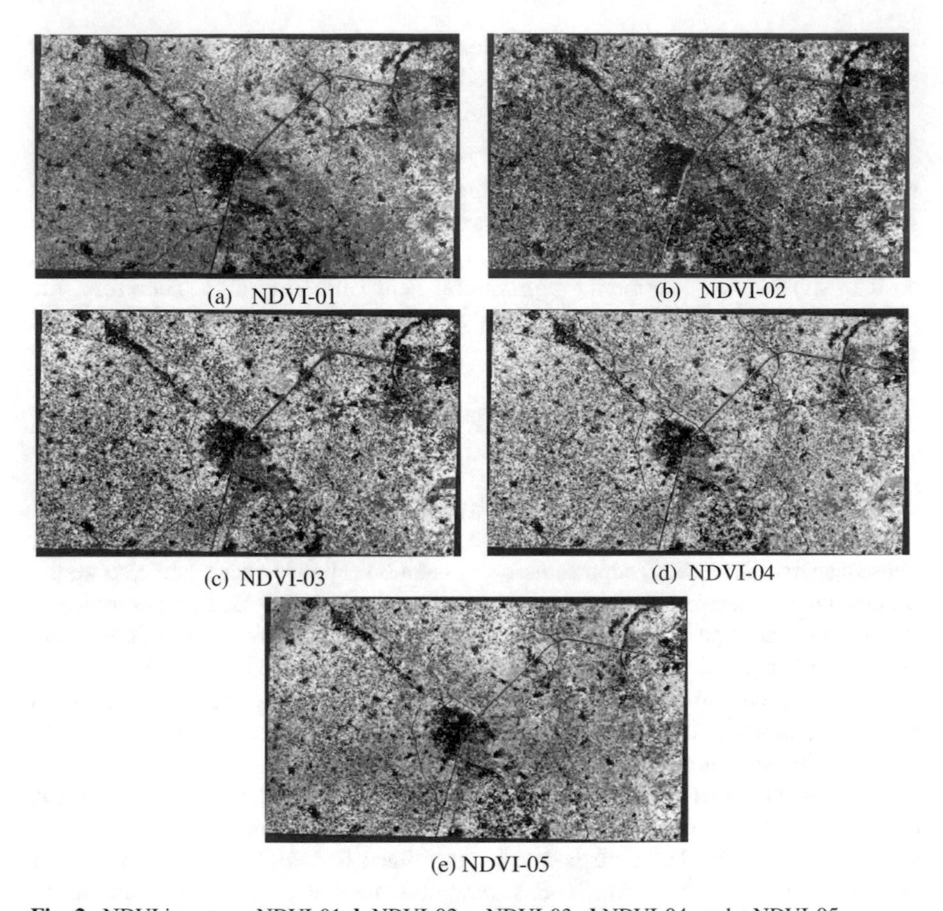

Fig. 2 NDVI images, **a** NDVI-01, **b** NDVI-02, **c** NDVI-03, **d** NDVI-04, and **e** NDVI-05

The water quality of vegetation is associated with the NDWI index. It's commonly used in forest health monitoring, precision agriculture, and other applications where vegetation's condition is sensitive to changes in water quality. It ranges from −1 to 1, with values ranging from 0.02 to 0.6 for green stable vegetation.

3.3 Analysis of Crop and Drought Stress in the Roorkee Region

This paper examines the strengths and shortcomings of mapping cultivated areas and cropping patterns during the Rabi (winter) season in Roorkee and its nearby agriculture fields. To monitor the winter wheat pattern, the NDVI image band also combined with other bands (i.e. NIR and Green) of the Sentinel-2 images to form a color composite image, which provides the distinguishing information of the various types

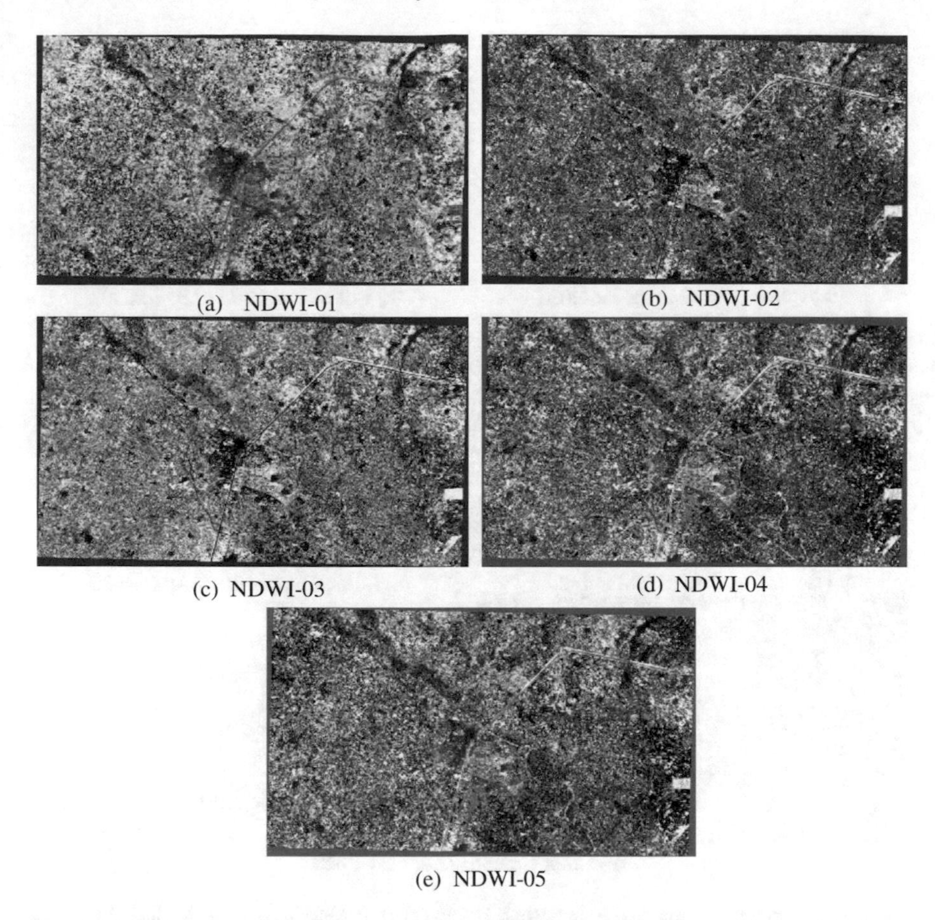

Fig. 3 NDWI images, **a** NDWI-01, **b** NDWI-02, **c** NDWI-03, **d** NDWI-04, and **e** NDWI-05

of vegetation present in the study region. In the present study, the color composite images are formed using the following color assignment: Red color is assigned to NIR Band, Green color is assigned to NDVI Band, and Blue color is assigned to the Green band, respectively. Color composite images formed using Image-01, Image-02, Image-03, Image-04, and image-05 are shown in Fig. 4a–e. The growing cycle of winter crop, especially wheat, can be easily monitored by making color composite images as shown in Fig. 4a–e, where green color represents dense vegetation with a closed canopy, bright yellow represents shrugs or less dense vegetation, golden yellow represent the grass, while blue and magenta colors represent the non-vegetated areas (i.e. Roorkee city's urban areas, Solani River and Ganga Canal).

To have proper monitoring of winter crops, approximately 300 samples have been collected for five different agriculture fields with the help of a ground truth survey and Google Earth. NDVI image provides the greenness information about the crop presented in the study region, while moisture content information can be obtained by

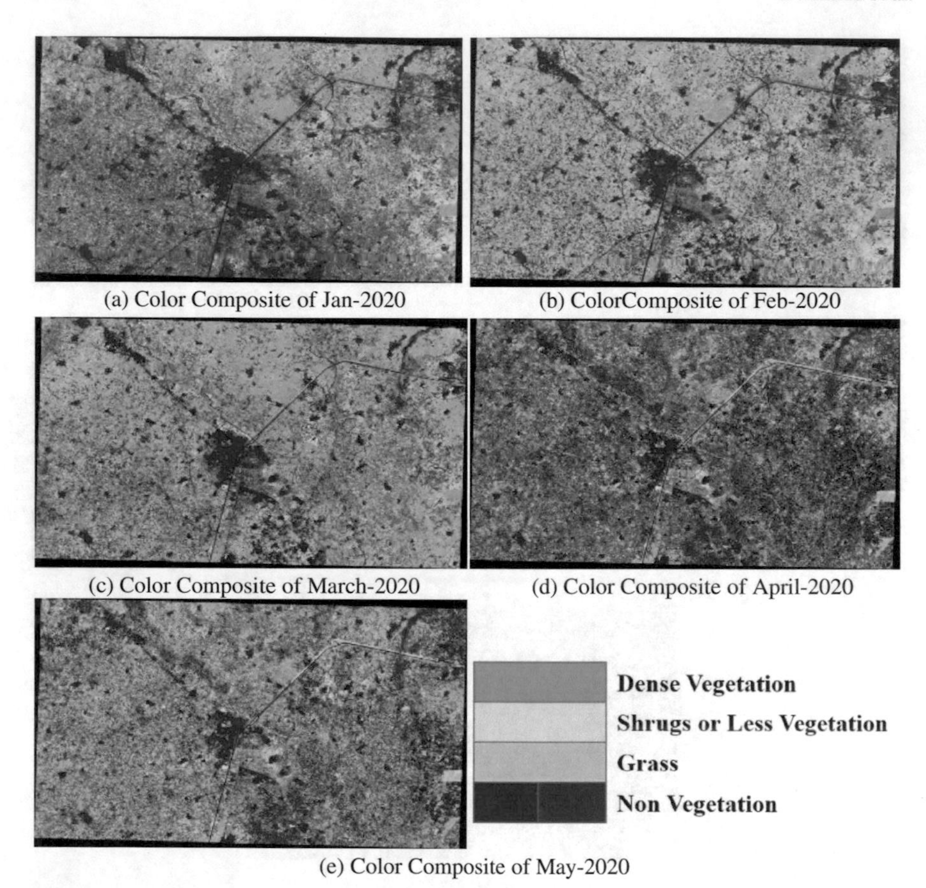

Fig. 4 Color composite images, **a** Jan-2020, **b** Feb-2020, **c** March-2020, **d** April-2020, and **e** May-2020

NDWI image. Moisture information can be utilized to monitor the drought stress of the crop. A comparative analysis has been carried out to analyze the crop and drought stress by obtaining the NDVI and NDWI values for the above-mentioned agriculture fields from the NDVI-01, NDVI-02, NDVI-03, NDVI-04, NDVI-05, NDWI-01, NDWI-02, NDWI-03, NDWI-04, and NDWI-05 images, respectively; all the values are plotted all together in the scatter plot, as shown in Fig. 5a and b.

From Fig. 4a, it is found that in January month dense vegetation is observed in quite a high amount along with sufficient soil moisture, while in February month as shown in Fig. 4b, greenness and sufficient water content have been increased which represents the progress of winter crop sowing in effectively and efficiently as shown in Fig. 5a and b. In the March month image, as shown in Fig. 4c, some non-vegetated region has been observed which suggests that due to the increase in temperature, some regions may dry up. From Fig. 4d, it can be seen that non-vegetated regions have increased, which indicates that April month is the time of winter month crop

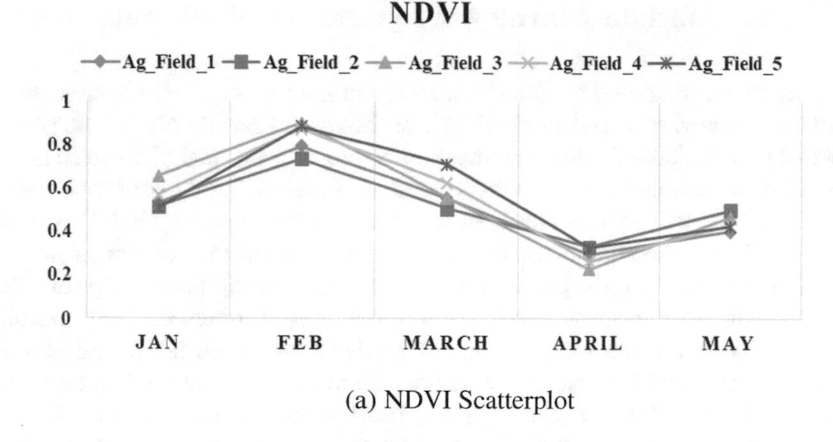

(a) NDVI Scatterplot

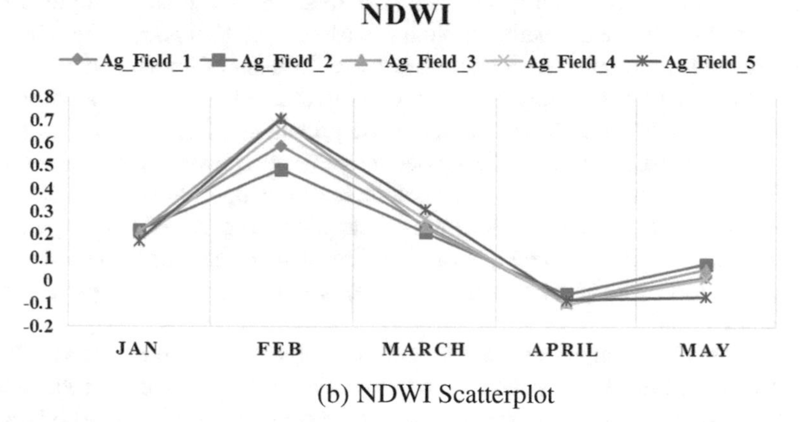

(b) NDWI Scatterplot

Fig. 5 Scatter plot of NDVI and NDWI values

maturity and wheat harvesting is also done during this period. After the harvesting of the wheat crop, vegetated regions are converted into non-vegetated regions, therefore, the magenta color is highly present in the April month image and water content is observed very low as shown in Fig. 5b. The last color composite image belongs to the May month as shown in Fig. 4e, in which it has been observed that greenness and water content were increased in some areas. The reason behind the increment of greenness in May month would be the sowing of new crops like corn, sorghum, sunflowers, etc. The crop stress and drought stress information obtained could be used to develop a machine learning-based smart agriculture monitoring system.

4 Role of Machine Learning in Agriculture Monitoring

For precision agriculture, building a smart agriculture system is the basic requirement for crop protection and monitoring. Satellite images can be utilized using image processing techniques for the classification of crops, pests, and yield forecasts. A model based on a decision tree can be used for determining the amount of water required for the field by monitoring the moisture, temperature, and humidity of the soil [20]. Smart agriculture-based systems are now using the concept of Internet of Things for elevating the production of food and reducing the wastage of water through deep reinforcement learning algorithms [21]. Through different machine learning algorithms, not only parameters related to crop are taken into account for quality products but also the selection of the best suitable crop to be grown is predicted [22]. Machine learning (ML) has become the most trending field of Artificial Intelligence, which gives the power to systems without being externally programmed to learn and develop through experience [4]. The prime idea is to allow computers to learn on their own without any involvement of humans. It enables the system to take action according to the situation. In the same context, programming is used for automation, and ML is automating the procedure of automation. Numerous machine learning algorithms are developed every year to provide real-time solutions [23]. Figure 6 illustrates the different machine learning algorithms for unsupervised learning through clustering and exploring techniques, supervised learning through regression and classification techniques and reinforcement learning through decision-making. The most widely used machine learning algorithms for smart agriculture are listed in the figure.

The three main components of all machine learning algorithms are as follows: (I) **Representation**: It is a technique to showcase the knowledge with the help of neural networks, support vector machines, model ensembles, instances, graphical models, sets of rules, decision trees, and others, (II) **Evaluation**: For evaluating any kind of hypothesis that is prediction, accuracy, probability, and so on, and (III) **Optimization**: What model will be created to optimize the given solution?

Supervised learning is the most advanced, well-researched, and widely used type of ML algorithm. It identifies the relationships that exist between dataset values and a target variable. Mostly predictive analytics models are referred to as supervised machine learning models because of their ability to forecast the future based on the past. Labeled data work as the input for supervised machine learning. This means that the model will be trained with historical data along with the correct answer labeled on them. To make a prediction, it tries to find a relationship between the predictors and the answer. The concept is that a data observation will appear in the future, and all we will know are the predictors. The predictors must then be used by the model to make an accurate prediction of the response value.

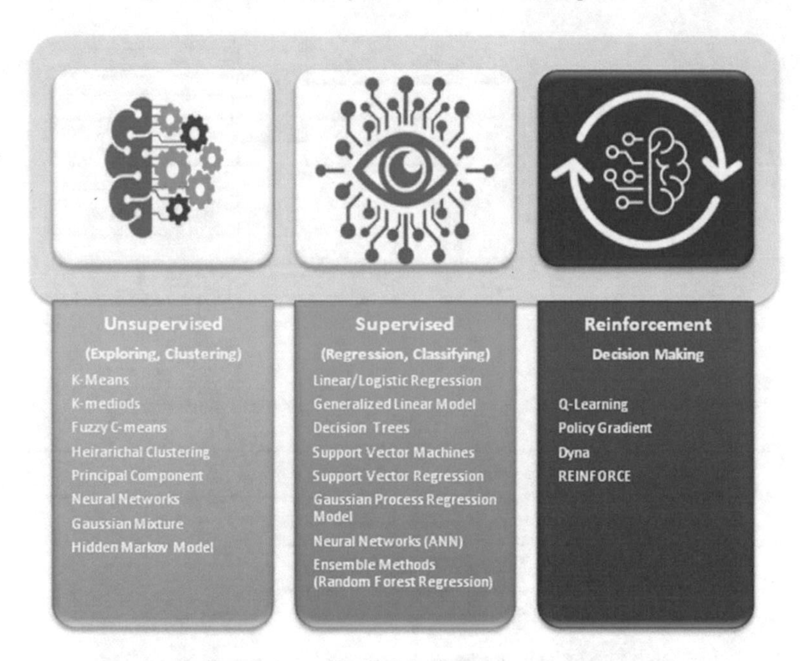

Fig. 6 Key elements of machine learning

4.1 Proposed Framework of Smart Agriculture System

Artificial and Deep Neural Networks (ANNs and DL) and Support Vector Machines (SVMs) are the most common models in agriculture among all the techniques accessible under supervised machine learning for implementing precision agriculture, which provides agriculture information at any stage of the process like for managing plant species, soil health monitoring, crop health management, or livestock management. In this paper, a framework for a smart agriculture system is proposed that requires a supervised machine learning algorithm to provide predictions for irrigation patterns based on crop health, soil health, and water level as shown in Fig. 7. Step-by-step execution of machine learning-based smart agriculture system is given below, which is as follows:

Step 1: Sentinel-2 images covering the study region need to be downloaded on a temporal basis, though proper crop monitoring can be done from sowing to harvesting.

Step 2: Pre-processing has been applied to resample the spatial resolution and retrieve the surface reflectance values.

Step 3: Calculation of NDVI and NDWI indices to obtain and analyze the crop health, crop stress, drought stress, and moisture map.

Step 4: Implementation of a machine learning algorithm to train and validate the smart agriculture system through crop information obtained from NDVI and NDWI images.

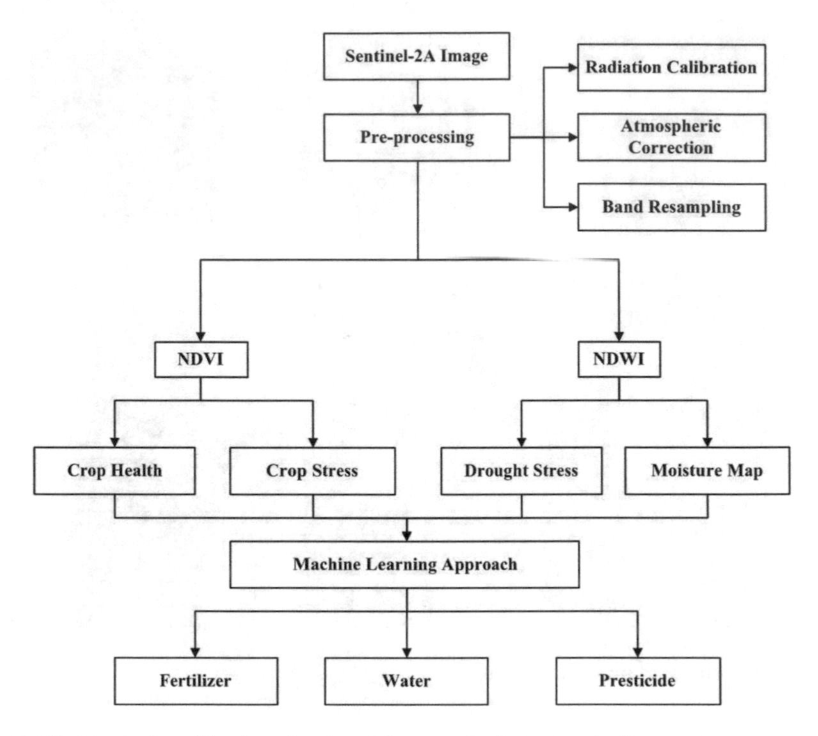

Fig. 7 Flowchart of machine learning-based framework of smart agriculture system

Step 5: Time series data will be used to understand the crop growth cycle, and will be helpful to monitor the crop stress. A developed smart agriculture system will be able to suggest the amount of fertilizer, water, and pesticides required for maintaining crop health.

Satellite images will also aid in the management of the natural ecosystem on and near fields, which is critical as farmers search for more productive farming practices. Farmers can fence off environmentally vulnerable sites and reduce the chance of inter-species disease transmission by using spatial mapping to help understand riparian zones and areas of natural cover for livestock and wild animals. Farmers can theoretically change boundary lines by analyzing how specific areas of a paddock operate, eliminating the need to transport cattle from field to field. This not only reduces tension for the livestock but also reduces potential demand on grazing fields, ensuring that they are preserved for future generations. Satellite images allow for more geospatial research, which contributes to the protection of natural flora and fauna, as well as a reduction in carbon emissions and increased operational productivity.

To train a smart agriculture system, previously labeled data is given to the machine with known answers so that it can learn the patterns involved. It examines various types of data and the answers to each problem to discover a pattern. The more the data there is, the more precise the findings would be. The data is then tested, in which a problem is provided to the computer, to solve, and the machine, knowing the pattern of solving the problem and the possible answers, which provides the most appropriate outcomes. The two most important steps in any grouping are learning and prediction. In the learning stage, the model is built using the available training data. In the prediction stage, the model should forecast the results based on the training data. Though, SVM classification could be performed to compare the results generated by the image processing for vegetation and water indices and further generate the result for the amount of fertilizer, water, and pesticides required for maintaining crop health. Based on the pre-processed dataset, the system calculates the crop's water, fertilizer, and pesticide requirements. The framework uses a machine learning approach to examine real-world data that has been fed into the machine learning system for further analysis. Following this approach, the ML algorithm compares the obtained results to the standard level and then determines whether or not water, fertilizers, and pesticides should be supplied. Moreover, the framework can be aligned with a real-time application providing all the necessary information along with report generation and notifications to users for the supply of water, fertilizers, and pesticides on the field. The agri-tech industry's use of satellites and emerging technology has the potential to change existing farm management activities. Access to real-time data on land, water, and crop status would improve organizational preparation and decision-making, allowing for significant productivity increases in the farming sector. This would result in the introduction of precise farming innovations, which are critical to the global agriculture industry.

5 Conclusion

A new generation of satellites with upgraded sensors could assist agronomists in maximizing and maintaining crop yield in the face of climate change. They lead to a deeper understanding of crop cultivar reaction to water deficit with the knowledge they have. The data they provide helps researchers better understand how different crop cultivars respond to water shortages. Farms based on machine learning are already evolving into artificial intelligence systems, even though they are still in their infancy. Machine learning solutions presently handle specific concerns, but agricultural methods will grow into knowledge-based agriculture as more automated data capture, data processing, machine learning, and decision-making are merged into an integrated framework, which will be capable of increasing production levels and product quality. This form of data-driven farm management relies on data to improve productivity while minimizing resource waste and contamination of the atmosphere. In this paper, after analyzing NDVI and NDWI indices for each image, the observation was that February has high NDVI and NDWI values compared to March, April, and May months. The reason behind these higher values is that this time crop has high greenness and high moisture content. The values that have decreased in March and April months indicate that the crop growth cycle is completing and the crop is maturing as April is the harvesting time of these winter crops. In future, satellite image-based smart agriculture monitoring systems can be used to monitor crop inputs based on the previous year's results, to ensure prudent lending decisions, or to assess vegetation risks using NDVI and other indices. Crop monitoring allows you to store all of your data in one location and perform an accurate and systematic analysis of environmental patterns, plant growth periods, the right amount and timing for seeding or fertilizer applications, and more. Investors and farm equipment vendors may assess the scale of the market and forecast demand trends in particular areas.

References

1. Guerrini F (2015) The future of agriculture? Smart Farming. Forbes
2. Abobatta WF (2014) Precision agriculture: a new tool for development. Encycl Earth Sci Ser 515–517 (2014). https://doi.org/10.1007/978-0-387-36699-9_132
3. Kadya V, Karani KP (2021) An implementation of IoT and data analytics in smart agricultural system—a systematic literature review (2021). https://doi.org/10.5281/zenodo.4496828
4. Talaviya T, Shah D, Patel N, Yagnik H, Shah M (2020) Implementation of artificial intelligence in agriculture for optimisation of irrigation and application of pesticides and herbicides. Artif Intell Agric 4:58–73. https://doi.org/10.1016/j.aiia.2020.04.002
5. Sishodia RP, Ray RL, Singh SK (2020) Applications of remote sensing in precision agriculture: a review. Remote Sens 12:1–31. https://doi.org/10.3390/rs12193136
6. Basavaraj P, Chetan HT (2017) Role of remote sensing in precision agriculture. Marumegh 1–8
7. Singh P, Pandey PC, Petropoulos GP, Pavlides A, Srivastava PK, Koutsias N, Deng KAK, Bao Y (2020) Hyperspectral remote sensing in precision agriculture: present status, challenges, and future trends. LTD
8. Segarra J, Buchaillot ML, Araus JL, Kefauver SC (2020) Remote sensing for precision agriculture: Sentinel-2 improved features and applications. Agronomy 10:1–18. https://doi.org/10.3390/agronomy10050641
9. Gumma MK, Tummala K, Dixit S, Collivignarelli F, Holecz F, Kolli RN, Whitbread AM (2020) Crop type identification and spatial mapping using Sentinel-2 satellite data with focus on field-level information. Geocarto Int 37(7):1833–1849. https://doi.org/10.1080/10106049.2020.1805029
10. Shukla R, Dubey G, Malik P, Sindhwani N, Anand R, Dahiya A, Yadav V (2021) Detecting crop health using machine learning techniques in smart agriculture system, 80:699–706
11. Cogato A, Pagay V, Marinello F, Meggio F, Grace P, De Antoni Migliorati M (2019) Assessing the feasibility of using medium-resolution imagery information to quantify the impact of the heatwaves on irrigated vineyards. Remote Sens 11:1–19
12. Swain S, Wardlow B, Narumalani S, Tadesse T, Callahan K (2011) Assessment of vegetation response to drought in nebraska using Terra-MODIS land surface temperature and normalized difference vegetation index. GIScience Remote Sens 48:432–455. https://doi.org/10.2747/1548-1603.48.3.432
13. Composites. https://gsp.humboldt.edu/OLM/Courses/GSP_216_Online/lesson3-1/composites.html
14. Kobayashi N, Tani H, Wang X, Sonobe R (2020) Crop classification using spectral indices derived from Sentinel-2A imagery. J Inf Telecommun 4:67–90. https://doi.org/10.1080/24751839.2019.1694765
15. Skakun S, Vermote E, Roger J-C, Franch B (2017) Combined use of Landsat-8 and Sentinel-2A images for winter crop mapping and winter wheat yield assessment at regional scale. AIMS Geosci 3:163–186. https://doi.org/10.3934/geosci.2017.2.163
16. Sonobe R, Yamaya Y, Tani H, Wang X, Kobayashi N, Mochizuki K (2018) Crop classification from Sentinel-2-derived vegetation indices using ensemble learning. J Appl Remote Sens 12:1. https://doi.org/10.1117/1.jrs.12.026019

17. Jelínek Z, Kumhálová J, Chyba J, Wohlmuthová M, Madaras M, Kumhála F (2020) Landsat and Sentinel-2 images as a tool for the effective estimation of winter and spring cultivar growth and yield prediction in the Czech Republic. Int Agrophys 34:391–406. https://doi.org/10.31545/INTAGR/126593
18. Ahmed T, Singh D (2020) Probability density functions based classification of MODIS NDVI time series data and monitoring of vegetation growth cycle. Adv Sp Res 66:873–886. https://doi.org/10.1016/j.asr.2020.05.004
19. Why does NDVI, NDBI, NDWI ranges from −1 to 1?—GIS resources. https://www.gisresources.com/ndvi-ndbi-ndwi-ranges-1-1/
20. Pratyush Reddy KS, Roopa YM, Kovvada Rajeev LN, Nandan NS (2020) IoT based smart agriculture using machine learning. In: Proceedings of the 2nd international conference on inventive research in computing applications, ICIRCA 2020, pp 130–134. https://doi.org/10.1109/ICIRCA48905.2020.9183373
21. Bu F, Wang X (2019) A smart agriculture IoT system based on deep reinforcement learning. Futur Gener Comput Syst 99:500–507. https://doi.org/10.1016/j.future.2019.04.041
22. Abhishek L, Rishi Barath B (2019) Automation in agriculture using IoT and machine learning. Int J Innov Technol Explor Eng 8:1520–1524
23. Mekonnen Y, Namuduri S, Burton L, Sarwat A, Bhansali S (2020) Review—machine learning techniques in wireless sensor network based precision agriculture. J Electrochem Soc 167:037522. https://doi.org/10.1149/2.0222003jes

Chapter 28
A Review on Deep Learning Algorithms for Real-Time Detection of Multiple Vehicle-Based Classes: Challenges and Open Opportunities

Iosun Stephen Shima, Abdulsalam Ya'u Gital, Abdullahi Madaki Gamsha, Mustapha Abdulrahman Lawal, Kwaghe Obed Patrick, and Okoro Christian Chukwuka

1 Introduction

When it comes to video-based systems, one of the many applications is with regard to the surveillance of vehicle traffic. In several cases, it helps in vehicle identification and presents information of changing lanes, proving valuable in automated levy systems, as previous vehicle detection systems were unable to determine charges for different vehicle classes [1]. Vehicle recognition systems have recently made significant contributions in vehicle detection and even for detecting lanes with heavy traffic, with the added benefit of classifying vehicles on highways be it cars, motorcycles, or heavy-duty vehicles. However, a drawback of the normal vehicle systems is that their accuracy tends to decline due to weather and general occlusion by obstacles in the background such as other vehicles, vegetation, and even road signs. This forces the systems to be highly dependent on traffic image analysis which can be trusted to detect, track, and present accurate vehicle classification [2].

Vehicle detection is an invaluable feature when it comes to networks ranging from autonomous driving systems to real-time vehicle surveillance systems. In terms of Autonomous driving, locating other vehicles has greater priority over other objects and with an ever-rising number of vehicles in the urban world. When dealing with Traffic surveillance systems, it is vital that the detecting system be quick in detecting vehicles in real time [3].

Owing to the emergence of new technologies such as GPU and TPU, there has been a drop-in training time in Deep learning networks over the use of CPU computations.

I. S. Shima (✉) · A. Y. Gital · A. M. Gamsha · M. A. Lawal · K. O. Patrick · O. C. Chukwuka
Department of Mathematical Sciences, Abubakar Tafawa Balewa University, Bauchi, Bauchi State, Nigeria
e-mail: stephen_iosun@yahoo.com

G. Mathur et al. (eds.), *Proceedings of 3rd International Conference on Artificial Intelligence: Advances and Applications*, Algorithms for Intelligent Systems,
https://doi.org/10.1007/978-981-19-7041-2_28

This has made Machine learning models less attractive in the field of object detection, stemming from their complex nature. Vehicle detection models fall into one of the three types which are motion-based models, feature-based models, and convolutional neural network-based models [4].

Background subtraction as well as optical flow, and frame subtraction is an important aspect of motion-based models. Prediction for the purpose of background subtraction within the network is done using auto-detection by means of the GMM [4]. The working process of the GMM involves differentiating the foreground from the background in order to make it possible to detect vehicles. However, its downside is that it is not ideal for the detection of vehicles which are stationary. On the other hand, in comparison to Feature-based models which are HOG, Haar, and SURF. Previously, deformable part-based models (DPM) have been observed to perform better during detection utilizing features of HOG as well as classifiers such as Adaboost and Support vector machines [4].

As a result of the fast-tracking improvement of deep neural networks, the performance of object detectors has seen a rapid improvement and owing to this, deep learning-based detection techniques have benefitted from active study over the past several years [5]. Object localization deals with identifying targeted visuals which lie within a picture or image or video and taking steps to locate their exact positions. With the aid of deep learning techniques [6], this technique employed for object identification and localization has allowed for computer vision to reach its peak. Owing to unwanted issues such as viewpoint inconsistency, lighting directions and lighting positions, and even dimensions, it has made it quite difficult to effectively achieve proper identification of objects.

Two types of object detection algorithms exist. The first one is the Object detection algorithms region proposals. For this one, the RCNN [7], Fast RCNN, and Faster-RCNN, etc., fall under it. These make it possible to create region proposal networks (RPN) which are divided thereafter. The second one is the Object detection algorithms using regression. The SSD and YOLO [8], etc., all fall under this one. These methods are also capable of generating region proposal networks (RPN), but contrary to the previous, the major difference is that they divide these region proposals into categories immediately after they are generated. The YOLO is enhanced by YOLO9000 to take into consideration targets which are above 9000 categories. However, this arranges them hierarchically. In comparison, YOLOv3 [8] implements multilevel classification. By the approach, there is no need for the estimation of the value function. The benefit of this is improvement when it comes to differentiating targets of a small size.

The potential of deep learning should not be overlooked especially in terms of object detection. Of all object detection approaches in recent times, the ones built on deep learning models have shown to achieve results which have been found to be promising, especially when it comes to the finding an object in images [9]. SSD which is a single shot detector presents the benefit of effectively balancing speed with result accuracy of by introducing a CNN-based model just once to the picture which serves as the input. This makes it possible to compute the feature map. The SSD also makes use of anchor boxes quite similar to like faster RCNN. However, these anchor boxes come in different aspect ratios. This operation is carried out across several CNN layers. These layers work on different scales and utilize several feature maps which makes it possible to detect targets of varying sizes. Experimentally, SSD is much more accurate on a wide range of datasets irrespective of how small the pictures are, in comparison to the opposing single stage methods [8].

Thus, this research work aims at reviewing deep learning approaches for vehicle detection that utilize deep learning techniques and focuses more on single stage detectors. The following sections in this paper are split into sub-sections of which Sect. 2 discusses the merits and Motivation for the study, Sect. 3 presents a review of a number of the deep learning methods covered within the literature, Sect. 4 provides the appliance of deep learning to the studies of auto-detection, and eventually, Sect. 5 discusses the review and attempts to explore existing gaps and research directions for future work.

1.1 Merits and Motivation

The crucial role of Vehicle detection cannot be downplayed, as well as tracking application especially within the field of civilian and military. These can be seen in the surveillance, control and management of traffic, as the present benefits range from vehicle categorization, identification and even analysis of traffic, to vehicle speed check, count and tracking. All this is irrelevant to environmental changes.

2 Literature Review

In this section, an attempt will be made to review literature which are relevant and at the same time current, going further to make extensive analysis of different theoretical frameworks and methodologies implemented in vehicle detection modeling. Different approaches used have been outlined and critically analyzed with regard to strengths, downsides, and gaps in preceding research works. The review expands upon various object detection methods which have had the most representation and have been seen as the most pioneering, and with prospects for future research. First, and foremost, their methodologies will be examined and their downsides explained. We will start with a quick review of the traditional base model for vehicle detection. There are three main classes when it comes to vehicle detection models. They are classified as follows.

2.1 *Motion-Based Models*

Motion-based models have to do with the process of background subtraction. This model employs the use of frame subtraction as well as optical flow, with the detection carried out using the GMM. However, as stated previously, it falls short when it comes to the detection of vehicles which are stationary. In the case of the optical flow method, it works on the principle of locating the pixels of the features which can be found on the object preceded by the tracking of the pixels of these vectors on the models. This process takes a lot of time, and its problem of complexity is not desirable. For the process of frame subtraction, the differences which lie between two or more frames are calculated. This process makes it possible to detect non-stationary vehicles. However, it is not recommended for movements which are passive and swift [4].

2.2 *Feature-Based Models*

These are HOG, Haar, and SURF. The HOG is able to group the pixels according to gradients, as it possesses a significantly high descriptive nature. The pixels are located on the object, after which their orientation is extracted for detection by the HOG. The Haar then points out the desired region after which the feature detection is carried out to spot features of interest such as vehicle parts and recognizable objects thereby allowing the SURF to process the classification. The SURF has the benefit of being much more robust, as well as efficient, having the best performance compared to the SIF. The SURF boasts of having an elevated level of effectiveness when it comes to object labeling [10].

2.3 *Convolutional Neural Network-Based Models*

Convolutional Neural Network (CNN)

The proposal for convolutional neural networks (CNN) for the purpose of image processing was made by LeCun. CNNs have been extensively implemented within a large number of computer applications with success. The general working principle of CNNs involves learning features which are mainly abstract. To achieve this, convolution layers and the pooling layers are alternated and stacked, respectively. The Convolutional neural network combines the raw input data and multiple local filters in order to produce translation, non-changing local features. Combining this with features which have been extracted by the pooling layers [11]. A Convolutional neural network is made up of layers which are placed in series in which case each layer describes a specific computation. This is as represented in Fig. 1.

From Fig. 1 the convolution layers of the CNN are depicted as the foremost layer in the network. The learning and working process of CNN takes place in two stages.

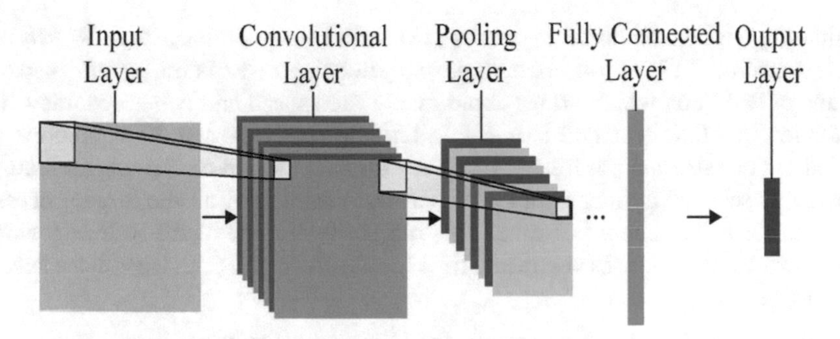

Fig. 1 Architecture of convolutional neural network [12]

These stages are (a) network training, (b) feature extraction, and classification. The first stage has two parts which are the forward part and the backward part. Looking at the forward part, the images which serve as the input flow into the network thereby obtaining an abstract representation. This in turn is used to compute the loss cost with regard to the given ground truth labels. As a result of the loss cost, the backward part is responsible for computing each parameter's gradients within the network, after which these parameters are updated as a result of the gradients in order to prepare for the next cycle of forward computation. When a satisfactory number of training iterations, within the second stage is reached, the network which has been trained can now be employed to extract deep features as well as classify images which are unknown [13].

Recurrent Convolutional Neural Network (RCNN)

The RCNN detector [14] first begins with generating a proposed region by means of an algorithm such as Edge Boxes [15]. The proposed regions are then cropped out of the base image after which resizing is done. The CNN then classifies the regions which have been cropped and resized. In the Final stage, the bounding boxes for proposed regions are then refined by means of the support vector machine (SVM) which undergoes training by means of CNN features as shown in Fig. 2.

In the case of [14], the goal of the bounding-box object detection involves the handling of a more convenient number of regions for candidate object [16, 17] and

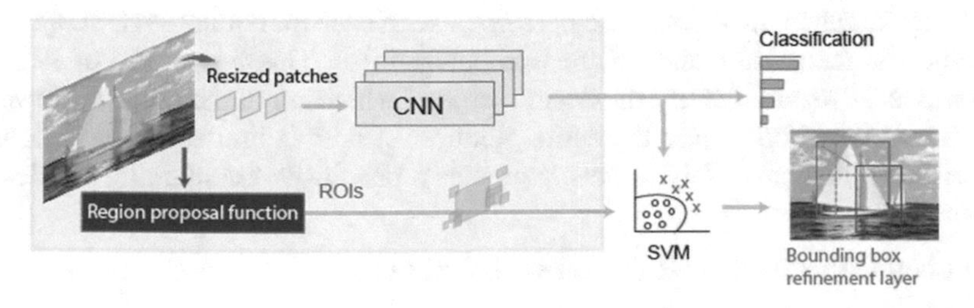

Fig. 2 Representation of RCNN approach

evaluating the convolutional networks [18] individually on each RoI. RCNN has been drawn-out [17, 19] in order to permit attending to RoIs on feature maps by means of RoI Pool, which brings about an elevated speed and better accuracy. The limitation of RCNN method based on a tailored search is that 2000 sections are pulled out or extracted per image. For every domain located on the image, features have to be selected by means of CNN. The combination of all the three processes takes a substantial amount of time. This raises the runtime of the RCNN method. Therefore, RCNN requires significant time for the prediction of the result, for several new images [8].

Fast Recurrent Convolutional Neural Network (Fast RCNN)

Instead of employing three different models of RCNN, the Fast RCNN utilizes one model to pull out or extract desired characteristics from the various domains. These domains are then distributed into numerous categories attributed to on extracted features, and all recognized divisions attributed to the boundary boxes come back together. The Fast RCNN utilizes spatial pyramid pooling as a means of calculating a single CNN representation for the entire image. It proceeds to move a single domain for an individual picture to a specific convolutional network model. It does this by substituting three discrete models for extraction of desired characteristics, which are distributed into divisions, leading to the creation of bounding boxes [8].

Downside—Fast RCNN method makes use of a selective search method in order for it to identify concerned regions. This is a time demanding approach; as in an ideal situation, this entire procedure takes close to two seconds per picture. Compared to the RCNN, the speed of the Fast RCNN is quite good, but taking into consideration real-life datasets, the speed of this approach is still lacking [20].

Faster Recurrent Convolutional Neural Network (Faster-RCNN)

The Faster RCNN [21] detector presents a region proposal network (RPN) which aims at introducing region proposals straight into the network opposed to utilizing an external algorithm like Edge Boxes. The Region proposed network makes use of Anchor Boxes for the detection of objects. In this case, introducing region proposals to the network is much faster and better suited for input data. A typical representation of Faster-RCNN is shown in Fig. 3.

Downside—Faster-RCNN method—This process requires multiple passes for a single picture in order to extract all the targets. Due to the fact that various systems are forced to run in a sequence, the performance of each individual system operation is subject to the performance of the one before it [8]. This makes use of RPNs in order to localize and identify the desired objects which are in the picture. However, this is not carried out across the entire picture instead it is limited to the regions of the picture which are likely to have a greater probability of containing the desired targets [20].

You Look Only Once (YOLO) Network (YOLO)

Based on one stage detector, for YOLO, the picture which serves as the input is subdivided into a domain represented by S × S grids. In this situation, the ability to

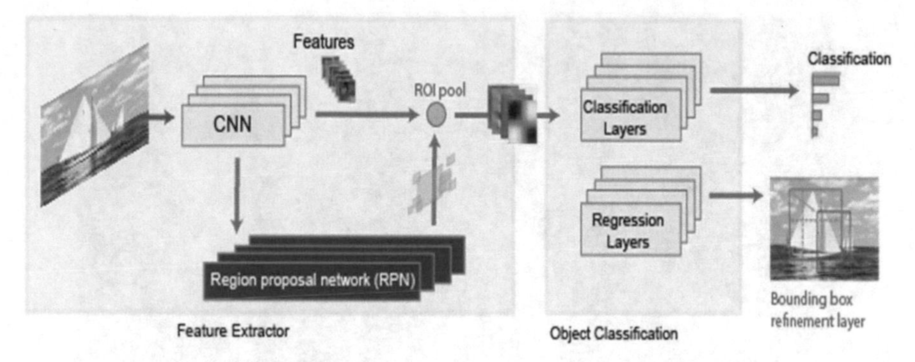

Fig. 3 Representation of Faster-RCNN

detect the target object is highly dependent on every grid which falls inside the input image. At this point, Grid cells are introduced into the region of interest in order to make predictions of the targets which fall within inside boundary boxes. There are five elements which are to be predicted are predicted for every boundary box and they are denoted as i, j, k, l, and c. At the center of the picture, there lies a target inside the box which is denoted by coordinates 'i' and 'j'. In this case, 'k' represents the height, 'l' the width, while 'c' represents the score for confidence, respectively. Relatively 'c' represents the measured probability of holding the target within the boundary box. Figure 4 shows a grid representation of YOLO.

YOLO v1 is not without its own limitations, which are many. Due to this, it is constrained to some extent. Its limitations are attributed to how close the objects appear in the picture. In a situation where the objects fall in a cluster, the YOLO v1

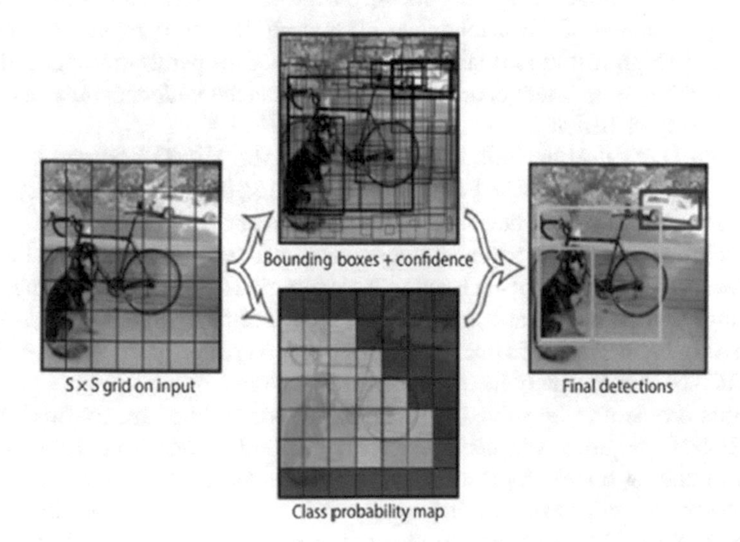

Fig. 4 S × S Grid representation of YOLO

Fig. 5 A sample drawback of YOLO V1

will be unable to detect the small objects. In another situation whereby the size of the object differs from the image which was used for the training of the data, this causes the localization and detection of objects to be difficult [8]. Figure 5 shows an example of the downside of YOLO v1 where it sometimes, fails by identifying the man as an airplane.

Redmon and Farhadi [22] are credited with proposing YOLO v2 and YOLO 9000 In 2016. YOLO v2 object detection deep network comprises a feature extraction network as well as a detection network as is shown in Fig. 5. (ResNet-50) which is a feature extraction network is also a deep transfer learning model. Yolo v2 surpasses Yolo as it does a great job of balancing the running time and the accuracy. The accuracy of Yolo v2 is raised by introducing batch normalization; this aids to augment 2% in map by means of the attachment of it to each layer of convolution. A further step is to add a high-resolution classifier which allowed to operate thoroughly through the modification of its filters in order to give an enhanced understanding of network time for it to work better.

The YOLO-V3 [22] network is an evolution of the YOLO [22] and YOLO-V2 networks. In Comparison to the Faster RCNN network, the YOLO network proceeds to transform the problem in detection into a problem of regression, as it is not necessary for it to have a proposal region thus it proceeds to create coordinates for bounding box as well as likelihoods of each individual class directly by means of regression. This brings about an increase in the detection speed compared to Faster RCNN. Though a benefit of YOLO is that it provides a much greater speed in comparison to Faster-RCNN, unfortunately its detection error is large.

For this problem to be solved, YOLO-V2 introduces the 'anchor box' into the Faster RCNN as it utilizes k-means clustering method in order to generate suitable a priori bounding boxes. As a result of this, the number of anchor boxes which are necessary in order to attain the same result of intersection over union (IoU) declines. YOLO-V2 hence presents a benefit of improving the network structure as it uses a convolution layer to replace the fully connected layer in the output layer of YOLO. YOLO-V3 is an improvement over YOLO-V2, as it is capable of using multi-scale prediction. It uses this feature to identify the final target, and also it has the benefit of having a more complex network structure in comparison to YOLO-V2. YOLO-V3 makes prediction of bounding boxes on varying scales, as a result, multi-scale prediction makes YOLO-V3 more effective for detecting small targets than YOLO-V2 [23]. The YOLOv3-tiny network is another version, which is a miniaturized version of the YOLOv3 object detection network. Both the YOLOv3 and the YOLOv3-tiny aim at multiple class detection opposed to detecting a single class [4].

SSD

SSD is a single shot detector which does a good job of balancing speed with result accuracy, and it achieves this by applying a CNN-based model just once to the input picture for the computation of the feature map. In addition to that, it utilizes anchor boxes just like faster RCNN at varying aspect ratios, followed by learning the offset instead of defining the box. This operation occurs on so many layers of CNN, in which case all the layers function on a wide range of scale and utilize multiple feature maps. Due to this, it is possible to detect targets of several sizes. On an experimental level, SSD possesses a much better accuracy when exposed to different datasets even on inputs pictures which are smaller size in comparison to the other single stage methods [8].

Table 1 depicts the comparison among the various techniques with practical results.

From Table 1, deep learning has achieved remarkable sway on how the world is adjusting to AI in recent years. Among the popular object detection techniques as discussed within the preceding section are (RCNN), Faster-RCNN, (SSD), and (YOLO). Among these, the SSD have shown improved accuracy, while on the other hand, YOLO achieves better performance when speed has higher priority over accuracy. Thus, deep learning based on SSD can combine with Mobile Nets to achieve effective application for detection and tracking. This algorithm carries out effective object detection while at the same time not degrading performance.

Table 1 Performance comparison among popular algorithm for vehicle detection

Technique	Learning paradigm	Regional proposal	Detection rate (%)	Computation time
CNN [24]	BP	Selective search	76.67	–
RCNN [25]	SGD	RPN	70	245 ms
FRCNN [25]	SGD	RPN	73.2	116 ms
Yolo [4]	SGD, BP	Selective search	77.44	5 ms
SSD [26]	SGD	–	98.68	362 ms

2.4 Related Work

Over time, several scholars have developed various kinds of face recognition algorithms, which include Sparse Coding (SC) algorithm, Local Binary Pattern (LBP) algorithm, Histograms of Oriented Gradients (HOG) algorithm, Linear Discriminant Analysis (LDA) algorithm, and Gabor feature algorithm. All these stated algorithms provide between 50 and 76% accuracy rates. These results are too low making these techniques obsolete compared to the recently developed deep learning algorithms.

In 2016, [27] proposed Vehicle Detection from 3D Lidar by means of Fully Convolutional Network followed by evaluating it on KITTI dataset. Experiments carried out using the KITTI dataset show the best available performance of the proposed method. However, with the addition of more training data as well as designing a deeper network, it is possible to further improve the detection performance.

In 2017, [24] proposed lightweight convolutional neural networks for object detection and evaluated it on the popular vehicle detection data class DETRAC and Pascal VOC, results showed that the vehicle detection models have the benefit of accuracy, speed and are therefore suitable for embedded visual applications. For the differences in detection accuracy, it is arguable that this is as a result of the variance in the number of classes across these datasets.

In 2018, [28] proposed Robust vehicle detection under various weather conditions. By means of MIPM with testing done on KITTI and DETRAC dataset, the obtained results have indicated that the algorithms which have been proposed have proven to be robust, showing higher accuracy in comparison to others. This is more evident when during the encountering of occlusions, variations in lighting conditions, as well as weather conditions. However, future work should focus on making the proposed framework to become robust when confronted with different atmospheric/environmental conditions, with emphasis on night testing, as well as proposing a method for improved detection results of vehicles passing through tunnels, traveling on winding as well as steep uphill roads. In addition to that, future work should look towards generalizing the framework to 3D information extraction.

Similarly, in 2018 [29] proposed Unauthorized Amateur UAV Detection which is based on WiFi Statistical Fingerprint analysis and evaluate it on UAV class. Findings show that this innovative detection technique has been tested in a wide range of real-life scenarios, and has proved very capable of effectively detecting and identifying intruder drones in all the considered experimental setups. However, extending the research to a larger set of traffic data obtained from different UAV types would likely strengthen the classification model and additionally increase the accuracy of amateur drone detection.

Also, in 2018, [26] proposed a Real-Time Object Detection and Tracking Using Deep Learning and OpenCV and test it on 21 objects only. Findings from the research show that this algorithm performs efficient object detection while not compromising on the performance; however, Future work should focus on reducing the cost of computation even further.

Moreover, in 2018, [25] proposed an Enhanced Region Proposal Network for the purpose of object detection by means of deep learning method and evaluating it on PASCAL VOC and COCO data classes, it was observed that, for the VGG-16 model, the ERPN obtained 78.6% mAP on VOC 2007 dataset, 74.4% mAP on VOC 2012 dataset, and 31.7% on the COCO dataset.

Recently, in 2020, [30] proposed A lightweight vehicle detection and tracking technique which can be used for systems focused on advanced driving assistance and evaluated it on KITTI dataset class. The deep learning algorithms have higher performance than the RT VDT, however, at the expense of much higher computational cost (high-end GPUs). More recently, in 2021 [31] proposed Pedestrian- and Vehicle-Detection Algorithm Based on Improved Aggregated Channel Features and evaluated the results with Caltech and KITTI datasets. The technique showed to improve the AP by 0.27% on average for Mv-ACF vehicle detectors. However, with respect to the detection of small objects more accurately; this is a challenge that remains for future work.

In 2021, [4] proposed A Delicate Real-Time Vehicle-Detection Algorithm based on YOLO-V3. The proposed network was carried out by introducing Spatial pyramid pooling into the network. The researcher used the Mean square error (MSE) as well as the Generalized intersection over union (GIoU) loss function for the bounding box regression for the purpose of raising the performance of the network. The network which was used for the system training included vehicle-based classes which were obtained from PASCAL VOC 2007, 2012, and MS COCO 2014 datasets. The Little YOLO-SPP network first of all proceeds to detect the vehicle in real-time, having the benefit of high accuracy irrespective of video frame and current condition of the weather. However, a drawback to the algorithm is that it still failed to detect some vehicles and was observed to have made a few mislabeling too.

As a result of this, according to the author [4] it is recommended that future research look into tackling these issues with a wider range of vehicle classes. Table 2 presents in chronological order, authors, data class, and the state-of-the-art literatures which have been reviewed in terms of vehicle detection.

3 Discussions and Open Issues

With the more advanced and effective visual object detectors in the fields of security, transportation, military, etc., the utilization of object detection is therefore experiencing a sharp increase. Despite all these advancements, there is still much room for additional development. It is not easy to make a clear comparison between different object detection methods. In most real-life applications, choices are made to balance accuracy with speed. As a result, it is important to be familiar with other characteristics that have a noteworthy impact on performance [8]. Among the popular object detection techniques as discussed within the preceding section are (RCNN), Faster-RCNN, (SSD), and (YOLO). However, SSD have shown to be more accurate, while YOLO achieves better performance when speed has more priority over accuracy.

Table 2 Summary of related work by method, datasets, findings, and limitation

Authors	Proposed approach	Datasets	Findings	Limitation
[27]	3D Lidar using CNN	KITTI dataset	Outperforms conventional models	Uses few trainings sample
[24]	Lightweight CNN	DETRAC and Pascal VOC	Its suitable for embedded visual applications	Classes disparity between the datasets
[28]	MIPM	KITTI and DETRAC	Achieved higher accuracy at varying conditions	Lacks generalization
[25]	An enhanced region proposal network	PASCAL VOC and COCO data classes	Obtained an excellent mAP for the datasets	Requires addition of more data classes
[26]	Deep learning and OpenCV	21 objects only	Achieved high detection rate	High cost of computation
[29]	UAV detection based on WiFi	UAV class	Efficiently detect intruder drones in all experimental setups	Requires larger set of traffic data derived from different UAV types
[30]	A vehicle detection and tracking technique for advanced driving assistance systems	KITTI dataset class	The deep learning algorithms have higher performance than the RT VDT	At the expense of much higher computational cost (high-end GPUs)
[32]	Vehicle detection-and-tracking method for autonomous driving	Actual road images	Achieved a fairly accurate and robust detection	Complex shadow patterns exist
[31]	Improved aggregated channel features	Caltech and KITTI datasets	Improve the AP by 0.27% on average for Mv-ACF vehicle detectors	The detection of small objects more accurately is a challenge
[4]	A delicate real-time vehicle detection algorithm	Vehicle-based classes from PASCAL VOC 2007, 2012	Achieve a higher mAP on the test datasets	Fail to address the issues of more vehicle classes

Thus, deep learning based on SSD can achieve effective application of detection and tracking. This algorithm has been demonstrated to accomplish effective object detection while not degrading performance.

Provided below are some of the latest trends in this domain for facilitating future research in the field of visual object detection with deep learning. From Table 1, we observed the following challenges and research opportunities.

i. By means of increasing the amount of training data and designing a deeper network, the detection performance of the aforementioned deep learning methods can be even further improved.
ii. It was also observed that for the differences in detection accuracy it is argued that this lapse is due to the variance in the number of classes which fall between these datasets, hence the need to train these models with multiple classes simultaneously still remains an open issue.
iii. Improving the proposed framework to become robust so as to get a better footing against conditions atmospheric/weather as well as bad lighting notably when carried out at night is an open issue. Also, they need to propose a method to improve detection rates of vehicles traveling within tunnels, and road situations such as winding and steep uphill traversing, is vital research the generalization of framework to 3D information extraction need to be looked into, hence a notable goal for future research should involve tackling these issues with a wider range of data classes.
iv. Reducing the cost of computation of the existing work even further is an area future work should focus on.
v. It was also observed that extending the existing research to a larger set of traffic data derived from different UAV types would most likely strengthen the classification model and thereby improving the accuracy of amateur drone detection.
vi. Future work should focus on augmenting the existing technique with the detection and tracking of pedestrians and cyclists.
vii. The detection of small objects more accurately is a challenge that remains open; hence the goal of future research should include addressing these issues with more vehicle classes.

Acknowledgements This study was supported by the Tertiary Education Trust Fund (TETFund) National Research Fund 2020 (NRF, 2020), through the directorate of Research and Innovation of Abubakar Tafawa Balewa University, Bauchi Nigeria.

References

1. Lai AH, Fung GS, Yung NH (2001) Vehicle type classification from visual-based dimension estimation. In: ITSC 2001. 2001 IEEE intelligent transportation systems. Proceedings (Cat. No. 01TH8585). IEEE.
2. Hadi RA, Sulong G, George LE (2014) Vehicle detection and tracking techniques: a concise review. arXiv preprint arXiv:1410.5894
3. Kumaran SK, Dogra DP, Roy PP (2019) Anomaly detection in road traffic using visual surveillance: a survey. arXiv preprint arXiv:1901.08292
4. Rani E (2021) LittleYOLO-SPP: a delicate real-time vehicle detection algorithm. Optik 225:165818
5. Wang Z et al (2021) WearMask: fast in-browser face mask detection with serverless edge computing for COVID-19. arXiv preprint arXiv:2101.00784

6. Kuang P et al (2018) Real-time pedestrian detection using convolutional neural networks. Int J Pattern Recognit Artif Intell 32(11):1856014
7. Chen X, Gupta A (2017) An implementation of faster rcnn with study for region sampling. arXiv preprint arXiv:1702.02138
8. Adarsh P, Rathi P, Kumar M (2020) YOLO v3-Tiny: object detection and recognition using one stage improved model. In: 2020 6th international conference on advanced computing and communication systems (ICACCS). IEEE
9. Loey M et al (2021) Fighting against COVID-19: a novel deep learning model based on YOLO-v2 with ResNet-50 for medical face mask detection. Sustain Cities Soc 65:102600
10. Bay H, Tuytelaars T, Van Gool L (2006) Surf: speeded up robust features. In: European conference on computer vision. Springer.
11. Zhao R et al (2017) Learning to monitor machine health with convolutional bi-directional LSTM networks. Sensors 17(2):273
12. Peng M et al (2016) Nirfacenet: a convolutional neural network for near-infrared face identification. Information 7(4):61
13. Li Y et al (2018) Deep learning for remote sensing image classification: a survey. Wiley Interdisc Rev: Data Min Knowl Discov 8(6):e1264
14. Girshick R et al (2014) Rich feature hierarchies for accurate object detection and semantic segmentation. In: Proceedings of the IEEE conference on computer vision and pattern recognition
15. Zitnick CL, Dollár P (2014) Edge boxes: locating object proposals from edges. In: European conference on computer vision. Springer
16. Uijlings JR et al (2013) Selective search for object recognition. Int J Comput Vision 104(2):154–171
17. Girshick R (2015) Fast R-CNN. In: Proceedings of the IEEE international conference on computer vision
18. Krizhevsky A, Sutskever I, Hinton GE (2012) Imagenet classification with deep convolutional neural networks. In: Advances in neural information processing systems
19. He K et al (2015) Spatial pyramid pooling in deep convolutional networks for visual recognition. IEEE Trans Pattern Anal Mach Intell 37(9):1904–1916
20. Zhao Z-Q et al (2019) Object detection with deep learning: a review. IEEE Trans Neural Netw Learn Syst 30(11):3212–3232
21. Ren S et al (2015) Faster R-CNN: towards real-time object detection with region proposal networks. In: Advances in neural information processing systems
22. Redmon J, Farhadi A (2017) YOLO9000: better, faster, stronger. In: Proceedings of the IEEE conference on computer vision and pattern recognition
23. Tian Y et al (2019) Apple detection during different growth stages in orchards using the improved YOLO-V3 model. Comput Electron Agric 157:417–426
24. Anisimov D, Khanova T (2017) Towards lightweight convolutional neural networks for object detection. In: 2017 14th IEEE international conference on advanced video and signal based surveillance (AVSS). IEEE
25. Chen YP, Li Y, Wang G (2018) An enhanced region proposal network for object detection using deep learning method. PLoS ONE 13(9):e0203897
26. Chandan G, Jain A, Jain H (2018) Real time object detection and tracking using deep learning and OpenCV. In: 2018 international conference on inventive research in computing applications (ICIRCA). IEEE
27. Li B, Zhang T, Xia T (2016) Vehicle detection from 3d lidar using fully convolutional network (2016). arXiv preprint arXiv:1608.07916
28. Yaghoobi Ershadi, N., J.M. Menéndez, and D. Jiménez, *Robust vehicle detection in different weather conditions: Using MIPM*. PloS one, 2018. **13**(3): p. e0191355.
29. Bisio I et al (2018) Unauthorized amateur UAV detection based on WiFi statistical fingerprint analysis. IEEE Commun Mag 56(4):106–111

30. Farag W (2020) A lightweight vehicle detection and tracking technique for advanced driving assistance systems. J Intell Fuzzy Syst (Preprint):1–18
31. Hua J et al (2021) Pedestrian-and vehicle-detection algorithm based on improved aggregated channel features. IEEE Access 9:25885–25897
32. Wael F (2020) A comprehensive vehicle-detection-and-tracking technique for autonomous driving. Int J Comput Digit Syst 9(4):567–580

Chapter 29
Optimization of Resource Management for Workload Allocation in Cloud Computing

N. Senthamarai

1 Introduction

The foremost goal of optimal resource allocation for any cloud person or cloud issuer may be both to improve the QoS parameters and resource utilization. The hybrid energy-efficient resource allocation technique can allocate resources dynamically and enhance the strength performance of the data center. This technique is used to consolidate and rearrange the allocation of resources in an efficient manner [1]. The queuing model is used for optimal resource allocation. The optimal allocation is to reduce the cost of providers under the condition of customer demand. It can provide high quality service [2].

The VM consolidation hassle is treated as the classical online bin packing optimization hassle, which considers VMs as items and PMs as bins, and in which the objective is to percent those objects in as few containers as viable. The VMs have exceptional aid demands and the PMs (packing containers) have distinct aid capacities. The trouble with this allocation is that special aid demands are assigned to special resource capacities.

Real-time application of virtual coaching requires an efficient method to constantly compare the pupil's overall performance. The majority of the tutorial system might be hosted in cloud surroundings. So, the cloud issuer is to satisfy the anticipated first-rate carrier and operational costs. The probabilistic resource allocation method is especially designed for educational establishments [3].

The over-provisioning and under-provisioning of resources for varying user requests may not only result in wastage of resources; but also may result in the need for high computational infrastructure. Hence, optimal resource allocation is

N. Senthamarai (✉)
Department of Networking and Communications, SRM Institute of Science and Technology, Chennai, India
e-mail: senthamn@srmist.edu.in

G. Mathur et al. (eds.), *Proceedings of 3rd International Conference on Artificial Intelligence: Advances and Applications*, Algorithms for Intelligent Systems,
https://doi.org/10.1007/978-981-19-7041-2_29

also an important concept to be considered when balancing the workload among servers. Therefore, the principle objective of the powerful useful resource allocation for any cloud user or cloud provider is to decorate the QoS parameters and resource usage within the cloud surroundings.

2 Related Work

One of the most effective ways to reduce the amount of power consumed in cloud facts center is energy management through the use of virtualization. Cloud carriers offered among cloud vendors and cloud customers should be dependable. But from time to time, wrong useful resource control leads to bad providers. So, prediction and optimization techniques are crucial to keep away from inefficient resource management. But making a prediction is very hard in green resource control in the cloud records center.

Energy conservation is specially targeted on the optimization of migration method which reduces the quantity of destiny migrations. In their work, the processing node velocity is calculated primarily based on the number of jobs finished in keeping with unit time, and the allocation is based on the minimum number of migration nodes [4].

The fuzzy set approach is used as a rating mechanism which includes decomposition, precedence, and aggregation. Using this technique, the gold standard cloud company is chosen [5]. Memory allocation is used as a critical function to select appropriate reminiscence vicinity in step with the character requirements. Thus, the memory region forecast procedure was completed to anticipate the detached memory region [6].

The processor workload prediction approach was proposed by way of Jabalin for a multitenant environment. The migrated information of the VM has to be moved in a secure way [7]. A Minimum Cost Maximum Flow (MCMF) algorithm is implemented for dynamic workloads and glide variations. This set of guidelines is combined with a prediction for reaching the best overall performance [8].

In Deng et al. [9], the author used a fog computing technique. Here, this idea is used to broaden a scientific framework to investigate the power consumption put off trouble in a fog–cloud computing machine. It can broaden the workload allocation problem and break up the primal trouble into 3 sub-issues approximately, which can be solved inside the corresponding subsystems. The work can offer steerage on reading the interaction and cooperation between the fog and cloud. It saves the verbal exchange bandwidth and reduces transmission latency. It moreover improves the general performance of cloud computing.

A job shop scheduling problem is proposed based totally on the constraint of gadget availability to restrict makespan. This paper proposes an operating time window algorithm that includes running period and breaking duration. This proposed method is executed to genuine problem however intricacy is more when contrasted with common cycle continue to plan issue [10].

Sun et al. [11] introduced a modeling framework called ROAR (Resource Optimization, Allocation, and Recommendation system) to simplify, optimize, and automate useful resource allocation. ROAR uses a DSL (Domain Specific Language) called GROWL (Generic Resource Optimization for Web applications Language) to hide the low-degree configuration and analysis information of load checking out for the complex multi-tier net packages. GROWL provides an optimized useful resource configuration to fulfill the QoS dreams of the specific cloud provider.

Ya-Hui Jia provided layer allocated CC (dCC) architecture with adaptive computing useful resource allocation for huge-scale optimization. The first layer is the dCC version which takes the fee of calculating the significance of subcomponents, and for this reason, allocating property. An effective allocating set of rules is designed that could adaptively allocate computing sources based mostly on a periodic contribution calculating method. The 2D layer is the pool version which takes the fee of making true usage of imbalanced beneficial aid allocation. Within this sediment, two super conformance guidelines are designed to help optimizers use the assigned computing resources efficiently [13].

Xudong presented a task technique, situating part figuring basically based most certainly energy IoT, which limits organization cast off. The job optimization assignment version is hooked up, and the most suitable job assignment orientated on put off amongs a couple of regional nodes is in addition found out on the idea of computing aid optimization inside the unmarried component node. The balanced initialization, useful resource allocation, and task allocation (BRT) set of rules are proposed [12].

The useful resource control proposed three optimization fashions (i.e., GMO, SP1O, and SP2O) that allow an IoT issuer company to find the maximum appropriate deployment of gateways, the most beneficial useful resource allocation for carrier capabilities, and the most beautiful routing consistent with a price function with an overall performance constraint in a NIoT machine [14].

Green strength-saving verbal exchange is one of the critical study topics for the improvement of the 5G era. In order to beautify the power efficiency performance of the cross-layer broadband wireless verbal exchange network machine, this takes a look at combining the evaluation of the broadband wireless communication device to optimize the electricity basic overall performance of the pass layer wireless communiqué gadget [15].

This paper addresses the preserve scheduling hassle with series-set up, set-up instances and activity lag times, which developments are essential in contemporary and manufacturing structures. Experiments are finished on instructions of facts set generated from well-known benchmark instances. The computational consequences show that HGA-TS outperforms other heuristic algorithms and might locate better bounds than MILP, proving its effectiveness in fixing flexible pastime shop scheduling trouble. It is used to restrict makespan [16].

3 Proposed System

Resource optimization is a vital concept in making selections and in studying problems. In mathematical terms, optimization trouble is one which finds the fine solution from many of the sets of all possible solutions. Job shop scheduling algorithm is used for resource optimization in which idle machines are assigned to resources at a particular time. Four different jobs of various processing instances are considered, which need to be scheduled on m machines with various processing electricity.

3.1 Enhanced Job Shop Optimization

Job shop scheduling consists of variables and constraints for the problem. The variables denote the beginning time of the task $t_{i,j}$. The constraints are divided into two categories: conjunctive constraint and disjunctive constraint. Conjunctive constraint derives from the condition of any two continuous tasks executed in the same job which means the second can be executed after completing the first one. For example, tasks (1, 2) and tasks (1, 3) are continuous tasks for job1. The execution time for task (1, 2) is 2; task (1, 3) must start at least after 2 units of time. So, the result of the constraint is

$$\mathbf{t}_{1,2} + 2 \le \mathbf{t}_{1,3} \tag{1}$$

Disjunctive constraint derives from the condition that a machine cannot work on two tasks at the same time. For example, task (1, 2) and task (2, 3) are processed on the same machine. So, the result of the constraint is

$$\mathbf{t}_{1,2} + 2 \le \mathbf{t}_{2,3} \quad \text{(if task (1, 2) is scheduled before task (2, 3))} \tag{2}$$

This algorithm is designed by considering conjunctive and disjunctive constraints. Both constraints are used to calculate finish time during processing the task and minimize the makespan.

Algorithm for Enhanced Job shop optimization

Input: number of systems, number of jobs and processing times.
Output: Optimized allocation
Step 1: Declare the variables
Step 2: Assign Systems=range (0, m_count) where m_count=3
 Assign Jobs=range (0, p_count) where p_count=4
Step 3: Define number of systems and processing times of each system
 $t_{p,q}$ be begin time of job j that is performed on system p,
 $f_{p,q}$ be finish time of job j that is performed on system p,
 $E_{p,q}$ be execution time of job j that is performed on system p,
 d_q be deadline
 M_{max} be makespan (completion time of job).
Step 4: if $t_{p,q}$- $t_{p,r}$>=$E_{p,q}$ or $t_{p,q}$- $t_{p,q}$>=$E_{p,q}$
 Only one job can be processed on a machine based on prediction table
Step 5: if $t_{i,j}$-$E_{i,j}$<=d_j
 Jobs must be finished based on prediction table
Step 6: $M_{max} \geq t_{ij} + E_{ij}$
 Minimize the makespan

This algorithm defines the number of machines and the processing time of each machine for processing the job based on the prediction table. Check if $t_{i,j} - t_{i,k} >= E_{i,k}$ or $t_{i,k} - t_{i,j} > = E_{i,j}$ is satisfied, and select the execution time from the prediction table according to the job size. Check if $t_{i,j} - E_{i,j} < = d_j$ is satisfied, it executes the job based on the prediction table. This model checks whether the makespan value is greater than the starting time and processing time, thereby reducing the makespan value.

Figure 1 depicts the comparative analysis of migrations with job shop optimization and without this model. The Comparative analysis of migrations are listed in Table 1. Job shop optimization is used to reduce unnecessary migration. The proposed method shows a clear improvement in medium-small jobs, medium-large jobs, and large jobs. It avoids unnecessary migrations.

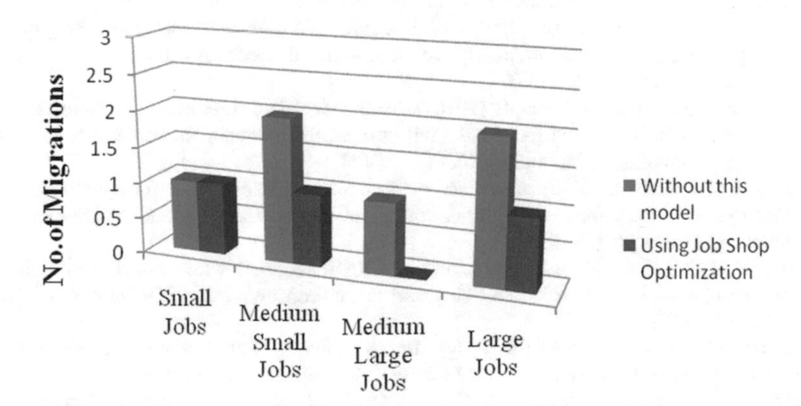

Fig. 1 Comparative analysis of migrations

Table 1 Comparative analysis of migrations

Jobs	Without this model	Using job shop optimization
Small jobs	1	1
Medium small jobs	2	1
Medium large jobs	1	0
Large jobs	2	1

4 Conclusion

The proposed method is useful for resource management in cloud computing and also used to reduce the makespan. Job shop scheduling algorithm is used for resource optimization to locate the satisfactory answer from the various set of all viable solutions. The main limitations in this module are the clients' expectation of phenomenal quick stacking administrations and readiness.

References

1. Pangotra N, Sharma M (2016) A pragmatic approach to optimize energy efficient resource allocation technique in cloud computing data center. Int J Adv Res Comput Commun Eng 5(2):184–191
2. Zhang N, Li R (2016) Resource optimization with reliability consideration in cloud computing. In: Reliability and maintainability symposium (RAMS). IEEE
3. Koch F, Assunção MD, Cardonha C, Netto MAS (2016) Optimising resource costs of cloud computing for education. Fut Gen Comput Syst 55:473–479
4. Sekhar VS, Joseph N (2014) Optimizing the virtual machine migrations in cloud computing systems by using future prediction algorithm. Int J Eng Res Technol (IJERT) 3(8):366–369
5. Aruna L, Aramudhan M (2016) Framework for ranking service providers of federated cloud architecture using fuzzy sets. Int J Technol 7(4):643–653
6. Jarlin Jeincy G, Shaji RS, Jayan JP (2016) A secure virtual machine migration using memory space prediction for cloud computing. In: International conference on circuit, power and computing technologies–ICCPCT
7. Jabalin Reeba P, Shaji, RS, Jayan, JP (2016) A secure virtual machine migration using processor workload prediction method for cloud environment. In: International conference on circuit, power and computing technologies–ICCPCT. IEEE
8. Hadji M, Zeghlache D (2012) Minimum cost maximum flow algorithm for dynamic resource allocation in clouds. In: IEEE fifth international conference on cloud computing. IEEE Computer Society, pp 876–882
9. Deng R, Lu R, Lai C, Luan TH, Liang H (2016) Optimal workload allocation in fog-cloud computing towards balanced delay and power consumption. IEEE Internet of Things J 3(6):1171–1181
10. Ploydanai K, Mungwattana A (2010) Algorithm for solving job shop scheduling problem based on machine availability constraint. Int J Comput Sci Eng 2(05):1919–1925
11. Sun Y, White J, Eade S, Schmidt DC (2016) ROAR: a QoS-oriented modeling framework for automated cloud resource allocation and optimization. J Syst Softw 116:146–161
12. Niu X, Shao S, Xin C, Zhou J, Guo S, Chen X, Qi F (2019) Workload allocation mechanism for minimum service delay in edge computing-based power Internet of Things, 7:83771–83784

13. Jia Y-H, Chen W-N, Gu T, Zhang H, Yuan H-Q, Kwong S, Zhang J (2019) Distributed cooperative co-evolution with adaptive computing resource allocation for large scale optimization. IEEE Trans Evol Comput 23(2):188–202
14. Pham T-M, Nguyen T-T-L (2020) Optimization of resource management for NFV-enabled IoT systems in edge cloud computing. IEEE Access 8:178217–178229
15. Wang Y, Zhu Q (2021) A hybrid genetic algorithm for flexible job shop scheduling problem with sequence-dependent setup times and job lag times. IEEE Access 9:104864–104873
16. Dong Z, Wei J, Chen X, Zheng P (2020) Energy efficiency optimization and resource allocation of cross-layer broadband wireless communication system. Green Commun Wirel Netw 8:50740–50754

Chapter 30
Evaluating Morphometric Feature Variability of Handwritten Numerals Among Malaysian Malays Using Self-organizing Maps

Loong Chuen Lee, Nur Fatin Syuhada Binti Roslee, and Hukil Sino

1 Introduction

Forensic document examination is one of the subfields of forensic sciences dealing with determining the authorship of questioned documents and ink or paper analysis [1]. The determination of authorship is often much more challenging than the ink or paper analysis because the process is labor intensive and experience demanding compared to ink or paper analysis which can be performed by various chemical instruments [2]. Meanwhile, the former analysis can only be accomplished via a comparison analysis and relies on the fact that no two people can have the exact same handwriting due to a combination of genes and environmental and anatomical factors [3]. Forensic document examiners will compare the questioned documents with known documents as references to identify or eliminate suspects as the source of handwriting. In other words, identification of the author based on handwriting requires an exemplar, i.e., known handwriting from a suspect. Hence, there is no way to determine the writer of a questioned document in case no suspect is available for the case.

In this context, it is feasible to narrow down the search of potential suspects to a particular ethnic origin or nationality based on handwriting. Kapoor and Saini [4] found that culture and ethnicity influence a person's handwriting, e.g., regularity, neatness, counterclockwise rotations and posture. Meanwhile, Cheng et al. [5] attempted to identify the class characteristics of the English handwriting written by

L. C. Lee (✉) · N. F. S. B. Roslee · H. Sino
Forensic Science Program, CODTIS, Faculty of Health Sciences, Universiti Kebangsaan Malaysia, 43600 Bangi, Selangor, Malaysia
e-mail: lc_lee@ukm.edu.my

L. C. Lee
Institute IR4.0, Universiti Kebangasaan Malaysia, 43600 Bangi, Selangor, Malaysia

G. Mathur et al. (eds.), *Proceedings of 3rd International Conference on Artificial Intelligence: Advances and Applications*, Algorithms for Intelligent Systems,
https://doi.org/10.1007/978-981-19-7041-2_30

three racial groups (Chinese, Malay, and Indian) in Singapore. Six class characteristics were determined to be influenced by the writers' native language. Malaysia is a multicultural country and the three main ethnic groups are Malay, Chinese, and Indian. On the other hand, Mohamed et al. [6] studied that the angularity of letters was affected by race and the education system in Malaysia. The authors found that Indian writers tend to have rounded handwriting, while Chinese writers prefer angular handwriting. Eventually, the authors concluded that variations in the education system in Malaysia cause differences in the handwriting of the three main ethnic groups.

Strictly speaking, handwriting includes both words and numerals. However, according to Bhardwaj et al. [7], numerals' characteristics are least studied compared to words for forensic investigation. Investigating the variability of the numerals' handwriting is just equally important since numerals are always involved in cases like forged or counterfeit cheques. Forger tends to alter the size of the numeral or by addition or elimination of additional strokes when disguising numerals. Despite the fact that several works have attempted to identify the class characteristics of handwriting, work considered Malaysians as a studied population is still very few [8–10]. Hence, this paper aims to explore the morphometric feature variability of numerals among Malaysian Malays. The inter-subject and intra-subject variations were carefully evaluated based on self-organizing maps (SOM). The insights of the variations could contribute to the discrimination of individuals or the classification of a writer by nationality or ethnic origin based on numerals among Malaysians.

2 Methodology

2.1 Handwritten Numeral Data

Five Malaysian Malays have provided their handwritten numerals voluntarily for this study. The five subjects fulfilled the following inclusion factors: (a) attended local primary and secondary education at a national school, (b) Malaysian, (c) aged between 20 and 30 years old, and (d) right-handed. Table 1 presents the background details of the five subjects. All the subjects were provided with the same type of paper (IK Yellow, white copy paper, 70 gsm, 210 mm $\times$ 297 mm) and a pen (gel pen, Faber Castell, 0.7 mm). They were required to consecutively write numbers 0–9 six times a day over 14 days. Eventually, each subject presented 840 numerals (10 numerals $\times$ 6 sets per day $\times$ 14 days).

A digital microscope (DINO-LITE Handheld Microscope USB 2.0, AMA4515-ZT-EDGE) was used to measure at least two morphometric features in each numeral. The selection was made in such a way that the features: (a) must be seen in the majority of the samples and (b) measurable. Qualitative features, e.g., shape and presence of hook, were not considered in this work. Figure 1 shows examples of numerals with the measured features. Basically, all the features were a distance between two points.

Table 1 Background details of the subjects

Code	Sex	Age	Education level	Occupation
S1	Female	26	Bachelor degree	R&D chemist
S2	Female	22	Bachelor degree	Student
S3	Female	22	Bachelor degree	Student
S4	Male	22	Bachelor degree	Student
S5	Female	22	Bachelor degree	Student

Fig. 1 A total of 33 morphometric features measured from the ten numerals via a digital microscope

Generally, the tiny entry of all the numerals presents a certain width after zooming in under the microscope. And it is expected that the variation between subjects would be in the range of millimeters. Therefore, to minimize bias, all the features were measured from the outermost or innermost line of the ink entry and the distance between the points shall be that of maximum value. In case one of the two markers was missing, such features cannot be measured and would be labeled as a missing value.

2.2 *Statistical Analysis*

The final data consists of 33 features and 4200 samples with 16 features presenting varying missing values. Thus, mean imputation methods were first performed to replace the missing values. Then, the data were converted into ratios by the numerals in order to alleviate size correlations from the morphometric data, as detailed in Table 2. Then, the data were processed by the self-organizing maps (SOMs) algorithm.

Table 2 Descriptions of 44 morphometric features

Numeral	Code	Definition	Numeral	Code	Definition
0	F1	f0a/f0b		F23	f5a/f5c
1	F2	f1a/f1b		F24	f5a/f5d
2	F3	f2a/f2b		F25	f5b/f5c
	F4	f2a/f2c		F26	f5b/f5d
	F5	f2b/f2c		F27	f5c/f5d
3	F6	f3a/f3b	6	F28	f6a/f6b
	F7	f3a/f3c	7	F29	f7a/f7b
	F8	f3a/f3d		F30	f7a/f7c
	F9	f3b/f3c		F31	f7a/f7d
	F10	f3b/f3d		F32	f7a/f7e
	F11	f3c/f3d		F33	f7b/f7c
4	F12	f4a/f4b		F34	f7b/f7d
	F13	f4a/f4c		F35	f7b/f7e
	F14	f4a/f4d		F36	f7c/f7d
	F15	f4a/f4e		F37	f7c/f7e
	F16	f4b/f4c		F38	f7a/f7b
	F17	f4b/f4d	8	F39	f8a/f8b
	F18	f4b/f4e		F40	f8a/f8c
	F19	f4c/f4d		F41	f8b/f8c
	F20	f4c/f4e	9	F42	f9a/f9b
	F21	f4d/f4e		F43	f9a/f9c
5	F22	f5a/f5b		F44	f9b/f9c

SOM is a type of neural network algorithm that adopts the unsupervised learning approach [11]. It segregates the data into a number of nodes (the dimension of nodes has to be optimized by the researcher) presented in a mapping plot [12]. The contribution of the variables can be inspected through the corresponding SOM plot. In order to ease the interpretation of SOM results, the 44 features were split into seven sets so that the most number of features to be modeled by SOM was only 10. Then, the segregation of the 420 numerals by the five subjects was evaluated based on the mapping plot, and the most discriminated morphometric feature was determined from the respective SOM plot.

All the statistical analyses were accomplished in the R statistical software, v.3.6.2 [13]. SOM was performed using the function som available by the kohonen package.

3 Results and Discussions

Figure 2 shows the SOM plot (left) and the corresponding mapping plot (right) computed for clustering the five subjects according to the 44 numeral features. The dimension of nodes has been carefully optimized, so each node shall be filled by at least one sample and presented meaningful segregation by the subjects.

Despite the segregation of the five subjects not seen in any plot, it is noted that several nodes of mapping plots estimated from numerals 0–2, and 4–6 were dominated by a particular subject (see Fig. 2a, c and d). Particularly in Fig. 2d, four of the five subjects were, respectively, dominated by one of the six nodes. Similar observations were also seen in Fig. 2c, though the homogeneity of nodes was not as evident as in Fig. 2d. Meanwhile, other numerals have not presented any apparent clustering by the subjects.

Next, the relative similarities among subjects could be inferred based on the distances of the nodes representing the subjects. Nodes close to each other denote high similarity. Based on Fig. 2c, S3 seems more similar to S5 than the other subjects. However, S3 was located far away from the node dominated by S5 when modeled using features deriving from the numeral 5 (Fig. 2d). In other words, S3 is best to be discriminated with S5 based on numerals 5 and 6. It is worth mentioning that S2 never dominates a node. This could be due to the fact that the subject's writing skill is less mature than others, thus showing minimal individualistic features.

Recently, Nasrul [8] performed a study to classify individuals based on geometric morphometric numerals among Malaysians. The researcher extracted various statistical features from the images of selected numerals. Based on the score plots of principal component analysis (PCA), Nasrul [8] concluded that the numeral '6' is useful in discriminating the ten subjects. However, this work found that 6 was useful only in discriminating S3 from the others. The discrepancies could be due to the two rationales.

Firstly, Nasrul [8] deployed image segmentation techniques in extracting multiple features from '6'. Meanwhile, this work adopted a manual approach to measure the morphometric features. Although the image segmented technique allows faster measurement than the manual approach, it is much more rigid since measurements rely on pre-selected landmarks. Since handwritten numerals can be very dynamic, thus this work adopted manual measurement assisted by digital microscopes.

Moreover, the current work concerned only a particular ethnic group of Malaysian, i.e., Malays, that was different from [8] considered Malaysian subjects regardless of ethnicity. As noticed by Mohamed et al. [6], the education system played a vital role in formulating the writing habit of a person. Four main streams of the primary school system are provided in Malaysia, i.e., Malay, Chinese, Indian, and English-based. Hence, the high discriminative capability of the numeral 6 in [8] could be partly contributed by the inhomogeneity of the subjects.

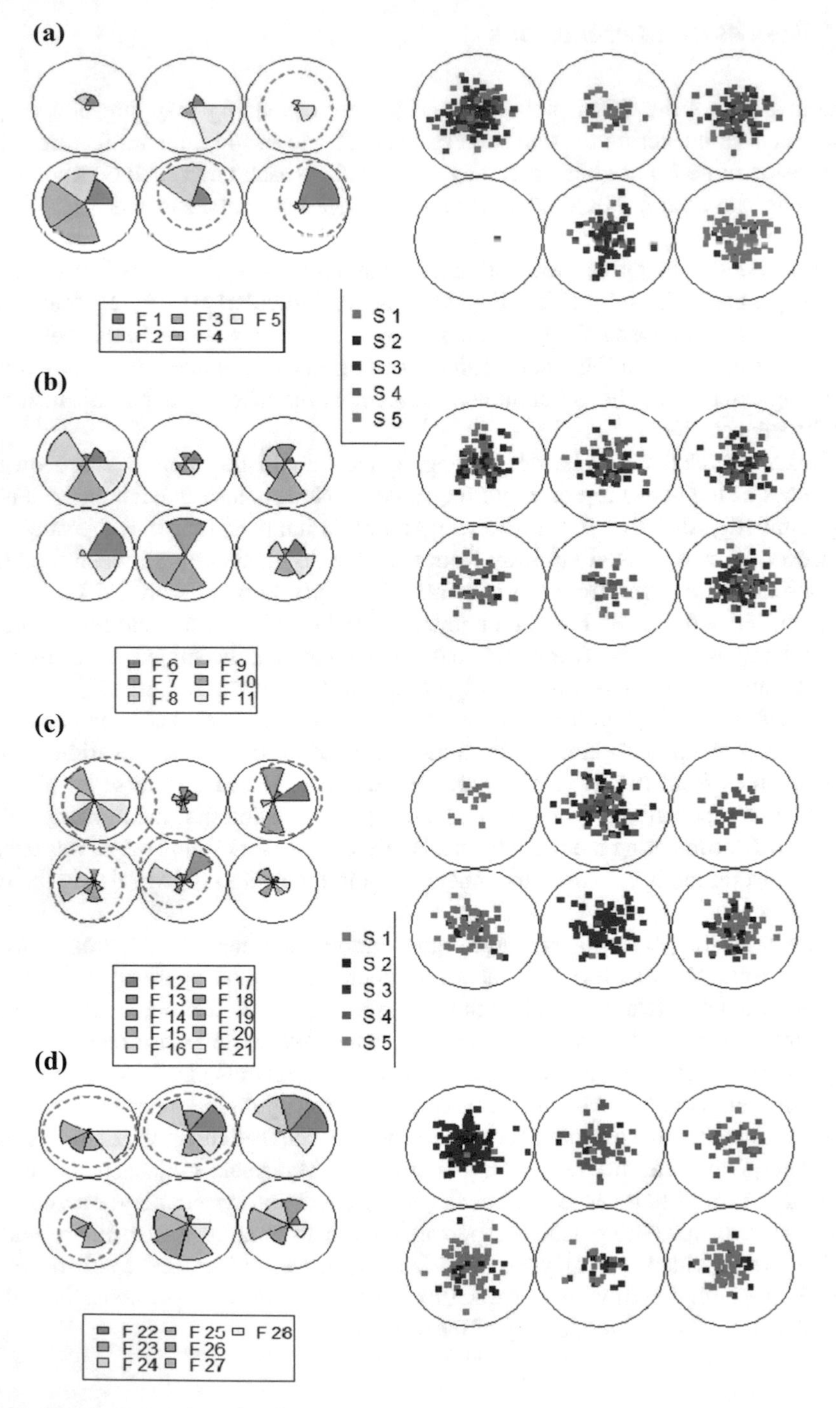

Fig. 2 SOM plots (left) and the corresponding mapping plots (right) computed from the morphometric features measured from numerals **a** 0–1–2, **b** 3, **c** 4, **d** 5–6, **e** 7, **f** 8, and **g** 9

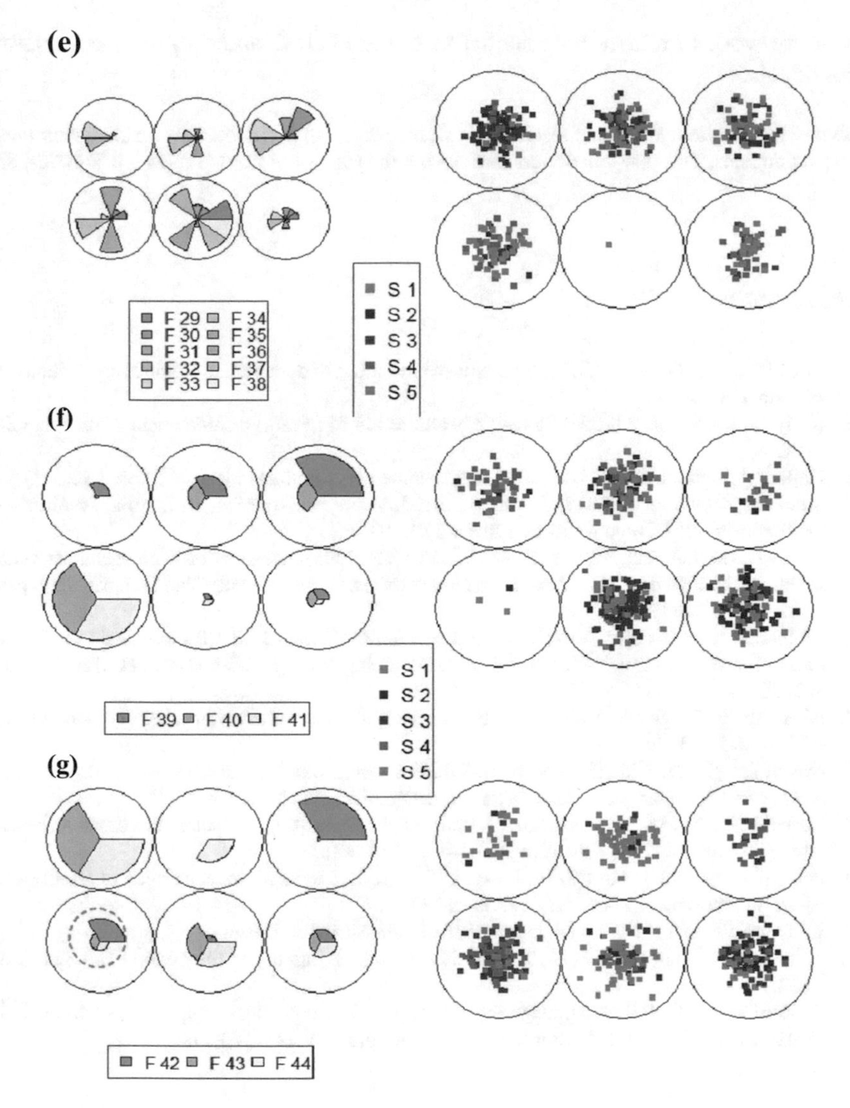

Fig. 2 (continued)

4 Conclusion

This study presents the first pilot work evaluating the variability of morphometric features among Malaysian Malays. The most discriminated numerals were identified based on self-organizing maps' outputs. It was found that numerals 0–2 and 4–6 demonstrated good potential to discriminate against Malays writers. However, this work has employed rather small data, and thus more samples shall be acquired in future work to affirm the potential of numerals 0–2 and 4–6 in discriminating individuals.

Acknowledgements We would like to thank all the subjects for voluntarily providing their handwritten numerals. This research was conducted with the support of CRIM-UKM (GUP-2020-085).

References

1. Ellen D, Day S, Davies C (2018) Scientific examination of documents: methods and techniques, 4th edn. CRC Press
2. Kelly JS, Lindblom B (2006) Scientific examination of questioned documents, 2nd edn. CRC Press
3. Kapoor AK, Saini M (2017) Handwriting as a means of cultural identity. J Foren Sci 3(1):1–2
4. Saini M, Kapoor AK (2015) Impact of heredity and environment in familial similarity of handwriting. Int J Comput Electron Res 4(1):1–10
5. Cheng N, Lee GK, Yap BS, Lee LT, Tan SK, Tan KP (2005) Investigation of class characteristics in English handwriting of the three main racial groups: Chinese, Malay, and Indian in Singapore. J. Foren. Sci. 50(1):1–8
6. Mohamed R, Hazir NM, Yong WK, Ahmad UK, Mohamad I (2010) Statistical examination of common characteristics for disguised handwriting amongst Malaysians. Health Environ J 1(2):30–37
7. Bhardwaj B, Kwatra H, Harne P (2019) Individuality of numerals in disguised handwriting. Int J Sci 2(1):13–20
8. Nasrul HI (2020) Classification of individuals using handwritten numeral with geometric morphometric techniques. Master's thesis, Universiti Sains Malaysia
9. Harne P, Mishra MK, Sodhi GS (2018) Analysis of handwriting characteristics based on diverse ethnic distribution. Int J Adv Trends Comput Appl 5(2):1–6
10. Peter KA, Arvind T, Sai SS, Hari VP (2018) Study on class characteristics of handwriting based on emotions. J Foren Sci Crim Invest 11(1):1–4
11. Kohonen T (2001) Self-organizing maps, 3rd edn. Springer, Berlin
12. Wehrens R, Kruisselbrink J (2018) Flexible self-organizing maps in Kohonen 3.0. J Stat Softw 87(7):1–18
13. R Core Team (2019) R: A language and environment for statistical computing. R version 3.6.2 (2019–12–12). R Foundation for Statistical Computing, Vienna, Austria

Chapter 31
Hesitant Fuzzy Sets Based TSK Model for Sentiment Analysis

Makrand Dhyani, Sanjay Kumar, and Govind Singh Kushwaha

1 Introduction

Emotions are considered psychological states brought on by neurophysiological changes, variously associated with feelings, thoughts, and a degree of pleasure or displeasure whereas sentiment is an attitude toward something. Sentiment analysis (SA) is the process of obtaining information related to ones emotions whenever expressed in conversations, questions, requests, or comments through techniques like natural language processing and Machine Learning. With social media and micro blogging websites booming at an exponential level, it is important to analyze the vast data being generated on a daily basis as it consists of valuable information regarding a person's views belief's likes dislikes, etc. towards a brand or a topic of social importance. SA depends mainly on information to classify expressions as neutral, negative, or positive.

Researchers are continuously putting efforts to determine the attitude/behavioral characteristics of an individual using linguistic data generally present in the form of questionnaires. Various models to determine the personality characteristics of an individual include models developed by Eysenck [1], Price et al. [2], Goldberg [3], and Sinha [4]. Detecting and analyzing personalities using questionaries' proved to be an effective method for researchers & psychologists. Advancement in technology and advent of artificial intelligence aided researchers to use a different approach that could be effective to determine the personality traits characteristics of an individual.

M. Dhyani (✉) · S. Kumar
Department of Mathematics, Statistics and Computer Science, G. B. Pant University of Agriculture and Technology, Pantnagar, India
e-mail: makdhyani23@gmail.com

G. S. Kushwaha
Department of Social Sciences and Humanities, G. B. Pant University of Agriculture and Technology, Pantnagar, India
e-mail: drgovindsingh@yahoo.co.in

G. Mathur et al. (eds.), *Proceedings of 3rd International Conference on Artificial Intelligence: Advances and Applications*, Algorithms for Intelligent Systems,
https://doi.org/10.1007/978-981-19-7041-2_31

Tools and techniques such as pattern recognition and corpus based Natural Language Processing toolkits are nowadays being used vastly to recognize and determine general attitude of an individual or groups of individual at community and global levels.

Pak and Paroubek [5] performed a linguistic analysis of a corpus of over 3 lakh text posts from twitter to build a sentiment classifier. Over the time various techniques have been developed for SA. According to Vinodhini and Chandrasekaran [6], Naive Bayes, Maximum Entropy (MaxEnt), and Support Vector Machines (SVMs) respectively are three text-based categorizations which have found a prominent application in the field of SA. Feldman [7] gives a detailed literature on types of SA namely aspect-based, comparative, sentence-level, and document-level sentiment analysis along with sentiment lexicon acquisition. Unigram model, tree kernel model, and feature-based model were used by Sahayak et al. [8] for SA of twitter data. The application of various deep learning techniques such as ANN, CNN, RMDN & RNTN and their comparison on different datasets in SA was reviewed by Ain et al. [9]. Correa et al. [10], Hughes et al. [11] have also anticipated few psychological criteria using data of SNSs of twitter and Facebook with help of questioners.

Decision tree, random forest, and SVM are also gained attention for classified and analysis of twitter data to understand the sentiments related to farmers protest at the international level Neogi et al. [12]. Corpus and dictionary-based methods were also used by Gautam et al. [13] to determine the semantic orientation of the opinion words in tweets using Naives Bayes classifier to perform SA of twitter data. Recently, Naseem et al. [14] used COVID-19 twitter data to develop an information policy to curb the negative sentiment growth on social media platforms. Algorithms like Convolution neural network model (CNN) and Long Short Term Memory (LSTM) were also used for SA by Gandhi et al. [15]. Recently, Kumar and Vardhan [16] used Plutchik's wheel of emotion which contains eight basic emotions to classify the sentiments of various tweets using Rule Based Emotion Classification (RBEM) algorithm.

Fuzzy set theory and fuzzy logic also known as the science of uncertainty has proven itself to handle uncertainty, vagueness, and non-determinism in real life models. Inference models given by Mamdani and Assilian [17], Sugeno and Kang [18] enabled users to directly obtain output using a simple knowledge base (usually referred to as rule base) where there is a non-linear relationship between input and output variables. Kushwaha and Kumar [19], Devi et al. [20], Pandey et al. [21] developed various fuzzy rule base approaches to predict the anxiety of students.

Considering real-life situations of making decisions or choices, people are usually irresolute and hesitant for one thing or another thus making it difficult to reach a final agreement. To deal with such cases, Torra and Narukawa [22] introduced the concept of a hesitant fuzzy set (HFS). The HFS, as one of the extensions of the fuzzy set introduced by Zadeh [23], allows the membership degree of an element to be a set represented by several possible values and can express the hesitant information more comprehensively than other extensions of the fuzzy set. Xia and Xu [24] defined various aggregation operators to aggregate hesitant fuzzy information. In the present study we propose a Hesitant fuzzy set-based Takagi-Sugeno-Kang fuzzy inference system to analyze the attitude of web users using data of a micro blogging web site called Twitter.

2 Preliminaries

2.1 *Fuzzy Set*

Let $X = (x_i : i = 1 \text{ to } n)$ be a finite crisp reference set. A FS A on X is a mathematical object of the form $A = \{x, \mu_A(x) : \forall x \in x\}$. Here $\mu_A(x)$ is the grade of membership of x in A and $\mu_A : X \rightarrow [0, 1]$.

2.2 *Hesitant Fuzzy Set*

HFS H over the reference set $Z = \{z_1, z_2, z_3 .... z_n\}$ is defined by a mathematical object of the form $H = \{\langle z, h_H(z)\rangle | \forall z \in z\}$ where h_H is membership function that returns the elements of U into possible subset of [0, 1] i.e. h_H: $U \rightarrow P[0, 1]$. Here, $P[0, 1]$ is collection of subsets of [0, 1].

Example: If $Z = \{z_1, z_2, z_3\}$ is the reference set, and $h_A(z_1) = \{0.3, 0.5, 0.7\}$ and $h_A(z_3) = \{0.2, 0.4, 0.6\}$, $h_A(z_3) = \{0.5, 0.6\}$ are the possible membership grade of $x_i(i = 1, 2, 3)$ to a set A, respectively. Then A can be considered as HFS and is represented as

$$A = \{<z_1, \{0.3, 0.5, 0.7\}>, <z_2, \{0.2, 0.4, 0.6\}>, <z_3, \{0.5, 0.6\}>\}.$$

Below is a representation of some operations on HFS.

1. Union—$h_1 \cup h_2 = \{\alpha_1 \vee \alpha_2 | \alpha_1 \in h_1, \alpha_2 \in h_2\}$.
2. Intersection—$h_1 \cap h_2 = \{\alpha_1 \wedge \alpha_2 | \alpha_1 \in h_1, \alpha_2 \in h_2\}$.
3. Complement—$h_1^c = \{1 - \alpha_1 | \alpha_1 \in h_1\}$.

Here $\wedge$ & $\vee$ represent min and max operators.

2.3 *Aggregation Operator for HFS*

Let H be an element whose elements h HFEs are determined by a function $h_H : X \rightarrow P([0, 1])$ then $H^{\cdot} = \{<x, O(h_H(x))>\} \forall x \in X$ is a fuzzy set and membership grades are calculated using the mapping $h_H : P([0, 1]) \rightarrow [0, 1]$ such that

$$O(\{x_1, x_2, \ldots.x_n\}) = 1 - \coprod_{i=1}^{n} (1 - x_i)^{w_i} \quad (1)$$

where n is the number of elements in the subset of [0, 1] and w_i is the weight of x_i s.t $\sum x_i = 1$ Bisht and Kumar [25].

3 Materials and Methods

3.1 Collection of Data and Pre-processing

From a site named vicinetas.com, 2000 tweets about the protest of farmer bills in various languages were collected. 100 tweets out of them in the English language were selected anonymously after eliminating retweets, links, date, time, and other non-meaningful symbols. Selected tweets were analyzed using "TextBlob" tool of a Natural language toolkit (NLTK) for their polarity and subjectivity which are taken as input variables in the proposed IFIS. A brief description of polarity and subjectivity associated with tweets in terms of psychology is given as follows.

Polarity: Polarity is usually defined in terms of negative or positive connotations that people have with an individual word. Fundamentally, polarity is a mixed construct referring to the fact that in natural language, some words are negative while some are positive.

Subjectivity: Subjectivity of a sentence expresses some personal feelings, views, or beliefs. It is important to note that a purely objective sentence is sentiment free (without any sentiment), while a sentence purely subjective in nature leans towards a negative or a positive sentiment. Subjectivity numerically lies in the interval [0 1] while polarity lies in [−1 1].

3.2 Hesitant Fuzzy Set Based Sugeno Fuzzy Inference System

The process of developing HFS-based TSK model includes the following steps.

1. Fuzzification of input and output data.
2. Construction of Hesitant Fuzzy Sets.
3. Construction of rule base and defuzzification.

Stepwise procedure of the proposed HFS-based TSK model for prediction of positivity is explained in the following steps.

Step 1. Fuzzification

After analyzing tweets using "TextBlob" tool of a natural language toolkit (NLTK), raw scores of polarity (x), subjectivity (y) are computed. Raw score of positivity

(z) associated with tweets is computed using "NaïveBayesClassifier" available in "TextBlob". These raw scores of polarity, subjectivity are mentioned in Tables 2 and 3 respectively. Raw scores of polarity and subjectivity with equal intervals are fuzzified using triangular fuzzy sets namely A_i (i = 1–5) & B_j (j = 1–4) with the following parameters.

Step 2. Construction of Hesitant Fuzzy Sets

After construction of triangular fuzzy sets, we use CPDA to determine the parameters of triangular membership grades for unequal intervals. Inputs namely polarity and subjectivity are then fuzzified to triangular membership functions. Parameters for unequal length so obtained using the cumulative probability distribution approach (CPDA) used by Gangwar and Kumar [26] are shown in Table 1. The hesitant fuzzy sets so obtained are presented in Tables 2 and 3 respectively.

Two membership grades thus obtained for equal and unequal intervals for polarity and subjectivity to construct an HFS as mentioned in Definition 1. The two membership grades are then aggregated using Eq. (1) with weights (w_i) = 0.5 to get the following aggregated memberships of polarity and subjectivity respectively. Tables 4 and 5 represent the aggregated memberships obtained from the above-mentioned hesitant fuzzy sets respectively.

Step 3. Construction of Rule Base and Defuzzification

The proposed HFS-based Takagi-Sugeno-Kang fuzzy inference system (TSK FIS) uses the rule base which consists of IF-Then rules. IF-THEN rules formulate the conditional statements that comprise intuitionistic fuzzy logic. Since numbers of IFSs for polarity (x) and subjectivity (y) are 5 and 4 respectively, a total of 20 rules of the following form can be constructed. As the expression of a fuzzy model, we use the implications and fuzzy reasoning method suggested by Takagi and Sugeno [18]. A fuzzy implication is of the following form.

$$R\text{: If } x_1 \text{ is } A_1 \text{ and } x_2 \text{ is } A_2 \text{ and} \ldots\ldots\ x_k \text{ is } A_k \text{ then}$$

Table 1 Parameters for triangular fuzzy sets with equal and unequal length

Interval for polarity		Interval for subjectivity	
Equal length	Unequal length	Equal length	Unequal length
$A_1 = [-1\ -1\ -0.5]$	$A_1^* = [-1\ -0.5\ -0.078]$	$B_1 = [0\ 0\ 0.5]$	$B_1^* = [0\ 0.09\ 0.19]$
$A_2 = [-1\ -0.5\ 0]$	$A_2^* = [-0.22\ -0.1\ 0.021]$	$B_2 = [0.25\ 0.5\ 0.75]$	$B_2^* = [0.074\ 0.21\ 0.35]$
$A_3 = [-0.5\ 0\ 0.5]$	$A_3^* = [-0.04\ 0.045\ 0.14]$	$B_3 = [0.5\ 0.75\ 1]$	$B_3^* = [0.27\ 0.40\ 0.52]$
$A_4 = [0\ 0.5\ 1]$	$A_4^* = [0.08\ 0.18\ 0.28]$	$B_4 = [0.75\ 1\ 1]$	$B_4^* = [0.43\ 0.71\ 1]$
$A_5 = [0.5\ 1\ 1]$	$A_5^* = [0.2\ 0.60\ 1]$		

Table 2 Hesitant fuzzy sets for polarity

S. no.	Input (x)	HA_1	HA_2	HA_3	HA_4	HA_5
1	0.25	0, 0	0, 0	0.5, 0	0.5, 0.3	0, 0.125
2	0.1	0, 0	0.7167, 0	0.2833, 0	0, 0.6167	0, 0
3	−0.15	0, 0.1706	0.3, 0.5833	0.7, 0	0, 0	0, 0
4	0.1416667	0, 0	0.7167, 0	0.2833, 0	0, 0.6167	0, 0
5	0	0, 0	0, 0	1, 0.4706	0, 0	0, 0
6	0.1473214	0, 0	0, 0	0.7054, 0	0.2946, 0	0, 0
7	0.119047	0, 0	0, 0	0.7619, 0.22	0.2381, 0.390	0, 0
8	0.15	0, 0	0, 0	0.7, 0	0.3, 0.7	0, 0
9	0.2	0, 0	0, 0	0.6, 0	0.4, 0.8	0, 0
10	0.1501	0, 0	0, 0	0.7, 0	0.3, 0.7	0, 0
11	0.0952380	0, 0	0, 0	0.809, 0.4712	0.1905, 0.152	0, 0
12	0.119047	0, 0	0, 0	0.7619, 0.22	0.238, 0.3905	0, 0
13	0.225	0, 0	0, 0	0.55, 0	0.45, 0	0, 0.0625
14	0.29765	0, 0	0, 0	0.4047, 0	0.5953, 0	0, 0.2441
15	0.5	0, 0	0, 0	0, 0	1, 0	0, 0.75

Table 3 Hesitant fuzzy sets for subjectivity

S. no.	Input (y)	HB_1	HB_2	HB_3	HB_4
1	0.2	0.6, 0	0, 0.9265	0, 0	0, 0
2	0.4	0.2, 0	0.6, 0	0, 1	0, 0
3	0.205	0.59, 0	0, 0.9632	0, 0	0, 0
4	0.66833	0, 0	0.2667, 0	0.7333, 0	0, 0.4
5	0.75	0, 0	0, 0	1, 0	0, 0.45
6	0.741072	0, 0	0.0357, 0	0.9643, 0	0, 0.8929
7	0.309523	0.381, 0	0.2381, 0	0, 0.304	0, 0
8	0.55	0, 0	0.8, 0	0.2, 0	0, 0.4286
9	0.8	0, 0	0, 0	0.8, 0	0.2, 0.68
10	0.35	0.3, 0	0.4, 0	0, 6154	0, 0
11	0.178571	0.649, 0.114	0, 0.7689	0, 0	0, 0
12	0.309523	0.381, 0	0.2381, 0.2891	0, 0.304	0, 0
13	0.425	0.15, 0	0.7, 0	0, 0.7917	0, 0
14	0.641666	0, 0	0.4333, 0	0.5667, 0	0, 0.756
15	0.5	0, 0	1, 0	0, 0	0, 0.25

Table 4 Aggregated membership grades for polarity

S. no	Polarity	HP_1	HP_2	HP_3	HP_4	HP_5
1	0.25	0	0	0.292893	0.408392	0.064586
2	0.1	0	0.467741	0.153419	0.380888	0
3	−0.15	0.089286	0.459917	0.452277	0	0
4	0.141667	0	0.467741	0.153419	0.380888	0
5	0	0	0	1	0	0
6	0.147321	0	0	0.457229	0.51987	0
7	0.119048	0	0	0.569216	0.318547	0
8	0.15	0	0	0.452277	0.541742	0
9	0.2	0	0	0.367544	0.65359	0
10	0.15	0	0	0.452277	0.541742	0
11	0.095238	0	0	0.68261	0.171669	0
12	0.119048	0	0	0.569216	0.318547	0
13	0.225	0	0	0.32918	0.25838	0.031754
14	0.297656	0	0	0.228443	0.36384	0.130575
15	0.5	0	0	0	1	0.5

Table 5 Aggregated membership grades for subjectivity

S. no.	Subjectivity	HS_1	HS_2	HS_3	HS_4
1	0.2	0.367544	0.728891	0	0
2	0.4	0.105573	0.367544	0	0
3	0.205	0.359688	0	0	0
4	0.668333	0	0.143671	0.48357	0.225403
5	0.75	0	0	1	0.25838
6	0.741071	0	0.018012	0.811056	0
7	0.309524	0.213234	0.28478	0.165734	0
8	0.55	0	0.552786	0.105573	0.24409
9	0.8	0	0	0.552786	0.501763
10	0.35	0.16334	0.225403	0.379839	0
11	0.178571	0.437609	0.519271	0	0
12	0.309524	0.213234	0.264042	0.165734	0
13	0.425	0.078046	0.452277	0.543601	0
14	0.641667	0	0.247205	0.341745	0.506036
15	0.5	0	1	0	0

$$y = p_0 + p_1 x_1 + p_2 x_2 \ldots\ldots p_k x_k$$

The proposed model has two inputs so the above implication takes the form:

$$\text{R: If } x \text{ is } \mathrm{P}_1 \text{ and } y_2 \text{ is } \mathrm{B}_2 \text{ then, } \quad z = ax + by + c.$$

The rule base used in the proposed model is presented in Table 6.

Defuzzification is the process of producing a quantifiable result and maps an fuzzy number to a crisp value. In the proposed model defuzzification process is influenced by the aggregated memberships of the inputs. If for a corresponding input, i number of rules is fired then the final defuzzified output i.e. positivity is obtained as

$$\mathrm{Z} = \frac{\alpha_1 z_1 + \alpha_2 z_2 \ldots. + \alpha_i z_i}{\alpha_1 + \alpha_2 + \ldots. + \alpha_i} \tag{2}$$

To calculate α for an individual rule we apply the following algorithm.

Table 6 Fuzzy rule base for the proposed HFS-based TSK model

R1	If x is A_2 and y is B_1 then $z = 9.684x + 19.05y + 1.638$	R9	If x is A_4 and y is B_1 then $\mathrm{z} = -6.243x + 7.596y + 1.656$
R2	If x is A_2 and y is B_2 then $z = 6.878x + 23.18y - 7.298$	R10	If x is A_4 and y is B_2 then $z = -3.262x + 15.79y - 5.436$
R3	If x is A_2 and y is B_3 then $z = 5.155x + 6.965y - 7.274$	R11	If x is A_4 and y is B_3 then $z = -3.71x + 16.44y - 9.701$
R4	If x is A_2 and y is B_4 then $z = -0.0643x + 0.7144y + 0.8573$	R12	If x is A_4 and y is B_4 then $z = -0.05642x + -2.257\mathrm{y} + -2.821$
R5	If x is A_3 and y is B_1 then $z = -1.935x + 0.4995y + 0.7125$	R13	If x is A_5 and y is B_1 then $z = 1.845x + 1.364y + 3.341$
R6	If x is A_3 and y is B_2 then $z = 1.776x - 3.177y + 1.643$	R14	If x is A_5 and y is B_2 then $z = 5.309x + 3.561\mathrm{y} + 9.771$
R7	If x is A_3 and y is B_3 then $z = -3.721\mathrm{x} + -6.98\mathrm{y} + 6.259$	R15	If x is A_5 and y is B_3 then $z = -1.595x - 1.728y - 2.658$
R8	If x is A_3 and y is B_4 then $z = -1.211x + 0.663y + 0.6264$		

Step 1

Select inputs x (polarity) & y (subjectivity).

Step 2 If $x \in A_i\,(0<i\leq 5)$ *and* $y \in B_j\,(0<j\leq 4)$

R_k *(number of rules fires)* = $i*j = n$ *(say)*

Step 3

For R_1

$$\alpha_1 = \{\min(\mu(x_i), \mu(y_j)\ \textit{where}\ x_i \in HP_i, y_j \in HS_j\}$$

Step 4

For all R_k *(k = 2 to n), go to step 3.*

End for.

4 Results and Discussion

Raw scores of polarity and subjectivity obtained from the Natural Language Processing toolkit were fuzzified into triangular membership functions with equal and unequal intervals. The parameters of unequal intervals were determined using CPDA. Two membership grades thus obtained corresponding to equal and unequal intervals for every single input were used to construct a hesitant fuzzy set. The two grades of membership of polarity and subjectivity are then aggregated using the hesitant fuzzy aggregation operator given by Bisht and Kumar et al. [25]. The aggregated membership grade serves as firing strength of individual rules in the proposed hesitant fuzzy set-based TSK model. The crisp output (positivity) obtained after the process of defuzzification for various inputs is presented in Table 7. Results so obtained are compared with results from the traditional TSK model with equal and unequal length for parameters as mentioned in Table 1 with the same rule base.

To confirm the outperformance of the proposed HFS-based TSK-FIS, we use the error measure of root mean square error (RMSE) and compare it with TSK-FIS with fuzzy sets of equal and unequal interval length. The low value of RMSE (1.1338754) confirms that the proposed HFS-based TSK-FIS outperforms over TSK-FIS with fuzzy sets of unequal interval length. Statistical analysis of the results obtained is presented in Table 8. Paired 2 tailed t test (Table 8) also confirms that the proposed model is significantly differs with this model at the confidence level of 5%. However the performance of the proposed HFS-based TSK-FIS in terms of both RMSE and statistical analysis is found more over similar to the TSK-FIS with fuzzy set with equal intervals, capability of the proposed model of handling hesitation due to multiple fuzzification makes it more efficient and reliable.

Table 7 Predicted score of positivity using the proposed HFS-based TSK model

Inputs			Predicted positivity		
S. no.	Polarity	Subjectivity	Proposed HFS-based TSK model	TSK with fuzzy set of unequal interval length	TSK with fuzzy set of equal interval length
1	0.25	0.2	0.4876	1.29	0.972
2	0.1	0.4	3.9298	1	0.76
3	−0.015	0.205	3.12095	−0.852	0.98
4	0.141	0.66833	0.629	−4.4	1.07
5	0	0.75	1.024	1.2	1.02
6	0.147	0.74107	1.3469	−4.61	1.01
7	0.119	0.3095	0.882096	−1.1	0.946
8	0.15	0.55	−0.74022	−4.5	0.95
9	0.2	0.8	−1.3045	−4.77	0.605
10	0.15	0.35	0.950708	−4.5	0.864
11	0.095238	0.1785	1.124521	0.334	0.96
12	0.119048	0.3095	0.85311	−1.11	0.946
13	0.225	0.425	0.803613	−3.57	0.838
14	0.297656	0.64166	0.9729	1	1.08
15	0.5	0.5	2.3325	−1.13	0.828
		RMSE	**1.338754**	**3.450629**	**0.194229**

Table 8 Statistical analysis

S. no.	Comparison with	Statistical parameters		
		SD	*t*-value	*p*-value
1	TSK FIS with fuzzy sets of equal interval length	1.2971	0.514	0.615
2	TSK FIS with fuzzy sets of unequal interval length	2.11905	5.134[a]	0.00

[a]Denotes the significance at 5%

5 Conclusion

Sentiment analysis has become a source of information pool with the increasing growth internet and social media. HFS-based TSK models as the name suggest uses Hesitant fuzzy sets as its base rather than traditional fuzzy sets along with hesitant aggregation operators. HFS-based TSK model proposed in this research study for sentiment analysis is a novel and improvised approach to handle Takagi-Sugeno-Kang systems. The major advantage of the proposed is that it enables users to incorporate the parameter of hesitancy when dealing with real-life problems. The study finds the general indication of positive attitude of tweeting individuals. As NLTK does not include parameters such as Sarcasm and Ironies into the account while calculating polarity and subjectivity which suggests that the system by default contains hesitancy at its core. The deviation of the result's in terms of RMSE could be attributed to the hesitancy involved which could not be incorporated in previous models that use Natural Language processing toolkits and other techniques used in Sentiment Analysis. The low value of RMSE validates the proposed model for sentiment analysis and indicates that the model may be well suited to handle real-life problems where there is a clash of opinion or hesitancy involved.

References

1. Eysenck HJ (1959) Learning theory, behaviour therapy. J Mental Sci 105(438):61–75
2. Price DD, Barrell JE, Barrell JJ (1985) A quantitative-experiential analysis of human emotions. Motiv Emot 9(1):19–38
3. Goldberg LR (1990) An alternative "description of personality": the big-five factor structure. J Pers Soc Psychol 59(6):1216
4. Sinha AKP (1995) Manual for Sinha's comprehensive anxiety test (scat). National Psychological Corporation, Agra
5. Pak A, Paroubek P (2010) Twitter as a corpus for sentiment analysis and opinion mining. In: LREc, vol 10, no 2010, pp 1320–1326, May 2010
6. Vinodhini G, Chandrasekaran RM (2012) Sentiment analysis and opinion mining: a survey. Int J 2(6):282–292
7. Feldman R (2013) Techniques and applications for sentiment analysis. Commun ACM 56(4):82–89
8. Sahayak V, Shete V, Pathan A (2015) Sentiment analysis on twitter data. Int J Innov Res Adv Eng (IJIRAE) 2(1):178–183
9. Ain QT, Ali M, Riaz A, Noureen A, Kamran M, Hayat B, Rehman A (2017) Sentiment analysis using deep learning techniques: a review. Int J Adv Comput Sci Appl 8(6):424
10. Correa T, Hinsley AW, De Zuniga HG (2010) Who interacts on the Web?: the intersection of users' personality and social media use. Comput Hum Behav 26(2):247–253
11. Hughes DJ, Rowe M, Batey M, Lee A (2012) A tale of two sites: Twitter vs. Facebook and the personality predictors of social media usage. Comput Hum Behav 28(2):561–569
12. Neogi AS, Garg KA, Mishra RK, Dwivedi YK (2021) Sentiment analysis and classification of Indian farmers' protest using twitter data. Int J Inform Manag Data Insights 1(2):100019
13. Gautam J, Atrey M, Malsa N, Balyan A, Shaw RN, Ghosh A (2021) Twitter data sentiment analysis using naive Bayes classifier and generation of heat map for analyzing intensity geographically. In: Advances in applications of data-driven computing. Springer, Singapore, pp 129–139
14. Naseem U, Razzak I, Khushi M, Eklund PW, Kim J (2021) COVIDSenti: a large-scale benchmark Twitter data set for COVID-19 sentiment analysis. IEEE Trans Comput Soc Syst 8(4):1003–1015
15. Gandhi UD, Malarvizhi Kumar P, Chandra Babu G, Karthick G (2021) Sentiment analysis on twitter data by using convolutional neural network (CNN) and long short term memory (LSTM). Wirel Pers Commun 1–10
16. Kumar P, Vardhan M (2022) PWEBSA: Twitter sentiment analysis by combining Plutchik wheel of emotion and word embedding. Int J Inform Technol 1–9
17. Mamdani EH, Assilian S (1975) An experiment in linguistic synthesis with a fuzzy logic controller. Int J Man Mach Stud 7(1):1–13
18. Sugeno M, Kang GT (1988) Structure identification of fuzzy model. Fuzzy Sets Syst 28(1):15–33

19. Kushwaha GS, Kumar S (2009) Role of the fuzzy system in psychological research. Europe's J Psychol 5(2):123–134. https://doi.org/10.5964/ejop.v5i2.271
20. Devi S, Kumar S, Kushwaha GS (2016) An adaptive neuro fuzzy inference system for prediction of anxiety of students. In: 2016 eighth international conference on advanced computational intelligence (ICACI), pp 7–13. https://doi.org/10.1109/ICACI.2016.7449795
21. Pandey DC, Kushwaha GS, Kumar S (2020) Mamdani fuzzy rule-based models for psychological research. SN Appl Sci 2:913. https://doi.org/10.1007/s42452-020-2726-z
22. Torra V, Narukawa Y (2009) On hesitant fuzzy sets and decision. In: 2009 IEEE international conference on fuzzy systems. IEEE, pp 1378–1382
23. Zadeh LA (1965) Fuzzy set, information and control. In: Zadeh LA (ed), pp 338–353. https://doi.org/10.1142/9789814261302_0021
24. Xia M, Xu Z (2011) Hesitant fuzzy information aggregation in decision making. Int J Approx Reason 52(3):395–407
25. Bisht K, Kumar S (2019) Hesitant fuzzy set based computational method for financial time series forecasting. Gran Comput 4(4):655–669
26. Gangwar SS, Kumar S (2014) Probabilistic and intuitionistic fuzzy sets–based method for fuzzy time series forecasting. Cybern Syst 45(4):349–361

Chapter 32
Toward a Better Model for the Semantic Segmentation of Remote Sensing Imagery

Muazu Aminu Aliyu, Souley Boukari, Abdullahi Madaki Gamsha, Mustapha Lawal Abdurrahman, and Abdulsalam Yau Gital

1 Introduction

This process is called satellite imagery which is also in another clime called earth observation. The process has applications in disaster and resource management and agriculture to mention a few [1]. Satellite imagery is data collected by a range of sensors from the ground which is also called remote sensing [2]. A few years ago, manual computation of satellite imagery was feasible and easy due to not many available earth observations, however, that is not the case now due to the proliferation of data known today as big data. Due to the high volume of data available today, performing analyses becomes a problem [5]. One of the major parts of the problem is labeling [6]. Where it is responsible for recognizing the structural pattern of the imagery acquired in the satellite imagery. Research in the computer vision industry has been deploying different techniques to address this problem of automating the analysis of big data in different ways, by implication if any condition of the data changes, it may result in the state-of-the-art framework not being feasible again due to change and will require rewriting it from scratch which is time-consuming, inefficient, and expensive [7]. The gap leads to academic research in the field to come up with better framework and effective approaches [8].

2 Methodology

This paper will deploy the use of a new 3D CNN architecture that will test a new approach different from the ones done before [9]. 3D-Unet processes, at the same

M. A. Aliyu (✉) · S. Boukari · A. M. Gamsha · M. L. Abdurrahman · A. Y. Gital
Department of Mathematical Science, Abubakar Tafawa Balewa University Bauchi, Lushi, Nigeria
e-mail: muazugeetal@gmail.com

G. Mathur et al. (eds.), *Proceedings of 3rd International Conference on Artificial Intelligence: Advances and Applications*, Algorithms for Intelligent Systems,
https://doi.org/10.1007/978-981-19-7041-2_32

time, the spectral and spatial component of the imagery with 3D convolutions to give out better results and products of the training with little trained parameters [3]. This work will introduce and study the use of big data for satellite imagery and representation of imagery. Individual imagery will be treated as $n * n * f$. This model ad architecture follows the traditional CNN architecture and will also apply 3D-Unet convolutions other than using a 1-layer convolution. Convolution layers are stacked on top of one another so as to layer a deep architecture model [10]. A 3D layer is inputted to conform to 3D voxels where each of the layer has a number of large kernels that concurrently analyze the convolutions in 3D of the inputs [11]. In the convolution stack, the 3D convolution is followed by a 1D convolution layer that expels the spatial neighbor and other connected layers. To put it simply, this model inputs 3D voxels and produces 3D maps that are later winnowed to a 1D feature vector layer among another layer [12, 13].

2.1 Proposed Model

3D-Unet contains 19,069,955 total parameters. This work looks forward to the reduction of bottlenecks by times the number of channels already before maximum pooling.

$n * n * f$ voxel tile of the image containing 7 channels. We will get a result in the last layer as $n * n * f$ voxels in x, y, and z dimensions. Having a voxel size of $1 : 76 * 1 : 76 * 2 : 04 * m^3$ approximatively receptive field are $155 * 155 * 180xm$ each voxel in the predicted segmentation. Individual voxel has more than enough context to learn [14, 15].

We shall introduce batch normalization (BN) for individual ReLU. During training, the individual batch is normalized with standard deviation, and the mean is reconditioned using these values. With a trailing layer for bias and scale. During test time, we normalize using computed global statistics. We Set the weights of unlabeled pixels to zero in order to ensure learning from only the labeled and not the entire data.

2.2 Training of Network

Coming up with a new 3D-Unet Layer, the work will explore deploying a different 3D-Unet network. U-Net [10] the first series are interspersed with max pooling layers. The work changes the Unet to zero padding in the convolution layers so that where there is need the size of the layers are the same. This work creates Unet with a meager preselect hyper-parameters [9].

2.3 Dataset

The size of the data set is 30 Gigabyte obtained using a drone. The High-resolution multispectral imagery captured Hamline Beach State Park of New York State [16].

2.4 Metric

This work will adopt accuracy, precision, and recall matrix to evaluate the performance of the model as follows:

$$\text{Accuracy} = \frac{TP + TN}{TP + FP + FN + TN} \tag{1}$$

$$\text{Precision} = \frac{TP}{TP + FP} \tag{2}$$

$$\text{Recall} = \frac{TP}{TP + FN} \tag{3}$$

3 Implementation

To implement the proposed model, this work adopts the use of high-resolution multispectral data in order to train the network. The imagery was captured using a drone over Hamlin Beach State Park, NY.

The imagery file has in it the validation, annotation training, and test sets, encompassing 18 object class labels as shown in Fig. 1. Size of the Data File is 0 3.0 GB.

The arrangement of the multispectral imagery is as numChannels-by-width and 0 by height arrays. The different tiles are infrared channels together with mask channel and after trained RGB image. The labeled image is a classified image of 18 object class labels.

Here we have RGB color channels in this order 1st, 2nd, and 3rd image channels as seen in Fig. 2 showing each of the color components of the training, validation, and test images as a montage.

Infrared Imagery (Fig. 3) displays the infrared emission of the test images, training, and validation as a montage.

Mask in Fig. 4 is used to enforce boundaries for training image (left), validation image (center), and test image (right).

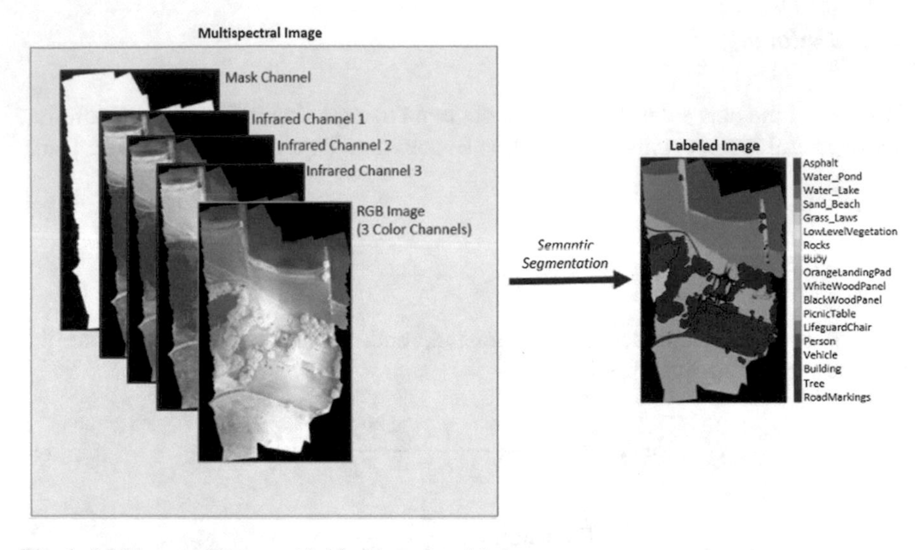

Fig. 1 Multispectral image with 18 object class labels

Fig. 2 Validation image and test image, RGB component of training image

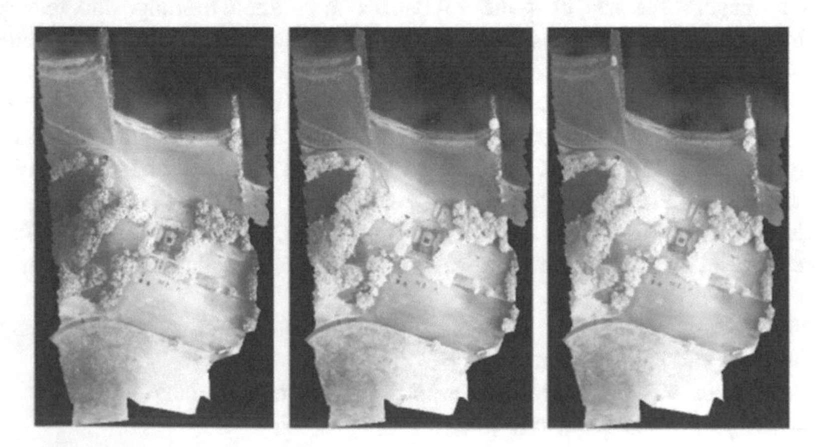

Fig. 3 IR channel 1, 2, and 3 test images

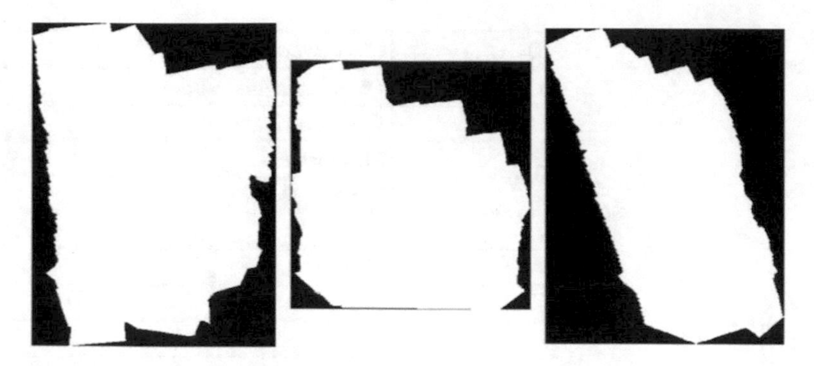

Fig. 4 Mask of test images, training, and validation

Labeled imagery is the representation of the ground truth of the segmented imagery with individual pixel assigned to one of the 18 classes. Table 1 shows the 18 classes and their IDs (Fig. 5).

The aim of this work is to calculate the vegetation cover extent in the multispectral image by dividing the number of vegetation pixels by the number of valid pixels; see Table 2 for settings of hyperparameter.

Now, we can adopt the new 3D-Unet for the semantic segmentation of the imagery.

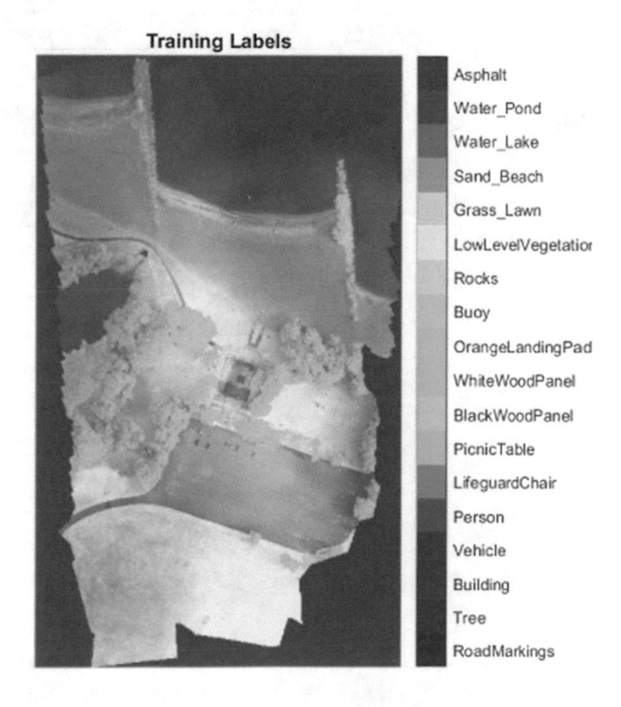

Fig. 5 Training labels for Rits18 datasets

Table 1 Image classes and IDs for the rits18 datasets

IDs	Class name
0	Other Class/Image Border
1	Road Markings
2	Tree -paths
3	Building blocks
4	Vehicle
5	Human
6	Chair
7	Table
8	Black Wood Panel
9	White Wood Panel
10	Orange Landing Pad
11	Buoy
12	Stones
13	Other Vegetation
14	Grass
15	Sand
16	Water Lake
17	Water Pond
18	Asphalt (Parking Lot/Walkway)

Table 2 Settings of Parameter

Parameters	Settings
Initial learning rate	0.05
Max epochs	150
Mini batch size	16
l2reg	0.0001
Momentum	0.9
Learn rate schedule	Piecewise
Shuffle	Every-epoch
Gradient threshold method	l2 norm
Gradient threshold	0.05
Verbose frequency	20

3.1 Result Presentation

To perform the forward pass on the trained network and also perform segmentation on the imagery patches using the semantics function. The typical sample of the segmented image is presented in Fig. 6.

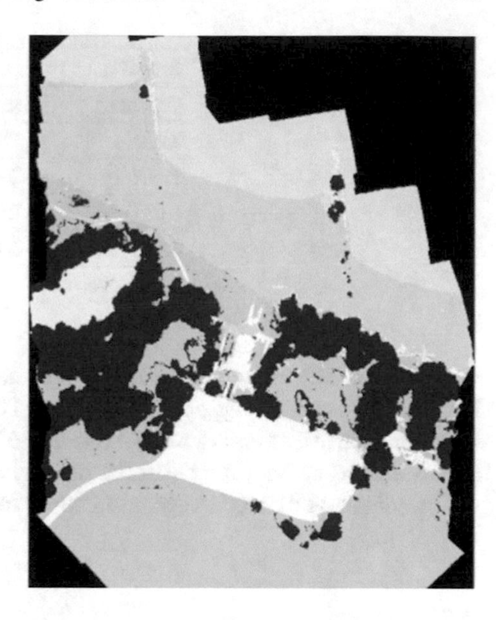

Fig. 6 Segmented image

The result of the segmented imagery and ground truth label is saved as PNG files. It is to be used later for the computation of accuracy metrics by overlaying the segmented image on the histogram-equalized RGB validation image as shown in Fig. 7.

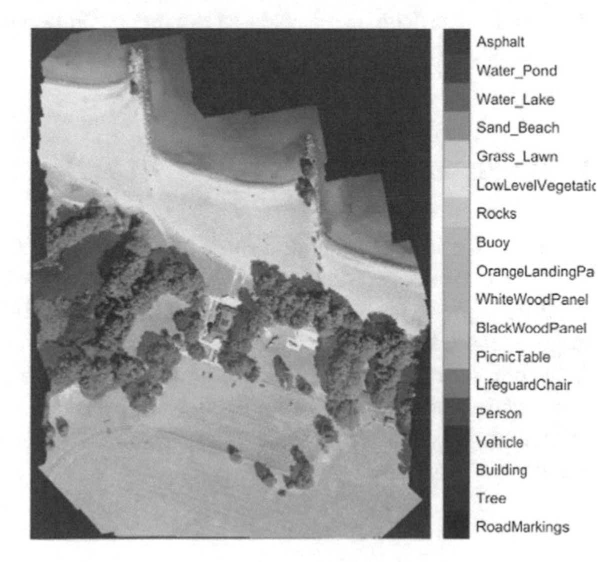

Fig. 7 Label validation image

Table 3 Performance comparison based on accuracy against existing algorithms

Model	Accuracy (%)	Precision	Recall
Proposed	90.698	94.321	96.756
MLP	30.4	40.337	49.701
KNN	27.7	31.613	28.993
SVM	29.6	38.556	30.662
Sharp Mask	57.3	67.999	61.224
Refine-Net	59.8	58.331	66.535

In Fig. 6, the model performance and dataset during training are shown, and the training accuracy was evaluated after 100 iterations. The classification accuracy obtained after training was 90.698%. For classification accuracy, the higher the classification accuracy the better the model has been built.

For this work, the total vegetation cover from the segmented image is 51.72%

3.2 *Result*

This Table 3 hereunder presents the accuracy of the RIT-18 test set evaluated against the existing state-of-the-art studies.

From the Table 3, it is apparent that the proposed deep learning algorithm outperforms the conventional state-of-the-art algorithms in terms of accuracy 3 It is quite obvious that the deep learning algorithm outperformed the conventional approach in terms of accuracy, precision, and recall.

4 Discussion

As seen in the in-classification task, a precision and recall value within the range of 90–100% can be concluded as an excellent performance that can support the accuracy score by a given classification model. Thus, From Table 3, it is noticed that the proposed model achieved a precision value of 94.321% and a recall value of 96.756%. Therefore, we can conclude that the proposed model has significantly improved the classification performance from existing approaches used in previous state of the art studies. It is proven that this approach can help develop better frameworks for superior performance.

References

1. Slavkovikj V, Verstockt S, De Neve W, Van Hoecke S, Van de Walle R (2015) Hyperspectral

image classification with convolutional neural networks. Proceedings of the 23rd Annual ACM conference on multimedia conference, 1159–1162, https://doi.org/10.1145/2733373.2806306
2. Wu Z, Gao Y, Li L, Xue J, Li Y (2019) Semantic segmentation of high-resolution remote sensing images using fully convolutional network with adaptive threshold. Connect Sci 31(2):169–184
3. Praveena S, Singh SP (2015) Hybrid clusteing algorithm and Neural Network classifier for satellite image classification. 2015 international conference on industrial instrumentation and control (ICIC), pp 1378–1383, https://doi.org/10.1109/IIC.2015.7150963
4. Shivaprakash M (2016) Semantic segmentation of satellite images using deep learning. Master's Thesis. Czech Technical University in Prague & Luleå University of Technology
5. Yang Z, Mu X-D, Zhao F-A (2018) Scene classification of remote sensing image based on deep network grading transferring. Optik 168:127–133
6. Çiçek Ö, Abdulkadir A, Lienkamp SS, Brox T, Ronneberger O (2016) 3D U-Net: learning dense volumetric segmentation from sparse annotation. In: Ourselin S, Joskowicz L, Sabuncu M, Unal G, Wells W (eds) Medical image computing and computer-assisted intervention—MICCAI 2016. MICCAI 2016. Lecture Notes in Computer Science, vol 9901. Springer, Cham
7. You J, Liu W, Lee J (2020) A DNN-based semantic segmentation for detecting weed and crop. Comput Electron Agric 178:105750
8. Yuan X, Shi J, Gu L (2020) A review of deep learning methods for semantic segmentation of remote sensing imagery. Expert Syst Appl, 114417
9. Simonyan K, Zisserman A (2014) Very deep convolutional networks for large-scale image recognition. arXiv preprint arXiv:1409.1556
10. Wurm M, Stark T, Zhu XX, Weigand M, Taubenböck H (2019) Semantic segmentation of slums in satellite images using transfer learning on fully convolutional neural networks. ISPRS J Photogramm Remote Sens 150:59–69
11. Pritt M, Chern G (2017) Satellite image classification with deep learning. 2017 IEEE applied imagery pattern recognition workshop (AIPR), pp 1–7, https://doi.org/10.1109/AIPR.2017.8457969
12. Eleyan A (2012) Breast cancer classification using moments. 2012 20th Signal processing and communications applications conference (SIU), pp 1–4
13. Bre F, Gimenez JM, Fachinotti VD (2018) Prediction of wind pressure coefficients on building surfaces using artificial neural networks. Energy Build 158:1429–1441
14. Ronneberger O, Fischer P, Brox T (2015) U-Net: convolutional networks for biomedical image segmentation. In: Navab N, Hornegger J, Wells W, Frangi A (eds) Medical image computing and computer-assisted intervention—MICCAI 2015. MICCAI 2015. Lecture Notes in Computer Science(), vol 9351. Springer, Cham. https://doi.org/10.1007/978-3-319-24574-4_28
15. Yamashita R, Nishio M, Do RKG, Togashi K (2018) Convolutional neural networks: an overview and application in radiology. Insights Imaging 9(4):611–629
16. Kemker R, Salvaggio C, Kanan C (2018) Algorithms for semantic segmentation of multispectral remote sensing imagery using deep learning. ISPRS J Photogramm Remote Sens 145:60–77

Chapter 33
Sentimental Segregation for Social Media Using Lexicon Technology

Pallavi Sapkale, Payal Bansal, Moresh Mukhedkar, and Sandhya Sharma

1 Introduction

Twitter is one of the huge social networks that enables the users to create short posts called "tweets". It has more than 187 million daily active users which tells us that an abundant number of tweets are generated every day. Twitter has a treasure of "Data". It is an open space for people to put forward their thoughts, views, and experiences about almost everything leading to it becoming a very useful platform for analyzing public opinions. In our project, we hunt for tweets in which the products launched are praised or criticized by the public and hence keep a track of reviews of people using the data obtained from Twitter. There are various software options available for analyzing the Twitter data one of which is R. R has a huge variety of tools which makes it more appealing to work on. Despite that, we opted for R studio since it is a little easier with scripts as compared to R. Now we have the unstructured text data which requires classification. The approach we follow is objects sentiment analysis in which a collection of words is used to decide the sentiment of the input text. It refers to the task of determining whether the attitude behind a particular text is positive, negative, or neutral. This can help in determining whether a product is viewed positively or negatively by the user. This is beneficial not just to the customers but also to the product or service-providing companies and organizations helping them understand

P. Sapkale (✉)
RAIT, Nerul Navi, Navi Mumbai, India
e-mail: pme932@rait.ac.in

P. Bansal
Poornima College of Engineering, Jaipur, India

M. Mukhedkar
DYP Ambi, Talegaon Dabhade, India

S. Sharma
Suresh Gyan Vihar University Jaipur, Jaipur, India

G. Mathur et al. (eds.), *Proceedings of 3rd International Conference on Artificial Intelligence: Advances and Applications*, Algorithms for Intelligent Systems,
https://doi.org/10.1007/978-981-19-7041-2_33

user sentiment and user behavior by knowing the response over their products hence saving them the huge amount of money that they might have invested in getting the feedback.

2 Motivation

According to the various text types and application fields, sentiment analysis is called evaluation extraction, and emotion prediction in Cambria et al. [1, 2]. Considering the various conditions of sentiment analysis algorithms, this paper divides the methods into the following categories, existing work, R packages, Twitter scrapping, analysis, and data pre-processing. This gave us the idea to make something that will overcome this drawback and will gather information from a social network that is widely used by people of all ages. So after, doing research on this issue, we found that the most used social networking sites were Instagram and Twitter, but the information passed by the users of Instagram was not fitting the criteria of our problem. Twitter has millions of users with billions of information to share with almost no boundary in ideas to share among people. This helped us to finalize Twitter as our source of project to collect data.

3 Need for Sentiment Analysis

Sentiment analysis is one of the important methods, because of the fast growth of Internet technology and social media Ravi and Ravi [3]. Such analysis can be used in all kinds of services. A few of them are shortlisted here as follows: • Sports: In the sports field, everyone wants to react to each action of sports so we can analyze which team is reaching high. • Business: In the business world, it's very important to know about the market value of a particular product so with sentiment analysis we can focus on proper things. • Politics: This analysis will help to find the winning team also during election time. • Public Actions: Sentiment analysis is also utilized to monitor and analyze social wonders, for the spotting situations that are possibly dangerous and deciding the general state of mind of the bloggers.

4 Analysis of Sentiments

In Kharde and Sonawane [4] author defined sentiment analysis based on a process that developed with attitude, views, and emotion-based tweets. And such texts are positive, negative, or neutral. Following word classification with opinion and views. Opinion: Its output varies with the person. View: subjective text. Belief: accepted thought Sentiment: text related to emotion and feelings. These are the various types of sentiments.

5 Existing Work and Methods Used with Twitter Datasets

Sentimental research analysis was started by Hearst in time span 1992. Mainly pair of techniques have been introduced for the classification of sentiments. The first method was lexical and another one was machine learning-based analysis. In the first-mentioned method, a dictionary-based system is considered manually developed by an origin. These sets are used to pretend the words meaning so that classification could be done easily. Researcher Deng et al. focused on domain-specific sentiment class in 2017. The devised technique was calculated through two huge corpora that included one million tweets associated with political issues and more than 7 lac of tweets regarding the stock market in Jain and Katkar [5]. Kharde and Sonawane [4] author devised the approach for analyzing the sentiments of tweets. They highlighted data mining using Naive Bayes and Bayes Net classifiers. In [6] researchers proposed the pattern-based method which spots the skepticism on Twitter. To search for derision, they utilize a pattern-based method with the help of Parts of Speech (PoS) and for classification, a machine learning approach. The main goal achieved by them was (a) Sentiment-based, (b) punctuation-based, (c) syntactic and semantic-based, and (d) pattern-based. The resulting part decided whether the text was sarcastic or not.

6 Proposed Method

We have gone through various literature and websites from which we have collected some gaps here. • There is a limitation in the boundary of the topic the user wants to research. • The reach of such sites is limited to users. • The accuracy of results was found to be fake in some cases. • The display available was not many users friendly. Several methods and techniques developed to segregate tweets into positive and negative types. Few of them Xia et al. [2–6] work with machine learning and few are with the lexicon method Musto et al. [7]. Both the methods can be used to differentiate the sentiments from huge datasets. But the work is limited. Hence we need to develop some new techniques. In our proposed part, we work with R shiny which is a freely available source for data and easily operates with windows and any operating system. Maximum 14 days' time span in which we can access tweets in windows. Tweets past that can't be harvested (Fig. 1).

	score	text
1	1	RT @sanket_jack: @waatho AAP ko happy makar sankrant! #AAP #Kejriwal cartoon --> #Jackartoon http://t.co/kIeaxBgNGt
2	0	If #Kejriwal would have passed Kanduri's (Uttarakhand) lokayukt in Delhi then his Law Minister Sobhnath Bharti would be in jail.
3	0	RT @IsaiditOrg: Media's Ignorance of @narendramodi Will Prove to be a Blessing http://t.co/ipZZ90dEgS #narendramodi #kejriwal #AAP #NaMo #M…
4	1	RT @IsaiditOrg: Media has shunned @narendramodi but it could work in his favor http://t.co/ipZZ90dEgS #narendramodi #Modi #AAP #kejriwal #p…
5	-1	Didn't #Kejriwal earlier say that there was no conflict in #AAP n Binny never asked for a post in the Cabinet? http://t.co/Ry3n7jbtnS
6	0	Y #Kejriwal ws silent on 25th dec whn thr was issue with #Binny , its time nation should understand difference between #Honesty n #Arrogance
7	-1	RT @IshaSG: http://t.co/fpHLAiS188 Now even #AAP members agree that thr is difference between before & after..saying & doing.. #Fraud #Kejr…
8	0	[illegible] #Kejriwal [illegible]
9	-1	http://t.co/fpHLAiS188 Now even #AAP members agree that thr is difference between before & after..saying & doing.. #Fraud #Kejriwal
10	0	RT @vibhask1: [illegible] @manoharparrikar [illegible] [illegible] .#kejriwal http://t.co/Y32i23Pohh
11	0	RT @2212akshay: Did anyone noticed that in @aajtaknews against #AAP nd #kejriwal side mein ad TOSHIBA ki Advtsmnt aa rhi h . Japanese firm…
12	0	Did anyone noticed that in @aajtaknews against #AAP nd #kejriwal side mein ad TOSHIBA ki Advtsmnt aa rhi h . Japanese firm !!!!
13	[illegible]	RT @vibhask1: [illegible] @manoharparrikar [illegible] [illegible] http://t.co/Y32i23Pohh
14	0	RT @vibhask1: [illegible] @manoharparrikar [illegible]

Fig. 1 An expected score of each Tweet

Even after the cleaning phase, a special character still remains, and the tweet won't be harvested. In the current work, we use the available data then we remove the unwanted things from it then we use propose lexicon and then manipulate it. Figure 2 shows the proposed method which describes the steps for the proposed work. Initially, the suitable environment is prepared by authorizing a Twitter API connection for that there is a need for the tweeter to develop an account and then it helps loading the given packages. Then it follows the process: First, the user provides input in the search term, and the shiny web application searches for the keywords in the whole database of Twitter and searches for the relevant tweets. If the number of tweets harvested is less than the default tweets value, it will simply throw an error. To avoid this, slider input is provided to regulate the number of tweets. After the tweets are harvested, the database undergoes a cleaning process and further lexical analysis is performed. The output is expressed visually and can be determined by looking at the visualizations. It was devised in four parts which are as follows: First data gathering then removing unwanted things, then applying the proposed Lexical method. After that finally, classify and calculate the score.

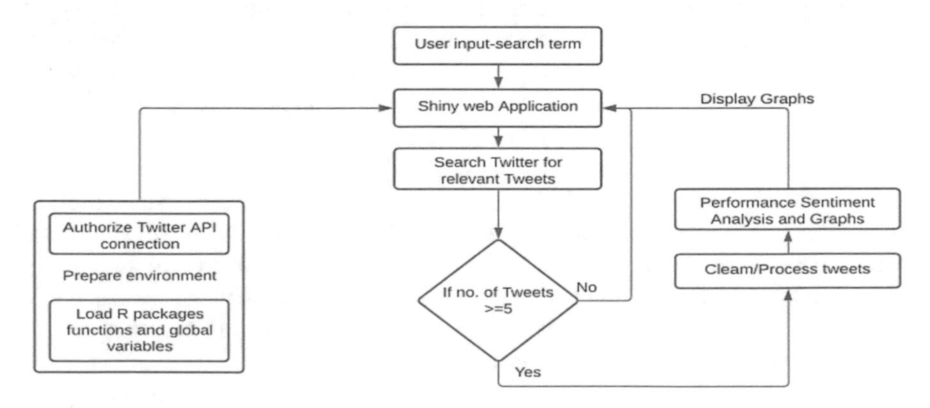

Fig. 2 Proposed method

7 Twitter Selection

Twitter is a social network that allows its users to send and read short messages that are known as tweets from Kouloumpis et al. [8], which do not exceed a length of more than 140 characters [8]. It easily promotes your research, by providing links to blog stories, journal articles, and news items. It also helps one reach new audiences seek feedback from others, contribute to discussions on events, and likewise many more tasks. All these qualities of Twitter help us to reach a large audience and get information about what we are looking for Twitter in the present-day has about 192 million active users and still counting which means that this social media platform has a large reach to the people with about billions of tweets around the world of about millions of topics being tweeted which indirectly makes Twitter a hub of information Alqaryouti et al. [9]. Whereas other social media platforms like Facebook, Instagram, etc. are more concentrated on entertainment purposes which narrows the main motive of fetching information one is looking for; in the case of Twitter, it is totally the opposite of it. So this makes Twitter the best case for us to further work on. Figure 3 shows the results of tweet parameters.

7.1 Harvesting

For harvesting the tweets, we need to link the R studio to Twitter. This is done by inserting the API key of the Twitter developer's account in R Studio. This allows refining the language, time, username, and Geolocation of the harvested tweet. The default number of tweets that can be scraped is 25. On the other hand, a significant number of tweets that are harvested are duplicates or "retweets". Twitter only allows tweets to harvest within the 14 days time span.

Fig. 3 Tweets results with parameter

7.2 Data Cleaning

The next step is data cleaning, to get the text in a clean condition. Data cleaning is used to identify incomplete, incorrect, inaccurate, or irrelevant parts of the data and then replace, modify, or delete the dirty or coarse data. Here, in this work, we fetched more than a thousand tweets from Twitter for analyzing the sentiments of Twitter user's toward social media. So, R or any other analytic tool is able to read and decode the text. Even one unknown symbol can stop the whole process. There are thousands of different symbols including emojis, URLs, and punctuation which require different cleaning methods. After removing all the unwanted characters like emojis, URLs, Symbols, etc., we can start to explore the cleaned texts and will try to fetch the data from Twitter for further modifications. The process has two cleaning phases: • Phase i: After the scraping of data to convert it into a standard format. • Phase ii: Sentiment function to unload the tweet which throws an error. Data collection and data cleaning is the very helpful or beneficial part of sentiment analysis.

7.3 Twitter Scraping and Performing Analysis

To harvest the tweets, we use the command search Twitter () and pass in the next parameters or filters. For example, the demonstrated search for apple iPhone was posted by Udemy. Here, the time and number of tweets were also defined. As a result, tweets with similar parameters were harvested. From Fig. 3, tweet parameters are described. This command will get 100 tweets related to Apple + iPhone. The function "search Twitter"is used to download tweets from the timeline. Now we want to convert this list of one thousand tweets into the information frame, so we are able to work thereon. Then finally we tend to convert the information frame into -CSV file.

7.4 Sentiment Lexicons

Sentiment Lexicons are simply a collection of words to which the text is compared against. The hits are counted and summed up to get a score. Once we have the tweets, we just need to apply some functions to convert these tweets into some useful information. The main working principle of sentiment analysis is to find the words in the tweets that represent positive sentiments and find the words in the tweets that represent negative sentiments. For this, we need a list Of words that contains positive and negative sentiment words. The lexicons are readily available on the Internet Vashishtha [13]. After downloading the list, save it in your working directory. The sentiment analysis uses two packages plyr and stringr for manipulation. The sentiment function calculates a score for each individual tweet. It first calculates the positive score by comparing words with the negative words list and then calculates the negative score by comparing words with the negative words list. So, after this process, each tweet is assigned with a certain sentiment score. Refined tweets into a data frame are done and exported as a.CSV file. When we import this CSV file, a dataset file is formed within the operating directory. The next step is to get the tweets, and this could be done by making a separate CSV file that contains the score of every tweet. The photograph of the score file shows the expected score of every tweet as an associate number before each tweet. The next step is to visualize the tweets by holograms. This can be done by using thc hist function. We use the packages ggplot and gganimate to play with colors and animations to make the visualization interesting [10].

8 Result

After doing the analysis, the result shows various forms which are described as follows:

A. **Word Cloud**

To know which words are mostly associated with the desired tweet, we use the word cloud visualization. In this, the most occurring words associated with that tweet comes in a large size than the one which has not occurred much. So with the help of this, we can come to know what is the action going on Twitter of the desired top. Figure 4 shows the word cloud.

B. **Histogram**

To know about the Positive and Negative sentiment score of the desired Tweet, we can use the visualization of Histogram which gives us the sentiment score histogram of both positive and negative Tweets which is cleared with Figs. 5 and 6.

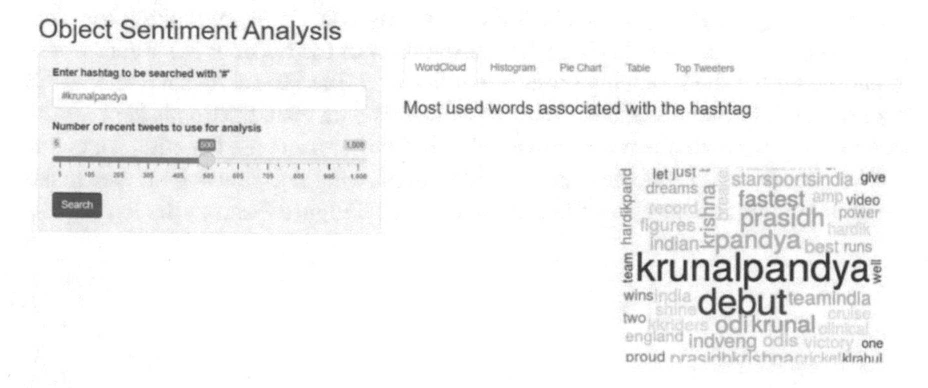

Fig. 4 Word cloud

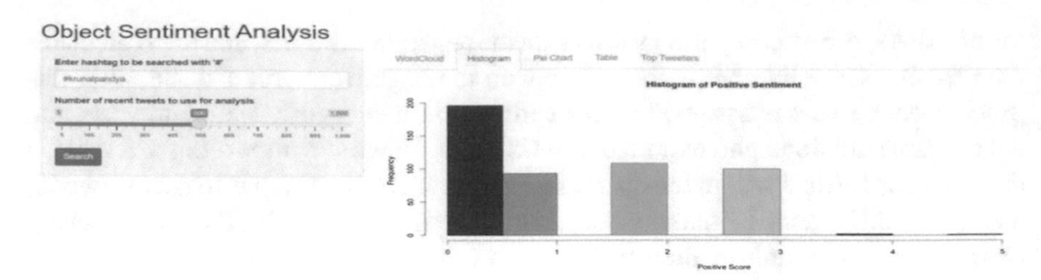

Fig. 5 Positive histogram

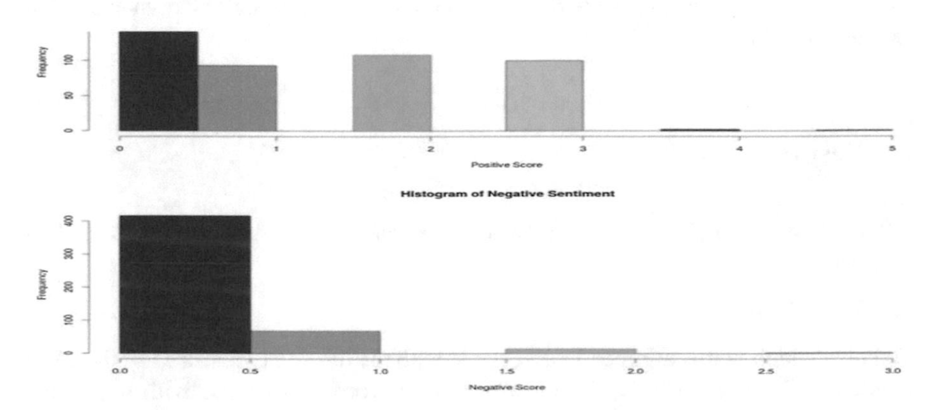

Fig. 6 Negative histogram

C. Pie Chart

To get a much better visualization of the positive and negative sentiment scores of the desired tweets, we use the Pie Chart visualization which is very easy to understand by everyone and is also a quick look view to make the decision of a few objects in some cases. The pie chart is depicted in Fig. 7. To know where a user is tweeting about the desired tweet, we can use the table which also gives us the positive percentage as well as the negative percentage about the following tweets we are looking for. This table can be also used for checking the transparency of the tweet or in simpler words to know whether the following tweet is genuine or a spam tweet by someone which is shown in the following table. For top tweets, this data visualization is kept for the user to know what are the maximum number of Twitter users are tweeting about our desired tweets. This gives the user some idea about what decision should they make as per the maximum number of tweets tweeted [11]. Figure 8 shows the top tweets.

Fig. 7 Pie chart

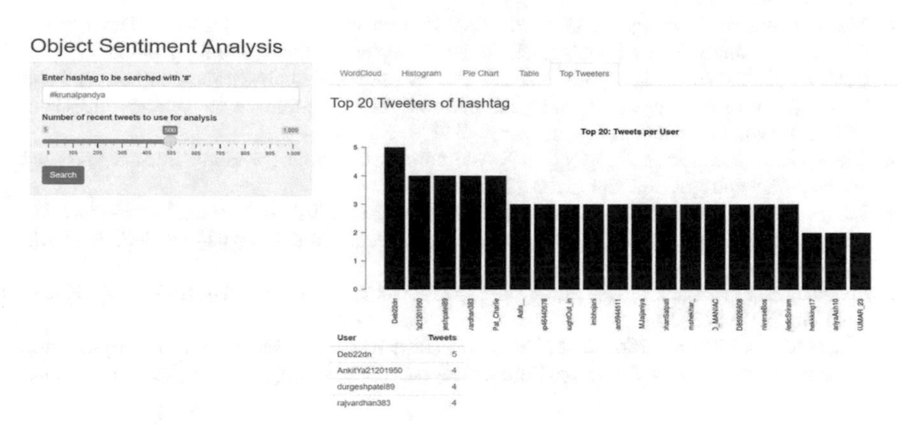

Fig. 8 Top Tweets

9 Conclusion

The research on Object sentiment analysis has been improved successfully where we introduced sentiments and their various types which depends on the importance of the sentiment analysis. Then we presented a taxonomy of sentiment analysis. We have explored the initial sentiment analysis datasets. The work on Object sentiment analysis can be further restudied and improved and we have worked with the same department to make the work easier and implemented a few ideas which have been elaborated. In this research, we have provided various data visualizations like Word cloud, Histogram, Pie Chart, Table, and Top Tweeters whose functions have been provided in their respective sections. This object sentiment analysis will be more effective in various domains such as implicit sentiment detection, determination of the quality of a product, and review system for pre-analysis of a particular object.

References

1. Cambria E, Schuller B, Xia Y, Havasi C (2013) New avenues in opinion mining and sentiment analysis. IEEE Intell Syst 28(2):15–21
2. Xia R, Xu F, Zong C, Li Q, Qi Y, Li T (2015) Dual sentiment analysis: considering two sides of one review. IEEE Trans Knowl Data Eng 27(8):2120–2133

3. Ravi K, Ravi V (2015) A survey on opinion mining and sentiment analysis: tasks, approaches and applications. Knowl Based Syst Elsevier 89:14–24
4. Kharde VA, Sonawane PS (2016) Sentiment analysis of twitter data: a survey of techniques, arXiv:1601.06971. LNCS Homepage, http://www.springer.com/lncs. Accessed 21 Nov 2016
5. Jain AP, Katkar VD (2015) Sentiments analysis of Twitter data using data mining. In: Information processing (ICIP), 2015 international conference on. IEEE, pp 807–810
6. Bouazizi M, Ohtsuki T (2016) A pattern-based approach for sarcasm detection twitter. IEEE Access 4:5477–5488
7. Musto, Semeraro C, Polignano G, Marco (2014) A comparison of lexicon-based approaches for sentiment analysis of microblog. CEUR Workshop proceedings, 1314, pp 59–68
8. Kouloumpis E, Wilson T, Moore J (2011) Twitter sentiment analysis: the good the bad and the omg!. In: Proceedings of the fifth international AAAI conference on weblogs and social media, Barcelona, Catalonia, Spain, pp 538–541
9. Xie X, Ge S, Hu F, Xie M, Jiang N (2019) An improved algorithm for sentiment analysis based on maximum entropy. Soft Comput 23(2):599–611
10. Kariya C, Khodke P (2020) Twitter sentiment analysis. 2020 International conference for emerging technology (INCET), Belgaum, India, pp 1–3, https://doi.org/10.1109/INCET49848.2020.9154143
11. Schouten K, Frasincar F (2016) Survey on aspect-level sentiment analysis. IEEE Trans Knowl Data Eng 28(3):813830
12. Alqaryouti O, Siyam N, Monem A, Shaalan K (2019) Aspect-based sentiment analysis using smart government review data. Appl Comput Inform. https://doi.org/10.1016/j.aci.2019.11.003

Chapter 34
Apply Rough Set Methods to Preserve Social Networks Privacy—A Review

B. S. Panda, M. Naveen Kumar, and Satyabrata Patro

1 Introduction

In society, social networks need continuously existed in various forms. A social network is a social construction that connects people and organizations. It shows how they are linked through a variety of social links, shows from casual acquaintance to tight kinship. Social networks can be used to model traffic, email, disease transmissions, and criminal behavior.

With the growth of computer record-keeping power and internet connectivity, equally the interactions within and the measure of these social networks are fetching visible as public institutions evolve. Web 2.0 programs, Instagram, Cake Financial, Daily Strength, Disaboom, Epernicus, Facebook, Twitter, Passport stamp, IDSA's Terrorism Tracker (T2), and others are some of the most popular social networks.

1.1 Issues of Privacy in Social Network

Scientists and researchers who collect data for public use are frequently forced to choose among two undesired outcomes. They can either issue data for others to examine, even if that study poses serious privacy risks, or they can withhold data due to privacy problems, preventing further research.

If extensively published, knowledge extraction of personal life can be a detrimental element. The public broadcast of a person's movements over time can have major privacy implications. Furthermore, the rise in violent terrorist acts in recent years has heightened a sense of insecurity that has persisted for years. As a result, the social

B. S. Panda (✉) · M. Naveen Kumar · S. Patro
Deparment of CSE, Raghu Engineering College, Visakhapatnam, India
e-mail: panda.bs@raghuenggcollege.in

G. Mathur et al. (eds.), *Proceedings of 3rd International Conference on Artificial Intelligence: Advances and Applications*, Algorithms for Intelligent Systems,
https://doi.org/10.1007/978-981-19-7041-2_34

network analysis (SNA) is concerned about maintaining individual privacy. Many social network data have been made widely available in recent years, [1, 2] it is an important concern that the data should be published to preserve privacy.

Example 1 During 2001 "the Enron Corporation's bankruptcy", the legal processes complete 500,000 email messages public, which researchers analyzed [3]. This information has benefited email communication, organizational arrangement, and social network study significantly, but it has also expected to result in significant privacy destructions for those engaged.

As a result, partly as a result of many web 2.0 applications, a growing amount of social network information has been made openly accessible and evaluated in some form. In comparison to substantial investigations of relational situations, we identify new obstacles in privacy-preserving publishing of social network information and investigate possible problem formulation in several dimensions such as privacy, background information, and data convenience.

1.2 Contributions and Paper Outline

The basic definition of Social networks and their representations are discussed in Sect. 2; we derive the basic issues of privacy in terms of Problem formulation. Here the rich rough set theory (RST) in Sect. 3, is referring to associated definitions and notations. Section 4 describes various privacy preservation models with anonymization techniques. Lastly, Section five adds the conclusion.

2 Social Network

Modeling the social network is one of the structural analysts' most popular concerns. So far, graphs and matrices have been the most widely adopted. In a social network, the major goal of employing mathematical and graphical tools is to describe network descriptions compactly and systematically. We can use computers to analyze network data, thanks to mathematical representations. Furthermore, network processing techniques and mathematical rules themselves suggest what we should look for in our data.

2.1 Represent Graph as Social Relations

A social network (Fig. 1) is made up of objects (or nodes) that denote arcs and lines (or edges) that denote relationships or relations in graph representations. Sociologists

Fig. 1 The social network

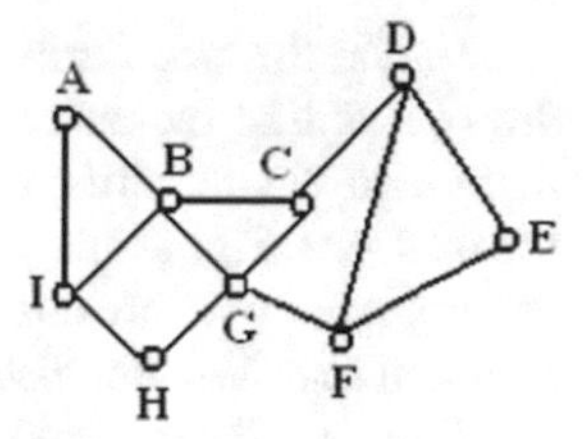

termed their graphs "sociograms" after borrowing this method of graphing from mathematics. Even though there are many variations on the concept of socio-grams, they always have one thing in common. The graph representation may be sufficient for a small graph, but the data and/or research problems are frequently too complex for this relatively simple method. A social network [4] can also be defined as a graph G = (V, E), with V = v1, v2, vn representing the set of vertices and E = e1, e2,….en representing the set of edges linking pairs of vertices. The edges of the network may have different functions depending on the network's goal.

2.2 *Represent Social Relations as Matrices*

The best common way to represent a social network is with a matrix, which has several rows and columns such as there are actors in the data collection (Table 1). The links between the actors are represented by elements. The most basic and often used matrix is binary, in which a one is inserted in a field if there is a tie, and a zero is imputed otherwise. Because it indicates who is close to or adjacent to whom in the "social space" delineated by the relations, the matrix resembles an "adjacency matrix."

Table 1 Social relation using matrix

	A	B	C	D	E	F	G	H	I
A	–	1	0	0	0	0	0	0	1
B	1	–	1	0	0	0	0	0	1
C	0	1	–	1	0	0	0	0	0
D	0	0	1	–	1	0	0	0	0
E	0	0	0	1	–	1	0	0	0
F	0	0	0	1	1	–	1	0	0
G	0	0	1	0	0	1	–	1	0
H	0	0	0	0	0	0	1	–	1
I	1	1	0	0	0	0	0	1	–

Despite the popularity of Social networks, attention draws on some common drawbacks like owner-controlled information access those bounds to keep information to a site, difficulties to export social profiles into machine-processable formats, etc. The developments of the collaborative social network, semantic web technologies, etc., are answers to them. However, in all cases, privacy needs to play a crucial role. We feel discussing these aspects will increase the length of the paper and limit our discussions on the basic issues of privacy and how to preserve it.

2.3 Problem Definition

With the availability of the social network in public and having little local knowledge about individual vertices, an opponent may attack the privacy of some sufferers. As a real example, consider a combined social network of "close-association to B" viewed in Fig. 2.

To privacy preserve, it is not enough to delete complete identities of individual nodes (Fig. 2b). If an adversary has little and more knowledge of nearly the neighbors of each node, the privacy may quite be escaped.

Even though the information is a public good that spreads within its immediate vicinity, there are significant linguistic, social, and cognitive hurdles to knowledge diffusion across social groupings. These hurdles can prevent knowledge from being passed on to those who require it. These social obstacles are significantly more difficult to overcome than those that exist in the interior spaces of organizations and institutions, due to society's immense complexity and diversity. Furthermore, most individuals, including policymakers and development program planners, are often blind to these limitations.

To formulate into a problem if a social network is mistakenly at large to the public, the privacy of knowledge may be compromised. We've seen how, when an individual, business, or social group innovates successfully, the knowledge that underpins that advancement becomes evident, at least in part, in the immediate vicinity. As time passes, such improvement is recognized and emulated. A systematic approach for anonymizing social network data before it is released is required.

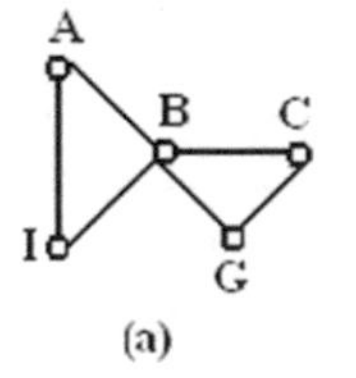

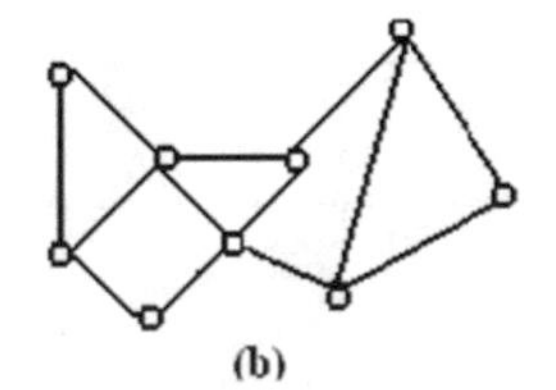

Fig. 2 **a** The 1-neighborhood of B **b** Privacy-preserved anonymous network

3 Rs (Rough Sets)

3.1 Notations and Definitions

RST is a principle that handles uncertainty in datasets. In this segment, here elaborated some fundamental notations of the Rough set theory [5–7].

- RS contains the fixed set of objects $U(\neq \phi)$
- The equivalence is the *indiscernibility relation* R over U.
- RS distinguishes between two or more values.
- The matched couple is called an approximation space A = (U, R).
- RS denotes the equality $[x]_R$ class or R containing an element $x \in U$.
- For any subset P$(\neq \phi) \subseteq \Re$, the intersection of every equivalence relations i P s denoted by IND(P) and is termed the *indiscernibility relation* P.
- Fundamental sets in A – the equality classes of R.
- Definable set in A–Any finite union of basic sets in A.
- For each $X \subseteq U$ and an equality relation $R \in IND(K)$, there associate both the subsets.

The lower approximation notation of X in A is the set $\underline{R}X = \cup\{Y \in U/R : Y \subseteq X\}$.

The elements $\underline{R}X$ are those elements of U which can be certainly classified as elements X with the knowledge of R.

The upper approximation notation of X in A is the set $\overline{R}X = \cup\{Y \in U/R : Y \cap X \neq \phi\}$.

$\overline{R}X$ is the pair of data elements X that can be possibly categorized as elements of X employing knowledge of R.

The boundary of X $\overline{R}X - \underline{R}X$ the elements of $\underline{R}X$ are those elements of U, which can certainly be classified as elements of X, and the elements of $\overline{R}X$ are those elements of U, which can be classified as elements of X, employing knowledge of R.

The marginal region is the un-decidable part of the universe as shown in Fig. 3.

Here assume that X is *rough* concerning R if and only if $\underline{R}X \neq \overline{R}X$, equally $BN_R(X) \neq \phi$. X is said to be R-*definable* if and only if $\underline{R}X = \overline{R}X$, or $BN_R(X) = \phi$.

3.2 Rough Set Usages and Applications

Rough sets deals with more applications of RST [2, 3, 6–9]. Some brief applications of RST are as follows:

a. Presentation of ambiguous or incomplete information.
b. Empirical knowledge and experience-based information gain.
c. Information evaluation.
d. Conflict resolution.

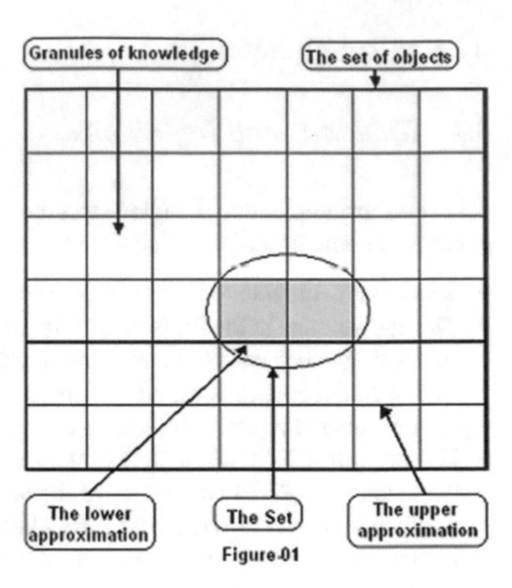

Fig. 3 Rough set approach

e. Assessment of the information's quality in terms of reliability and the occurrence or absence of recurring data frames.
f. Identifying and assessing data needs.
g. Classification of pattern based on a rough estimate.
h. Making decisions in the face of uncertainty.
i. Data reduction that preserves information.

3.3 *Equivalence Classes and Relations of Rough Sets in Social Network*

The rough set principle is founded on the concept of equality relations, as mentioned in the definitions and notations. This term refers to divisions made consisting of indistinguishable object classes with similar upper and lower approximations. Classical approximation spaces may be used to assess incomplete data. A global set and an incomplete relation, which is a completely equality relation, make up such parts. Equivalence classes are information granules denoted by these parts. The equivalency relation separates the universe U into equivalence classes of objects, which are pair-wise disjointing subsets. For an object x, that is. The equivalence class of information holding x is denoted by [P] = {P'IPE(V)p'}.

According to Fiksel [4], a social network consists of dynamic nodes partitioned into certain equivalence classes. "Individuals of such a social network may be divided into equivalence classes so that the original network may be represented by a reduced network containing one node for each equivalence class". The aggregation technique is used to get an associated group.

4 Privacy Preservation Models

4.1 Existing Systems (Anonymization)

To address the basic security issues, prior work in the database management domain [10] has been expended. The majority of works come with an emphasis on developing strategies for effectively anonymizing data so that detailed results can be published and shared. The main technique explored by the research literature follows the method of domain generalization. The ultimate goal is to conceal each individual into an appropriately constructed group, in a way that an attacker cannot easily reason about the participation of individuals in the group.

Definition 1 (*Principle of k-Anonymity*) A table fulfills k-anonymity if for each tuple $s \in S$ there occurs $k-1$ other tuples $s_{i1}, s_{i2}, \ldots, s_{i\,k-1} \in S$ such that $s[X] = s_{i1}[X] = s_{i2}[X] = \ldots = s_{i\,k-1}[X]$ for all $X \in Q_{I}$.

In present days, a novel meaning of privacy called k-anonymity [11] gained a reputation in the database management and data mining community. Recently, Zhou and Pei [12] proposed a method for preserving privacy in social networks against neighborhood attacks using the model of k-anonymity.

Formula 1 (k-Anonymity): Assume P be a social network data and P' anonymization of P. If P' is k-anonymous, after performing the neighborhood background information, any vertex in P cannot be re-identified in P' with confidence bigger than 1/k.

Definition 2 (*Principlek-anonymity p-sensitive property*) The masked microdata meets the p-sensitive k-anonymity state if it satisfies k-anonymity and the number of different attributes for each confidential variable is minimum *p* within the similar group of records with the equal combination of key variables that exist in covered microdata.

In the case of homogeneity and background knowledge attacks, k-anonymous social network data and associate principles may still outflow privacy. k-anonymity cannot avoid attribute disclosure, according to several writers. Machanavajjhala et al. [13] proposed a new concept of privacy known as *l-diversity*, which states that each equivalence class's distribution of sensitive attributes must have minimum well-represented values for every complex data attribute.

Definition 3 (*Principle of l-diversity*) If there are at minimum l "well-represented" standards for the sensitive property, an equality class is said to have *l*-diversity [3]. If every equivalence class in a table has l-diversity, the table is said to have the *l*-diversity principle.

Recently, l-diversity also has found some limitations to prevent attribute disclosure. Some attacks like skewness attacks and similarity attacks become insufficient cases to prevent attribute disclosure. Moreover, practically it is highly difficult and unnecessary to achieve l-diversity under the usage of entropy l-diversity with an l-set of small values. A new model "t-closeness" [3] is published, satisfying the above limitations and dominating as the latest model in keeping the privacy of database management and data mining systems.

Definition 4 (*Principle of t-closeness*) If the space between the sharing of a sensitive data attribute in this method of class and the sharing of the attribute in the entire table of data is less than a threshold t, an equivalence class is said to have t-closeness. If all equability classes in a table have t-closeness, the table is supposed to have t-closeness.

4.2 Preserving Privacy in Social Networks by k-Anonymity with a Single Neighborhood

The method of Zhou and Pei [12] focuses on privacy preservation issues in one neighborhood having two broad steps as summarized below.

Definition 5 In a social network data G, the neighborhood of u ϵ V(G) is the made sub-graph of the neighbors of u, represented by $N_u = \{v|(u, v) \epsilon E(G)\}$ where $neighbor_G(u) = G(N_u)$.

Step 1: Extracting neighborhoods and organizing vertices: Each vertex's neighborhood is removed, and distinct components are segregated. Isomorphism checks are performed because all graphs in the similar group must be anonymized to only one graph. The following processes are performed for each component of the vertex for this purpose.

In addition, the lowest DFS code is chosen. The component is claimed to be represented by this code. The property "Two graphs G and G0 are isomorphic if and only if DFS(G) = DFS (G0)" is satisfied by the minimum DFS code (G0). Then, to obtain a single code for one vertex, the neighboring component code order is employed.

Step 2: Anonymization is accomplished by combining vertices from the same group. If no match is found, the cost factor is utilized to determine which pair of vertices should be built.

4.3 Privacy Preserving in Social Networks with Binary Neighborhoods by k-anonymity

We modify the above approach and extend it to binary neighborhoods.

Step 1: Assume that u, v, V(G), u, and v live in the same neighborhood. The labels are then generalized or kept unaltered, resulting in isomorphic neighborhoods for u and v. In addition, the vertices' labels are identical in both neighborhoods.

Step 2: If the neighborhoods are not identical, the value is determined, and the vertices with the lowest value are chosen.

Step 3: To make them look alike, the essential edges are added.

Step 4: The vertices pair is subjected to the same procedure as in Step 1.

4.4 Achieving Social Network Anonymity Using Rough Set Based Structural Equivalence

According to Fiksel [4], little can be forecast about the actions of the network, if every node has a different transition rule. State changes can chaotically propagate throughout the network, with no clear pattern emerging. This can only be achieved if the nodes may be partitioned into equality classes, so that equivalent nodes share certain structural properties. Moreover, equivalent nodes have identical transition rules.

Definition 6 (*Principle of Structural equivalence*) It refers to how closely two nodes are linked to one another, i.e., how similar their social surroundings are. It's a common assumption that architecturally equivalent nodes will also be similar in other respects, such as attitudes, behaviors, or performances.

Nodes of a social network that look basically related may be fuzzy to an adversary, despite external knowledge. A complete form of structural comparison between nodes is termed automorphic equivalence.

5 Conclusion

In this article, we discussed the issues related to preserving privacy in the social network data. The relevance of the equivalence class of social network and equivalence relation of rough set theory is promising for integration. We have discussed the flexibility of rough set theory in different applications including equivalence relations. Further, we proposed anonymization techniques in social networks with the extension of k-anonymity. This work further has the potential to extend using other anonymization techniques like t-closeness and l-diversity.

The model of Zhou and Pei levels the nodes of social networks for anonymization. However, one can also extend the other possibility to label the edges instead. Because some edges between nodes are sensitive and need to be hidden from public access which may lead to a concrete approach to prevent link re-identification.

References

1. Doran J (1985) The computational approach to knowledge, communication and structure in multiactor system. In: Nigel-Gilbert G, Heath C (eds) Social action and ArtificialIntelligence, Gower, England
2. Tripathy BK, Panda GK (2009) On covering based approximations of classifications of sets. Accepted to publish In: LNAI series, Int J Springer
3. Zhou B, Pei J (2008) Preserving privacy in social networks against neighborhood attacks. In: Proceedings of the 24th international conference on data engineering (ICDE '08), Cancun, Mexico
4. Fiskel J (1980) Dynamic Evolution in Societal networks. J Math Soc, 27–46
5. Pawlak Z (1986) Rough Sets J Inf Comp Sc II:341–356
6. Pawlak Z (1991) Rough sets-theoretical aspects of reasoning about data. Kluwer Acad Publ
7. Pawlak Z, Skowron A (2007) Rough sets some extensions. J Inf Sci 177(1):28–40
8. Wasserman S, Faust K (1994) Social network analysis: methods and applications. Cambridge University Press
9. Panda BS, Abhishek R, Gantayat SS (2012) Uncertainty classification of expert systems—a rough set approach is published in ISCON proceedings with IJCA, ISBN: 973-93-80867-87-0
10. Cormode G, Srivatava D (2009) Anonymized data: generation, models, usage. In: SIGMOD'09, Providence, Rhode Island, USA and ACM
11. Sweeney L (2002) K-anonymity: a model for protecting privacy. Int J Uncertainty Fuzziness Knowl Syst 10(5):557–570
12. Machanavajjhala, Gehrke J, Kifer D, Venkitasubramaniam M (2006) L-diversity: privacy beyond k-anonymity. In: Proceedings 22nd international conference data engineering (ICDE), p 24
13. Li N, Li T, Venkatasubramanian S (2004) t-closeness: privacy beyond k-anonymity and l-diversity

Chapter 35
System Development to Analyze Recruitment Process and Eligibility Criteria Using Machine Learning Algorithms

Megharaj Sonawane, Aditya Borse, Hrishikesh Sonawane, Aashish Mali, and Prachi Rajarapollu

1 Introduction

College recruitment/Placement are nothing but the outstanding platform where students can apply their skill practically, which is majorly related to their domain of interest. Primarily, student's intention of achieving better placement in reputed organizations or MNC's, to successfully reach their aim after completion of college. Thangavel et al. [1] stated the importance of machine learning based analyzers for the recruitment process. For any educational institutions/universities, students are a major contributor and student's performance really affects the country's future growth. Any technical field serves technical knowledge to the students, which makes them more technically sound and in today's era of globalization, competition is going to increase, it's more in the technical field so making 100% student placements it's become a critical and challenging task for institutions. Hence, before presenting students in front of recruitment companies, they should be well skilled and technically fulfilled but to ensure that, institutions should know about them. So, dataset with various attributes including soft skill also need to be taken into consideration (Pre-evaluation) while evaluating models for students campus placement and prediction. Anvesh et al. [2] described the process followed for scrutiny of students based on skills for placement.

Many recruiting agencies are always looking for fresher scholars and their talent, which plays a vital role for society development therefore candidates have to do excellent work through which the recruiter's expectation may fulfill. Harinath et al. [3] states the use of an efficient algorithm of machine learning which is useful in data analysis and prediction process of placement. Generally, while in placement

M. Sonawane · A. Borse · H. Sonawane · A. Mali · P. Rajarapollu (✉)
School of Electrical Engineering, MIT Academy of Engineering Pune, Pune, India
e-mail: prrajarapollu@mitaoe.ac.in

G. Mathur et al. (eds.), *Proceedings of 3rd International Conference on Artificial Intelligence: Advances and Applications*, Algorithms for Intelligent Systems,
https://doi.org/10.1007/978-981-19-7041-2_35

process, recruiting agencies doesn't focus on particular domain of interest of candidate because there are again number of branches with various job roles are available like job roles are different for software engineer, different for electronics engineer, Mechanical and civil engineer, so at the time of recruitment process, recruiters try to analyzes candidates all areas and then after overall evaluation they provide convenient job role to the particular candidate. Evaluating candidates based on only their aptitude skill, technical questions, verbal and reasoning skills it's not the right way of analysis done by many third party online platforms for the candidate recruitment process. Attributes like student certifications, internships they are done with, soft skills/behavior and so on are also taken into consideration while student screening.

To make models with proper and accurate predictions with higher efficiency by considering all parameters/attributes, using machine algorithms are best fitted for candidate selection and make predictions which cannot be possible with normal algorithms.

2 Literature Review

Manvitha and Swaroopa [4] had described the use of supervised machine learning techniques for efficient and accurate placement analysis. This paper presents an approach to improve the performance of students during the campus placement activities. The model is built using the Random forest and Decision Tree algorithms. The accuracy of the algorithm for both the Decision tree and the Random Forest is 84% and 86% Respectively. So here we can see random forest can be a better option to predict the placement results.

Student Placement Analyzer: A Recommendation System Using Machine Learning by Rao et al. [5], proposed system uses historical data collected by the previous year's students to predict their future placement. It then rates the chances of them getting placed into the institution according to their knowledge and skills. The concept of the system is to cluster the students according to their characteristics, which includes knowledge, skills, and attitude. Maurya et al. [6] described the method of student separation and placement prediction based on the academic performance. Reddy et al. [7] focused on the total investment on recruitment process and also elaborated on resources involvement for placement. It is very important to have an error free and accurate process of selection to avoid failure. Nurhopahip et al. [8], describes the k-NN-Linear Interpolation technique used in the SMOTE method. Karnakar et al. [9] in their paper proposed a system; Applicant Personality Prediction System Using Machine Learning (APPS) will be predicting the personality of the candidate based on the candidate's resume details and also by using some personality tests. Sudha et al. [10], based on the resume, how personality prediction can be done is the major focus of the paper. Arora et al. [11], in this paper the author had focused on performance parameters of faculties in higher education systems and their placement importance for improving the education quality. Yusuf and Lhaksmana [12] had discussed the comparative results of KNN and SVM algorithms and proved that for set problems, SVM is giving the better results. Kumar et al. [13], in their paper the author had worked for supervised and unsupervised learning and tried to get best results for placement activity. The system has met the objectives, which are to predict the placement status of the students in final year. Kumar et al. [14], author tried to focus on prediction of placement so that this prediction will help the students in better preparation and work on weak points in view of placement. Giri et al. [15], author focused on placement of students in the IT sector and their requirements so that students can get well prepared.

3 System Architecture

To make intelligent decisions, Machine learning is one of the growing fields in data science and a subset of artificial intelligence, to make more accurate predictions and outcomes by using machine learning algorithms.

Typical implementation phases in machine learning are Data Gathering, Data preprocessing, Training and validating dataset of model, testing model and prediction. The steps are shown in Fig. 1.

3.1 Data Gathering

In machine learning, data gathering is the most important phase which defines potential usefulness of our model with accuracy. To gather data, we need to first identify sources from where we can college data and use it for our model for further

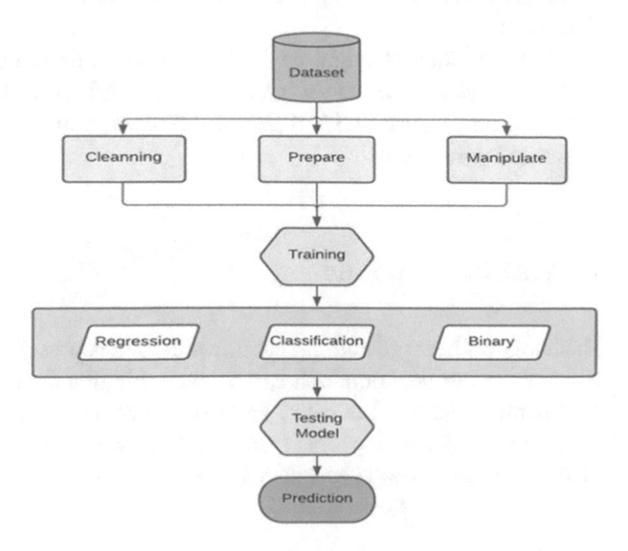

Fig. 1 System Architecture for implemented algorithm

processing on it. In the case of making predictions about student placement, the source is college/institutions, Universities etc. And after collection of data student data we need to aggregate the collected data into a single dataset. Based on the collected dataset as an input, algorithms predict output for our model. Machine learning algorithm output accuracy and efficiency totally depends on the selected dataset. Collected data set containing all the parameters and attributes which are used in prediction of student placement like academics marks, work experience, soft skill attributes, internship and certifications, student interest area etc. and also containing some random information like university, Board of exam, gender etc. which is really not useful for processing. So, we need to deal with those parameters and do processing accordingly.

3.2 Data Pre-processing

Data collected from the real world generally contain unusable parameters, noises, some data may be not in suitable form for our machine learning model which is required to be converted into another form. So, the process of making collected raw dataset into required machine dataset is known as data pre-processing. To obtain higher accuracy, it is necessary to clean the dataset and keep it in formatted form while creating machine learning models.

As we consider all required parameters in our dataset, it does also contain some values which are not a number and it's not good for our dataset so, we have to deal with these values. After taking a look from these values, we have to impute these values with zeros, which might be present in the form of characters or numbers and our system is also unable to understand the meaning of these kinds of values in the dataset. As we have to replace these values with zeros we used fill any method to deal with.

Collected dataset might be containing some unnecessary parameters which are really redundant to us so we have to reduced this column by using dropmethod and obtained new dataset with replaced unknown values with zeros and by reducing unwanted parameters.

4 Outlier Handling

Basically outliers are nothing but unusual values present in our dataset which totally differ from our data points. It can be available in a dataset while collecting data due to misinterpretation. Let's take an example of covid vaccination for adults in which they need to deal with people from age 18 onwards so, here in this dataset the outlier will be any one who is less than 18 year old. If we consider a person whose age is less than 18 then it deviates our model accuracy, so we have to handle this outlier and it also depends on the condition of our dataset. There are also various methods available for outlier handling through which we are using the IQR Method.

4.1 IQR Method

InterQuartile Rangc Method majorly used to find outliers from dataset and variation in it. In the given IRQ diagram as shown in Fig. 2, all the main values lie in the given box where Q1 is 25 percentile and Q3 is 75 percentile. The maximum value that is allow for outlier is Q3 + 1.5*IQR and minimum values is Q1–1.5*IQR. So, anything that is outside these values are outliers.

After execution of piece of code we have obtained some graph with outlier values as.

Figures 3 and 4 shows the BoxPlot based on the Secondary, Higher secondary and Undergraduate percentage with its effect on employability or selection chances using the concept of Quartile as discussed earlier, while Fig. 5 shows boxplot without a single outlier.

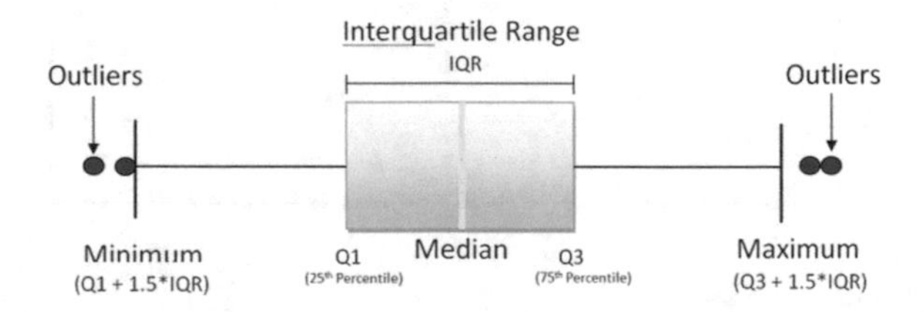

Fig. 2 Interquartile range

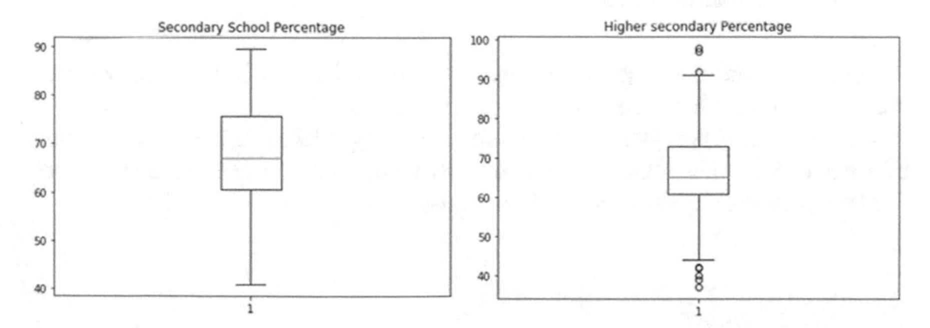

Fig. 3 BoxPlot for Education percentage comparison of Secondary and higher secondary

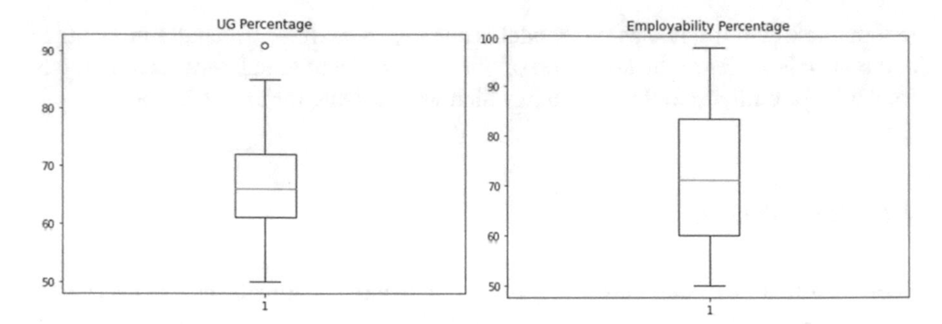

Fig. 4 BoxPlot for Education percentage comparison of UG percentage and Employability percentage

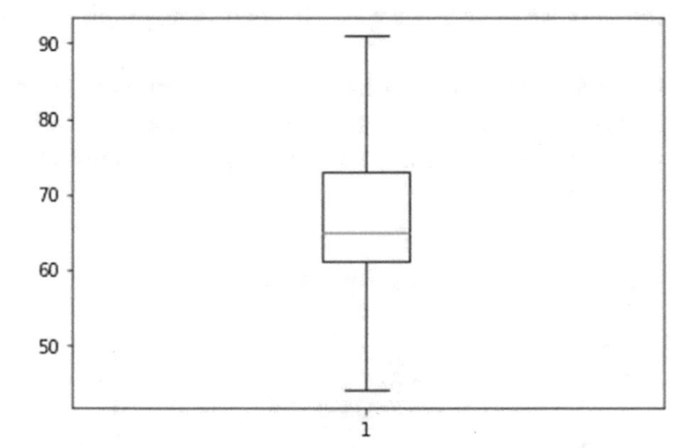

Fig. 5 BoxPlot without outliers

4.2 Data Visualization

Based on the dataset used graphical representation is done for better effect. Figure 6 presents the charts for stream selection by the candidate.

Figure 7 shows the distribution plot for the salary offered. If there is any gender bias while offering a salary we can go for a violin plot and can make inferences. Figure 8 presents the violin chart for the same.

4.3 Data Encoding

Machine learning model cant interpret with data which is not in numerical form and in case of our dataset there may be some parameters like gender, specialization, work experience are in the form of characters but our machines can interpret with

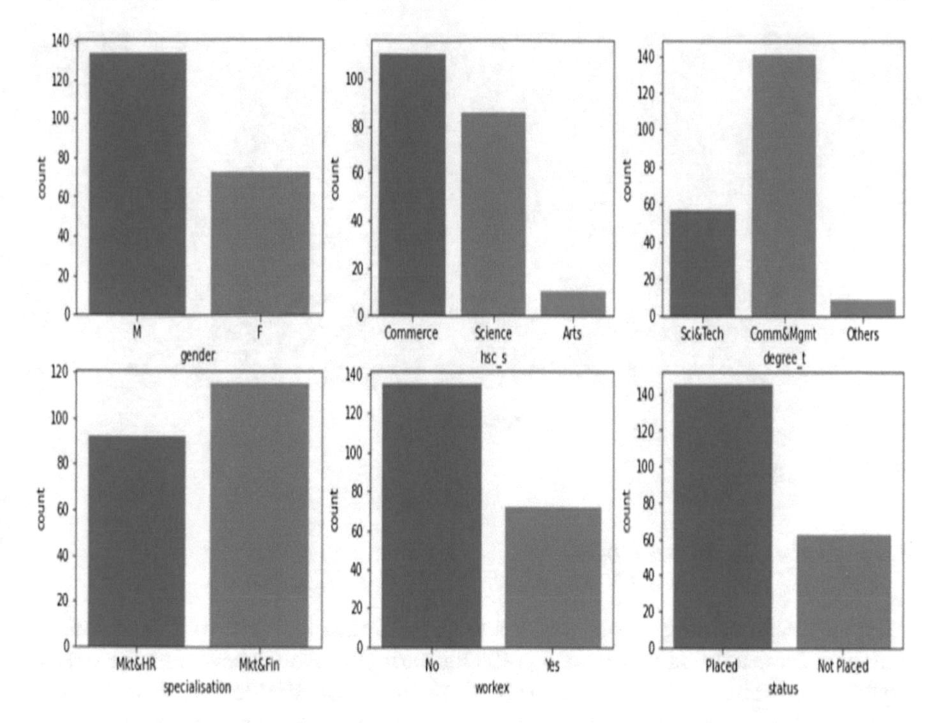

Fig. 6 Count plot for stream selection

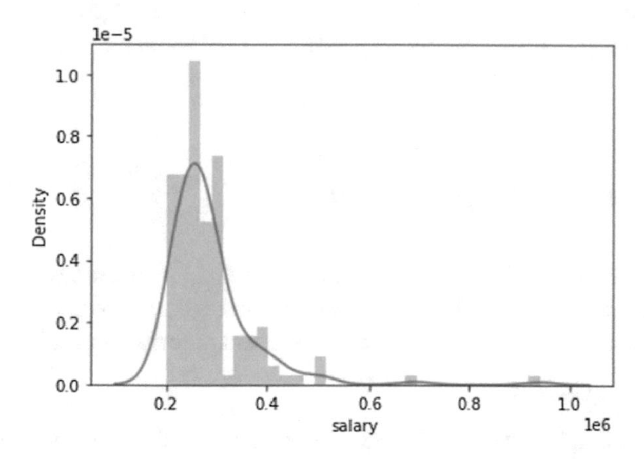

Fig. 7 Distribution plot based on the salary offered

categorical data so, the process of converting this categorical data into numerical data is known as data encoding. In this study, we used two encoding technique as

- Label encoding
- OneHot Encoding

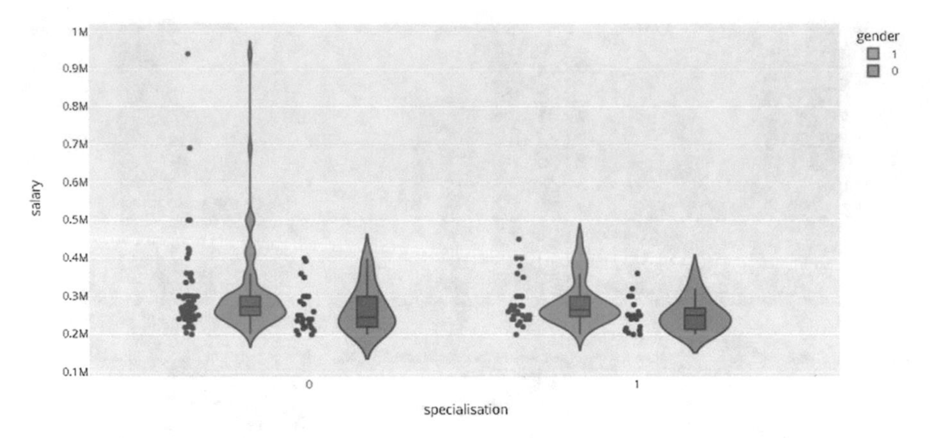

Fig. 8 Violin plot

4.4 *Train, Test and Split*

After all pre-processing is over we need to go for Train, Test and Split the dataset in which we have to assign target and predictor variables. Now to evaluate the overall performance of our machine learning algorithm we need to divide the dataset into two subset through which the first subset is used as a training dataset to fit the model and another one is referred to as a test data set. As we divide the dataset into two parts, training and testing dataset in which training part is always greater and generally 80% of data assigned for training and 20% [4] {2019Nov} is assigned for testing purpose.

4.5 *Machine Learning Algorithms*

To build an actual machine learning model with high accuracy for a gathered dataset, algorithm selection is very important and as per literature, there is very little research/studies that are conducted on student placement prediction [6].

1. **Logistic Regression**

Logistic regression is a statistical model that gives a parabolic view of classification [1] using logistic function to model binary dependent variables. Basically, it is used to modal probability of certain classes like placed or not placed. The mathematical formula for the same is given in Eq. 1.

$$\text{Logistic Function} = \frac{1}{1 + e^{-x}} \tag{1}$$

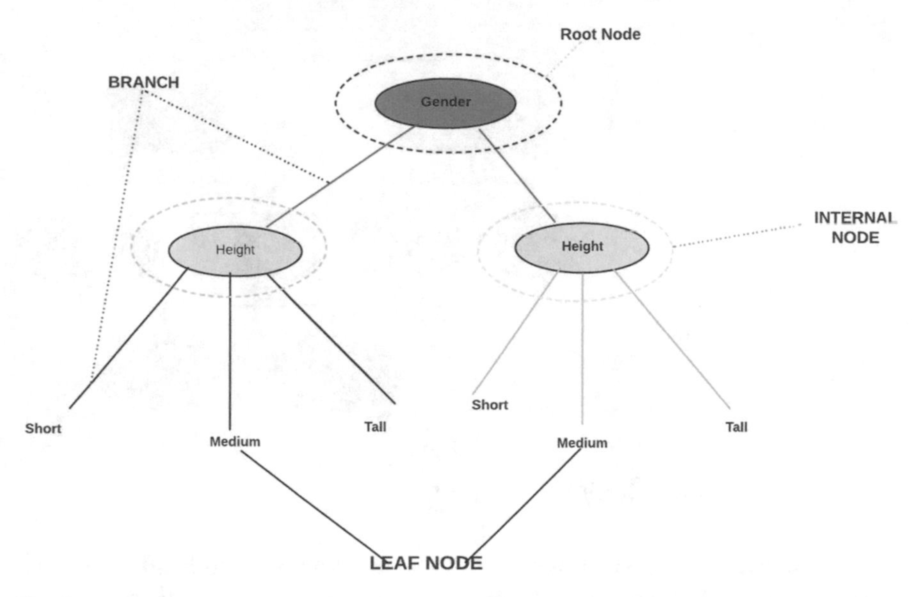

Fig. 9 Decision tree

2. **Decision Tree**

The machine learning model where according to certain attributes data gets split continuously and forms a tree like structure model of decision branches interpret decision rules and leaf shows outcome of decision tree. The decision tree for the problem statement is shown in Fig. 9. To measure uncertainty or impurity of decision tree, concept of entropy is used.

3. **Random Forest**

Random forest is nothing but a machine learning algorithm generally called as a bunch of decision trees put together or by combining many classifiers to make solutions on complex problems and also solve overfitting problems for dataset. Figure 10 shows the random forest diagram of the problem statement. Since one tree can't predict accurate result we are using aggregation of trees to produce more accurate results by using Eq. 2.

$$H(S) = \sum_{xeX} p(x) \log_2 \frac{1}{p(x)} \tag{2}$$

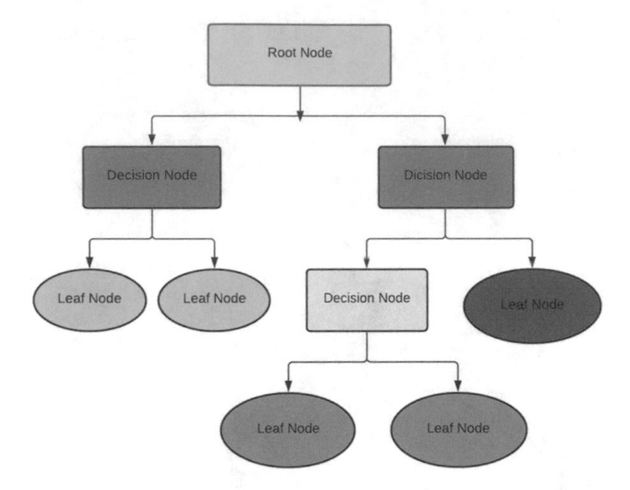

Fig. 10 Random forest

5 Result and Conclusion

The algorithms are applied on the data set and attributes used to build the model. The accuracy obtained after analysis for Decision tree is 73% and for the Random Forest is 78%.and using Logistic regression we are getting 83%. Hence, from the above said analysis and prediction it's better if the Logistic Regression algorithm is used to predict the placement results. In this paper.

Three different machine learning algorithms are examined; they are Logistic Regression, Decision Tree, and Random Forest. The obtained results show that Logistic Regression is the best performing algorithm for predicting student placement. It has outperformed all the other methods and showed an accuracy of 83%.

References

1. Thangavel SK, Divya Bharathi P, Sankar A (2017) Student placement analyzer: a recommendation system using machine learning. International conference on advanced computing and communication systems (ICACCS -2017), Jan. 06–07, 2017, Coimbatore, India
2. Anvesh K, Satya Prasad B, Venkata Sai Rama Laxman V, Satya Narayana B (2019) Automatic student analysis and placement prediction using advanced machine learning algorithms. Int J Innov Technol Exploring Eng (IJITEE) 8 (12), ISSN: 2278–3075
3. Harinath S, Prasad A, Suma HS, Suraksha A, Mathew T (2019) Student placement prediction using machine learning. Int Res J Eng Technol (IRJET) 6(4)
4. Manvitha P, Swaroopa N (2019) Campus placement prediction using supervised machine learning techniques. Int J Appl Eng Res ISSN 14(9):0973–4562
5. Apoorva Rao R, Deeksha KC, Vishal Prajwal R, Vrushak K, Nandini MS (2018) Student placement analyzer: a recommendation system using machine learning 4(3) IJARIIE-ISSN(O)-2395–4396

6. Maurya LS, Hussain MS, Singh S (2021) Developing classifiers through machine learning algorithms for student placement prediction based on academic performance. Appl Artif Intell 35(6):403–420
7. Jagan Mohan Reddy D, Regella S, Seelam SR (2020)Recruitment prediction using machine learning. 2020 5th international conference on computing, communication and security (ICCCS), 2020, pp 1–4
8. Nurhopipah A, Ceasar Y, Priadana A (2021) Improving machine Leaming accuracy using data augmentation in recruitment recommendation process. 2021 3rd east indonesia conference on computer and information technology (EIConCIT), pp 203–208
9. Karnakar M, Rahman HU, Santhosh ABJ, Sirisala N (2021) Applicant personality prediction system using machine learning. 2021 2nd global conference for advancement in technology (GCAT), pp 1–4
10. Sudha G, Sasipriya KK, Nivethitha D, Saranya S (2021)Personality prediction through CV analysis using machine learning algorithms for automated e-recruitment process. 2021 4th international conference on computing and communications technologies (ICCCT), pp 617–62
11. Arora S, Kawatra R, Agarwal M (2021) An empirical study—the cardinal factors towards recruitment of faculty in higher educational institutions using machine learning. 2021 8th international conference on signal processing and integrated networks (SPIN), pp 491–497
12. Yusuf M, Lhaksmana KM (2020) An automated interview grading system in talent recruitment using SVM. 2020 3rd international conference on information and communications technology (ICOIACT), pp 34–38
13. Kumar VU, Krishna A, Neelakanteswara P, Basha CZ (2020) Advanced prediction of performance of a student in an university using machine learning techniques. 2020 international conference on electronics and sustainable communication systems (ICESC), pp 121–126
14. Kumar N, Singh AS, TK Rajesh E (2020) Campus placement predictive analysis using machine learning. 2020 2nd international conference on advances in computing, communication control and networking (ICACCCN), pp 214–216
15. Giri A, Bhagavath MVV, Pruthvi B, Dubey N (2016) A placement prediction system using k-nearest neighbors classifier. 2016 second international conference on cognitive computing and information processing (CCIP), pp 1–4

Chapter 36
Broadband SIW Cavity-Backed High Gain Slot Antenna

Shivoy Pandey, Samrat Mehta, Sachin Kumar Sahu, and Ankit Sharma

1 Introduction

Because of its significant features such as elevated gain, unidirectionality of the radiation and superior isolation, wireless communication systems use cavity-backed slot antennas. However, because of its tremendous visibility, typical cavity back-end slot antennas are challenging to incorporate on planar circuits. However, by incorporating the SIW structure into the planar cavity back-end slot antenna, the problem can be solved. Because of the SIW cavity's high Q factor, the SIW cavity-supported slot antenna's impedance bandwidth is limited.

To enhance the SIW cavity-supported slotted antenna's impedance bandwidth, many approaches were suggested. When we tried to remove the substrate under or next to the slot, the proposed antenna's bandwidth limited itself to 5% [1, 2]. Because of the coupling between the slots of varying lengths, Ref. [3] was able to attain 8.5%, respectively. To improve the width of broadband, hybrid modes were used in Refs. [4, 5]. Using a rectangular slot, which formed by a hybrid mode in [4], a 6.3% operating bandwidth was reached. By converting the rectangular slot into a bow-tie slot, the impedance bandwidth is increased by 9.4%. A twin resonant slot patch antenna with the perturbative TM11 mode has an impedance bandwidth of 17.32% [6]. Reference [7] achieves impedance bandwidth of 15.2% and 17.5% by linking the lower and higher modes. A 28 GHz, a broadband dual mode (Modes TE102 and TE201) TCRS antenna (triangular-complimentary-split-ring-slot) published in [8] attains the bandwidth of 16.7%, although the antenna has a high level of cross-polarization and is limited at lower frequencies gain. The coupled feeding element in the multi-layer

S. Pandey · S. Mehta · S. K. Sahu · A. Sharma (✉)
Galgotia's College of Engineering and Technology, Uttar Pradesh, Knowledge Park I, Greater Noida 201310, India
e-mail: ankit.deli@gmail.com

G. Mathur et al. (eds.), *Proceedings of 3rd International Conference on Artificial Intelligence: Advances and Applications*, Algorithms for Intelligent Systems,
https://doi.org/10.1007/978-981-19-7041-2_36

structure shown in [8, 9] also serves in order to broaden the impedance bandwidth. Bow-tie slot antenna was able to achieve a 21.8% impedance bandwidth thanks to the meta-mushroom structure and coupled feeding mechanism in [9]. Reference [10] talks about an empty SIW (ESIW) cavity-backed antenna. The ESIW configuration brings the wide measured bandwidth to 40.8% and the maximum gain to 11.1 dB. The final reference talks about a SIW antenna with T-shaped slots in addition to the modified cross slots and two shorting vias. The mentioned antenna achieves a peak gain of 3.58 dB [11].

In this case study, a cavity-backed slot antenna with a broad bandwidth with the large gain characteristics has been created by integrating a modified slot with the modes of higher order in the SIW cavity. The suggested antenna has a 0.66 GHz impedance bandwidth and a maximum gain of 11.54 dB, according to testing. The following is the breakdown of the structure of the paper: The arrangement and field of the antenna are explained in Sect. 2, as well as the impact of various parameters on the simulated |S11|.

The simulated results are shown in Sect. 3. The paper comes to a close with Sect. 4.

2 Configuration and Design Analysis

As illustrated in Fig. 1, the suggested structure is totally constructed on one layer of the substrate of the material Rogers-Duriod 5580 with 0.787 mm of width. The square Substrate Integrated Waveguide cavity is made up of metallic vias layered around the parameter and the top and bottom of the substrate is of metal layers.

On the top metallic plate, the redesigned slot is engraved and positioned "D_s" from the cavity's short end. As a feeding element, a 50 Ω grounded coplanar waveguide (GCPW) with an additional titled slot is used to quickly develop higher-order modes and integrate with the planar circuit in the cavity. At the same time, the named slots separate the cavity of the SIW into two sub-cavities, creating the needed field distribution differences. The GCPW is also connected to a 50Ω microstrip feeding line for easy measurement.

The impedance of the SIW cavity-backed slot antenna bandwidth is controlled by the alteration of the slots and the cavity modes. As a result, analysis of slots and modes are done separately in order to have a good comprehension of the broadband theory of antenna proposed.

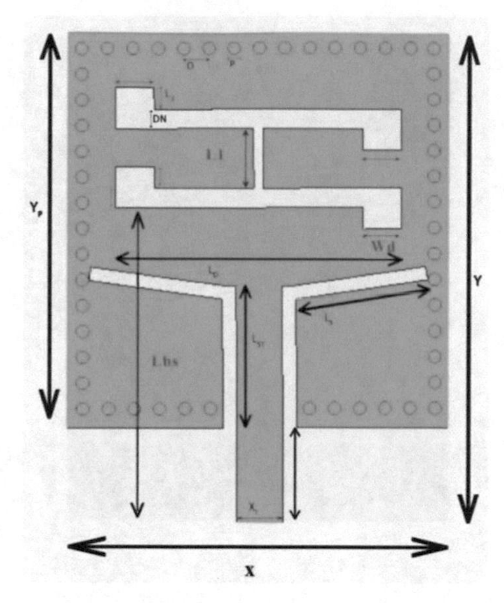

Fig. 1 Y = 25, Y_p = 20, D = 0.65, P = 1.3, L_2 = 1.095, D_n = 1.L_D = 14.7, L_s = 6.5, L_{st} = 7, X_T = 2.4, X = 20, L1 = 4, wd = 2 Lbs = 16.1

3 Evolution of the Proposed Structure

The purpose of using the evolution has resulted in a redesigned slot of the slot for radiation, which converted to a bow-tie from a slot shape to allow for a wider impedance bandwidth. Three basic variants of the intended antenna slot have been realized. First, a lower slot has been added to the antenna, resulting in a peak gain of 11.45 dB at 26.65 GHz when simulated. When simulated, the second slot, often known as the upper slot, produces the maximum peak gain of 9.98 dB at 26.55 GHz. The combination of the two variants is the third variation introduced in the slot. This has a peak gain of 11.65 dB at 26.65 GHz when measured.

The final slot in our design is to add a strip vertically that connects the upper and lower slots, as seen in Fig. 2d. At 26.6 Hz, the max gain achieved in this slot is 11.54 (Fig. 3).

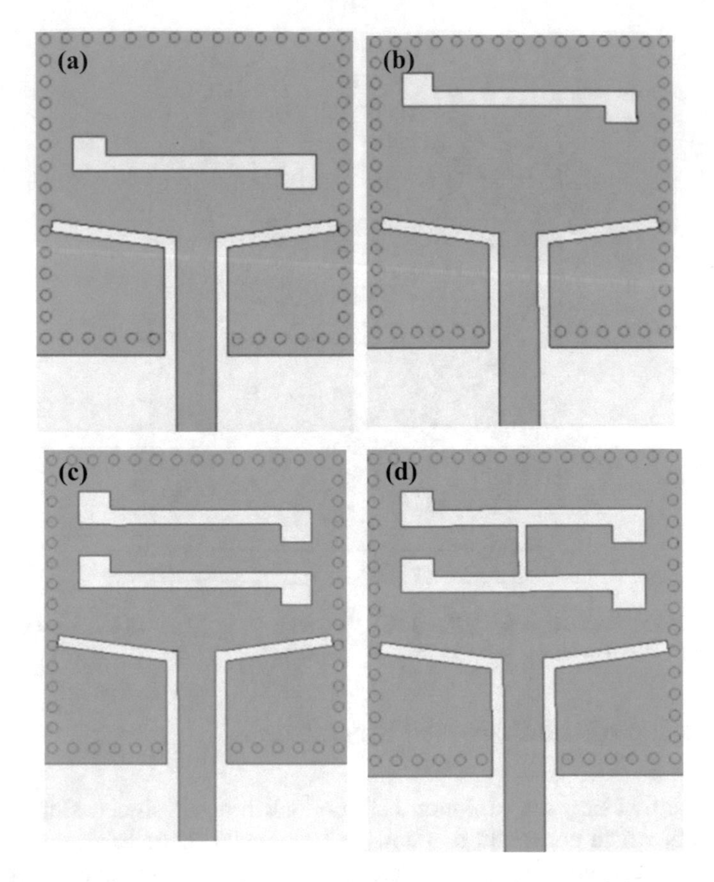

Fig. 2 The evolution of modified slot **a** Ant 1, **b** Ant 2, **c** Ant 3, **d** Proposed slots

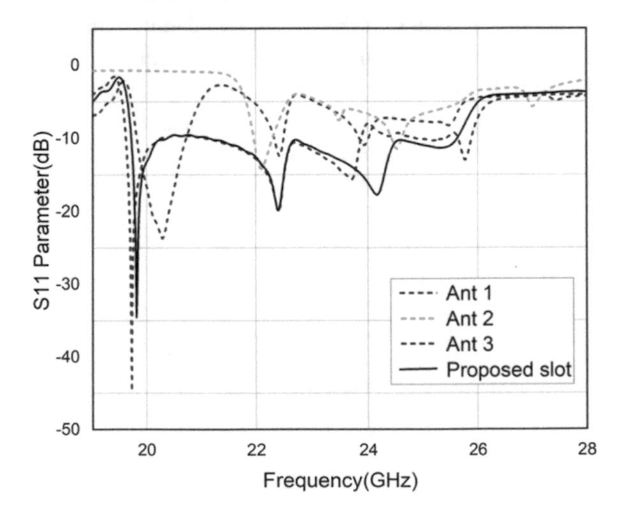

Fig. 3 The Simulated |S11| of Ant 1–4

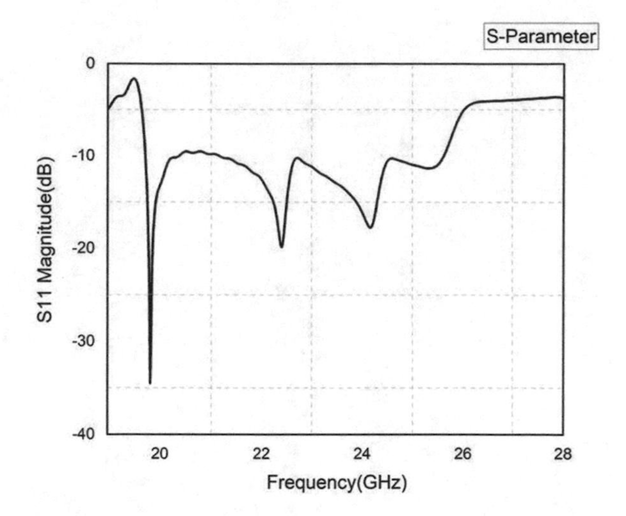

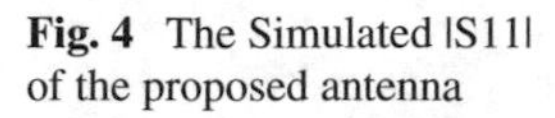

Fig. 4 The Simulated |S11| of the proposed antenna

4 Result and Discussion

For illustrating the results of the proposed antenna, a model is created. Figure 4 depicts a picture of the proposed antenna's S parameter (Figs. 5 and 6).

The differences between previously published single-layered antennas and the suggested antenna are summarized in Table 1. When these antennas are compared, the recommended antenna clearly offers the advantage of a large impedance bandwidth, high gain and low cross-polarization level.

5 Conclusion

The GCPW feeding structure is used to activate wideband SIW cavity-backed improved slot antenna that is put forward in the paper. The interplay between the slot and feeding structure exits high-order modes and bringing the high-order modes near together produces broadband impedance bandwidth. The antenna's gain is also increased because of the revised slot and modes of higher order.

Furthermore, its low cross-polarization level for the modified slot is influenced by the prominent placement of the transverse slot. The recommended antenna is better suitable for a wide range of broadband, low cross-polarized and high gain applications due to the advantages.

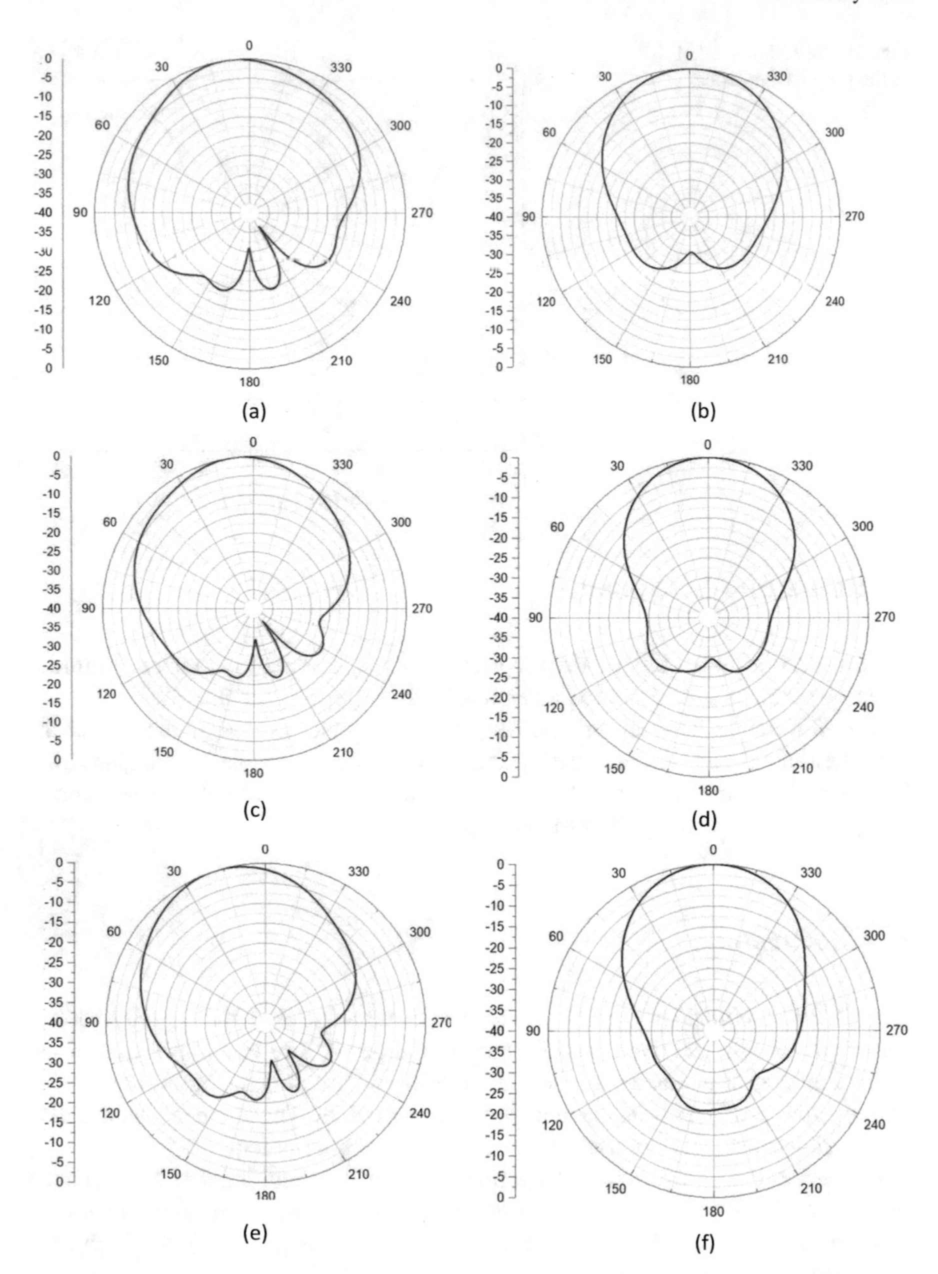

Fig. 5 Simulated Radiation Patterns: **a**, **c**, **e** shows E-plane at 19 GHz, 21 GHz, 23 GHz and **b**, **d**, **f** shows H-plane at 19 GHz, 21 GHz, 23 GHz, respectively

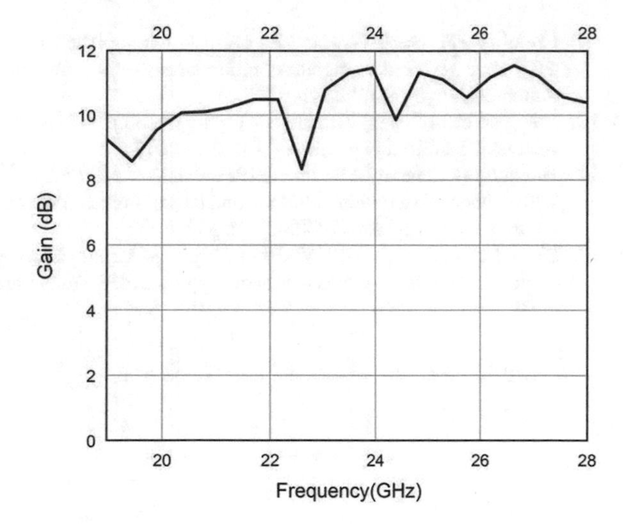

Fig. 6 The measured gains of the proposed antenna

Table 1 Suggested antenna

References	Type	Imp.BW (10 dB)%	Max Gain (dB)
[6]	Bow-tie slot	9.4	3.7
[7]	Slot-patch	17.32	7.79
[12]	Quad-resonant	17.5	7.27
[9]	TCSRS	16.67 at 28 GHz 22.2 at 45 GHz	<9
This work	Modified-slot shape	26.05	11.54

References

1. Yun S, Kim DY, Nam S (2012) Bandwidth enhancement of cavity backed slot antenna using a via-hole above the slot. IEEE Antennas Wirel Propag Lett 11:1092–1095
2. Yun S, Kim DY, Nam S (2012) Bandwidth and efficiency enhancement of cavity-backed slot antenna using a substrate removal. IEEE Antennas Wirel Propag Lett 11:1458–1461
3. Farrall AJ, Young PR (2004) Integrated waveguide slot antennas. Electron Lett 40(16):974–975
4. Mbaye M, Hautcoeur J, Talbi L, Hettak K (2013) Bandwidth broadening of dual-slot antenna using substrate integrated waveguide (SIW). IEEE Antennas Wirel Propag Lett 12:1169–1171
5. Luo GQ, Hu ZF, Li WJ, Zhang XH, Sun LL, Zheng JF (2012) Bandwidth-enhanced low-profile cavity-backed slot antenna by using hybrid SIW cavity modes. IEEE Trans Antennas Propag 60(4):1698–1704
6. Mukherjee S, Biswas A, Srivastava KV (2014) Broadband substrate integrated waveguide cavity-backed bow-tie slot antenna. IEEE Antennas WirelPropag Lett 13:1152–1155
7. Kim DY, Lee JW, Lee TK, Cho CS (2011) Design of SIW cavitybacked circular-polarized antennas using two different feeding transitions. IEEE Trans Antennas Propag 59(4):1698–1704
8. Kang H, Park SO (2016) Mushroom meta-material based substrate integrated waveguide cavity backed slot antenna with broadband and reduced back radiation. IET Microw Antennas Propag 10(14):1598–1603

9. Choubey PN, Hong W, Hao ZC, Chen P, Duong TV, Mei J (2016) A wideband dual-mode SIW cavity-backed triangular-complimentarysplit- ring-slot (TCSRS) antenna. IEEE Trans Antennas Propag 64(6):2541–2545
10. Wang R et al (2021) Broadband high-gain empty SIW cavity-backed slot antenna. IEEE Antennas Wirel Propag Lett 20(10):2073–2077
11. Hemanth A, Charan BSS, Teja GSR, Sahanth P, NS (2021) Compact SIW cavity slot antenna with enhanced bandwidth. 2021 second international conference on electronics and sustainable communication systems (ICESC), pp 592–595
12. Shi Y, Liu J, Long Y (2017) Wideband triple- and quad resonant substrate integrated waveguide cavity-backed slot antennas with shorting vias. IEEE Trans Antennas Propag 65(11):5768–5755

Chapter 37
Private Browsing Does Not Affect Google Personalization: An Experimental Evaluation

Maheen Ashraf, Syeda Fatima, Misbah Fatma, Sidratul Muntaha, Umme Rooman, Shahab Saquib Sohail, and Ahmad Sarosh

1 Introduction

Search engines have become an indispensable part of our lives. They essentially act as filters for the abundance of information available online. They allow users to find information of their interest almost immediately, without having to wade through various irrelevant search results. Though there are numerous search engines on the internet, Google takes the top spot with a striking 91.42%[1] of the market share. Most of us don't even bother picking another search engine, we simply "Google" words on the internet.

But not all users look for the same information online even if the keywords used by them are the same, this led to the potential deterioration of the quality of the search results [1–4]. To address this issue, search engines came up with the concept of "personalized search" [5, 6]. Google came forth with the idea of personalization back in 2004 which was later implemented in 2005 to understand the user's requirements better, filtering ambiguous search queries and retrieving relevant results for them [7].

---

[1] https://www.reliablesoft.net/top-10-search-engines-in-the-world/.

M. Ashraf (✉) · S. Fatima · M. Fatma · S. Muntaha · U. Rooman · S. S. Sohail
Department of Computer Science and Engineering, SEST, Jamia Hamdard, New Delhi 110062, India
e-mail: maheen.ashraf137@gmail.com

S. S. Sohail
e-mail: shahabssohail@jamiahamdard.ac.in

A. Sarosh
Department of Mathematics, College of Science, Jouf University, P.O. Box 2014, Sakaka, Saudi Arabia

G. Mathur et al. (eds.), *Proceedings of 3rd International Conference on Artificial Intelligence: Advances and Applications*, Algorithms for Intelligent Systems,
https://doi.org/10.1007/978-981-19-7041-2_37

While search results were traditionally based on website ranking, personalized search offered users results that were based on their search history [8–10]. Google with the help of machine learning used sophisticated algorithms to track users' activity online to find out areas they were most interested in and offered search results influenced by them.

While most of us were busy enjoying this privilege of a less time-consuming and more efficient search experience, we overlooked an important factor called "privacy" [11]. As algorithms Eagle-Eyed the user's activity online, this eventually made people feel it was invading their privacy. Again, to solve this issue, Google offered the "incognito" window in their browser Google Chrome which offered a non-personalized search and a way to bypass these behavioral identifiers [12]. Soon enough other browsers came up with their private browsing windows as well.

In this paper we aim to put this battle of privacy versus personalization to a test, to assess the longitudinal impact of personalization on user's privacy by providing state-of-the-art knowledge of the degree of personalization in private versus normal browsing. We analyzed the personalization of search results in the Google search engine in five different browsers to find out whether or not Google search offers personalization in their normal window and no personalization in their private windows. Our findings are both surprising and disappointing concerning the claims made by googling [13–16]. Our research aims to offer a better understanding of the personalization of search results on Google search while contributing further to the study of this field.

Road Map: Our paper is conventionally divided into 4 sections, in Sect. 2, we have elaborated our robust methodology to test our subject with quantified inputs and mathematical formulas that helped us derive our results, in Sect. 3, we have portrayed our results in written and tabulated forms for better understanding of the findings, and finally Sect. 4 unfolds our conclusion and thoughts about the results hence derived from the activities performed.

2 Methodology

In this section, we have detailed the methodology selected for our study. Required data collection, methods of sample selection, planned research design, and the analysis of the results of our study will be explained in this section. We delineate our datasets and methods to test Google's claim that its search engine offers personalized results in a normal window and no personalization in a private window.

2.1 Dataset

To collect datasets for our study, we asked five volunteers to perform an activity. These volunteers belong to different locations, age groups, and occupations. We picked India as the country from which we picked volunteers from remotely far locations. We did not include other countries as different countries have different search indexes which could alter our results. Each of our volunteers was given five different keywords. Since our study majorly focuses on tracking the nature of Google's algorithms to give personalized link recommendation that ultimately leads to filter bubbles, we deliberately tried to manipulate the keywords selection where a few of them were related to the volunteer's area of interest. Professionally, the first volunteer is a poet, the second user is an avid traveler. The third user is a college professor with an interest in books and academics. The fourth user is a college student with an interest in food and sports. At length, our fifth user is a middle-aged corporate worker who fancies accessories and jewelry. Table 1 depicts the user's area of interest, the browser, and keywords allotted, for all 5 volunteers.

2.2 Procedure

For our dataset collection, we planned activity to visvisualizee the amount of personalization Google offers in its normal window versus the amount of personalization in a private window. We picked five users with different browsers and different IP addresses having equal search engines, i.e., Google. We allotted them five keywords each and then asked them to search them on google on their respective browsers.

In our activity, each volunteer was told to first search their chosen keywords in a normal browsing window while signing in with their Google accounts and then search the same keywords in the private mode with their google accounts logged out. Later, we recorded the results for all 25 keywords both in normal and private mode and analyzed the results. Figure 1 represents the step-by-step illustration of how our methodology worked.

2.3 Measuring Personalisation

After drawing all the results, we asked all five volunteers to mention the top 8 links displayed on the Search Engine Results Page (SERP) while searching each keyword [17]. We only considered the raw links and ignored sections like maps, top stories, and infoboxes [18]. We have numbered all the links in the tabular form from each user for a better understanding of the results and calculated the correlation coefficient between the results of normal versus private windows for each keyword.

Table 1 Details of all volunteers with their respective keywords

Participants	Browser used	Area of interest	Keyword 1	Keyword 2	Keyword 3	Keyword 4	Keyword 5
User 1	Brave	Poetry	Pigeon	Serendipity	Gold	Basketball	Story
User 2	Google Chrome	Travel/Excursion	Song	Footwear	Weather	Train	Parachute
User 3	Mozilla Firefox	Books/Science	Atrocity	Planner	Sunshine	Invention	Experiment
User 4	Microsoft Edge	Food/Sports	Apple	Bear	Football	Jeans	Anxiety
User 5	Opera	Jewelry	Watch	Accessories	Silver	Environment	Electronics

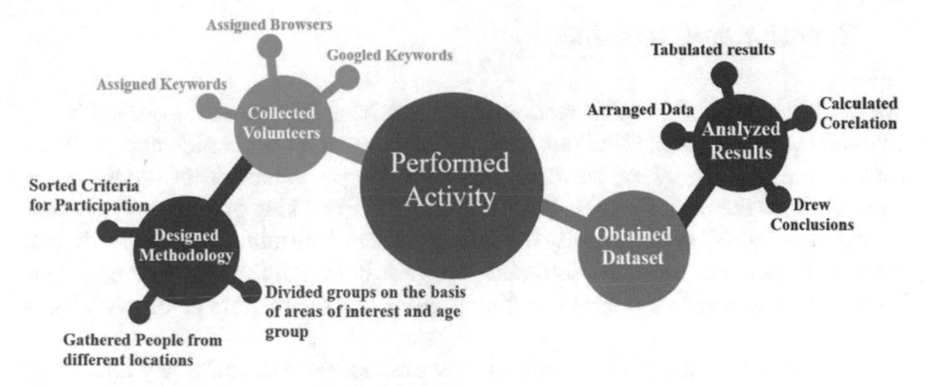

Fig. 1 Quintessential illustration of methodology

Calculating Correlation Coefficient

When calculating a correlation coefficient, we have to keep in mind the following representations:

- (x(i), y(i)) = a pair of data
- $\underline{X}$ = the mean of x(i)
- $\underline{Y}$ = the mean of x(i)
- s(x) = the standard deviation of the second coordinates of y(i)

Steps to calculate the correlation coefficient:

1. Determine data sets.
2. Calculate the standardized value for x variables.

$$(Z(\mathrm{x}))(\mathrm{i}) = (\mathrm{x}(\mathrm{i}) - \underline{\mathrm{x}})/\,\mathrm{s}(\mathrm{x}) \tag{1}$$

3. Calculate the standardized value for y variables.

$$(Z(\mathrm{y}))(\mathrm{i}) = (\mathrm{y}(\mathrm{i}) - \underline{\mathrm{y}})/\,\mathrm{s}(\mathrm{y}) \tag{2}$$

4. Multiply and find the sum.

$$(\mathrm{z}(\mathrm{x}))(\mathrm{i}) * (\mathrm{z}(\mathrm{y}))(\mathrm{i}) \tag{3}$$

   After multiplying the values, add them together to find the sum.
5. Divide the sum and then calculate the correlation coefficient.

From Coefficient Correlation, we find out that the value of correlation is exactly 1 for all 25 results.

3 Experimental Results

We derived a total of 200 link recommendations from 25 results for our study. By calculating coefficient correlations of all the results using the coefficient correlation formula and as alleged above we concluded our findings. For the sake of simplicity, we have tabulated the Search Engine Results Page (SERP) ranking of different keywords for user 1 in Table 2 systematically to analyze the data. The table given below depicts the search results of user 1 for different keywords both in normal as well as private browsing. Similarly, we prepared different tables for other users and keywords as well.

In our study for private browsing to increase, decrease or make any alterations in the personalization of the search results, we anticipated different search results in the private browsing window in comparison to the results normal browsing window. But after calculating the correlation of all the queries, we got the correlation value of 1 which leads us to the conclusion that all the results are strongly correlated with each other. In other words, the results in both windows were the same. This led us to believe that private browsing does not affect the personalization trends of Google's search engine.

4 Conclusion and Discussion

Personalization is a concept that refers to using additional information about the user along with the query to increase the relevancy of the search results. Users receive different search results for the same query based on their preferences, community, location, or browsing history. It allows search engines to give visitors customized experiences suited to their preferences, rather than giving a single, generic experience. Personalization isn't a new concept by any means. People like hairdressers or waiters commonly greet their returning customers by their name and even know what their "regular" is. The idea of personalization is so widespread in the offline world and that is exactly what search engines aimed to offer online with the help of algorithms.

Personalization in search results has made filtering content online easier and saved a ton of effort for the users, however, all good things come with a price and so has personalization. Personalized search offered zero user privacy which was unacceptable to many. In this paper, we made contributions toward addressing the given problem.

Our research in this paper has shown that Google's claim to offer personalized search results in normal mode is incorrect to a high extent. In our experiment, we barely noticed any personalization in normal mode. We purposely picked five individuals with remotely different preferences and age groups knowing the content they search online is entirely different. If algorithms were to pick up on their choice of content, they should have offered results influenced by the same.

Table 2 Search results of User 1

Keyword 1		Keyword 2		Keyword 3	
Normal	Private	Normal	Private	Normal	Private
pigeonindia.in	pigeonindia.in	Merriam-webster.com	Merriam-webster.com	goldprice.org	goldprice.org
stovekraft.com	stovekraft.com	imdb.com	imdb.com	wikipedia.org	wikipedia.org
wikipedia.org	wikipedia.org	dictionary.com	dictionary.com	gold.org	gold.org
britannica.com	britannica.com	wikipedia.org	wikipedia.org	kitco.com	kitco.com
allaboutbirds.org	allaboutbirds.org	dictionary.cambridge.org	dictionary.cambridge.org	moneycontrol.com	moneycontrol.com
flipkart.com	flipkart.com	vocabulary.com	vocabulary.com	britannica.com	britannica.com
amazon. in	amazon. in	collinsdictionary.com	collinsdictionary.com	m.economictimes.com	m.economictimes.com
pigeon.com	pigeon.com	rottentomatoes.com	rottentomatoes.com	goldreturns.in	goldreturns.in

Keyword 4		Keyword 5	
Normal	Private	Normal	Private
wikipedia.org	wikipedia.org	oxfordlearnersdictionaries.com	oxfordlearnersdictionaries.com
britannica.com	britannica.com	dictionary.cambridge.org	dictionary.cambridge.org
nba.com	nba.com	Merriam-webster.com	Merriam-webster.com
olympics.com	olympics.com	dictionary.com	dictionary.com
flipkart.com	flipkart.com	wikipedis.org	wikipedis.org
amazon. in	amazon. in	collinsdictionary.com	collinsdictionary.com
Live score. in	Live score.in	storysaver.net	storysaver.net
usab.com	usab.com	one-story.com	one-story.com

Also, what Google states about the private mode is that the activity data isn't kept on our device nor do the algorithms track the users' search history when logged out of their Google accounts.[2] So, if one searches something in incognito mode without logging in to their Google account, the results won't show up in their browsing history, and since there is no activity tracking there, there should be zero personalization in the search results.

So, to validate their claims, google must offer a personalized search result in normal mode with a google account signed in, and a varied old-school un personalized result in private mode while the accounts are logged out. But to our surprise, the resulting scenario was entirely different from what we imagined. We found that the results were firmly correlated and hence were the same. From the links to their alignment, everything was the same in normal and private mode for all 25 keywords.

In this paper, we majorly focused on finding out the existence of personalization in both normal and private windows and to the extent of our knowledge, we believe that the claims made by google search engine were incorrect. As for now if google does offer personalization that went unnoticed by us, we believe that Google should work on making personalization more transparent for the users. However, we do plan on studying more in the field of personalization and web searches to get a better view of this issue but we will leave that up for future exploration.

References

1. Salehi S, Du JT, Ashman H (2018) Use of Web search engines and personalisation in information searching for educational purposes. Inf Res Int Electron J 23(2):n2
2. Santally MI, Alain S (2006) Personalisation in web-based learning environments. Int J Distance Educ Technol (IJDET) 4(4):15–35
3. Schewe KD, Thalheim B (2007) Personalisation of web information systems–a term rewriting approach. Data Knowl Eng 62(1):101–117
4. Smyth B, Balfe E (2006) Anonymous personalization in collaborative web search. Inf Retr 9(2):165–190
5. Krafft TD, Gamer M, Zweig KA (2019) What did you see? A study to measure personalization in Google's search engine. EPJ Data Sci 8(1):38
6. Spink A, Khopkar Y, Shah P, Debnath S (2003) Search engine personalization: an exploratory study. First Monday
7. Gao M, Liu K, Wu Z (2010) Personalisation in web computing and informatics: theories, techniques, applications, and future research. Info Syst Front 12(5):607–629
8. Teevan J, Dumais ST, Horvitz E (2010) Potential for personalization. ACM Trans Comput Human Interaction (TOCHI) 17(1):1–31
9. Terren L, Borge-Bravo R (2021) Echo chambers on social media: a systematic review of the literature. Rev Commun Res 9:99–118
10. Yoganarasimhan H (2020) Search personalization using machine learning. Manage Sci 66(3):1045–1070

[2] https://support.google.com/chrome/answer/9845881?hl=en&ref_topic=9845306#zippy=%2Chow-incognito-mode-works%2Chow-incognito-mode-protects-your-privacy.

11. Dahlgren PM (2021) A critical review of filter bubbles and a comparison with selective exposure. Nordicom Rev 42(1):15–33
12. Fan W, Gordon M, Pathak P (2000) Personalization of search engine services for effective retrieval and knowledge management
13. Hannak A, Sapiezynski P, Molavi Kakhki A, Krishnamurthy B, Lazer D, Mislove, A, Wilson C (2013) Measuring personalization of web search. In: Proceedings of the 22nd international conference on world wide web, pp 527–538
14. Hawalah A, Fasli M (2015) Dynamic user profiles for web personalisation. Expert Syst Appl 42(5):2547–2569
15. Keenoy K, Levene M (2003) Personalisation of web search. In: IJCAI workshop on intelligent techniques for web personalization. Springer, Berlin, Heidelberg, pp 201–228
16. Kliman-Silver C, Hannak A, Lazer D, Wilson C, Mislove A (2015) Location, location, location: the impact of geolocation on web search personalization. In: Proceedings of the 2015 internet measurement conference, pp 121–127
17. Stamou S, Ntoulas A (2009) Search personalization through query and page topical analysis. User Model User-Adap Inter 19(1):5–33
18. Feuz M, Fuller M, Stalder F (2011) Personal web searching in the age of semantic capitalism: diagnosing the mechanisms of personalization. First Monday

Chapter 38
Spherical Fuzzy Parameterized Soft Set-Based Multi-criteria Decision-Making Method

Laxmi Rajput and Sanjay Kumar

1 Introduction

In social science, economics, engineering, medicine, and human resource development, MCDM has proven to be an effective tool for resolving a wide range of real-world decision-making issues. Since traditional mathematical tools are less adaptable to dealing with non-probabilistic uncertainty. As a result, Zadeh [1] devised a fuzzy set (FS) to represent the non-probabilistic uncertainty induced by imprecise and ambiguous data. After the development of the fuzzy set, many extensions are developed. Atanassov [2] generalized FS and defined an Intuitionistic fuzzy set (IFS) to include the non-determinacy due to a single membership function (MF) of an element in FS. Later on, Yager [3, 4] and Coung [5, 6] defined the Pythagorean fuzzy set (PyFS) and picture fuzzy set (PFS) as an extension of IFS. Ashraf et al. [7] recently introduced spherical fuzzy sets (SFSs), which are more rational and practical than IFS, PyFS, and PFS. All of these ideas, however, have an inherent limitation of the insufficiency of the parametrization tool connected with them.

Moltodtsov [8] created soft set theory (SST) to handle the parameterizations component of the analysis. Maji et al. [9] merged soft sets (SS) and FS. Later, Maji et al. [10] also defined an intuitionistic fuzzy soft set (IFSS). A number of academics have provided methods and techniques for dealing with DM issues in the SS environment. Soft matrix theory was created by Çağman and Enginoğlu [11]. After that, the soft discernibility matrix and its application were proposed by Feng and Zhou [12]. Soft expert sets (SESs), according to Alkhazaleh and Salleh [13], are a concept in which the viewer may grasp the ideas of all specialists in one model and solve problems. Alkhazaleh and Salleh [14] also discussed the fuzzy soft experts

L. Rajput (✉) · S. Kumar
Department of Mathematics, Statistics and Computer Science, G. B. Pant University of Agriculture and Technology, Pantnagar 263145, Uttarakhand, India
e-mail: laxmi31rajput@gmail.com

G. Mathur et al. (eds.), *Proceedings of 3rd International Conference on Artificial Intelligence: Advances and Applications*, Algorithms for Intelligent Systems,
https://doi.org/10.1007/978-981-19-7041-2_38

set (FSESs) and showed how to apply it in decision-making issues. Thus, there are several extensions of a fuzzy soft set (FSS). Yang et al. [15] developed an Adjustable soft discernibility matrix based on PFSS and DM issues. Peng et al. [16] developed PyFSS. Garg and Arora [17] developed a nonlinear-programming methodology for MCDM with interval-valued IFSSs information. Garg and Arora [18] proposed IFSS with its applications. Akram et al. [19] proposed hesitant fuzzy N-soft sets. Khalil et al. [20] created new operations on interval-valued PFS and interval-valued PFSS. Perveen et al. [21] created SFSSs with their uses in MCDM issues.

Çağman et al. [22] created the idea of fuzzy parameterized soft sets (FPSS). Further, several applications of FPSS are investigated in a variety of real-world challenges. Deli and Çağman [23] created the IFPSS theory. Joshi et al. [24] proposed the IFPSS theory with its application. Aydın and Enginoğlu [25] proposed Interval-valued IFPSS with its application in DM. Khan et al. [26] proposed bi-parametric distance and similarity measures of PFSs and applications in medical diagnosis. PA and John [27] created SF-SESs.

In this research article, we mention the limitation of IFPSS and propose the use of SFPSS to overcome that limitation. We also extend the construction method of Singh et al. [29] and propose a new method to construct SFS from IFS. We have also proposed a simple MCDM method using SFPSS. The rest of the paper is organized as follows: Sect. 2 contains definitions and limitations of IFPSS. Section 3 contains a novel construction method for SFPSS from IFS. Section 4 contains SFPSS-based MCDM application. Section 5 contains comparison analysis and Sect. 6 contains the conclusion.

2 Preliminaries

This part covers the definition of a SS, fuzzy soft set (FSS), IFS, IFPSS, SFS, and SFPSS, as well as the IFPSS limitation. In this paper, $\tilde{M}$ represents the universal set (US), S set of parameters, and $P(\tilde{M})$ is a set of all subsets of $\tilde{M}$, membership function (MF), neutral membership function (NeMF), and non-membership function (NMF).

Definition 1 Molodtsov [8] Let a subset T of set S and a mapping $\tilde{i}$ given by $\tilde{i} : T \rightarrow P(\tilde{M})$ then a SS over set $\tilde{M}$ define by pair ($\tilde{i}$, ,T).

Definition 2 Çağman et al. [22] Let a subset T of set S and a mapping $\tilde{i}$ given by $\tilde{i} : T \rightarrow P(\tilde{M})$, where $P(\tilde{M})$ is the set of all fuzzy subsets of $\tilde{M}$ then a FSS over $\tilde{M}$ define by pair ($\tilde{i}$, ,T).

Definition 3 Atanassov [2] A IFS over set $\tilde{M}$ is defined as $E = \left\{ < \tilde{m}, x_E(\tilde{m}), y_E(\tilde{m}) >: \tilde{m} \in \tilde{M} \right\}$. Here $x_E : \tilde{M} \rightarrow [0, 1]$ and $y_E : \tilde{M} \rightarrow [0, 1]$ [0, 1] represent MF and NMF, respectively, and satisfy the condition $0 \leq x_E(\tilde{m}) + y_E(\tilde{m}) \leq 1, \forall \tilde{m} \in \tilde{M}$.

Definition 4 Deli and Çağman [23] An IFPSS set over $\tilde{M}$ is defined as $\Lambda_E = \left\{\left(< \tilde{m}, x_E(\tilde{u}), y_E(\tilde{u}) >, \tilde{i}_E(T)\right) : t \in T\right\}$, where E is IFS over T and $x_E : T \to [0, 1]$, $y_E : T \to [0, 1]$ are MF and NMF, respectively, and $\tilde{i}_E$ is a mapping $\tilde{i}_E : T \to P(\tilde{M})$ such that $\tilde{i}_E(t) = \phi$ if $x_E(t) = 0$ and $y_E(t) = 1$.

Note If $x_E(t) = 0$ and $y_E(t) = 1, \forall$ t $\in$ T, then IFPSS becomes empty IFPSS and if $x_E(t) = 1$, $y_E(t) = 0$ and $\tilde{i}_E(t) = \tilde{M}, \forall$ t $\in$ T, then IFPSS is called universal IFPSS.

Definition 5 Ashraf et al. [7] A SFS over set $\tilde{M}$ is defined as E = { $(x_E(\tilde{m}), z_E(\tilde{m}), y_E(\tilde{m})) : \tilde{m} \in \tilde{M}\}$, where $x_E : \tilde{M} \to [0, 1]$, $z_E : \tilde{M} \to [0, 1]$, and $y_E : \tilde{M} \to [0, 1]$ are the MF, NeMF, and NMF, respectively, and satisfy the condition $0 \leq x_E^2(\tilde{m}) + z_E^2(\tilde{m}) + y_E^2(\tilde{m}) \leq 1, \forall \tilde{m} \in \tilde{M}$.

Definition 6 A SFPSS over $\tilde{M}$ is defined as $\Lambda_S = \left\{\left(< \tilde{m}, x_S(\tilde{m}), z_S(\tilde{m}), y_S(\tilde{m}) >, \tilde{i}_S(P)\right) : t \in T\right\}$, where S is SFS over T and $x_S : T \to [0, 1]$, $z_S : T \to [0, 1]$, and $y_S : T \to [0, 1]$ are MF, NeF, and NMF, respectively, and $\tilde{i}_S$ is a mapping $\tilde{i}_S : T \to P(\tilde{M})$ and satisfies the condition $0 \leq x_S^2(\tilde{m}) + z_S^2(\tilde{m}) + y_S^2(\tilde{m}) \leq 1, \forall \tilde{m} \in \tilde{M}$.

2.1 Limitations of IFPSS

In this segment, we discuss the limitations of the IFPSS using the following example.

Example 1 Let a decision-maker select a restaurant from a list of five restaurants based on five parameters {Tasty food, good services, Reasonable cost, Range of beverages, convenient location}.

Let $\{o_1, o_2, o_3, o_4, o_5\}$ be set of restaurant and $\{q_1, q_2, q_3, q_4, q_5\}$ set of parameters, where q_1 = "Tasty food", q_2 = "good services", q_3 = "reasonable cost", q_4 = "Range of beverages", and q_5 = "Convenient location".

Let $\{< q_1, 0.7, 0.1 >, <q_2, 0.4, 0.4 >, <q_3, 0.5, 0.3 >, <q_4, 0.6, 0.3 >, < q_5, 0.7, 0.1 >\}$ be an IFS, then IFPSS is given by

$$\begin{pmatrix} (< q_1, 0.7, 0.1 >, \{o_1, o_2\}), (<q_2, 0.4, 0.4 >, \{o_1, o_2, o_4\}), (<q_3, 0.5, 0.3 >, \{o_3, o_5, o_4, o_2\}), \\ (<q_4, 0.6, 0.3 >, \{o_3, o_2, o_4\}), (< q_5, 0.7, 0.1 >, \{o_1, o_3, o_4, o_5\}) \end{pmatrix}.$$

So, using the methodology given by Deli and Çağman [23], we reduced IFPSS to a reduced fuzzy set.

$$\{< o_1, 0.32 >, < o_2, 0.34 >, < o_3.0.31 >, < o_4, 0.34 >, < o_5, 0.22 >\}.$$

Hence, according to the above reduced fuzzy set, the best alternatives are o_2 and o_4 but we cannot say which one is better between o_2 and o_4. We did not get the desired result.

3 A Novel Construction Method for SFS from IFS

We propose a construction method from IFS to SFS. It is an extension of the novel construction method of IFS from FS proposed by Singh et al. [29]. IFS consists of MF and NMF but in the construction method of SFS, we calculate NeMF without affecting MF and NMF. It would be more convincing to modify the degree of the hesitancy of an element for NeMF. The method is presented in form of prepositions as follows:

Preposition 1 Let the mapping $G : [0, 1] \times [0, 1] \rightarrow [0, 1] \times [0, 1] \times [0, 1]$ is defined by$G(s, t) = (G_x(s, t), G_z(s, t), G_y(s, t))$, where $G_x(s, t) = s$,$G_z(s, t) = h(1 - s^2 - t^2)$ and $G_y(s, t) = t$ and $h : [0, 1] \times [0, 1]$ is any mapping with the property $h(0) = 0,\ h(s) \leq s$.

Preposition 2 Let $\Lambda_{IFS} = \{(\tilde{m}_i, x_{\Lambda_{IFS}}(m_i), y_{\Lambda_{IFS}}(\tilde{m}_i)) : \tilde{m}_i \in \tilde{M}\}$ be an IFS and $h : [0, 1] \times [0, 1]$ be any mapping. Then, $\Lambda_{SFS} = \{(\tilde{m}_i, G_x(x_{\Lambda_{IFS}}(\tilde{m}_i), y_{\Lambda_{IFS}}(\tilde{m}_i)), G_z(x_{\Lambda_{IFS}}(\tilde{m}_i), y_{\Lambda_{IFS}}(\tilde{m}_i)), G_y(x_{\Lambda_{IFS}}(\tilde{m}_i), y_{\Lambda_{IFS}}(\tilde{m}_i)) : \tilde{m}_i \in \tilde{M}\}$ is an SFS, where $G_x(x_{\Lambda_{IFS}}(\tilde{m}_i), y_{\Lambda_{IFS}}(\tilde{m}_i)) = x_{\Lambda_{IFS}}(\tilde{m}_i), G_z(x_{\Lambda_{IFS}}(\tilde{m}_i), y_{\Lambda_{IFS}}(\tilde{m}_i)) = h_{SFS}(1 - x^2_{\Lambda_{IFS}} - y^2_{\Lambda_{IFS}})$,$G_y(x_{\Lambda_{IFS}}(\tilde{m}_i), y_{\Lambda_{IFS}}(\tilde{m}_i)) = y_{\Lambda_{IFS}}(\tilde{m}_i)$.

Hence, h can be used to compute NeMF or the degree of the hesitancy of SFS, having property $h_{SFS}(0) = 0,\ h_{SFS}(s) \leq s$. So, h can be assumed as a NeMF or hesitancy function according to the purpose of defining.

Preposition 2 Let $h : [0, 1] \rightarrow [0, 1]$ be a mapping for the neutral membership function or hesitancy function for SFS be defined as

$$h(s_i) = \left\{ \begin{array}{l} \dfrac{s_i}{(a+1)} \times \bar{s};\ 0 \leq \text{s}_i \leq 0.5 \\ \dfrac{(1-s_i)}{(1+a)} \times \bar{s};\ 0.5 \leq \text{s}_i \leq 1 \end{array} \right\},\ \text{where } a = \max\{s_i\},\ s_i \in \tilde{M}, \text{ and } \bar{s} = \frac{\sum s_i}{\text{no. of elements } \tilde{M}}.$$

3.1 Proposed SFPSS-Based MCDM Method

The following are various steps that are used to construct SFPSS from a given IFS.

Step 1. Take an IFS and construct SFS using prepositions 1,2,3.

Step 2. After the construction of SFS, construct SFPSS over $\tilde{M}$.

Step 3. After the construction of SFPSS, reduce SFPSS to SFS using the given equations:

Let Λ_S be a SFPSS and reduced SFS is defined as follows:

$\Lambda_{rsf} = \{< \tilde{m}_i, x_{\Lambda_{rfs}}(\tilde{m}_i), z_{\Lambda_{rfs}}(\tilde{m}_i), y_{\Lambda_{rfs}}(\tilde{m}_i) >: \tilde{m}_i \in \tilde{M}\}$, where $x_{\Lambda_{rsf}} : \tilde{M} \to [0, 1]$, $z_{\Lambda_{rsf}} : \tilde{M} \to [0, 1]$, and $y_{\Lambda_{rsf}} : \tilde{M} \to [0, 1]$. $x_{\Lambda_{rsf}} = \frac{1}{|\tilde{M}|} \sum_{q \in E, t_i \in \tilde{M}} x_i(q)\chi_f(t_i)$, $z_{\Lambda_{rsf}} = \frac{1}{|\tilde{M}|} \sum_{q \in E, t_i \in \tilde{M}} z_i(q)\chi_f(t_i)$, $y_{\Lambda_{rsf}} = \frac{1}{|\tilde{M}|} \sum_{q \in E, t_i \in \tilde{M}} y_i(q)\chi_f(t_i)$, and $\chi_f(t_i) = \left\{ \begin{matrix} 1; t_i \in g_{\Lambda_{rsf}}(t_i) \\ 0; t_i \notin g_{\Lambda_{rsf}}(t_i) \end{matrix} \right\}$ where $x_{\Lambda_{rsf}}$, $z_{\Lambda_{rsf}}$, and $y_{\Lambda_{rsf}}$ set as operators of Λ_{rsf}..

Step 4. Reduced SFS to IFS.

$\Lambda_{rIF} = \{< \tilde{m}_i, x_{\Lambda_{rIF}}(\tilde{m}_i), y_{\Lambda_{rIF}}(\tilde{m}_i) >: \tilde{m}_i \in \tilde{M}\}$, where $x_{\Lambda_{rIF}} : \tilde{M} \to [0, 1]$ and $y_{\Lambda_{rIF}} : \tilde{M} \to [0, 1]$.

$x_{rIF}(\tilde{m}_i) = x(1 - y)$ and $y_{rIF}(u_i) = z(1 - y)$.

Step 5. Reduced IFS to FS using the following equation.

$$x_{FS} = x(1 - y)$$

4 Application of Proposed SFPSS-Based MCDM Method

Applying the developed technique to solve the limitation of IFPSS given by example 1 to find the desired result. So, we present stepwise computation as follows:

Step 1 The following SFS from IFS is constructed using prepositions 1,2,3:

$$S = \{(o_1, 0.7, 0.17, 0.1), (o_2, 0.4, 0.11, 0.4), (o_3, 0.5, 0.12, 0.3), (o_4, 0.6, 0.15, 0.3) \\ (o_5, 0.7, 0.17, 0.1)\}$$

Step 2 The following SFPSS over $\tilde{M}$ is constructed.

$$\Lambda_{SFPSS} = \{ (< q_1, 0.70, 0.17, 0.1 > \{o_1, o_2\}), (<q_2, 0.4, 011, 0.4 > \{o_1, o_2, o_4\}), \\ (<q_3, 0.5, 0.12, 0.3 > \{o_3, o_5, o_4, o_2\}), (<q_4, 0.6, 0.15, 0.3 > \{o_3, o_2, o_4\}) \\ (<q_5, 0.7, 0.17, 0.1 > \{o_1, o_3, o_5, o_4\}) \}$$

Step 3 SFPSS is reduced to the following SFS.

$$\Lambda_{rSF} = \{ (o_1, 0.36, 0.09, 0.12), (o_2, 0.36, 0.11, 0.22), (o_3, 0.36, 009, 0.14), \\ (o_4, 0.44, 0.11, 0.22), (o_5, 0.24, 0.058, 0.08) \}$$

Step 4 Reduced SFS is converted into IFS.

$$\Lambda_{rIFS} = \{ (o_1, 0.32, 0.08), (o_2, 0.28, 0.09), (o_3, 0.31, 0.08), (o_4, 0.34, 0.09), (o_5, 0.22, 0.05) \}$$

Step 5 Reduced IFS is converted into the following FS.

$$\Lambda_{rFS} = \{(o_1, 0.29),\ (o_2, 0.25),\ (o_3, 0.29),\ (o_4, 0.31),\ (o_5, 0.21)\}.$$

Hence, the maximum membership grade (MG) of o_4 is 0.31. So, the best alternative is o_4. Thus, o_4 is a restaurant that satisfies all parameters of the DM.

If step 4 is completed using fuzzification of IFS Ansari et al. [28] to convert IFS into the fuzzy set, we get the following results.

1. We get the following fuzzy set on assigning non-determinacy to maximum grade.

$$\tilde{A} = \{(o_1, 0.92, 0.08), (o_2, 0.91, 0.09), (o_3, 0.92, 0.08), (o_4, 0.91, 0.09), (o_5, 0.95, 0.05)\}$$

   In Ã the maximum MG of o_5 is 0.95. Thus o_5 is the best alternative.
2. We get the following fuzzy set on assigning equally non-determinacy to both membership grade (MG) and non-membership grade (NMG).

$$\tilde{A} = \{(o_1, 0.62, 0.38), (o_2, 0.60, 0.41), (o_3, 0.62, 0.39), (o_4, 0.63, 0.38), (o_5, 0.59, 0.42)\}$$

   In Ã the maximum MG o_4 is 0.63. Thus o_4 is the best alternative.
3. We get the following fuzzy set on assigning non-determinacy in proportionate allocation to both MG and NMG.

$$\tilde{A} = \{(o_1, 0.80, 0.20), (o_2, 0.76, 0.24), (o_3, 0.48, 0.21), (o_4, 0.79, 0.21), (o_5, 0.81, 0.19)\}$$

   In Ã the maximum MG of o_5 is 0.81. Thus o_5 is the best alternative.

5 Comparison Analysis

We compare our proposed MCDM approach to an existing technique Çağman et al. [22] in this section.

Firstly, we do fuzzification of the IFPS set using the following three different methods of Ansari et al. [28] and obtain three different FPSSs.

1. Allocating a MG to the degree of indeterminacy.
2. By giving both MG and NMG the same amount of non-determinacy.
3. Assigning non-determinacy to both MG and NMG in proportionate allocation.

$$\Lambda^1_{FSS} = \left\{ \begin{array}{l} (q_1, < \frac{0.9}{o_1}, \frac{0.9}{o_2}, \frac{0}{o_3}, \frac{0}{o_4}, \frac{0}{o_5} >), (q_2, < \frac{0.6}{o_1}, \frac{0.6}{o_2}, \frac{0}{o_3}, \frac{0.6}{o_4}, \frac{0}{o_5} >), \\ (q_3, < \frac{0}{o_1}, \frac{0.7}{o_2}, \frac{0.7}{o_3}, \frac{0.7}{o_4}, \frac{0.7}{o_5} >), (q_4, < \frac{0}{o_1}, \frac{0.7}{o_2}, \frac{0.7}{o_3}, \frac{0.7}{o_4}, \frac{0}{o_5} >), \\ (q_5, < \frac{0.9}{o_1}, \frac{0}{o_2}, \frac{0.9}{o_3}, \frac{0.9}{o_4}, \frac{0.9}{o_5} >) \end{array} \right\}$$

$$\Lambda^2_{FSS} = \left\{ \begin{array}{l} (q_1, < \frac{0.8}{o_1}, \frac{0.8}{o_2}, \frac{0}{o_3}, \frac{0}{o_4}, \frac{0}{o_5} >), (q_2, < \frac{0.5}{o_1}, \frac{0.5}{o_2}, \frac{0}{o_3}, \frac{0.5}{o_4}, \frac{0}{o_5} >), \\ (q_3, < \frac{0}{o_1}, \frac{0.6}{o_2}, \frac{0.6}{o_3}, \frac{0.6}{o_4}, \frac{0.6}{o_5} >), (q_4, < \frac{0}{o_1}, \frac{0.65}{o_2}, \frac{0.65}{o_3}, \frac{0.65}{o_4}, \frac{0}{o_5} >), \\ (q_5, < \frac{0.8}{o_1}, \frac{0}{o_2}, \frac{0.8}{o_3}, \frac{0.8}{o_4}, \frac{0.8}{o_5} >) \end{array} \right\}$$

$$\Lambda^3_{FSS} = \left\{ \begin{array}{l} (q_1, < \frac{0.87}{o_1}, \frac{0.87}{o_2}, \frac{0}{o_3}, \frac{0}{o_4}, \frac{0}{o_5} >), (q_2, < \frac{0.5}{o_1}, \frac{0.5}{o_2}, \frac{0}{o_3}, \frac{0.5}{o_4}, \frac{0}{o_5} >), \\ (q_3, < \frac{0}{o_1}, \frac{0.62}{o_2}, \frac{0.62}{o_3}, \frac{0.62}{o_4}, \frac{0.62}{o_5} >), (q_4, < \frac{0}{o_1}, \frac{0.66}{o_2}, \frac{0.66}{o_3}, \frac{0.66}{o_4}, \frac{0}{o_5} >), \\ (q_5, < \frac{0.87}{o_1}, \frac{0}{o_2}, \frac{0.87}{o_3}, \frac{0.87}{o_4}, \frac{0.87}{o_5} >) \end{array} \right\}$$

In all Λ^1_{FSS}, Λ^2_{FSS} and Λ^3_{FSS} MG is associated with the restaurant o_4. Hence, o_4 is the best restaurant that satisfies the parameter set by the DM. After comparing our proposed technique with the already existing technique, we get the same result as in the proposed SFPSS-based MCDM method.

6 Conclusions

In this research, we have extended the construction method of Singh et al. [29] and proposed a new approach to construct SFPSS from IFS. We mention the limitation of IFPSS in the form of a decision-making problem and show how SFPSS can be used to solve them. Comparing the proposed method to the already existing method and obtaining similar results, the validation of the proposed construction approach is confirmed.

References

1. Zadeh LA (1965) Fuzzy sets. Inf Control 8(3):338–353
2. Atanassov KT (1986) Intuitionistic fuzzy sets. Fuzzy Sets Syst 20(1):87–96
3. Yager RR (2013) Pythagorean fuzzy subsets. In: 2013 joint IFSA world congress and NAFIPS annual meeting (IFSA/NAFIPS). IEEE, pp 57–61
4. Yager RR (2013) Pythagorean membership grades in multicriteria decision making. IEEE Trans Fuzzy Syst 22(4):958–965
5. Cuong BC (2013a) Picture fuzzy sets-first results. Part 1, seminar neuro-fuzzy systems with applications. Institute of Mathematics, Hanoi

6. Cuong BC (2013b) Picture fuzzy sets-first results. Part 2 Seminar. Neuro-fuzzy systems with applications. Institute of Mathematics, Hanoi, Preprint 04/2013
7. Ashraf S, Abdullah S, Mahmood T, Ghani F, Mahmood T (2019) Spherical fuzzy sets and their applications in multi-attribute decision making problems. J Intell Fuzzy Syst 36(3):2829–2844
8. Molodtsov D (1999) Soft set theory—first results. Comput Math Appl 37(4–5):19–31
9. Maji PK, Biswas RK, Roy A (2001) Fuzzy soft sets
10. Maji PK, Biswas R, Roy AR (2001) Intuitionistic fuzzy soft sets. J Fuzzy Math 9(3):677–692
11. Çağman N, Enginoğlu S (2010) Soft matrix theory and its decision making. Comput Math Appl 59(10):3308–3314
12. Feng Q, Zhou Y (2014) Soft discernibility matrix and its applications in decision making. Appl Soft Comput 24:749–756
13. Alkhazaleh S, Salleh AR (2011) Soft expert sets. Adv. Decis Sci 2011:757868–757871
14. Alkhazaleh S, Salleh AR (2014) Fuzzy soft expert set and its application. Appl Math
15. Yang Y, Liang C, Ji S, Liu T (2015) Adjustable soft discernibility matrix based on picture fuzzy soft sets and its applications in decision making. J Intell Fuzzy Syst 29(4):1711–1722
16. Peng XD, Yang Y, Song J, Jiang Y (2015) Pythagorean fuzzy soft set and its application. Comput Eng 41(7):224–229
17. Garg H, Arora R (2018) A nonlinear-programming methodology for multi-attribute decision-making problem with interval-valued intuitionistic fuzzy soft sets information. Appl Intell 48(8):2031–2046
18. Garg H, Arora R (2018) Generalized and group-based generalized intuitionistic fuzzy soft sets with applications in decision-making. Appl Intell 48(2):343–356
19. Akram M, Adeel A, Alcantud JCR (2019) Hesitant fuzzy N-soft sets: a new model with applications in decision-making. J Intell Fuzzy Syst 36(6):6113–6127
20. Khalil AM, Li SG, Garg H, Li H, Ma S (2019) New operations on interval-valued picture fuzzy set, interval-valued picture fuzzy soft set and their applications. IEEE Access 7:51236–51253
21. Perveen PAF, Sunil JJ, Babitha KV, Garg H (2019) Spherical fuzzy soft sets and its applications in decision-making problems. J Intell Fuzzy Syst 37(6):8237–8250
22. Cagman N, Enginoglu S, Citak F (2011) Fuzzy soft set theory and its applications. Iranian J Fuzzy Syst 8(3):137–147
23. Deli I, Çağman N (2015) Intuitionistic fuzzy parameterized soft set theory and its decision making. Appl Soft Comput 28:109–113
24. Joshi BP, Kumar A, Singh A, Bhatt PK, Bharti BK (2018) Intuitionistic fuzzy parameterized fuzzy soft set theory and its application. J Intell Fuzzy Syst 35(5):5217–5223
25. Aydın T, Enginoğlu S (2021) Interval-valued intuitionistic fuzzy parameterized interval-valued intuitionistic fuzzy soft sets and their application in decision-making. J Ambient Intell Humaniz Comput 12(1):1541–1558
26. Khan MJ, Kumam P, Deebani W, Kumam W, Shah Z (2021) Bi-parametric distance and similarity measures of picture fuzzy sets and their applications in medical diagnosis. Egypt Inf J 22(2):201–212
27. PA FP, John SJ (2020) On spherical fuzzy soft expert sets. In: Narayanamoorthy, S. (eds.) AIP conference proceedings, Vol 2261, No. 1, p 030001, AIP Publishing LLC
28. Ansari AQ, Philip J, Siddiqui SA, Alvi JA (2010) Fuzzification of intuitionistic fuzzy sets. Int J Comput Cognition 8(3)
29. Singh A, Joshi DK, Kumar S (2019) A novel construction method of intuitionistic fuzzy set from fuzzy set and its application in multi-criteria decision-making problem. In: Mandal J, Bhattacharyya D, Auluck N (eds) Advanced computing and communication technologies. advances in intelligent systems and computing, vol. 702. Springer, Singapore

Chapter 39
Enhanced Error Correction Algorithm for Streaming High Definition Video Over Multihomed Wireless Networks (MWN)

S. Vijayashaarathi and S. Nithya Kalyani

1 Introduction

The ever-increasing demand for users and the development of modern communication technologies have led to the transformation of communication networks from the 1st Generation (1G) network to various 4G networks. In addition, 4G with multinetwork space will provide features such as, "Stay Connected", "Anytime Anywhere" and seamless communication [1]. Due to various features of various networks such as bandwidth, delays, costs, coverage and Service Quality (QoS) there are a few open and unresolved issues including traffic management, network management, security etc. [2, 3]. Wireless networks with IP are all very challenging issues for 4G networks. IPv6 for mobile (MIPv6) developed by the Internet Engineering Task Force (IETF) has packet management for switching devices for the same wireless networks [4]. In addition, the network management of the same networks depends on the network-related parameter i.e., Signal strength Received (RSS) [5, 6].

The movement management of various networks, however, depends not only on network-related parameters, but also on terminal-velocity, battery power, location information, user profile and preferences and service QoS service capabilities etc. [7, 8]. Designing the full-fledged motion management—IP while processing issues such as network contexts, terminals, user and services is a major concern for the industry and researchers in the current era [6, 9].

S. Vijayashaarathi (✉)
Sona College of Technology, Salem, India
e-mail: vijayashaarathi.s@sonatech.ac.in

S. N. Kalyani
K.S.R College of Engineering, Tiruchengode, India

G. Mathur et al. (eds.), *Proceedings of 3rd International Conference on Artificial Intelligence: Advances and Applications*, Algorithms for Intelligent Systems,
https://doi.org/10.1007/978-981-19-7041-2_39

2 Methodology

The work involves a Forward Error Correction technique for fixing errors and avoiding frame loss, and it has been shown to be the most effective method for avoiding frame loss. In just about all cases, a high frame error rate decreases the quality of video streaming. As a result, FEC is utilized to rectify those mistakes. FEC's key advantage is that it doesn't require interaction with the video encoder; therefore it may be used with any video coding scheme and on both stored and live video. This is FEC mechanism's basic operation. Every K video packet is safeguarded by (N–K) FEC packets in the FEC approach. The first K packets are the original data packets, whereas the following (N-K) data packets are known as parity data (refer Fig. 1).

Furthermore, to provide seamless video streaming services over heterogeneous wireless networks, a handover-aware video streaming method was developed. According to the large bandwidth variance, this changes the sending rate and quality level of the transmitted video streams in the system. This approach uses an explicit notification message that tells the streaming server of a client's handover occurrence to analyze the reaction of bandwidth variance owing to a handover. A simulation environment for a vertical handover between wireless local area networks and cellular networks was designed to test the performance.

A sub-frame-level (SFL) scheduling approach divides huge video frames into smaller sub-frames and sends each one to its own wireless network. In terms of PSNR, the SFL scheduling indicates an improvement in frame-level delays and video quality. SFL improves the average video quality in terms of PSNR, reduces the average end-to-end video frame latency, and ensures that more than 98% of video frames arrive at the client within the 200 ms deadline even though bandwidth is severely limited [10].

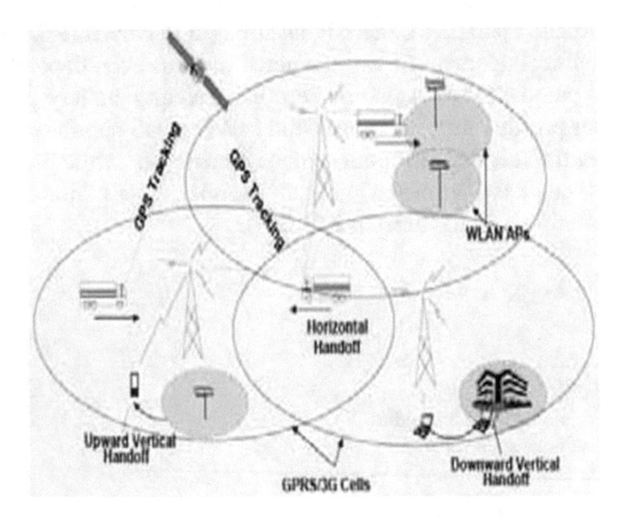

Fig. 1 Handoff

The video stream allocation based on frame splitting is handled on the server side. The video streaming rate, video frame size, available bandwidth, and fixed delay are all used to determine the stream allocation ratio (including the propagation delays of both wired and wireless networks).

The available bandwidth and path propagation delays of each wireless network are the feedback information from the client side. A header was inserted to each sub-frame, containing the sequence number, the actual video frame size, an absolute byte position within the frame, and the sub-frame size. The sequence number represents the frame order that is used for inter-frame homologous recombination, whereas the other three numbers in the header are used for intra-frame reassembling. Finally, the video packet transmitter is in charge of sending video packets to various wireless networks. The data transmission employed in this study is UDP.

In real-time video applications, each video frame has a decoding deadline connected with it. As a result, the first action when a video is brought to the table at the destination node is to see if it has passed the decoding deadline. If the packet is already beyond its due date, it will be discarded. The receiver buffer will be emptied of all received sub-frames with the same sequence number. In order to recover the original video streaming, intra-frame reassemble and inter-frame re-sequence are performed. Following that, sub-frames received from various wireless network interfaces will be reconstructed depending on the packet header's absolute byte offset. The rebuilt frames may arrive at the client out of order due to network heterogeneity using the following steps:

Step 1: Adding redundancy bits on compressed source bits to enable error detection and correction.

Step 2: Add a parity check bit at the end of a block of Video stream which can detect all single bit errors.

Step 3: For every k source bits add l channel bits to create $n = k + l$ bits, channel coding rate $r = k/n$.

3 Simulation Results

An analysis of three common video delivery methods at the frame level over various wireless networks, namely, EDPF, LBA and PMT was performed. The first EDPF algorithm was used on the side of the video server by considering the available bandwidth, link delay and frame size of the video. Then, arrival time is estimated based on the above metrics.

End-to-end video delays at the end of the EDPF are set under various bandwidth limits [11]. In general, low bandwidth leads to longer transmission, which in turn results in greater end-to-end delays (refer Fig. 2). The time split and reassembly took time to calculate the independent delay from end to end. Therefore, the split and

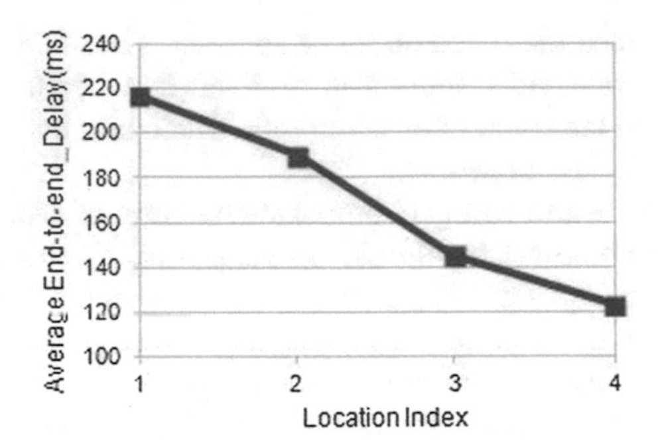

Fig. 2 Location index versus average end to end delay

reassembly time is insignificant compared with the end-to-end frames of each video [12, 13].

Upload the rating algorithm (LBA).In the LBA algorithm; video frames can be rated to reach customers at a time when they will be discarded on the server side.

PSNR value is better for LBA compared to EDPF and PMT when in Position 3 (refer Fig. 3). In PMT, the opportunity-generating function (PGF) and z-transform method are used to detect PGF video package delays and any careless time. The supply options for each wireless network are set based on the PGF.

In places 1 and 2, the ratios of framed frame-home client are indicated. The SFL simulation results show significant improvements in frames that are very low. In Areas 3 and 4, the delivery of all video frames has exceeded the deadline within the maximum delay limit which is why all independents met the coding deadline [14]. The rating of video frames that has passed the coding deadline indicates the potential for disruption seen by end users (refer Fig. 4).

Finally the growing expansion function (CDF) of each individual PSNR was created. PSNR values 78% of video frames in SFL are 31 dB. This leads to high median PSNR values (refer Fig. 5).

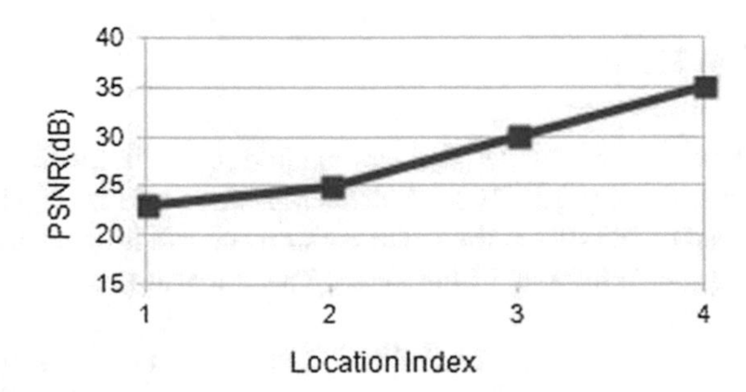

Fig. 3 Location index versus PSNR

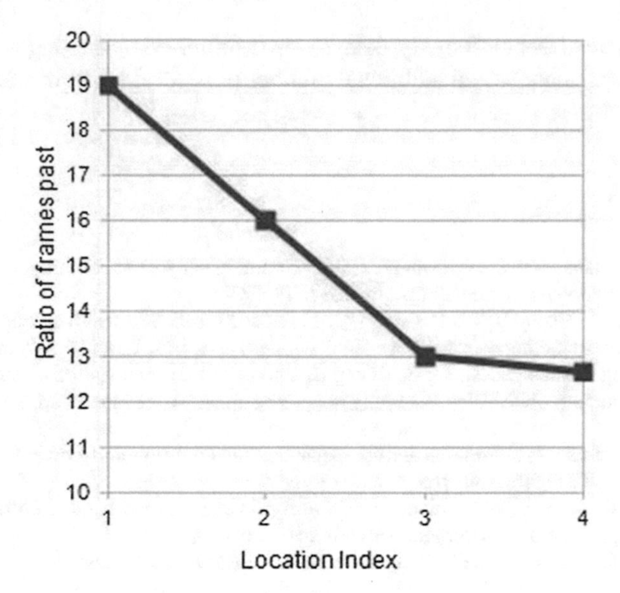

Fig. 4 Location index versus ratio of frames past

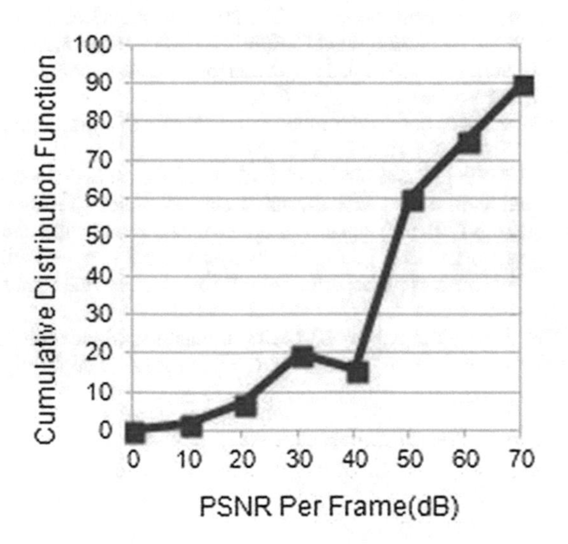

Fig. 5 PSNR versus CDF

4 Conclusion

The approach is different where large-format video frames are subdivided into sub Frames (SFL) standard frames and organize each of them into separate wireless networks on multi delays from end-to-end and improved video quality. In line with the SFL Advanced Debt Recovery, delays were minimized by dealing with a handoff algorithm. As a future job an additional number of networks will be considered.

References

1. Oliveira T, Mahadevan S, Agrawal DP (2011) Handling network uncertainty in heterogeneous wireless networks. In: Proceedings of IEEE INFOCOM
2. Wu J, Shang Y, Cheng B, Wu B, Chen J (2014) Loss tolerant bandwidth aggregation for multi homed video streaming over heterogeneous wireless networks. Wirel Pers Commun
3. Wu J, Shang Y, Huang J, Zhang X, Cheng B, Chen J (2013) Joint source-channel coding and optimization for mobile video streaming in heterogeneous wireless networks. Wirel Commun Netw
4. Sabeenian RS (2014) Estimation and compensation of video motion—a review. J Convergence Inf Technol (JCIT) 9(6):164–169
5. Chebrolu K, Rao R (2007) University of California at SanDiego bandwidth aggregation for real-time applications in heterogeneous wireless networks
6. Yooon J, Zhang H, Banerjee S, Rangarajan S (2012) MuVi: a multicast video delivery scheme for 4G cellular networks
7. Kamiyama N, Kawahara R, Mori T, Harada S, Hasegawa H (2011) Parallel video streaming optimizing network throughput. Comput Commun 34(10):1182–1194
8. Ci S, Sharif H (2005) Improving goodput in IEEE802.11 wireless LANS by using variable size and variable rate(VSVR) schemes. Wirel Commun Mobile Comput 5(3):329–342
9. Choudhury S, Gibson JD (2007) Payload length and for multimedia communications in wireless LANs. IEEE J Sel Areas Commun 25(4):796–807
10. Geun KJ, Krunz MM (2000) Bandwidth allocation in wireless networks with guaranteed packet-loss performance. IEEE/ACM Trans Netw 8:337–349
11. Zhihai H, Hongkai X (2006) Transmission distortion analysis for real-time video encoding and streaming over wireless networks. IEEE Trans Circuits Syst Video Technol 16:1051–1062
12. Maani E, Katsaggelos A (2010) Unequal error protection for robust streaming of scalable video over packet lossy networks. IEEE Trans Circuits Syst Video Technol 20:407–416
13. Bajic IV (2007) Efficient cross-layer error control for wireless video multicast. IEEE Trans Broadcast 53:276–285
14. Deng DJ, Ke CH, Huang YM, Chen HH (2008) Contention window optimization for IEEE 802.11 DCF access control. IEEE Trans Wirel Commun 7(12):5129–5135

Chapter 40
Tumordc.AI: A Comprehensive Deep Learning-Based Brain Tumor Detection and Classification System

Saurav Telge, Ryan Rodricks, Mrunmayee Waingankar, Adarsh Singh, and Ranjan Bala Jain

1 Introduction

Brain tumors are one of the significant causes of death in children and adults below 40. Over 12,000 people are diagnosed with a primary brain tumor each year, including 500 children and young people, which is around 33 people every day [1]. Brain tumors can begin in one's brain (primary brain tumors), or cancer can begin in other parts of one's body and spread to the brain as secondary (metastatic) brain tumors. Many different types of brain tumors exist. Some of them are noncancerous (benign), and some are cancerous (malignant).

Depending on how abnormal the cancer cells look under a microscope and how quickly the tumor is likely to grow and spread, brain tumors can be classified into 4 grades according to the World Health Organization (WHO) tumor grading system-grade I (lower grade), grade II, grade III, grade IV (higher grade) [2]. Grade I and II tumors rarely spread to nearby tissues. They can be cured but have a possibility of recurring. Grade III and IV tumors can quickly spread to the neighboring tissues and usually cannot be completely removed by surgery.

Doctors can detect lower grade tumors easily but identifying grade III and IV tumors and classifying them require a time-consuming proper medical procedure to confirm their type and grade. Thus, it is very important to develop and use a computer-aided system that will not only precisely detect a tumor but also will predict its type or grade accurately.

S. Telge (✉) · R. Rodricks · M. Waingankar · A. Singh · R. B. Jain
Vivekanand Education Society's Institute of Technology, Chembur, Mumbai, India
e-mail: 2018.saurav.telge@ves.ac.in

G. Mathur et al. (eds.), *Proceedings of 3rd International Conference on Artificial Intelligence: Advances and Applications*, Algorithms for Intelligent Systems,
https://doi.org/10.1007/978-981-19-7041-2_40

In this work, a cross-platform website is developed capable of detecting a tumor through MRI images of the brain. Later the type and grade of the tumor, if present, are also predicted. This system will help the doctors diagnose the tumor and start the treatment at an early stage, which will consequently help decrease the threat of the tumor.

2 Related Work

As detecting the tumor manually and classifying it might result in inaccurate predictions, a computer-aided system is necessary for accurate detection. Plenty of research has been done on brain tumor detection and classification, and here are some citations relevant to the research proposed in this paper.

A method was proposed by Mohsen et al. [3] using the DNN learning architecture for classification of brain tumors using brain MRIs. The dataset for model training, which comprised of 66 actual MRI scans of the human brain with 22 normal and 44 aberrant images. The Fuzzy C-Means clustering algorithm was employed during the pre-processing step, and after feature extraction using the Discrete Wavelet Transform (DWT) and reduction using the Principal Component Analysis technique, the classification using DNN was done. This DNN architecture gave the highest accuracy of about 96.97%, with 0.97 precision and 0.97 recall.

In order to overcome the lack of training data and labeling noise, Lu et al. [4] introduced a tumor classification method based on Magnetic Resonance Spectroscopy (MRS) data that blends deep neural networks with a novel data distillation and augmentation. The network takes into account both conspicuous aspects in the data that are frequently employed in clinical settings and features that are uncommon but have a significant impact on the model's performance. The accuracy of the deep learning network is 76.15%, with sensitivity and specificity of 73.15% and 76.30%.

A method was proposed by Hossain et al. [5] using Fuzzy C-Means clustering algorithm to detect brain tumors from 2D MRI images followed by traditional classifiers Convolutional Neural Networks (CNN). The dataset used was the BRATS dataset for binary classification, containing over 217 MRI images (187 for tumor and 30 for non-tumor). Among the traditional classification methods, Support Vector Machine gave the highest accuracy, 92.42%. CNN based methods provided an accuracy of 97.87% when the dataset was split in the ratio of 80:20 training and testing sets, respectively.

Sultan et al. [6] have developed two models for classification. The first model classifies the tumor into Meningioma, Glioma, and Pituitary tumor. The dataset used includes 3064 MRI images acquired from 233 patients from Nanfang Hospital and General Hospital, Tianjin Medical University, China, from 2005 to 2010, available online on various sites like Kaggle. The accuracy of the model was roughly 96.13%. The second approach assigned grades II, III, and IV to brain tumors. The cancer imaging archive (TCIA) public access repository provided 516 MRI pictures from 73 patients, which were utilized in the study. They used a proprietary deep neural network topology to create the model. Starting with the input layer, which retains the preprocessed pictures, and proceeding through the convolution layers and their activation functions, the proposed network has 16 layers. They also employed data augmentation, which helped them obtain higher results, and they ended up with an overall accuracy of 98.7%.

Anaraki et al. [7] proposed a method for noninvasively classifying different stages of Glioma using magnetic resonance imaging that employs evolutionary algorithms on top of CNN (MRI). In addition, the prediction error variance is reduced using an ensemble technique. For categorizing three Glioma grades, the findings obtained are 90.9% accurate. Glioma, Meningioma, and Pituitary tumor types were also correctly classified 94.2% of the time. Automatic feature extraction is a benefit of this strategy over shallow machine learning methods. The dataset consists of roughly 600 MR images of healthy people (without any lesions).

Irmakl et al. [8] have used pre-trained models to train their models. But to increase the accuracy and other metrics, they trained another CNN model by considering Hyper-Parameter Optimization. After training this model, they got an overall accuracy of 92.66%, 97.85% accuracy for classifying Glioma, 97.60% accuracy for Meningioma, 97.47% for metastatic, 95.44% for a healthy brain, and 96.96% for the Pituitary type of tumor which is grade classification. The main advantage of this method is that considering different parameters and using them in the training model can improve the model's overall accuracy.

It is observed that different techniques are available in the literature for the detection of brain tumors, but many of them lack pre-processing or are built on a limited dataset. Most of the models proposed either detect the brain tumor or show its type. There is no model proposed for combined work. Hence, this research focuses on a cross-platform website capable of detecting the tumor, its type, and its grade. The watershed algorithm is used for pre-processing. The Watershed algorithm can precisely position image edges and single-pixel width, closing and accurate regional boundaries with fast calculations. Visual Geometry Group-16 (VGG-16), a CNN architecture, is used as a classification model to detect the tumor. To train other models, customized CNN sequential models are used. These models are very effective in reducing the number of parameters without affecting the quality of the images, which helps in faster computation.

The rest of the paper is organized as follows. Section 3 describes the proposed solution, the methodology applied, the dataset used, the architecture of the proposed model, and other classification details. Section 4 discusses the results obtained and the inferences drawn. Section 5 concludes the paper followed by references at the end.

3 Proposed Work

In this section, the proposed system, pre-processing steps involved, CNN architecture, and the classification process is shown and described.

Figure 1 describes the flow of how the system processes the images and gives the required output. The proposed system is a web-based application called *'Tumordc.AI'* that is developed using the Django framework. It accepts a person's MRI image of the brain as an input and displays if the tumor exists or not, and if it exists, what its type and grade are. The input MRI image is preprocessed and segmented using the watershed algorithm. Then comes the first classification, which is used to check if the tumor exists or not. The model is trained using VGG-16 architecture. If the tumor exists, it is rechecked for the type (Meningioma, Glioma, or Pituitary) and grade (II, III, or IV) of the tumor simultaneously. For this classification, two similar customized CNN sequential models are used, consisting of Conv2D, averagePooling2D, Dropout, flatten, and Dense layers. The final result of the MRI is displayed accordingly. Measures such as accuracy and precision are calculated, and performance is evaluated.

3.1 Dataset

Having a large dataset is a critical aspect of the experiments for having firm results. Datasets available online were not large enough for the model to give good accuracy; hence a combined dataset was created with proper labeling. The dataset was acquired from BRATS 2020 [9], Kaggle [10], Radiopaedia [11, 12], and The Cancer Imaging Archive (TCIA) [13, 14] websites.

Further, the images had to be converted from DICOM format to JPEG format to train the neural network models. For this purpose, a python code was utilized to convert the images. The dataset used in this work consists of 1,193 images, that is, 400 images each of grades II, III, and IV. The dataset containing YES and NO labels for the tumor consist of 3060 images, with 1530 images of NO label and 1530 images of YES label. The dataset includes 2475 images of Meningioma, Pituitary, and Glioma tumors.

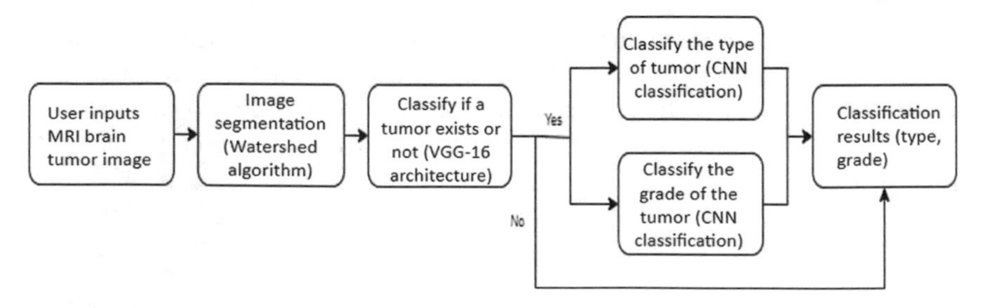

Fig. 1 Proposed System

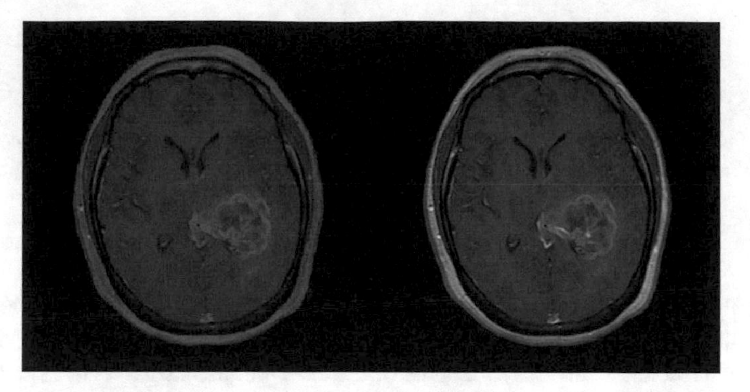

Fig. 2 Comparison of watershed-generated image (left) and normal image (right)

3.2 Pre-processing

To enhance the neural network's feature extraction, the Watershed algorithm is used. It improves the image by segmenting and highlighting the required part from the background. The watershed algorithm treats pixel values as a local topography (elevation) starting from user-defined markers. The algorithm floods basins from the markers until watershed lines connect basins allocated to distinct markers. Markers are frequently chosen as the image's local minima, from which basins are flooded [15].

Figure 2 compares a watershed-generated image of a grade III brain tumor and a regular image. As the watershed algorithm peaks, the pixels present on the boundary of a transition, and the tumor region is emphasized compared to the other areas, which helps in better feature extraction for the CNN model described ahead.

3.3 CNN Architecture

The proposed system has used CNN algorithms in models. CNNs comprises of fully connected feed-forward neural networks. Features can be extracted from images using CNN. It makes the architecture well suited to processing 2D data, such as images as it convolves learned features with input data and then uses the 2D convolutional layers. Using this, we don't need to select the required features from the images to classify them, eliminating the need to manually extract the features.

In this system, three models are used. First, the classification model is built using transfer learning (VGG-16 architecture). If it determines the presence of a tumor, then next, a customized CNN sequential model predicts the grade of the tumor ranging from II to IV. Then the final model predicts the types, which include Meningioma, Pituitary, and Glioma.

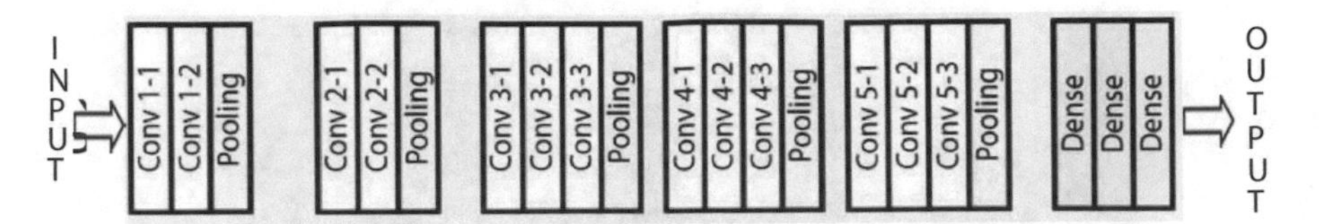

Fig. 3 VGG-16 architecture

We have used a VGG-16 model that has 16 different layers within it. While training a pre-trained version of the network trained on more than a million images from the ImageNet database is used. The details of these layers are as given in Fig. 3.

Convolutional layers: The first two layers are convolutional layers with 3*3 filters, with a stride of 1. And in total employ 64 filters, resulting in a volume of 224*224*64.

Pooling layer: Following convolutional layer, a pooling layer with a max-pool of 2*2 size and stride 2 is employed to lower the volume from 224*224*64 to 112*112*64.

Next Convolutional layers: After Pooling layers, there are two further convolution layers with 128 filters. As a result of this new dimension obtained is 112*112*128. With this, the volume decreases further to 56*56*128. Two further 256-filter convolution layers are added, followed by a downsampling layer that shrinks the size to 28*28*256.

Max-pool layer: It separates two additional stacks with three convolution layers, each with volume 7*7*512 is flattened into a Fully Connected (FC) layer. After the final pooling layer, a total of 4096 channels and 1000 softmax output classes are obtained.

Figure 4 describes the CNN model used. The CNN model has two Conv2d layers. The first one takes in filters as parameters with a minimum value of 32 and a maximum value of 128, with a step size of 32. Filters are a collection of kernels. Kernels are used to extract features from the image. The kernel advances horizontally from the image's top left corner, then shifts down and moves horizontally again. The output matrix is the dot product of the image pixel value and the kernel pixel value. Initially, the kernel value initializes randomly, and it's a learning parameter. The value for the kernel size ranges from 3 to 5 with a RELU activation function. The activation function is responsible for transforming the total weighted input from nodes into the activation of the node or output for that input. The second Conv2d layer uses the same filter and step size but uses a kernel size ranging from 2 to 4 [16].

The system compresses into sizes of 150 $\times$ 150 pixels and passes these images into two layers. After this, an average pooling layer with a pool size of 2 $\times$ 2 is used. The pooling layer's function is to reduce the spatial size of the representation to reduce the network complexity and computational cost. In average pooling, an average value of all the pixels in the batch is selected, and a stride of 1 is used, which helps the kernel to move within the image. Also, the data was flattened to feed it into the dense layer using the RELU activation function. In this model, the RELU

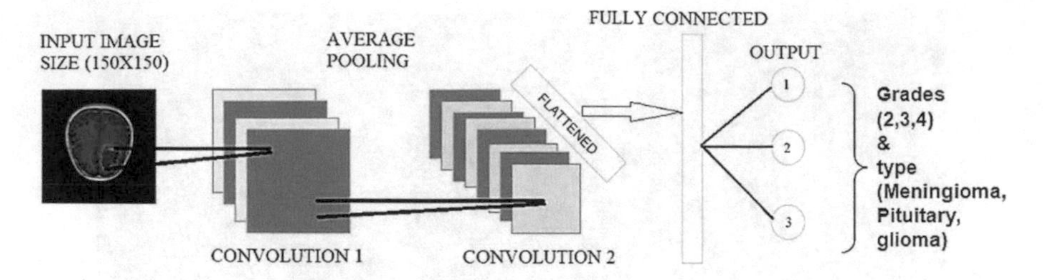

Fig. 4 The architecture of the CNN sequential model

activation function is used to solve the issue of the vanishing gradient problem. The backpropagation algorithm causes the vanishing gradient problem in recurrent neural networks. In this problem, the weights of the shallower neurons are impacted by gradients calculated at a deep stage of the network. Thus, stopping the neural network from further learning. The data into a 1-dimensional array for inputting to the next layer is converted by flattening. Specifically, it creates a single long feature vector and connects to the final classification model. Thus, all the pixel data are aligned in one line and make connections with the Fully Connected (FC) layer. The final layer is the SoftMax layer, which is used to convert the vector of values into probabilities which help us classify multiple grades and types of tumors.

4 Results and Discussion

In this section, the results of the models based on the proposed system are illustrated. All the analysis and computation were performed on Google Colab and Django was used to integrate the model with the website. The datasets have been split into 80% for training and 20% for testing. The 80% part is further divided into 80% for training and 20% for validation of the results.

Figure 5 shows the result obtained for classifying yes or no after training for 30 epochs with a training accuracy of 98.26%, validation accuracy of 98.25%, and the testing accuracy is 97.91%. On further training of the model over 30 epochs, it started overfitting the data and the graph reached a saturation point. Hence, the optimal results were obtained after 30 epochs.

Figure 6 illustrates classifying the type of tumor which includes Meningioma, Glioma, and Pituitary tumor, after training for 12 epochs has a training accuracy of 99.90%, and validation accuracy of 97.27%, and the testing accuracy is 94.70%. Here as well, the model started overfitting the data; hence the model wasn't trained further.

Figure 7 highlights the result obtained for classifying the grade of tumor, which includes grades II, III, and IV, after training for four epochs has a training accuracy of 99%, validation accuracy of 90%, and the testing accuracy is 80%.

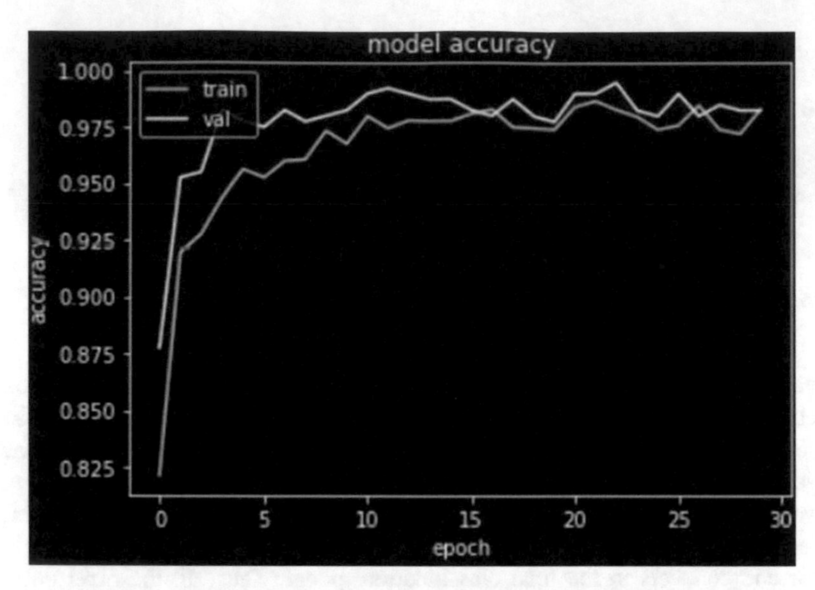

Fig. 5 Training and validation accuracy for the VGG-16 model

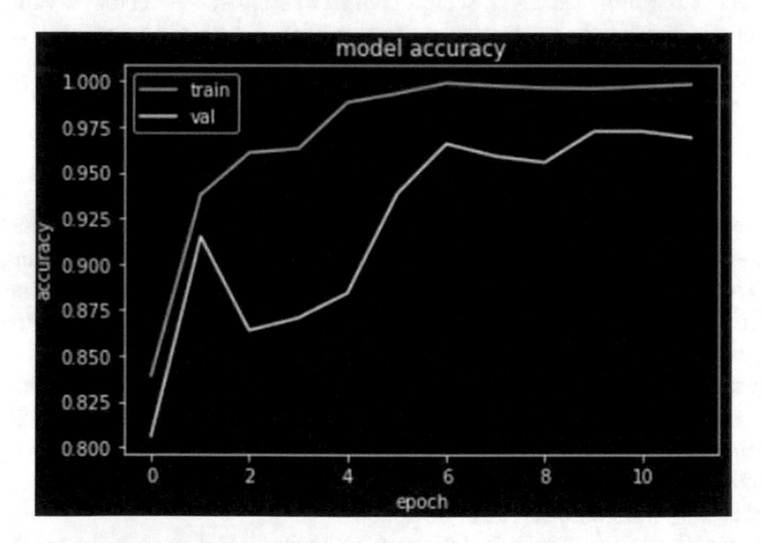

Fig. 6 Training and validation accuracy for type classification

Figure 8 shows the website describing the steps to be followed by the user to get the results.

Figure 9 shows the website application created for testing brain tumor images. On uploading an image, the system classifies if the patient has a tumor or not. If yes, then its grade and type are also displayed.

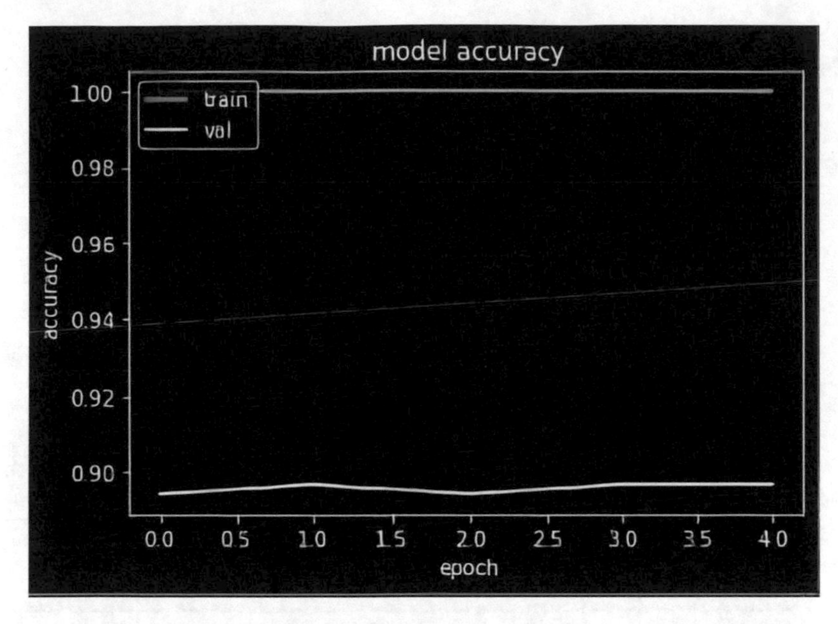

Fig. 7 Training and validation accuracy for grade classification

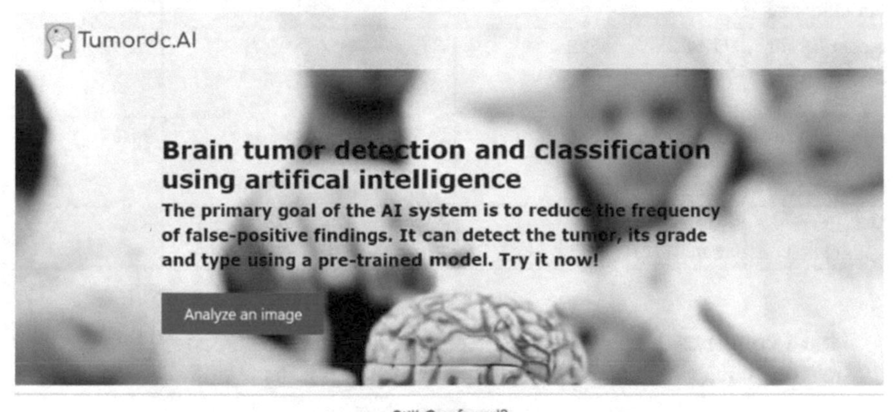

Fig. 8 Dashboard of the website

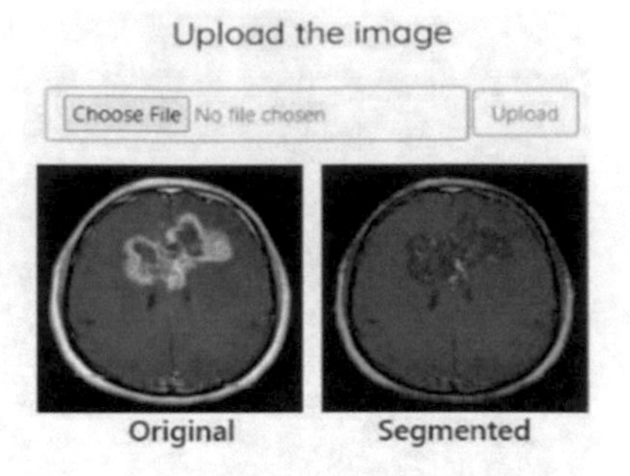

Fig. 9 *TumorDC.AI* Website for testing images

Table 1 Comparison of the work with other works

Evaluation measure	Proposed Solution *Tumordc.AI*	Mohsen et al. [3]	Hossain et al. [5]	Sultan et al. [6]	Anaraki et al. [7]
Accuracy for yes/no classification	*98.25% (VGG-16)*	96.97%	97.87%	NA	NA
Accuracy for grade classification	*90% (CNN Sequential)*	NA	NA	98.7%	90.4%
Accuracy for type classification	*97.27% (CNN Sequential)*	NA	NA	96.13%	94.2%
Website application	*Yes (Django framework)*	No	No	No	No

Table 1 compares the proposed system with other methods [3, 5–7] based on the different accuracies obtained, and the availability of a website portal.

5 Conclusion

It is vital to have a comprehensive system to detect brain tumors and alert the user of its possible outcomes. In recent years, deep learning models have provided promising results in the medical image analysis field. A simple, cost-effective, and robust system is proposed that includes pre-processing phases such as image processing and segmentation (Watershed algorithm) and classification based on deep learning algorithms. Due to this, feature extraction is bolstered, resulting in improved accuracy. Also, the proposed model's testing and training include large datasets such as Brats

2020, Radiopaedia, and Cancer Imaging Archive websites which is an added advantage over other solutions. This solution is first-of-a-kind since it encompasses and detects all types of tumors possible in one system. Finally, a web-based application *Tumordc.AI* is developed for detecting the tumor and its grade.

The proposed system provides better accuracy than other existing brain tumor identification and classification systems. This allows the system to be used in computing techniques for the early detection of brain tumors. It will serve as a helping hand to doctors.

References

1. The brain tumor charity site. https://www.thebraintumourcharity.org/get-involved/donate/why-choose-us/the-statistics-about-brain-tumors/. Accessed 12 Nov 2021
2. NCI. https://www.cancer.gov/types/brain/patient/adult-brain-treatment-pdq. Accessed 11 Nov 2021
3. Mohsen H, El-Dahshan E-S, El-Horbaty E-S, Salem A-B (2018) Classification using deep learning neural networks for brain tumors. Future Comput Inf J 3(1):68–71
4. Lu D, Polomac N, Gacheva I, Hattingen E, Triesch J (2021) Human-expert-level brain tumor detection using deep learning with data distillation and augmentation. In: ICASSP 2021–2021 IEEE international conference on acoustics, speech and signal processing (ICASSP). IEEE, pp 3975–3979
5. Hossain T, Shishir FS, Ashraf M, Al Nasim MA, Muhammad Shah F (2019) Brain tumor detection using convolutional neural network. In: 2019 1st international conference on advances in science, engineering and robotics technology (ICASERT), pp 1–6. https://doi.org/10.1109/ICASERT.2019.8934561
6. Sultan HH, Salem NM, Al-Atabany W (2019) Multi-classification of brain tumor images using deep neural network. IEEE Access 7:69215–69225. https://doi.org/10.1109/ACCESS.2019.2919122
7. Anaraki KA, Ayati M, Kazemi F (2019) Magnetic resonance imaging- based brain tumor grades classification and grading via convolutional neural networks and genetic algorithms. Biocybern Biomed Eng 39.1: 63–74
8. Irmak E (2021) Multi-classification of brain tumor mri images using deep convolutional neural network with fully optimized framework. Iran J Sci Technol Trans Electric Eng 1–22
9. BRATS dataset. https://www.med.upenn.edu/cbica/brats2020/data.html. Accessed 15 Nov 2021
10. Kaggle dataset. www.kaggle.com/abhranta/brain-tumor-detection-vgg16/data. Accessed 15 Nov 2021
11. Radiopedia. https://radiopaedia.org/cases/diffuse-astrocytoma-nos-protoplasmic-1. Accessed 16 Nov 2021
12. Radiopedia. https://radiopaedia.org/articles/anaplastic-astrocytoma. Accessed 15 Nov 2021
13. CancerImaging. https://wiki.cancerimagingarchive.net/display/Public/QIN-BRAIN-DSC-MRI. Accessed 16 Nov 2021
14. CancerImaging. https://wiki.cancerimagingarchive.net/pages/viewpage.action?pageId=50135264. Accessed 17 Nov 2021
15. Scikit. https://scikit-image.org/docs/dev/auto_examples/segmentation/plot_watershed.html. Accessed 16 Nov 2021
16. Zhuang JX, Tao W, Xing J, Shi W, Wang R, Zheng WS (2021) Understanding of Kernels in CNN models by suppressing irrelevant visual features in images. arXiv:2108.11054

Chapter 41
Simulation of Density Based Traffic Control System Using Proteus 7.1 Professional

Kaushik Ghosh, Utkarsh Pandey, Anshumaan Pathak, and Surajit Mondal

1 Introduction

Road network is one of the over-stretched infrastructures in certain parts of the world and this has led to an increase in road traffic in those areas. Regardless of the way that traffic lights have reliably been used for controlling the congestion of traffic, managing the traffic in numerous metropolitan areas around the world has continued being a concerning matter [1]. Accordingly, to get rid of these issues or possibly diminish them to a significant level, newer plans need to be executed by acquiring sensor based automation procedures in the area of traffic signalling systems [2]. A traffic light controls the traffic stream at road crossing points. It comprises of three essential lights which incorporate red, yellow and green. Red signal is utilized to stop traffic from continuing; yellow signal cautions vehicles for brief stop and green signal alarms vehicles for procession in the demonstrated directions [3]. The formatter will need to create these components, incorporating the applicable criteria that follow. Traffic congestion is a circumstance that occurs when vehicles travel slower than expected due to an increase in their numbers than the normal capacity of that particular road, at any given point of time. Otherwise known as traffic jams, traffic congestions may result from roads being impeded, terrible roads, legitimate traffic light system to framework vehicles, unseemly driving by road users etc. [4]. Metropolitan communities began to make traffic precepts to limit crashes, while

K. Ghosh
Department of Computer Science, University of Petroleum and Energy Studies, Dehradun, India
e-mail: kghosh@ddn.upes.ac.in

U. Pandey · A. Pathak · S. Mondal (✉)
Department of Electrical and Electronics Engineering, University of Petroleum and Energy Studies, Dehradun, India
e-mail: smondal@ddn.upes.ac.in; surajitmondalee@gmail.com

G. Mathur et al. (eds.), *Proceedings of 3rd International Conference on Artificial Intelligence: Advances and Applications*, Algorithms for Intelligent Systems,
https://doi.org/10.1007/978-981-19-7041-2_41

Fig. 1 Arduino connected with IR sensors

traffic flags were used to organize option to continue at major metropolitan convergences. Traffic control development in metropolitan streets has supported simple movement and utilization of automobiles in megacities [5]. Most of the significant roads have viable traffic control framework, which has encouraged easy progression of vehicles in metropolitan regions. Railways were invented filling as a temporary solution to the creating issue of street traffic control. However, this makes congestion at terminals inside metropolitan communities [6]. The infrared sensors (IR) are the main segments of this task. The sensors demonstrate like a switch as it controls the exchanging of the LEDs. The IR sensors have been applied to a few traffic frameworks. The IR system is planned in such a way that its transmitter and receiver are mounted on one or the other of the road with the end goal activated whenever a vehicle passes between the two sensors. The sensors empower this framework to be automated except it is a normal traffic control system which has been delivered insufficiently in densely populated territories. They decide whether there is heavy traffic on one path and allow the permit of traffic in inclination to other less dense lanes [7–10]. Figure 1 below shows the connection of IR sensor with Arduino [10].

Proteus is a complete hardware design simulation platform and embedded system software, Proteus ISIS is an intelligent schematic input system, framework plan and simulation of the fundamental platform to accomplish the mix of single-chip microcomputer simulation and pspice circuit re-enactment. It has the elements of analog circuit simulation, computerized circuit simulation, system simulation made out of single-chip microcomputer and its fringe circuit, I2C debugger, SPI debugger, RS232 dynamic simulation, keyboard and LCD framework simulation, and various virtual instruments [11]. Software product structure of proteus is shown in Fig. 2 [11].

In this paper, we have proposed a density based traffic control system using Proteus 7. The rest of the paper has been structured as follows: in Sect. 2 we have discussed the existing literature and in Sect. 3 we have described our proposed work along with the operational model. Section 4 has the necessary results and finally, we have concluded in Sect. 5.

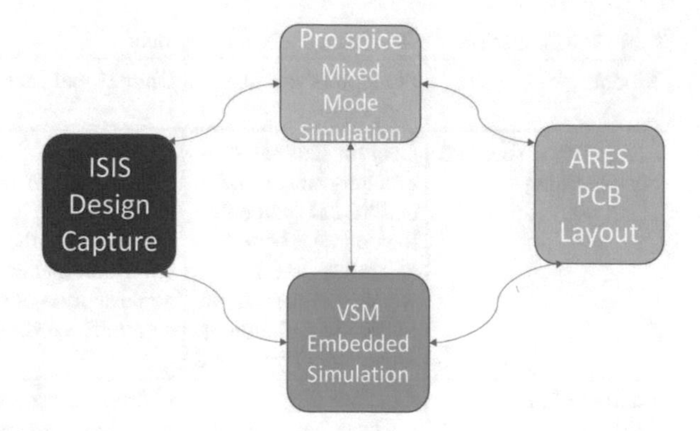

Fig. 2 Software product structure of proteus

2 Literature Survey

Previously used techniques for traffic control have their demerits such as induction loop in calculating number of vehicles on the road. To count number of vehicles on the road, different techniques have been proposed [12, 13]. However, if spacing between vehicles is small then the calculation of number of vehicles may be false too [14]. Here, proposed a traffic control system based on image processing that operates by measuring the area covered by cars on the road. Image capture, grayscale transformation, and image enhancement are the three basic processes used by the system. The camera must identify green lines placed on the road at random intervals. There is a level of traffic on the road if sections of the lines are not recognized, and the traffic lights are modified to fit the circumstance. MATLAB software was used to build the system [15]. Cameras for image detection, a storage device for the pictures discovered, an object identification algorithm, background object subtraction and shadow removal, blob segmentation, and object classification are all part of the system named camera-based autonomous road surveillance system (ARSS). The system was created with the goal of being able to work with current video surveillance systems [16]. Observed pictures are analyzed for each lane, and the blob ratio is utilized to establish the type of vehicle detected. The traffic signals are modified to relieve congestion based on the number of cars detected in all directions. Camera-based traffic control systems need more storage space and processing power to store and process pictures [17].

3 Comparison with Latest Work

Proposed work is compared with other models such as image processing and deep learning [18] and internet of things [19] in Table 1 below.

Table 1 Comparison of IR sensors with other models

Model	Solution of traffic congestion	Operational ease	Robustness
Image Processing and deep learning	Over the darknet problem, tensor flow and YOLO (you only look once) are used and to validate the working of the system SUMO (simulation of urban mobility) is used	On the basis of two scenarios; a road system that is less complex and the condition that exists in a real world network simulation proposed	Performance increased by 67.68% in comparison with real world situation
Internet of things (IOT)	At a four-direction roadway junction, an optimized regression approach is used to collect multi-path data and calculate single-point nifty decisions using waiting vehicle density	Arduino is used to access a smart traffic framework for a four-direction junction along with taking current infrastructure into consideration	By decreasing the waiting time for green light shift and with using the recorded vehicles images to track high speed vehicles this system ease the traffic condition
Proximity infrared sensors	Infrared sensors collects data and send to Arduino Uno to take decision based on density of vehicles in a particular junction	Four IR sensors mounted on four sides of junction and control delay is used for shifting of signals	System able to ease the traffic flocking

4 Proposed Work

The working of the proposed system is just as a normal traffic light system. The feature that sets this system apart from conventional system is, this system works on the basis of number of vehicles or pedestrians accumulated on a specific path of the road. On this condition, the sensors placed on that specific path turns out as low if accumulation of traffic or pedestrians are high, else it reads high signal. The traffic jam is controlled by the signals provided by Infrared sensors which are placed on every side of the path. When we get low signal from any of the sensor, then the traffic signal will turn green. Figure 3 shows the block diagram of Density based traffic control system.

A. Details of no. of components used

See (Table 2).

(1) *Animated LEDs*- LED is a SC light source that gives away light when current is passed through it. Electrons in it recombine with electron openings, giving away energy as positive ions. The color of the led light is controlled by the energy required for electrons to cross the band gap of the SC. In this circuit diagram,

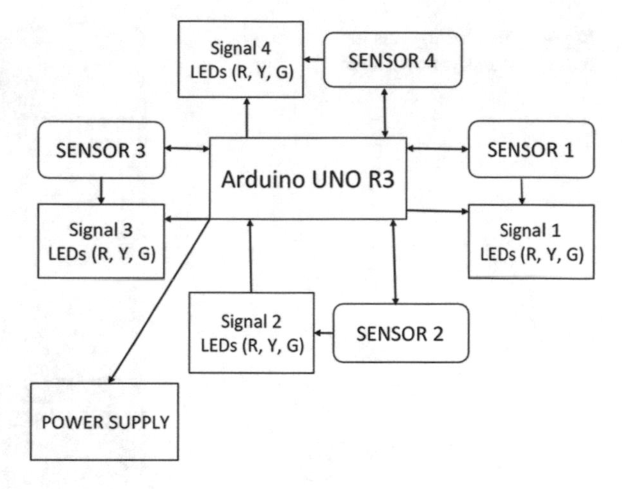

Fig. 3 Lock diagram of density based traffic control system

Table 2 No. of components used and their quantity

Components used	Quantity
Animated LEDs	12 (4 red, 4 yellow and 4 green)
Arduino Uno R3	1
Proximity infrared obstacle sensor	4
Battery	4
Logic state	4
Control delay	4

we are using 12 animated LEDs (4 red, 4 yellow and 4 green) as a substitute for the traffic signals. Figure 4 shows the Circuit diagram of three types of LEDs.

(2) *Arduino Uno R3-* In this circuit diagram we are using a single Arduino Uno R3. It is an open source microcontroller device which is used for small scale university projects and other small projects. It has 6 analog pins and 14 digital pins. It takes input from different sensors and gives the output on LEDs and other devices and shows the output on LCDs. It has a reset pin and other pins namely A0-A5 and 0-1. In the connections, we do not use the 0, 1 pins.

(3) *Proximity infrared obstacle sensor-* Total 4 number of IR sensors are required in this circuit diagram. An infrared sensor emanates and additionally recognizes infrared radiation to detect its environmental factors. The basic idea of an Infrared Sensor which is used as Obstacle finder is to transmit an IR sign, this IR sign bounces from the outer surface and the sign goes to the IR collector. Figure 5 shows the Circuit diagram of infrared sensor.

(4) *Battery-* A battery is a device which produces current by converting chemical energy into electrical energy and produces DC current. Here, we have connected

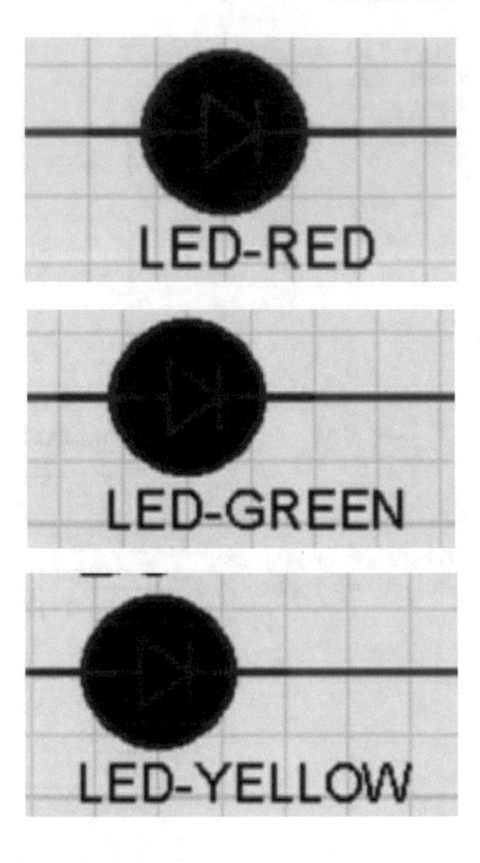

Fig. 4 Circuit diagrams of red, green and yellow LEDs

4 batteries of 12V each to 4 Infrared Sensors for giving them a supply to operate. Figure 6 shows the Circuit diagram of a 12V battery.

(5) *Logic state-* It is an indicator which monitors the circuit without affecting it. In this circuit, 4 logic state indicators are connected with the test pins of the IR sensor. Figure 7 shows the Circuit diagram of logic state.

(6) *Control delay-* Control delay is the portion of the complete delay credited to the traffic signal. Control delay incorporates developments at more slow speeds and stops on crossing point approaches as vehicles move in line position or hinder upstream of an intersection. 4 control delays are connected with each of the traffic signal lights. Figure 8 shows the Circuit diagram of control delay.

B. Operational model

This system is working as the input given to the sensor state by the operator to turn ON the IR sensor in the junction where traffic has been accumulated in a larger amount more than in comparison to other junctions. Then IR sensor gives the command to microcontroller to open the traffic by turning ON the green and yellow lights for vehicles and pedestrians respectively and restraining the traffic in other junctions. When traffic will reduce in that junction then the operator gives the input to other sensor

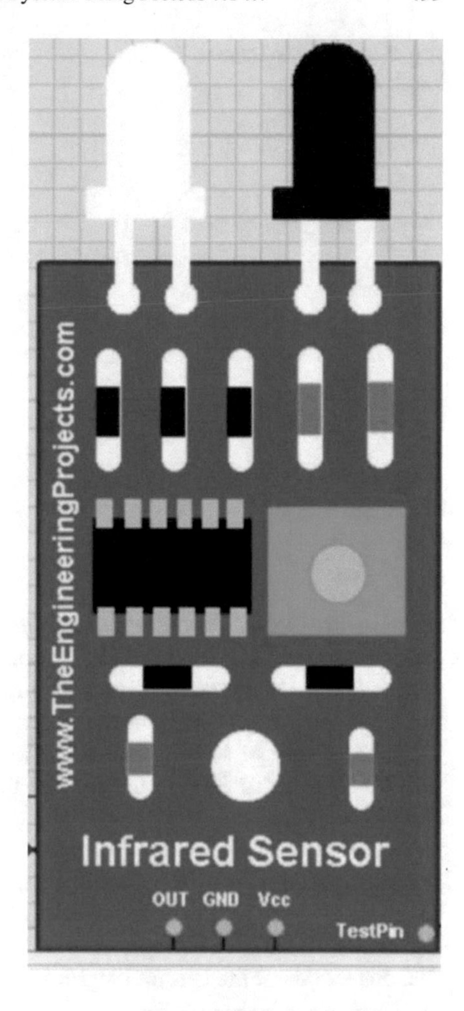

Fig. 5 Circuit diagrams of proximity infrared sensor

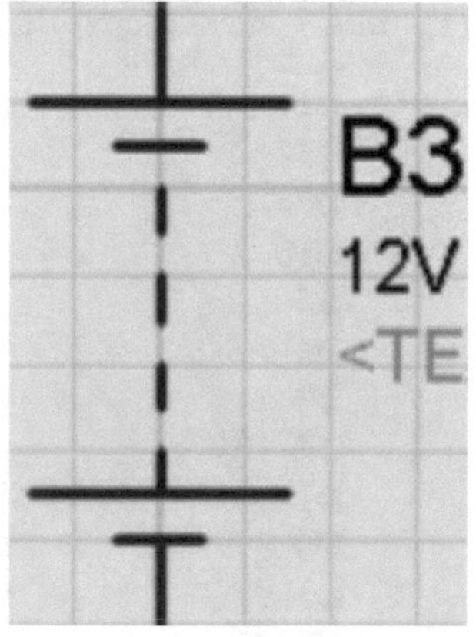

Fig. 6 Circuit diagrams of 12 V battery

Fig. 7 Circuit diagrams of Logic state

$$Kc. \frac{1 + Ti.p}{1 + d.Ti.p}$$

Fig. 8 Circuit diagrams of control delay

states to turn ON the IR sensor and this process will work in the loop. So, in whichever junction traffic will be accumulated the operator would have to give a command to IR sensor to open that junction. Transfer function for delaying is provided in the circuit before connection of LEDs and after every green LED command in the coding section to better in case of shifting of LEDs from one junction to another. If the delaying function is not provided in the system then microcontroller will have a hard time shifting the signals. A battery is provided to give the supply to IR sensors when logic state is turned ON, when the logic state will be OFF so do the battery. Battery is 12v multi-cell connected with all the Infrared sensors. Circuit diagram is shown in Fig. 9.

C. Experimental Setup

**Connections of the Circuit diagram are as follows–

(1) LEDs with Arduino Uno R3

D11–A0
D3–A1
D7–A2
D10–A3
D4–A4
D6–A5
D8–11
D2–12
D12–13
D5–8
D1–9
D9 – 10

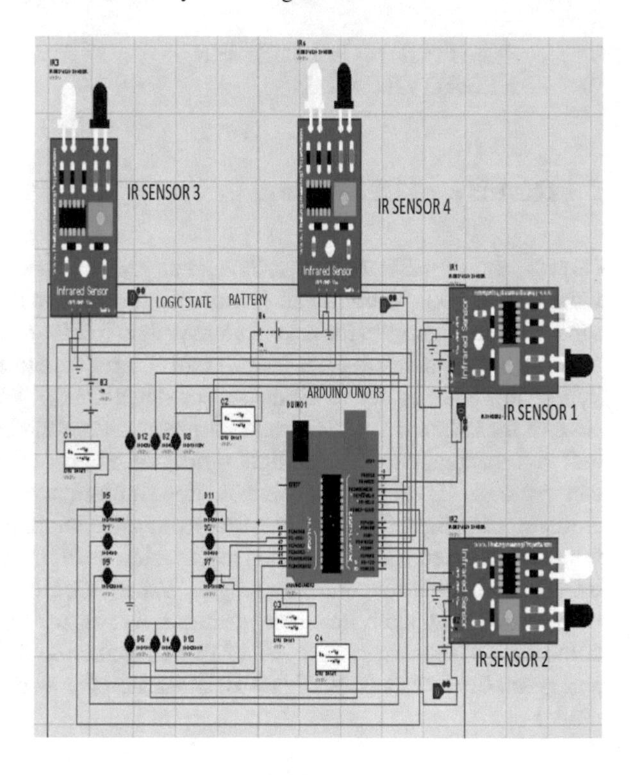

Fig. 9 Circuit diagrams of density based traffic control system

(2) *IR sensor Output pin with Arduino Uno R3*

IR1 – Pin 2
IR2 – Pin 3
IR3 – Pin 4
IR4 – Pin 5

(3) Logic state with IR sensor.

IR1–TEST PIN
IR2–TEST PIN
IR3–TEST PIN
IR4–TEST PIN

(4) Vcc of each IR sensor is connected to a 12 v battery namely B1, B2, B3 and B4.
(5) Control Delay with IR sensor and Yellow LED.

POSITIVE C1–IR3 TEST PIN
NEGATIVE C1–D5
POSITIVE C2–IR4 TEST PIN
NEGATIVE C2–D8
POSITIVE C3–IR1 TEST PIN
NEGATIVE C3–D7

POSITIVE C4–IR2 TEST PIN
NEGATIVE C4–D6

5 Result and Discussion

Circuit has 12 LED lights (yellow, red, green respectively), 4 Infrared Sensors mounted on each junction and sensor state control to manually start and stop each sensor. Green lights are for vehicles and yellow lights are for pedestrian to go through. Initially, at the starting of simulation all the red lights are open and all green and yellow lights are closed as all the sensor states are off. When we turn ON the sensor state of the Infrared Sensor 1, then the green and yellow lights of Junction 1 turn ON and in the other Junctions red light turns ON. So, cars and pedestrians on junction 1 can move on. There is a transfer function implemented for delaying the process of shifting green light from one Junction to another and delaying the shifting will make the whole system works well. It is represented as digitalWrite (G1, HIGH) followed by a delay(100) in Junction 1 and digitalWrite (G2, HIGH) followed by a delay(100) respectively and same for other junctions. When we turn OFF the logic state then command if (sensorstatus1 == LOW) will make red light on the Junction turn ON and green light turns OFF. Figure 10 illustrates the working simulation figure of the model.

6 Conclusion

The proposed system is working on the basis of data given by the operator to sensors and for this logic state has been put on the side of every sensor. With field utilization of this innovation, the level of traffic can be adequately put on a line by circulating the schedule timeline depending on the value of the vehicles in specific paths of multi-intersection crossing. We have effectively performed the simulation on the proteus software so that the lights will shift according to the density of vehicles and this will be followed on each and every side of the traffic lights. As one of the solutions, we have presented in our project that whenever there are large no. of vehicles accumulated on the junction so traffic on all the other junctions stopped in order to free the junction where cars have been accumulated in large no. otherwise, as the current traffic rules the junction moves slowly and cause massive traffic jam.

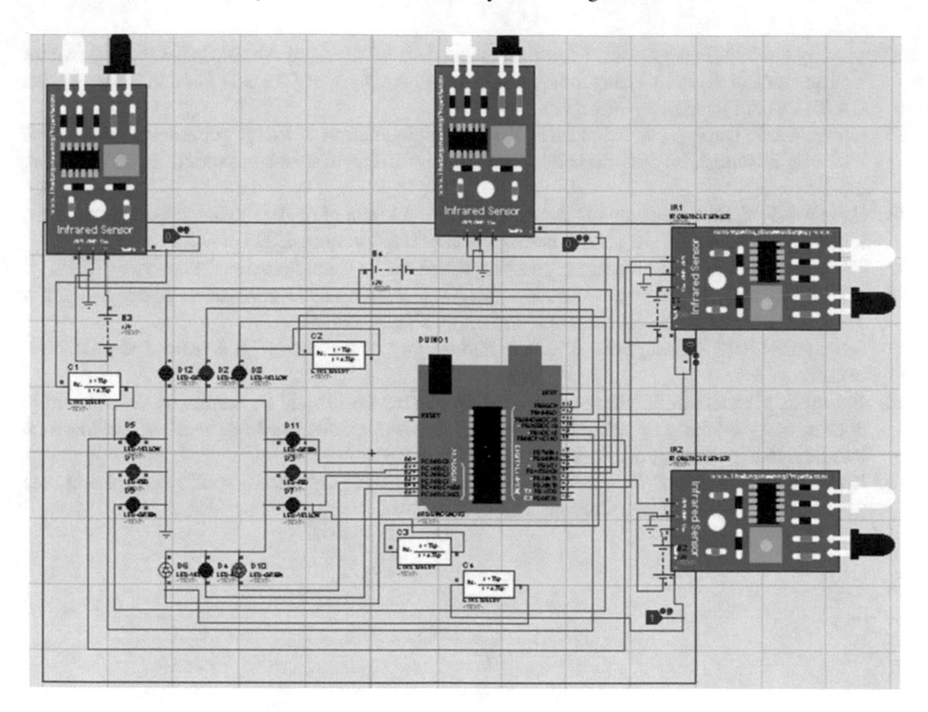

Fig. 10 Working simulation of model

References

1. Udoakah1 YN, Okure IG (2017) Design and implementation of a density-based traffic light control with surveillance system. Nigerian J Technol (NIJOTECH) 36(4):1239–1248
2. Shruthi KR, Vinodha K (2012) Priority based traffic lights controller using wireless sensor networks. Int J Electron Signals Syst (IJESS) 1(4), ISSN: 2231-5969
3. Kham NH, Nwe CM (2014) Implementation of modern traffic light control system. Int J Sci Res Publ 4(6):1–6
4. Traffic Control Systems Handbook (2005) Prepared for federal highway administration by Dunn Engineering Associates in association with Siemens Intelligent Transportation Systems
5. Kell JH, Fullerton IJ (1998) Manual of traffic signal design. Institute of Transportation Engineers, Prentice-Hall, Inc., pp 138
6. Sinhmar P (2012) Intelligent traffic light and density control using IR sensors and microcontroller. Int J Adv Technol Eng Res (IJATER) 2(2):30–35
7. Nwoye CD, Usikalu MR, Babarimisa IO, Achuka JA, Ayara WA (2017) Construction of an automatic power switch using infrared motion sensor. Journal of Informatics and Mathematical Sciences 9(2):331–337
8. Ayara WA, Omotosho TV, Usikalu MR, Singh MS, Suparta W (2017) Development of a solar charged laboratory bench power supply. J Phys Conf Ser 852(1):012044
9. Usikalu MR, Shittu AH, Obafemi LN (2018) Construction of an intelligent and efficient light control system. Int J Mech Prod Eng Res Dev (IJMPERD) 8(4):1057–1066
10. Dakhole A, Moon M (2013) Design of intelligent traffic control system based on ARM. Int J Adv Res Comput Sci Manag Stud 1(6):76–80
11. Zhou S, Jaji S, Wang CH (2015) Construction and practice of virtual simulation experimental teaching center. Comput Teach 2015(9):5–11

12. Vismay Pandit, Jinesh Doshi, Dhruv Mehta, Ashay Mhatre and Abhilash Janardhan, "Smart Traffic Control System Using Image Processing. Int J Emerg Trends Technol Comput Sci (IJETTCS) 3(1). (January–February 2014)
13. Choudekar P, Banerjee S, Muju MK (2011) Implementation of image processing in real time traffic light control. In: 3rd international conference on electronics computer technology, April, 2011
14. Semertzidis T, Dimitropoulos K, Koutsia A, Grammalidis N (2010) Video sensor network for real-time traffic monitoring and surveillance. Inst Eng Technol 4(2):103–112
15. Gaikwad OR et al (2014) Image processing based traffic light control
16. Hazrat Ali MD, Kurokawa S, Shafie AA (2013) Autonomous road surveillance system: a proposed model for vehicle detection and traffic signal control
17. Sinhmar P (2012) Intelligent traffic light and density control using IR sensors and microcontroller
18. Baroni R, Premebida S, Martins M, Oliva D, Freitas de Morais E, Santos M (2021) Traffic control using image processing and deep learning techniques. In: Metaheuristics in machine learning: theory and applications. Springer, Cham, pp 319–335
19. Kuppusamy P, Kalpana R, Venkateswara Rao PV (2019) Optimized traffic control and data processing using IoT. Clust Comput 22(1):2169–2178

Chapter 42
Security Enhancement Using Cryptography in Cloud-Based Education Portals

Sonia Rani, Sangeeta Rani, and Vikram Singh

1 Introduction

Cloud computing is playing a significant role in web-based online education delivery systems. Cloud computing has been considered as a model to enable convenient as well as on-demand network access to a shared pool of configurable computing resources. These resources could be networks, servers, storage, applications, and services. These resources could be allotted with minimum management effort. Thus cloud computing and distance learning are frequently integrated. A review of distance learning and cloud computing literature has been performed to understand the benefits and issues faced during the integration of cloud computing in distance learning. This section has explained cloud computing and the need for the cloud in distance learning.

The need for cloud-based distant learning is increasing every day. In the topic of online education systems for remote learning in cloud computing, there has been a variety of studies. These studies, however, have certain drawbacks. Some of the cloud-based education solutions for distant learning that have been previously introduced have insufficient security. However, some researchers have chosen to strengthen security, but there is still a performance problem since securing the data takes a long time. The goal of this study is to look at concerns like performance and security in a cloud-based education system for distant learning. There is a need for a system that can safeguard educational information without compromising performance. The cloud system and the necessity for an online education system are discussed in this study, as well as prior research on cloud-based online education systems, including methods and limits. The necessity for improvements in current research to retain security and performance, as well as the breadth of such a system, is then outlined.

S. Rani · S. Rani · V. Singh (✉)
Department of Computer Science & Engineering, Chaudhary Devi Lal University, Sirsa, India
e-mail: vikramsinghkuk@yahoo.com

G. Mathur et al. (eds.), *Proceedings of 3rd International Conference on Artificial Intelligence: Advances and Applications*, Algorithms for Intelligent Systems,
https://doi.org/10.1007/978-981-19-7041-2_42

2 Literature Review

Patil et al. [1] have studied cloud-based e-Learning solutions for setting up a virtual classroom in online distance education environments. During the training phase of their mechanism, they have employed Cloud Computing. A pre-existing e-learning system was taken into account while working on this study. Their research has focused on how to use cloud computing into a virtual classroom.

According to the work of Bouyer et al. reported in Bouyer et al. [2], online education "shall" more and more move on cloud computing platforms. A dynamic scalability of cloud computing has been the key component of their research claim and proposal. With its mobile Internet connectivity, Cloud Computing may provide a variety of online services. Virtual technologies are becoming more significant in online education as a result of technological advancements. The benefits of online training have also been listed by researchers.

Korucu and Atun [3] have outlined the characteristics and features of cloud systems utilized in online education. Because of advances in technology, it has been shown that the variety and value of data utilized in education are expanding. Web technologies and their contributions to a distant learning system were examined in a study. Consideration was also given to mobile systems, which are frequently used in remote education. Access to web-based technology has become much more convenient as a result. Individuals may now access info from the online without being constrained by geography and time thanks to web technology.

"When it comes to academics, cloud computing has both advantages and disadvantages," claimed and argued Mary & Rose, in Mary and Rose [4]. A number of advantages have been brought about by cloud computing in the academic arena. Using the cloud to store and process sensitive data has its drawbacks.

Meslhy et al. [5] have reported a data security model which was developed to ensure the safety of cloud applications. Using a single default gateway, this research study proposes a data security approach that protects sensitive user information across a variety of public and private cloud services. Encrypting sensitive data before sending it to the cloud is possible without crashing cloud apps with this gateway platform. A new encryption technique that is both fast and secure has been developed by researchers.

Saife et al. [6] have presented research findings on how to implement performance analysis of cloud-dependent web services necessary for virtual Learning Environment Systems. Cloud-based web services may be used in diverse contexts, according to a study. A wide range of high-quality services are being offered by protocols.

DNA cryptography was applied by Pandey [7] to protect the cloud app. The Huffman algorithm has been utilized for compression in research. In order to transmit data between sender and receiver programs, the author used socket programming. The cloud has been utilized in research to safeguard compressed data. The performance of the system has been impacted by 13% due to the mechanism.

With the use of the RSA algorithm, Suresh [8] has worked toward protecting the cloud environment. AES, DES, RSA, and other encryption and decryption algorithms have been examined in the context of security research. RSA has been implemented in this study using an asymmetric key algorithm.

Research on RSA-based data security for cloud applications was reported by Singh et al. [9]. It has been determined that the RSA algorithm's performance is dependent on three factors, which the author has investigated. Throughput, Space Complexity, and Timing Completion are the three main factors to consider. A method called RSA has been used to encrypt data in this study to ensure that only authorized people may access it. Before uploading data to the cloud, it is encrypted.

Cybersecurity for the e learning education system was the main theme of Bandara et al. [10]. For the most part, the concept of "cybersecurity" has been conceptualized as a set of guidelines for safeguarding electronic data. It's becoming more difficult to keep an e-learning system secure. A strategy to cybersecurity monitoring and management for e-Learning systems is described by the author in this study.

Kumar and Chelikani [11] have investigated the security issues arising in cloud e-learning systems and have come out with suggestions to improve the security. Authors have conducted both the theoretical and empirical studies on data collected from e-learning platforms hosted on cloud.

Mishra et al. [15] have pondered upon the applications of cloud services in educational sectors. They have concluded that cloud services may play a vital role in elevating the interest of students in a variety of educational disciplines. Cloud computing may help come over the heterogeneous coverage of computing resources over geographies.

Balobaid and Debnath [16] have proposed a new cloud-based model for delivery of education in distance mode. Authors have showcased how the deprived students in third world countries might benefit from the proposed model.

Zhihong et al. [17] have demonstrated that cloud-based education portals have a positive contribution toward distance education in terms of cutting efforts on repetitive CAI modules. Authors have proposed a general cloud computing model for increased personalization of remote teaching-learning.

Most research reviewed as a part of present endeavor have focused on issues related to adhering to cloud computing for faculty, staff, and students in the education sector. Researchers have also focused on security issues and risk divisions. Areas where cloud computing might influence the education field are considered. Security management and challenges faced by the education sector in developing countries are major issues. Security issues involve the hacking and cracking activities by intruders. Another issue is 24-hour availability. Presently students need to access required information at any time at any place using cloud space. Some of the researches explored the challenge of reducing the cost of distance learners. Providing online cloud-based education to underprivileged students and children in an underdeveloped country is a complicated task. However, some existing research provided security to the cloud system using RSA, DNA cryptography, and other security protocols. But there remains an issue of performance. Existing research that has provided security has reduced the overall performance from 7 to 20%. There have been several factors that are influencing the performance of the cloud such as packet size, time to encrypt the data, time taken to filter the data by a firewall, or malware detection mechanism.

Thus there is a need to introduce a mechanism that should be capable of enhancing security along with performance.

3 Security Enhancement in Cloud-Based Education Portal

Present research endeavor was aimed at developing a secure and high-performance model to address performance and security challenges that have been reported but not addressed in earlier studies. An experimental research approach was employed in the current study. The proposed work has been implemented in Java, and the application was created using NetBeans Integrated Development Environment. The proposed work aims to establish cloud-based infrastructure for distance education. A few other sorts of studies in related domains have been conducted with particular restrictions, though. The main issue is the slow speed of the network and the large data volume that need to be carried for seamless streaming of educational content. High-speed networks are needed to transport instructional information between client and server locations and between different levels of servers. During transmission, the instructional materials have to be encrypted and compressed. In order to transport instructional information via the cloud, a more secure and high-performance data transmission technique is needed. For the purpose of comparing and contrasting old and new methods of education, this project proposes to integrate the new mechanism into an educational module, evaluate its efficiency and safety, and then report on the results.

Compression followed by the encryption of data at the source and decryption followed by decompression at the destination are the primary focus of the proposed research (refer Fig 1). Before it is sent from the server to the client, the instructional information is compressed and encrypted. The data security to the data is provided by applying the exclusively-OR-based encryption mechanism. Thereafter, the data is sent to the client station. Decryption and decompression are performed here in order to restore the original instructional material.

4 Results and Discussion

The "Netbean IDE 6.1" environment was used to construct the transmitter and receiver modules of the network application. Port number, file location, and token to decode data are listed on the recipient end. The data has been encrypted using an exclusively-OR algorithm. To ensure the security of data transmission, the user-specified port has been used. It's because ports 1 through 1023 have already been designated for established protocols. To save the received material in a text file, the file path name text field is filled in. To utilize the exclusively-OR operation on incoming data, the user must give the port number, file name, and token on the receiving end. Afterward, the user clicks on the "enable upload" option to activate the receiver. The

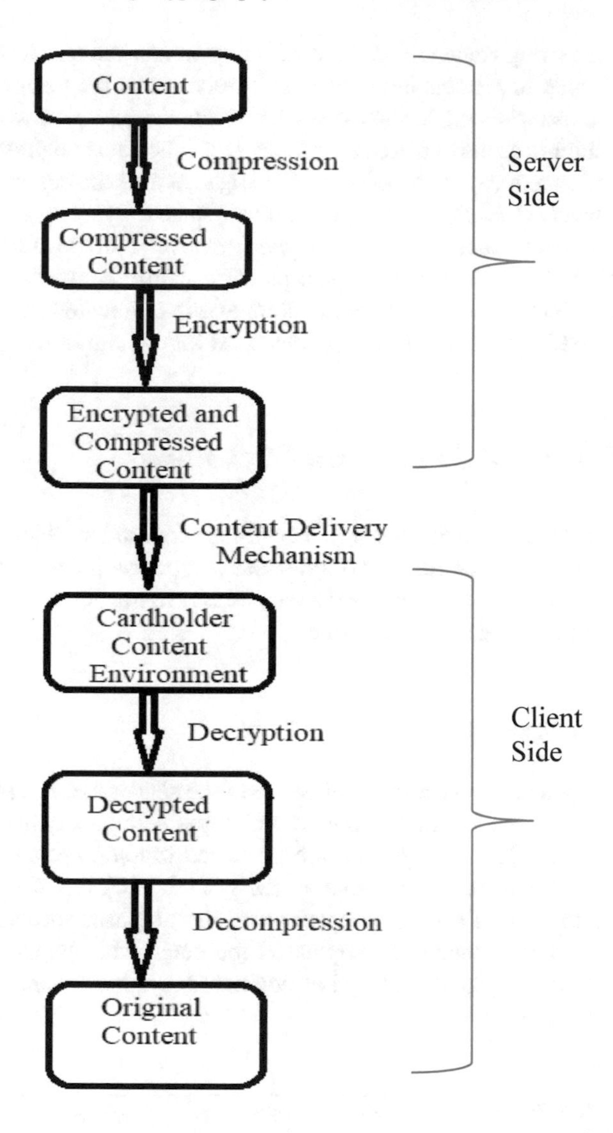

Fig. 1 Process flow of proposed work

module is used to transport data during the sender implementation. The path of the file, the server's IP address, and the token used to encrypt the code are all specified here. As a result of this, the sender would enter a port number that is identical to the client's port number. A file path and name are then entered by the user. You may enter the receiver's IP address in the IP address field. The code to encrypt data would be sent in the token box by the sender. The receiver's path to the file, port number, and decoding code are all established during the receiver's implementation. Sender to recipient would exchange a text file. An exclusively-OR program would replace the enormous string content of this file with tiny words and encode the data. During server-side code implementation, the port number must be greater than 1023. The port must be user-defined during the sender module's execution. It must also be the same port number as the one used by the receiver. The data compression method has been utilized to reduce the size of packets being sent over the network. Smaller sized material replaces the larger sized stuff in this instance.

In order to construct the transmitter and receiver module, Java was used as a programming language on the Netbean platform. Different packet counts are used to track how long a certain task took in comparison to how long it might take in the future. A MATLAB environment has been used for simulations.

4.1 Efficiency of the Proposed Mechanism

For the purpose of evaluating performance of the proposed mechanism, three parameters, namely, encryption time, error rate, and compression ratio, have been considered. The amount of time it takes for a data packet to travel between the sender and receiver modules is taken into account.

4.1.1 Encryption Time

Time taken for a data packet to travel between the sender and receiver does directly depend on the encryption time. In the present experiment, encryption times for three schemes, namely, RSA, DNA, and the proposed cryptography mechanisms have been simulated and placed in Table 1. RSA- and DNA-based techniques take a longer time to encrypt data in comparison to the proposed mechanism. Moreover earlier investigations have not compressed the data before transmission. Thus the time consumption is plainly reduced as compared to others owing to the lower size of the data packets.

Table 1 Comparison of encryption time

Number of packets	RSA cryptography	DNA cryptography	Proposed mechanism
10	1	0.9	0.85
20	1.7	1.6	1.3
30	2.4	2.2	1.7
40	3.1	2.9	2.4
50	4	3	2.7
60	4.7	4.4	3.5

Table 2 Comparison of error rate

Number of packets	RSA cryptography	DNA cryptography	Proposed mechanism
10	0.9	0.7	0.6
20	1.4	1.2	0.9
30	2.4	2.2	1.7
40	2.6	2.3	1.9
50	4	3	2.5
60	4.2	3.8	3.6

4.1.2 Error Rate

There always remain chances of errors during data transmission. But if the packet size is reduced and the packet remains for less time on the network then the probability of error is minimized. There remain less chances of error because the size string is reduced using a replacement mechanism. But RSA and DNA cryptography [12–14] mechanisms used in previous research have not reduced the size of packets. Thus, the proposed mechanism reduces the error rate indirectly, i.e., by diminishing the packet size. Table 2 has shown a comparative analysis of the error rate for RSA, DNA, and the proposed cryptographic schemes.

4.1.3 Packet Size

Replacement mechanism used in proposed work has reduced the content volume. Therefore, it can be said that packets to be transmitted consume lesser bandwidth and channel resources. So, the proposed mechanism compresses the original data into smaller packets as compared to RSA and DNA mechanisms. Previous research that made use of RSA and DNA cryptography [12–14] did not compress data before sending it off to the transmission lines. Comparative analysis of data volumes for RSA, DNA, and the proposed cryptography schemes have been placed in Table 3.

Table 3 Comparison of message size

Number of packets	RSA cryptography	DNA cryptography	Proposed mechanism
10	9	7	6
20	14	11	9
30	21	18	12
40	26	18	16
50	31	25	20
60	42	32	26

4.2 Security of the Proposed Mechanism

This subsection presents the impact of the proposed work on security aspects of the network and compares it with the RSA- and DNA-based mechanisms. In case of the proposed work, the number of packets affected is less as the number of attacks increases. From previous research, it has been found that DNA cryptography [12] is better as compared to basic RSA [13] and advanced RSA [14]. But the proposed work is better than DNA cryptography. Tables 4, 5, 6, 7, 8, and 9 have presented experimental evidence of the fact that a lesser number of packets are affected in the case of proposed work as compared to RSA- and DNA-based cryptography approaches.

Table 4 Comparison of response to man-in-the-middle attack

Number of attacks	RSA cryptography	DNA cryptography	Proposed mechanism
10	10	8	5
20	14	10	7
30	21	12	10
40	26	16	12
50	30	20	15
60	33	28	18

Table 5 Comparison of response to brute-force attack

Number of attacks	RSA cryptography	DNA cryptography	Proposed mechanism
10	10	8	6
20	14	9	7
30	21	11	8
40	26	16	12
50	30	20	15
60	33	28	20

Table 6 Comparison of response to denial of service

Number of attacks	RSA cryptography	DNA cryptography	Proposed mechanism
10	9	7	4
20	13	9	6
30	21	12	8
40	27	14	11
50	32	20	15
60	34	28	22

Table 7 Comparison of response to application-level attack

Number of attacks	RSA cryptography	DNA cryptography	Proposed mechanism
10	9	7	6
20	14	9	8
30	21	13	10
40	28	17	14
50	30	20	15
60	34	29	18

Table 8 Comparison of response to attack by malicious insider

Number of attacks	RSA cryptography	DNA cryptography	Proposed mechanism
10	10	7	5
20	14	10	6
30	21	13	9
40	28	16	11
50	30	20	15
60	35	29	18

Table 9 Comparison of response to attack on cloud services

Number of attacks	RSA cryptography	DNA cryptography	Proposed mechanism
10	10	8	6
20	13	9	8
30	22	13	10
40	25	16	13
50	31	20	16
60	38	28	20

4.2.1 Man-In-The-Middle Attack

Impact on the man-in-the-middle attacks on data packets in the case of RSA, DNA, and the proposed cryptographic schemes have been simulated and the results have been shown in Table 4.

4.2.2 Brute-Force Attack

A brute-force attack involves guessing login information via trial and error. Encryption keys and a hidden web page are also used. Comparative analysis of these attacks is shown in Table 5.

4.2.3 Denial of Service Attack

A denial-of-service attack is a kind of cyberattack that attempts to prevent people from accessing a computer or network resource. Due to reduced size of packet and less time taken during transmission over network the probability of denial of service gets reduced. Thus the impact of denial of service is less in case of proposed work. Table 6 presents a comparative analysis of denial of service.

4.2.4 Application-Level Attack

The attacks made at application level have been reduced by providing a special user interface to users. The chances of sending and receiving data without using that user interface are negligible. But previous research work has not provided a special user interface. Table 7 presents the comparative analysis of traffic hijacking in the case of RSA, DNA, and the proposed cryptography.

4.2.5 Attack by Malicious Insider

The possibility of malicious insider attack has been reduced by allowing different keys to encode and decode data. Previous research has used the same key in different sessions. Thus the chances of attack by malicious insiders were more in case of previous research. Table 8 presents the comparative analysis of attacks by malicious insiders in the case of RSA, DNA, and the proposed work.

4.2.6 Attack on Cloud Services

Use of exclusive OR after compression of data and user-defined port number has reduced the chances of different attacks on cloud services. Table 9 presents the comparative analysis of the attack on cloud service in the case of RSA, DNA, and the proposed cryptography mechanisms.

5 Conclusions

The integrated technique was examined, in which the content replacement mechanism reduced the size of the data packet while the exclusively-OR operation offered encryption security. The proposed research has assured that educational cloud systems are secure and work well. Simulation findings show that the proposed cloud-based education system outperforms traditional options. It is secure and speedy since data is compressed first and then encrypted on the sender side. At the receiving end, the data is decrypted and decompressed. Because the quantity of data being sent is reduced, the issues of error rate and latency are no longer an issue. In addition, the packet dropping ratio falls. Compared to traditional security systems, the proposed method is more resistant to a variety of attacks such as man in the middle, denial of service, brute-force attack, and attack on cloud services, as well as attacks from hostile insiders and application layer attacks. The recommended solution is more secure than RSA and DNA cryptography-based approaches and the proposed mechanism.

The future work may provide a better compression mechanism. The security could be increased in upcoming research. Future research might provide better performance along with reduced error rate by integrating advanced cloud services and optimization mechanisms. The use of soft computing techniques could improve the reliability and quality of services. To improve the dependability of the cloud in distant learning, researchers may examine its high availability and zero downtime. Nowadays, cloud computing as well as e-learning is rising speedily. It is playing a vital and powerful role in the area of education and learning. These are supporting smartphone mobile users to implement operations effectively at less cost. Such systems are utilizing cloud-dependent applications that have been provided by various cloud service providers. Upcoming research should be capable of providing key technology for the safety of cloud applications for distance education. Moreover, it should capable of increasing the quality of service and it is supposed to be eligible for teaching resource management. Moreover, future research should be capable of providing intelligent service management.

References

1. Patil P (2016) A study of e-learning in distance education using cloud computing. Int J Comput Sci Mobile Comput 5(8):110–113. (August 2016)
2. Bouyer A, Arasteh B (2014) The necessity of using cloud computing in educational system, CY-ICER 2014, 1877–0428. Elsevier
3. Korucu AT, Atun H (2016) The cloud systems used in education: properties and overview. World Acad Sci Eng Technol Int J Educ Pedag Sci 10(4)
4. Mary TAC, Rose PJAL (2020) The impact of graduate student's perceptions towards usage of cloud computing in higher education sectors. Univers J Educ Res 8(11):5463–5478. https://doi.org/10.13189/ujer.2020.081150
5. Meslhy E, Hatem AE, Sherif E (2013) Data security model for cloud computing. J Commun Comput 1047–1062
6. Saife O, Mohammed EA, Saad A (2016) Performance analysis of cloud-based web services for virtual learning environment systems integration. Int J Innov Sci Eng Technol 3
7. Pandey GP (2019) Implementation of DNA cryptography in cloud computing and using Huffman algorithm, socket programming, and new approach to secure cloud data
8. Suresh P (2016) Secure cloud environment using RSA algorithm. Int Res J Eng Technol 03(02). (Feb 2016)
9. Singh SK, Manjhi PK, Tiwari RK (2016) Data security using RSA algorithm in cloud computing. Int J Adv Res Comput Commun Eng 5(8):11–16
10. Bandara I, Ioras F, Maher K (2014) Cybersecurity concerns in e-learning education. In: Proceedings of ICERI2014 conference, 17–19 Nov 2014, Seville, Spain

11. Kumar G, Chelikani A (2011) Analysis of security issues in cloud based e-learning. Dissertation, University of Borås/School of Business and IT. http://urn.kb.se/resolve?urn=urn:nbn:se:hb:diva-20868
12. Ali A, Bajpeye A, Srivastava AK (2015) e-learning in distance education using cloud computing. Int J Comput Techn 2(3). (May–June 2015)
13. Sharma SK, Goyal N, Singh M (2014) Distance education technologies: using e-learning system and cloud computing. Int J Comput Sci Inf Technol 5(2):1451–1454
14. Shi Y, Yang HH, Yang Z, Wu D (2014) Trends of cloud computing in education. In: Cheung SKS et al (eds) ICHL 2014, LNCS 8595, pp 116–128, 2014. © Springer International Publishing, Switzerland, 2014
15. Mishra JP, Panda SR, Pati B, Mishra SK (2019) A novel observation on cloud computing in education. Int J Recent Technol Eng 8(3). (Sept 2019).
16. Balobaid A, Debnath D (2016) A novel proposal for a cloud-based distance education model. Int J e-Learn Secur 6(2). (Sept 2016)
17. Zhihong X, Junhua G, Yongfeng D, Jun Z, Yan L (2013) Expand distance education connotation by the construction of a general education cloud. In: International conference on advanced information and communication technology for education (ICAICTE 2013)

Chapter 43
Feature Ranking Merging: FRmgg. Application in High Dimensionality Binary Classification Problems

Alberto F. Merchán, Alba Márquez-Rodríguez, Paola Santana-Morales, and Antonio J. Tallón-Ballesteros

1 Introduction

Data mining term [15] has been coined in recent decades and its meaning is the process of extracting useful information from databases. A data mining system usually allows one to collect, store, access, process and ultimately, describe and visualize data sets [35]. Databases are growing in size making traditional analysis and visualization techniques stop working.

There are many approaches within data science. Just to mention a few of them we briefly review the most common ones. Data mining and Knowledge Discovery in Databases (DM and KDD) processes are responsible for extracting patterns and models from huge databases. KDD refers to the overall process of discovering useful knowledge from data with the difference that data mining refers to a particular step in this process [11]. The CRISP-DM (Cross Industry Standard Process for Data Mining) methodology [36] is the successor from KDD [21]. Data mining may be coped with multiple artificial intelligence branches such as computational intelligence, which encompasses neural networks, fuzzy sets, and evolutionary computation [35].

A. F. Merchán · A. Márquez-Rodríguez · P. Santana-Morales
University of Huelva, Huelva, Spain
e-mail: alberto.fernandez320@alu.uhu.es

A. Márquez-Rodríguez
e-mail: alba.marquez139@alu.uhu.es

P. Santana-Morales
e-mail: paola.morales@alu.uhu.es

A. J. Tallón-Ballesteros (✉)
Department of Electronic, Computer Systems and Automation Engineering,
University of Huelva, Huelva, Spain
e-mail: antonio.tallon.diesia@zimbra.uhu.es

G. Mathur et al. (eds.), *Proceedings of 3rd International Conference on Artificial Intelligence: Advances and Applications*, Algorithms for Intelligent Systems,
https://doi.org/10.1007/978-981-19-7041-2_43

Generative Adversarial Networks (GANs) are on the rise although the data preparation stage at the feature level is especially important and sometimes may do the machine learning stage unaffordable [29]. Once we have applied data mining, we will use machine learning. In the field of machine learning there are several strategies [2] like decision trees, rule-based systems, etc. Many ingredients are indispensable to create a knowledge-based system like feature selection, association rules network, and theory building.

There are two fundamental matters like over-fitting, which is undesirable in the training procedure of any machine learning algorithm [4], and data quality, which should be tackled with as a prior step to any model training [24].

2 Background

Feature selection is the procedure for selecting a reduced subset of features that is necessary and sufficient to describe the target concept and discarding the attributes that are irrelevant and do not provide useful information [19]. It can be an arduous process because it could be that there exists complex relationships between a set of features and the target [5]. For instance, the genome-wide prediction field involves complex traits [13]. Feature subset selection [14] has been successfully applied in the domain of software engineering [40]. The scalability of any data preparation (DP) method is a major issue. Therefore, some computational intelligence techniques like genetic algorithms have been tried to achieve a remarkable efficiency [16]. For high-dimensionality problems, there are reviews that even being 8-year-old works are still valid and up to date [3]. The typical step after the DP phase is machine learning [37]. In some cases, DP and the machine learning stage are done in a tandem [23].

There are four methods used for feature selection: filter [27], embedded, semi-wrapper [34], and wrapper. On the one hand, a filter method attempts to find predictive subsets of the features using statistical measures from the empirical distribution [39]. Embedded approaches make use of an inner measure, which is mandatory to create the learning model without evaluating the model by means of the algorithm. The semi-wrappers are in-between methods, which compute the performance of the feature set using a supervised machine learning method that is different from the target learner. On the other hand, wrapper-based feature selection approaches use the same algorithm that will be applied to build the final classifier. It searches in the space of subsets of features, optionally using cross-validation to compare the performance of a given classifier on each tested subset. The wrapper methods perform better than filters; they require more computational time [39], are limited to any concrete learning approach and are not very convenient for transfer learning.

3 Proposal

High-dimensionality feature selection many times involves a penalization procedure in order to limit the number of selected features [25]. For low-medium feature spaces, discarding a certain fixed number of them may be enough to achieve a good prediction [31] or even an exact number taking into account the number of attributes and classes [33].

This paper tries to refine a previous proposal that establishes a percentage in order to set a cut-off point in the context of feature ranking-based feature selection [28]. In particular, the introduced approach combines a couple of feature ranking methods using a common percentage from those features that range from a weak to a strong influence in the class label. These methods are run in an independent way and merge the solutions achieved by every ranking-based feature selection. The proposed framework has been called Feature Ranking merging (FRmgg). Basically, any percentage may be used with the current proposal; nonetheless if the percentage is very close to a 10% the classification model will be trained earlier than in the case of a high percentage like a 90%. This paper applies feature ranking methods with a positive percentage lower than 100% from the attributes with a positive incidence to force that a subset is chosen.

4 Experimentation

There are many tools for data mining like Shogun [30], Keras [26] (recently it has been coded for C language, [8]) and MLPack [9] (and its newly extension for R language [17]) to cite a few of them. This paper makes use of a visual tool [6] which is Weka framework. Attribute selection is applied to the training set of three databases (Table 1) which are submitted to a hold-out cross-validation with stratification [20] with 75:25 percentages for training and test, respectively, and also, we report the performance of three classifiers. The data distribution is balanced and hence we do not need to apply any rebalancing procedure [22].

Four scenarios have been considered to assess the proposal. On the one hand, we have evaluated the base case that we have named Full and, on the other hand, three situations where the number of selected attributes is constrained to a percentage as

Table 1 Table of databases

Data set	No. instances	No. features
Arcene	200	10000
Dexter	600	20000
Gisette	7000	5000
Average	2600.0	11667.7

Table 2 Hellinger Tree's parameters

Name of parameter	Value
BinarySplits (B)	True
minNumObj (M)	2
Unprunned (U)	True
Use laplace (A)	True
maxEqual (E)	0.0

well as the merging of both solutions. Full represents all the features of the data set when no selector is applied. The features are selected by Info Gain or Gain Ratio. The main difference between both selectors is that Info Gain evaluates the value of an attribute by measuring the information gain with respect to the class, while the Gain Ratio classifier evaluates the value of the attribute by measuring the gain ratio with respect to the class. Finally, the Feature Ranking merging (FRmgg) combines the attributes of Info Gain and Gain Ratio in one list where the first attributes are those that have been selected by both selectors and are followed by the attributes which have been chosen only once doing a one-by-one insertion. That is, in a turn, we pick up an attribute from one feature ranking method and, in the next, an attribute from the counterpart feature ranking method. As we can observe, this proposal incorporates an improved feature sorting procedure [32] which we call graded multiple feature sorting (GMFS).

As we have mentioned, the proposal of the contribution is to take a database in such a way that a percentage of the attributes that has a positive incidence in the data set is taken. For the experimentation, we have defined a percentage of a 90% and hence this proposal applies the methods InfoGain (90%) and GainRatio (90%) and then combine both solutions for the FR merging scenario.

For evaluating the performance of these classifiers, we will take into account the accuracy and the maximum partial FP Rate. The accuracy refers to the percentage of observation that has been correctly classified, while the false positive rate (FPRate) represents the probability of classifying an observation in a wrong class [12].

We have used three classification methods: Hellinger Tree (HTree), K-nearest neighbors (IBk) and Naive Bayes (NB) [38]. In the following, we will explain each of these methods and the parameters we have used in the experiments.

HTree generates a Hellinger distance decision tree that uses this distance as the splitting criteria. It has advantages over other alternatives like entropy and it works, especially well, on imbalanced data and it has good performance in balanced data sets [7]. The parameters used for this classifier are shown in Table 2.

The parameter "*BinarySplits*" means whether to use binary splits on nominal attributes when building the trees and the "*minNumObj*" is the minimum number of instances per leaf on the tree.

Table 3 Ibk's parameters

Name of parameter	Value
*k*NN (K)	1
windowSize (W)	0

Table 4 NaiveBayes' parameters

Name of parameter	Value
useKernelEstimator (K)	False
useSupervisedDiscretization (D)	False

As the parameter "*Unprunned*" is true, it means that the tree has not been pruned. If "*Use Laplace*" is true, it indicates that counts at leaves are smoothed based on Laplace [7].

The Instance-Based algorithm (IBk) is derived from the nearest neighbor pattern classifier with some editions that reduce storage requirements with small losses in the accuracy [1]. The parameters that we used on this classifier are shown in Table 3.

The "*kNN*" parameter represents the number of nearest neighbors that are being taken into account. While, on the other hand, the parameter "*windowSize*" indicates the maximum number of instances allowed in the training pool [1].

Naive Bayes is one of the most used algorithms in data mining [38]. The Bayesian classifier is optimal when features are independent of the class. It shows well performance in domains with attribute dependencies [10]. The Naive Bayesian classifier provides an easy approach to the representation, use, and learning of probabilistic knowledge. It is designed for supervised learning [18]. The parameters that we have used for this classifier are shown in Table 4.

The option *useKernelEstimator* is false, so the kernel estimator for numeric attributes will not be used. Instead, it will use a normal distribution. If we had established *useSupervisedDiscretization* to true, there would be converted numeric attributes to nominal ones. But since it is set to false, the numeric attributes remain numeric.

5 Results

The test results that we have obtained are shown in Table 5.

For the Arcene database, the performance of the classifier considering the full attribute space is superior to the reduced attribute space.

We can notice that the predictive capacity of FRMerging is the same, in this case, as the predictive capacity of GainRatio. This shows that although we have more

Table 5 Test results

		Classifiers					
Data set		*HTree*		IBk		NB	
	Scenario	Performance Metric					
		Accuracy	*Max Partial FP Rate*	*Accuracy*	*Max Partial FP Rate*	*Accuracy*	*Max Partial FP Rate*
Arcene	Full	*76*	*0.318*	*80*	*0.273*	*78*	*0.250*
	Gain Ratio	*74*	0.545	76	0.409	70	0.357
	Info Gain	*64*	0.364	80	0.364	68	0.429
	FR merging	*74*	0.545	78	0.364	70	0.393
Dexter	Full	91	0.093	63	0.427	85	0.160
	Gain Ratio	*91*	*0.107*	83	0.200	91	0.133
	Info Gain	92	*0.093*	80	0.333	91	0.120
	FR merging	92	*0.093*	79	0.387	91	0.133
Gisette	Full	95	0.057	95	0.065	92	0.142
	Gain Ratio	*95*	*0.053*	97	0.038	91	0.155
	Info Gain	*95*	*0.053*	97	0.043	92	0.138
	FR merging	*96*	*0.046*	97	0.045	91	0.154
Mean ± SD	Full	87.45 ± 10.09	0.16 ± 0.14	79.33 ± 16.33	0.26 ± 0.18	84.96 ± 6.78	0.18 ± 0.06
	Gain Ratio	86.84 ± 11.29	0.23 ± 0.27	85.21 ± 10.71	0.22 ± 0.19	84.18 ± 12.28	0.22 ± 0.12
	Info Gain	83.81 ± 17.24	0.17 ± 0.17	85.56 ± 9.63	0.25 ± 0.18	83.83 ± 13.72	0.23± 0.17
	FR merging	87.22 ± 11.59	0.23 ± 0.28	84.43 ± 10.57	0.40 ± 0.05	83.98 ± 12.11	0.23 ± 0.14

attributes, it does not represent a limitation. Moreover, without information from the problem domain the performance matches the best approximation. For the future, the recommendation would be to obtain the attribute spaces of Info Gain and Gain Ratio, combine them and only run the classification algorithm with FRMerging.

On the Dexter database, we can appreciate that the best results are obtained with Info Gain and FR merging scenarios regardless of the classifier used in the experiments. So, it can be said that results are improved with respect full scenario in all cases. And that overall, HTree classifier gives the best results.

Table 6 Comparison with a state-of-the-art feature selector

		Classifiers					
Data set		*HTree*		IBk		NB	
	Feature selector	Performance metric					
		Accuracy	*Max partial FP rate*	*Accuracy*	*Max partial FP rate*	*Accuracy*	*Max partial FP rate*
Arcene	CFS	*78*	*0.318*	*80*	*0.364*	*74*	*0.286*
	FR merging	*74*	0.545	78	0.364	70	0.393
Dexter	CFS	90	0.107	83.33	0.227	85.33	0.187
	FR merging	*92*	*0.093*	79	0.387	91	0.133
Gisette	CFS	93.37	0.071	94.34	0.059	92.74	0.009
	FR merging	*96*	*0.046*	97	0.045	91	0.154

Finally, in the Gisette database, we perceive that, in general, we have the best classifiers performance and that it is very consistent as the results do not vary much between the different classifiers and scenarios.

It can be noticed that although the performance does not vary a lot, for the IBk classifier we obtain better results. While with the NB classifier we obtain the worst of the experiments for this data set as the performance does not improve with any of the scenarios with respect Full scenario. It can be also considered that Gain Ratio, Info Gain and FR merging scenarios are the ones with the most similar forecasting capability.

The main advantage is time reduction and, what is more, the possibility of being able to obtain the classifier in a reasonable time and allow keeping the data set in memory in order to do any further pre-processing.

As a final milestone, a comparison between a state-of-the-art feature subset selector, CFS (this is a very well-known method in the data pre-processing community which is based on the correlation concept), and the proposal introduced in this paper, FR merging, which is an advanced feature ranking method, is reported on Table 6. From a qualitative analysis, it can be asserted that CFS wins 9 times, FR merging (our proposal) wins 8 times, and once was a tie; from this point of view our proposal achieves competitive results. From a computational efficiency perspective, FR merging is more efficient than CFS given that the average time to conduct the feature selection varies from a couple of seconds (Arcene) to about 20 seconds (Gisette), whereas CFS requires between five minutes (Gisette) and ten minutes (Arcene and Dexter). An important difference is that CFS does not need to set up any parameter related to the amount of attributes to select; in the case of FR merging, we have only specified a percentage of attributes with positive incidence (in this paper a 90%).

6 Conclusions

This paper introduced Feature Ranking merging (FRmgg) framework to feature selection which is a two-step methodology; firstly, it defines a percentage in order to set a cut-off point in the context of feature ranking-based feature selection; secondly, it merges the solutions achieved a couple of feature ranking methods using a common percentage from those features which have from a weak to strong influence in the class label, which run in an independent way. The proposed methodology has been tested with three high-dimensionality binary classification problems which have been trained with three classifiers and two measures have been reported. The test results are acceptable given that the feature space reduction is convenient compared to the raw data set and the results in some cases are better. The proposed approach may be handy for high-dimensionality problems with a few classes or medium-dimensionality data set where the number of class labels is high or very high, when there is an important requirement in terms of completion times (the sooner the better, although keeping an acceptable quality).

Acknowledgements This work has been partially subsidized by the project US-1263341 (*Junta de Andalucía) and FEDER funds.*

References

1. Aha DW, Kibler D, Marc KA (1991) Instance-based learning algorithms. Mach Learn 6(1):37–66
2. Bishop CM (1999) Pattern recognition and feed-forward networks. In: The MIT encyclopedia of the cognitive sciences, vol 13. MIT Press
3. Bolón-Canedo V, Sánchez-Marono N, Alonso-Betanzos A, Benítez JM, Herrera F (2014) A review of microarray datasets and applied feature selection methods. Inf Sci 282:111–135
4. Cawley GC, Talbot NLC (2010) On over-fitting in model selection and subsequent selection bias in performance evaluation. J Mach Learn Res 11:2079–2107
5. Chawla S (2010) Feature selection, association rules network and theory building. In: Feature selection in data mining. PMLR, pp 14–21
6. Cho S-B, Tallón-Ballesteros AJ (2017) Visual tools to lecture data analytics and engineering. In: International work-conference on the interplay between natural and artificial computation. Springer, pp 551–558
7. Cieslak DA, Ryan Hoens T, Chawla NV, Philip Kegelmeyer W (2012) Hellinger distance decision trees are robust and skew-insensitive. Data Mining Knowl Discov 24(1):136–158
8. Conlin R, Erickson K, Abbate J, Kolemen E (2021) Keras2c: a library for converting Keras neural networks to real-time compatible c. Eng Appl Artif Intell 100:104182
9. Curtin RR, Edel M, Lozhnikov MI, Mentekidis Y, Ghaisas S, Zhang S (2018) mlpack 3: a fast, flexible machine learning library. J Open Source Softw 3(26):726
10. Domingos P, Pazzani M (1997) On the optimality of the simple Bayesian classifier under zero-one loss. Mach Learn 29(2):103–130
11. Fayyad U, Stolorz P (1997) Data mining and kdd: promise and challenges. Future Gener Comput Syst 13(2–3):99–115
12. Flach P (2012) Machine learning: the art and science of algorithms that make sense of data. Cambridge University Press

13. González-Recio O, Rosa GJM, Gianola D (2014) Machine learning methods and predictive ability metrics for genome-wide prediction of complex traits. Livest Sci 166:217–231
14. Guyon I, Elisseeff A (2003) An introduction to variable and feature selection. J Mach Learn Res 3(Mar):1157–1182
15. Han J, Pei J, Kamber M (2011) Data mining: concepts and techniques. Elsevier
16. Hong J, Cho S (2006) Efficient huge-scale feature selection with speciated genetic algorithm. Pattern Recognit Lett 27(2):143–150
17. Hothorn T (2022) Cran task view: machine learning & statistical learning
18. John GH, Langley P (2013) Estimating continuous distributions in Bayesian classifiers. arXiv:1302.4964
19. Kira K, Rendell LA et al (1992) The feature selection problem: traditional methods and a new algorithm. Aaai 2:129–134
20. Kohavi R et al (1995) A study of cross-validation and bootstrap for accuracy estimation and model selection. In: Ijcai, vol 14. Montreal, Canada, pp 1137–1145
21. Larose DT, Larose CD (2014) Discovering knowledge in data: an introduction to data mining, vol 4. Wiley
22. Li J, Wu Y, Fong S, Tallón-Ballesteros AJ, Yang X-S, Mohammed S, Wu F (2022) A binary pso-based ensemble under-sampling model for rebalancing imbalanced training data. J Supercomput 78(5):7428–7463
23. Li L, Darden TA, Weingberg CR, Levine AJ, Pedersen LG (2001) Gene assessment and sample classification for gene expression data using a genetic algorithm/k-nearest neighbor method. Combin Chem High Throughput Screen 4(8):727–739
24. Liu J, Li J, Li W, Wu J (2016) Rethinking big data: a review on the data quality and usage issues. ISPRS J Photogramm Remote Sens 115:134–142
25. Ma S, Huang J (2008) Penalized feature selection and classification in bioinformatics. Brief Bioinform 9(5):392–403
26. Manaswi NK (2018) Understanding and working with Keras. In: Deep learning with applications using python. Springer, pp 31–43
27. Sánchez-Maroño N, Alonso-Betanzos A, Tombilla-Sanromán M (2007) Filter methods for feature selection–a comparative study. In: International conference on intelligent data engineering and automated learning. Springer, pp 178–187
28. Santana-Morales P, Merchán AF, Márquez-Rodríguez A, Tallón-Ballesteros AJ (2022) Feature ranking for feature sorting and feature selection: Fr4(fs)^2. In: International work-conference on the interplay between natural and artificial computation. Springer
29. Schlegl T, Seeböck P, Waldstein SM, Schmidt-Erfurth U, Langs G (2017) Unsupervised anomaly detection with generative adversarial networks to guide marker discovery. In: International conference on information processing in medical imaging. Springer, pp 146–157
30. Sonnenburg S, Rätsch G, Henschel S, Widmer C, Behr J, Zien A, de Bona F, Binder A, Gehl C, Franc V (2010) The shogun machine learning toolbox. J Mach Learn Res 11:1799–1802
31. Tallón-Ballesteros AJ, Cavique L, Fong S (2019) Addressing low dimensionality feature subset selection: Relieff (-k) or extended correlation-based feature selection (ecfs)? In: International workshop on soft computing models in industrial and environmental applications. Springer, pp 251–260
32. Tallón-Ballesteros AJ, Fong S, Leal-Díaz R (2019) Does the order of attributes play an important role in classification? In: International conference on hybrid artificial intelligence systems. Springer, pp 370–380
33. Tallón-Ballesteros AJ, Riquelme JC (2017) Low dimensionality or same subsets as a result of feature selection: an in-depth roadmap. In: International work-conference on the interplay between natural and artificial computation. Springer, pp 531–539
34. Tallón-Ballesteros AJ, Riquelme JC, Ruiz R (2019) Semi-wrapper feature subset selector for feed-forward neural networks: applications to binary and multi-class classification problems. Neurocomputing 353:28–44
35. Wang L, Fu X (2006) Data mining with computational intelligence. Springer Science & Business Media

36. Wirth R, Hipp J (2000) Crisp-dm: towards a standard process model for data mining. In: Proceedings of the 4th international conference on the practical applications of knowledge discovery and data mining, vol 1. Manchester, pp 29–40
37. Witten IH, Frank E, Hall MA, Pal CJ, Mining Data (2005) Practical machine learning tools and techniques. In: Data Mining, vol 2, p 4
38. Wu X, Kumar V (2009) The top ten algorithms in data mining. CRC Press
39. Xing EP, Jordan MI, Karp RM et al (2001) Feature selection for high-dimensional genomic microarray data. In: Icml, vol 1. Citeseer, pp 601–608
40. Xu Z, Li S, Luo X, Liu J, Tao Z, Tang Y, Xu J, Yuan P, Keung J (2019) Tstss: a two-stage training subset selection framework for cross version defect prediction. J Syst Softw 154:59–78

Chapter 44
Correlated Features in Air Pollution Prediction

Farheen and Rajeev Kumar

1 Introduction

Transport and industrial activities contribute to air pollution. It results in health hazards and higher mortality as well as economic loss. A particulate matter (PM2.5 & PM10) is the main constituent of air pollution. Specifically, PM2.5 impacts the cardiovascular system and other internal organs. Due to its adverse effects on lungs, respiratory, and nervous system, it is the most common cause of lung cancer, asthma, and stroke [10, 12, 14].

Hence, monitoring and predicting this particulate matter concentration is crucial. Several factors affect PM2.5 concentration. A dataset with multiple factors is referred to as a multivariate dataset. Multiple features can improve prediction results, but they can also cause increased dimensionality problem. There are two approaches available to resolve it:

1. decomposition into pattern subspaces (e.g., Kumar and Rockett [7, 8]), and
2. reducing feature sets by reducing correlated features and eliminating irrelevant features.

In this work, we consider the second approach and reduce the feature set using correlation. There are different ways of selecting features. It can be categorized into filter method, wrapper method, and embedded method. We use the Pearson correlation method, which comes under the filter method [2].

In this paper, we have analyzed two datasets from the UCI data repository [3]. We analyze in two parts:

Farheen · R. Kumar (✉)
Data to Knowledge (D2K) Lab, School of Computer and Systems Sciences, Jawaharlal Nehru University, New Delhi 110 067, India
e-mail: rajeevkumar.cse@gmail.com

G. Mathur et al. (eds.), *Proceedings of 3rd International Conference on Artificial Intelligence: Advances and Applications*, Algorithms for Intelligent Systems,
https://doi.org/10.1007/978-981-19-7041-2_44

1. Using Pearson correlation, we find correlated features and based on our analysis, we identify the reduced set of features, and
2. A Long Short-Term Memory (LSTM) model is used to predict PM2.5.

We use the previous 24 h data to predict the next hour. The root mean square error (RMSE) and mean absolute error (MAE) of the complete feature set are compared with the reduced feature set.

This article is arranged as follows. Section 2 reviews the literature on related topics. Section 3 discusses the methodology. Section 4 contains the experimental results and discussion. The conclusion of this paper is in Sect. 5.

2 Related Work

We included two subsections in this literature review: one discusses dimensionality reduction techniques and the second discusses PM2.5 prediction based on correlation analysis.

2.1 Curse of Dimensionality

In recent years, analyzing high-dimensional data has become a challenge. Literature provides a variety of techniques for eliminating irrelevant and redundant features. Dimension reduction techniques help to enhance processing speed and reduce the time and effort required to extract valuable information [1]. There are two approaches to reducing dimensionality: decomposition into pattern subspaces (e.g., Kumar and Rockett [7, 8]) and feature selection. There are different approaches to feature selection. We are using here the pearson correlation method [4]. Ibrahim et al. [6] used the correlation and principal component analysis method for feature selection to diagnose breast cancer.

2.2 Feature Selection in Prediction of Air-Pollution

Air pollution prediction is an active area of research. It is necessary to anticipate it in order to control it. Here, we review some relevant work on PM2.5 prediction using the correlation method.

Tao et al. [13] proposed a deep learning model built upon 1D convolutional neural network and bi-directional Gated Recurrent Units (CBGRU) to predict air pollution. They obtained the dataset from the UCI repository, which includes PM2.5 and meteorological data. To examine the relationship between various influencing factors and PM2.5, the correlation coefficient method was applied. A deep learning CBGRU model created by them was used to predict air pollution with a reduced set of features. Lastly, they compared their results with various models, including Support Vector Regression (SVR), Gradient Boosted Regression (GBR), Multinomial Regression (MR), Decision Tree Regressor (DTR), Simple Recurrent Neural Network (RNN), Long Short-Term Memory Networks (LSTM), Gated Recurrent Unit (GRU), and Bidirectional Gated Recurrent Unit (BGRU). Compared to other models, they analyzed that the CBGRU model gives better prediction results.

Li et al. [9] proposed a deep learning model called Long Short Term Memory Neural Network Extended (LSTME) model. Air quality data were collected from 12 stations in Beijing. Correlation coefficients were computed amongst these stations. All the stations were highly correlated with PM2.5. Hence, a single model was used instead of separate models for each station. They provided predictions for various time slices, e.g., for 1 h, 2 h, 3 h, 4–6 h, 7–12 h, and 13–24 h; their performance was found satisfactory even for 13–24 h predictions. They compared their results of the LSTME model with the Spatiotemporal Deep Learning (STDL) model, Time-Delay Neural Network (TDNN) model, Autoregressive Moving Average (ARMA) model, Support Vector Regression (SVR) model, and Traditional Long Short Term Memory Neural Network (LSTM NN) model. Compared to other models, the LSTME model was found to be more accurate.

Mahajan et al. [11] examined pollutants that significantly impacted the air quality index of different regions and identify patterns for those features. Information was taken from the central pollution control board. Air quality measurements were taken for a five year period from July 2015 to July 2020 of various cities. There are twelve pollutants in the dataset and an air quality index (AQI). Pollutants are PM2.5, PM10, NO, NO_2, NO_x, NH_3, SO_2, O_3, CO, Benzene, Toluene, and Xylene. They used the correlation coefficient method to analyze the data. In their analysis, particulate matter (PM2.5, PM10) and NO_2 were strongly correlated with AQI, so these pollutants were selected as major pollutants. They analyzed the trends across these pollutants on different dimensions, like regions, months, weekdays, weekends, and holidays to identify the patterns.

3 The Proposed Methodology

3.1 Correlation for Feature Selection

The Pearson correlation method for feature selection determines strength amongst features. We have used a heat map to analyze the correlation strength of various influencing factors with the target value (PM2.5) and in between the influencing factors. In the Beijing dataset, we have analyzed that some meteorological features such as temperature and dew point are strongly related to each other. Pressure is also strongly correlated with them. Hence, we have dropped the dew point and

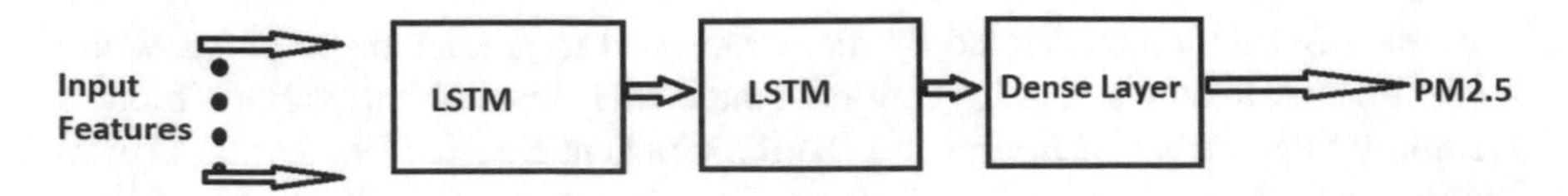

Fig. 1 LSTM model

pressure from this dataset. The analysis for meteorological variables presented above is consistent across all datasets.

3.2 Proposed LSTM Architecture

Hochrieter and Schmidhuber [5] proposed the LSTM cell to deal with long-term dependencies. They introduced a gate into the RNN cell for improving its capacity to memorize. In comparison to a Simple Recurrent Neural Network, each neuron in LSTM functions as a memory cell. There are three gates in a neuron: an input gate, a forget gate and an output gate. Internal gates help to solve the problem of long term dependency. Internal gates help to overcome long-term dependence.

Figure 1 shows an LSTM model that we used in our work. We use two LSTM layers and a dense layer. We input various influencing factors with PM2.5 in our model and predict the target (PM2.5) series.

4 Experimental Results

4.1 Dataset

In this paper, we use two benchmark datasets from the UCI repository. Datasets contain missing values. To deal with them, we use backward filling. We use a label encoder to convert the categorical variables.

Beijing PM2.5 Dataset Data considered is from Beijing, China. It includes PM2.5 concentrations and meteorological features such as Dew Point, Wind Direction, Temperature, Snow, Pressure, and Wind Speed. It is an hourly dataset for a four year period from January 2010 to December 2014. Total records in the dataset are 43824.

We divided the data into three categories: training, validation, and testing. Training data contains the first 35000 records, and validation data contains 7000 records. Test data is for the period from 18-10-2014 to 31-12-2014.

Beijing Multi-Site Air-Quality Dataset Air-quality data were collected from twelve sites in Beijing, China. It is an hourly dataset of four year period from March 2013 to February 2017. It includes multiple features such as Particulate Matter (PM2.5,

PM10) concentration, SO_2, NO_2, CO, O_3, Temperature, Pressure, Dew Point, Rain, Wind Direction, and Wind Speed.

We divided the data into three categories: training, validation, and testing. Training data includes the first 28000 records, and validation data contains 6008 records. Test data is for the period from 17-01-2017 to 28-02-2017.

4.2 Correlation Guided Feature Selection

We have plotted heat maps for various datasets. Figure 2 illustrates the correlation plot for the Beijing dataset. Figures 3, 4 and 5 display the correlation plot for all the twelve sites of the multi-site air-quality dataset. In the next step, we examined these plots. We found a high correlation between a few meteorological features (Temperature, Dew Point, and Pressure) across all datasets. Based on our findings, we removed some features from our dataset. Table 1 shows the complete set of features and the reduced set for both datasets.

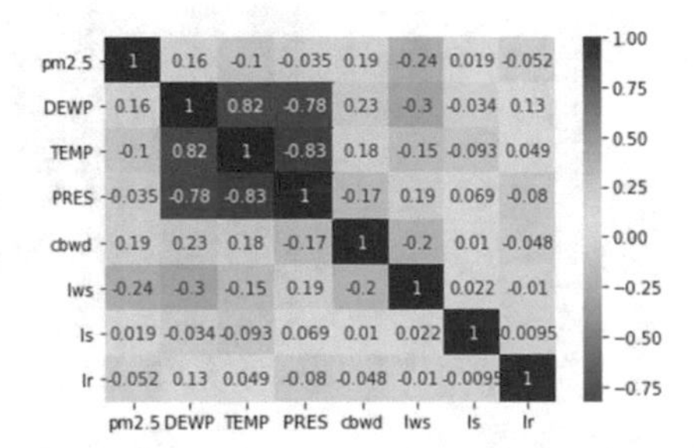

Fig. 2 Heat map for Beijing PM2.5 dataset

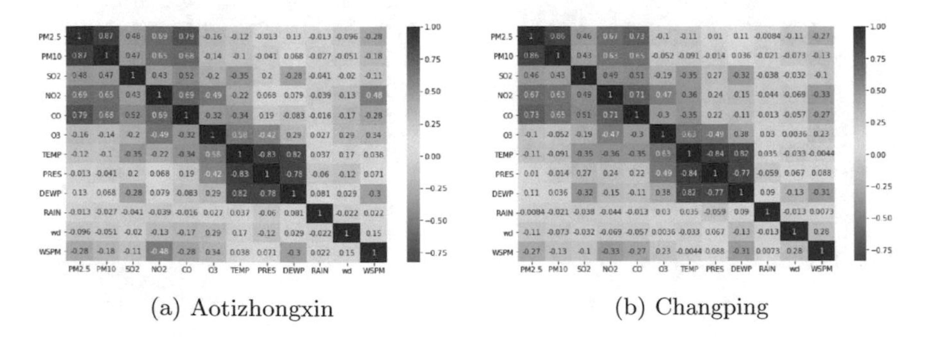

(a) Aotizhongxin (b) Changping

Fig. 3 Heat maps for Beijing multi-site air-quality dataset

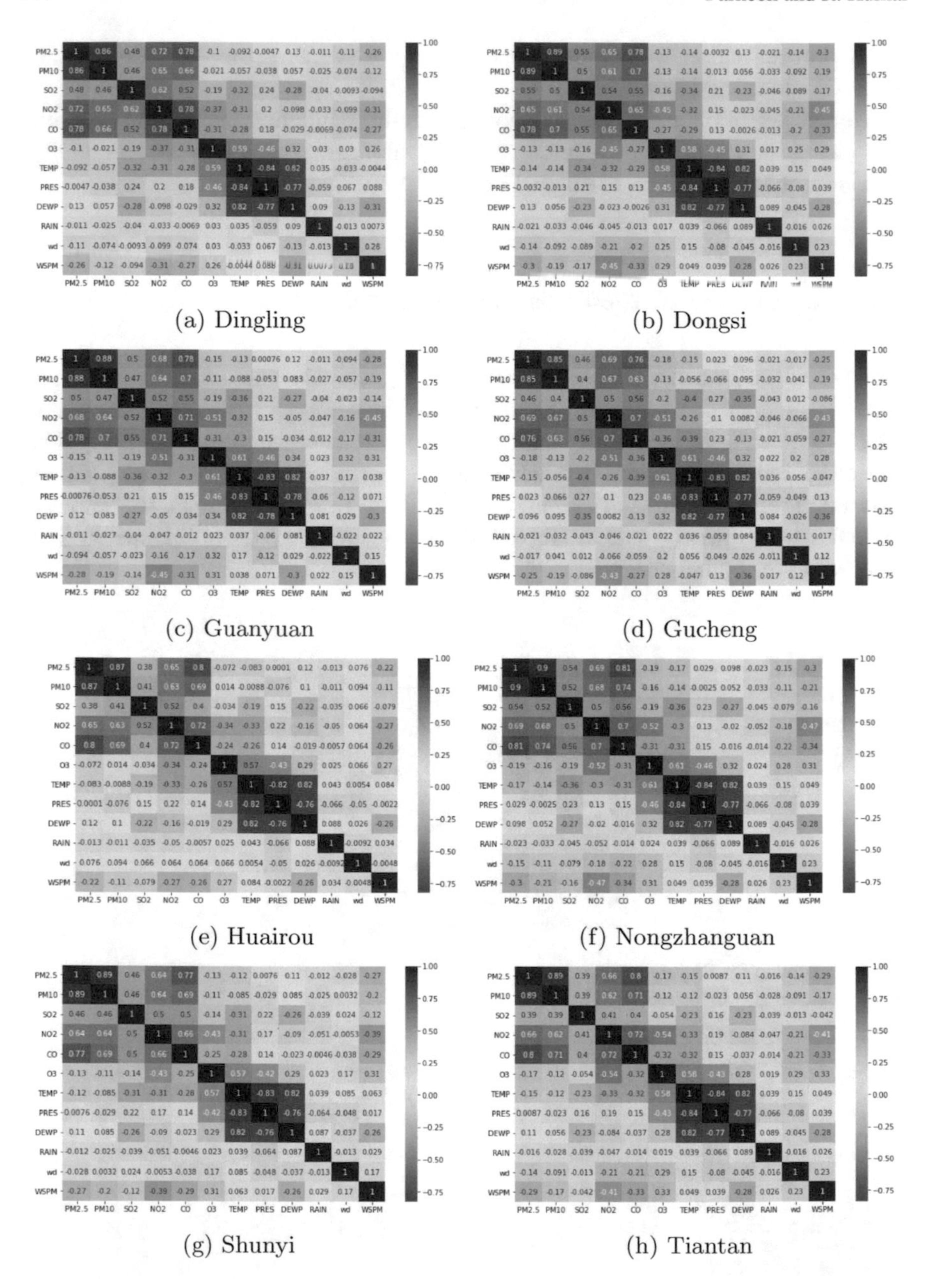

(a) Dingling (b) Dongsi

(c) Guanyuan (d) Gucheng

(e) Huairou (f) Nongzhanguan

(g) Shunyi (h) Tiantan

Fig. 4 Heat maps Beijing multi-site air-quality dataset

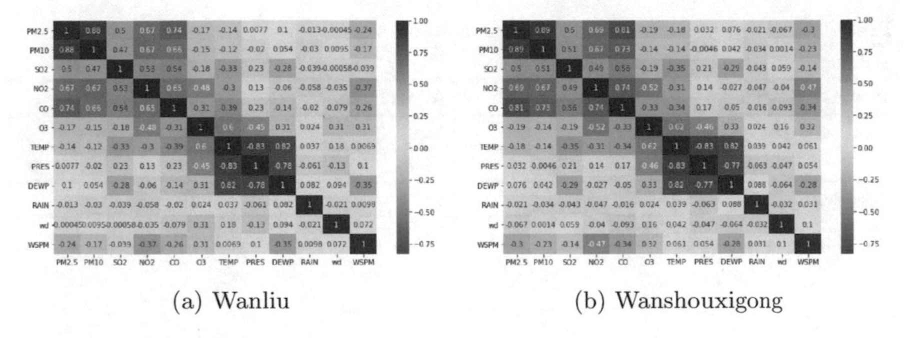

(a) Wanliu (b) Wanshouxigong

Fig. 5 Heat maps for Beijing multi-site air-quality dataset

Table 1 Feature description for both datasets

Feature table		
Data set	All features	Reduced features
UCI Beijing PM2.5	PM2.5, Dew Point, Wind Direction, Temperature, Snow, Pressure, Rain and Wind Speed	PM2.5, Wind Direction, Temperature, Snow, Rain and Wind Speed
Multi-Site Air Quality Beijing	PM2.5, PM10, SO_2, NO_2, CO, O_3, Temperature, Pressure, Dew Point, Rain, Wind Direction, and Wind Speed	PM2.5, PM10, SO_2, NO_2, CO, O_3, Temperature, Rain, Wind Direction, and Wind Speed

Table 2 Accuracy measure for full and reduced feature set

Beijing dataset		
Accuracy measure	All features	Reduced feature
MAE	13.56	12.53
RMSE	22.51	22.26

4.3 Prediction with LSTM

After reducing the number of features, the LSTM model as described above is used to predict. This model predicts the next hour by using data from the previous 24 hours. The results of RMSE and MAE for the complete feature set and the reduced feature set are shown in Tables 2 and 3. We are only sharing the results of the Beijing PM2.5 dataset and two sites (Aotizhongxin & Changping) from all the 12 sites of the multi-site air quality data. Tables 2 and 3 show that RMSE and MAE values for the reduced feature set are not degraded even after removing some features from both datasets.

Table 3 Accuracy measure for full and reduced feature set

Beijing multi site air quality dataset			
Name of the site	Accuracy measure	All features	Reduced feature
Aotizhongxin	MAE	11.48	11.34
	RMSE	23.48	22.96
Changping	MAE	10.34	10.07
	RMSE	19.32	19.14

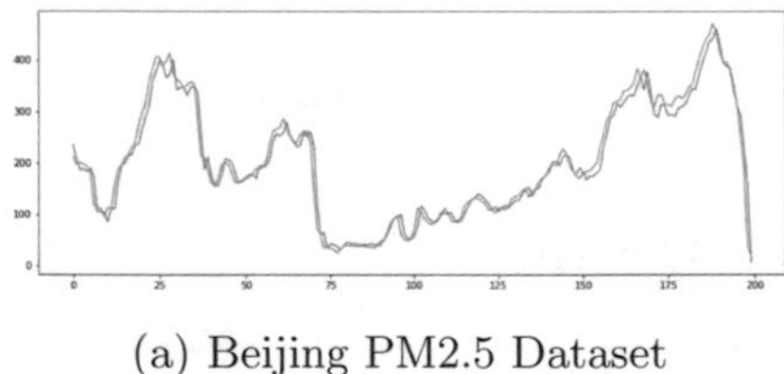

(a) Beijing PM2.5 Dataset

Fig. 6 Prediction plots for reduced feature set

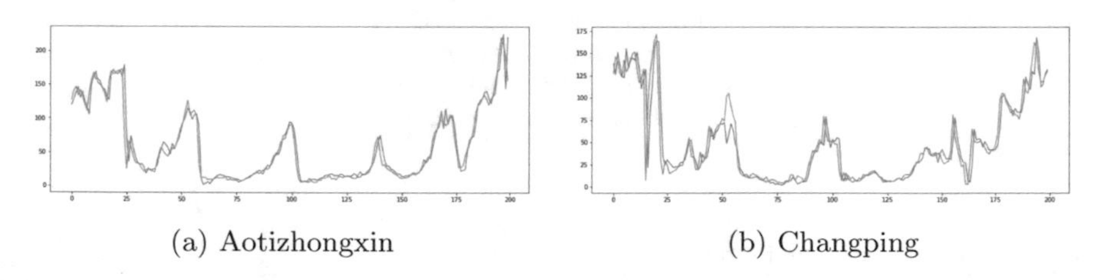

(a) Aotizhongxin (b) Changping

Fig. 7 Prediction plots for reduced feature set (Multi-site air-quality dataset)

Figures 6 and 7 illustrate the actual and predicted test results. The first 200 values are plotted from the test dataset, and these plots are shared only for a reduced feature set.

4.4 Discussion

We examined correlation plots for various datasets and found that a few meteorological features across all datasets highly correlated with each other. In the Beijing PM2.5 dataset, meteorological elements such as temperature, dew point, and pressure are highly correlated. Further, we examined the same for all 12 sites of the multi-site air quality dataset to confirm our findings and found consistent results.

The meteorological variables temperature, dew point, and pressure show strong correlations across all datasets. PM10, CO, and NO_2 are strongly correlated with the target value (PM2.5), and also these three have a strong inter-correlation with each other.

5 Conclusion

In this work, we have used the correlation method to identify the strength amongst features. We analyze that certain meteorological variables were strongly correlated. We eliminated such correlated features and made predictions based on the complete feature set and the reduced feature set based on our findings. We created an LSTM model to predict a single step. We compared the RMSE and MAE values of the complete and reduced feature sets and found that the results were not degraded. In future, we wish to extend these experiments with several other appropriate deep learning models.

References

1. Ayesha S, Hanif MK, Talib R (2020) Overview and comparative study of dimensionality reduction techniques for high dimensional data. Inf Fusion 59:44–58
2. Chandrashekar G, Sahin F (2014) A survey on feature selection methods. Comput Electric Eng 40:16–28
3. Dua D, Graff C (2017) Uci machine learning repository
4. Guyon I, Elisseeff A (2003) An introduction to variable and feature selection. Mach Learn Res 3:1157–1182
5. Hochreiter S, Schmidhuber J (1997) Long short-term memory. Neural Comput 9:1735–80
6. Ibrahim S, Nazir S, Velastin SA (2021) Feature selection using correlation analysis and principal component analysis for accurate breast cancer diagnosis. Imaging 7:225
7. Kumar R, Rockett P (1998) Multiobjective genetic algorithm partitioning for hierarchical learning of high-dimensional pattern spaces: a Learning-follows-Decomposition strategy. IEEE Trans Neural Netw 9(5):822–830
8. Kumar R, Rockett PI (1996) ANCHOR-a connectionist architecture for hierarchical nesting of multiple heterogeneous neural nets. In Proceedings of AAAI workshop integrating multiple learning models (IMLM 96), pp 59–65
9. Li X, Peng L, Yao X, Cui S, Yuan H, You C, Chi T (2017) Long short-term memory neural network for air pollutant concentration predictions: method development and evaluation. Environ Pollut 231:997–1004
10. Lu F, Xu D, Cheng Y, Dong S, Guo C, Jiang X, Zheng X (2015) Systematic review and meta-analysis of the adverse health effects of ambient PM2.5 and PM10 pollution in the Chinese population. Environ Res 136:196–204
11. Mahajan M, Kumar S, Pant B, Tiwari UK, Khan R (2021) Feature selection and analysis in air quality data. In: Proceedings of 11th international conference on cloud computing, data science & engineering (Confluence), pp 280–285
12. Ostro B, Lipsett M, Reynolds P, Goldberg D, Hertz A, Garcia C, Henderson KD, Bernstein L (2010) Long-term exposure to constituents of fine particulate air pollution and mortality: Results from the California Teachers Study. Environ Health Perspect 118(3):363–9

13. Tao Q, Liu F, Li Y, Sidorov D (2019) Air pollution forecasting using a deep learning model based on 1D convnets and bidirectional GRU. IEEE Access 7:76690–76698
14. World health organization, WHO air quality guidelines for particulate matter, ozone, nitrogen dioxide and sulphur dioxide.global update (2005)

Chapter 45
Deep Learning-Based Condition Monitoring of Insulator in Overhead Power Distribution Lines Using Enhanced Cat Swarm Optimization

J. Jey Shree Lakshmi, J. Subalakshmi, J. Joyslin Janet, B. Vigneshwaran, and M. Sivapalanirajan

1 Introduction

Insulation coordination on power equipment needs more concentration in the power grid due to transmission and distribution advancements. Recently, intelligent techniques are commonly used in high-power apparatuses for better diagnosis and tracking before breakdown. There are billions of high- voltage insulators in service Sampedro et al. [1]. Field inspection activities by helicopters or walking patrols are essential for medium and high-voltage insulators. Human experts' field inspection activities cause safety issues for those experts. In continuation to that, there may be an incorrect diagnosis about the status of the HV insulators. Therefore, it indicates that artificial intelligence technique-based assessment is attractive for assessing power apparatus status in live lines Prates et al. [2].

In common, it is tough to predict the insulator components from the insulator images effectively because the individual insulator's shape, color, and textures may look different. The background clutters the real-time insulator images with a series of other insulators. Besides, the insulator gives blur images due to jitter during the inspection of the robot's movement Gao et al. [3].

Recently, CNN architecture has been increasing its depth to learn more unique features from image datasets. Wang et al. [4] VGGnet and Ray et al. [5] GoogLeNet have more than ten layers in their architecture. Deep learning networks are tremendously hard because of vanishing gradients and degradation. Recently, Liu et al. [6] Highway networks and Kumar et al. [7] ResNet proposed to include connections between the adjacent convolutional layers, which have the property to improve

J. Jey Shree Lakshmi · J. Subalakshmi · J. Joyslin Janet · B. Vigneshwaran ·
M. Sivapalanirajan (✉)
National Engineering College, Kovilpatti, India
e-mail: sivapalanirajan@gmail.com

G. Mathur et al. (eds.), *Proceedings of 3rd International Conference on Artificial Intelligence: Advances and Applications*, Algorithms for Intelligent Systems,
https://doi.org/10.1007/978-981-19-7041-2_45

the vanishing gradient issue. However, optimizing these deep learning networks is complicated because of the massive quantity of parameters and the amount of database memory. An alternating network, integrating the existing layers in the deep network rather than deepening by adding new layers, is proposed by Liu et al. [8] and Shengtao and Jianying [9]. They are the updated version of the networks with multi-level intermediate layers and added new side branches. Thus, deep fusion integrates the power of individual layers and produces the maximum recognition rate. Though these deep fusion networks have attained talented recitals, they motionlessly need many extra parameters required for generating the side branches.

Traditionally available optimization techniques like Particle Swarm Optimization (PSO) by Ababneh and Bataineh [10], Bee Colony Optimization (BCO) by Karaboga [11], Ant Colony Optimization (ACO) by Kannan et al. [12], etc., The above-said methods provide higher results by using the details from previous iterations without knowing the features. These methods failed to provide valid information about the global optimum based on exploration capability, convergence speed, and solution quality. The proposed algorithm practices a modified meta-heuristic evolutionary optimization algorithm from Cat Swarm Optimization (CSO) Shu and Tsai [13] called ECSO.

This proposed work performs condition monitoring of high-voltage insulators from OPDLs using CFN with ECSO to predict the materials and insulator defects with real-time backgrounds. CFN architecture is designed based on the plain Convolution Neural Network (CNN) model in the first stage. CFN is a fusion architecture that integrates the in-between layers with adaptive weights. Our proposed algorithm consists of salient features like efficient side outputs made in the CFN by using 1×1 convolution and global average pooling Piao et al. [14] to make side branches from in-between layers. Secondly, locally connected layers in the network act as a fusion module. Finally, the CFN module's extraction features are optimized using ECSO for evaluating the high-level components, and those features are input to Multi-class NLSVM. These approaches abridged the workload of the data acquisition process, which improved the recognition rate of the proposed model.

2 Data Collection

An image compilation contains information about the shape and type of materials, defects, and interference for the data-gaining stage. The image dataset collected from a studio has a smooth and controlled environment in terms of external background information. But the drawback behind such a dataset is the lack of natural environment backgrounds, including streets, infrastructure, and objects that originate in the city. The overlap of conductors with the environment and the difference in the image brightness, dimensions, and angles. In this proposed work, four different configurations (C1–C4) of insulator types are considered for recognition, as shown in Table 1.

In this proposed work, four different types of distribution HV insulators are chosen, which function in 11 kV. Image Collection Station (ICS) is used for data

Table 1 Different configurations of insulator

Configuration	Types of insulators
C1	Ceramic Insulator without defect
C2	Ceramic Bicolour Insulator without defect
C3	Polymeric Insulator without defect
C4	Glass Insulator without defect

acquisition from a series of different insulator images totaling 800 shots, out of which 200 of each type of insulator, 100 photo collection relates to intact images (ten for each angle of θ, with equal intervals of 10°). The remaining 100 consists of defective images of insulators with different defects and tips. For each insulator, 520 shots were collected to make 2080 images. The 520 image collection of a component has 400 intact parts and 120 defective components. It results in an approximate ratio of 4 photographs of intact components for each defective component. The dataset contains 2880 images in total and acts as the database.

3 Methodology

Traditional approaches follow a standard format of feature descriptors and classifiers to make the feature extraction from aerial insulators. Such images of the insulator material make it difficult to differentiate and detect defective insulators. It extracts only shallow image features from the image dataset for the recognition of defective insulators. The extraction of superficial features with complex background images reduces the classification rate.

3.1 Convolution Neural Network (CNN)

Traditional CNN can learn gradable image features from aerial images. Once CNN is deeper, it considers global and local information from the image dataset. The significant drawbacks behind the extracted features from the CNNs are not dependent upon the insulators' size, position, and direction. Traditional CNN approaches only classify the image pixels and consider the image blocks around the pixels as CNN's input for training and testing.

This proposed work presents a standard CNN model with ten convolutional layers, four max-pooling layers, one average pooling layer, and at last, ReLU and softmax layer. The Max-pooling layer takes part after the pair of convolution layers. The initial four-pair convolution layer consists of the max-pooling layer, and the last layer is the average pooling layer. Except for the last convolution layer, every layer consists of the ReLU layer. After two consecutive convolutional layers, it produces a high-dimensional vector that increases computational time. A CFN algorithm reduces the

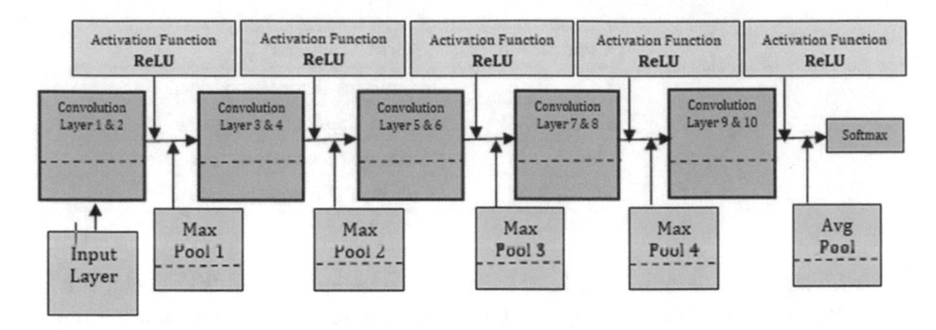

Fig. 1 Deep convolutional neural network architecture

computation time with an increment in recognition rate by extracting the unique and informative features from the input patterns. The proposed plan network is shown in Fig. 1.

3.2 Deep Fusion Network

In recent years, multilayer fusion networks in the current literature present the importance of intermediate outputs openly in deep networks [15–18]. In-depth, the researchers created side branches from the in-between layers, and finally they fused the extracted features for prediction. Cheng et al. [19] proposed DAG CNNs models add all the multi-scale outputs from in-between layers of CNN. The drawback behind the algorithm is that adding many parameters in the CNN failed to consider the importance of the side branches. CFN can learn adaptive weights for fusing side branches in our proposed algorithm while adding a few parameters.

3.3 Convolution Fusion Networks (CFN)

CFN has new side branches in between the layers and combines them into a locally connected (LC) fusion module through a basic CNN model, as in Fig. 2. Conventional methods add a new fully-connected layer in the side branch, which causes an increase in the number of branches as an alternative CFN can add side branches effectively in-between the layer with few parameters. Initially, the side branches are taken from the pooling layer of each convolution layer and connected to the 1×1 Convolution layer as the main branch. It is essential to determine all the 1×1 Convolution layers have the same number of input channels for integration. The global average pooling (GAP) will convert the feature vector into one-dimensional. The main branch of 1×1 Conv layers follows the convolution layer and not the pooling layer.

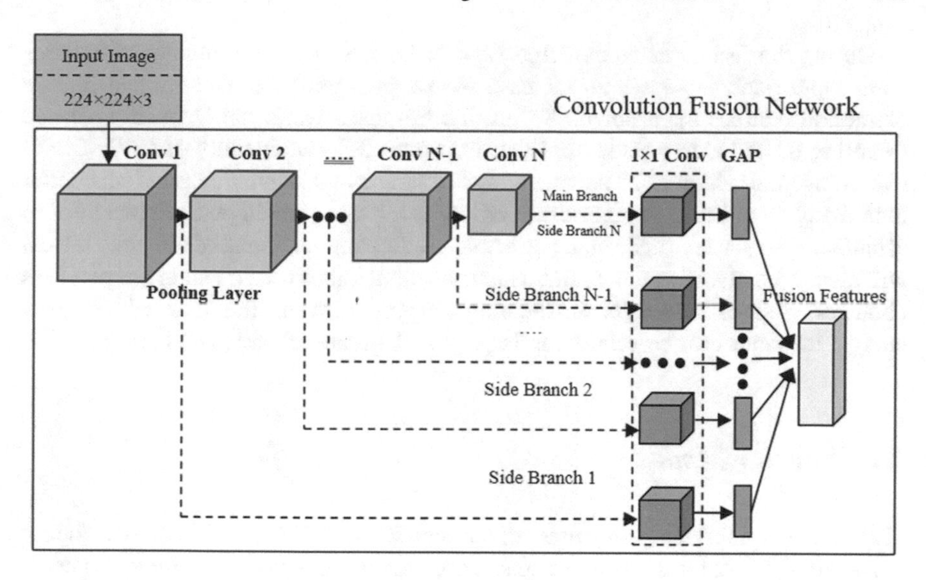

Fig. 2 The general pipeline of the proposed CFN

Let us assume there are N side branches in the proposed network for condition monitoring of insulator status. The last side branch, the Nth branch, is called the main branch, directly connected to the convolution layer. For the Nth side branch, the input for the 1 × 1 Conv layer is $v_{i,j}^{(n)}$, where $n = 1,2,\ldots N$ and (i, j) is the spatial location of feature maps. $H^{(n)}$ and $W^{(n)}$ from the Nth 1 × 1 Conv layers denote the actual height and width of the feature vectors. Then GAP is performed to make a single-dimensional feature vector, and it is calculated by Eq. (1)

$$G_k^{(n)} = \frac{1}{H^{(n)}W^{(n)}} \sum_{i=1}^{H^{(n)}} \sum_{j=1}^{W^{(n)}} f_{i,j,k}^{(n)} \quad (1)$$

where $G_k^{(n)}$ is the kth element in the nth GAP feature vector. $G^{(n)}$ denotes the GAP feature from the main branch.

4 Selection of Feature Using Enhanced Cat Swarm Optimization

DCNN is used to extract the features from the input patterns. In continuation to that, selecting appropriate features is the next stage for recognizing the insulator status. Feature selection is a procedure of finding the most pertinent and optimal features in the image without experiencing damage to information. The proposed work can improve the search capability within the workspace, which helps in extracting the unique features from the DCNN.

During the feature extraction algorithm in DCNN, a greater number of features from input patterns causes an increase in computation cost. The optimal feature extraction rectifies the problem. So, In this proposed work, the Opposition-Based Learning (OBL) algorithm is used that improves the searchability of CSO. ECSO overcomes the influence of parameter convergence and stagnation problems better than many meta-heuristic algorithms. ECSO is a Swarm Intelligence (SI) algorithm established to portray cats' usual performance. In common, most of the time, the cat will always be in resting mode. But it also has great attentiveness and inquisitiveness about the objects around its surroundings. Apart from inactive time which saves energy, it spends time searching for its prey with greater attention and energy.

4.1 Seeking Mode

This model explains the resting times of the cats. Even though in seeking mode, the cat moves to different locations in the space, nonetheless, it remains attentive. It understands the solutions as a local search. It is essential to denote four important factors in the seeking mode: seeking memory pool (SMP), counts of dimension to change (CDC), seeking a range of the selected dimension (SRD), self-position considering (SPC), and Mixture Ratio (MR). SMP determines the number of simulation cats as a copy. SRD defines the cat dimension difference between old and new for mutation. CDC explains the counts of the cat at different dimensions experienced for mutation. MR determines that the cats spend most of their time resting and observing. The following steps are involved in the seeking mode of the CSO algorithm:

1. Delineate the number of copies (Q) of the jth cat.
2. Add or subtract SRD values randomly from the current position of cats and restore the previous values for all copies.
3. Calculate the fitness for all copies. Identify the finest solution and implement it at the position of jth cat. Equation (2) determines the chosen likelihood of each candidate.

$$Prob_j = \frac{|fitness_j - f|}{fitness_{max} - fitness_{min}} \tag{2}$$

where $Prob_j$ defines the chosen likelihood of candidate j, $fitness_{max}$ is the maximum fitness value, $fitness_{min}$ is the minimum fitness value, $fitness_j$ is a fitness value for the candidate j. When the aim is to find the minimum fitness, f is $fitness_{max}$.

4.2 Tracking Mode

The following steps are involved in the tracking mode:

1. Initially, each cat's fitness should measure for finding the best cat in the tracking mode
2. The velocity of each cat varies according to Eq. (3)

$$C_{new} = C_{k,d}^{m} + A \times t_1 \times \left(y_{best,d}^{m} - y_{k,d}^{m}\right) \quad (3)$$

where $C_{k,d}^{m}$ is the Cat_k velocity, t_1 is a constant value.
A is a random value that lies between 0 and 1.
d is the dimension that varies between 1 and D.
m is the current iteration's total $y_{best,d}^{m}$ is the best position of the cat during iteration m.
$y_{k,d}^{m}$ denotes the Cat_k position, C_{new} determines the new velocity.

3. The position of each cat is changed as per Eq. (4)

$$y_{new} = y_{k,d}^{m} + C_{new} \quad (4)$$

where y_{new} denotes the new Cat_k position.

5 Results and Discussion

The results obtained throughout the assessment of the different training optimizer models on the test set composed of 520 datasets for material classification and 1600 dataset classification are shown in Tables 2, 3, and 4.

This proposed work uses SVM, NLSVM with RBF, and NLSVM with sigmoidal function for feature recognition. The average accuracy rate of SVM is 85.96% for

Table 2 Confusion matrix of the proposed system with support vector machine

		Predicted class				
		C1	C2	C3	C4	RR (%)
Actual class	C1	452	25	30	13	86.92
	C2	38	461	11	10	88.65
	C3	16	36	445	23	85.57
	C4	37	24	29	430	82.69
Accuracy						85.96

Table 3 Confusion matrix of the proposed system with NLSVM–RBF function

		Predicted class				
		C1	C2	C3	C4	RR (%)
Actual class	C1	465	21	11	23	89.42
	C2	27	471	16	4	90.58
	C3	19	23	460	18	88.46
	C4	15	18	7	480	92.31
Accuracy						90.19

Table 4 Confusion matrix of the proposed system with NLSVM–Sigmoidal function

		Predicted class				
		C1	C2	C3	C4	RR (%)
Actual class	C1	500	6	4	10	96.15
	C2	8	503	6	3	96.73
	C3	3	7	510	0	98.08
	C4	1	0	1	518	99.62
Accuracy						97.64

material classification and 87.40% for insulator defect classification. The results show the data augmentation and a structural architecture to make CFN. It also plays a significant role in image classification using DL. Results presented in Table 3 expose the influence of NLSVM with RBF function.

The results summarized in the table reveal that the proposed work achieves better results compared to the existing methodologies. The best classification rate is 97.64% for NLSVM with a sigmoidal kernel function to predict the insulator materials. The results obtained in this sequence shows the proposed system conveys a valuable solution for the power line inspection automation process. It also helps greatly in reducing the workload faced by manual inspection with specialized human inspectors to diagnose defects. Convinced image features, such as a destructive change in shape, texture, and overlaps, make misclassifications.

6 Conclusion

For a better recognition rate, we propose a talented methodology for the automatic examination of insulators found in OPDL using a deep fusion network combined with ECSO. The proposed method discusses the prediction of material type and its defect in the HV insulator using CFN built on a plain CNN. The architecture evaluates discriminative features from the input images for better classification. For experimental analysis, initial photographs were taken in a studio using controlled conditions considered for training. At the same time, another set of images are collected from a specially made distribution lines looking real. These types of CFN models are not very deep, but they can challenge both deep (e.g., 10 layers) and much more profound (e.g., 100 layers). CFN inherits the generalization capabilities of CNN.

References

1. Sampedro C, Rodriguez-Vazquez J, Rodriguez-Ramos A, Carrio A, Campoy P (2019) Deep learning-based system for automatic recognition and diagnosis of electrical insulator strings. IEEE Access 7:101283–308
2. Prates RM, Cruz R, Marotta AP, Ramos RP, Simas Filho EF, Cardoso JS (2019) Insulator visual non-conformity detection in overhead power distribution lines using deep learning. Comput Electric Eng 78:343–55
3. Gao Z, Yang G, Li E, Shen T, Wang Z, Tian Y, Wang H, Liang Z (2019) Insulator segmentation for power line inspection based on modified conditional generative adversarial network. J Sens 1–8
4. Wang X, Zhang J, Yan WQ (2019) Gait recognition using multichannel convolution neural networks. Neural Comput Appl 32:14275–14285
5. Ray S, Ganguly B, Dey D (2020) Identification and classification of stator inter-turn faults in induction motor using wavelet kernel based convolutional neural network. Electric Power Compon Syst 48:1421–1432
6. Liu H, Hussain F, Shen Y, Morales-Menendez R, Abubakar M, Junaid Yawar S, Arain HJ (2019) Signal processing and deep learning techniques for power quality events monitoring and classification. Electric Power Compon Syst 47:1332–1348
7. Kumar D, Samantaray SR, Kamwa I, Sahoo NC (2014) Reliability-constrained based optimal placement and sizing of multiple distributed generators in power distribution network using cat swarm optimization. Electric Power Compon Syst 42:149–164
8. Liu Y, Guo Y, Georgiou T, Lew MS (2018) Fusion that matters: convolutional fusion networks for visual recognition. Multimed Tools Appl 77:29407–29434
9. Li S, Li J (2017) Condition monitoring and diagnosis of power equipment: review and prospective. High Volt 2(2):82–91
10. Ababneh JI, Bataineh MH (2008) Linear phase FIR filter design using particle swarm optimization and genetic algorithms. Digit Signal Process 18(4):657–668
11. Karaboga N (2009) A new design method based on artificial bee colony algorithm for digital IIR filters. J Frankl Inst 346(4):328–348
12. Kanan HR, Faez K, Hosseinzadeh M. (2007). Face recognition system using ant colony optimization-based selected features. In: Proceedings of the 2007 IEEE symposium on computational intelligence in security and defense applications (CISDA 2007). IEEE, pp 57–62
13. Shu CC, Tsai FW (2007) Computational intelligence based on the behavior of Cats. Int J Innov Comput Inf Control 3(1):163–173
14. Piao J, Chen Y, Shin H (2019) A new deep learning based multi-spectral image fusion method. Entropy 21:1–16
15. Zhai Y, Chen R, Yang Q, Li X, Zhao Z (2018) Insulator fault detection based on spatial morphological features of aerial images. IEEE Access 6:35316–35326
16. Chen Q, Yan B (2019) Research on aerial insulators convolution neural network detection and explosion detection. J Electron Meas Instrum 31(6):942–953

17. Liao S, An J (2016) Aerial detection of damaged insulators on transmission lines. J Syst Simul 33(4):176–179
18. Jiang Y, Han J, Ding J (2017) Glass insulator identification and blasting defect diagnosis based on multi-feature fusion. China Electric Power 50(5):52–58
19. Cheng J, Wang L, Xiong Y (2018) Modified cuckoo search algorithm and the prediction of flashover voltage of insulators. Neural Comput Appl 30:355–370

Chapter 46
Music Generation and Composition Using Machine Learning

Akanksha Dawande, Uday Chourasia, and Priyanka Dixit

1 Introduction

Music is a universal language that has many formalizations, but there is no one guiding theory for music composition. Moreover, each person has unique music preferences. For the last 2 decades, musical generation via deep learning models seems to have been a hot issue. Music presents a distinct difficulty than visuals in terms of three primary dimensions: To begin with, music is temporal, having a hierarchical system and cross-temporal relationships. Secondly, music is made up of a variety of interconnected tools that evolve over time. Finally, music is divided into chords, arpeggios, and melodies, resulting in several outcomes for every time-step [1, 2].

Harmony occurs when numerous notes or voices are played at the same time to create a new sound. The harmonies' blended tones are complementary and agreeable to the ear.

Harmony may be seen in chords and chord progressions. 3 or even more notes play at the very same time in a chord. The melody is supported or complimented by the chords and chord progressions in a musical piece.

Harmony is also achieved by combining voice components. A choir's united vocals are an excellent illustration. A choir's numerous voices mix together to create a pleasing sound.

Melody is a musical phrase made up of a series of notes or voices. The melody is frequently a most distinctive and recognized aspect of a song.

A. Dawande (✉) · U. Chourasia · P. Dixit
RGPV University, Bhopal, India
e-mail: akankshadawande15@gmail.com

G. Mathur et al. (eds.), *Proceedings of 3rd International Conference on Artificial Intelligence: Advances and Applications*, Algorithms for Intelligent Systems,
https://doi.org/10.1007/978-981-19-7041-2_46

Instruments or voices may be used to generate melodies. They consist of two or even more artistically appealing notes in a series. The majority of compositions are made up of numerous melodies which repeat. Pitch and rhythm are indeed the two most important aspects of a melody:

- The aural vibration generated by an equipment or speech is known as pitch. It's the pitch of a note, whether it's high or low. A melody is created by combining various tones in a succession. The amount of time every pitch will sound is referred to as Rhythm or Duration. Full notes, ½ notes, 1/4 notes, triplets, and other beat subdivisions are used to split these durations.
- Rhythm is a crucial component of music that has several meanings. Consider the following scenario: A recurrent motion of notes and pauses (silences) in time is referred to as rhythm. It has to do with how people perceive time. A pattern of weak and strong notes or sounds which repeats throughout one song is sometimes referred to as rhythm. Drums, percussion, instruments, and voices may all be used to produce these patterns.

Audio data, on the other hand, contains a number of characteristics that are similar to what is often investigated in DL (computer vision and NLP). The music's continuous structure reminded us of natural language processing, something one can do with RNN. There are several other audio 'channels' (in terms of tones and instruments) that resemble visuals that CNN may be utilized for. Deep generative networks, on the other hand, are an interesting new field of study with the potential to generate realistic synthetic data. Variation Auto encoders (VAEs) and Generative Adversarial Networks (GANs), and also language models in NLP, are a few examples. In the family of neural networks, the RNN is a member of feedback neural networks (a subfamily of neural networks) with feedback connections. The ability to send information over time-steps makes the RNN different from the other members of the family of the neural network. When we talk about the working style of RNN, it considers sequential information as input and provides information in sequential form instead of accepting stable input and providing stable output. The output provided by the layers inside the networks re-enters into the layer as the input which helps in computing the value of the layer and by this process the network makes itself learn based on the current data and previous data together. Talking about the architecture of the basic RNN, we can say that an RNN consists of many copies of the neural network connected and working in a chain. As we have discussed the capabilities of the RNN are that it can work with sequential data as input and output Audio data is likewise sequentially data collected in the sense that it may be thought of as a signal that modulates with timing, analogous to time series information wherein data sets are recorded in a sequence with time values.

2 Literature Review

Mao et al. [1] proposed a framework for gaining knowledge about musical style utilizing DNN and conclusively proved a method for influencing the prototype to start generating music with such a given mix of artist styles using a distributing representation of style. The Biaxial modeling was modified by incorporating loudness and style that ultimately increased quality of the created music. Nevertheless, the produced music's lack of long-term framework and common point is an issue that needs to be addressed. DeepJ is an end-to-end probabilistic graphical model competent of creating music focusing on a particular mix of musician styles, as described by the author.

Mason et al. [2] investigated the use of unit selecting as well as sequences as a method of producing music that used a ranking-based process, in which a unit is defined as an arbitrary length multitude of musical measures. The unit database's width and volume, as well as the unit length, are both critical to the program's effectiveness. The capacity to recreate never-before-seen music by selecting units from a database was demonstrated using an autoencoder. The authors first investigate if a unit selection method limited to a finite size component librarian is adequate for covering a broad range of music. The author accomplishes this by creating a deep auto–encoder which encodes a musical input and reconstructs it using a library of choices. The authors then characterize a generative method that incorporates a deep structured semantic model (DSSM) and an LSTM to estimate the next unit, with units consisting of four, two, and one musical measures, respectively.

Jhamtanie et al. [3] proposed a generative model for music generation focusing on self-repetition. The authors use a Generative Adversarial Network formulation to learn a model which can generate compositions with long-term repetition structures similar to those found in training data. The authors propose to represent self-repetition in a composition using a self-similarity matrix constructed by computing similarity between pairs of measures. To avoid optimization issues due to the discrete nature of notes in musical compositions, and to provide more flexibility in identifying similarity between measure pairs, the authors propose to encode measures into low dimensional measure.

Jean-Pierre et al. [4] established the domain, examined earlier and pioneer approaches, and proposed an analytical model to aid in the analysis and classification of the great range of programs and operations documented in the literature, with many examples that illustrate it. Todd's Time-Windowed and conditioned recurrent architectures were the first attempts to employ ANN to make music. This proposed model is intended to aid in the examination of diverse viewpoints that led to the construction of different DL-based music generating systems. It consists of five basic characteristics that describe distinct methods of using DL algorithms to produce the musical material, as well as the categorizations that correspond to each dimension. Purpose, Interpretation, Structure, Requirements, and Strategy are all terms that may be used to describe anything. Feed-forward neural architectural has a Feedforward approach, as does Todd's Time-Windowed architecture, which is dependent on the

Iterative strategy. The recurrent architecture Strategy of recursion is the foundation of the well-known LSTM (Long Short-Term Memory) architecture. A bidirectional RNN is made up of two RNNs, whereas the Recurrent neural network-RBM architecture [1] is made up of a Recurrent neural network and an RBM structure. A variational autoencoder (VAE) that is based on refinement include the GLSR-VAE architecture, as well as a variational recurrent autoencoder (VRAE). A stacked autoencoder design, such as the DeepHear design, and a recurrent autoencoder structures, such as the MusicVAE architecture, are two types of autoencoder architectures. The anticipation-RNN architecture, C-RNN-GAN architecture, MidiNet architecture, and DeepBach architecture are all examples of this kind of architecture. This paper continues with a lesson on deep learning-based music creation before moving on to a conceptual framework for analyzing the many ideas and dimensions involved.

Garca et al. [5] proposed a methodology for autonomously producing and finishing musical compositions. The model is based on ML and DL generative learning techniques like RNN. Related efforts treat sound as a document, enabling the network to fully understand the sheet music's grammar and symbol relationships. This requires a lot of training and, in many situations, may lead to overfitting. Business Knowledge, Information Understanding, Information Processing, Formulation, Validation, and Implementation are the six steps of this process. This work adds by removing the most difficult constraints from the information, enabling the musical substance to be isolated from the language. A website app depending on the training set is also shown. Beginners may use the programme to create automated music form start or from a specified songbooks fragment. This study explains how DL technologies related to language modelling approaches may be used to generate music automatically. RNN using Gated Recurrent Units (GRUs) and embedding input layers achieves a 44.50% accuracy.

Hewahi et al. [6] suggested a concept for using a long short-term memory neural network to generate music parts. The proposed approach accepts midi files, turns them to song files, then encodes them such that they may be used as NN inputs. Before feeding data into the neural network, an augmentation step is carried out, which involves splitting the file into multiple keys. The file is then put into the neural network, which is then trained. The last stage is to create music. The primary goal was to provide a random note to the neural network, which would then progressively reconstruct it until it produced a nice piece of music. Several tests have been carried out to determine the optimal parameter values that may be used to generate nice music. With certain files, the results were incredible, as the created music pieces were perfectly in tune in terms of rhythm and harmony.

Jaques et al. [7] established a methodology for sequentially learning in which the sequential prediction is filtered and improved while keeping the strong predictive qualities learnt from the information by maximizing certain imposed incentive functions. They investigated the applicability of their method in the context of music creation. To anticipate the next note in a musical sequence, an LSTM is trained on a huge corpus of 30,000 MIDI tunes. The reward function for this Note-RNN is a mix of rewards based on music theory principles, and the results of some other learned Note-RNN, which is then improved via reinforcement learning. The results showed that combining machine learning with reinforcement learning could not only generate more attractive tunes, but also dramatically minimize undesirable RNN behaviors and failure modes.

Manan et al. [8] collected information regarding the advances in GANs that had shown interesting outcomes. Adding layers after preceding one have convergence is proved to aid in achieving better convergence and stabilization of the network and also decreasing the time for training by a significant amount. As a result, they apply this training approach to gradually validate the system in the time and pitch domains. They used a layer of determinism binary neuron at the ending of the generators to produce binaries value outcomes rather than fraction between 0 and 1, since deterministic binary neurons have been shown to assist improved performance in several earlier proposed frameworks.

Dua et al. [9] provided an upgraded version of the existing sheet music system. The utilization of recurrent neural networks and lstm was critical to the success of the project. In order to attain the aim, two modules were employed specifically, since the end outcomes with these modules was superior than that of the one used previously.

Mangal et al. [10] created a music suite using a fully trained model. Experimentation and modeling training sessions were carried out on Google Colab, on Google Cloud Platform, using deep learning and Keras code. Their article proposes the development that may be used to autonomously compose music and melodies without the need for human intervention. Using a single layered LSTM model, the machine can remember past dataset information and synthesize polyphonic music.

Chen et al. [11] suggested a producing framework that takes the Generative Adversarial Network architecture to produce note sequencing. The usage of a CNN, which was optimized as per the peculiarities of musical notes, was crucial. The optimization approach aided the CNN in concentrating on understanding all of the music qualities while also speeding up the experimental phase. On the classic piano dataset, Kong et al. [12] built a DNN model for various music categorizations.

Cheng et al. [13] suggested a CNN approach to classify music genres. On the GTZAN dataset, the author retrieved Mel spectrum as a feature and passed it into Convolution Neural Network for additional training, achieving an accuracy of 84.0%.

For music genre categorization, Shi et al. [14] suggested a CNN model based on VGG-16. The chroma feature, which reflects the time domain and frequencies domain of musical characteristics and takes into account the presence of harmony, was retrieved.

3 Methodology

The goal of this study is to identify various forms of music for song writing. Figure 1 shows a block schematic of the suggested technique. In three phases, the suggested technique is explained. The initial phase is gathering music data. The next stage was to extract features from the music signals. The next stage is to divide music

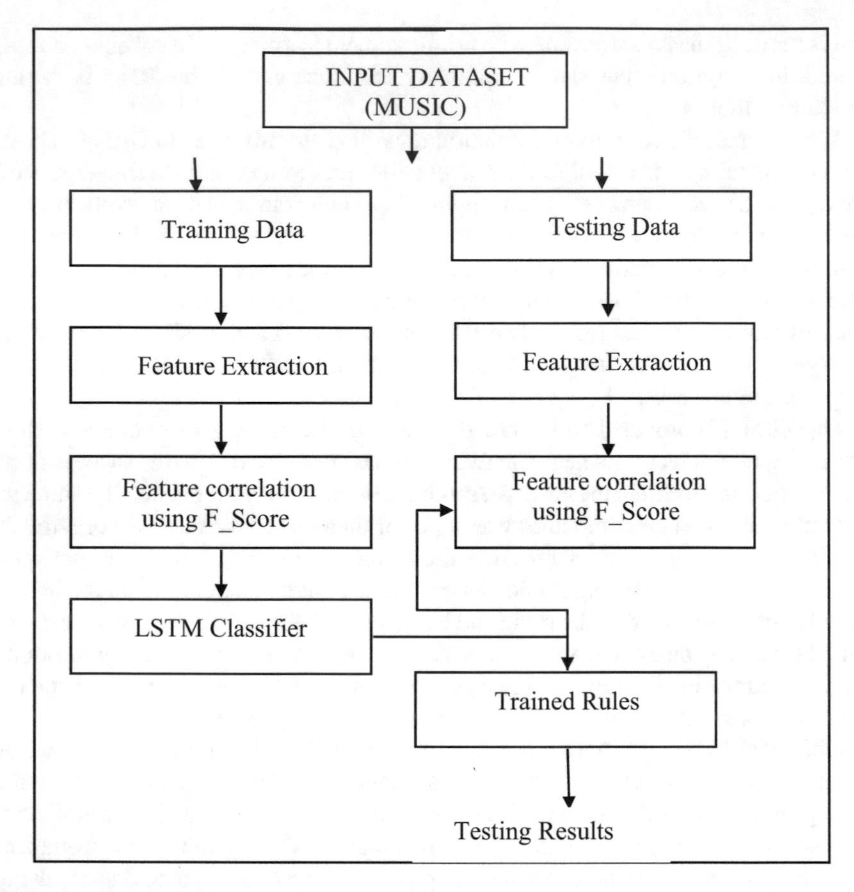

Fig. 1 Block diagram of proposed methodology

into several genres. Figure 2 depicts the suggested technique, which is addressed in further depth in the next section.

3.1 Signal Gathering

The data is taken from the classic piano midi collection. The dataset was created on a digital piano using a MIDI-based sequencer and then transferred to audio formats. There are also some scores and audios available that display the scores as they are being played.

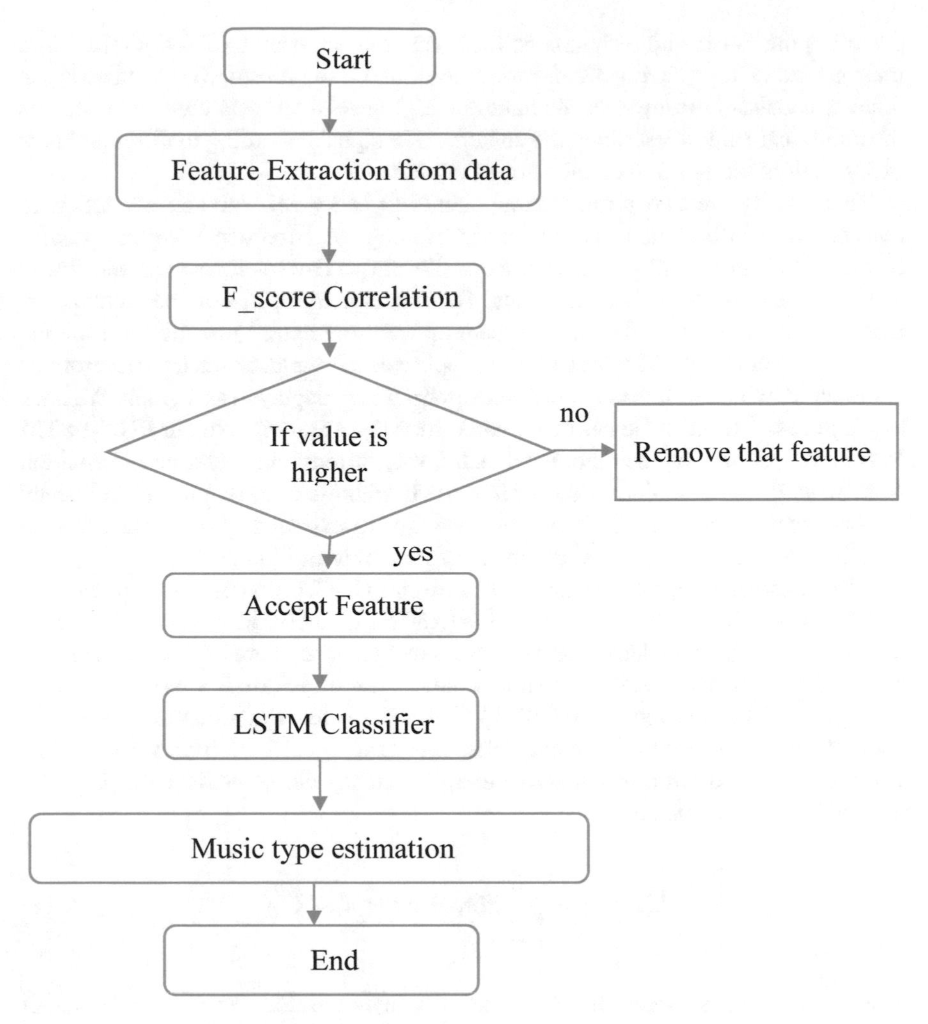

Fig. 2 Flowchart of proposed methodology

3.2 Feature Extraction

The process of computing a compact or concise numerical representation that is used to typify a fragment of audio is referred to as feature extraction. The aim of feature extraction is to represent a music piece or fragment into a compact and descriptive way. Suitable machine learning algorithms or deep learning algorithms are then used to classify the audio signals into the desirable outputs (such as genre) using the significant features extracted. Feature extraction is the foremost procedure of pattern recognition systems. Content- based features and text-based features are the two major features that can be extracted from the music audio signals. The content-based features involve the extraction of features that are embedded in the audio file that describes the music audio signal and these features are used to classify music into their genres or identify important information about the music. The content-based features consist of the low-level features and high-level features. Low-level feature is the numerical values describing the contents of a signal according to different kinds of inspection: temporal, spectral, perceptual, etc.

There really are two primary components to every ML project: extraction of features from information and modelling training. Mel-frequency cepstral coefficients (MFCCs) are often retrieved from the music dataset for sound and music extracting features for ML applications. The model is trained using these characteristics. MFCC feature extraction is a technique for extracting just the most crucial data from an audio file. The MFCCs use a mel scale to extract characteristics from an audio input, which results in a mel spectrogram when displayed as a graph. We have to preprocess the audio file before we can utilize this as an input to our RNN-LSTM NN. Preprocessing, on the other hand, is the way of removing relevant information from an audio stream. Since it determines the brightness of a sound, Mel-frequency cepstral coefficient (MFCC) is one of the ways to extract important data from a signal. It can also be used to determine the sound's timbre (quality).

Other characteristics are retrieved in addition to MFCC and are listed below:

STFT: The Short-time Fourier transform (STFT) is a Fourier-related transformation that can be used to detect sine frequency and phase content of tiny sections of a signal as they change over time. A short-time Fourier transform is a series of Fourier transforms of a windowed signal (STFT). The STFT offers time-localized frequency components whenever the frequency elements of a signal vary from time to time, while the standard Fourier transform offers frequency components averaging over the entire signal's time period.

$$\mathrm{X}[\mathrm{k}, \mathrm{n}_0] = \sum_{n=0}^{Nft=1} x[no + n].w(n).e^{-j\frac{2\pi kn}{Nft}} \tag{1}$$

where, k = frequency index, $\mathrm{n_o}$ = time index, w(n) = window function $e^{-j\frac{2\pi kn}{Nft}}$ = DFt kernel

X[no + n] = $\mathrm{N_{FT}}$ points of x starting at n.

There really are two primary components to every ML project: extraction of features from information and modelling training. Mel-frequency cepstral coefficients (MFCCs) are often retrieved from the music dataset for sound and music extracting features for ML applications. The model is trained using these characteristics. MFCC feature extraction is a technique for extracting just the most crucial data from an audio file. The MFCCs use a mel scale to extract characteristics from an audio input, which results in a mel spectrogram when displayed as a graph. We have to preprocess the audio file before we can utilize this as an input to our RNN-LSTM NN. Preprocessing, on the other hand, is the way of removing relevant information from an audio stream. Since it determines the brightness of a sound, Mel-frequency cepstral coefficient (MFCC) is one of the ways to extract important data from a signal. It can also be used to determine the sound's timbre (quality).

Other characteristics are retrieved in addition to MFCC and are listed below:

STFT: The Short-time Fourier transform (STFT) is a Fourier-related transformation that can be used to detect sine frequency and phase content of tiny sections of a signal as they change over time. A short-time Fourier transform is a series of Fourier transforms of a windowed signal (STFT). The STFT offers time-localized frequency components whenever the frequency elements of a signal vary from time to time, while the standard Fourier transform offers frequency components averaging over the entire signals time period.

$$\mathrm{X[k, n_0]} = \sum_{n=0}^{Nft=1} x[no+n].w(n).e^{-j\frac{2\pi kn}{Nft}} \tag{2}$$

where, k = frequency index, n_o = time index, w(n) = window function $e^{-j\frac{2\pi kn}{Nft}}$ = DFt kernel

X[no + n] = N_{FT} points of x starting at n.

Spectral Centroid: The spectral centroid is a metric used to describe a spectrum in digital signal processing. It shows where the spectrum's center of mass is situated. It has a strong perceptual link with the perception of a sound's intensity. A spectral power distribution's "center of mass" is represented by the spectral centroid. It is computed as the weighted mean of the frequencies contained in a signal, as determined by a Fourier transform, with the magnitude of the frequency serving as weights:

$$\text{Centroid, } \mu = \frac{\sum_{i=1}^{N} fi \cdot mi}{\sum_{i=1}^{N} mi} \tag{3}$$

where m_i represents the magnitude of bin number i, and f_i represents the center frequency of that bin.

Harmony: Harmony is the sound of two or more notes played at the same time in music. In fact, this wide definition may also cover certain instances of notes being played in succession. The ear generates its own simultaneity in the very same way as the eye experiences movement in a motion picture if the consecutively played notes remind you of the tones of a known chord (a group of notes sounded together).

Mean: The focus/center of a group of values is referred to as the mean. For every sub-band signal, the Mean is taken into account.

$$Mean = 1/N \sum_{i=1}^{n} Xi \tag{4}$$

Median: The median is the middle value that separates the top and lower halves of a data sample/population, or a probability distribution. It may be thought of it as the "intermediate" values of an information gathering in simple relationships.

Variance: The sum of the squared deviations from the Mean.

$$\sigma = (x + a)^n = \sum_{i=1}^{n} (X - \mu)^2 / N \tag{5}$$

Standard deviation: The standard deviation is a straightforward measurement of an information set's variability. The RMS departure of its data from the mean is called the standard deviation.

$$std = \sqrt{(\sum_{i=1}^{n} (Xi - X)^2))/N - 1} \tag{6}$$

Skewness: Skewness is a measure of the asymmetry of a real-valued random variable's probability distribution around its mean. The value of skewness might be positively or negatively, or it can be indefinite.

$$skewness = E(X - \mu\sigma)^3 \tag{7}$$

Kurtosis: Kurtosis refers to a distribution's relative peakedness or flattening in comparison to the normal distribution. This is provided by:

$$kurtosis = \mu 4/\sigma^4 \tag{8}$$

where, $\mu 4$ is the fourth moment about the mean and б is the standard deviation.

3.3 F_score Correlation

After feature extraction, the feature vector matrix of datasets are further send to the F-Score correlation algorithm to determine co-related features. F-Score correlation is an algorithm which is used to determine the direct or indirect relation among data values. Suppose the dataset contained n features then F_score determines n*n relation among them.

Let's considers:

Input dataset (x_n), where n = 1, 2… k.

Number of classes = c, where ($c >= 2$).

Then F-Score correlation of ith feature with jth feature is determined as in Eq. (i):

$$F_i = \frac{\sum_{j=1}^{c}\left(\overline{x}_i^j - \overline{x}_i\right)^2}{\sum_{j=1}^{c}\frac{1}{n_j - 1\sum_{n=1}^{n_j}(x_{n,i}^j - x_i^j)^2}} \tag{9}$$

where $\overline{x_i}$ = mean of ith feature of entire dataset.

$\overline{x_i}^j$= mean of ith feature of the jth dataset.

$\overline{x_{n,i}}^j$= ith feature of the nth instance in the jth dataset.

The numerator of above-mentioned equation, Eq. (9), represents the discrimination among classes in the dataset. Whereas the denominator represents the discrimination within each of the classes in the dataset. If this F-Score value is smaller among feature set then those features are not related to each other whereas if the value is higher, then that feature is highly related and can be added to the feature subset.

3.4 Feature Classification

Once the characteristics have been acquired, they are merged into a vector that may be used. The supplied data set is initially divided into two groups: training and testing. After the data has been divided, the classifier is applied to the training phase for every proportion to create classifier rules. The data was then classified using the test dataset. LSTM is used for categorization. LSRMs are a sort of RNN that may acquire ordering dependency in sequence prediction challenges. This is a need in a variety of complicated issue areas, including machine translation, voice recognition, and others. LSTMs work including both LTMs and STMs, and they employ the notion of gates to make the computations efficient and simple.

Very much like a straightforward RNN, a LSTM additionally has a secret state where H(t–1) addresses the secret condition of the past timestamp and Ht is the secret condition of the current timestamp. Notwithstanding that LSTM likewise have a cell state addressed by C(t–1) and C(t) for past and current timestamp separately.

Here the secret state is known as Short term memory and the cell state is known as Long term memory. Please refer Fig. 3.

4 Result Analysis

4.1 Parameters for Evaluating Performance

Accuracy: This metric represents the recognition accuracy of every known test inputs as a proportion of the total training data and is calculated as follows:

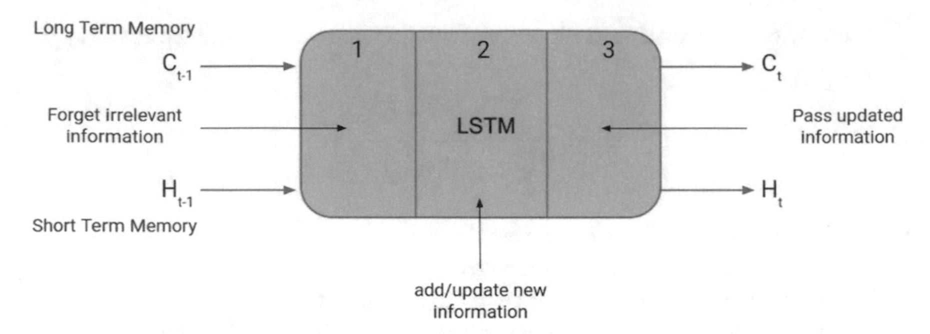

Fig. 3 LSTM architecture [9]

$$Accuracy = (TP + TN)/TP + TN + FP + FN \tag{10}$$

Precision: It's calculated by dividing the number of properly detected motions in each class by the total number of correctly recognized movements in all courses:

$$Precision = TP/(TP + FP) \tag{11}$$

Recall: The amount of correct result divide it by the number of outcomes that should've been given is referred to as recall.

$$Recall = TP/(TP + FN) \tag{12}$$

F-Measure: The F-Measure is the combined accuracy rate and recall merit. This component was used to assess the implementation's performance in terms of proper outcomes, i.e., without taking into account incorrect recognition data, and the outcome was provided by:

$$F_{measure} = \frac{2 * (precision * recall)}{(Precision + Recall)} \tag{13}$$

4.2 Result Analysis

This section proposed music generation that is based on machine learning approach such as LSTM. The result was evaluated on two datasets i.e., classic piano dataset and GTZAN dataset. Their results are illustrated individually. When training the LSTM model for music detection, the number of iterations is observed by the loss value. Figure 4 shows the training loss vs number if iterations graph, where Y-axis depicts the loss and X-axis depicts the number of iterations while learning on the classic piano dataset. Similarly, Fig. 5 illustrates the training loss on the GTZAN dataset. It is clear from the graph that as the number of iterations increases the loss decreases.

In this section, error rates are analyzed on 5 different testing samples and their relationship with different errors like MSE as well as RMSE. Table 1 shows the

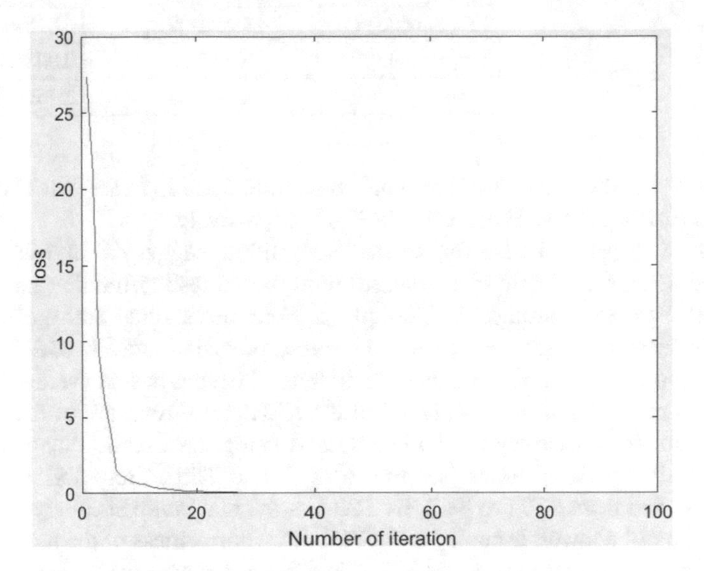

Fig. 4 Training loss versus number of iteration on classic piano dataset

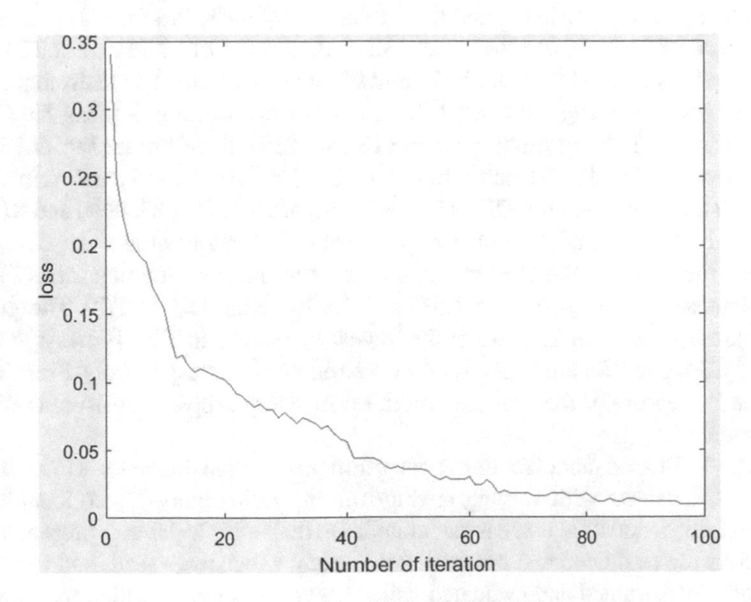

Fig. 5 Training loss versus number of iteration on GTZAN dataset

Table 1 MSE and RMSE evaluation with varying training/testing ratios

Testing samples	MSE	RMSE
Test_Sample_1	1.33E-05	0.003651
Test_Sample_2	3.01E-06	0.001734
Test_Sample_3	8.67E-06	0.002946
Test_Sample_4	4.65E-09	6.82E-05
Test_Sample_5	8.32E-06	0.0029
Average	6.66E-06	2.26E-03

variation of the MSE and RMSE with different training ratio in LSTM. The average MSE and RMSE are 6.66E-0s6 and 2.26E-03 respectively.

The LSTM-based model for music recognition suggested in this study is computed using the Classic Piano dataset with varied testing/training ratios designated Test_Sample_1 through Test_Sample_5. Four metrics are tested and compared across numerous datasets presented: accuracy, precision, recall, and f-measure. According to experimental data, the recommended design has an average accuracy of 85.56%, precision of 92.08%, recall of 80.47%, and F-measure of 83.76%. The suggested model's accuracy (in %) is assessed using the Classic Piano Dataset at different training/testing ratios (see Fig. 6). (TS1 to TS5). From TS1 to TS5, the accuracy varied from 82.71 to 88.72%. TS3 has the maximum accuracy of 88.72%, while TS2 has the lowest accuracy of 82.71%. The correctness of the recommended model is clear, with an average of 88.72%. Figure 7 shows the proposed model's precision using the Classic Piano dataset at various training ratios (TS1 to TS5). From TS1 to TS5, precision varied from 88.21 to 94.89%. Precision is best at TS3, with a score of 94.89%, and lowest at TS2, with a score of 88.21%. For TS1, TS4, and TS5, precision is 91.24%, 93.76%, and 92.30%, respectively. On average, precision is 92.08%. The suggested model's recall is shown in Fig. 8 using the Classic Piano dataset at different training ratios (TS1 to TS5). Recall ranged from 73.17 to 84.70% from TS1 to TS5. Recall is lowest in TS1, at 73.17%, and greatest in TS3, at 84.70%. The recall rates for TS2, TS4, and TS5 are 79.53%, 83.78%, and 81.16%, respectively. Recall is 80.47% on average. Figure 9 illustrates the F-measure of the proposed model using the Classic Piano dataset at different training ratios (TS1 to TS5). F-measure ranged from 78.05% to 87.89% from TS1 to TS5. The greatest F-measure is 87.89% in TS3, while the lowest is 78.05% in TS1. F-measure values for TS2, TS4, and TS5 are 81.58%, 86.86%, and 84.42%, respectively. From TS1 to TS2, the F-measure of the proposed model noticeably drops. F-measure is 83.37% on average.

The LSTM-based model for music detection proposed in this study is trained using the GTZAN dataset with varying testing/training ratios named Test_Sample_1 to Test_Sample_5, and its average is calculated. On the multiple datasets presented, four parameters are evaluated and compared: accuracy, precision, recall, and f-measure. The model was trained and evaluated using LSTM as a classification model on the GTZAN dataset. The suggested design has an average accuracy of 96.40%, precision

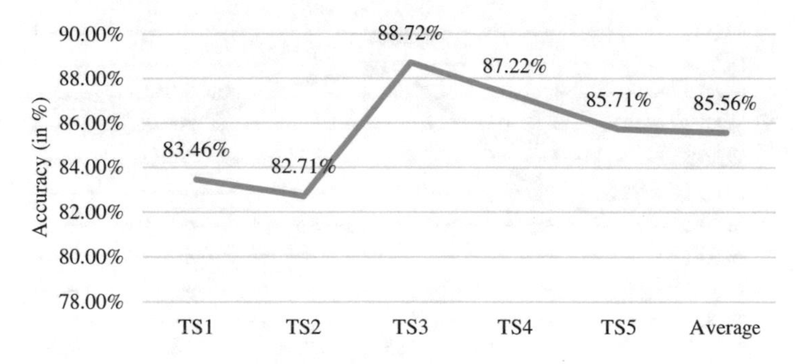

Fig. 6 Accuracy evaluation on classic piano dataset

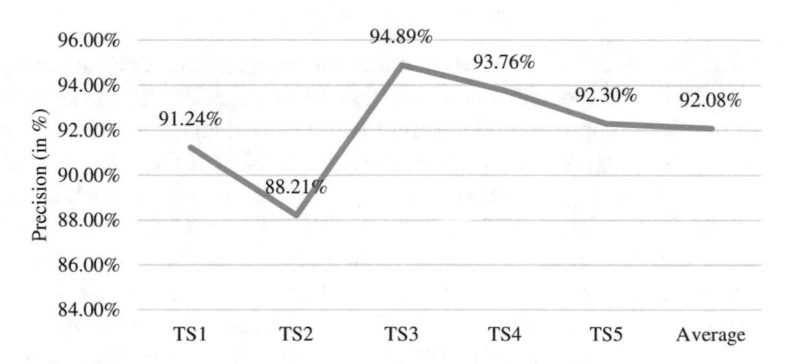

Fig. 7 Precision evaluation on classic piano dataset

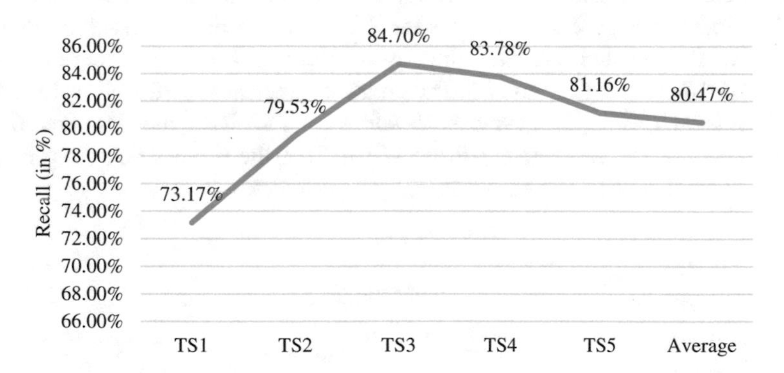

Fig. 8 Recall evaluation on classic piano dataset

of 96.70%, recall of 96.51%, and F-measure of 96.54%, according to experimental data.

Figure 10 depicts the proposed model's accuracy (in %) assessment using the GTZAN dataset at various training ratios (TS1 to TS5). The accuracy ranged from 94 to 98% from TS1 to TS5. The accuracy is highest at TS1, which is 98%, and

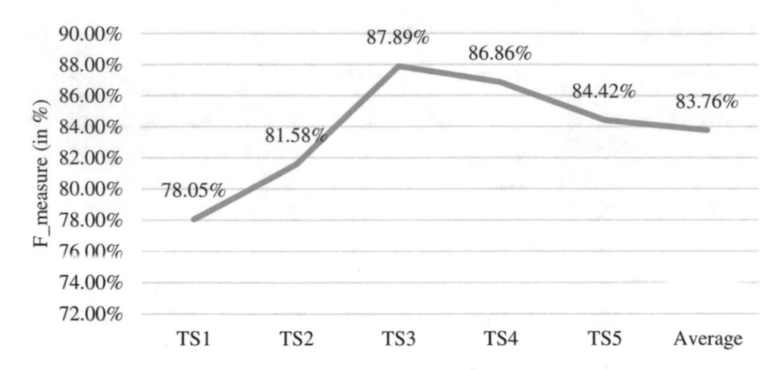

Fig. 9 F_measure evaluation on classic piano dataset

lowest at TS5, which is 94%. It is evident that the suggested model's accuracy falls from TS1 to TS2. The accuracy rate is 96.40% on average. Figure 11 demonstrates the precision of the proposed model using the GTZAN dataset at different training ratios (TS1 to TS5). Precision ranged from 98.35% to 94.56% from TS1 to TS5. The precision at TS1 is the highest, at 98.35%, while at TS5, it is the lowest, at 94.56%. Precision is 96.55%, 97.46%, and 96.60% for TS2, TS3, and TS4. It is evident that the suggested model's precision declines from TS1 to TS2. Precision is 96.70% on average. Figure 12 shows the proposed model's Recall using the GTZAN dataset at various training ratios (TS1 to TS5). From TS1 to TS5, Recall varied from 98.08% to 94.24%. TS1 has the highest Recall, at 98.08%, while TS5 has the lowest, at 94.24%. For TS2, TS3, and TS4, Recall is 96.60%, 97.66%, and 95.97%, respectively. The Recall of the recommended model clearly decreases from TS1 to TS2. On average, Recall is 96.51%. Figure 13 illustrates the F-measure of the proposed model using the GTZAN dataset at different training ratios (TS1 to TS5). F-measure ranged from 98.16% to 94.26% from TS1 to TS5. The greatest F-measure is 98.16% in TS1, while the lowest is 94.26% in TS5. F-measure values for TS2, TS3, and TS4 are 96.53%, 97.52%, and 96.21%, respectively. From TS1 to TS2, the F-measure of the proposed model noticeably drops. F-measure is 96.54% on average.

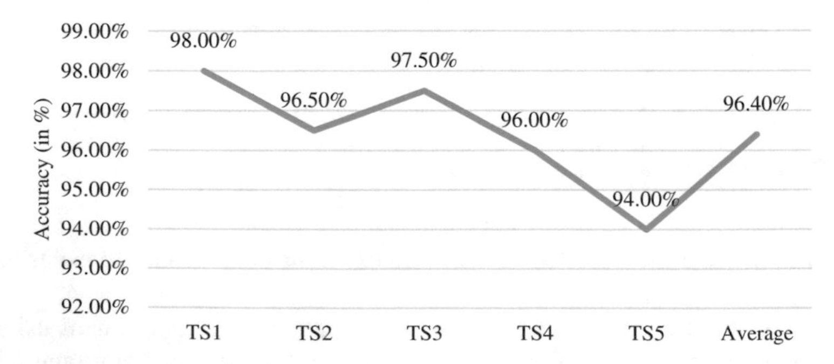

Fig. 10 Accuracy evaluation on GTZAN dataset

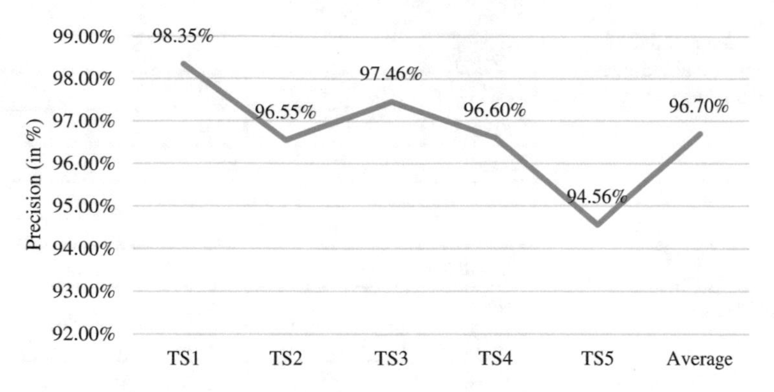

Fig. 11 Precision evaluation on GTZAN dataset

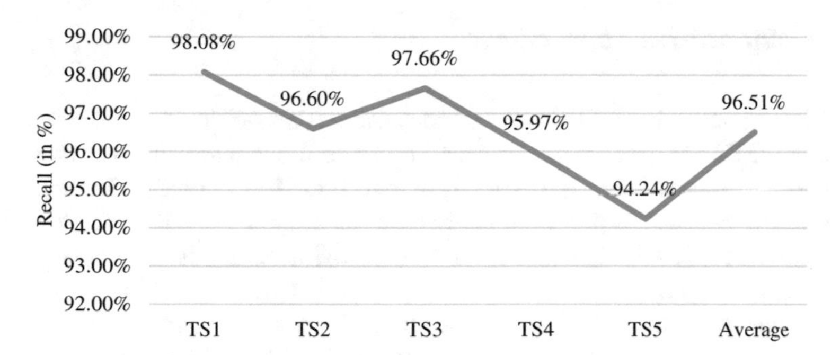

Fig. 12 Recall evaluation on GTZAN dataset

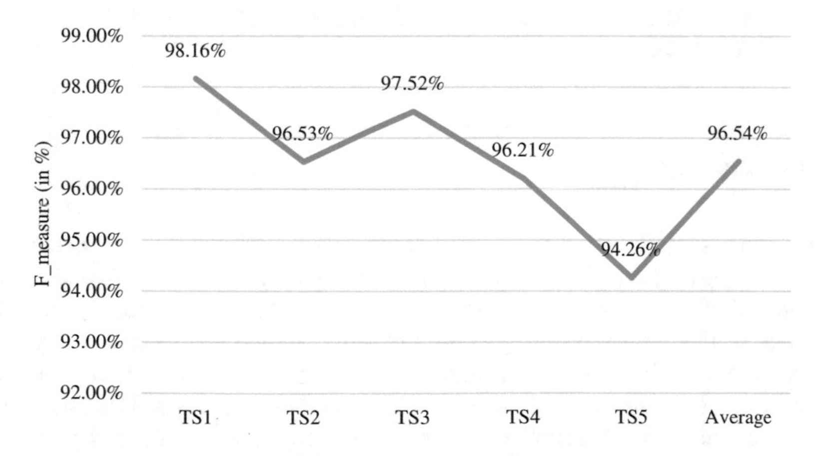

Fig. 13 F_measure evaluation on GTZAN dataset

Table 2 Comparative performance evaluation on classic piano dataset

Algorithms	Accuracy (%)
CNN [12]	66.9
CRNN [12]	73.9
LSTM	85

Table 3 Comparative performance evaluation on GTZAN dataset

Algorithms	Accuracy (%)
CNN [13]	83.30
CNN(VGG16) [14]	92.1
LSTM	96

4.3 Comparative State-of-Art

Music is detected using LSTM on two independent datasets in this research. Classic Piano and GTZAN datasets, to be specific. Table 2 compares the accuracy of the various algorithms CNN [12], CRNN [12], and LSTM [proposed]. When using the classic piano dataset, our proposed method has an accuracy of 85%, which is higher than CNN and CRNN. Table 2 shows that LSTM accuracy is 18.1% higher than CNN and 11.1% higher than CRNN. Similarly, Accuracy of different algorithms is compared using the GTZAN dataset. Table 3 shows comparative Performance Evaluation on the GTZAN dataset. When using the GTZAN dataset, our proposed method has an accuracy of 96%, which is higher than CNN and CRNN. Table 3 shows that LSTM accuracy is 12.7% higher than CNN [13] and 3.9% higher than CNN-VGG16 [14].

5 Conclusion

Music genre categorization may assist users, particularly specialized artists and novices, locate the music they're looking for. Since they are new to music and inexperienced with diverse music types, finding a certain kind of music through streaming media takes a long time, resulting in inefficiencies. Whenever an artist is exploring out music in a certain genre, if he listens to it for a long time and evaluates it, the sound will grow fatigued and the judgement will fail, and they'll also spend too much time looking for music. As a result, for these individuals, a music genre categorization tool is a time-saving solution. This work is dedicated to design a machine learning tool for music genre classification and generation. For this, the model is divided into three steps, feature extraction, feature correlation and classification. For feature extraction audio signals characteristics are extracted such as STFT, mean spectral centroid, mean harmony, mean MFCC, and other signal statistical features are used. The features are relevant for music classification are evaluated using correlation analysis. In this work f_score correlation analysis was performed. Then correlated features are fed into LSTM for music genre classification. The work was evaluated on two datasets, classic piano and GTZAN dataset. It was observed from the result analysis that an average MSE and RMSE for music generation was approx. 10^{-6} and 10^{-3} respectively. While the model achieved approx. 85% accuracy on the classic piano dataset and approx. 96% accuracy on the GTZAN dataset. The result was compared with some existing works and it was observed the proposed model shows on an average of 15% improvement on the classic piano dataset. Similarly, on the GTZAN dataset, it shows improvement of 7% on an average. In future, this work can be extended on real-time music note prediction and composition of new music or song with the help of the machine learning approach.

References

1. Mao HH (2018) DeepJ: style-specific music generation. In: Proceedings of 12th IEEE international conference semantic computing ICSC 2018, vol 2018 Jan, pp 377–382. https://doi.org/10.1109/ICSC.2018.00077
2. Bretan M, Weinberg G, Heck L (2017) A unit selection methodology for music generation using deep neural networks. In: Proceedings of the 8th international conference computing creativity ICCC 2017, pp 1–13
3. Jhamtani H, Berg-Kirkpatrick T (2019) Modeling self-repetition in music generation using generative adversarial networks. http://ifdo.ca/
4. Briot JP (2021) From artificial neural networks to deep learning for music generation: history, concepts and trends. Neural Comput Appl 33(1):39–65. https://doi.org/10.1007/s00521-020-05399-0
5. García JC, Serrano E (2019) Automatic music generation by deep learning. Adv Intell Syst Comput 800:284–291. https://doi.org/10.1007/978-3-319-94649-8_34
6. Hewahi N, AlSaigal S, AlJanahi S (2019) Generation of music pieces using machine learning: long short-term memory neural networks approach. Arab J Basic Appl Sci 26(1):397–413. https://doi.org/10.1080/25765299.2019.1649972
7. Jaques N, Gu S, Turner RE, Eck D (2016) Generating music by fine-tuning recurrent neural networks with reinforcement learning. Thesis, pp 410–420. http://repositorium.ub.uni-osnabrueck.de/bitstream/urn:nbn:de:gbv:700-2008112111/2/E-Diss839%7B_%7Dthesis.pdf%5Cn, http://arxiv.org/abs/1611.02796
8. Oza M, Vaghela H, Srivastava K (2020) Progressive generative adversarial binary networks for music generation. Adv Intell Syst Comput 1087:181–192. https://doi.org/10.1007/978-981-15-1286-5_16
9. Dua M, Yadav R, Mamgai D, Brodiya S (2020) An Improved RNN-LSTM based novel approach for sheet music generation. Procedia Comput Sci 171:465–474. https://doi.org/10.1016/j.procs.2020.04.049
10. Mangal S, Modak R, Joshi P (2019) LSTM based music generation system. Iarjset 6(5):47–54. https://doi.org/10.17148/iarjset.2019.6508
11. Chen H, Xiao Q, Yin X (2019) Generating music algorithm with deep convolutional generative adversarial networks. In: 2019 2nd international conference electronics technology ICET 2019, pp 576–580. https://doi.org/10.1109/ELTECH.2019.8839521
12. Kong Q, Choi K, Wang Y (2020) Large-scale MIDI-based composer classification. http://arxiv.org/abs/2010.14805

13. Cheng YH, Chang PC, Kuo CN (2020) Convolutional neural networks approach for music genre classification. In: Proceedings of 2020 international symposium on computer, consumer and control, IS3C 2020, pp 399–403. https://doi.org/10.1109/IS3C50286.2020.00109
14. Shi L, Li C, Tian L (2019) Music genre classification based on chroma features and deep learning. In: 10th International conference intelligence control information processing ICICIP 2019, no 61901356, pp 81–86. https://doi.org/10.1109/ICICIP47338.2019.9012215

Chapter 47
Tumor Visualization Model for Determining Pathway in Radiotherapy

Garima Iyer, Nidhi Panchal, Pranav Pandya, and Shruti Dodani

1 Introduction

Radiotherapy is a form of cancer treatment that uses high-energy X-ray beams that are delivered to stop the cancer cells from growing. The radiation beam affects the cell's ability to multiply. However, it can affect both the cancer cells as well as the normal tissues. As a result, an ideal method to prevent the normal tissues from getting damaged is by shaping the field of radiation to match the tumor and making the treatment as accurate as possible, which is done by choosing the right pathway and beam angle to send the radiation based on the distribution of radiation's dosage.

A survey was conducted to understand the awareness of radiotherapy among the people and concur with the relevance and need of our proposed model. As per the data of the survey, we can comprehend, from Fig. 1a that 74% of responders were aware of the negative consequences that the radiation has on healthy tissues which further led to an 83.1% positive response to the proposition of building a program that can accurately detect the tumor and find an effective pathway through which radiation can be sent, as shown in Fig. 1b.

The procedure for radiotherapy involves locating the tumor and organs at risk on CT scans and then the treatment beams are designed to deliver enough radiation to the tumor while limiting the radiation dose to the nearby healthy tissues. These treatment plans are designed by the oncologists manually—which means that they manually contour the tumor on the CT scans and decide the beam angle for sending the radiation dose. According to the survey's findings, as shown in Fig 1c and d, a

G. Iyer · N. Panchal · P. Pandya (✉) · S. Dodani
Department of Biomedical Engineering, D. J. Sanghvi College of Engineering, Mumbai, India
e-mail: pranavpandya.work@gmail.com

S. Dodani
e-mail: shruti.savant@djsce.ac.in

G. Mathur et al. (eds.), *Proceedings of 3rd International Conference on Artificial Intelligence: Advances and Applications*, Algorithms for Intelligent Systems,
https://doi.org/10.1007/978-981-19-7041-2_47

Q. Did you know the high energy x-ray, which is used in radiotherapy, can affect healthy tissues as well?

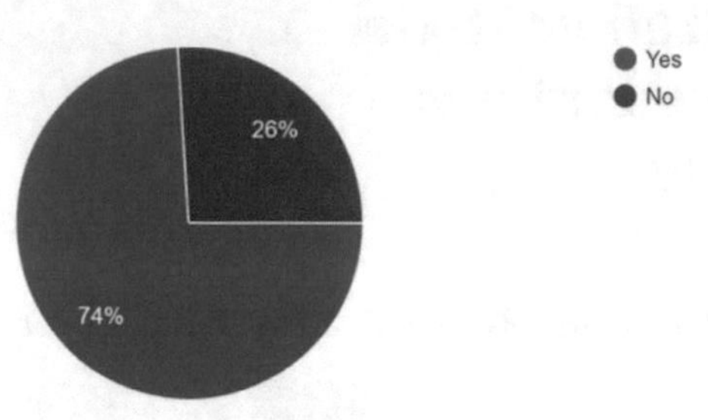

Q. How essential do you think a proper treatment plan is for minimal damage to healthy tissues?

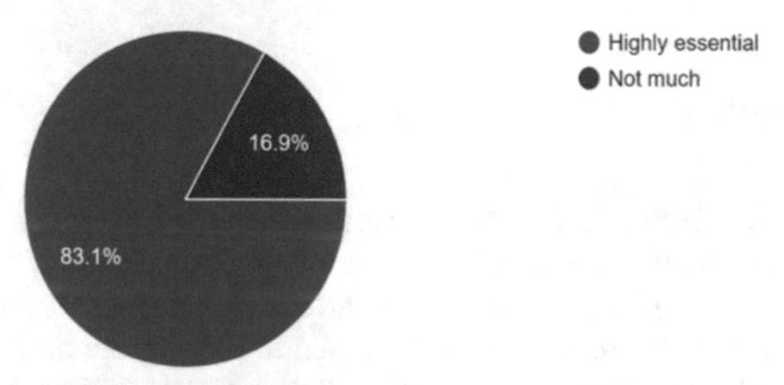

Q. Are you aware of the errors involved with treatment planning?

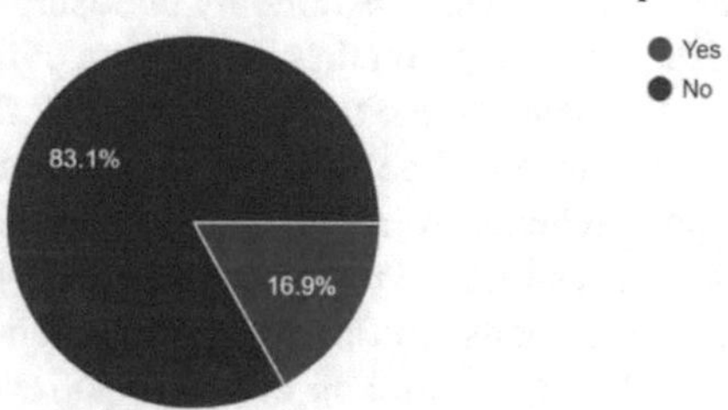

Q. Do you think the manual process will be time-consuming with open areas for errors?

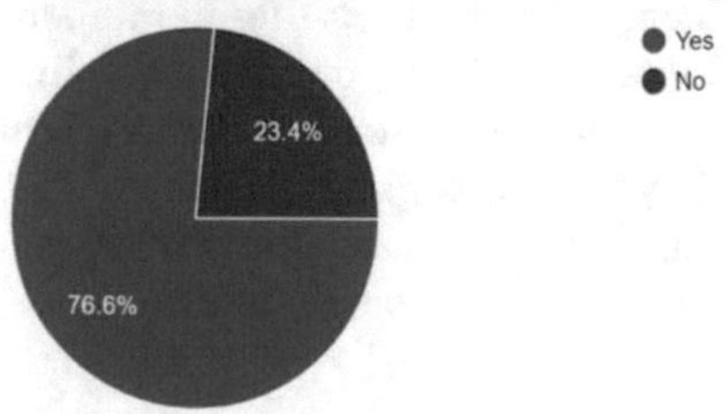

Fig. 1 Survey outcome

majority of respondents agree that the manual approach for treatment planning can prove to be time-consuming and error-prone.

Additionally, software like Eclipse, Monaco, Xio, etc., which are used in treatment planning and give information on the amount of radiation tend to be very expensive, and are thus not equipped in third-world countries which cannot afford the systems.

Based on the statistics [1], Head and neck cancers account for 30–40% of all malignancies in India. Mouth cancers, throat cancers, laryngeal cancers, nasal cancers, sinus cancers, and salivary gland cancers can all be discovered in the oral cavity, throat, larynx, nasal cavity, sinuses, and salivary glands. They can also be seen in the head and nerves and muscles of the neck. Head and neck cancers have a high mortality rate, with around 10% of patients dying within 6 months after diagnosis [1]. Furthermore, more than 500 new cases of brain tumors are diagnosed every day around the world, with a significant number of them being children [2]. The paper focuses on the above type of cancers because of their increase in numbers, high mortality rate, and complexity in locations.

Mortality rates in children and infants who are affected by cancer are also more than that in adults, because of the small area that gets affected by radiation during radiotherapy, making it necessary to build a system that can accurately detect the tumor and the pathway through which radiation is to be sent.

As a result, to limit minimal damage to the organs at risk and avoid any errors, we proposed a model to automatically detect and contour the tumor, and build a threedimensional model based on its density and surface areas, which can aid in detecting the radiation pathway for minimal damage to the surrounding tissues and overall treatment planning.

2 Literature Survey

The procedure was divided into two parts: (i) Tumor detection and segmentation and (ii) 3D modeling and Visualization of the tumor. To determine the most effective approach, research was carried out in these two distinct parts.

In all, 4 papers were reviewed to decide the best approach.

2.1 Tumor Detection and Segmentation

Mohanarathinam [3] has proposed the processing of brain tumor image filtration and segmentation by passing the data through an anisotropic filter and then enhancing it using threshold-based segmentation and bounding box method. The morphological operations performed on the image reduce the contrast between consecutive pixels which can give information about the data. After resizing and converting it into grayscale, the tumor is segmented using the bounding box method. As this process

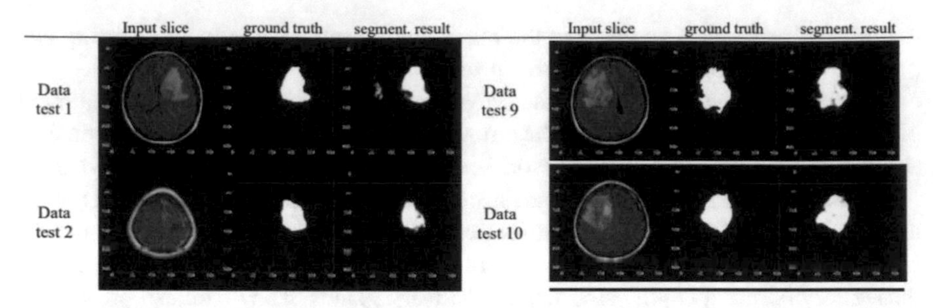

Fig. 2 Segmentation results of testing data [4]

is not performed on the tumor-less region, it improves the process of detection. The software used for the detection is MATLAB.

Pravitasari et al., in their paper [4], have reviewed several deep learning models for tumor segmentation and concluded that the proposed model, namely UNet-VGG16 with Transfer Learning is running well on the computer with a processor of Intel Core i7, 32 GB RAM, 128 GB SSD, and without GPU and VRAM. The proposed model has great performance compared to the UNet model (in four scenarios) since it has the minimum value of loss and maximum value of accuracy. The segmentation results under the proposed model tend to approach the ROI target of each brain tumor MRI image very well. The results of segmentation from testing data were obtained with a CCR value of 95.69%. The sample result of 4 test Data is shown in below Fig. 2.

In this paper, Wozniak et al. [5] proposed a new correlation learning mechanism (CLM) for deep neural network topologies that blends convolutional neural networks (CNN) with traditional design. The support neural network aids CNN in determining which files are best for pooling and convolution layers. As a result, the main neural classifier improves its learning speed and efficiency. Their findings suggest that the CLM model can achieve 96% accuracy, 95% precision, and 95% recall. They explained their hypothesized method and analyzed numerical data in order to draw inferences and demonstrate future work.

Hence, comparing both approaches in [3–5] for tumor detection and segmentation, we have chosen UNet-VGG16 with Transfer Learning as the main approach of our research, as python has several advantages over MATLAB and the model gives greater accuracy. However, we have implemented this deep learning model on CT DICOM images.

2.2 2D to 3D Modeling

In the paper [6], Nguyen et al., focuses on 3 main algorithms for 3D reconstruction, they are—Marching cube, Ray casting, and Texture-based rendering as shown in Fig. 3. C++ is used to implement the algorithm. The application is based on the VTK

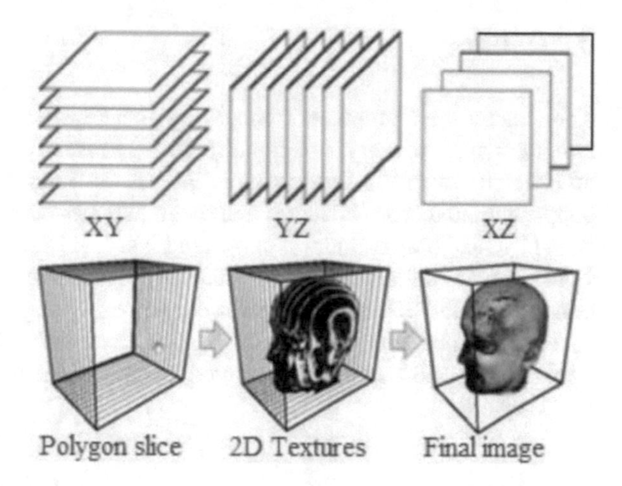

Fig. 3 Texture-based volume rendering [5]

Library. The Marching cube is flexible for modification and can extract necessary surfaces. It also has fast-renders, no limitation in memory, and is good for surface reconstruction. Ray casting offers a high-quality image, a flexible extension, and ease of use. However, because we have to compute one ray for each pixel, the operation is time expensive. The fastest algorithm is texture-based volume rendering, and the image quality is maintained. It cannot, however, be used for scans with many slices due to its limitations. The figure below shows the basic steps involved in texture-based volume rendering.

In the paper [7], Mamdouh et al., proposed a model that is implemented using three different tools, Autodesk 3D Maya, Autodesk Mesh mixer, and Python and CV2. Using python and open CV, the brightness and contrast of the DICOM are increased as it is easy to interact with the 3D model and the noise is reduced. The algorithm is divided into 3 phases. In the first and second steps, the DICOM files are analyzed, noise is removed, and the files are converted to a 3D file format. The third phase involves converting DICOM files to low-poly 3D models that may be used with any virtual or simulation engine.

BIG DATA is a massive collection of information (structured, semi-structured, and unstructured) stored in the world's archives. Huge data generates a store incentive and prepares for larger volumes of computerized data that can't be broken down using traditional processing processes. Numbers, dates, and strings aren't the only things that make up a large amount of data. Geospatial information, 3D information, medical image information, sound, video, unstructured stuff, including log papers, and web networking are all part of the massive data.

As we are going to work on a large set of DICOM images acquired from the Nanavati Hospital Database, we need an algorithm that works for a bigger amount of data efficiently. Hence, Big Data algorithms are most suitable.

3 Dataset

The data for the Computed Tomography scans were acquired through the Siemens Biograph mCT scanner. It is a multi-slice CT scanner that can detect, characterize, and monitor even the tiniest cancer lesions with reproducible quantification, thus supporting more cost-effective treatment. We were given CT scans for 10 patients with HNC and brain tumors as shown in Figs. 4 and 5, respectively; by the Advanced Centre for Radiation Oncology (ACRO) department at the Nanavati hospital for constructing the 3D model of the tumor and executing the pathway determination.

For the CNN model, the data was acquired through Kaggle uploaded by Mr. Amit Kumar, and was later augmented to increase the dataset size for better training accuracy.

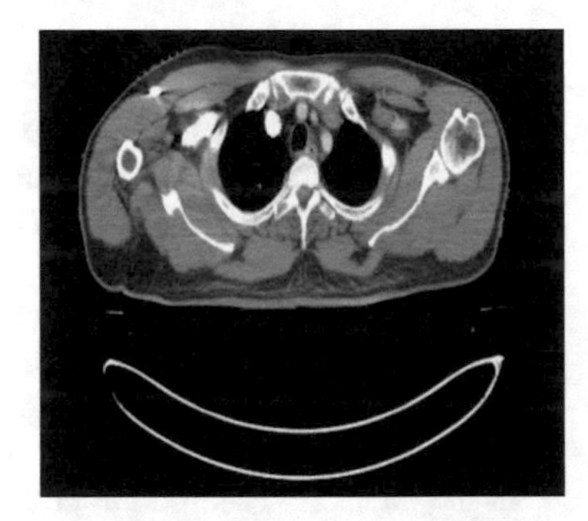

Fig. 4 HNC CT image

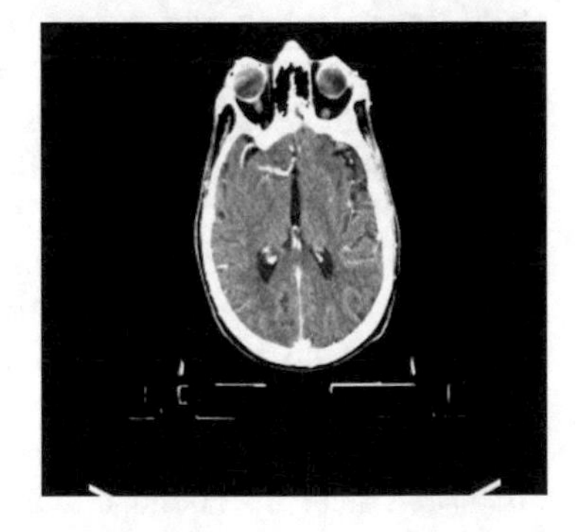

Fig. 5 Brain CT image

4 Methodology and Implementation

4.1 Tumor Detection

The block diagram, as shown in Fig. 6 is divided into 2 parts: Tumor Detection and Tumor Segmentation and 3D modeling.

Block diagram (1): The initial part of the program is tumor detection. The dataset acquired from Kaggle, of around 250 CT DICOM images, is converted into a jpeg format. This data is used for building the model using CNN, where data is divided into train and test groups. We got a Boolean output on running the model based on whether a tumor is present or not. The data is then separated based on their Boolean output. The data, together with their Boolean results, is recorded in a CSV file.

Block diagram (2): After the 1st step, we have exclusive CT slices that have tumors. For 2D contouring, individual contouring of each slice is done, and after that, the contoured images are stacked to build a 3D model of the tumor.

The first goal of this study was to create a CNN model that could determine whether or not a subject had a tumor based on a CT scan. We used the VGG16 model [4] architecture and weights to train the model for this binary problem. Accuracy was used as a metric to justify the model performance. The formula for accuracy can be defined as shown in Eq. (1)

$$\text{Accuracy} = \frac{Number\ of\ correctly\ predicted\ images}{Total\ number\ of\ tested\ images} \times 100\% \tag{1}$$

One of the deep learning algorithms used in semantic segmentation is the fully convolutional network (FCN). FCN evaluates pixel-to-pixel mapping and determines the pixel class using ground truth. The FCN is a variation of the traditional convolutional neural network (CNN). The major layers of CNN are convolution, pooling, and fully connected, with the fully connected layer being replaced by the convolution layer in the FCN. As a result, FCN can identify each pixel in the image and offer them the ability to generate predictions on inputs of any size.

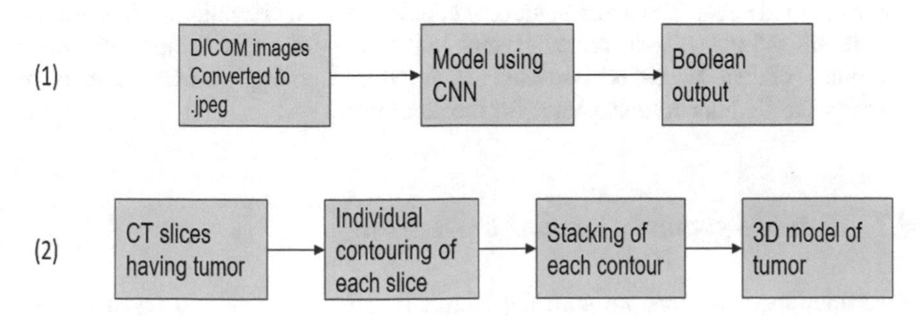

Fig. 6 Block diagram of the proposed algorithm

The convolution neural network (CNN) model is made up of different layers described as follows:

1. **Convolution layer**: Two-dimensional convolution operations are performed on the images by employing the kernel matrix thereby giving a convoluted feature matrix as result.
2. **Activation layer**: In the nodes, the Rectified linear unit (ReLU) is employed as an activation function. ReLU is the most widely used activation function. The mathematical equation for ReLU is shown in Eq. (2)

$$f(x) = a = max(0, x) \quad (2)$$

   The main advantage of using the ReLU function over other activation functions is that it does not activate all the neurons at the same time. The negative components are made zero in the matrix to avoid summing up values to zero. If the input is negative it will convert it to zero and the neuron does not get activated.
3. **Average pooling layer**: Multiple times Average pooling function can be used to reduce the convoluted feature matrix thereby improving the computation power.
4. **Fully connected layer**: In this layer two functions are performed. Firstly, flatten is used to convert a 2D matrix to a ID array. It reduces the dimension and computation complexity. Secondly, the Dense function is used which takes numerous inputs to give one output.
5. **Batch Normalization layer**: It is used to improve the speed, stability, and performance of the model. It reduces the generalized error.

Aside from CNN layers, illustrated in Fig. 7, a variety of optimizers are employed to speed up the gradient descent process. Here, we employed the [9] ADAM optimizer. Adaptive Moment Estimation is an algorithm for optimization technique for gradient descent. The method is efficient when working with a large problem involving a lot of data or parameters as it requires less memory and is efficient. Intuitively, it is a combination of the "gradient descent with momentum" algorithm and the "RMSP" algorithm [8].

Gradient Descent is an iterative optimization process for determining a function's minimal value. The fundamental concept is to start with random values for the parameters and then take incremental steps in the direction of the "slope" with each iteration. Gradient descent is often used in supervised learning to minimize the error function and find the optimal values for the parameter.

4.2 Tumor Segmentation and Contouring

For tumor segmentation, we used the inbuilt functions provided in OpenCV and Matplotlib libraries.

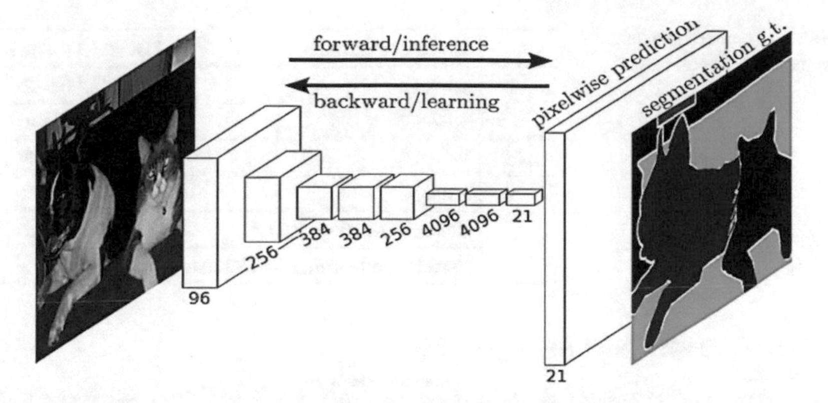

Fig. 7 Structure of fully connected convolutional neural network [8]

OpenCV is a Python open-source toolkit for computer vision in applications such as artificial intelligence, machine learning, and face recognition. The CV is an abbreviated form of computer vision in OpenCV, which is defined as a branch of research that assists computers in understanding the content of digital pictures such as photographs and videos.

Matplotlib is a Python package that allows you to create static, animated, and interactive visualizations.

The goal of computer vision is to figure out what's going on in the images. It extracts the description from the images, which may be an object, a text description, a threedimensional model, or something else entirely.

5 Results

5.1 Tumor Detection

On comparing with the performance of different models displayed in Table 1, our model using UNet-VGG16 has given a better accuracy with the CCR (correct classification ratio) value of 97.55% obtained from training data.

Figure 8 shows the result obtained by giving an Epoch of 50. Different architectural or convolutional block scenarios could be obtained for future research to obtain additional alternative models. The validation accuracy which is the number of accurately predicted images divided by the total number of images; stood at 86.96%.

Table 1 The performance comparison of each model

	Model comparison	Loss	Accuracy
	Model 1	0.124	0.942
	Model 2	0.083	0.951
UNet	Model 3	0.085	0.953
	Model 4	0.244	0.938
	UNet- VGG16 with TL	0.054	0.961
	Correlation learning mechanism	0.083	0.95

```
Epoch 00049: val loss did not improve from 0.58573
Epoch 50/50
=============) - 165 25/step - loss: 0.0749 accuracy: 0.9755 val loss: 0.7052 - val accuracy: 0.8696
```

Fig. 8 Acquired Result for Training Accuracy

5.2 *Tumor Segmentation*

The steps to segment and contour the tumor from the DICOM image are as follows:

1. By using FOR loop, all Digital Imaging and Communications in Medicine format (.dcm) images in a single patient's series were converted to Joint Photographic Experts Groups (.jpeg) format.
2. All the jpeg images are cropped such that only the maximal of the brain image is being processed, as shown in Fig. 9.

   The image in Fig. 9 is cropped in such a way that only the useful part of the image, that is the Cerebral Cortex, and its underlying regions remain in view. This way we reduce the number of pixels accounted for and make the tumor contouring process faster.
3. The histogram, as shown in Fig. 10, is acquired using the inbuilt function in the NumPy library on python and plotted using matplotlib.

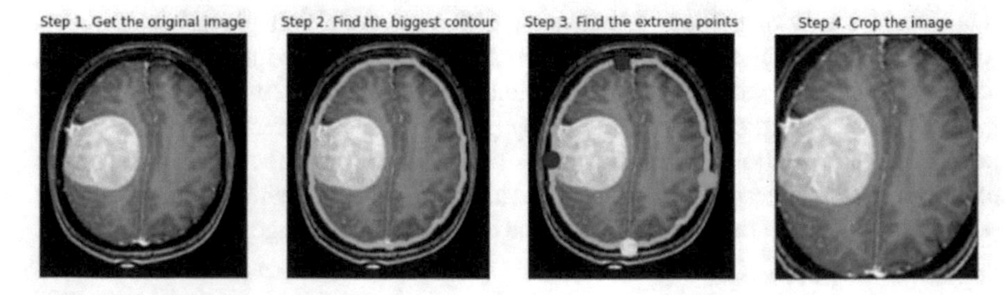

Fig. 9 Maxima crop of DICOM image

Figure 10 shows the occurrence of gray level values in the image. The threshold values for contouring are selected according to the gray level peaks obtained from the histogram.

4. Suitable threshold values are selected, and the image is subjected to multilevel thresholding which converts all pixels outside the threshold to black or white depending on the selected values of the threshold, as shown in Fig. 11.

   In Fig. 11, it can be seen that the gray levels between the threshold values are unaffected and the ones beyond them change to either black if the value is lower than the lower threshold or white if the value is higher than the higher threshold.

5. The functioning of the Canny edge detector is now utilized. The canny edge detector is an edge detection operator that detects a wide range of edges in images using a multi-stage approach.

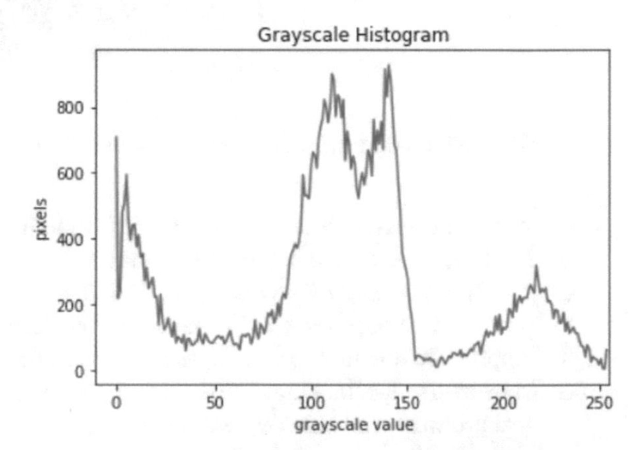

Fig. 10 Grayscale histogram of the image

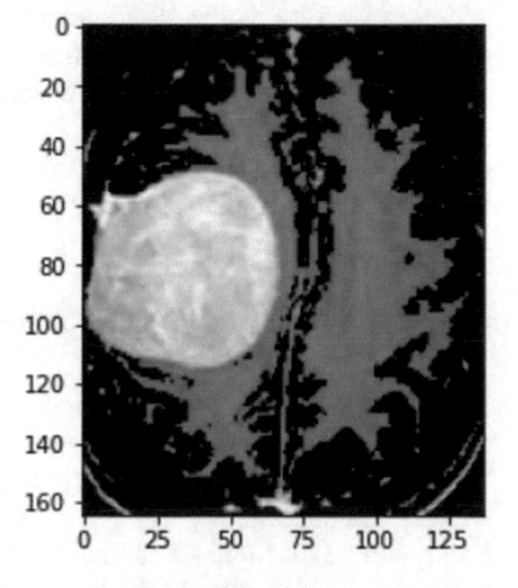

Fig. 11 Multilevel thresholding

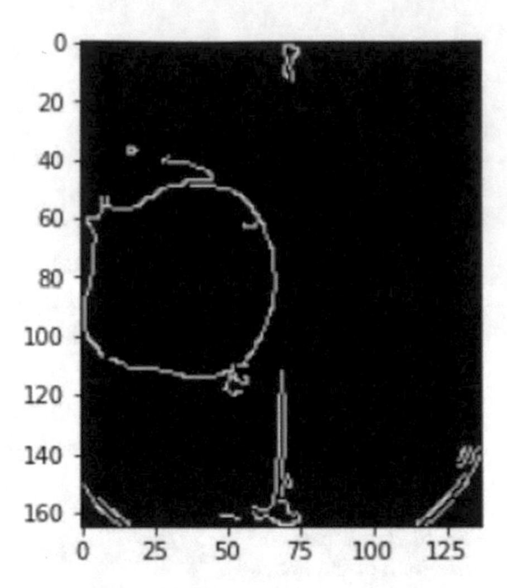

Fig. 12 Tumor contour

The Canny edge detection technique can be broken down into five distinct steps:

a. Apply a Gaussian filter to the image to smooth it out and remove the noise.
b. Find the intensity gradients of the image.
c. To eliminate spurious edge detection responses, use gradient magnitude thresholding or lower bound cut-off suppression.
d. Apply a double threshold to determine potential edges.
e. Hysteresis edge tracking: Finish edge detection by suppressing all other edges that are weak and not coupled to strong edges.

Since we are using an inbuilt function to carry out the operation, all the mentioned steps are included in the process. The user needs to manually enter the dual thresholds depending on the histogram for the image.

In Fig. 12, the outline of the contoured tumor is achieved using the canny edge detection feature.

6. Now for all such slices of the CT scan, all the detected contours are placed on top of each other, thus forming a multidimensional array.

6 Conclusion and Future Scope

The model proposed in the paper lays out a base model that can be used for Radiotherapy treatment planning. The tumor detection model, obtained using the UNet-VGG16 algorithm, has an accuracy of 97.55%, following which the tumor contouring module was successfully implemented using image processing techniques such as Canny Edge Detection. After stacking these contoured scans, we successfully built a 3D visualization model which, if need be, can be used for radiation's pathway determination based on the correlation of the various parameters such as density, surface area, and dosage distributions.

Even though this research is limited only to the HNC and brain tumors, it can be trained for other types of cancers as well. A further extension for automatically determining the pathway is in works, but the 3D model can also help in navigation during surgeries. Since this is an open-source project, it cannot replace the software that are being used in hospitals. But our project can be used for training inexperienced oncologists, cross-checking the results, and for research purposes.

Acknowledgements We would like to thank Dr. Nagraj Huilgol, Mr. Anand Parab, and Ms. Shubhangi Barrsing at the Advanced Center for Radiation Oncology department in Nanavati Max Super Specialty Hospital for providing us with the CT DICOM images and teaching us various concepts in radiotherapy. Along with this, we would also like to thank Kaggle as we acquired another set of data from them. We are thankful for the altruistic cooperation, help, advice, and guidance of all these people.

References

1. Kulkami MR (2013) Head and neck cancer burden in India. Int J Head Neck Surg 4(1):29–35. https://doi.org/10.5005/jp-journals-10001-1132
2. Zahid, Parul (2016) World brain tumor day. National Health Portal India, June 06, 2016. https://www.nhp.gov.in/world-brain-tumor-day_pg
3. Mohanarathinam A (2020) Enhanced image filtrationusing threshold based anisotropic filter for brain tumor image segmentation. In: 2020 3rd international conference on intelligent sustainable systems (ICISS), pp 308–316. https://doi.org/10.1109/ICISS49785.2020.9315924
4. Pravitasari A et al (2020) UNet-VGGI 6 with transfer learning for MRI-based brain tumor segmentation. In: TELKOMNIKA telecommunication, computing, electronics & control, vol 18, no 3, pp 1310–1318. (June 2020). https://doi.org/10.12928/telkomnika.vl8i3.14753
5. Wozniak M, Silka J, Wieczorek M (2021) Deep neural network correlation learning mechanism for CT brain tumor detection. Neural Comput Appl. https://doi.org/10.1007/s00521-021-0584l-x
6. Nguyen VS, Tran MH, Quang Vu HM (2016) A research on 3D model construction from 2D DICOM. In: 2016 international conference on advanced computing and applications (ACOMP), 2016, pp 158–163. https://doi.org/10.1109/ACOMP.2016.031
7. Mamdouh R et al (2020) Converting 2D-medical image files "DICOM" into 3D-models, based on image processing, and analysing their results with python programming. WSEAS Trans Comput 19:10–20. https://doi.org/10.37394/23205.2020.19.2
8. Shelhamer E, Long J, Darrell T (2017) Fully convolutional networks for semantic segmentation. IEEE Trans Pattern Anal Mach Intell 39(4), 640–651. (I April 20 I7). https://doi.org/10.1109/TPAMI.2016.2572683
9. Prakhar (2020) Intuition of Adam optimizer. GeeksforGeeks. 24 Oct 2020. https://www.geeksforgeeks.org/intuition-of-adam-optimizer/

Chapter 49
Emotional AI-enabled Interview Aid

Tejas Dhopavkar, Omkar Ghagare, Onkar Bhatlawande, and Sujata Khedkar

1 Introduction

The ability to speak is the distinguishing factor between mankind and other animal species. With the help of speech, one is able to communicate with others, share experiences, and spread knowledge and information. Sometimes, speech serves as a medium for passing down traditions and rituals from one generation to another.

Hendrix and Morrison [1] proposed that a human being experiences various emotions depending on the situation. Such emotions can be anger, surprise, disgust, happiness, and many other emotions. Expressing one's emotions is a part of human communication. These emotions tend to affect verbal communication skills including public speaking, communication with peers, negotiations, etc., which are an important part of professional as well as personal life.

According to Lieskovská et al. [2], non-verbal communication is the backbone of the communication process. It enhances the effectiveness of verbal communication. Simple hand gestures, good body language, and facial expressions help the orator to convey a message better. For a message to be delivered correctly, it is necessary that the facial expressions and body language match with the verbal message being

T. Dhopavkar (✉) · O. Ghagare · O. Bhatlawande · S. Khedkar
Department of Computer Engineering, Vivekanand Education Society's Institute of Technology, Mumbai, India
e-mail: 2018.tejas.dhopavkar@ves.ac.in

O. Ghagare
e-mail: 2018.omkar.ghagare@ves.ac.in

O. Bhatlawande
e-mail: 2018.onkar.bhatlawande@ves.ac.in

S. Khedkar
e-mail: sujata.khedkar@ves.ac.in

G. Mathur et al. (eds.), *Proceedings of 3rd International Conference on Artificial Intelligence: Advances and Applications*, Algorithms for Intelligent Systems,
https://doi.org/10.1007/978-981-19-7041-2_49

delivered. In case there is a mismatch between the facial expression and the verbal message, there is a chance that the receiver of the message will interpret the message incorrectly.

The Human–Computer Interaction focuses on creating an effective and natural communication interface between humans and computers to enhance user experience, help in human development, and so on. Emotions in speech can be captured, analyzed, and interpreted for various applications like anger detection for drivers, stress detection for airline pilots and in the gaming industry as well, as proposed by Abbaschian et al. [3]. Kerkeni et al. [4] presented a study that researchers and scholars have developed and used various machine learning models and algorithms for speech emotion recognition. Some of them are the hidden Markov model, convolutional neural networks, LSTM and GRU networks, Recurrent Neural Networks, and so on.

One of the important stages in the life of a student is his or her first job. Naturally, a candidate will be quite nervous during an interview. Due to such nervousness, a candidate might not be able to speak properly or project his emotions correctly. Prior practice can reduce the negative impact of nervousness. In order to make the candidate become more confident, we propose Emotional AI-Enabled Interview Aid, which helps capture, analyze, and interpret the underlying emotion in the input given by the user. The input can be in the form of audio or text. The user can then visualize the emotion in the given input and can make the necessary corrections in the tone and writing style. This will help the user make a better impact during the interview and become the best fit for the company.

2 Literature Survey

Iqbal and Barua [5] proposed a model where live recorded speech data is given as input to the system. The algorithms used were Gradient Boosting, Support Vector Machine (SVM), and K-Nearest Neighbor (KNN). A total of 34 audio features are extracted: Zero crossing rate (ZCR), energy, entropy of energy, spectral centroid, spectral spread, spectral entropy, spectral flux, spectral rolloff, 13 MFCCs, 12 chroma vectors and chroma deviation. The emotion has been classified into 4 states: Anger, Happiness, Sadness, and Neutral. The proposed model of the paper classifies emotion only into 4 categories. And the model only takes voice signals as input which is also one of the limitations of the system. The system lacks noise reduction. Noise affects the accuracy of the system during testing so it can be worked upon in the future.

The model proposed by Yoon et al. [6] uses a deep dual recurrent encoder model that utilizes both text data and audio signals. MFCC values are utilized to extract speech information from audio sources. Also, The F0 frequency, the voicing probability, and the loudness contours were among the 35 prosodic features that were used. The OpenSMILE toolkit is used to extract features from the data. The model classifies emotions into four categories: happy, sad, angry, and neutral with an accuracy of 68.8–71.8%. It's not a real emotion recognition system as it should use visual features along with the audio. The model proposed incorrectly classifies the neutral class as the happy class.

Kwon [7] has made use of novel adaptive noise thresholding. To achieve this, the author have found an energy–amplitude relationship. There exists a direct relation between energy and amplitude. Higher the energy, higher the amplitude. After reading the input with a sampling rate of 16,000; after computing the energy–amplitude relationship, the maximum amplitude for that particular frame is found. After passing each frame through a suitable threshold, the noise and silent portions are removed from the spectrogram. The spectrogram itself is generated using STFT. Further, the authors pass these cleaned spectrograms as input to the Deep Stride CNN model. This model eliminates the need for pooling layers by using a deep stride mask of dimension 2*2. The deep stride mask is responsible for down-sampling the input spectrogram. This method can obtain an accuracy of 81%.

Principal Component Analysis is a technique used for reducing the dimensionality of a dataset while minimizing information loss. Zheng et al. [8] performed PCA with 60 components on the input spectrogram. This leads to the creation of a PCA whitened spectrogram. The advantage of this approach is that the PCA whitened spectrogram has visible differences in terms of texture and color for different emotions. The PCA whitened spectrogram is further given as an input to the Deep Convolutional Neural Network which can classify 4 emotions with an accuracy of 40%.

Kerkeni et al. [4] used algorithms such as SVM, RNN, and Multivariate linear regression after extracting features and applying feature selection to the input. The datasets used are Berlin and Spanish databases. SER reported the best recognition rate of 94% on the Spanish database using the RNN classifier. For the Berlin database, all of the classifiers achieve an accuracy of 83%. The output is a set of 7 emotions. RNN often performs better with more data and it suffers from the problem of very long training times, thus making it a less considerable option practically as compared to SVM and MLR.

Sajjad et al. [9] proposed a model where useful segments from the audio file have been extracted using the K-means clustering algorithm where each segment is converted into a spectrogram. These spectrograms are fed to the CNN model to extract high-level features and fed to bi-directional LSTM for it to output an emotion. Accuracy for IEMOCAP dataset is 72.25%, 85.57% for EMO- DB dataset, and for RAVDESS dataset, it is 77.02%. However, the K-Means clustering algorithm is used to get the important sequences from the given input speech but may not always be able to select the efficient sequence from the speech, which may lead to poor accuracy of the model.

Our contributions are stated as follows:

- The proposed system is one of a kind, as it provides an opportunity for individuals to hone their skills as it provides an opportunity for them to practice interview questions while taking benefit of quality feedback.
- To build the system, various models had to be tried out like CNN, Logistic Regression, Decision Tree, Random Forest, and SVM. While trying to optimize these

models, we gained some insight into preprocessing techniques like GloVe embeddings and how they result in increasing the accuracy. The same has been discussed in the paper.
- The proposed system is a dual emotion recognition system, which means the user has the option to give an input in the form of text as well as in the form of audio. This ensures a holistic analysis of the individual's caliber.
- To the best of our knowledge, the proposed approach to emotion recognition performs better than other approaches in terms of both audio and text input. The same has been discussed in Sect. 4.

3 Methodology

Figure 1 refers to the proposed solution which uses supervised learning. The dataset when created has to be preprocessed before it is fed to the model for training and testing. Once satisfactory accuracy is achieved, the model then has to be integrated with the web application. Once the web application is ready, it can be deployed on the web. The steps mentioned in the block diagram have been discussed in detail in Sects. 3.1, 3.2, 3.3 and 3.4.

3.1 Dataset

Text Dataset. The Big Five personality traits dataset, created by Pennebaker and King [10] is a huge dataset of 2400 stream-of-consciousness texts labeled with personality. The Big Five personality traits are a taxonomy, or grouping, of personality traits that are also known as the five-factor model (FFM) and the

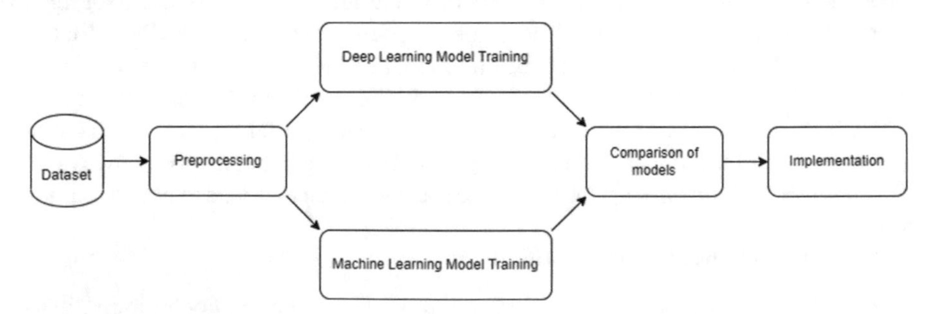

Fig. 1 Block diagram

OCEAN model. The following are the five factors: Openness to experience (inventive/curious vs. consistent/cautious), Conscientiousness (efficient/organized vs. easy-going/careless), Extraversion (outgoing/energetic vs. solitary/reserved), Agreeableness (friendly/compassionate vs. detached), and Neuroticism (sensitive/nervous vs. secure/confident).

Another dataset that was used and combined with the Big Five dataset is the Myers–Briggs Type Indicator (MBTI) dataset proposed by Briggs and Myers [11]. It's a self-report questionnaire that reveals people's psychological preferences for how they perceive the world and make judgments. The four traits of MBTI are: Intuition/Sensing, Feeling/Thinking, Perception/Judging, and Introversion/Extraversion, these four traits are similar to the four in Big Five with neuroticism being the exclusion.

Audio Dataset. The Ryerson Audio-Visual Database of Emotional Speech and Song (RAVDESS) proposed by Livingstone and Russo [12] is one of the most widely used datasets when the task of speech emotion recognition is considered. It consists of 7356 audio files recorded by 24 different actors, with each audio file representing one emotion out of the eight which are neutral, calm, happy, sad, angry, fearful, disgusted, and surprised. Each speech audio file has a duration of 3 s. RAVDESS follows a specific naming convention for naming the files. With the help of this, video and song files can be distinguished from speech files. For the proposed system, only the speech files have been considered.

The audio files are digital in nature. They are a representation of the words spoken in the amplitude domain. Amplitude is a measure of the loudness of the sound. However, the frequency domain provides a better understanding of the properties of a signal. MFCC, Mel Scale, and Chroma are concepts that rely deeply on the frequency domain.

3.2 Preprocessing

Preprocessing for text dataset. In the proposed system, 2 different datasets will be utilized. Machine learning and Deep Learning models will be used on both datasets as well as the combination of these two datasets to ascertain which gives better accuracy.

The text from the datasets is cleaned by a certain set of methods. All text is changed to lowercase, and all characters other than ASCII letters, digits, exclamation marks, and single and double quotation marks are removed. Mohammad and Turney [13] presented a list of English words and their associations with eight basic emotions. The words that have no emotional category are removed as they are of no use with the help of the NRC emotion lexicon which contains 14,182 unigrams.

The MBTI dataset needs to be converted into Pennebaker and King dataset format and then merged so as to increase the data for training. Furnham [14] gave a correlation between the traits where Openness to experience (correlates with Intuition), Agreeableness (correlates with Feeling), Conscientiousness (correlates with Judging), Extraversion (correlates with Extraversion), and there is no related emotion for Neuroticism in MBTI. The resultant dataset has a total of 11,142 rows and 4 labels.

Word embedding is the representation of words for the purpose of text analysis, typically as a vector of real-valued values that indicate the meaning of the words so that the words as near as possible in the vector space are expected to have similar meanings. In the first approach to extract features from the text, CountVectorizer is used on the preprocessed Pennebaker and King dataset. It helps to transform a given text into a vector on the basis of the frequency of each word that occurs in the entire text. The second approach uses GloVe pre-trained word vectors as proposed by Pennington et al. [15] for the purpose of word embedding on the preprocessed Pennebaker and King dataset, and the third approach also uses GloVe pre-trained word vectors but on the merged dataset. Stanford researchers have developed an unsupervised learning algorithm for generating word embeddings based on global word-word co-occurrence matrices from the corpus. For each utterance in our dataset, pre-trained GloVe embeddings with 300-dimensional vectors have been used.

Preprocessing for audio dataset. MFCC has been widely used for the task of SER. To derive MFCC from the audio signal, a Fourier transform needs to be applied to the frame of the audio signal under consideration. This helps to convert the audio signal in the time/space domain to the frequency domain. However, the Fourier transform assumes that the data is infinite. To counter this, windowing is done. A power spectrum of the frame under consideration can be generated by applying N-point DFT to it. The next step is to take the logarithm of the power spectrum, which was mapped to the mel scale. Logarithm helps us to obtain the count of a factor that is repeated. Thus, a logarithm of the frequency mapped to the mel scale will help us obtain the occurrences of certain frequencies, which can be unique to different emotions. The DCT is an energy compression technique that converts the signal to the frequency domain as studied by Licciardi [16]. Applying the DCT to the logarithms of the frequency mapped to the mel scale helps us obtain the mel spectrum cepstral coefficients.

The Mel Scale represents the pitch, which shares a direct relationship with frequency. The mel spectrograms in Fig. 2a–h represent changing frequencies according to pitch classes, for 8 emotions, over 3 s. The formula for converting the frequency from Hertz to Mel Scale is given in Eq. 1

$$m = 2595\left(1 + \frac{f}{700}\right) \tag{1}$$

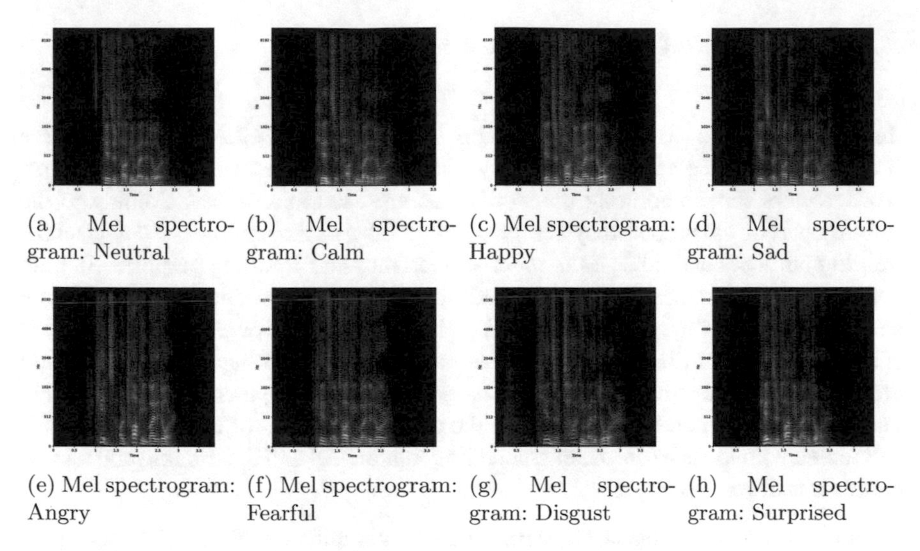

(a) Mel spectrogram: Neutral (b) Mel spectrogram: Calm (c) Mel spectrogram: Happy (d) Mel spectrogram: Sad

(e) Mel spectrogram: Angry (f) Mel spectrogram: Fearful (g) Mel spectrogram: Disgust (h) Mel spectrogram: Surprised

Fig. 2 Mel spectrograms for 8 emotions

Chroma bins are used to segregate the pitch into 12 different classes as proposed by Kattel et al. [17] and they are represented using chromagrams. This type of classification is very important as it can be utilized to identify chroma signatures unique to each of the eight emotions present in RAVDESS. The chromagrams for 8 emotions present in RAVDESS are represented in Fig. 3a–h, where each subfigure's color scheme represents the classification of the pitch into 12 different classes over a period of 3 s.

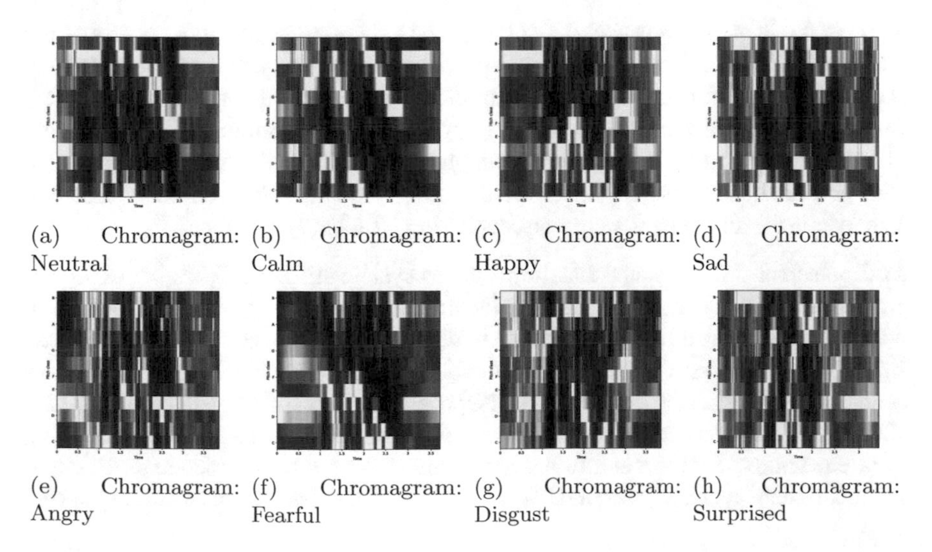

(a) Chromagram: Neutral (b) Chromagram: Calm (c) Chromagram: Happy (d) Chromagram: Sad

(e) Chromagram: Angry (f) Chromagram: Fearful (g) Chromagram: Disgust (h) Chromagram: Surprised

Fig. 3 Chromagrams for 8 emotions

3.3 Training and Testing Datasets

Text Dataset. After forming word vectors with the help of GloVe embeddings the preprocessing phase ends and for every essay in the dataset, we use these formed word vectors for representing the words that are used in an essay. Coming to the training part of the proposed system, we utilize 80% of the preprocessed dataset for training purposes and 20% of the preprocessed dataset for testing purposes. In both training and testing sets that are formed, the dependent variable 'y', which is the traits columns having binary values indicating the presence or absence of a trait, is split into 5 given traits, i.e., 5 dependent variables representing the 5 traits. These are later used to construct a model for every trait by passing them as the dependent variable to the model. For all words in the essays, the mean of the word vectors is calculated with the help of 'mean embedding vectorizer' and the training and testing data is transformed using it.

Audio Dataset. The proposed system utilizes a combination of MFCCs, mel spectrograms, and chroma as features for training purposes. For each of the audio files, a combination of these three features is generated and stored in an array that consists of the labels as well. For training purposes, this array of features and labels is divided into the ratio of 80:20. This ensures that sufficient features and labels are fed to the model for training purposes while maintaining enough data for testing purposes.

3.4 Model Construction

Text. In the Sect. 3.2, it is stated that the proposed system uses count vectors or the GloVe embedding vectors based on the approach. Accordingly, in the first and second approaches, a total of 1973 training samples were provided for training to SVM, Decision Tree, Naive Bayes, Logistic Regression, and Random Forest classifiers. In the third approach, 8913 training samples were provided for training to SVM, Decision Tree, Naive Bayes, Logistic Regression, and the Random Forest classifier. The results for all the approaches are depicted in Tables 3, 4, 5, 6 and 7.

Audio. In Sect. 3.2, it is stated that the proposed system utilizes MFCCs, Mel Spectrogram, and Chroma features for training the model. Accordingly, 2304 audio samples, with a total of 180 features each, were provided for training to the Decision Tree Classifier, Random Forest Classifier, and Convolutional Neural Network. CNN has also been utilized for the task of SER. The CNN model consists of 4 layers with 256, 128, 128, and 128 neurons, respectively. The output layer consists of 8 neurons generating a flattened output for the 8 emotions. A dropout of 0.1 has been added in the first and second hidden layers. The second hidden layer also consists of a max-pooling kernel

Table 1 Performance metrics for audio dataset

Random forest classifier values	
Accuracy	91.31
Precision	93
Recall	91
F1 Score	91

of dimensions 8*8 to downsample the input vector. All the layers except the output layer use ReLU as the activation function, while the output layer utilizes Softmax as the activation function. The results of the Random Forest Classifier Model, which gives the highest accuracy, are depicted in Table 1.

4 Results

The results of the models described in Sect. 3.4 are depicted in Tables 1, 2, 3, 4, 5, 6 and 7. From Table 1, it can be said that the Random Forest Classifier gives an accuracy of 91.31% for detecting emotions from audio. From Tables 3, 4, 5, 6 and 7, it can be observed that three approaches have been implemented which are CountVec, GloVe, and GloVeMerged; along with five different models. Out of these three approaches, GloVeMerged gives the highest accuracy for almost all the models. CountVec and GloVe use CountVectorizer and GloVe word embeddings, respectively, for preprocessing. GloVeMerged combines the text dataset and preprocesses it using GloVe word embeddings. For the GloVeMerged approach, Logistic Regression gives the highest accuracy of 70.18% for detecting emotions from text input as depicted in Table 2. Therefore, for implementing the system, Random Forest Classifier and Logistic Regression were considered audio and text input, respectively. However, for the text input, the combined dataset cannot be trained for neuroticism using the GloVeMerged approach as that emotion is not present in [15]. Thus, we have implemented Logistic Regression using approach 2 which uses GloVe pre-trained word vectors on the preprocessed Pennebaker and King dataset, for detecting neuroticism in the proposed system.

Table 2 Performance metrics for text dataset

Logistic regression values	
Accuracy	70.18
Precision	68.5
Recall	70
F1 score	67.25

Table 3 Results for openness

OPN	SVM	DT	NB	LR	RF
CountVec	54.66	52.02	52.02	56.28	59.92
GloVe	59.92	54.06	59.72	61.13	57.69
GloVeMerged	77.52	71.20	75.95	80.30	79.50

Table 4 Results for agreeableness

AGR	SVM	DT	NB	L	RF
CountVec	52.02	48.79	52.02	51.01	52.22
GloVe	54.86	55.47	52.22	54.66	52.33
GloVeMerged	68.06	57.65	62.23	68.19	67.12

Table 5 Results for conscientiousness

CON	SVM	DT	NB	LR	RF
CountVec	55.26	53.44	50.40	55.26	54.05
GloVe	56.88	50	55.47	56.88	52.83
GloVeMerged	58.23	53.03	56.30	60.25	59.22

Table 6 Results for extraversion

EXT	SVM	DT	NB	LR	RF
CountVec	51.82	51.62	50.40	53.04	54.86
GloVe	52.02	48.38	54.45	52.43	51.41
GloVeMerged	71.83	60.48	68.64	71.96	71.69

Table 7 Results for neuroticism

NEU	SVM	DT	NB	LR	RF
CountVec	52.22	47.98	51.62	52.83	52.83
GloVe	56.48	53.04	53.64	57.09	55.26
GloVeMerged	NA	NA	NA	NA	NA

5 Implementation

As seen from the results, Random Forest and Logistic Regression give the best accuracy for the audio dataset and the text dataset, respectively. Thus, the proposed system utilizes these two models for implementation. The web application has been designed using technologies like HTML, CSS, JavaScript, and Flask. Figures 4, 5, 6, 7, and 8 represent the application design. Figure 4 is the home page of the application, where the user has a choice to select either text or audio-based interviews. Based upon the option selected, the user is prompted to give the answer to the question which is put forth as depicted in Figs. 5 and 7. After each question, the user can check the analysis for the answer, which is generated by the models. The analysis pages are depicted in Figs. 6 and 8.

Fig. 4 User interface of the proposed system

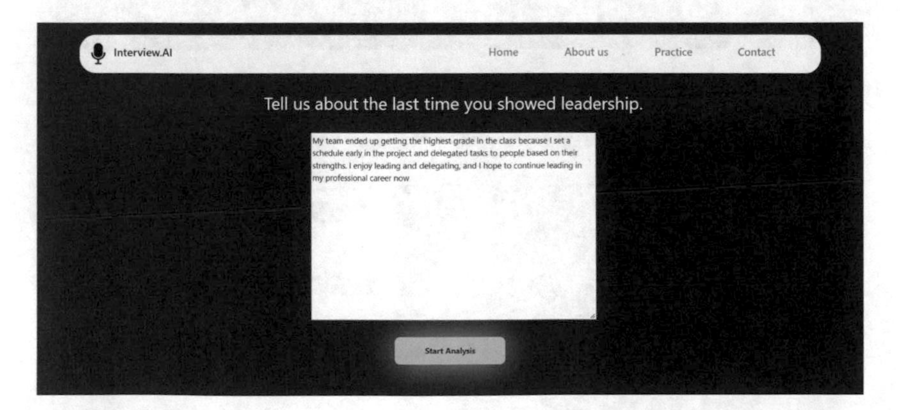

Fig. 5 Text input

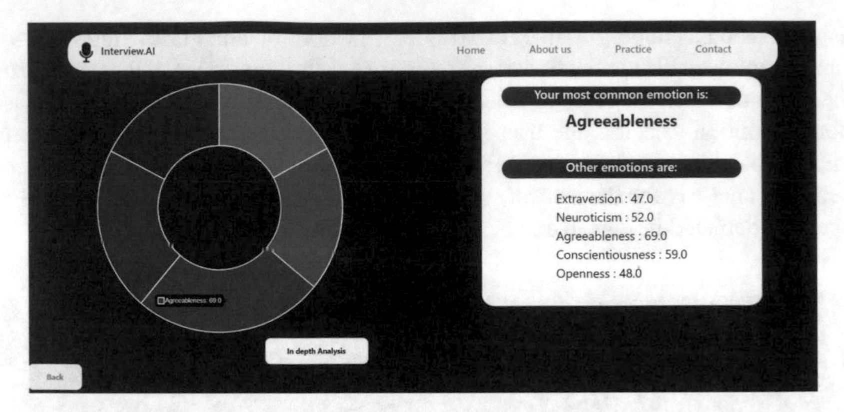

Fig. 6 Text analysis

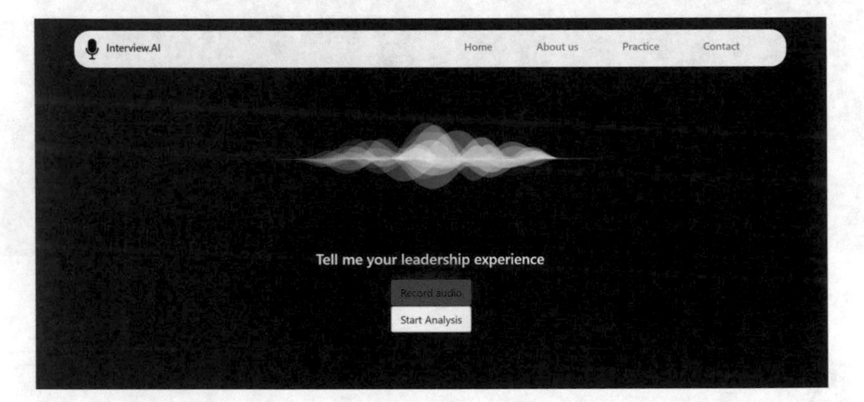

Fig. 7 Audio input

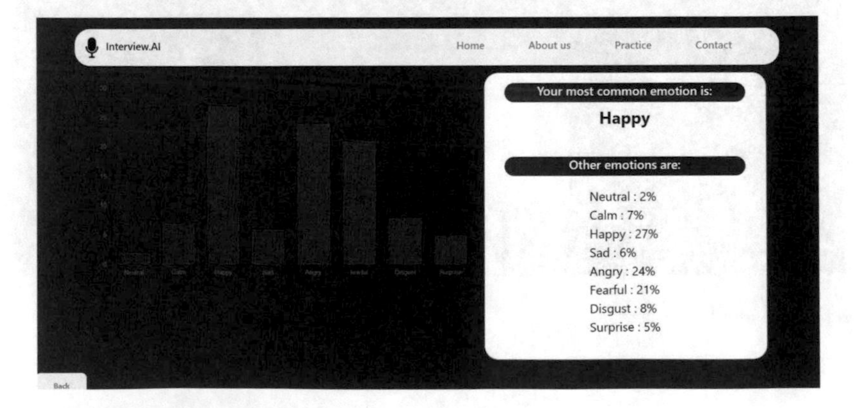

Fig. 8 Audio analysis

6 Conclusion and Future Work

For the goal of recognizing emotions in user input, the proposed system developed three approaches for detecting emotions from the audio input and six for detecting emotions from text input. In order to solve the problem of data scarcity, we combined two datasets to create the text model. Accuracy was enhanced using GloVe embeddings on the combined dataset. The most accurate models for text emotion recognition are Logistic Regression and Random Forest for audio emotion recognition. As a result, they were incorporated into the proposed system. The system suggested was built using Flask, HTML, CSS, and Bootstrap. The users can see the results of the analysis in the form of bar graphs and pie charts which were implemented with the help of Chart.js.

The amount of text data available for 'neuroticism' is currently limited, which has impacted the model's accuracy. As a result, if new data for this trait becomes available in the future, the accuracy of the model to detect this trait could improve. Visual components, in addition to text and audio, may be incorporated in the future to improve the system, resulting in better and more accurate emotion recognition of the person.

References

1. Hendrix RE, Morrison CC (2020) Student emotional responses to different communication situations. J Appl Commun 104(3):1–20
2. Lieskovsk´a, E., Jakubec, M., Jarina, R., & Chmul´ık, M. (2021) A review on speech emotion recognition using deep learning and attention mechanism. Electronics 10(10):1163
3. Abbaschian BJ, Sierra-Sosa D, Elmaghraby A (2021) Deep learning techniques for speech emotion recognition, from databases to models. Sensors 21(4):1249
4. Kerkeni L, Serrestou Y, Mbarki M, Raoof K, Mahjoub MA, Cleder C (2019) Automatic speech emotion recognition using machine learning. In: Social media and machine learning. IntechOpen
5. Iqbal A, Barua K (2019) A real-time emotion recognition from speech using gradient boosting. In: 2019 international conference on electrical, computer and communication engineering (ECCE). IEEE, pp 1–5
6. Yoon S, Byun S, Jung K (2018) Multimodal speech emotion recognition using audio and text. In: 2018 IEEE spoken language technology workshop (SLT). IEEE
7. Kwon S (2019) A CNN-assisted enhanced audio signal processing for speech emotion recognition. Sensors 20(1):183
8. Zheng WQ, Yu JS, Zou YX (2015) An experimental study of speech emotion recognition based on deep convolutional neural networks. In: 2015 international conference on affective computing and intelligent interaction (ACII). IEEE, pp 827–831
9. Sajjad M, Kwon S (2020) Clustering-based speech emotion recognition by incorporating learned features and deep BiLSTM. IEEE Access 8:79861–79875
10. Pennebaker JW, King LA (1999) Linguistic styles: language use as an individual difference. J Pers Soc Psychol 77(6):1296
11. Briggs KC, Myers IB (1977) The myers-briggs type indicator: form G. Consulting Psychologists Press

12. Livingstone SR, Russo FA (2018) The Ryerson Audio-Visual Database of Emotional Speech and Song (RAVDESS): a dynamic, multimodal set of facial and vocal expressions in North American English. PLoS ONE 13(5):e0196391
13. Mohammad SM, Turney PD (2013) Crowdsourcing a word–emotion association lexicon. Comput Intell 29(3):436–465
14. Furnham A (1996) The big five versus the big four: the relationship between the Myers-Briggs Type Indicator (MBTI) and NEO-PI five factor model of personality. Personality Individ Differ 21(2):303–307
15. Pennington J, Socher R, Manning CD (2014) Glove: global vectors for word representation. In: Proceedings of the 2014 conference on empirical methods in natural language processing (EMNLP), pp 1532–1543
16. Licciardi GA (2020) Hyperspectral compression. In: Data handling in science and technology, vol 32, pp 55–67. Elsevier
17. Kattel M, Nepal A, Shah AK, Shrestha D (2019) Chroma feature extraction. In: Conference: chroma feature extraction using fourier transform, vol 20

Chapter 50
Skin Cancer Detection Using Convolutional Neural Networks and InceptionResNetV2

Deepika Vodnala, Konkathi Shreya, Maduru Sandhya, and Cholleti Varsha

1 Introduction

In general, cancer begins in cells, which are the structural components of tissues. The skin and other organs of the body are made up of tissues. Normal cells divide and expand to generate new cells in response to the requirements of the body. When old or damaged cells die, new ones often replace them. When new cells develop when the body does not need them, this natural process goes wrong. These additional cells join together to produce a mass of tissue known as a tumor or growth. Skin growths can be benign (not cancerous) or malignant (cancerous). Benign tumors are not as dangerous as malignant tumors.

Benign growths are seldom life-threatening and can be removed. Benign tumors, in general, do not develop, do not infiltrate the tissues around them, and don't disseminate to other body parts. Malignant growths can be dangerous to one's life; they can be removed, but they can reappear. Melanoma [1] can infiltrate and damage nearby organs and tissues, as well as spread to other areas of the body where treatment is very challenging.

Melanoma is curable if detected early. Clinicians typically screen for skin cancer by performing a visual examination. Clinicians search for moles as well as other regions that vary in color from healthy skin during cancer screening. A rule for spotting cancer [2] is the ABCDE rule.

D. Vodnala (✉)
Department of Computer Science and Information Technology, CVR College of Engineering, Hyderabad, India
e-mail: deepuvodnala19@gmail.com

K. Shreya · M. Sandhya · C. Varsha
Department of Information Technology, Vignana Bharathi Institute of Technology, Hyderabad, India

G. Mathur et al. (eds.), *Proceedings of 3rd International Conference on Artificial Intelligence: Advances and Applications*, Algorithms for Intelligent Systems,
https://doi.org/10.1007/978-981-19-7041-2_50

- A—asymmetry (a lack of equality or equivalence between different sections or elements of something). Ex: Ameoba.
- B—border irregularities (The mole's edges are jagged, notched, or fuzzy).
- C—color (The mole's color varies in intensity from tan to brown to black, and is not uniform).
- D—Diameter that is greater than one inch (about the size of an eraser).
- E—Visual screening of physicians for skin cancer is evolving (the mole changes over time) and cannot ensure 100% detection and may occasionally cause harm. Unnecessary operations such as skin biopsy or excision for lesions that do not turn out to be malignant or lesions that are ignored and do not go for biopsy, ending in mortality, are examples of potential damage.

The following is how this paper is organized. Section 2 discusses the importance of detecting skin cancer. In Sect. 3, related work covers recent existing systems for skin cancer detection and their drawbacks. In Sect. 4, the overview of the proposed system and its methodology have been discussed. In Sect. 5, results are provided and the paper is concluded in Sect. 6.

2 Problem Statement

In general, skin cancer may or may not lead to the loss of human life. It is noticed that skin cancers are increasing day by day for both men and women. In certain circumstances, a skin cancer that is not found in the early stages can be fatal.

CNN Classification can be used to identify this sort of skin cancer in its initial phases. The dataset was initially divided into training and testing sets. The training process should be initiated on the test dataset, and evaluation should be done on the test dataset. Before initiating actual training and testing, dataset images need to be preprocessed. The model determines whether an image from the testing dataset is benign or malignant when it is uploaded to it.

3 Related Work

According to Pham et al. [3], one of the most dangerous diseases on the globe is skin cancer. The Hybrid technique was presented in this study to address the class imbalance in skin disease categorization. There are four parts: Balanced Mini-Batch logic, Real-time Image Augmentation, CNN, and Custom Fully Connected Layers. The Balanced Mini-Batch logic described above differs from standard Batch Mini-Batch logic. The training procedure is to choose pictures for mini-batch training, then train images to identify features using CNN, and finally train custom fully connected layers using the previously chosen features through CNN. To reduce the class imbalance of skin lesion detection, the Hybrid technique combines the algorithm-level method with real-time picture augmentation.

Nie et al. [4] proposed that to forecast melanoma classes, one of the layers of machine learning is Deep learning. The majority of publicly available datasets in this study are unbalanced and gathered in an unethical manner. There are two benign and two malignant samples in the dataset that was gathered. To classify whether it is benign or cancerous, DCNN is trained using a five-step validation process. The goal of this study is to use the DCCN and K-Fold cross-validation to classify melanoma skin cancer in images taken with a tiny dermoscopy. 760 photos were trained using fivefold cross-validation in this experiment, yielding a new test set with a 63.35 percent accuracy.

Hemsi et al. [5] proposed a convolutional neural network ensemble by using a sample time regularly space-shifting approach. The proposed system aims to improve convolutional deep classification models that combine class metadata into an ensemble and perform various test image iterations. Then, shift vectors are dispersed across a lattice. Then, using ensemble scores, two subsets are fused using CNN. This proposed system's future work will include testing more deep networks [6] and lattice topologies. However, it does not apply to medical image classification problems. Image features are critical for understanding every class, as well as generating an appropriate classification model for skin images. The unequal distribution of cancers in the dataset is its main disadvantage, which affects training, as well as the extreme specialization of the class.

CNN-based classification models have allegedly emerged as the optimal solution for melanoma diagnosis in recent years. According to Naeem et al. [7], Skin cancer images are categorized by CNN-based classification models in the same way that dermatologists do, allowing for life-saving identification immediately. In this paper, the most recent studies on CNN-based melanoma categorization are analyzed in detail. Images from datasets are used to remove extraneous features like dark spots, hairs, and wrinkles to produce the best results, and training and testing are defined in this step. The outcomes of the aforementioned techniques are compared to achieve the best possible result.

According to Albahli et al. [8], YOLOv4 is one of the deep neural networks capable of overcoming artifacts and generating decision boundaries for melanoma illness. Active Contour Segmentation is employed to enhance things, and infected melanoma areas are removed. Dermo scope photos are used to produce ground truth annotations. During the skin refinement step, training images are converted to hairs, gel bubbles, and other picture improvements. After that, the refined and ground pictures were sent to YOLOv4. YOLOv4 is a fantastic melanoma detector. YOLOv4 accurate normal skin class, Melanoma lesion, and melanoma segregation using prediction result, prediction over many dimensions to forecast melanoma lesion, and the YOLOv4 localized melanoma detection are analyzed using Active Contour Segmentation.

4 Proposed System

The Proposed System of Skin Cancer Detection Using Convolutional Neural Networks and InceptionResNetV2 works based on Image Processing. The process of transferring an image to a digital format and then performing different operations on it to extract valuable information. When specified signal processing methods are used, the image processing system typically interprets all pictures as 2D signals. The proposed system is implemented using CNN, K-fold cross-validation, and inception-ResNetV2 techniques. It has the following modules, i.e., Preprocessing, training, and testing on datasets, prediction, calculating mean accuracy, and standard deviation and calculating accuracy.

Figure 1 depicts a block diagram for detecting skin cancer. Initially, the whole dataset needs to be split into two parts, i.e., train and test datasets. Preprocessing will be done on both datasets. In the preprocessing step, flipping, resizing, rescaling, reshaping, and extra noise will be removed. After that, CNN is built by adding convolution and pooling layers and it is initialized.

Convolutional Neural Networks (CNNs)

CNNs [9] are neural networks with one or more convolutional layers that are used primarily for image processing, classification, segmentation, and other auto-correlated data. Convolution is essentially the process of sliding a filter over the input. CNN has five five layers that are illustrated below. Figure 2 depicts CNN's architecture.

a. Convolution layer: Convolution is the initial layer that is utilized to fetch various attributes from the source images. The mathematical operation of convolution between the input image and a filter of size M $\times$ M is carried out by this layer. The dot product between the filter and the regions of the source data that is

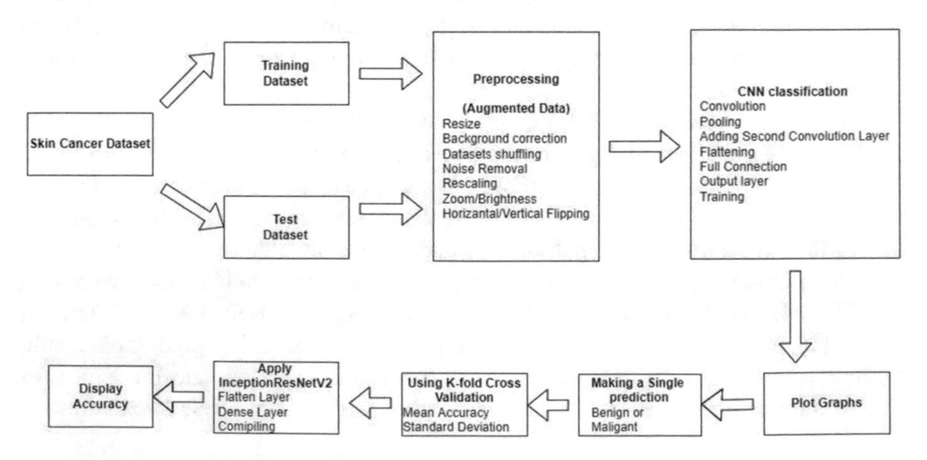

Fig. 1 Block diagram for detecting skin cancer

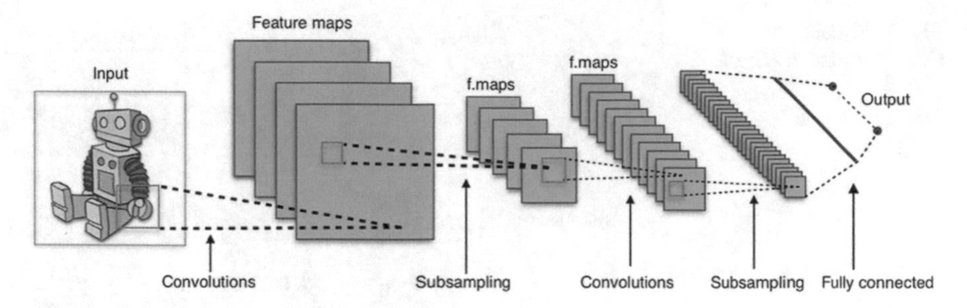

Fig. 2 CNN architecture

proportionate to the filter's size is calculated by swiping the filter over the input image (M × M).

b. Pooling Layer: The main objective of this layer is to minimize the dimension of the preprocessed feature map in order to minimize processing costs. It can be achieved by decreasing the connections across layers and functioning separately for every feature map. Pooling activities are classified into several kinds based on the approach utilized.
c. Fully connected layer: Neurons from several layers are connected using the Fully Connected (FC) layer, which includes weights and biases. All those layers, which constitute a couple of levels of the architecture of CNN, are frequently positioned well before the output layer.
d. Drop Out: When all characteristics are coupled to the FC layer, the training dataset frequently exhibits overfitting. When a particular classifier operates accurately on a training set which has a detrimental impact on the classifier's efficiency once try with new samples, this is known as overfitting.
e. Activation Function: A crucial component of the CNN paradigm is the activation function [10]. They are employed to discover and calculate any kind of intricate and continuous relationship between network variables. In other words, it decides what classifier information should be sent to thc network end and also what should not.

The CNN is utilized to train the training dataset. Evaluation will be done on the testing dataset. Accuracy graphs are plotted using train and test data accuracies. For loss using train and test data loss, a single prediction of the cancer is done when the image is uploaded. Mean accuracy and standard deviation are calculated by utilizing k-fold cross-validation.

4.1 K-Fold Cross-Validation

A resampling technique called cross-validation is used to evaluate machine learning algorithms on a short sample data. The number of clusters through which a specific

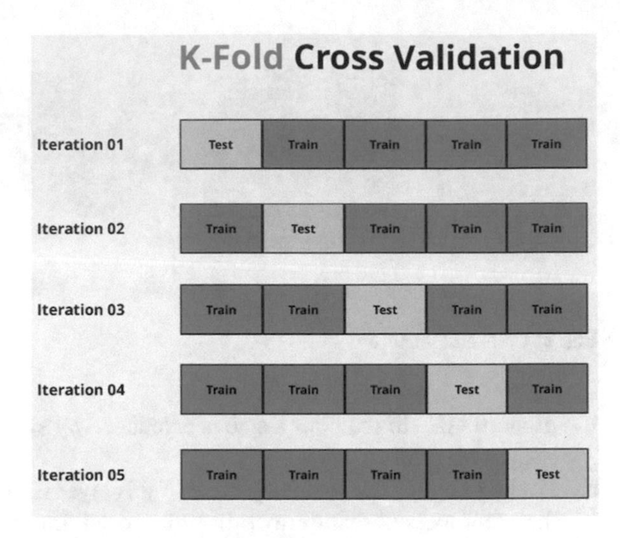

Fig. 3 K-Fold Cross-Validation Iterations

dataset is to be partitioned is determined by the process's sole parameter, k. The procedure is hence commonly known as k-fold cross-validation. Because it is simple to understand and generates a less distorted or realistic estimate of model ability than other strategies, like a straightforward train/test split, it is a common strategy. Figure 3 displays iterations in K-Fold Cross-Validation.

4.2 InceptionResNetV2

A convolutional neural network called InceptionResNetV2 is used to train millions of pieces of datasets. The 164-layer network can categorize images into 1000 different item types. The next step InceptionResNetV2 is built using flattened and Trampled Layers. After building, the classifier compilation is done and the accuracy is displayed. Figure 4 displays the InceptionResNetV2's layers.

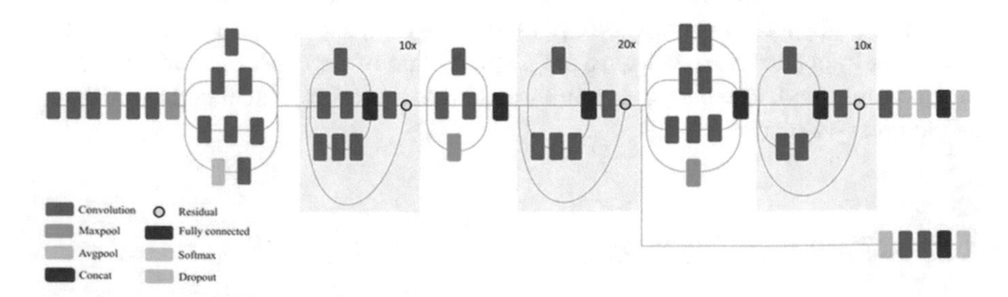

Fig. 4 Layers of InceptionResNetV2

5 Results

Skin Cancer Detection Using Convolutional Neural Networks and InceptionResNetV2 is implemented using CNN, K-Fold Cross-Validation, and InceptionResNetV2 Algorithms. The proposed system is implemented using Google Colab Notebook. Initially, it performs preprocessing steps on training and testing datasets, and Dataset classification is shown in Fig. 5. The training dataset is used to initialize and train the CNN, and the testing dataset is used to evaluate its performance.

The accuracy graph is plotted using training data accuracy and testing data accuracy where the number of epochs and accuracy is represented on the X- and Y-axis which is shown in Fig. 6.

The model loss graph is plotted using training data loss and testing data loss where the X- and Y-axis represents the number of epochs and the loss as depicted in Fig. 7. Selected Image from the Skin Cancer Dataset is shown in Fig. 8.

Once the image is uploaded to the model, it detects the kind of skin cancer which is shown in Fig. 9.

The dataset is divided into Splits, and for each split, Accuracy is calculated. Mean Accuracy and Standard Deviations are calculated by using a k-fold cross-validation algorithm as depicted in Fig. 10.

InceptionResNetV2 algorithm is used for improving accuracy and it is shown in Fig. 11.

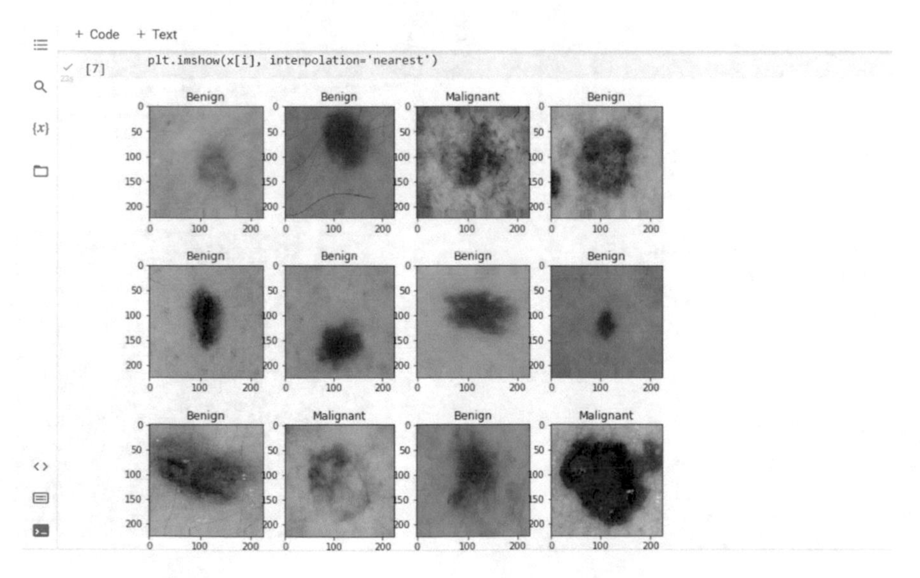

Fig. 5 Dataset classification

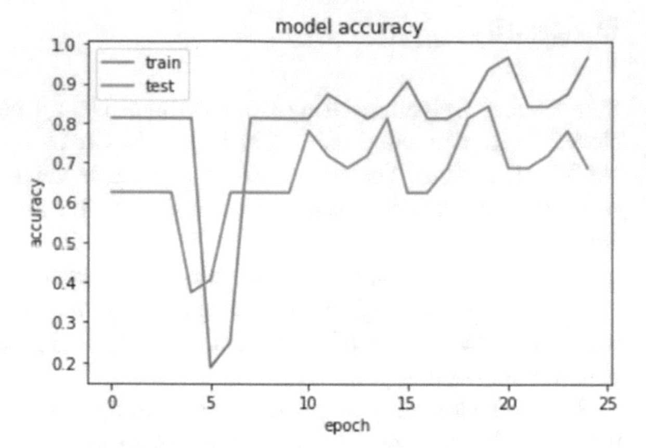

Fig. 6 Accuracy graph

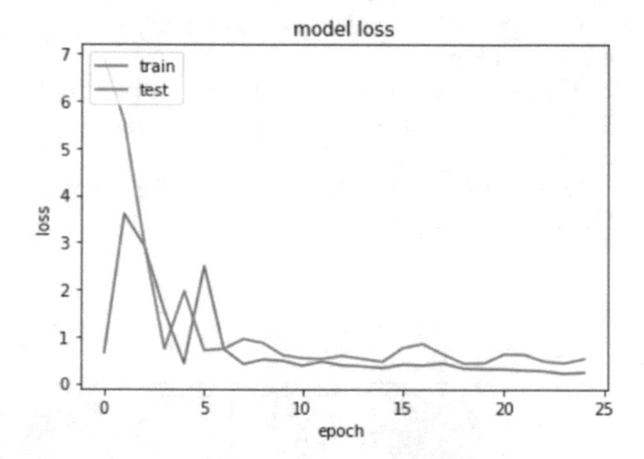

Fig. 7 Loss graph

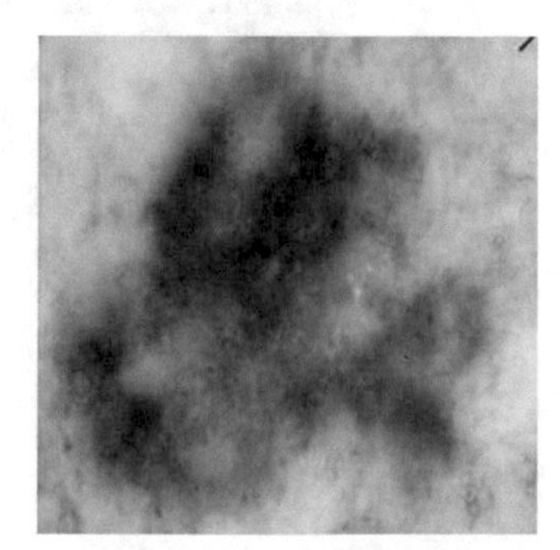

Fig. 8 Selected image from the skin cancer dataset

Fig. 9 Detection of cancer

```
+ Code   + Text                                   RAM / Disk    Editing
[32] print("Mean Accuracy %.2f%% (+/- %.2f%%)" % (np.mean(cvscores), np.std(cvscores)))

accuracy: 100.00%
accuracy: 50.00%
accuracy: 100.00%
accuracy: 66.67%
accuracy: 66.67%
accuracy: 66.67%
accuracy: 66.67%
accuracy: 100.00%
accuracy: 100.00%
accuracy: 100.00%
Mean Accuracy 81.67% (+/- 18.93%)
```

Fig. 10 Mean accuracy and standard deviation

```
[31] modelss.compile(loss = 'categorical_crossentropy', optimizer='adam', metrics=['accuracy'])
     r2 = modelss.fit(x = training_set, validation_data = test_set, epochs =1)
     x=r2
     #history = cnn.fit(x = training_set, validation_data = test_set, epochs =5 )

     59/59 [==============================] - 2134s 36s/step - loss: 0.9300 - accuracy: 0.7928 - val_loss: 1431.6261 - val_accuracy: 0.5530
```

Fig. 11 Accuracy for the model using InceptionResNetV2

6 Conclusion

The proposed system, i.e., Skin Cancer Detection Using Convolutional Neural Networks and InceptionResNetV2 detects the type of skin cancer whether it is Malignant or Benign. The type of skin cancer is detected by using algorithms such as Convolutional Neural networks and K-Fold Cross-Validation and InceptionResNetV2. The proposed system performs the prediction of the type of skin cancer by using the CNN algorithm, and to calculate Mean Accuracy and Standard Deviation, it uses K-fold Cross-Validation. And to yield better and constant Accuracy, it uses the InceptionResNetV2 algorithm. Further, the research work can be enhanced by adding various hybrid algorithms to detect skin cancer more effectively in advance.

References

1. Razmjooy N, Sheykhahmad FR, Ghadimi N (2018) A hybrid neural network – world cup

optimization algorithm for melanoma detection. Open Med 13(1), 9–16
2. Yu L, Chen H, Dou Q, Qin J, Heng P-A (2017) Automated melanoma recognition in dermoscopy images via very deep residual networks. IEEE Trans Med Imag 36(4):994–1004
3. Pham T-C, Doucet A, Luong C-M, Tran C-T, Hoang V-D (2020) Improving skin-disease classification based on customized loss function combined with balanced mini-batch logic and real-time image augmentation. IEEE Access vol 8
4. Nie Y, De Santis L, Carratù M, O'Nils M, Sommella P, Lundgren J (2020) Deep melanoma classification with K-fold cross-validation for process optimization, In: IEEE International Symposium on Medical Measurements and Applications (MeMeA), (2020).
5. Thurnhofer-Hemsi K, López-Rubio E, Domínguez E, Elizondo DA (2021) Skin lesion classification by ensembles of deep convolutional networks and regularly spaced shifting. IEEE, vol 9
6. Esteva A, Kuprel B, Novoa RA, Ko J, Swetter SM, Blau HM, Thrun S (2017) Dermatologist-level classification of skin cancer with deep neural networks. Nature 542(7639):115–118
7. Naeem A, Farooq MS, Khelifi A, Abid A (2020) Malignant melanoma classification using deep learning: datasets, performance measurements, challenges and opportunities. In: IEEE access, vol 8
8. Albahli S, Nida N, Irtaza A, Yousaf MH, Mahmood MT (2020) Melanoma lesion detection and segmentation using YOLOv4-DarkNet and active contour. In: IEEE access, vol 8
9. Nida N, Irtaza A, Javed A, Yousaf MH, Mahmood MT (2019) Melanoma lesion detection and segmentation using deep region-based convolutional neural network and fuzzy C-means clustering. Int J Med Inform 124:37–48
10. Vadrevu A, Rajeshwari R, Pabbathi L, Sirimalla S, Vodnala D (2022) Image forgery detection using metadata analysis and ELA processor. In: Saini HS, Sayal R, Govardhan A, Buyya R (eds) Innovations in computer science and engineering, vol 385. lecture notes in networks and systems. Springer, Singapore, pp 579–586

Chapter 51
Development of Thermocycler for On-Site Detection of Meat Authenticity and Microbial Contamination

Sunil Rajora, S. M. Annapurna, Yoginder Kumar, V. Sai Krishna, and Navjot Kumar

1 Introduction

The recent advances in genomic technology and precision in testing have vast potential to exploit in food technology. The genetic component of meat, the DNA, cannot be altered by processing, cooking, and any other technique. They are the best suitable markers for food authentication, adulteration, and spoilage [1]. The compact instrumentation with result-ready technology will enable regulators to test random samples in the field with high accuracy.

The identification of admixing in commercially processed and raw meats is the most challenging task for regulatory agencies as well to identify the species authentication as the country has huge consumption and export of different meat species ranging from chicken, goat, buffalo, and pig. India has exported 10,85,619.92 MT of buffalo meat products to the world for the worth of Rs. 23,460.38 Crores/ 3126.75 USD Millions during the year of 2020–21 [2].

DNA-based technologies have been implied for meat authentication and adulteration for long and evolved with the technology. Earlier where Restriction fragment length polymorphism (RFLP) and fingerprinting techniques were used, now highly precise mitochondrial DNA, 16S ribosomal subunit, and other highly conserved genes are amplified for accurate quantification of different meat varieties. Technological evolution has made it possible even to trace as low as 0.01% of adulterated

S. Rajora (✉) · S. M. Annapurna · Y. Kumar · V. S. Krishna · N. Kumar
CSIR-Central Electronics Engineering Research Institute, Pilani, Jhunjhunu 333031, India
e-mail: sunil.rajora3@gmail.com

V. S. Krishna
e-mail: sai@ceeri.res.in

N. Kumar
e-mail: navjot@ceeri.res.in

G. Mathur et al. (eds.), *Proceedings of 3rd International Conference on Artificial Intelligence: Advances and Applications*, Algorithms for Intelligent Systems,
https://doi.org/10.1007/978-981-19-7041-2_51

meat samples. However, the bottleneck in the DNA-based detection system is a requirement for lab setup, highly precise instrumentation, and the cost of the test.

Existing detection approaches for authenticity and admixing in non-vegetarian food items primarily rely on DNA analysis with few available protein-based analytical methods. The authentic nature of the meat products also employs protein-based analytical methods with the inclusion of mass spectrometry [3]. The limitations of protein-based analytical methods are non-specificity, high levels for the limit of detection, and processed meat products due to denatured proteins. Levin et al. [4] performed the study in relation to the authenticity measurement of the food items based on DNA approaches and various PCR protocols. Similarly, pathogenic contamination of meat products is majorly done by microbiological methods, which include bacterial growth and analysis [5]. Recently advancement of DNA-based methods is replacing these time-consuming microbial testing methods. DNA-based analyses are simple, rapid, sensitive, and accurate but require laboratory setup and trained persons and are time-consuming. The highly specific DNA-based methods can detect as small as 0.01% of adulteration among raw and processed food samples and identify pathogen contamination accurately. Keeping in view the demand of the huge meat industry worldwide, a number of private companies and laboratories offer meat testing services using these technologies. Private research labs have modified the procedures for more accurate results, but the test kits are expensive and require a sophisticated lab setup.

The research work includes the development of the thermocycler system integrating PID control strategy, temperature sensor, heating and cooling equipment, i.e., Peltier module, heater driver circuitry, and a 4 × 4 array copper-made sample holder for the placement of samples. The rapid change in the temperature (heating and cooling) is obtained, thereby allowing the process to alternate between primer annealing (55 °C), DNA amplification (70 °C), and strand melting cycles (95 °C). Each cycle mainly includes these three steps for denaturation, primer annealing, and extension, after initial preheating of the sample at 60 °C. The device will be suitable for use by a non-technical person with simplified sample processing steps and on-the-spot results related to meat authentication, admixing, and microbial contaminants detection.

2 Materials and Methods

2.1 4 x 4 Array Sample Holder

The 4 × 4 array sample holder made of copper was successfully designed, optimized, and fabricated. The newly fabricated 4 × 4 sample holder and its side view are shown in Figs. 1 and 2. At a time 16 cuvettes can be placed in the designed holder and the heat sin, exhaust fan assembly ensures the proper heat transmission. The chemical as well as physical properties of copper justify it as electrical conductive and high

Fig. 1 Newly fabricated 4 × 4 sample holder

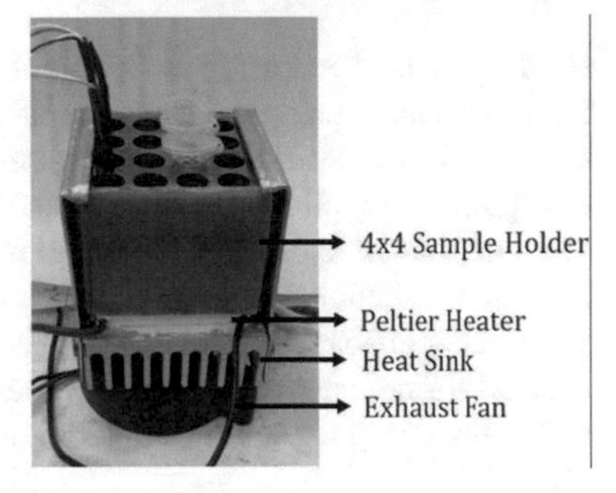

Fig. 2 4 × 4 sample holder side view

thermal conductive nature [6]. The resistance to the electron flow is observed at the room temperature, the reason for which is basically the scattered electrons on the lattice surface [7]. The study says that above some maximum permissible-current density, the excessive heating of copper occurs. So the 4 × 4 sample holder of (50 mm × 50 mm × 30 mm), mechanical chassis was designed, optimized, and fabricated so as to achieve the efficient heat transmission generated by the Peltier heater to the sample placed in the sample holder made of copper.

2.2 Development of Thermocycler

Figure 3 shows the block diagram of the thermocycler system. As shown in Fig. 3, the system is categorized into four subsystems which are the heating chamber, controller, temperature sensor, and power management unit. In the initial step to begin with the experiment, four different types of temperature sensors have been tested for fast response and accuracy of the measurement. Among these four temperature sensors, the thermocouple is finalized. Further, the error value is calculated. The error correction is performed and tuned automatically and the generated PWM signal adds to the temperature control of the thermocycler [8]. The sample is placed in the cuvette

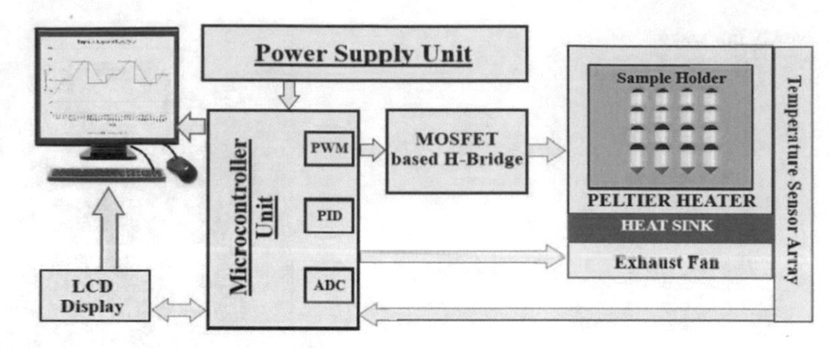

Fig. 3 Block diagram of developed thermocycler system

residing in the 4 × 4 array sample holder as shown in Fig. 1. The temperature can be increased and decreased by the thermal electric module. The thermal electric module includes 5A Peltier Heater which features 127 semiconductor couples making it easy to use for this application. The thermal electric module acts as a heat pump that can increment and decrement the sample temperature. Four different temperature sensors provide the feedback to the microcontroller to maintain the desired set temperature value. Proportional–Integral–Derivative Controllers (PID) tuning methodology is applied for controlling the heater module driven by the MOSFET-based H-bridge and maintaining the required value of temperature.

2.3 PID Temperature Controller

Proportional–Integral–Derivative Controllers (PID) are mainly used in the process control mechanism for several industrial applications in automation system design [9, 10]. This paper deals with the temperature control of a thermocycler using a PID controller.

The thermocycler employs precise temperature control and rapid temperature changes based on the heating mechanism to conduct the polymerase chain reaction, PCR. Ineffective temperature control methodology can cause overshooting or undershooting of the trace point resulting in the desecration of the processed sample and employing inaccurate results.

The achievement of the three thermal points with the initial preheat of the sample at around 60–70 °C and further the melting, annealing, and amplification process are completed. The parameters: Kp—the controller path gain, Ki—the controller's integrator time constant, and Kd—the controller's derivative time constant [11] are tuned to achieve the desirable point with accurate results.

3 Results and Discussions

The lab-level setup of the developed thermocycler is shown in Fig. 4. Three different temperature steps were used to study the ramp rate characteristics of the developed thermocycler which were initial preheating treatment at 60 °C incrementing toward 95 °C, 95 °C to 55 °C, and finally achieving 70 °C. The thermocycler is initially tested at the different values for the proportional gain and further other coefficients Ki and Kd values were fixed for the optimum temperature control. Each temperature input step is tested in the thermocycler under different proportional gains, Kp value, fixed integral gain, Ki value, and derivative gain Kd value.

The control algorithm methodology analyzes the error in the measurement and reduces it by the further calculation of the new weighted sum value. The weighted sum is calculated as follows:

$$x(t) = K_p e(t) + K_i \int_0^t e(\tau)d\tau + K_d \frac{de(t)}{d(t)} \quad (1)$$

where x is the calculated weighted sum value, e: error value, and t: Instantaneous time. The control signal x(t) is thus the sum of three positive constants: Kp, Ki, and Kd.

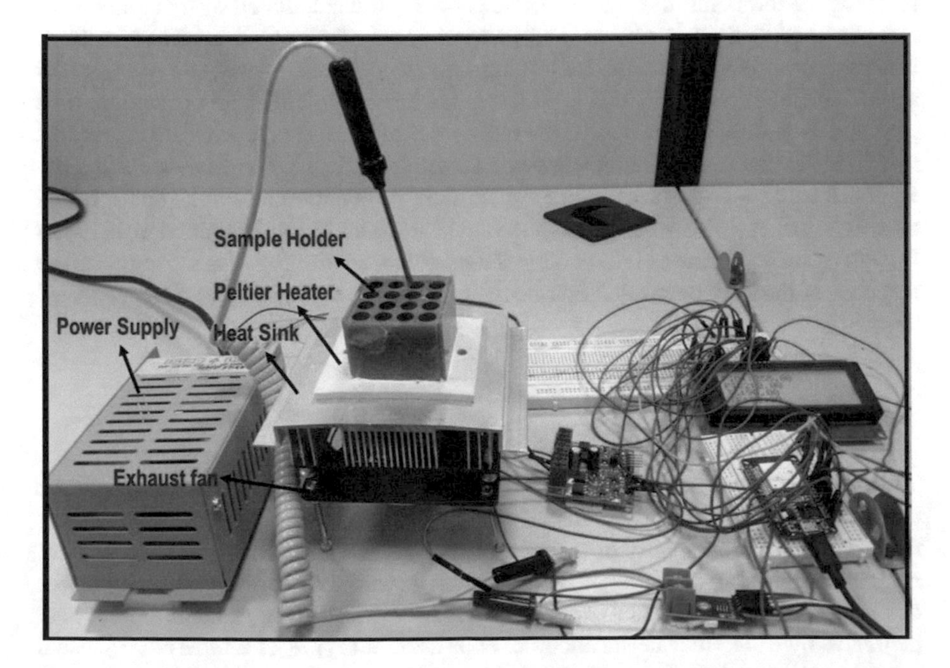

Fig. 4 Lab-level setup for thermocycler to be used for the execution of RT-PCR-based test

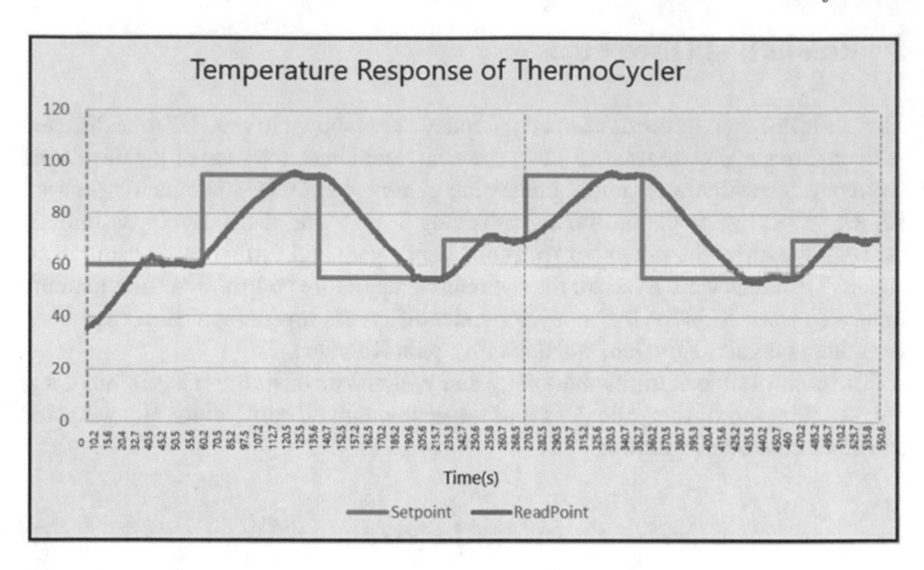

Fig. 5 Temperature response of the developed thermocycler at different set points

The initial thermal point has been set to 60 °C through the warm-up mechanism during the initialization. During the first cycle after the preheat the rise time for reaching the temperature value of 95 °C is 60.3 s, the calculated overshoot is 3 °C, and the settling time is 65.2 s. The thermal value of 55 °C is achieved in 50 s, undershoot value is 1 °C, and the settling time is 67.3 s. Similarly, the rise time for achieving the set point of 70 °C is 60.4 s, the overshoot is 2 °C, and the settling time is 65.3 s. To obtain the set points, the two cycles of the thermocycler are completed in 550.6 s (9.16 min). The developed PID-controlled thermocycler evaluates the feedback using a fixed or set point temperature to generate an error signal. Based on this value, it calculates the output value x(t). The output obtained will swing frequently in the region of the set temperature value. Figure 5 shows the temperature response of the PID-controlled thermocycler at different temperature stages.

4 Conclusion

The design and development of the RT-PCR-based thermocycler in order to track the set temperature profiles rapidly and accurately which can be used for the on-site detection of the authenticity of meat products and microbial contamination in them has been implemented. The 4 × 4 array mechanical chassis for holding the sample was designed, optimized, and fabricated to be used for the efficient transmission of heat generated by the Peltier heater module. Four different types of temperature sensors were tested out of which a thermocouple has been finalized for the temperature measurement of the sample to provide the feedback to the controller in order to measure and maintain the temperature of the PCR block. The developed thermal cycler is shown in Fig. 4. The temperature response of the developed thermocycler shows that by initial preheating of the sample to about 60 °C, the three temperature stages were obtained within 4.5 min (270 s) and the two cycles were completed in 9.16 min (550.6 s). The result obtained was saved and recorded for the further analysis of the samples.

Acknowledgements The authors would like to express their sincere gratitude to the Director, CSIR-Central Electronics Engineering Research Institute, Pilani, for his patient guidance and encouragement and to all the members of Dairy and Food Instrumentation Division, CSIR-CEERI, IIH-Jaipur Centre for their support and cooperation during the completion of this research work.

References

1. Sheikha AFE, Mokhtar NFK, Amie C, Lamasudin DU, Isa NM, Mustafa S (2017) Authentication technologies using DNA-based approaches for meats and halal meats determination. Food Biotechnol 31:281–315
2. APEDA agriXchange product profile page, https://agriexchange.apeda.gov.in/Home.aspx. Last Accessed 16 Feb 2022
3. Montowska M, Pospiech E (2010) Authenticity determination of meat and meat products on the protein and DNA basis. Food Rev Intl 27(1):84–100
4. Levin RE, Ekezie FGC, Sun DW (2018) DNA-based technique: polymerase chain reaction (PCR), pp 527–616
5. Kumar Y (2020) Isothermal amplification-based methods for assessment of microbiological safety and authenticity of meat and meat products
6. Hammond CR (2000) The elements, in handbook of chemistry and physics, 81st edn. CRC Press
7. Trigg GL, Immergut EH (1992) Encyclopedia of applied physics. Vol. 4: combustion to diamagnetism. VCH Publishers, pp 267–272
8. Bu M, Nielsen IRP, Jørgensen KS, Skov J (2013) A temperature control method for shortening thermal cycling time to achieve rapid polymerase chain reaction (PCR) in a disposable polymer microfluidic device. J Micromech Microengin
9. Dinca M, Gheorghe M, Galvin PT (2009) Design of a PID controller for a PCR micro reactor. IEEE Trans Educ 116–125
10. Jiang W, Jiang XC (2012) Design of an intelligent temperature control system based on the fuzzy self-tuning PID, Proc Engin
11. Singhal R, Padhee S, Kaur G (2012) Design of fractional order PID controller for speed control of DC motor. Int J Sci Res Publ 2:2250–3153

Chapter 52
Augmented Reality-Based Smart Mobile Application to Make ASD Children Accustomed to Daily Challenges

Jahanvi Singh, Pranita Ranade, and Tanmoy Goswami

1 Introduction

Autism is a complex neurological condition identified when a person's social relationships, speech, and conduct are significantly impaired, as defined by American Psychiatric Association, 2018 [1].

Problems in social relatedness have been the most common problem of the disease since Kanner's (1943) initial conception of the most recent Diagnostic and Statistical Manual of Mental Disorders (DSM-IV-TR). Individuals with ASD struggle to begin conversations, utilise proper introductions, and maintain shared attention. All social-behavioural adjustments in academic, personal, vocational, and communal contexts are based on these deficiencies. Additionally, people with ASD may avoid or shun social contact with their peers, limiting their chance to practice proper social behaviours. Eventually it results in deficits in social relatedness which could last for lifetime. As an individual ages, the noticeable abnormalities from typical social interactions may become more severe. The cumulative effects of social deficiencies over a lifetime necessitate effective evidence-based best practices early in a child's development to reduce obstacles to learning and smooth utilisation of social interactions [2].

The Rehabilitation Council of India clearly explains that Autism is a condition that can be managed rather than a disease that can be cured. As a result of the vast range of conditions, no single "technique" will work for all autistic children. However, there is substantial evidence that computers may benefit autistic children [3]. In this paper, a study has been conducted on AR enable solution for the skill development of children with ASD.

J. Singh (✉) · P. Ranade · T. Goswami
Symbiosis International (Deemed University), Symbiosis Institute of Design, Pune, Maharashtra, India
e-mail: jahnvisingh2000@gmail.com

G. Mathur et al. (eds.), *Proceedings of 3rd International Conference on Artificial Intelligence: Advances and Applications*, Algorithms for Intelligent Systems,
https://doi.org/10.1007/978-981-19-7041-2_52

2 Methodology

At the initial stage, primary and secondary research was conducted to understand the user behaviour, needs, and requirements. A thorough literature analysis was done based on autism, immersive technology, early intervention, Applied Behaviour Analysis (ABA) Therapy, and so on. Reliable research portals such as Scopus, Researchgate, and Google Scholar were chosen for acquiring legitimate research articles and case studies. Following secondary research, user interviews and surveys were conducted to understand better autistic children's experiences, which techniques are most effective for them, and what activities may trigger their fear. This part of the study was conducted on special educators and parents of autistic children based in Pune, India. During personal interviews, preliminary questions were asked to get the interviewees comfortable. Direct questions later followed this; some examples include "How did you sense that your child has autism?", "What measures are you taking/have you taken to be able to support your child?", "Do you think his/her condition has improved from when you started therapy to now?", and "Do you think game therapy/inclusive technology can improve this situation?". Significant insights were gained from these interactions, and some common suggestions given by parents include "autistic children may function differently, but if you let them stay on their path, they will eventually get to their destination. If you force them into something, then they will probably cause damage to their growth and themselves" and "autism's strength is visual, they are very visual learners, so if you show them anything on the screen or by picture card or menu card or any similar type of game, it works wonders for them". Later, an AR-based innovative application was designed to assist Autistic children.

3 Literature Review

3.1 Various Types and Symptoms

Another paper describes the five types of Autism as follows:

- Asperger's Syndrome—The DSM-5 diagnostic handbook has categorised it as a level 1 autism spectrum disorder. While communicating, Asperger's syndrome is commonly used in autistic groups and is used more frequently than level 1 spectrum condition.
- Rett Syndrome—Rett syndrome is an uncommon neurodevelopmental disease in early childhood. Although it is more common in girls, it may also be diagnosed in boys. Rett syndrome poses difficulties in nearly every element of a child's life.
- Childhood Disintegrative Disorder (CDD)—CDD, also known as "Heller's syndrome or disintegrative psychosis, is a neurodevelopmental condition marked

by the emergence of developmental difficulties in language, motor abilities, or social function at a later age."

- Kanner's Syndrome—Kanner's syndrome was first identified in 1943 by John Hopkins University psychiatrist Leo Kanner, who classified it as infantile autism. According to doctors, the disease is also a classic autistic disorder.
- Pervasive Developmental Disorder-Not Otherwise Specified (PDD-NOS) is a minor form of autism that manifests itself in various ways. Challenges in social and linguistic development were the most prevalent signs [4]

As per the National Institute of health, the possible indicators to check if the child has Autism Spectrum Disorders are as follows: [5]

- By the age of one year, they have not babbled, pointed, or made significant motions.
- By the age of sixteen, they have not spoken a single word.
- It is not possible to divide two words by two years.
- Doesn't reply when called by name.
- Becomes illiterate or socially inept.
- Limited eye contact.
- Doesn't appear to understand how to play with toys.
- Toys or other objects are stacked excessively.
- Is devoted to a single toy or thing and does not grin.
- At times, he appears to be deaf.

3.2 Problems Regularly Faced by ASD Children

Out of the many difficulties Autistic people face, National Institute on Deafness and other communication disorders broadly identifies the following as the major issues encountered by Autistic people [6].

- Face trouble interacting and involving with others.
- Find things like high intensity of lights or loud noises unfriendly or painful.
- Get agitated or nervous in new settings or social interactions.
- Take longer to comprehend information or repeatedly do or believe the same things.

3.3 Interactive System for Children with ASD

Caregivers of Autistic children face a difficult task in assisting them in achieving their aim of having a happy and fulfilling life. Because there is no known treatment for autism, education is critical to accomplishing this aim. It's crucial to be aware of autism as a type of mental impairment requiring specialised instruction and supervision rather than a psychiatric condition requiring therapy. Interactive systems for children with ASD are characterised as systems that reply in real time and are customised based on each child's behaviour. This software-based approach piques children's attention with a well-designed system that focuses on specific behaviours. Interactive smart applications have been demonstrated to benefit children with ASD at this stage. The findings indicate that using these platforms allows afflicted youngsters to connect, socialise, communicate, and learn in innovative ways. Several studies have been conducted on employing interactive systems to teach various abilities to children with ASD. Most of the research in the evaluations above showed that these children's learning improved after utilising the interactive systems [7]. This project aimed to develop specialised heuristics for the formative assessment of interactive solutions for children who have ASD to construct an engaging system [8]. The research found that new-age technologies such as virtual reality, augmented reality, virtual agents, and sensors in educational games helped children with ASD and showed the importance of more research in the development of teaching solutions using the advancement of technologies to improve the practical skills of children with ASD.

3.4 Use of AR/VR in Education

One of the studies conducted in 2020 explores the use of a virtual world as in virtual reality (VR) for the interaction of children with ASD and also with a physical world as in augmented reality (AR), or a combination of both as in mixed reality (MR), and embraces the confidence that technology is changing with the changes in learning experiences [9]. Interacting with the virtual world may involve using a specific VR headset, which many children with ASD may find challenging to wear or utilise. In the context of a study, the AR application may be used with a tablet or smartphone and provides a more ubiquitous method for individual and ASD intervention. AR draws the attention of children with ASD, according to evidence-based research. The AR-based applications give children with ASD a multimodal interaction to acquire multiple abilities as part of intervention or therapy sessions. In the coming years, the AR industry is anticipated to develop faster than the VR market [10]. In another study conducted in 2021, gamification was discussed for ASD children to improve their memory skills [11].

Gamification

Gamification features are a significant and valuable aspect of ASD learning methodologies [12]. Children with ASD in affluent countries rely heavily on technology in today's world. By definition, gamification encompasses not only games but also the entire psychological environment [13]. Individuals can be encouraged to compete with others and achieve the given tasks and goals if game-playing is adequately developed and implemented. Furthermore, it motivates people to maintain their activity and self-improvement to improve and break records.

3.5 Various Ways to Introduce a Holistic AR Experience

- **Imagination:** The goal is to develop a graphical environment that replicates the imagination that a youngster would typically have. Children can express their creativity via play, but autistic children encounter several challenges in this area. Their behaviours frequently appear to be meaningless and repetitious. Despite this, these children want to play and be accepted by their classmates, and how their peers react to them significantly impacts their feelings of isolation [14].
- **Attention:** Autistic children tend to focus on some items in their environment while excluding others. Although not a diagnostic symptom of autism, it is one of the early signs of the disorder. Eye movement recordings have shown promise in the detection and treatment of autism. Compared to traditional educational techniques, computer-based technology can help improve focus and, on occasion, improve learning [15].
- **Social skills:** Social skills are attitudes that predict substantial social consequences in children and teenagers in similar settings. The application of AR to meet the needs of people with social skill impairments is still a challenge. Making eye contact and giving embraces were the subject of previous research. Some of these studies have shown that they can assist children with autism acquire social skills and a more natural social attitude, allowing them to engage more effectively with their peers. Adult and peer-involved treatments, peer formation and beginning by an autistic child, class-extent teaching or engagement, and the use of scripts are all examples of these therapies [16].
- **Emotion:** AR is a method that allows for variable adaptation in this regard. The goal is for the individual to exhibit appropriate replies quickly and allow for breaks for re-evaluation and intention contacting to enhance response time. Children with autism have difficulties feeling emotions like neurotypical individuals, such as joy, grief, amazement, rage, hatred, and fear. AR's purpose in this field is to help children with autism understand these emotions since it has already been shown that they can identify these feelings in others and themselves following training [17].
- **Navigational abilities:** Humans employ navigational skills to get from one place to another. Autistic children's quality of life and happiness levels rise with these talents. It permits them, for example, to take public transportation when necessary. To do this, mobile devices with navigation software might be utilised. These devices are readily available, have a high acceptance rate among individuals with impairments, and may be used for various tasks other than navigation [18].

3.6 ASD in India

"Prevalence of autism spectrum disorder in Indian children: A systematic review and meta-analysis", a paper that talks about how India has a population of about 1.3 billion people, with children under the age of 15 accounting for roughly a third of the population. In India, an estimation has been done that more than 2 million people have ASD. Most of the research on ASD that has been published is based on hospital data. Therefore, there is no information on the prevalence of this disease in India. Only a few researchers have looked at its prevalence in community settings. Furthermore, the lack of universal use of adequately established and translated autistic diagnostic methods makes estimating the precise incidence of ASD challenging. It has also been predicted that there was a delay in ASD diagnosis at an early age, and there is also under-recognition of the disease [19].

4 User Study

4.1 Telephonic Interviews

Several parents and physicians were interviewed. They shared their experiences with autistic children and their encountered issues. Before the telephonic discussions, a comprehensive questionnaire was created to obtain real insights that would later be employed in the proposed framework design. Insights of interviews are that each child is unique, and their symptoms vary, need for patience while teaching necessary, and visual learners. Special educators and physicians discussed a few strategies for teaching motor skills and improving speech in ASD youngsters. According to doctors, there is a significant information gap among parents regarding symptoms, treatments, and numerous therapies that may be used to help their autistic children better.

4.2 Survey

A Google form was prepared and then shared in various groups that consisted of parents of the autistic child. The questions were designed to determine the average age at which parents learn that their child is autistic and what measures or actions they take to assist their children. The findings revealed that delayed speech, not answering their name, being alone, having a short attention span, experiencing Echolalia, and high sensitivity to touch, light, and sound were all typical ASD symptoms. More than 50% of parents opted for special schools and home-schooling for their autistic children and had a belief that their child is not able to cope with the curriculum of a typical School and neither do they teach daily activities that a child has to go through on their own and how to tackle the difficulties that they face.

4.3 Conceptualisation of the System

From the user research, it has been found that autistic individuals experience anxiety while going out and handling everyday problems, resulting in a lack of confidence and dependency on others. Several treatments and therapies are available to address various issues, but none have addressed the anxiety concerns and feelings of reliance that people face daily. This study focuses on a method for lowering autistic children's anxiety and dread of unfamiliar locations by increasing their attention spans and giving them a sense of independence. The paper explores the possibilities of Augmented Reality (AR) and how it may be presented in a fun and artistic way to keep children engaged. Also, as it is easily accessible on phones and tablets, children can start learning and exploring independently.

5 Ideation

5.1 Defining the Audience

Three concepts were brainstormed and assessed to determine which one addressed the most pain points while still entertaining enough for autistic children to participate. Artificial intelligence and augmented reality are used to power the solution. The system will use AR to educate the kid about a specific situation in a step-by-step tutorial and interaction at different places, after which the child will be given a brief quiz. The system then evaluates the performance using artificial intelligence, based on which the learning process is curated to fit the child's learning ability. In Fig. 1, the overall information architecture framework has been explained.

5.2 System Framework

Once the final concept is validated, parents fill up the child's information, such as name, age, and level of Autism. This onboarding allows students to store their progress and return to it at any time to see where they left off. The process begins when a child is handed a jigsaw puzzle to solve with a scenario printed on it and a special QR unique scenario that may be of point while still entertaining enough for autistic children to participate.

Artificial intelligence and Augmented Reality are used to power the solution. The system will use AR to educate the child in a specific situation in a step-by-step tutorial along with interaction at different places, after which the child will be given a brief quiz. The system then evaluates the performance using artificial intelligence, based on which the learning process is curated to fit the child's learning ability. In Fig. 1, the overall information architecture framework has been explained.

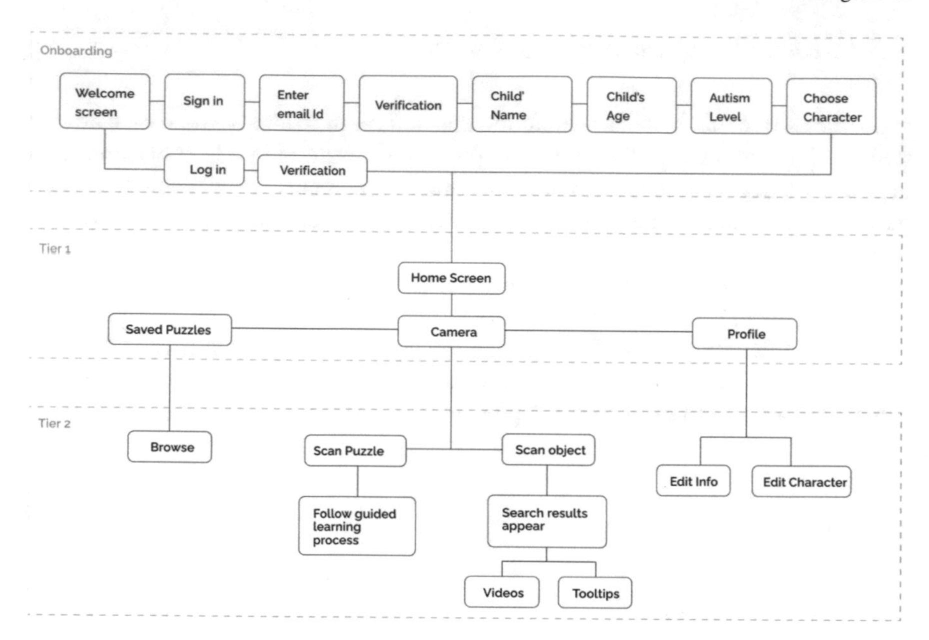

Fig. 1 Information architecture of the proposed system

5.3 *Proposed Design Solution*

All of the features were selected and implemented systematically after thoroughly analysing the approaches and strategies provided by special educators and physicians. This will ultimately assist children in learning their day-to-day activities in a comical storyline style. The first step is to onboard the system, which will be done by their parents, as shown in Fig. 2. Once validated, they can fill up the child's information, such as name, age, and level of Autism.

This onboarding allows children to store their progress and return to it at any time to see where they left off. The process, represented in Fig. 3, begins when a child is handed a jigsaw puzzle to solve, which will have a scenario printed on it along with a unique QR code; this scenario may be of anything from confined places such as a kitchen or classroom to open locations with a large crowd such as a bus station, shop, or playground. When the child solves the puzzle, the system will direct them and

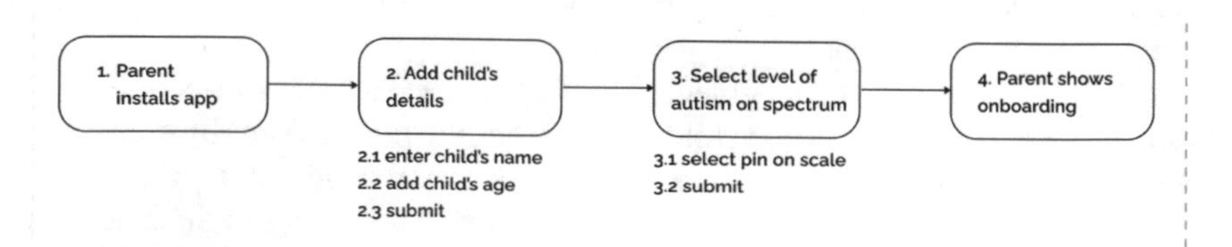

Fig. 2 User flow 1 of proposed system—onboarding

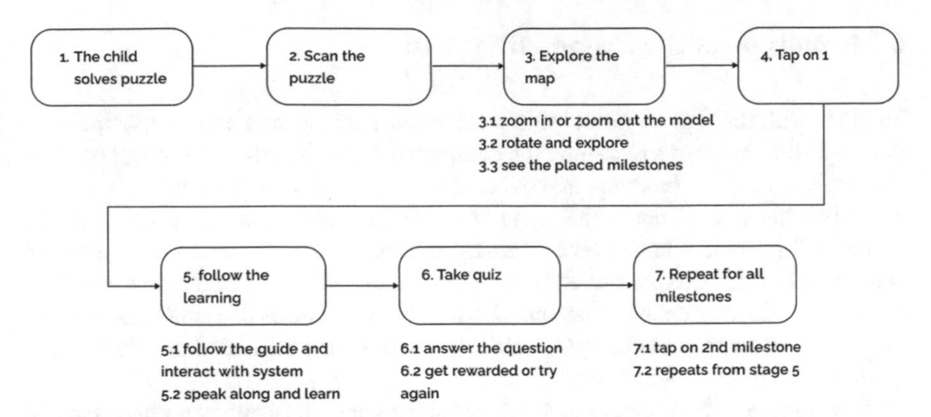

Fig. 3 User flow 2 of proposed system—learning

inform them that they may scan it for further information and take tests. As soon as the child understands how to study, the system will create a 3D model of the situation on top of the puzzle, with two or three milestones. These milestones represent stages the kid must pass through to gain experience in that scenario. The 3D model may be interacted with by zooming in, zooming out, rotating, and even putting it in other locations.

The device also incorporates another scanning feature combining machine learning and artificial intelligence, Fig. 4. This functionality can scan everyday items seen in the real world. The kid must point the camera towards the thing they want to watch and then press the shutter button. The system will then overlay search results on top of that object, describing what it does and its different components. The child can also look at the numerous learning videos and the facts about the thing. This feature helps children learn about everyday items and how to utilise them without fear.

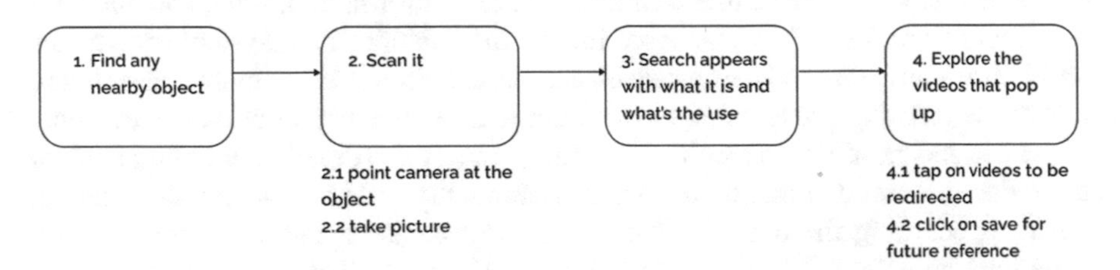

Fig. 4 User flow 3 of proposed system—exploring

6 Prototype of the Proposed System

To begin with the digital prototype, a bus station scenario was chosen and incorporated into the prototype. Scanning the completed puzzle yields a 3D model of a bus station, having a bus, bus stop, and ticket counter, as well as a few people. The child may also choose to listen to the scene or begin the learning process. The listening option will provide a breakdown of all the noises and sounds that a person would hear at a bus stop; this feature allows the system to aid in reducing anxiety in ASD children by allowing enabling become familiar with the noises in a controlled setting. Individual noises may be changed, and the intensity can be regulated to help the child gradually become used to them.

The next stage is to purchase a ticket, during which an animated character will talk and educate about each step and refer to the child by their name to help them connect with their own name. The process of purchasing a ticket is broken down into three stages: requesting access, paying for the ticket, and collecting the ticket. The animated character also asks them to initiate and participate in the conversation by speaking along with a few sentences. AI evaluates these lines on the backend to determine pronunciation correctness and how the child responds to the characters; the data is also utilised to generate reports. After the child has understood the procedure of purchasing a ticket, the animated character instructs them on how to inspect the ticket and what elements are included. The child may now go to the quiz, where the system will construct questions according to the Autism level specified during onboarding. The child must correctly answer these questions to earn stars that can be used to personalise their Avatar, which can later be shown on the 3D models they view. Negative marking is not used in the system to encourage the child to try again. If the child answers a question incorrectly, they will be allowed to try again or to look up the solution using the stars they have earned in earlier questions. Weekly or daily reports are generated based on the child's performance; this also enables the system in adjusting to adjust to the learning process and quiz; this can also be managed manually in the profile, where they can access their personal information, stars, and purchase new items to make their avatar more like them. This incentive system encourages children to learn more and participate in quizzes. Children become more concentrated due to the learning process and examinations; their memory improves; they know about new locations they haven't been before; anxiety and dread of unfamiliar places are reduced. After completing the first milestone, the child is guided to the second, which is boarding the bus. The technology will create a new augmented reality situation where the child can see inside the bus. When they first arrive on this screen, an animated character will invite them to pick a seat based on their preferences. While they do so, the animated character will teach them about safety rules for selecting the safest seat on a bus. The child will also be able to engage with other characters, including the conductor, driver, and passenger. This way, they will learn how to begin discussions, engage, and behave appropriately. This is where they will solve a new problem and repeat the procedure. Some prototype screens have been shown in Fig. 5.

Fig. 5 High fidelity prototype: The AR model of the bus when the puzzle is scanned and the Animated character guiding and teaching about buying a ticket

7 Usability Testing

A screener document was created to perform the usability testing, and 6–12 years old autistic children and their parents or caregivers were chosen. The only skill necessary for participation in the usability test was the ability to operate a smartphone or tablet. This was an in-person usability lab test with the child and their parents. Towards the end of the trial, personal interaction was held to understand the parents' perspectives better and see if anything could be added or deleted from the current solution. Before the test, parents were given a summary of what the test entailed and were informed that it was the solution being tested, not the child. They were also given a solution to explore further. Once parents were satisfied, their children were given the full opportunity to explore the app to see whether they were ready to proceed with the test. When it was confirmed that the child was prepared to move, they were given the puzzle to solve and instructions on how to scan. And as the AR model formed in front of their eyes, they were free to explore and engage with the system flow. When the child encounters a problem, they are permitted to seek assistance to determine where the system is failing and what may be improved. Parents and children have expressed their admiration for the app's colours and design and the innovative technology it incorporates. Parents noticed that their youngster spent more time than usual and was fascinated by the 3D models in the live environment. Many children were happy to give answers so that they could watch the reward animation play. Parents agreed that this is a positive start and that there are many possibilities for the future. More animation and vocal prompts were among the ideas suggested.

8 Conclusion and Future Scope

It was indicated that children would require some assistance learning how to use the app at first, but after they are comfortable with it, they would try to get started independently. As technology advances, these AR models may be created using earlier software versions while maintaining graphic quality, broadening the scope.

Special AR glasses could be developed specifically for autistic children with adequate budgets and technical advancements. The glasses would aim to scan and superimpose related information on the glass and give auditory feedback. Another option in the same field is Leap Motion Controller; it is an input device that detects hand and finger movements without direct contact with the touch screen; this technology would merge well with the proposed solution to make it more interactive.

References

1. American Psychiatric Association (2018) "What Is Autism Spectrum Disorder?". Accessed from https://www.psychiatry.org/patients-families/autism/what-is-autism-spectrum-disorder.
2. Boudreau J, Harvey MT (2013) Increasing recreational initiations for children who have ASD using video self-modelling. Educ Treat Child 36(1):49–60
3. Historical Overview of Autism (2017) Accessed from http://www.rehabcouncil.nic.in/writereaddata/autism.pdf
4. Carmigniani J, Furht B, Anisetti M, Ceravolo P, Damiani E, Ivkovic M (2011) Augmented reality technologies, systems and applications. Multimedia Tools Appl 51(1):341–377
5. U.S. Department of Health and Human Services, National Institutes of Health, National Institute of Mental Health. (March 2022). Autism Spectrum Disorder. Accessed from https://www.nimh.nih.gov/health/topics/autism-spectrum-disorders-asd
6. Baron-Cohen S (2001) Theory of mind in normal development and autism. Prisme 34(1):74–183
7. Hadwin J, Baron-Cohen S, Howlin P, Hill K (1996) Can we teach children with autism to understand emotions, belief, or pretence? Dev Psychopathol 8(2):345–365
8. Valencia K, Rusu C, Quiñones D, Jamet E (2019) The impact of technology on people with autism spectrum disorder: a systematic literature review. Sensors (Basel, Switzerland) 19(20):4485
9. Khowaja K, Banire B, Al-Thani D, Sqalli MT, Aqle A, Shah A, Salim SS (2020) Augmented reality for learning of children and adolescents with autism spectrum disorder (ASD): a systematic review. IEEE Access 8:78779–78807
10. Shukla D (2020) AR and VR market size likely to grow exponentially. In: Electronics for you, pp 84–87. Accessed from https://india.theiet.org/media/1315/efy-magazine-ar-vr-market-to-grow-exponentially.pdf
11. Goswami T, Arora T, Ranade P, Enhancing memory skills of autism spectrum disorder children using gamification. J Pharmac Res Int 33(34B):125–132
12. Khowaja K, Salim SS (2020) A framework to design vocabulary-based serious games for children with autism spectrum disorder (ASD). Univ Access Inf Soc 19(4):739–781
13. Dymora P, Niemiec K (2019) Gamification as a supportive tool for school children with dyslexia. In: Informatics, vol 6, No. 4. Multidisciplinary Digital Publishing Institute, p 48
14. Khowaja K, Salim SS (2015) Heuristics to evaluate interactive systems for children with autism spectrum disorder (ASD). PLoS ONE 2–3. https://doi.org/10.1371/journal.pone.0132187
15. McMahon DD, Smith CC, Cihak DF, Wright R, Gibbons MM (2015) Effects of digital navigation aids on adults with intellectual disabilities: comparison of paper map, Google maps, and augmented reality. J Spec Educ Technol 30(3):157–165
16. Michel P (2004) The use of technology in the study, diagnosis and treatment of autism. Final term Paper for CSC350: Autism and Associated Developmental Disorders, pp 1–26

17. Maskey M, Warnell F, Parr JR, Le Couteur A, McConachie H (2013) Emotional and behavioural problems in children with autism spectrum disorder. J Autism Dev Disord 43(4):851–859
18. Volkmar FR, Reichow B, McPartland J (2022) Classification of autism and related conditions: progress, challenges, and opportunities. Dialogues in clinical neuroscience
19. Chauhan A, Sahu JK, Jaiswal N, Kumar K, Agarwal A, Kaur J, Singh S, Singh M (2019) Prevalence of autism spectrum disorder in Indian children: a systematic review and meta-analysis. In: Neurology India, (January), pp 100–101. https://doi.org/10.4103/0028-3886.253970

Chapter 53
Performance Evaluation of Secured Containerization for Edge Computing in 5G Communication Network

Abubakar Saddiq Mohammed, Isiaka Olukayode Mosudi, and Suleiman Zubair

1 Introduction

The deployment of 5G communication standards is a precursor to the explosive evolution of Information and Communication Technology (ICT) innovations for mobile devices. Aggregation and integration of wide range of applications and operations such as: Machine-to-Machine (M2M) Communication, Internet of Things (IoT), emerging Vehicle Technologies, Virtual Reality (VR), Augmented Reality (AR), etc. are at its wake. With these comes a correspondingly large increase in the number of smart mobile devices to approximately over 50 billion. However, this number is small compared to the exponential growth in the volume of data generated by these powerful applications and feature-rich contents. This will create a hype for mobile data traffic and high computing requirements according to Skarpness [1]. As a result, constraints of computational resources and network resources are envisaged for cellular mobile communication User Equipment (UE).

To resolve these problems, the computational requirements of mobile applications can be offloaded to tethered external infrastructures with adequate resources for processing. Different interventions have been proposed, which include Cyber Foraging, Cloudlet, Mobile Cloud Computing (MCC) and Multi-access Edge Computing (MEC), Mosudi et al. [2].

A. S. Mohammed (✉) · I. O. Mosudi · S. Zubair
Telecommunication Engineering Department, Federal University of Technology, Niger State, Minna, Nigeria
e-mail: abu.sadiq@futminna.edu.ng

G. Mathur et al. (eds.), *Proceedings of 3rd International Conference on Artificial Intelligence: Advances and Applications*, Algorithms for Intelligent Systems,
https://doi.org/10.1007/978-981-19-7041-2_53

2 Containerization

Containerization is the process by which the Operating System (OS) kernel allows running of isolated user-space instances called containers. These are standard collection of software that bundles up the code and all its dependencies together as an abstraction at the OS application layer, thereby enabling the application to run quickly and reliably from one computing environment to another. Container images are typically tens of MBs in size. It can handle more applications and require fewer machines and OS, Adufu et al. [5]. Containers compared to Virtual Machines (VM) are more suitable for MEC for the sake of storage limitation and computing resources optimization. Containerization allows hardware resources to be decoupled from software, enabling packaged software to execute on multiple hardware architectures providing several benefits such as rapid construction, instantiation, and initialization of virtualized instances, Taleb et al. [3].

MEC resources can be allocated to containers for better isolation, performance and allowing for easy collaboration and deployment of applications across different mobile environments, Willis [6]. Orchestrated containerized MEC will provide efficient infrastructures needed for migration of monolith legacy applications onto 5G service platforms. This enables the breaking down of large applications into microservice deployable on a large number of interconnected MEC platforms, Alam et al. [4]. The containerization ecosystem has become so matured, presenting a whole lot of its orchestrators such as Docker Swan, Kubernetes (k8s), Marathon, and Amazon container engine. Google Container Engine (GKE), and Azure container service, Kata [7], Piparo et al. [8], Sanchez [9], Hoque et al. [10] and Augustyn and Warchal [11]. The test bed experiments in the paper made use of Kata-containers.

2.1 Kata Container

Kata containers run in dedicated kernels to provide isolation of network input/output and memory and can utilize hardware-enforced isolation with virtualization extensions. However, it is backward compatible with industry standards such as Open Containers Initiatives (OCI) container format, Kubernetes Container Runtime Interface (CRI), as well as legacy virtualization technologies while consistent with standard Linux containers in performance. Kata is based on the Kernel Virtual Machine (KVM) hypervisor with an option for Quick Emulator/Net Emulator (QEMU/NEMU). NEMU is actually a stripped-down version of QEMU by removing emulation not required thereby reducing the attack surface. It is more secure than a traditional container by replacing the default container runtime (runC) with Kata-runtime. Relying on Kata-agent, shim for I/O while running Kata runtime instead of runC container runtime as available in Docker. Kata containers are light and fast containers.

3 Design of MEC Deployment Scenario

In this section, the 5G network 3GPP and non-3GPP transport components specifications were evaluated, and models for MEC deployment scenarios for 5G network were designed. This was carried out to provide the platform to compare MEC application end-to-end transport latency in 5G and 4G deployments. This work leveraged on control/user plane separation (CUPS), Lower Layer Functional Splits (LLFS), and Higher Layer Functional Splits (HLFS) and 3GPP 5G Service-Based Architecture (SBA), ESTI [12], and distributed Common Compute Platform (CCP) which permits the location of Virtualized Network Functions (VNFs) in different parts of the network for management of different capabilities. MEC hosts were located at the Centralized Unit (CU) connected directly to the Packet Data Convergence Protocol (PDCP) thereby, reducing the estimated total UP latency for MEC deployment in 5G.

The end-to-end transport latency has an effect on determining the value of UP latency, and in combination with CP latency, determines the effective QoE. Higher Layer Functional Split (HLFS) option 2 for the mid-haul and Lower Layer Functional Split (LLFS) option 7 for front-haul will permit four RAN deployment scenarios and thus four MEC deployment scenarios (Fig. 1),

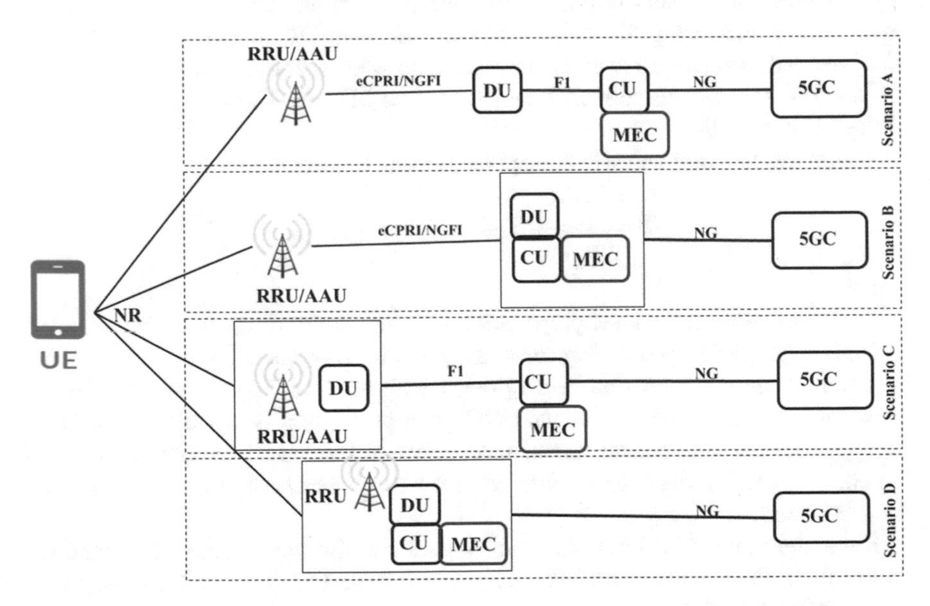

Fig. 1 5G MEC deployment models (SBA RAN)

3.1 Latency Evaluation

The latency figure, T can be obtained using the relationship for the total one-way UP latency for an application deployed on 4G/LTE,

$$T = T_{Radio} + T_{Backhaul} + T_{Core} + T_{Transport} \quad (1)$$

by modifying (1) for LTE/EPC, we obtain

$$T = T_{LTE} + T_{EPC} + T_{Transport} \quad (2)$$

where,

$\mathbf{T_{LTE}}$ = one-way packet propagation delay between UE and eNB, plus packet processing time; $\mathbf{T_{EPC}}$ = one-way packet propagation delay between eNB and EPC, plus processing delay with the core network; $\mathbf{T_{Transport}}$ = one-way packet propagation between the EPC and Packet Data Network (PDN). This might include propagation delay to the internet, if service requested by the UE has to be sourced from the Internet.

Latency values will vary from one MEC deployment scenario to another. Quantifying all the parameters is challenging due to differences in the performance of equipment. Considering 5G end-to-end network from DU to CU, and MEC host. However, we assumed a 1-way latency range between 5 and 8 ms between CU and DU and in essence, 8 ms network latency between CU and DU eases the co-location of the CU with MEC.

Therefore, the total one-way user plane latency becomes,

$$T = T_{NR} + T_{DU} + T_{CU} + T_{Transport} \quad (3)$$

where,

$\mathbf{T_{NR}}$ = the one-way packet propagation for New Radio (NR) delay between UE and DU, plus packet processing time; $\mathbf{T_{DU}}$ = one-way packet propagation delay between DU and CU, plus processing delay with the CU; $\mathbf{T_{CU}}$ = one-way packet propagation delay between CU and 5GC, plus processing delay within the 5GC; $\mathbf{T_{Transport}}$ = one-way packet propagation between the 5GC and DN. This might include propagation delay to the Internet if the service requested has to be sourced from the Internet.

Deployment of MEC in all the four scenarios in the model above provided the option for a direct connection between MEC and the CU. The total one-way user plane latency becomes,

$$T = T_{NR} + T_{DU} + T_{CU} \quad (4)$$

There is a need for a 5G-capable integrated development environment in the quest to investigate the deployment of MEC at the 5G CU, but we could not get a simulator for this purpose. Instead, leaning on Docker containers, Kata-runtime, and OS-builder -Kata containers guest OS building scripts, a sandbox application was built to gain insight into the advantages of computing at the network edge compared to at the remote cloud servers. In order to compare MEC versus MCC deployments of resource-intensive applications, a mobile web application was built, shipped in a secure container image, saved as a code and pushed to a repository. This combination of definition files were deployed on Docker engines hosted in the remote cloud and edge servers. The chosen target was the web application platform because of its capabilities of execution on a wide range of devices and mobile environments without modification of the application codebase.

3.2 Experimental Environment

Experimental environment incorporated Ubuntu server, Ubuntu Docker image and Docker container engine with its default runtime (runC), replaced with Kata-Runtime. The Ubuntu Kata-container image was created using OS-builder. Python programming language was used for application logic, dataset result generation, cleaning and graphing. The test application backend was based on Python Flask micro web framework while the front-end was built using HTML/CSS/JavaScript. Docker was used for application shipping and Locust framework for load testing. Publicly available Atlassian Bitbucket git and Docker Hub repositories were used for web application code base and container image repositories respectively. The containerized mobile application was deployed using Docker, but Kata-runtime replaced runC to ensure app isolation at the kernel level. This ensured the deployment of the MEC application at the speed of containers while maintaining the security available in VMs. The test mobile application was a memory and processor-intensive mobile web application that generates Rubik cubes images and provides a breakdown of the cube details. These details were total cubelets faces, cubelets and hidden cubelets. It also holds the generated graphics in the memory while rendering it on the end-user devices. This is comparable to graphics generation and rendering in mobile game applications.

4 Experimental Results and Discussion

There were two sets of results obtained. The first set was obtained from the application of 5G transport interface specifications on the four proposed MEC deployment models. Evaluating Eq. (4) for eMBB proposed four deployment scenarios shown in Fig. 1. by applying 5G specification values Mosudi et al. [2];

4.1 Deployment Scenario

Scenario A: Optimally this prototype produced an estimated round-trip time (RTT) value of 11.2 ms. Using Eq. (4),

$$T = T_{NR} + T_{DU} + T_{CU}$$

$$\begin{aligned} \text{Minimum } T &= 4000 + 100 + 1500\,\mu s \\ &= 5.6 \times 10^{-3}\,s \\ \text{Maximum } T &= 4000 + 100 + 10000\,\mu s \\ &= 14.1 \times 10^{-3}\,s \end{aligned}$$

Scenario B: Optimally, the RTT is 8.2 ms.

$$\begin{aligned} T &= 4000 + 100\,\mu s \\ &= 4.1 \times 10^{-3}\,s \end{aligned}$$

Scenario C: Optimal RTT is 11 ms.

$$\begin{aligned} \text{Minimum } T &= 4000 + 1500\,\mu s \\ &= 5.5 \times 10^{-3}\,s \\ \text{Maximum } T &= 4000 + 10000\,\mu s \\ &= 14 \times 10^{-3}\,s \end{aligned}$$

Scenario D: Optimal RRT is 8 ms.

$$T = 4000\,\mu s$$

$$= 4 \times 10^{-3}\,s$$

4.2 Load Test

The second set of results were obtained from the experimental load test of the secured containerized mobile web application deployed on remote cloud location and servers. Figure 2 depicts the MCC Test Scenario.

The results obtained from the experiments included; the total requests per second (req/s) made to the application deployed, request time stamps, requests failure per second (req/s), minimum, median and maximum application response time, average

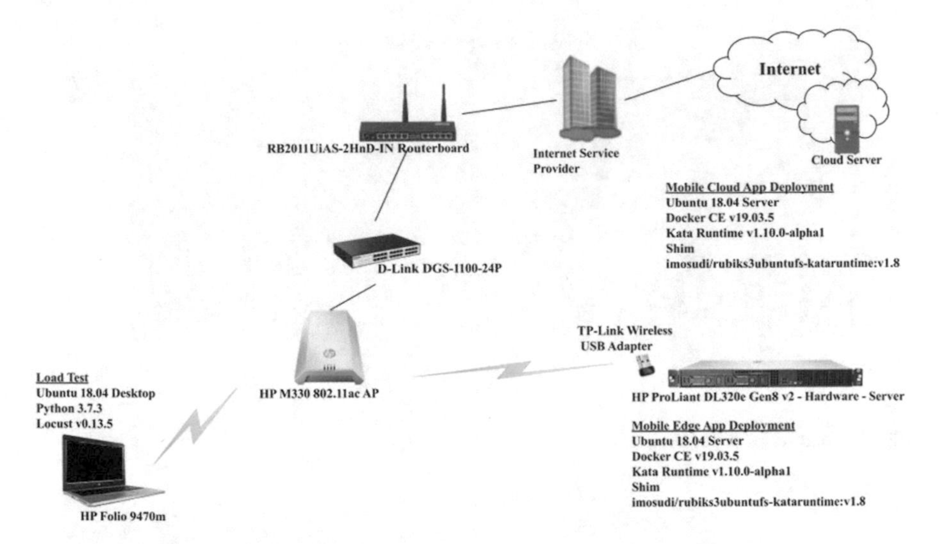

Fig. 2 MCC test scenario

application data download size, 50 percentile and 95 percentile application response time among other result parameters. All these were determined for both edge site and cloud application deployments respectively. Each time, the experiment lasted about 2 h. It was observed that the application response time and the amount of downloaded application data follow the same pattern corresponding to the size of the Rubik's cube being rendered. Likewise, failures were more prevalent with the MCC deployments compared to the relatively stable MEC deployments.

Request rate for both the edge site and cloud deployments are presented in the results. The intention to simulate a random Rubik's cube size between values of three and fourteen did pay off. The requested rate for both deployments were about the same as indicated in Fig. 3, which justifies a fair comparison. Figure 4 shows that failures were more prevalent in cloud deployments. Application maximum and median response time for both edge and cloud deployments were presented in Fig. 5.

The average application data download are presented in Fig. 6. The same application version was deployed for both scenarios and with the convergence in the application request rates as shown in Fig. 3. These had a significant effect on the amount of application data downloaded for both scenarios during the experiments. The application's average data download also converged, validating fair comparison. The application response for half of the experiment duration, 50 percentile, second quartile or median response time is plotted in Fig. 7.

The general application response time over a period of two hours for 50 percentiles is discussed below. The application response time for less than 95 per cent of the time span of the experiment and 95 per cent is presented in Figs. 7 and 8.

The round-trip times (RTT) from the MEC deployment are all within the latency requirements for VR and AR of 7-12 ms. Tactile Internet is < 10 ms, Vehicle-to-Vehicle is < 10 ms, Manufacturing and Robotic, Control/Safety Systems are between

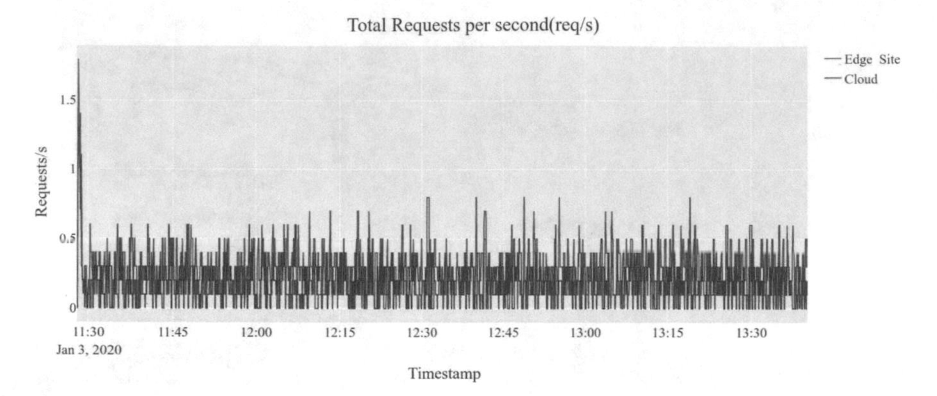

Fig. 3 Total request per sec

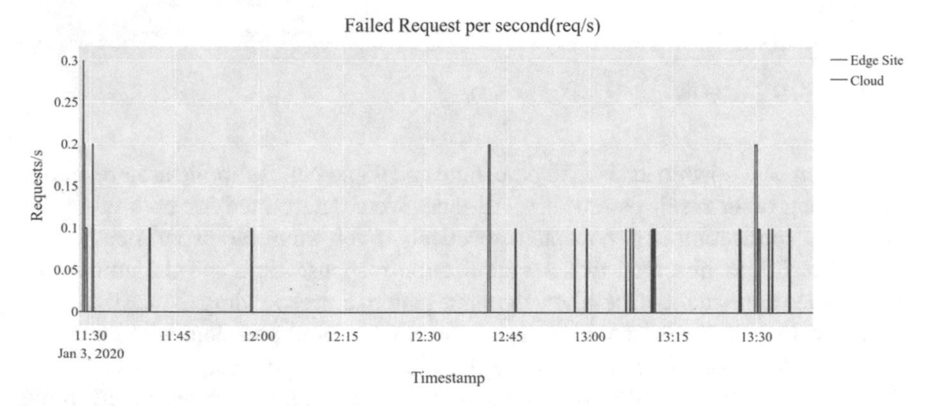

Fig. 4 Total failed request rate

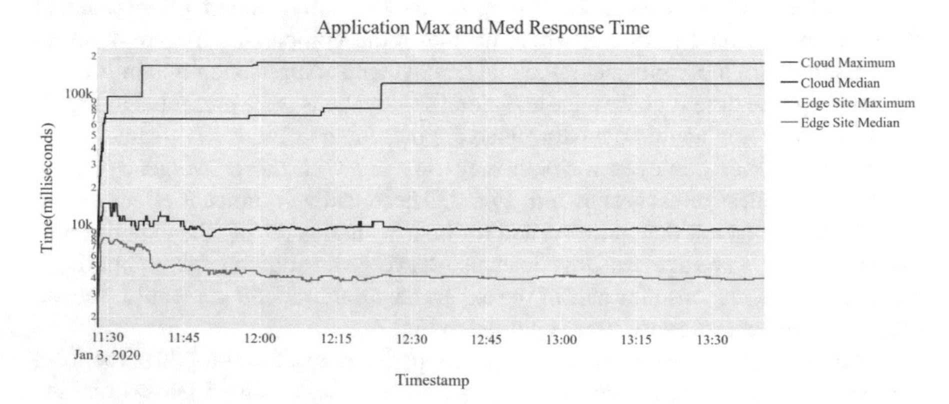

Fig. 5 Application max. and medium response time

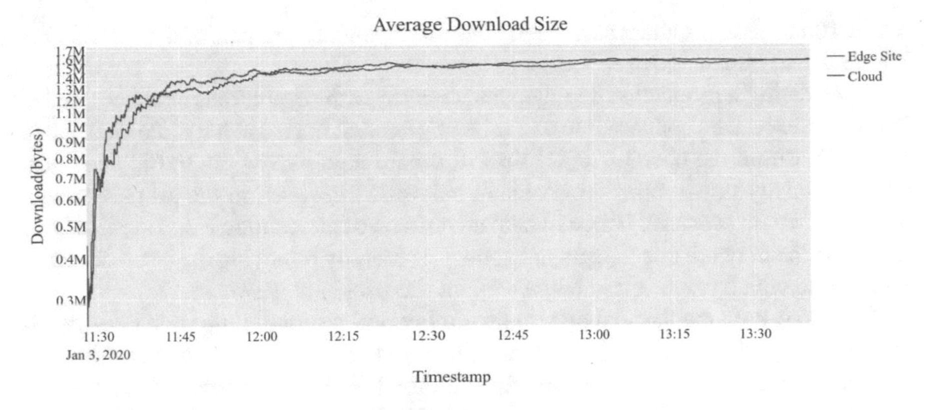

Fig. 6 Average application content download in bytes

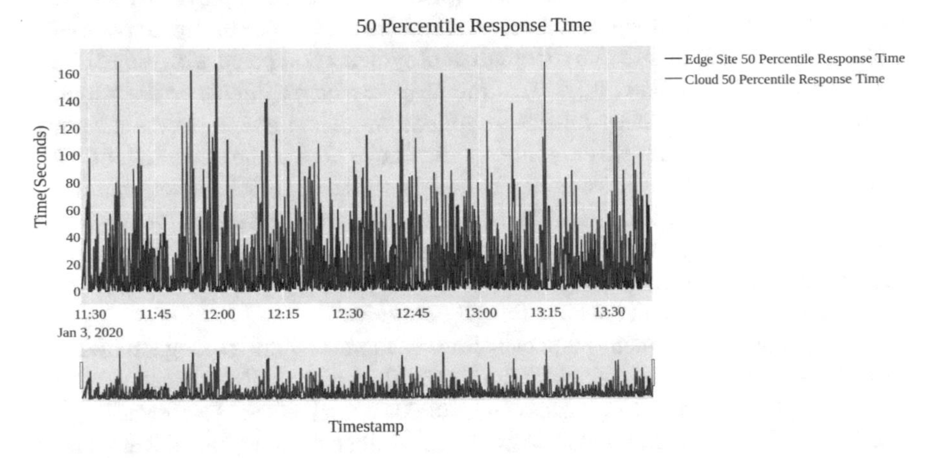

Fig. 7 Second quartile application response time

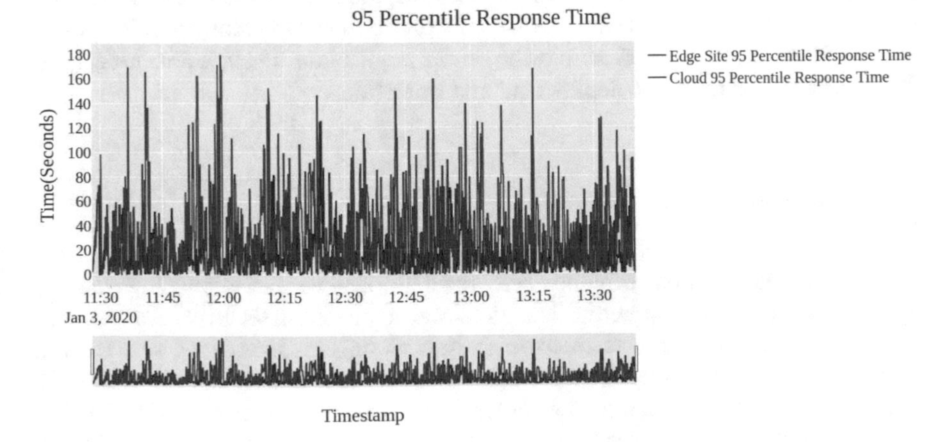

Fig. 8 95 percentile of response time

1 and 10 ms. The attendant savings in compute resources are part of the reasons for the choice of containers over VM.

Considering a production deployment scenario with several thousand geographically dispersed UEs connected to a mobile application hosted at a remote cloud data center or within the operator DN, this will create high bandwidth traffic and exert serious penalty on the operator backhaul network. However, in the proposed edge site deployment scenario, UEs will take advantage of consuming MEC applications hosted at the network edge, deployed at the CU thereby removing the issue of heavy data traffic which can result in bottleneck on the backhaul networks.

Secondly, both median and maximum application response time were considered for both edge sites and cloud deployments as indicated in Fig. 5. It was observed that in the maximum latency figures for the cloud deployment required for smooth running of mobile applications, there were initial failures reported for both deployment at application startup. In fact, the figure for the edge deployment was very poor, but failure finally disappeared. However, there is evidence that there is a higher application failure response rate for cloud deployments compared with the edge site deployment as indicated in Fig. 4. This might have adverse effects on the adoption of new and emerging latency-sensitive applications. The maximum edge response time for edge deployment was initially a little above 60 s compared to about 170 s for cloud deployment as shown in Fig. 6. This test was not a 5G network but it was obvious that deployment on a real 5G network with adequate MEC server resources can normalize the edge figures to more acceptable values. Furthermore, the comparison of minimum application response time confirms the proposal for containerized applications deployment at the edge for 5G networks.

Clearly, edge deployment response time is more visible looking around low latency values, unlike the cloud which dominates the graph skyline indicating consistent unacceptably high latency values. The fact that containers are secure and deployable for MEC infrastructures will increase the ability of enterprise developers by improving collaboration to quickly deliver scalable and reliable applications and services at paces required for 5G rollout not jeopardizing the security of the end-to-end network. Containers can provide the required DevOps ecosystem for developers and engineers to work across the entire application lifecycle, from designs, development, and testing to deployment and operations.

5 Conclusion

Secured containerization technique was used to achieve end-to-end low latency figures with MEC infrastructure isolation and application security. The secured containerized application was deployment in both the cloud and edge scenarios, it was validated that the edge scenario has lower User Plane latency figures, less backhaul traffic and a lower application failure rate. Secured containers using Docker and Kata containers provide most of the essential features required for MEC infrastructures making it suitable for resource-intensive mobile applications speed and isolation requirements to guarantee safety within the mobile ecosystem expected with massive deployment of 5G UEs. Container workflow orchestration provides automation capabilities for MEC to deliver just in time applications and services, enabling service providers to provide ubiquitous access to specified applications or services since applications infrastructures are pushed to repositories and can be deployed by downloading from the repository.

References

1. Skarpness M (2017) Beyond the cloud: edge computing. In: Keynote speech at Embedded Linux conference Europe, Prague, Czech Republic
2. Mosudi IO, Abolarinwa J, Zubair S (2019) Multi-access edge computing deployments for 5G networks. In: Proceeding, 3rd international engineering conference, pp 472–479
3. Taleb T, Samdanis K, Mada B, Flinck H, Dutta S, Sabella D (2017) On multi-access edge computing: a survey of the emerging 5G network edge cloud architecture and orchestration. IEEE Commun Surv Tutor 19(3):1657–1681
4. Alam M, Rufino J, Ferreira J, Ahmed SH, Shah N, Chen Y (2018) Orchestration of micro-services for IoT using docker and edge computing. IEEE Commun Mag 56(9):118–123
5. Adufu T, Choi J, Kim Y (2016) Is container-based technology a winner for high performance scientific applications. In: Proceedings of 7th Asia-Pacific network operations and management symposium (APNOMS), pp 507–510
6. Willis J (2015) Docker and the three ways of DevOps. Docker Blog. May 26, 2015. https://blog.docker.com/2015/05/docker-three-ways
7. Kata Containers (2019) Learn: an overview of the Kata Containers project. Accessed May 17, 2019, from https://katacontainers.io/learn/
8. Piparo D, Tejedor E, Mato P, Mascetti L, Moscicki J, Lamanna M (2018) SWAN: a service for interactive analysis in the cloud. Futur Gener Comput Syst 78:1071–1078
9. Sanchez C (2015) Scaling docker with Kubernetes. Website. http://www.infoq.com/articles/scaling-docker-with-kubernetes
10. Hoque S, Brito MSD, Willner A, Keil O, Magedanz T (2017) Towards container orchestration in fog computing infrastructures. In: 41st annual computer software and applications conference (COMPSAC), pp 294– 299
11. Augustyn DR, Warchał L (2010) Cloud service solving N-body problem based on windows azure platform. In: International conference on computer networks. Berlin, Heidelberg, pp 84–95
12. European Telecommunications Standards Institute (ETSI) (2018) TS 129 500 - V15.0.0–5G; 5G system; Technical realization of service based architecture. - ETSI. Retrieved May, 2019, https://www.etsi.org/deiver/etsi_ts/129500_129599/129500/15.00.00_60/ts_129500v50000p.pdf

Chapter 54
An Optimal Monte Carlo Approach Related to Multidimensional Artificial Intelligence

Venelin Lyubomirov Todorov and Ivan Tomov Dimov

1 Introduction

Lin [5] and Lin et al. [6] considered the difficult task of evaluating multidimensional integrals used in artificial intelligence. The first Lin type of integrals are

$$\int_{\Omega} p_1^{u_1}(x)\dots p_s^{u_s}(x)dx, \tag{1}$$

and the second Lin type integrals are

$$\int_{\Omega} e^{-Nf(x)}\phi(x)dx, \tag{2}$$

where $f(x)$ and $\varphi(x)$ are polynomials with an integer N. Up to now multidimensional Lin type integrals (1) and (2) are computed unsatisfactory with deterministic Watanabe [12] and algebraic methods Song et al. [11], and Monte Carlo (MC) methods Paskov [9], Pencheva et al. [10] outperforms the deterministic ones for which is typical "curse of dimensionality" Dimov [4].

The system model is based on evaluation of Lin type multidimensional integrals and the methodology of our work Owen [8] is to apply several stochastic techniques

V. L. Todorov (✉)
Institute of Mathematics and Informatics, Department of Information Modeling, Bulgarian Academy of Sciences, Acad. Georgi Bonchev Str., Block 8, 1113 Sofia, Bulgaria
e-mail: vtodorov@math.bas.bg; venelin@parallel.bas.bg

I. T. Dimov
Institute of Information and Communication Technologies, Department of Parallel Algorithms, Acad. Georgi Bonchev Str., Bulgarian Academy of Sciences, Block 25A, Sofia, Bulgaria
e-mail: ivimov@bas.bg

G. Mathur et al. (eds.), *Proceedings of 3rd International Conference on Artificial Intelligence: Advances and Applications*, Algorithms for Intelligent Systems,
https://doi.org/10.1007/978-981-19-7041-2_54

which will give a solution with very high accuracy even for high dimensional cases. For the first time a specific optimal MC method is used for evaluating multidimensional integrals in artificial intelligence and also the comparison between the three methods has never been performed before.

The paper is organized as follows. The definitions for the optimal MC approach is given in Sect. 2. The numerical study with multidimensional Lin type integrals are given in Sect. 3. Finally, the conclusions are described in Sect. 4.

2 The Optimal Stochastic Approach

We follow the idea of the method developed in Atanassov and Dimov [2]. For more clear explanation of the terms and the detailed description of the algorithm, see Atanassov and Dimov [2].

Let d and k be natural numbers, $d, k \geq 1$. Let us define

$$F_0 \equiv \mathbf{W}^k\left(\| f \|; U^d\right) \tag{3}$$

where f is defined over $\mathrm{U}^{\mathrm{d}} = [0; 1)^{\mathrm{d}}$, for which

$$\frac{\partial^r f(\boldsymbol{x})}{\partial x_1^{\alpha_1} \ldots \partial x_d^{\alpha_d}}, \quad \alpha_1 + \cdots + \alpha_d = r \leq k, \tag{4}$$

and $\| \cdot \|$ on W^{k} is

$$f = \left\{ \sup \left| \frac{\partial^k f(\boldsymbol{x})}{\partial x_1^{\alpha_1} \ldots \partial x_d^{\alpha_d}} \right|, \quad \alpha_1 + \ldots + \alpha_d = k, \quad \boldsymbol{x} \equiv (x_1, \ldots, x_d) \in U^d \right\}. \tag{5}$$

Now for $n, s, k \geq 1$we make a MC formula relies on $m \geq 1$ and $\binom{s+k-1}{s}$ points (ps) in $[0; 1]^{\mathrm{s}}$. Points $x^{(r)}$ are total $\binom{s+k-1}{s}$ and if for $P(x)$, $\deg P \leq k : P(x(r)) = 0$, then.

$P \equiv 0$. If $N = n^s$ for $n \geq 1$ we divide $[0; 1]^{\mathrm{s}}$ into n^{s} K_j, i.e.

$$[0, 1]^s = c_{i=1}^{n^s} K_j$$

and

$$K_j = \prod_{i=1}^{s} \left[a_i^j, b_i^j\right),$$

$$b_i^j - a_i^j = \frac{1}{n}, i = 1, \ldots, s.$$

For every cube K_j we evaluate the coordinates of $\binom{s+k-1}{s}$ points $y^{(r)}$, determined by

$$y_i^{(r)} = a_i^r + \frac{1}{n} x_i^{(r)}.$$

We choose m random ps $\xi_i(j, s) = (\xi_1(j, p), ..., \xi_s(j, p))$ from every K_j, so all $\xi_i(j, p)$ are independent udr points, and evaluate all $f\ (y^{(r)})$ and $f\ (\xi_i(j, p))$, and the Lagrange polynom of f in the point z. Let the obtained polynom will be $L_k\ (f, z)$. For every P, $\max\deg P = k - 1 : L_k(f; z) \equiv z$. We sum for every $j = 1, \ldots N$ and after some transformations consequently obtain:

$$\int\limits_{K_j} f(x)dx \approx \frac{1}{mn^s} \sum_{s=1}^{m} [(\xi(j, p)) - L_k(f, \xi(j, p))] + \int\limits_{K_j} L_k(f, x)dx. \tag{6}$$

$$I(f) \approx \frac{1}{mn^s} \sum_{j=1}^{N} \sum_{s=1}^{m} [(\xi(j, p)) - L_k(f, \xi(j, p))] + \sum_{j=1}^{N} \int\limits_{K_j} L_k(f, x)dx. \tag{7}$$

$$\int\limits_{K_j} f(x)dx \approx \frac{1}{mn^s} \sum_{s=1}^{m} [(\xi(j, p)) - L_k(f, \xi(j, p))] + \int\limits_{K_j} L_k(f, x)dx. \tag{8}$$

$$I(f) \approx \frac{1}{mn^s} \sum_{j=1}^{N} \sum_{s=1}^{m} [(\xi(j, p)) - L_k(f, \xi(j, p))] + \sum_{j=1}^{N} \int\limits_{K_j} L_k(f, x)dx. \tag{9}$$

Thus we obtain an optimal Monte Carlo approximation with an optimal order of convergence $\mathcal{O}\left(N^{-\frac{1}{2}-\frac{k}{d}}\right)$ for d-dimensional functions from the class W^k.

3 Numerical Examples

We will use the following notations: LHSMC = Latin Hypercube sampling Minasny and McBratney [7], SOBQMC = Sobol quasi-random sequence Antonov and Saleev [1] and Bratley and Fox [3], OPTMC = optimal approach. In the next Tables, the relative errors (RES) evaluated with the three approaches are given for the corresponding multidimensional integral (MI) and number of samples (NS) and preliminary given time in seconds (PGTS).

We study the following Lin multidimensional integrals (1) and (2):

Example 1 s = 3.

$$\int_{[0,1]^3} \exp(x_1x_2x_3)\mathrm{dx} \approx 1.14649907. \tag{10}$$

Example 2 s = 4.

$$\int_{[0,1]^4} x_1x_2^2e^{x_1x_2}\sin(x_3)\cos(x_4)\mathrm{dx} \approx 0.1089748630. \tag{11}$$

Example 3 s = 5.

$$\int_{[0,1]^5} \exp(-100x_1x_2x_3)(\sin(x_4)+\cos(x_5))\mathrm{dx} \approx 0.1854297367. \tag{12}$$

Example 4 s = 7.

$$\int_{[0,1]^7} e^{1-\sum_{i=1}^{3} sin(\frac{\pi}{2}.x_i)}.\arcsin\left(\sin(1)+\frac{\sum_{j=1}^{7} x_j}{200}\right)\mathrm{dx} \approx 0.75151101. \tag{13}$$

Example 5 s = 15.

$$\int_{[0,1]^{15}} \left(\sum_{i=1}^{10} x_i^2\right)\left(x_{11}-x_{12}^2-x_{13}^3-x_{14}^4-x_{15}^5\right)^2\mathrm{dx} \approx 1.96440666. \tag{14}$$

Example 6 s = 25.

$$\int_{[0,1]^{25}} \frac{4x_1x_3^2e^{2x_1x_3}}{(1+x_2+x_4)^2}e^{x_5+\cdots+x_{20}}x_{21}\ldots x_{25}\mathrm{dx} \approx 108.808. \tag{15}$$

Example 7 s = 30.

$$\int_{[0,1]^{30}} \frac{4x_1x_3^2e^{2x_1x_3}}{(1+x_2+x_4)^2}e^{x_5+\cdots+x_{20}}x_{21}\ldots x_{30}\mathrm{dx} \approx 3.244540. \tag{16}$$

For the 3-MI for NS N = 10^7 the best method is OPTMC—it gives a relative error 5.34e-9—see Table 1 and for PGTS the best method for 100 s is OPTMC—the relative error is 8.61e-8 in Table 2. For the 4-MI for a NS N = 10^7 the best method is OPTMC—it gives a relative error 8.16e-9—see Table 3 and for PGTS the best method for 20 s is OPTMC—the relative error is 6.54e-7 in Table 4. For the 5-MI for NS $N = 10^7$ the best method is again OPTMC—it gives a relative error 7.01e-8—see Table 5 and for PGTS the best for 20 s is again the optimal MC—the result is 8.37e-8 in Table 6. For the 7-MI for NS $N = 10^7$ the best MC is again OPTMC—it gives a relative error 1.45e-7—see Table 7 and for PGTS the best MC for 20 s is now the Sobol method SOBQMC—the relative error is 3.87e-6 in Table 8. For the 15-MI for NS $N = 10^6$ the best MC is again OPTMC—it gives a relative error 8.29e-6—see Table 9 and for PGTS the best MC for 100 s is again the Sobol approach—the relative error is 8.17e-7 in Table 10. For the 25-MI for NS $N - 10^6$ the best MC is again OPTMC—it gives a relative error 3.11e-5—see Table 11 and for PGTS the best MC for 20 s is now the optimal MC—the relative error is 3.13e-3 in Table 12. For the 30-MI for NS $N = 10^6$ the best MC is again OPTMC—it gives a relative error 2.11e-4—see Table 13 and for PGTS the best MC for 20 s is the optimal MC—the relative error is 4.63e-3 in Table 14. From all the results we can conclude that for PGTS the optimal OPTMC always outperforms the Sobol method SOBQMC, but the optimal MC is a computationally expensive algorithm, and sometimes for a PGTS the SOBQMC or even the Latin hypercube sampling approach LHSMC outperforms the optimal approach OPTMC.

Table 1 RES of 3-MI

N	SOBQMC	T	LHSMC	t	OPTMC	t
10^3	5×10^{-4}	0.4	6×10^{-3}	0.003	3×10^{-5}	0.79
10^4	2×10^{-4}	1.8	6×10^{-4}	0.05	2×10^{-6}	4.09
10^5	2×10^{-5}	15	1×10^{-4}	0.5	4×10^{-7}	31.4
10^6	7×10^{-6}	104	7×10^{-5}	5.2	6×10^{-8}	153
10^7	1×10^{-6}	929	1×10^{-5}	16	5×10^{-9}	1048

Table 2 RES of 3-MI for a fixed time

time(s)	SOBQMC	LHSMC	OPTMC
1 s	3×10^{-4}	5×10^{-4}	1×10^{-5}
5 s	8×10^{-5}	7×10^{-5}	1×10^{-6}
10 s	4×10^{-5}	4×10^{-5}	7×10^{-7}
100 s	7×10^{-6}	5×10^{-6}	8×10^{-8}

Table 3 RES of 4-MI

N	SOBQMC	T	LHSMC	t	OPTMC	t
10^4	2×10^{-5}	2.1	5×10^{-4}	0.05	1×10^{-5}	4.6
10^5	5×10^{-6}	17	3×10^{-4}	0.5	7×10^{-6}	44
10^6	1×10^{-6}	191	4×10^{-5}	4.7	2×10^{-7}	351
10^7	8×10^{-7}	1117	8×10^{-6}	47	8×10^{-9}	2645

Table 4 RES of 4-MI for a fixed time

time,s	SOBQMC	LHSMC	OPTMC
0.1 s	4×10^{-4}	4×10^{-4}	4×10^{-5}
1 s	3×10^{-5}	3×10^{-4}	2×10^{-5}
5 s	5×10^{-5}	4×10^{-5}	1×10^{-5}
10 s	6×10^{-6}	3×10^{-5}	7×10^{-6}
20 s	4×10^{-6}	2×10^{-5}	6×10^{-7}

Table 5 RES of 5-MI

N	SOBQMC	T	LHSMC	t	OPTMC	t
10^3	5×10^{-4}	0.03	9×10^{-3}	0.006	2×10^{-5}	1.5
10^4	1×10^{-4}	0.2	3×10^{-3}	0.06	7×10^{-6}	2.5
10^5	2×10^{-5}	2.6	2×10^{-3}	0.6	2×10^{-6}	6
10^6	6×10^{-6}	23	1×10^{-4}	6	5×10^{-7}	19
10^7	2×10^{-6}	241	2×10^{-5}	59	7×10^{-8}	104

Table 6 RES for the 5-MI for a fixed time

time,s	SOBQMC	LHSMC	OPTMC
0.1 s	1×10^{-4}	3×10^{-3}	1×10^{-4}
1 s	7×10^{-5}	8×10^{-4}	5×10^{-5}
5 s	1×10^{-5}	3×10^{-4}	1×10^{-6}
10 s	9×10^{-6}	8×10^{-5}	8×10^{-7}
20 s	7×10 -6	5×10^{-5}	5×10^{-7}

Table 7 RES of 7-MI

N	SOBQMC	T	LHSMC	t	OPTMC	t
10^4	2×10^{-4}	0.66	1×10^{-3}	0.1	2×10^{-4}	9
10^5	1×10^{-4}	7.35	2×10^{-4}	1.1	4×10^{-5}	39
10^6	4×10^{-5}	70	8×10^{-5}	10	1×10^{-6}	165
10^7	9×10^{-6}	687	1×10^{-5}	100	1×10^{-7}	590

Table 8 RES of 7-MI for a fixed time

time,s	SOBQMC	LHSMC	OPTMC
0.1 s	2×10^{-3}	1×10^{-3}	2×10^{-3}
1 s	6×10^{-4}	1×10^{-4}	3×10^{-4}
5 s	8×10^{-5}	9×10^{-5}	1×10^{-4}
10 s	1×10^{-5}	8×10^{-5}	8×10^{-5}
20 s	3×10^{-6}	5×10^{-5}	6×10^{-5}

Table 9 RES of 15-MI

N	SOBQMC	T	LHSMC	t	OPTMC	t
10^3	2×10^{-3}	0.88	1×10^{-2}	0.1	7×10^{-3}	24
10^4	2×10^{-4}	8	7×10^{-3}	1.0	6×10^{-4}	80
10^5	1×10^{-4}	90	1×10^{-3}	9	7×10^{-5}	239
10^6	1×10^{-5}	930	1×10^{-4}	97	8×10^{-6}	719

Table 10 RES of 15-MI for a fixed time

time, s	SOBQMC	LHSMC	OPTMC
1 s	1×10^{-3}	3×10^{-3}	3×10^{-2}
5 s	2×10^{-4}	7×10^{-4}	1×10^{-2}
10 s	9×10^{-5}	1×10^{-4}	9×10^{-3}
20 s	9×10^{-6}	4×10^{-5}	7×10^{-3}
100 s	8×10^{-7}	4×10^{-6}	9×10^{-5}

Table 11 RES for the 25-MI

N	SOBQMC	t	LHSMC	t	OPTMC	t
10^3	1×10^{-1}	0.3	7×10^{-1}	0.01	3×10^{-3}	2.01
10^4	5×10^{-2}	5.5	5×10^{-2}	0.12	3×10^{-3}	18
10^5	7×10^{-3}	32	2×10^{-2}	1	5×10^{-5}	177
10^6	2×10^{-3}	159	1×10^{-4}	8	3×10^{-5}	1203

Table 12 RES of 25-MI for a fixed time

time,s	SOBQMC	LHSMC	OPTMC
1 s	9×10^{-2}	2×10^{-2}	7 × 10 -2
5 s	5×10^{-2}	1×10^{-2}	8×10^{-3}
10 s	2×10^{-2}	9×10^{-3}	5×10^{-3}
20 s	8×10^{-3}	7×10^{-3}	3×10^{-3}

Table 13 RES of 30-MI

N	SOBQMC	T	LHSMC	t	OPTMC	t
10^3	1×10^{-1}	0.32	8×10^{-1}	0.01	2×10^{-2}	5
10^4	8×10^{-2}	4	6×10^{-2}	0.1	6×10^{-3}	13
10^5	1×10^{-2}	29	2×10^{-2}	1	1×10^{-3}	141
10^6	9×10^{-3}	165	9×10^{-3}	8	2×10^{-4}	1280

Table 14 RES of 30-MI for a fixed time

time,s	SOBQMC	LHSMC	OPTMC
1 s	1×10^{-1}	2×10^{-2}	4×10^{-1}
5 s	7×10^{-2}	1×10^{-2}	1×10^{-2}
10 s	5×10^{-2}	9×10^{-3}	8×10^{-3}
20 s	1×10^{-2}	7×10^{-3}	4×10^{-3}

4 Conclusion

In our work an optimal MC approach for computing multidimensional integrals of Lin type (1) and (2) which are important for artificial intelligence, from 3 to 30 dimensions, has been presented. In our case study the Sobol sequence, the Latin hypercube sampling algorithm and the optimal Monte Carlo approach have been compared on some case test Lin functions. The optimal Monte Carlo is one of the best MC for MI and one of the few possible approaches, because the deterministic methods are impractical for higher dimensions. Meanwhile, the optimal algorithm is suitable to deal with 30 MI for just a minute on a laptop. It is a key element because this may be fundamental in some control test examples in artificial intelligence. In the future, the scope of our work will be to develop other optimal Monte Carlo approaches for this very important computational problem in artificial intelligence.

Acknowledgements Venelin Todorov is supported by the Bulgarian National Science Fund (BNSF) under Project KP-06-M32/2—17.12.2019 "Advanced Stochastic and Deterministic Approaches for Large-Scale Problems of Computational Mathematics". The work is also supported by the BNSF under Project KP-06-N52/5 and the Bilateral Project KP-06-Russia/17.

References

1. Antonov I, Saleev V (1979) An economic method of computing LP-sequences. USSR Comput. Math. Phy. 19:252–256
2. Atanassov E, Dimov IT (1999) A new optimal monte carlo method for calculating integrals of smooth functions. J Monte Carlo Methods Appl 5(2):149–167. https://doi.org/10.1515/mcma.1999.5.2.149
3. Bratley P, Fox B (1988) Algorithm 659: implementing Sobol's Quasirandom sequence generator. ACM Trans Math Softw 14(1):88–100
4. Dimov I (2008) Monte Carlo methods for applied scientists. World Scientific, New Jersey, p 291p
5. Lin S (2011) Algebraic methods for evaluating integrals in Bayesian statistics," Ph.D. dissertation, UC Berkeley, May 2011
6. Lin S, Sturmfels B, Xu Z (2009) Marginal likelihood integrals for mixtures of independence models. J Mach Learn Res 10:1611–1631
7. Minasny B, McBratney B (2006) A conditioned Latin hypercube method for sampling in the presence of ancillary information. J Comput Geosci Arch 32(9):1378–1388
8. Owen A (1995) Monte Carlo and Quasi-Monte Carlo methods in scientific computing. Lecture notes in statistics, vol 106, pp 299–317

9. Paskov SH (1994) Computing high dimensional integrals with applications to finance, Technical report CUCS-023–94, Columbia University
10. Pencheva V, Georgiev I, Asenov A (2021) Evaluation of passenger waiting time in public transport by using the Monte Carlo method. In: AIP conference proceedings, vol 2321, No 1. AIP Publishing LLC, p 030028
11. Song J, Zhao S, Ermon S (2017) A-nice-mc: adversarial training for mcmc. In: Advances in neural information processing systems, pp 5140–5150
12. Watanabe S (2001) Algebraic analysis for nonidentifiable learning machines. Neural Com-put. 13:899–933

Chapter 55
Using Numerous Biographical and Enrolling Observations to Predict Student Performance

Mpho Mendy Nefale and Ritesh Ajoodha

1 Introduction

The research of forecasting student success using biographical history, pre-college observations, and enrolling data is introduced in this section. The problem statement, related work, purpose statement, overview of techniques, results in summary, and overall report structure will all be discussed in this section.

1.1 Problem Statement

For many students, student performance has been a major concern. For most students, getting a spot at a university is a life-changing experience, with many students picking a course only because they heard it pays better. The sad reality is that the majority of students who are accepted into the university programs do not complete their studies owing to a lack of academic ability or performance in their field of study, Ajoodha [1].

1.2 Purpose Statement

The major goal of this research is to create a system that is capable of accurately predicting a student's likelihood of success in a given set of disciplines based on var-

M. Mendy Nefale (✉) · R. Ajoodha
University of the Witwatersrand, School of Computer Science and Applied Mathematics, Johannesburg, South Africa
e-mail: nefalempho97@gmail.com

R. Ajoodha
e-mail: Ritesh.Ajoodha@wits.ac.za

G. Mathur et al. (eds.), *Proceedings of 3rd International Conference on Artificial Intelligence: Advances and Applications*, Algorithms for Intelligent Systems,
https://doi.org/10.1007/978-981-19-7041-2_55

ious background and enrollment data. Throughout their studies, students are anxious about the success of their degrees. Having a system that predicts student performance would help students determine how much effort they need to put in order to ensure their success, Ajoodha [1].

1.3 *Related Work*

Several studies have been conducted to predict student performance. A study conducted by Osmanbegovic [2] and Kabakchieva [3], used data mining technologies to predict student performance. A study conducted by Elbadrawy [4] used personal data to predict students performance by the application of linear logistic regression. A study conducted by Ajoodha [1] used features of the learner's background, personality, and school to predict student performance using several machine learning algorithms.

1.4 *Methodology*

In this study, several machine learning models were utilized to predict student performance. The synthetic data included biographical information, pre-college information, and enrollment information. Highest risk, high risk, lowest risk, and medium risk are the four risk categories for the response variables. The SVM, random forest, and decision tree were among the machine learning models utilized. Accuracy, confusion matrix, and precision were used to evaluate the models.

1.5 *Results*

After applying all the models for this study, the SVM outperformed all other models with an accuracy of 95.54% and precision of 96%.

1.6 *Contribution*

This study will add to a current knowledge by developing a model that can predict learner risk status before completing a certain course of study and comparing aspects that are more successful in predicting student performance.

2 Related Work

Students that participate in a students success courses have a higher chance of succeeding than students who do not participate in a students success courses, Tien and Armen [5]. Success courses help students move from their previous education or experience. These may be high school or college students. Cho [6] found that 68% of students who finished on time took student success courses, indicating that students enrolled in students success courses have a better chance of succeeding. However, some students succeed despite not being enrolled in students success courses. Most students who drop out are those who had high grades in high school but may have been encouraged to study certain courses by family members, Kember [7]. Students who switched majors after a poor performance in engineering, on the other hand, appeared to have high hopes for the course, which could have been affected by the possibility of financial benefits in post-graduate studies or jobs, Nicholls et al. [8]. First entering students are at the highest possibility of dropping out during the first semester of study or failing to complete their program/degree on time in all institutions of higher learning, Gray et al. [9]. The majority of students are between the ages of 30 and 40, with more than 68% of students being over 30. This age category is also associated with a higher likelihood of failing the course, with a rate of students failing the course of 37.7%, which is higher than the overall percentage of students failing the course in the student population (38%), Arnold et al. [10]. Engineering majors were more likely to be pursued by students that have a high average academic performance and quantitative score's test than their lower performing peers. Students with superior overall academic performance and standardized students with higher exam scores were more likely to pursue engineering majors than their less-achieved counterparts; students who studied engineering for the love of it were more likely to excel than those who pursued it for the sake of higher pay after graduation, Nicholls et al. [8]. Students with disabilities have a higher chance of failing than those who do not. Depending on their ethnic heritage, the percentage of students that successfully finished the course varies dramatically. Thirty three percent of the students on these courses were enrolled in bachelor's degree programs in applied sciences, Caruso and Salaway [11]. In comparison to students enrolled in bachelor's degree programs in business, they have a higher chance of failing the course. Finally, students who enroll in courses during the summer semester are more likely to fail than those who enroll during the fall semester who take it during the fall or spring semesters, Herrera [12].

The random forest algorithm is one of the well known supervised learning system. It creates a "forest" out of a collection of decision trees that are commonly trained using the "bagging" method, Oshiro et al. [13]. The bagging method's primary idea is that it combines many learning models enhances total output. In a random forest, the process of dividing a node analyzes only a subset of the features at random, Oshiro et al. [13]. A study conducted by Ndou [14] used random forest to predict the student performance and obtained 94.04% accuracy. Year of commencement, plancode, plan description, stremline, age in first year, school quintile, mathematics matric major, home province, rural or urban, life orientation, physics Chem, English first language, home country, additional mathematics, mathematics matric literacy, computer studies, and English first additional were all included in a study that yielded an accuracy of 80% when employing a Naive Bayes model, Cho and Karp [6].

Nghe [15] and Ajoodha et al. [1] used BNT Models to predict students performance and achieved 61.54% and 70% accuracy respectively.

3 Research Methodology

This section will provide a fast overview of the data collecting and the methods that has been used to analyze and model the data.

3.1 Data Collection and Sampling Methods

The study used a synthetic data based on the learned Bayesian network structure modeling. The value of the parent node is taken from their unconditional distribution, and the value of the child node is taken from the parent set. We repeated the sampling process until all node values are generated. The Gaussian distribution to model continuous variables was used. The Gaussian distribution, usually called the normal distribution, is a continuous function with a mean and standard deviation, provided that the data is normal. Negative values such as cumulative estimates and probabilities have negative values. Remove negative values from the data set, they could not be changed because it would change the network's distribution. The level of the factor is represented by tabular conditional probability density (CPD), which is used to model discrete variables. The data was not balanced, however, synthetic minority oversampling technique (SMOTE) was used to balance the data.

3.2 Features

The data includes biographical information, pre-college observations, and university enrollment observations. We have utilized this information to perform our research. Gender, race, first-year age, home language, home province, home country, and place of origin, whether rural or urban, are all biographical characteristics that has been employed in the study. In the pre-university observations, we utilized the school quintile to indicate school courage, with quintile 1 being the poorest and quintile 5 being the richest. Finally, for the observation of university enrollment, we used the year of commencement of studies, the description of the plan, professional history, the possibility of success in different branches of science (mathematics, physics, earth sciences, and life sciences), the totality of course grades, and the number of years that the completion of studies was spent.

3.3 Methods

3.3.1 Random Forest

Random forest is a simple machine learning technique that in most cases gives excellent results even without hyperparameter tuning and is widely used due to its simplicity and versatility, Biau and Scornet [16]. The random forest produces understandable predictions and can handle large data sets effectively. The random forest has been discovered to be more accurate in predicting outcomes, Biau and Scornet [16]. By expanding the number of trees, this strategy improves accuracy, Yates and Islam [17]. Random forest has proven to be useful in different kinds of problems, Shi and Horvath [18]. Random forest has been successfully applied in remote sensing, bioinformatics, predicting students' performance, analyzing customers' behavior, and many other predicting problems, Belgiu and Dragut [19]. Random forest generates and blends numerous decision trees to produce a more exact and dependable prediction. The random forest has the benefit of being able to be used for both classification and regression problems, both of which are common in modern machine learning systems. Random forest has nearly identical hyperparameters to decision trees and bagging classifiers. The precise information is that you do not want to mix a bagging classifier decision tree due to the fact that you may use random forest classifier-class, Paul et al. [20]. Random forest also can manage regression duties. Another great advantage of the random forest approach is that determining the proportional value of each feature on the prediction is a breeze. We used the Gini index when performing random forests on categorization data. To determine how nodes on a decision tree branch, we apply the formula below:

$$Gini = 1 - \sum_{i=1}^{c} (pi)^2 \tag{1}$$

The Gini of each branch on a node is calculated using the class and probability formula above. Entrophy can also be used to figure out how a decision tree's nodes branch.

$$Entrophy = \sum_{i=1}^{c} -pi * log_2 pi \tag{2}$$

The relative frequency of the class you're looking at in the dataset is represented by pi, and the number of classes is represented by c.

3.3.2 Decision Tree

One of the supervised machine learning algorithms is decision tree. It can be used to tackle problems related to regression and classification, Quinlan [21]. The decision tree method is often the method of choice for predictive modeling since it is both simple to learn and highly effective, Song and Ying [22]. The decision to make strategic splits has a significant impact on the correctness of a tree, Song and Ying [23]. Classification and regression trees have different decision criteria. Decision tree have been successfully applied in a range problems including business, medicine, computer science and many more, Lipyanina et al. [24] and it was found to give highly accurate results. Because the method clearly spell out the problem and allow all choices to be altered, decision tree is an effective technique of decision-making.

3.3.3 SVM

The SVM is a type of supervised machine learning model that can be used to solve classification problems, Noble [25]. We depict each data item as a point in n-dimensional space in this approach, where n is the number of features you have. The SVM approach is beneficial because it recognizes non-linearity in the data and produces an accurate prediction model, Pisner [26]. When the data is linearly or non-linearly separable, the SVM performs well in terms of accuracy. When the data is linearly separable, the SVM produces a separating hyperplane that optimizes the margin of separation between classes along a line perpendicular to the hyperplane. The SVM's primary role is to look for a hyperplane that can distinguish between the two classes, Osisanwo et al. [27]. The SVM was applied to predict membrane protein types and virulent protein in bacterial pathogens, Gai et al. [28]. The method was found to be accurate. Based on the related work, we can say that SVM is a good method for prediction problems, Gai et al. [28].

3.3.4 Evaluations

The goal of evaluation is to put a model through its paces on data that differs from what it was trained on. This gives an unbiased assessment of learning performance.

In this study, we employed the evaluation functions listed below:

- Accuracy
- Confusion matrix
- Precision.

4 Results and Discussion

The outcomes of our experiment are presented and discussed in this section.

4.1 *Data Analysis*

Python 3 was used to evaluate the bogus data. The findings were predicted using the machine learning models mentioned in the preceding section. When a student begins a program, the data displays four risk statuses: highest risk (students at this risk status end up dropping out), high risk(students at this risk status finish the degree in more than 5 years), medium risk (students at this risk status finish their degree in more 3 years), and low risk (Students at this risk status gets to finish their degree on time or in less than 3 years).

4.1.1 Feature Information Gain

This section investigates the contribution of each characteristic to classifying the risk status class variable. Using the IGR, the most contributing features were found as illustrated in the table below by order from the top one. Plan description, year started, language, home province, home country, gender, and rural or urban status were determined to be the most significant features.

4.1.2 Features Statistics

In this brief section, we will investigate how the significant features connect to other variables and the target variables.

4.2 *Classification*

The outcomes of the six fitted machine learning classification algorithms are examined in this section: SVMs, Decision tree, Random forest, Linear logistic regression, KNN, and Naive Bayes.

4.2.1 Confusion Matrix and Accuracies

Accuracy is the first metric we used and it is the most basic. It provides an answer to the question:

How often does the classifier get it right?

It is easily obtained by applying the following formulas:

$$Accuracy = \frac{\text{number of correctly classified items}}{\text{number of all classified items}} \tag{3}$$

Table 1 Confusion Matrix table

Risk status	High risk	Highest risk	Lowest risk	Medium risk
High risk	1637	29	21	19
Highest risk	9	1670	0	2
Lowest risk	28	0	1635	43
Medium risk	69	14	71	1593

Confusion Matrix is yet another indicator commonly used to assess the success of a classification algorithm. This metric was used in this study. If we were to utilize a confused matrix to forecast if an email is spam or not, we would have the following matrix::

	Predicted: RE	Predicted: SE
Actual: RE	TN	FP
Actual: SE	FN	TP

where: RE = real email, SE = spam email, TN = true negatives, TP = true positives, FN = false negatives, and TF = false positive

The anticipated classes in the matrix's columns, are represented. The actual classes, on the other hand, are represented in the matrix's rows. There are four cases:

- TP: when the classifier expected spam and The emails were unquestionably spam.
- TN: where the classifier expected "not spam" and therefore the emails were real
- FP: where the classifier expected "spam," though The emails were authentic.
- FN: where the classifier expected "not spam," however the emails were spam.

True or false suggests if the classifier appropriately expected the class, while positive or negative shows whether or not the classifier efficaciously expected the goal class. The accuracy is then predicted by the use of following formula:

$$Accuracy = \frac{TN + TP}{TN + TP + FN + FP} \tag{4}$$

The method of tenfold cross validation was used to assess accuracy.

The SVM outperformed all other algorithms with an accuracy of 95.56%, which means that the classifier got it right 95.56% times out of the entire testing data, followed by Random Forest with an accuracy of 94.23%, which means that the classifier got it right 94.23% of the entire testing data, and decision tree with an accuracy of 93%, which means that the classifier got it right 93% of the entire testing data.

Table 1 **shows the confusion Matrix for the SVM.**

Table 1 displays the following results:

High Risk: The actual high risk is 1706. Predicted high risk in actual high risk is 1637, out of 1706 actual high risk, 1637 are predicted correctly as high risk wheres 29, 21 and 19 were incorrectly predicted as highest, lowest and medium risk respectively.

Highest Risk: The actual highest risk is 1681. Predicted highest risk in actual highest risk is 1670, 1670 are predicted correctly as highest risk wheres 9 and 2 and incorrectly predicted as high risk and medium risk respectively.

Lowest Risk: The actual lowest risk is 1706. Predicted lowest risk in actual lowest risk is 1635, out of 1706 actual lowest risk, 1635 are correctly predicted wheres 28 and 43 are incorrectly predicted as high risk and medium respectively.

Medium Risk: The actual medium risk is 1747. Predicted medium risk in actual medium risk is 1593, out of 1747 actual medium risk, 1593 are correctly predicted wheres 69, 14 and 17 are incorrectly predicted as high, highest and lowest respectively.

Precision:
Aside from accuracy, the confusion matrix was used to construct a number of additional performance indicators. Precision provides an answer to the question:

How often does it get it right when it forecasts the outcome? This is accomplished through the application of the formula:

$$Precision = \frac{TP}{TP + FP} \tag{5}$$

When the goal is to reduce the quantity of FP, precision is typically used.

The SVM has outperfomed all other model with an Accuracy of 95.54% and 96% Precision.

4.3 Interpretation

SVM performs better mostly reason being that it offers superb consequences in phrases of accuracy whilst the information is linearly or non-linearly separable. When the information is linearly separable, the SVM end result is a isolating hyperplane, which maximizes the margin of separation among classes, measured alongside a line perpendicular to the hyperplane.

Random Forest is one the models that gave good results in this study mostly reason being that a large number of relatively independent models (trees) operating in committee will prevail over the individual constitutive models. The key is the low correlation between the models.

Decision tree had a high accuracy in this study, decision tree is one of the most used models due to the fact it wrecks down complicated records into extra practicable parts. Due to the fact it wrecks down complicated records into extra practicable parts.

5 Conclusion

Our research into classifying students into the appropriate risk profiles using biographical background, individual, and schooling variables revealed that biographical characteristics, followed by individual traits, have the greatest impact on student attrition or risk classification. Pre-college factors have little or no impact on determining student risk profiles. Similarly, the eight most significant contributing attributes are biographical and individual characteristics, according to the results. They serve a significant role in classifying students into the appropriate risk profiles, according to the conceptual model. When compared to models fitted with a controlled balanced class data set using the SMOTE technique, the results show that the fitted models performed well on an imbalanced class data set. This could be related to the fact that the SMOTE method does not take into account neighboring samples from different classes when generating synthetic samples. It can then lead to class overlap and the introduction of additional noise. The fitted machine learning algorithms were able to recognize the various risk profiles successfully. However, the positive class identification rate varies depending on the quantity of each class, Cho and Karp [6].

The findings show that the machine learning models used were able to estimate learner susceptibility based on the attributes provided. The SVMs outperformed all other models with an accuracy of 95.56% followed by random forest with 94.23%, and decision tree with 93%.

The least performing model on this study was found to be decision tree.

The observe concludes that scholar attrition is stricken by biographical and individual attributes, and consequently those elements have to be considered in the better schooling enrollment system.

Significance: This study was important in assisting students in determining which level of risk the course they are taking has based on their biographical and pre-college because many students do not consider their biographical and pre-college when choosing a course of study, and the majority of them do not complete their degree, which is unfortunate.

Future Work: This research can be expanded in a number of ways, including developing a model that will recommend the most appropriate course for a student's success, thereby reducing the number of students who drop out.

Acknowledgements This work is based on the research supported in part by the National Research Foundation of South Africa (Grant numbers: 121835). The student performance prediction system developed in this study can be adopted by students across the world because it is easily accessible, and if widely adopted, this system would have a great impact on the students' performance.

References

1. Ajoodha R, Jadhav A, Dukhan S (2020) Forecasting learner attrition for student success at a south african university. In: Conference of the South African Institute of computer scientists and information technologists 2020, pp 19–28
2. Osmanbegovic E, Suljic M (2012) Data mining approach for predicting student performance. Econ Rev: J Econ Busin 10(1):3–12
3. Kabakchieva D (2013) Predicting student performance by using data mining methods for classification. Cybern Inform Technol 13(1):61–72
4. Elbadrawy A, Polyzou A, Ren Z, Sweeney M, Karypis G, Rangwala H (2016) Predicting student performance using personalized analytics. Computer 49(4):61–69
5. Tien I, Armen DK, Reliability engineering system safety
6. Cho S-W, Karp M (2013) Student success courses in the community college. Commun Coll Rev 41:86–103
7. Kember D (1995) Open learning courses for adults: a model of student progress. Educational Technology (1995)
8. Nicholls GM, Wolfe H, Besterfield-Sacre M, Shuman LJ, Larpkiattaworn S, A method for identifying variables for predicting stem enrollment. J Eng Educ 96(1):33–44
9. Gray G, McGuinness C, Owende P, Hofmann M (2016) Learning factor models of students at risk of failing in the early stage of tertiary education. J Learn Anal 3(2):330–372
10. Arnold AS, Wilson JS, Boshier MG, Smith J (1998) A simple extended-cavity diode laser. Rev Sci Instrum 69(3):1236–1239. http://link.aip.org/link/?RSI/69/1236/1
11. Caruso JB, Salaway G (2007) The ECAR study of undergraduate students and information technology, 2007. Retrieved December, vol 8, p 2007
12. Herrera OL (2006) Investigation of the role of pre- and post-admission variables in undergraduate institutional persistence, using a markov student flow model. Conf Paper 1(16):06
13. Oshiro TM, Perez PS, Baranauskas JA (2012) How many trees in a random forest? In: International workshop on machine learning and data mining in pattern recognition. Springer, pp 154–168
14. Ndou N, Ajoodha R, Jadhav A (2020) Educational data-mining to determine student success at higher education institutions. In: 2nd international multidisciplinary information technology and engineering conference (IMITEC). IEEE, pp 1–8
15. Tran T-O, Dang H-T, Dinh V-T, Phan X-H et al (2017) Performance prediction for students: a multi-strategy approach. Cybern Inf Technol 17(2):164–182
16. Biau G, Scornet E (2016) A random forest guided tour. Test 25(2):197–227
17. Yates D, Islam MZ (2021) Fastforest: Increasing random forest processing speed while maintaining accuracy. Inf Sci 557:130–152
18. Shi T, Horvath S (2006) Unsupervised learning with random forest predictors. J Comput Graph Stat 15(1):118–138
19. Belgiu M, Drăguţ L (2016) Random forest in remote sensing: a review of applications and future directions. ISPRS J Photogramm Remote Sens 114:24–31
20. Paul A, Mukherjee DP, Das P, Gangopadhyay A, Chintha AR, Kundu S (2018) Improved random forest for classification. IEEE Trans Image Process 27(8):4012–4024
21. Quinlan JR (1996) Learning decision tree classifiers. ACM Comput Surv (CSUR) 28(1):71–72
22. Damanik IS, Windarto AP, Wanto A, Andani SR, Saputra W et al. (2019) Decision tree optimization in c4. 5 algorithm using genetic algorithm. J Phys: Conf Ser 1255(1). IOP Publishing, p 012012

23. Song Y-Y, Ying L (2015) Decision tree methods: applications for classification and prediction. Shanghai Arch Psychiatry 27(2):130
24. Lipyanina H, Sachenko A, Lendyuk T, Nadvynychny S, Grodskyi S (2020) Decision tree based targeting model of customer interaction with business page. In: CMIS, pp 1001–1012
25. Noble WS (2006) What is a support vector machine? Nat Biotechnol 24(12):1565–1567
26. Pisner DA, Schnyer DM (2020) Support vector machine. In: Machine learning. Elsevier, pp 101–121
27. Osisanwo F, Akinsola J, Awodele O, Hinmikaiye J, Olakanmi O, Akinjobi J (2017) Supervised machine learning algorithms: classification and comparison. Int J Comput Trends Technol (IJCTT) 48(3):128–138
28. Cai Y-D, Ricardo P-W, Jen C-H, Chou K-C (2004) Application of svm to predict membrane protein types. J Theor Biol 226(4):373–376

Author Index

G. Mathur et al. (eds.), *Proceedings of 3rd International Conference on Artificial Intelligence: Advances and Applications*, Algorithms for Intelligent Systems, https://doi.org/10.1007/978-981-19-7041-2

Printed in the United States
by Baker & Taylor Publisher Services